New Development of Neutrosophic Probability, Neutrosophic Statistics, Neutrosophic Algebraic Structures, and Neutrosophic Plithogenic Optimizations

New Development of Neutrosophic Probability, Neutrosophic Statistics, Neutrosophic Algebraic Structures, and Neutrosophic Plithogenic Optimizations

Editors

Florentin Smarandache
Yanhui Guo

MDPI • Basel • Beijing • Wuhan • Barcelona • Belgrade • Manchester • Tokyo • Cluj • Tianjin

Editors
Florentin Smarandache
Mathematics
University of New Mexico
Gallup
United States

Yanhui Guo
Department of Computer Science
University of Illinois
Springfield
United States

Editorial Office
MDPI
St. Alban-Anlage 66
4052 Basel, Switzerland

This is a reprint of articles from the Special Issue published online in the open access journal *Symmetry* (ISSN 2073-8994) (available at: www.mdpi.com/journal/symmetry/special_issues/Neutrosophic_Probability_Statistics_Algebraic_Structures_Plithogenic).

For citation purposes, cite each article independently as indicated on the article page online and as indicated below:

LastName, A.A.; LastName, B.B.; LastName, C.C. Article Title. *Journal Name* **Year**, *Volume Number*, Page Range.

ISBN 978-3-0365-2955-4 (Hbk)
ISBN 978-3-0365-2954-7 (PDF)

Contents

About the Editors

Florentin Smarandache

Florentin Smarandache is a professor of mathematics at the University of New Mexico, United States. He received his MSc in Mathematics and Computer Science from the University of Craiova, Romania, PhD in Mathematics from the State University of Kishinev, and Postdoctoral in Applied Mathematics from Okayama University of Sciences, Japan, and The Guangdong University of Technology, Guangzhou, China. He has been the founder of neutrosophy (generalization of dialectics), neutrosophic set, logic, probability and statistics since 1995 and has published hundreds of papers and books on neutrosophic physics, superluminal and instantaneous physics, unmatter, quantum paradoxes, absolute theory of relativity, redshift and blueshift due to the medium gradient and refraction index in addition to the Doppler effect, paradoxism, outerart, neutrosophy as a new branch of philosophy, Law of Included Multiple-Middle, multispace and multistructure, hypersoft sets, degree of dependence and independence between neutrosophic components, refined neutrosophic sets, neutrosophic over-under-off-sets, plithogenic sets / logic / probability / statistics, neutrosophic triplet and duplet structures, quadruple neutrosophic structures, extension of algebraic structures to NeutroAlgebras and AntiAlgebras, NeutroGeometry and AntiGeometry, Dezert–Smarandache Theory, and so on to many peer-reviewed international journals and many books and he has presented papers and plenary lectures at many international conferences around the world.

In addition, he has published many books of poetry, drama, children's stories, translations, essays, a novel, folklore collections, traveling memories, and art albums (http://fs.unm.edu/FlorentinSmarandache.htm).

Yanhui Guo

Yanhui Guo received his B. S. degree in Automatic Control from Zhengzhou University, China, M.S. degree in Pattern Recognition and Intelligence System from Harbin Institute of Technology, China, and Ph.D. degree in the Department of Computer Science, Utah State University, USA. He was a research follow in the Department of Radiology at the University of Michigan and an assistant professor in St. Thomas University. Dr. Guo is currently an assistant professor in the Department of Computer Science at the University of Illinois at Springfield. Dr. Guo has published more than 100 journal papers and 30 conference papers, completed more than 10 grant-funded research projects, and worked as an associate editor of different international journals, a reviewer for top journals and has been a part of many conferences. His research areas include computer vision, machine learning, big data analytics, and computer-aided detection/diagnosis.

Preface to "New Development of Neutrosophic Probability, Neutrosophic Statistics, Neutrosophic Algebraic Structures, and Neutrosophic Plithogenic Optimizations"

This Special Issue presents state-of-the-art papers on new topics related to neutrosophic theories, such as neutrosophic algebraic structures, neutrosophic triplet algebraic structures, neutrosophic extended triplet algebraic structures, neutrosophic algebraic hyperstructures, neutrosophic triplet algebraic hyperstructures, neutrosophic n-ary algebraic structures, neutrosophic n-ary algebraic hyperstructures, refined neutrosophic algebraic structures, refined neutrosophic algebraic hyperstructures, quadruple neutrosophic algebraic structures, refined quadruple neutrosophic algebraic structures, neutrosophic image processing, neutrosophic image classification, neutrosophic computer vision, neutrosophic machine learning, neutrosophic artificial intelligence, neutrosophic data analytics, neutrosophic deep learning, and neutrosophic symmetry, as well as their applications in the real world.

The neutrosophic extended triplet group (NETG) is a novel algebra structure studied here by Xin Zhou, Ping Li, Florentin Smarandache and Ahmed Mostafa Khalil in an article (Results on Neutrosophic Extended Triplet Groups Equipped with a Partial Order) presenting the concept of a partially ordered neutrosophic extended triplet group (po-NETG), considering the properties and structure features of po-NETGs. The authors propose the concepts of the positive cone and negative cone in a po-NETG, study the specificity of the positive cone in a partially ordered weak commutative neutrosophic extended triplet group (po-WCNETG), and introduce the concept of a po-NETG homomorphism between two po-NETGs.

In the next selected paper (Single-Valued Neutrosophic Ideals in Šostak Sense), Yaser Saber, Fahad Alsharari and Florentin Smarandache introduce the notion of single-valued neutrosophic ideals sets in Šostak's sense, and then the concept of a single-valued neutrosophic ideal open local function for a single-valued neutrosophic topological space, studying the basic structure, especially a basis for such generated single-valued neutrosophic topologies and several relations between different single-valued neutrosophic ideals and single-valued neutrosophic topologies. For the purpose of symmetry, the authors also define the single-valued neutrosophic relations.

Guansheng Yu, Shouzhen Zeng and Chonghui Zhang study the single-valued neutrosophic linguistic distance measures based on the induced aggregation method in their paper (-Valued Neutrosophic Linguistic-Induced Aggregation Distance Measures and Their Application in Investment Multiple Attribute Group Decision Making), suggesting a new extension of the existing distance measures based on the induced aggregation view, namely the single-valued neutrosophic linguistic-induced ordered weighted averaging distance (SVNLIOWAD) measure. Based on SVNLIOWAD, in order to eliminate the defects of the existing methods, the authors develop a novel induced distance for single-valued neutrosophic linguistic sets, called the single-valued neutrosophic linguistic weighted induced ordered weighted averaging distance (SVNLWIOWAD). Then, the relationship between the two proposed distance measures is explored, and a numerical example concerning an investment selection problem is constructed to prove the efficiency of the proposed method under a single-valued neutrosophic linguistic environment.

Ashraf Al-Quran, Hazwani Hashim and Lazim Abdullah extend, in the paper Hybrid Approach of Interval Neutrosophic Vague Sets and DEMATEL with New Linguistic Variable, the concept of Interval Neutrosophic Vague Sets (INVS) to the linguistic variable that can be used in the decision-making process. The advantages of the linguistic variable of INVS, which is a useful tool to deal with uncertainty and incomplete information, derives from allowing the greater range of value for membership functions. In addition, a case study on the quality of hospital service is evaluated to demonstrate the approach, and a comparative analysis to check the feasibility of the method is presented, showing that different methods produce different relations and levels of importance, which is due to the inclusion of the INVS linguistic variable.

The paper of Imaginative Play in Children Using Single-Valued Refined Neutrosophic Sets, by Vasantha W. B., Ilanthenral Kandasamy, Florentin Smarandache, Vinayak Devvrat and Shivam Ghildiyal, introduces the Single-Valued Refined Neutrosophic Set (SVRNS), a generalized version of the neutrosophic set consisting of six membership functions based on imaginary and indeterminate aspects, and hence it is more sensitive to real-world problems. Machine learning algorithms such as K-means, parallel axes coordinate, etc., are applied for a real-world application in child psychology. The study of imaginative pretend play of children in the age group from 1 to 10 years is analyzed using SVRNS, helping in detecting the mental abilities of a child on the basis of imaginative play. The authors conclude that SVRNS is better at representing these data when compared to other neutrosophic sets.

Aggregation operators are key features of decision-making theory, while the neutrosophic cubic set (NCS), as a generalized version of NS and INS, is a very effective choice when dealing with vague and imprecise data. Majid Khan, Muhammad Gulistan, Mumtaz Ali and Wathek Chammam intend, in the paper Generalized Neutrosophic Cubic Aggregation Operators and Their Application to Multi-Expert Decision-Making Method, to generalize these aggregation operators by presenting neutrosophic cubic-generalized unified aggregation (NCGUA) and neutrosophic cubic quasi-generalized unified aggregation (NCQGUA) operators. The authors employ the multi-expert decision-making method (MEDMM) to express this complex framework.

In the paper Single Valued Neutrosophic Data Envelopment Analysis: Application to Hospital Performance Measurement, Wei Yang, Lulu Cai, Seyed Ahmad Edalatpanah and Florentin Smarandache introduce a model of data envelopment analysis (DEA) in the context of neutrosophic sets, proposing an innovative process to solve it. Furthermore, the authors analyze the problem of healthcare system evaluation with inconsistent, indeterminate, and incomplete information using the new model. The triangular single-valued neutrosophic numbers are also employed to deal with the data, and the method is used to assess 13 hospitals of Tehran University of Medical Sciences of Iran. The results prove the efficiency of the suggested approach and emphasize that the model has practical outcomes for decision makers.

Another study (Novel Dynamic Multi-Criteria Decision Making Method Based on Generalized Dynamic Interval-Valued Neutrosophic Set, by Nguyen Tho Thong, Florentin Smarandache, Nguyen Dinh Hoa, Le Hoang Son, Luong Thi Hong Lan, Cu Nguyen Giap, Dao The Son and Hoang Viet Long) selected for this Special Issue introduces the generalized dynamic internal-valued neutrosophic sets, which are an extension of dynamic internal-valued neutrosophic sets. Based on this extension, the authors develop some operators and a TOPSIS method to deal with the change of both criteria, alternatives, and decision makers by time. For example, the method is applied to rank students according to attitude–skill–knowledge evaluation model.

The associative law reflects the symmetry of operation, and other various variation associative laws reflect some generalized symmetries. In the article Kind of Variation Symmetry: Tarski Associative Groupoids (TA-Groupoids) and Tarski Associative Neutrosophic Extended Triplet Groupoids (TA-NET-Groupoids), Xiaohong Zhang, Wangtao Yuan, Mingming Chen and Florentin Smarandache introduce a new concept of Tarski associative groupoids (or transposition associative groupoid (TA-groupoid)), presenting many examples, to determine their basic properties and structural characteristics, and as well to discuss the relationships among a few non-associative groupoids. Moreover, the researchers suggest a new concept of Tarski associative neutrosophic extended triplet groupoids (TA-NET-groupoid), examining related properties.

In the next paper (Components Semigroups and Multiset Neutrosophic Components Semigroups), Vasantha W. B., Ilanthenral Kandasamy and Florentin Smarandache define the usual product and sum operations of neutrosophic components (NC). Four different NCs are defined using the four different intervals: (0, 1), [0, 1), (0, 1] and [0, 1]. In the neutrosophic components, it is assumed that the truth value or the false value or the indeterminate value is from the intervals (0, 1) or [0, 1) or (0, 1] or [0, 1]. All the operations defined on these neutrosophic components on the four intervals are symmetric. In all four cases, the NC collection happens to be a semigroup under product, and all of them are torsion-free semigroups or weakly torsion-free semigroups. The NC defined on the interval [0, 1) is a group under addition modulo 1, while the NC defined on the interval [0, 1) is an infinite commutative ring under addition modulo 1. The authors also examine the multiset NC semigroup using the four intervals, and derive several interesting properties of these structures.

Abdul Alamin, Sankar Prasad Mondal, Shariful Alam, Ali Ahmadian, Soheil Salahshour and Mehdi Salimi focus on analyzing the homogeneous linear difference equation in a neutrosophic environment in the paper and Interpretation of Neutrosophic Homogeneous Difference Equation. The authors interprete the solution of the homogeneous difference equation with initial information, the coefficient and both as a neutrosophic number. The theoretical work is followed by numerical examples and an application in actuarial science, which shows the great impact of neutrosophic set theory in mathematical modeling in a discrete system to better understand the behavior of the system.

The grand theory of action of Parsons has an important place in social theories, but there are many uncertainties in this theory, for which classical logic is often insufficient. In the study Modeling of Talcott Parsons's Action and Decision-Making Applications for It, the researchers Cahit Aslan, Abdullah Kargın and Memet Şahin export, for the first time, the grand theory of action of Parsons in neutrosociology. The authors achieve a more effective way of dealing with the uncertainties in the theory of Parsons as in all social theories, proposing a similarity measure for single-valued neutrosophic numbers, showing, in addition, that this measure of similarity satisfies the similarity measure conditions, obtaining applications that allow finding the ideal society in the theory of Parsons within the theory of neutrosociology. Finally, the authors compare the data obtained in this study with the results of the similarity measures defined previously.

In the article of the Single-Valued Neutrosophic Fuzzy Set and the Soft Set with Applications in Decision-Making, Ahmed Mostafa Khalil, Dunqian Cao, Abdelfatah Azzam, Florentin Smarandache and Wedad R. Alharbi propose a novel concept of the single-valued neutrosophic fuzzy soft set by combining the single-valued neutrosophic fuzzy set and the soft set. Five types of operations (e.g., subset, equal, union, intersection, and complement) on single-valued neutrosophic fuzzy soft sets are presented for possible applications. Additionally, several theoretical operations of single-valued neutrosophic fuzzy soft sets are given. In addition, the first type for fuzzy decision making based on a single-valued neutrosophic fuzzy soft set matrix is constructed. To clarify the applicability, the authors scrutinize a numerical example using the AND operation of the single-valued neutrosophic fuzzy soft set for fuzzy decision making.

Fault diagnosis has become more and more important with increasing automation, and the factors that cause mechanical failures are becoming more and more complex. In order to contribute to a solution for the given problem, Shchur Iryna, Yu Zhong, Wen Jiang, Xinyang Deng and Jie Geng propose a single-valued neutrosophic set ISVNS algorithm for processing uncertain and inaccurate information in fault diagnosis in the paper -Valued Neutrosophic Set Correlation Coefficient and Its Application in Fault Diagnosis. In order to solve the fault diagnosis problem more effectively, the authors generate a neutrosophic set by triangular fuzzy number and introduce the formula of the improved weighted correlation coefficient. Experiments show that the algorithm can significantly improve the accuracy degree of fault diagnosis, and can better satisfy the diagnostic requirements in practice.

The paper New Multi-Sensor Fusion Target Recognition Method Based on Complementarity Analysis and Neutrosophic Set, by Yuming Gong, Zeyu Ma, MeWang, Xinyang Deng and Wen Jiang, investigates a multi-sensor fusion recognition method based on complementarity analysis and a neutrosophic set. The proposed method has two parts: complementarity analysis and data fusion. Complementarity analysis applies the trained multi-sensor to extract the features of the verification set into the sensor, obtaining the recognition result of the verification set. Based on the recognition result, the multi-sensor complementarity vector is obtained. Then, the sensor output the recognition probability and complementarity vectors are used to generate multiple neutrosophic sets, which are then merged within the group through the simplified neutrosophic weighted average (SNWA) operator. Finally, the neutrosophic set is converted into a crisp number, and the maximum value is the recognition result.

Yaser Saber, Fahad Alsharari, Florentin Smarandache and Mohammed Abdel-Sattar introduce the notion of r-single-valued neutrosophic connected sets in single-valued neutrosophic topological spaces, which is considered as a generalization of r-connected sets in Šostak's sense and r-connected sets in intuitionistic fuzzy topological spaces. In their paper (and Stratification of Single-Valued Neutrosophic Topological Spaces), they introduce the concept of r-single-valued neutrosophic separated and obtain some of its basic properties, also attempting to show that every r-single-valued neutrosophic component in single-valued neutrosophic topological spaces is an r-single-valued neutrosophic component in the stratification of it. For the purpose of symmetry, the authors conclusively define the single-valued neutrosophic relations.

Neutrosophy has been used in sentiment analyses of textual data, but it has not been used in speech sentiment analysis. Consequently, Kritika Mishra, Ilanthenral Kandasamy, Vasantha Kandasamy W. B. and Florentin Smarandache (in the paper Novel Framework Using Neutrosophy for Integrated Speech and Text Sentiment Analysis) suggest a novel framework that performs sentiment analysis on audio files by calculating their single-valued neutrosophic sets (SVNS) and clustering them into positive–neutral–negative, combining the results with those obtained by performing sentiment analysis on the text files of these audio files.

In another selected paper (Novel MCDM Method Based on Plithogenic Hypersoft Sets under Fuzzy Neutrosophic Environment), Muhammad Rayees Ahmad, Muhammad Saeed, Usman Afzal and Miin-Shen Yang advance the study of plithogenic hypersoft sets (PHSS), by investigating four classifications of PHSS that are based on the number of attributes chosen for application and the nature of alternatives or that of attribute value degree of appurtenance. The proposed classifications cover most of the fuzzy and neutrosophic cases with possible neutrosophic applications in symmetry. As an extension of the technique for order preference by similarity to an ideal solution (TOPSIS), the paper also suggests a novel multi-criteria decision making (MCDM) method that is based on PHSS. The proposed PHSS-based TOPSIS method can be employed to solve real MCDM problems precisely modeled by the concept of PHSS. As an application, a parking spot choice problem is solved by the proposed PHSS-based TOPSIS under a fuzzy neutrosophic environment and it is validated by considering two different sets of alternatives along with a comparison with fuzzy TOPSIS in each case, and the results prove that the method is able to be extended to analyze time series and in develop algorithms for graph theory, machine learning, pattern recognition, and artificial intelligence.

Dongsik Jo, S. Saleh, Jeong-Gon Lee, Kul Hur and Chen Xueyou introduce the new notion of interval-valued neutrosophic crisp sets, providing a tool for approximating undefinable or complex concepts in the real world in the paper Structures via Interval-Valued Neutrosophic Crisp Sets. The authors also propose an interval-valued neutrosophic crisp (vanishing) point and obtain some of its properties, and then define an interval-valued neutrosophic crisp topology, base (subbase), neighborhood, and interior (closure), respectively, and investigate each property, and give some examples. Finally, they define an interval-valued neutrosophic crisp continuity and quotient topology and study each property.

Fahad Alsharari aims in the paper £-Single Valued Extremally Disconnected Ideal Neutrosophic Topological Spacesto mark out new concepts of r-single valued neutrosophic sets, called r-single-valued neutrosophic £-closed and £-open sets. The definition of £-single-valued neutrosophic irresolute mapping is provided, discussing its properties, and then the concepts of £-single-valued neutrosophic extremally disconnected and £-single-valued neutrosophic normal spaces are established. A useful implication diagram between the r-single-valued neutrosophic ideal open sets is obtained.

Finally, in the last paper (Study on Some Properties of Neutrosophic Multi Topological Group) of this Special Issue, the researchers Bhimraj Basumatary, NWary, Dimacha Dwibrang Mwchahary, Ashoke Kumar Brahma, Jwngsar Moshahary, Usha Rani Basumatary and Jili Basumatary study some properties of the neutrosophic multitopological group, by introducing the definitions of a semi-open neutrosophic multiset, semi-closed neutrosophic multiset, neutrosophic multi regularly open set, neutrosophic multi regularly closed set, neutrosophic multi continuous mapping, and then investigating some of their properties. Since the concept of the almost topological group is new, the authors also provide the definition of the neutrosophic multi almost topological group, and for the purpose of symmetry, they use the definition of neutrosophic multi almost continuous mapping to define a neutrosophic multi almost topological group and examine its properties.

The fields of neutrosophic probability and neutrosophic statistics, neutrosophic algebraic structures, neutrosophic optimization, and neutrosophic applications in symmetry are receiving more and more attention, and their importance is hopefully proved by the papers selected for *Symmetry*'s Special Issue Development of Neutrosophic Probability, Neutrosophic Statistics, Neutrosophic Algebraic Structures, and Neutrosophic & Plithogenic Optimizations.

Florentin Smarandache, Yanhui Guo
Editors

symmetry

MDPI

Article

New Results on Neutrosophic Extended Triplet Groups Equipped with a Partial Order

Xin Zhou [1,*], **Ping Li** [1], **Florentin Smarandache** [2] and **Ahmed Mostafa Khalil** [3,4]

[1] School of Science, Xi'an Polytechnic University, Xi'an 710048, China; liping7487@163.com
[2] Departmet of Mathematics, University of New Mexico, 705 Gurley Ave., Gallup, NM 87301, USA; smarand@unm.edu
[3] College of Mathematics and Information Science, Shaanxi Normal University, Xi'an 710062, China; a.khalil@azhar.edu.eg
[4] Department of Mathematics, Faculty of Science, Al-Azhar University, Assiut 71524, Egypt
* Correspondence: sxxzx1986@163.com; Tel.:+861-599-170-3526

Received: 13 November 2019; Accepted: 10 December 2019; Published: 13 December 2019

check for updates

Abstract: Neutrosophic extended triplet group (NETG) is a novel algebra structure and it is different from the classical group. The major concern of this paper is to present the concept of a partially ordered neutrosophic extended triplet group (po-NETG), which is a NETG equipped with a partial order that relates to its multiplicative operation, and consider properties and structure features of po-NETGs. Firstly, in a po-NETG, we propose the concepts of the positive cone and negative cone, and investigate the structure features of them. Secondly, we study the specificity of the positive cone in a partially ordered weak commutative neutrosophic extended triplet group (po-WCNETG). Finally, we introduce the concept of a po-NETG homomorphism between two po-NETGs, construct a po-NETG on a quotient set by providing a multiplication and a partial order, then we discuss some fundamental properties of them.

Keywords: partially ordered neutrosophic extended triplet group; positive cone; homomorphism; quotient set

1. Introduction

Groups play a very important role in algebraic structures [1–3], and have been applied in many other areas such as chemistry, physics, biology, etc. The concept of neutrosophic set theory is proposed by Smarandache in [4], which is the generalization of classical sets [5], fuzzy sets [6], and intuitionistic fuzzy sets [5,7]. Neutrosophic sets have received wide attention both on practical applications [8–10] and on theory as well [11,12]. The main idea of the concept of a neutrosophic triplet group (NTG), is defined in [13,14]. For an NTG $(G,*)$, every element a in G has its own neutral element (denoted by $neut(a)$) satisfying $a*neut(a) = neut(a)*a = a$, and there exists at least one opposite element (denoted by $anti(a)$) in G relative to $neut(a)$ satisfying $a*anti(a) = anti(a)*a = neut(a)$. Here, $neut(a)$ is not allowed to be equal to the classical identity element as a special case. By removing this restriction, the concept of neutrosophic extended triplet group (NETG), is presented in [13]. Many significant results and several studies on NTGs and NETGs can be found in [15–20]. On the other hand, some algebraic structures are equipped with a partial order that relates to the algebraic operations, such as ordered groups, ordered semigroups, ordered rings and so on [21–28].

Regarding these developments, as the motivation of this article, we will consider what it is like to endow a NETG with a partial order and introduce the concepts of partially ordered NETGs and positive cones. Then we consider a question: is a subset P of a NETG G the positive cone relative to some compatible order on G if P satisfies some conditions? To solve this problem, we investigate structure features of partially ordered NETGs and try to characterize the positive cones. Finally, we study properties of homomorphisms and quotient sets in partially ordered NETGs, and discuss the relationships between homomorphisms and congruences. In particular, the quotient set equipped with a special multiplication and a partial order provides a way to obtain a partially ordered NETG. All these results lay the groundwork for investigation of category properties of partially ordered NETGs.

The rest of this paper is organized as follows. In Section 2, we review some basic concepts, such as a neutrosophic extended triplet set, a neutrosophic extended triplet group, a weak commutative neutrosophic extended triplet group and a completely regular semigroup, and several results were published in [16,19]. In Section 3, we define a partially ordered neutrosophic extended triplet group and partially ordered weak commutative neutrosophic extended triplet group. Several of their interesting properties of partially ordered neutrosophic extended triplet group and partially weak commutative neutrosophic extended triplet group are explained. The homomorphisms and quotient sets of partially ordered neutrosophic extended triplet group are shown in Section 4. Finally, conclusions are given in Section 5.

2. Preliminaries

In this section, we recall some basic notions and results which will be used in this paper as indicated below.

Definition 1. *([13]) Let G be a non-empty set together with a binary operation* $*$. *Then G is called a neutrosophic extended triplet set if for any* $a \in G$, *there exist a neutral of "a" (denoted by* $neut(a)$*) and an opposite of "a" (denoted by* $anti(a)$*), such that* $neut(a) \in G$, $anti(a) \in G$, *and*

$$a * neut(a) = neut(a) * a = a;$$

$$a * anti(a) = anti(a) * a = neut(a).$$

The triplet $(a,\ neut(a),\ anti(a))$ *is called a neutrosophic extended triplet.*

Definition 2. *([13]) Let* $(G, *)$ *be a neutrosophic extended triplet set. If* $(G, *)$ *is a semigroup, then G is called a neutrosophic extended triplet group (for short, NETG).*

Proposition 1. *([[16] Theorems 1 and 2]) Let* $(G, *)$ *be a NETG. The following properties hold:* $\forall a \in G$

(1) $neut(a)$ *is unique;*
(2) $neut(a) * neut(a) = neut(a)$;
(3) $neut(neut(a)) = neut(a)$.

Notice that $anti(a)$ *may be not unique for every element a in a NETG* $(G, *)$. *To avoid confusion, we use the following notations:*

$anti(a)$ *denotes any certain one opposite of a and* $\{anti(a)\}$ *denotes the set of all opposites of a.*

Proposition 2. *([[19], Theorem 1]) Let* $(G, *)$ *be a NETG. The following properties hold:* $\forall\ a \in G$, $\forall\ p,\ q \in \{anti(a)\}$

(1) $p * neut(a) \in \{anti(a)\}$;
(2) $p * neut(a) = q * neut(a) = neut(a) * q$;
(3) $neut(p * neut(a)) = neut(a)$;
(4) $a \in \{anti(p * neut(a))\}$;
(5) $anti(p * neut(a)) * neut(p * neut(a)) = a$.

Definition 3. *([16]) Let $(G, *)$ be a NETG. If $a * neut(b) = neut(b) * a$ $(\forall a \in G, \forall b \in G)$, then G is called a weak commutative neutrosophic extended triplet group (WCNETG).*

Proposition 3. *([[16], Theorem 2]) Let $(G, *)$ be a NETG. Then G is a WCNETG iff G satisfies the following conditions: $\forall a \in G$, $\forall b \in G$*

(1) $neut(a) * neut(b) = neut(b) * neut(a)$;
(2) $neut(a) * neut(b) * a = a * neut(b)$.

Proposition 4. *([[16], Theorem 3]) Let $(G, *)$ be a WCNETG. The following properties hold: $\forall a \in G$, $\forall b \in G$*

(1) $neut(a) * neut(b) = neut(b * a)$;
(2) $anti(a) * anti(b) \in \{anti(b * a)\}$.

Definition 4. *([29]) A semigroup $(S, *)$ will be called completely regular if there exists a unary operation $a \mapsto a^{-1}$ on S with the properties:*

$$(a^{-1})^{-1} = a, \; a * a^{-1} * a = a, \; a * a^{-1} = a^{-1} * a.$$

Proposition 5. *([[19], Theorem 2]) Let $(G, *)$ be a groupoid. Then G is a NETG iff it is a completely regular semigroup.*

Note 1. In semigroup theory, a^{-1} is called the inverse element of a and it is unique. However, in a NETG, $anti(a)$ is called an opposite element of a and it may not be unique. From Proposition 5, we get that for arbitrary element a of a NETG $(G, *)$, if we define a unary operation $a \mapsto a^{-1}$ by $a^{-1} = anti(a) * neut(a)$, then $(G, *)$ is a completely regular semigroup.

In the following, we will regard all NETGs as completely regular semigroups, in which $a^{-1} = anti(a) * neut(a)$ for arbitrary element a. Then by Proposition 2, we have in a NETG $(G, *)$, for each $a \in G$, $a^{-1} \in \{anti(a)\}$ and $a^{-1} * a = a * a^{-1} = neut(a)$.

3. Partially Ordered NETGs

An NETG is a special set endowed with a multiplicative operation. Assuming that we introduce a partial order which is compatible with multiplication in a NETG, we will get the definition of partially ordered NETGs as indicated below.

Definition 5. *Let $(G, *)$ be a NETG. If there exists a partial order relation $\leq$ on G such that $a \leq b$ implying $c * a \leq c * b$ and $a * c \leq b * c$ for all $a \in G$, $b \in G$, $c \in G$, then $\leq$ is called a compatible partial order on G, and $(G, *, \leq)$ is called a partially ordered NETG (for short, po-NETG).*

Similarly, if $(G, *)$ is a WCNETG and endowed with a compatible partial order, then $(G, *, \leq)$ is called a partially ordered WCNETG (po-WCNETG). Hence, po-WCNETGs must be po-NETGs.

Remark 1. *Obviously, the properties of NETGs and WCNETGs are holding in po-NETGs and po-WCNETGs, respectively.*

In the following, we give an example of a po-NETG.

Example 1. *Let $G = \{0, a, b, c, 1\}$ with the Hasse diagram as shown in Figure 1, in which 0 denotes the bottom element (mean the element is smallest element w.r.t. to partial order) and 1 denotes the top element (mean the element is largest element w.r.t. to partial order) of G. Then G is a partially ordered set.*

Define multiplication $*$ on G as shown in Table 1 , where a, b, c to label the elements in the po-NETG and the multiplication $*$ among these elements.

Table 1. Multiplication $*$ on G.

*	0	a	b	c	1
0	0	0	0	0	0
a	0	b	c	a	1
b	0	c	a	b	1
c	0	a	b	c	1
1	0	1	1	1	1

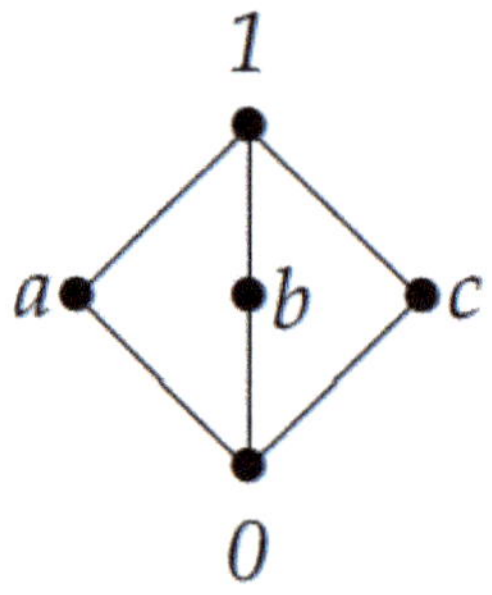

Figure 1. Hasse diagram.

We can verify that $(G, *)$ is a WCNETG. Moreover,

$$neut(0) = 0,\ \{anti(0)\} = \{0,\ a,\ b,\ c,\ 1\},\ 0^{-1} = 0;$$

$$neut(a) = c,\ \{anti(a)\} = \{b\},\ a^{-1} = b;$$

$$neut(b) = c,\ \{anti(b)\} = \{a\},\ b^{-1} = a;$$

$$neut(c) = c,\ \{anti(c)\} = \{c\},\ c^{-1} = c;$$

$$neut(1) = 1,\ \{anti(1)\} = \{a,\ b,\ c,\ 1\},\ 1^{-1} = 1.$$

It is easy to see that the partial order shown in Fig.1 is compatible with multiplication $*$. Hence, $(G, *, \leq)$ is a po-WCNETG.

Definition 6. *If $(G, *, \leq)$ is a po-NETG, then $a \in G$ is said to be a positive element if $neut(a) \leq a$; and a negative element if $a \leq neut(a)$. The subset P_G of all positive elements of G is called the positive cone of G, and the subset N_G of all negative elements the negative cone.*

Remark 2. *By Proposition 1, $\forall\ a \in G,\ neut(a) \in P_G \cap N_G$, so $P_G \cap N_G \neq \emptyset$.*

Lemma 1. *Let $(G, *)$ be an NETG. Then $\forall\ a \in G$,*

$$[neut(a)]^{-1} = neut(a) = neut(a^{-1}).$$

Proof. Let $a \in G$. Then

$$\begin{aligned}[neut(a)]^{-1} &= anti(neut(a)) * neut(neut(a)) \\ &= anti(neut(a)) * neut(a) \\ &= neut(neut(a)) \\ &= neut(a).\end{aligned}$$

On the other hand, by Proposition 2(3), we have $neut(a^{-1}) = neut(anti(a) * neut(a)) = neut(a)$. □

Remark 3. *If G is a po-NETG and $P \subseteq G$, we shall use the notation*

$$P^{-1} = \{a^{-1} : a \in P\}.$$

Proposition 6. *Let $(G, *, \leq)$ be a po-NETG. Then $P_G \cap P_G^{-1} = \{a \in G : a = neut(a) = a^{-1}\}$.*

Proof. ($\Longrightarrow$) Let $a \in G$. By Proposition 1 and Lemma 1, we have

$$neut(a) \in \{a \in G : a = neut(a) = a^{-1}\},$$

so $\{a \in G : a = neut(a) = a^{-1}\} \neq \emptyset$. By Lemma 1, it is clear that

$$\{a \in G : a = neut(a) = a^{-1}\} \subseteq P_G \cap P_G^{-1}.$$

($\Longleftarrow$) Let $b \in P_G \cap P_G^{-1}$, then $neut(b) \leq b$ and $\exists\, c \in P_G$ such that $b = c^{-1}$, so

$$b = c^{-1} = anti(c) * neut(c) \leq anti(c) * c = neut(c) = neut(b^{-1}) = neut(b),$$

that is, $b \leq neut(b)$, whence $b = neut(b)$. Hence,

$$c = b^{-1} = [neut(b)]^{-1} = neut(b) = b.$$

Then we can conclude that $b \in \{a \in G : a = neut(a) = a^{-1}\}$, and so

$$P_G \cap P_G^{-1} \subseteq \{a \in G : a = neut(a) = a^{-1}\}.$$

Thus, $P_G \cap P_G^{-1} = \{a \in G : a = neut(a) = a^{-1}\}$. □

Remark 4. *If $(G, *, \leq)$ is a po-NETG and $P \subseteq G$, then we shall use the notation*

$$P^2 = \{a * b : a,\, b \in P\}.$$

Proposition 7. *(1) If $(G, *, \leq)$ is a po-NETG, then $P_G \subseteq P_G^2$.*
(2) If $(G, *, \leq)$ is a po-WCNETG, then $P_G = P_G^2$.

Proof. (1) If $(G, *, \leq)$ is a po-NETG, then $\forall\, a \in P_G$, by $neut(a) \in P_G$, we have $a = a * neut(a) \in P_G^2$, and so $P_G \subseteq P_G^2$.

(2) If $(G, *, \leq)$ is a po-WCNETG, then $\forall\, a \in P_G$, $\forall b \in P_G$, by Propositions 3 and 4, we have $neut(a * b) = neut(b) * neut(a) = neut(a) * neut(b) \leq a * b$, and so $a * b \in P_G$, thus $P_G^2 \subseteq P_G$. Consequently, $P_G = P_G^2$. □

Proposition 8. *Let $(G, *, \leq)$ be a po-WCNETG. Then $\forall\, a \in G$, $aP_Ga^{-1} \subseteq P_G$.*

Proof. Let $a \in G$ and $b \in P_G$, then by Propositions 3 and 4, we have $neut(a * b * a^{-1}) = neut(a^{-1}) * neut(a * b) = neut(a * b) * neut(a^{-1}) = [neut(b) * neut(a)] * neut(a^{-1}) = neut(b) * [neut(a) * neut(a^{-1})] = neut(b) * neut(a^{-1} * a) = neut(b) * neut(neut(a)) = neut(b) * neut(a) = neut(b) * (a * a^{-1}) = [neut(b) * a] * a^{-1} = [a * neut(b)] * a^{-1} \leq a * b * a^{-1}$, thus $aba^{-1} \in P_G$. Therefore, $aP_Ga^{-1} \subseteq P_G$. □

Lemma 2. *Let $(G, *)$ be a WCNETG. Then $\forall\, a \in G$, $\forall b \in G$, $(a * b)^{-1} = b^{-1} * a^{-1}$.*

Proof. We know $a * b$ is an element of $G\ \forall\ a \in G$, $\forall b \in G$ and by Proposition 4, we have $anti(b) * anti(a) \in \{anti(a * b)\}$. Then using Propositions 1, 5 and Note 1 we get the following identities:

$$\begin{aligned} b^{-1} * a^{-1} &= [anti(b) * neut(b)] * [anti(a) * neut(a)] \\ &= anti(b) * [neut(b) * anti(a)] * neut(a) && \text{(Because the multiplication $*$ is associative)} \\ &= anti(b) * [anti(a) * neut(b)] * neut(a) && \text{(Because G is a WCNETG)} \\ &= [anti(b) * anti(a)] * [neut(b) * neut(a)] && \text{(Because the multiplication $*$ is associative)} \\ &= [anti(b) * anti(a)] * neut(a * b) && \text{(By Proposition 3)} \\ &= (a * b)^{-1}. \ \square \end{aligned}$$

Lemma 3. *Let $(G, *, \leq)$ be a po-NETG. Then $P_G = P_N^{-1}$ and $P_G^{-1} = P_N$.*

Proof. Let $a \in G$. If $a \in P_G$, then $neut(a) \leq a$, it follows by Lemma 1 that $a^{-1} = neut(a^{-1}) * a^{-1} = neut(a) * a^{-1} \leq a * a^{-1} = neut(a) = neut(a^{-1})$, and so $a^{-1} \in P_N$, whence $a = (a^{-1})^{-1} \in P_N^{-1}$. Hence, $P_G \subseteq P_N^{-1}$. Similarly, we can prove that if $a \in P_N$ then $a^{-1} \in P_G$, so $P_N^{-1} \subseteq P_G$. Consequently, $P_G = P_N^{-1}$. Similarly, $P_G^{-1} = P_N$. $\square$

Definition 7. *Let $(G, *)$ be a WCNETG. If $\forall a \in G$, $\forall b \in G$, $\forall c \in G$, $a * neut(c) = b * neut(c)$ implies $a = b$, then we say G satisfies neutrosophic cancellation law.*

Lemma 4. *Let $(G, *)$ be a WCNETG satisfying neutrosophic cancellation law and $P \subseteq G$ satisfy $\forall\ a \in P$, $a * a = a$. Then $\forall\ a \in G$, $\forall\ b \in G$, $a * neut(b) \in P$ implies $neut(a) = a = a^{-1}$.*

Proof. If $a * neut(b) \in P$, then $a * neut(b) = (a * neut(b)) * (a * neut(b)) = (a * a) * neut(b)$, and so $a * a = a$, whence $neut(a) = a\ \forall\ a \in G$, $\forall\ b \in G$. Then by Lemma 1, we get $a^{-1} = [neut(a)]^{-1} = neut(a) = a$. $\square$

Proposition 9. *Let $(G, *)$ be a WCNETG satisfying neutrosophic cancellation law and $P \subseteq G$ satisfy the following conditions:*

(1) $P^2 \subseteq P$;
(2) $P \cap P^{-1} = \{a \in G : neut(a) = a = a^{-1}\}$;
(3) $\forall\ a \in P$, $a * a = a$;
(4) $\forall\ a \in G$, $aPa^{-1} \subseteq P$,

then a compatible partial order on G exists such that P is the positive cone of G relative to it. Moreover, G is a chain with respect to this partial order if and only if $P \cup P^{-1} = G$.

Proof. Define the relation $\leq$ on G by

$$a \leq b \Leftrightarrow b * a^{-1} \in P.$$

By Proposition 1 and Lemma 1, we have $\forall\ a \in G$, $neut(a) \in P \cap P^{-1} \subseteq P$, and so $\leq$ is reflexive on G obviously.
If now $a \leq b$ and $b \leq a$, then $b * a^{-1} \in P$ and $a * b^{-1} \in P$. Since by Lemma 2 we know that

$$(a * b^{-1})^{-1} = (b^{-1})^{-1} * a^{-1} = b * a^{-1},$$

we conclude

$$b * a^{-1} \in P \cap P^{-1}.$$

It follows by (2) that $b * a^{-1} = neut(b * a^{-1})$. However, by Proposition 4 and Lemma 1,

$$neut(b * a^{-1}) = neut(a^{-1}) * neut(b) = neut(a) * neut(b),$$

thus

$b * neut(a) = b * a^{-1} * a = neut(b * a^{-1}) * a = [neut(a) * neut(b)] * a = neut(a) * [a * neut(b)] = [neut(a) * a] * neut(b) = a * neut(b)$, that is, $b * neut(a) = a * neut(b)$.

However, by Proposition 3, we have

$$b * neut(a) = neut(b) * neut(a) * b = neut(a * b) * b,$$

and similarly,

$$a * neut(b) = neut(a) * neut(b) * a = [neut(b) * neut(a)] * a = neut(a * b) * a,$$

therefore,

$$neut(a * b) * b = neut(a * b) * a,$$

and by neutrosophic cancellation law, consequently $a = b$. Hence, $\leq$ is anti-symmetric.

To prove that $\leq$ is transitive, let $a \leq b$ and $b \leq c$. Then

$$b * a^{-1} \in P \text{ and } c * b^{-1} \in P.$$

It follows by (1) that

$$P \supseteq P^2 \ni (c * b^{-1}) * (b * a^{-1}) = c * (b^{-1} * b) * a^{-1} = c * neut(b) * a^{-1} = (c * a^{-1}) * neut(b).$$

By (3) and Lemma 4, we have

$$neut(c * a^{-1}) = c * a^{-1} = (c * a^{-1})^{-1},$$

and so

$$c * a^{-1} \in P \cap P^{-1} \subseteq P,$$

that is, $c * a^{-1} \in P$. Thus, $a \leq c$. Therefore, $\leq$ is a partial order on G.

To see that it is compatible, let $x \leq y$. Then $y * x^{-1} \in P$ and it follows by (1) and (4) that, for every $a \in G$,

$$(a * y) * (a * x)^{-1} = (a * y) * (x^{-1} * a^{-1}) = a * (y * x^{-1}) * a^{-1} \in P,$$

$$(y * a) * (x * a)^{-1} = y * (a * a^{-1}) * x^{-1} = y * neut(a) * x^{-1} = (y * x^{-1}) * neut(a) \in P^2 \subseteq P,$$

which shows that

$$a * x \leq a * y \text{ and } x * a \leq y * a.$$

It follows that $\leq$ is compatible.

Finally, note that $\forall\, a \in G$,

$$neut(a) \leq a \Leftrightarrow a * [neut(a)]^{-1} \in P \Leftrightarrow a * neut(a) \in P \Leftrightarrow a \in P,$$

so P is the associated positive cone. Suppose now that $(G, \leq)$ is a chain, then for every $a \in G$, we have either

$$neut(a) \leq a \text{ or } a \leq neut(a).$$

It follows by Lemma 3 that

$$a \in P \text{ or } a \in P^{-1}.$$

Thus $G = P \cup P^{-1}$. Conversely, if $G = P \cup P^{-1}$, then for all $a,\ b \in G$, we have

$$a * b^{-1} \in P \text{ or } a * b^{-1} \in P^{-1},$$

that is,

$$a * b^{-1} \in P \text{ or } b * a^{-1} = (a * b^{-1})^{-1} \in P.$$

Hence, we have either $b \leq a$ or $a \leq b$. Therefore, $(G, \leq)$ is a chain. □

By the following example, we clarify the above proposition as:

Example 2. *Let $G = \{a,\ b,\ c\}$. Define multiplication $*$ on G as shown in Table 2, where a, b, c to label the elements in the po-NETG and the multiplication $*$ among these elements.*

Table 2. Multiplication $*$ on G.

*	a	b	c
a	a	b	c
b	b	c	a
c	c	a	b

It is easy to verify that $(G, *)$ is a WCNETG and $(G, *)$ satisfies neutrosophic cancellation law, in which

$$neut(a) = neut(b) = neut(c) = a,$$

$$\{anti(a)\} = \{a\},\ a^{-1} = a;$$

$$\{anti(b)\} = \{c\},\ b^{-1} = c;$$

$$\{anti(c)\} = \{b\},\ c^{-1} = b.$$

Let $P = \{a\}$, then P satisfies all conditions mentioned in Proposition 9. Define the relation $\leq$ on G by $x \leq y \Leftrightarrow y * x^{-1} \in P$, then $\leq$ is a partial order on G and $(G, \leq)$ is a antichain. Obviously, P is the positive cone of G with respect to this partial order $\leq$.

Proposition 10. *Let $(G, *)$ be a po-WCNETG. Then $\forall\ x \in G$, $\forall y \in G$, $x \leq y$ implies $y * x^{-1} \in P_G$.*

Proof. Let $\forall\ x \in G$, $\forall y \in G$. If $x \leq y$, then $neut(x) = x * x^{-1} \leq y * x^{-1}$, hence, by Proposition 4 and Lemma 1, we have $neut(y * x^{-1}) = neut(x^{-1}) * neut(y) = neut(x) * neut(y) \leq (y * x^{-1}) * neut(y) = neut(y) * (y * x^{-1}) = (neut(y) * y) * x^{-1} = y * x^{-1}$. Thus, $y * x^{-1} \in P_G$. □

4. Homomorphisms and Quotient Sets of po-NETGs

Definition 8. *Let $(G, *, \leq_1)$ and $(T, \cdot, \leq_2)$ be two po-NETGs. The map $f : G \to T$ is called a po-NETG homomorphism of po-NETGs, if f satisfies: $\forall a \in G$, $\forall b \in G$*

(1) $f(a * b) = f(a) \cdot f(b)$;
(2) $a \leq_1 b$ *implies* $f(a) \leq_2 f(b)$.

Proposition 11. *Let $(G, *, \leq_1)$ and $(T, \cdot, \leq_2)$ be two po-NETGs, and let $f : G \to T$ be a po-NETG homomorphism of po-NETGs. The following properties hold:*

(1) $\forall a \in G, f(neut(a)) = neut(f(a))$;
(2) $\forall a \in G, \{f(b) : b \in \{anti(a)\}\} \subseteq \{anti(f(a))\}$, *and if f is bijective, then* $\{f(b) : b \in \{anti(a)\}\} = \{anti(f(a))\}$;
(3) $\forall a \in G, [f(a)]^{-1} = f(a^{-1})$;
(4) $\forall a \in P_G, f(a) \in P_T$;
(5) $\forall a \in N_G, f(a) \in N_T$.

Proof.

(1) $\forall\, a \in G,\ \forall\, b \in \{anti(a)\}$, since

$$f(a) \cdot f(neut(a)) = f(a * neut(a)) = f(a) = f(neut(a) * a) = f(neut(a)) \cdot f(a),$$

$$f(a) \cdot f(b) = f(a * b) = f(neut(a)) = f(b * a) = f(b) \cdot f(a),$$

then we obtain $f(neut(a)) = neut(f(a))$.

(2) From the proof of (1), we can get that

$$\forall\, a \in G,\ \forall\, b \in \{anti(a)\},\ f(b) \in \{anti(f(a))\},$$

and so

$$\{f(b) : b \in \{anti(a)\}\} \subseteq \{anti(f(a))\}.$$

If f is bijective, then $\forall\, d \in \{anti(f(a))\},\ \exists\, c \in G$ such that $f(c) = d$. Since

$$f(c * a) = f(c) \cdot f(a) = d \cdot f(a) = neut(f(a)) = f(neut(a)),$$

we have $c * a = neut(a)$. Similarly, we can get $a * c = neut(a)$. Thus, $c \in anti(a)$ and so

$$d = f(c) \in \{f(b) : b \in \{anti(a)\}\}.$$

By the arbitrariness of d, we have

$$\{anti(f(a))\} \subseteq \{f(b) : b \in \{anti(a)\}\}.$$

Then,

$$\{f(b) : b \in \{anti(a)\}\} = \{anti(f(a))\}.$$

(3) Let $a \in G$ and $b \in \{anti(a)\}$. By (2), $f(b) \in \{anti(f(a))\}$. Then by (1), we have

$$[f(a)]^{-1} = anti(f(a)) \cdot neut(f(a)) = f(b) \cdot f(neut(a)) = f(b * neut(a)) = f(a^{-1}).$$

(4) Since $\forall\, a \in P_G,\ neut(a) \leq_1 a$, we have $neut(f(a)) = f(neut(a)) \leq_2 f(a)$, and so $f(a) \in P_T$.

(5) It is similar to (4).

□

Definition 9. *Let $(G, *, \leq)$ be a po-NETG and θ be an equivalence relation on G. If θ satisfies*

$$\forall a \in G,\ \forall b \in G,\ \forall c \in G,\ \forall d \in G,\ (a, b) \in \theta\ \&\ (c, d) \in \theta \Rightarrow (a * c, b * d) \in \theta,$$

then θ is called a congruence on G.

Obviously, $\theta_1 = \{(a, a) : a \in G\}$ and $\theta_2 = \{(a, b) : \forall\, a,\ b \in G\}$ are both congruences on G, and they are called identity congruence on G and pure congruence on G, respectively.

Definition 10. *Let $(G, *, \leq)$ be a po-NETG and θ be a congruence on G. A multiplication $\circ$ on the quotient set $G/\theta = \{[a]_\theta : a \in G\}$ is defined by*

$$[a]_\theta \circ [b]_\theta = [a * b]_\theta.$$

Proposition 12. *Let a relation* $\preceq$ *on* $(G/\theta, \circ)$ *be defined by*

$$\forall\ [a]_\theta \in G/\theta,\ \forall [b]_\theta \in G/\theta,\ [a]_\theta \preceq [b]_\theta \Leftrightarrow a \leq b.$$

Then, $(G/\theta, \circ, \preceq)$ *is a po-NETG.*

Proof. We can verify that $\circ$ is associative. Let $[a]_\theta \in G/\theta$ (see Definition 10), since

$$[neut(a)]_\theta \circ [a]_\theta = [neut(a) * a]_\theta = [a]_\theta = [a * neut(a)]_\theta = [a]_\theta \circ [neut(a)]_\theta,$$

and

$$[anti(a)]_\theta \circ [a]_\theta = [anti(a) * a]_\theta = [neut(a)]_\theta = [a * anti(a)]_\theta = [a]_\theta \circ [anti(a)]_\theta,$$

we conclude that $(G/\theta, \circ)$ is a NETG, in which $\forall\ [a]_\theta \in G/\theta,\ neut([a]_\theta) = [neut(a)]_\theta$ and $[anti(a)]_\theta \in \{anti([a]_\theta)\}$. Then it is easy to see that $\preceq$ is a partial order on $(G/\theta, \circ)$. Moreover, $\forall\ [a]_\theta \in G/\theta,\ \forall [b]_\theta \in G/\theta,\ \forall [c]_\theta \in G/\theta$, if $[a]_\theta \preceq [b]_\theta$, then $a \leq b$, so we have $a * c \leq b * c$, and $c * a \leq c * b$. Thus,

$$[a]_\theta \circ [c]_\theta = [a * c]_\theta \preceq [b * c]_\theta = [b]_\theta \circ [c]_\theta$$

and

$$[c]_\theta \circ [a]_\theta = [c * a]_\theta \preceq [c * b]_\theta = [c]_\theta \circ [b]_\theta.$$

Thus, $(G/\theta, \circ, \preceq)$ is a po-NETG. □

In the following, we give an example to illustrate Proposition 12.

Example 3. *Consider the po-NETG* $(G, *, \leq)$ *is given in Example 1. Now we define a relation* θ *on G by*

$$\theta = \{(0,0),\ (a,a),\ (b,b),\ (c,c),\ (1,1),\ (a,b),\ (b,a),\ (a,c),\ (c,a),\ (b,c),\ (c,b)\}.$$

Then we can verify that θ *is a congruence on G with the following blocks:*

$$[0]_\theta = \{0\},\ [a]_\theta = \{a,\ b,\ c\},\ [1]_\theta = \{1\}.$$

So the quotient set $G/\theta = \{[0]_\theta,\ [a]_\theta,\ [1]_\theta\}$. *By Proposition 12, we know* $(G/\theta, \circ, \preceq)$ *is a po-NETG, in which* $neut([0]_\theta) = [0]_\theta$, $neut([a]_\theta) = [c]_\theta = [a]_\theta$, $neut([1]_\theta) = [1]_\theta$, $\{anti([0]_\theta)\} = \{[0]_\theta,\ [a]_\theta,\ [1]_\theta\}$, $\{anti([a]_\theta)\} = \{[a]_\theta\}$, $\{anti([1]_\theta)\} = \{[a]_\theta,\ [1]_\theta\}$, *and then* G/θ *is a chain, because* $[0]_\theta \preceq [a]_\theta \preceq [1]_\theta$.

Proposition 13. *Let* $(G, *, \leq)$ *be a po-NETG and* θ *be a congruence on G. Then the natural mapping* $\natural_\theta : (G, *, \leq) \to (G/_\theta, \circ, \preceq)$ *given by* $\natural_\theta(a) = [a]_\theta$ *is a po-NETG homomorphism of po-NETGs.*

Proof. As $\natural_\theta(a * b) = [a * b]_\theta = [a]_\theta \circ [b]_\theta = \natural_\theta(a) \circ \natural_\theta(b)\ \forall\ a \in G,\ \forall b \in G$. If $a \leq b$, then $[a]_\theta \preceq [b]_\theta$ which implies $\natural_\theta(a) \preceq \natural_\theta(b)$. Thus, the natural mapping $\natural_\theta : (G, *, \leq) \to (G/_\theta, \circ, \preceq)$ is a po-NETG homomorphism of po-NETGs. □

Next, we give an example to explain Proposition 13.

Example 4. *From Example 3, we consider the natural mapping* $\natural_\theta : (G, *, \leq) \to (G/_\theta, \circ, \preceq)$. *Thus,* $\natural_\theta(0) = [0]_\theta$, $\natural_\theta(a) = \natural_\theta(b) = \natural_\theta(c) = [a]_\theta$, $\natural_\theta(1) = [1]_\theta$. *It is easy to verify that* $\natural_\theta$ *is a po-NETG homomorphism of po-NETGs.*

Proposition 14. *Let* $(G, *, \leq_1)$ *and* $(T, \cdot, \leq_2)$*be two po-NETGs and* $f : (G, *, \leq_1) \to (T, \cdot, \leq_2)$ *be a po-NETG homomorphism of po-NETGs. We shall use the notation*

$$Kerf = \{(a, b) \in G \times G : f(a) = f(b)\},$$

then we can get the following properties:

(1) $Kerf$ *is a congruence on* G;
(2) f *is a injective po-NETG homomorphism of po-NETGs if and only if* $kerf$ *is an identity congruence on* G;
(3) *There exists an injective po-NETG homomorphism of po-NETGs* $g : (G/Kerf, \circ, \preceq) \to (T, \cdot, \leq_2)$ *such that* $f = g \circ \natural_{Kerf}$.

Proof.

(1) Obviously, $Kerf$ is an equivalence relation on G. Let $\forall\, a \in G, \forall\, b \in G, \forall\, c \in G, \forall\, d \in G$, if $(a, b) \in Kerf$ and $(c, d) \in Kerf$, then $f(a) = f(b)$ and $f(c) = f(d)$. Since f is a po-NETG homomorphism of po-NETGs, we have $f(a * c) = f(a) \cdot f(c) = f(b) \cdot f(d) = f(b * d)$, and so $(a * c, b * d) \in Kerf$. Thus, $Kerf$ is a congruence on G.
(2) If f is an injective po-NETG homomorphism of po-NETGs and if $(a, b) \in kerf$ then $f(a) = f(b)$. Therefore, we get $a = b$. Hence, by the arbitrariness of (a, b), we obtain $kerf$ is an identity congruence on G.

   Conversely, suppose that $kerf$ is an identity congruence on G. $\forall\, a \in G, \forall\, b \in G$, if $f(a) = f(b)$, then $(a, b) \in kerf$, so $a = b$. Therefore, f is an injective po-NETG homomorphism of po-NETGs.
(3) We define a map $g : G/Kerf \to T$ by $\forall\, [a]_{Kerf} \in G/Kerf$, $g([a]_{Kerf}) = f(a)$, then g is injective. $\forall\, [a]_{Kerf}, [b]_{Kerf} \in G/Kerf$, we have $g([a]_{Kerf} \circ [b]_{Kerf}) = g([a * b]_{Kerf}) = f(a * b) = f(a) \cdot f(b) = g([a]_{Kerf}) \cdot g([b]_{Kerf})$, and if $[a]_{Kerf} \preceq [b]_{Kerf}$, then $a \leq_1 b$, thus, $f(a) \leq_2 f(b)$, that is, $g([a]_{Kerf}) \leq_2 g([b]_{Kerf})$. Hence, g is an injective po-NETG homomorphism of po-NETGs.

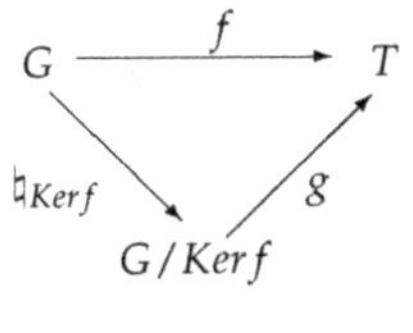

$\forall\, a \in G$, $(g \circ \natural_{Kerf})(a) = g(\natural_{Kerf}(a)) = g([a]_{Kerf}) = f(a)$, that is, $f = g \circ \natural_{Kerf}$.

□

In the following, we present an example to illustrate Proposition 14.

Example 5. *Consider* $(G, *, \leq_1)$ *be the po-NETG is given in Example 1, in which the partial order* $\leq_1$ *is the same as the partial order* $\leq$ *in Example 1. Assume that* $T = \{m, n, p, q, r\}$ *be a bounded lattice with a partial order* $\leq_2$ *with the Hasse diagram shown as in Figure 2 whose multiplication* $\cdot$ *is defined as* $\wedge$.

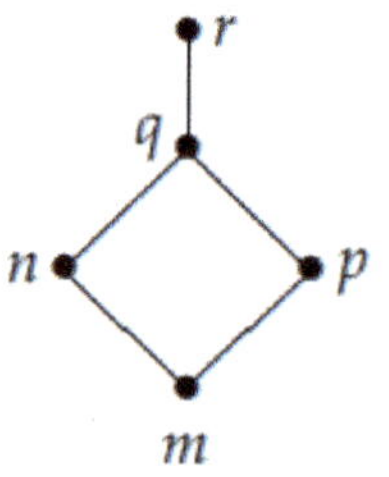

Figure 2. Hasse diagram.

We can verify that $(T, \cdot, \leq_2)$ is a po-NETG, in which $\forall\ x \in T$, $neut(x) = x$, $\{anti(m)\} = \{m, n, p, q, r\}$, $\{anti(n)\} = \{n, q, r\}$, $\{anti(p)\} = \{p, q, r\}$, $\{anti(q)\} = \{q, r\}$, $\{anti(r)\} = \{r\}$. Now, we define a map $f : G \rightarrow T$ by $f(0) = m$, $f(a) = f(b) = f(c) = f(1) = r$, then f is a po-NETG homomorphism of po-NETGs, and $Kerf = \{(0,0), (a,a), (b,b), (c,c), (1,1), (a,b), (a,c), (a,1), (b,a), (b,c), (b,1), (c,a), (c,b), (c,1), (1,a), (1,b), (1,c)\}$. Obviously, $Kerf$ is a congruence on G. f is not injective, and of course, $kerf$ is not an identity congruence on G.

5. Conclusions

In this paper, inspired by the research work in algebraic structures equipped with a partial order, we proposed the concepts of po-NETGs, deeply studied the relationships between po-NETGs and their positive cones, and characterized the positive cone of a WCNETG after defining a partial order relation on it. Moreover, we found that the quotient set of a po-NETG can construct another po-NETG by defining a special multiplication and a partial order on the quotient set, and we also achieved the interrelation of homomorphisms and congruences of po-NETGs. All these results are useful for exploring the structure characterization (for example, category properties) of po-NETGs. As a direction of future research, we will consider the application of the fuzzy set theory and the rough set theory to the research of algebraic structure of po-NETGs. Furthermore, we will discuss the relation between the homomorphisms and congruences of po-NETG and the morphisms of ordered lattice ringoids [30]. Finally, in the next paper, we will study sub-structures of po-NETGs and we give some examples using constructions such as central extensions or direct products related to sub-structures of po-NETGs.

Author Contributions: The authors contributed equally to writing this article. All authors read and approved the final manuscript.

Funding: This work was supported by the National Natural Science Foundation of China (Grant no. 11501435) and supported by the PhD research start-up fund (Grant no. BS1529).

Acknowledgments: Authors thank the editors and the anonymous reviewers for their insightful comments which improved the quality of the paper.

Conflicts of Interest: The authors declare no conflict of interest.

References

1. Dummit, D.S.; Foote, R.M. *Abstract Algebra*, 3rd ed.; John Viley & Sons Inc.: Hoboken, NJ, USA, 2004.
2. Herstein, I.N. *Topics in Algebra*; Xerox College Publishing: Lexington, KY, USA, 1975.
3. Surowski, D.B. The uniqueness aspect of the fundamental theorem of finite Abelian groups. *Amer. Math. Mon.* **1995**, *102*, 162–163. [CrossRef]
4. Smarandache, F. *Neutrosophy, Neutrosophic Probability, Set, and Logic;* American Research Press: Rehoboth, DE, USA, 1998.

5. Smarandache, F. Neutrosophic Set, a Generalization of the Intuitionistic fuzzy set. In Proceedings of the International Conference on Granular Computing, Atlanta, GA, USA, 10–12 May 2006; pp. 38–42.
6. Zadeh, L.A. Fuzzy sets. *Inf. Contr.* **1965**, *8*, 338–353. [CrossRef]
7. Atanassov, T.K. Intuitionistic fuzzy sets. *Fuzzy Sets Syst.* **1986**, *20*, 87–96. [CrossRef]
8. Peng, X.D.; Dai, J.G. A bibliometric analysis of neutrosophic set: two decades review from 1998 to 2017. *Artif. Intell. Rev.* **2018**, 1–57. [CrossRef]
9. Peng, X.D.; Liu, C. Algorithms for neutrosophic soft decision making based on EDAS, new similarity measure and level soft set. *J. Intell. Fuzzy Syst.* **2017**, *32*, 955–968. [CrossRef]
10. Peng, X.D.; Dai, J.G. Approaches to single-valued neutrosophic MADM based on MABAC, TOPSIS and new similarity measure with score function. *Neural Comput. Applic.* **2018**, *29*, 939–954. [CrossRef]
11. Zhang, X.H.; Bo, C.X.; Smarandache, F.; Dai, J.H. New inclusion relation of neutrosophic sets with applications and related lattice structure. *Int. J. Mach. Learn. Cyber.* **2018**, *9*, 1753–1763. [CrossRef]
12. Zhang, X.H.; Bo, C.X.; Smarandache, F.; Park, C. New operations of totally dependent-neutrosophic sets and totally dependent-neutrosophic soft sets. *Symmetry* **2018**, *10*, 187. [CrossRef]
13. Smarandache, F. *Neutrosophic Perspectives: Triplets, Duplets, Multisets, Hybrid Operators, Modal Logic, Hedge Algebras and Applications*; Pons Publishing House: Brussels, Belgium, 2017.
14. Smarandache, F.; Ali, M. Neutrosophic triplet group. *Neural Comput. Appl.* **2018**, *29*, 595–601. [CrossRef]
15. Zhang, X.H.; Wu, X.Y.; Smarandache, F.; Hu, M.H. Left (right)-quasi neutrosophic triplet loops (groups) and generalized BE-algebras. *Symmetry* **2018**, *10*, 241. [CrossRef]
16. Zhang, X.H.; Hu, Q.; Smarandache, F.; An, X. On neutrosophic triplet groups: basic properties, NT-subgroups and some notes. *Symmetry* **2018**, *10*, 289. [CrossRef]
17. Zhang, X.H.; Smarandache, F.; Liang, X.L. Neutrosophic duplet semi-group and cancellable neutrosophic triplet groups. *Symmetry* **2017**, *9*, 275. [CrossRef]
18. Zhang, X.H.; Wang, X.J.; Smarandache, F.; Jaíyéolá, T.G.; Lian, T.Y. Singular neutrosophic extended triplet groups and generalized groups. *Cogn. Syst. Res.* **2019**, *57*, 32–40. [CrossRef]
19. Zhang, X.H.; Wu, X.Y.; Mao, X.Y.; Smarandache, F.; Park, C. On neutrosophic extended triplet groups (loops) and Abel-Grassmann's groupoids (AG-groupoids). *J. Intell. Fuzzy Syst.* **2019**, *37*, 5743–5753. [CrossRef]
20. Ma, Y.C.; Zhang, X.H.; Yang, X.F.; Zhou, X. Generalized neutrosophic extended triplet group. *Symmetry* **2019**, *11*, 327. [CrossRef]
21. Blyth, T.S. *Lattices and Ordered Algebraic Structures*; Springer: Berlin, Germany, 2005.
22. Blyth, T.S.; Pinto, G.A. On ordered regular semigroups with biggest inverses. *Semigroup Forum* **1997**, *54*, 154–165. [CrossRef]
23. Certaine, J. Lattice-Ordered Groupoids and Some Related Problems. Ph.D. Thesis, Harvard University, Cambridge, MA, USA, 1945.
24. Darnell, M.R. *Theory of Lattice-Ordered Groups*; Marcel Dekker: New York, NY, USA, 1995.
25. Fuchs, L. *Partially Ordered Algebraic Systems*; Pergamon Press: Oxford, UK, 1963.
26. Hion, J.V. Archimedean ordered rings. *Uspechi Mat. Nauk.* **1954**, *9*, 237–242.
27. Xie, X.Y. *Introduction to Ordered Semigroups*; Science Press: Beijing, China, 2001.
28. Birkhoff, G. Lattice-orderd groups. *Ann. Math.* **1942**, *43*, 298–331. [CrossRef]
29. Howie, J.M. *Fundamentals of Semigroup Theory*; Oxford University Press: Oxford, UK, 1995.
30. Ludkowski, S.V. Skew continuous morphisms of ordered lattice ringoids. *Mathematics* **2016**, *4*, 17. [CrossRef]

symmetry

MDPI

Article

Single-Valued Neutrosophic Linguistic-Induced Aggregation Distance Measures and Their Application in Investment Multiple Attribute Group Decision Making

Guansheng Yu [1], Shouzhen Zeng [2,3] and Chonghui Zhang [4,5,*]

1 School of Economics, Fujian Normal University, Fuzhou 350117, China; yuguansheng@163.com
2 School of Business, Ningbo University, Ningbo 315100, China; zszzxl@163.com
3 School of Management, Fudan University, Shanghai 200433, China
4 School of Business, Sichuan University, Chengdu 610064, China
5 College of Statistics and Mathematics, Zhejiang Gongshang University, Hangzhou 310018, China
* Correspondence: zhangch1988@zjgsu.edu.cn

Received: 11 December 2019; Accepted: 18 January 2020; Published: 2 February 2020

Abstract: This paper studied the single-valued neutrosophic linguistic distance measures based on the induced aggregation method. Firstly, we proposed a single-valued neutrosophic linguistic-induced ordered weighted averaging distance (SVNLIOWAD) measure, which is a new extension of the existing distance measures based on the induced aggregation view. Then, based on the proposed SVNLIOWAD, a novel induced distance for single-valued neutrosophic linguistic sets, namely the single-valued neutrosophic linguistic weighted induced ordered weighted averaging distance (SVNLWIOWAD), was developed to eliminate the defects of the existing methods. The relationship between the two proposed distance measures was also explored. A multiple attribute group decision making (MAGDM) model was further presented based on the proposed SVNLWIOWAD measure. Finally, a numerical example concerning an investment selection problem was provided to demonstrate the usefulness of the proposed method under a single-valued neutrosophic linguistic environment and, then, a comparison analysis was carried out to verify the flexibility and effectiveness of the proposed work.

Keywords: single-valued neutrosophic linguistic set; distance measure; weighted induced aggregation; MAGDM; investment selection

1. Introduction

The growing uncertainties and complexities in multiple attribute decision making (MADM) make it increasingly difficult for people to judge their attributes accurately. Accordingly, how to measure such complex and uncertain information effectively has become a key issue during the process of decision making. Several tools, such as fuzzy set [1], intuitionistic fuzzy set (IFS) [2], picture fuzzy set [3,4], linguistic term [5], and neutrosophic set [6], have been introduced to deal with inaccurate and uncertain information. The single-valued neutrosophic linguistic set (SVNLS), introduced by Ye [7], is an up-to-date tool to measure uncertainty or inaccuracy of information by combining the advantages of single-valued neutrosophic set [8] and linguistic terms [5]. The basic element of the SVNLS is the single-valued neutrosophic linguistic number (SVNLN), which makes it more suitable for solving uncertain and imprecise information than the existing tools. Ye [7] extended the conventional the technique for order preference by similarity to ideal solutions (TOPSIS) [9] approach to SVNLS environment and explored its application in investment selection problems. Wang et al. [10] studied the operational laws for SVNLS and presented the SVNL Maclaurin symmetric mean aggregation operator. Chen et al. [11] studied the ordered weighted distance measure between SVNLSs. Wu et al. [12] studied the application of the SVNLS in a 2-tuple MADM environment. Kazimieras et al. [13] presented a weighted aggregated sum product assessment approach for SVN decision making problems. Garg and Nancy [14] proposed some SVNLS aggregation operators based on the prioritized method to solve the attributes' priority in MADM problems. Cao et al. [15] studied the SVNL decision making approach based on a combination of ordered and weighted distances measures.

Distance measure is one of the most popular tools to express the deviation degree between two sets or variables. Consequently, many types of distance measures have been investigated and proposed in the existing literature, such as the weighted distance (WD) measure [16], ordered weighted averaging distance (OWAD) measure [17], combined weighted distance (CWD) measure [18], and induced OWAD (IOWAD) measure [19]. Among them, the IOWAD measure is a widely used one, recently proposed by Merigó and Casanovas [19]. The key advantage of the IOWAD is that it summarizes the minimun and maximum distance measures and can use induced-ordering variables to depict the intricate attitudinal characteristics. Now, the IOWAD operator has been widely used in MADM problems and extended to accommodate several fuzzy environments, such as fuzzy IOWAD (FIOWAD) [20], fuzzy linguistic IOWAD [21], intuitionistic fuzzy IOWAD (IFIOWAD) [22], and 2-tuple linguistic IOWAD (2LIOWAD) [23].

However, as far as we know, there is no research on the application of the SVNLS with the IOWAD method. In accordance with the previous analysis, the SVNLS is an excellent method to describe fuzzy and uncertain information, while the IOWAD is a new tool that can be well integrated into the complex attitudes of decision makers. In order to develop and enrich the measure theory of SVNLS, this study explored the usefulness of the IOWAD measure in SVNL environments. For this purpose, the rest of the article is set out as follows: in Section 2, we briefly introduce some basic concepts. Section 3 firstly develops the single-valued neutrosophic linguistic induced ordered weighted averaging distance (SVNLIOWAD) operator, which is the extension of the IOWAD operator with SVNL information. Furthermore, the single-valued neutrosophic linguistic weighted induced ordered weighted averaging distance (SVNLWIOWAD) is then introduced to overcome the defects of the SVNLIOWAD operator and other existing induced aggregation distances. In Section 4, a MAGDM model based on the SVNLWIOWAD operator is formulated and a financial decision making problem is also provided to demonstrate the usefulness of the proposed method. Finally, Section 5 gives a conclusion for the paper.

2. Preliminaries

In this section, we mainly recap some basic concepts of the SVNLS and the IOWAD operator.

2.1. The Single-Valued Neutrosophic Set (SVNS)

Definition 1 *[24]. Let u be an element in a finite set U. A single-valued neutrosophic set (SVNS) A in U can be defined as in (1):*

$$A = \left\{ \langle u, T_A(u), I_A(u), F_A(u) \rangle \middle| u \in U \right\}, \tag{1}$$

where $T_A(u)$, $I_A(u)$, and $F_A(u)$ are called the truth-membership function, indeterminacy-membership function, and falsity-membership function, respectively, which satisfy the following conditions:

$$0 \le T_A(u), I_A(u), F_A(u) \le 1,\ 0 \le T_A(u) + I_A(u) + F_A(u) \le 3. \tag{2}$$

A single-valued neutrosophic number (SVNN) is expressed as $(T_A(u), I_A(u), F_A(u))$ and is simply termed as $u = (T_u, I_u, F_u)$. The mathematical operational laws between SVNNs $u = (T_u, I_u, F_u)$ and $v = (T_v, I_v, F_v)$ are defined as follows:

(1) $u \oplus v = (T_u + T_v - T_u * T_v, I_u * T_v, F_u * F_v)$;

(2) $\lambda u = (1-(1-T_u)^\lambda, (I_u)^\lambda, (F_u)^\lambda), \lambda > 0$;
(3) $u^\lambda = ((T_u)^\lambda, 1-(1-I_u)^\lambda, 1-(1-F_u)^\lambda), \lambda > 0$.

2.2. The Linguistic Set

Let $S = \{s_\alpha | \alpha = 1, \ldots, l\}$ be a finite and totally ordered discrete term set, where s_α indicates a possible value for a linguistic variable (LV) and l is an odd value. For instance, given $l = 7$, then a linguistic term set S could be specified $S = \{s_1, s_2, s_3, s_4, s_5, s_6, s_7\}$ = *{extremely poor, very poor, poor, fair, good, very good, extremely good}*. Then, for any two LVs, s_i and s_j in S, should satisfy rules (1)–(4) [24]:

(1) $s_i \le s_j \Leftrightarrow i \le j$;
(2) $Neg(s_i) = s_{-i}$;
(3) $\max(s_i, s_j) = s_j$, if $i \le j$;
(4) $\min(s_i, s_j) = s_i$, if $i \le j$.

The discrete term set S is also extended to a continuous set $\overline{S} = \{s_\alpha | \alpha \in R\}$ for reducing the loss of information during the operational process. The operational rules for LVs $s_\alpha, s_\beta \in \overline{S}$ are defined as follows [25]:

(1) $s_\alpha \oplus s_\beta = s_{\alpha+\beta}$;
(2) $\mu s_\alpha = s_{\mu\alpha}, \mu \ge 0$.

2.3. The Single-Valued Neutrosophic Linguistic Set (SVNLS)

Definition 2 *[7]. Let U be a finite universe set and $\overline{S}$ be a continuous linguistic set, a SVNLS B in U is defined as in (3):*

$$B = \left\{ \left\langle u, [s_{\theta(u)}, (T_B(u), I_B(u), F_B(u))] \right\rangle \middle| u \in U \right\}, \tag{3}$$

where $s_{\theta(u)} \in \overline{S}$, the truth-membership function $T_B(u)$, the indeterminacy-membership function $I_B(u)$, and the falsity-membership function $F_B(u)$ satisfy condition (4):

$$0 \le T_B(u), I_B(u), F_B(u) \le 1,\ 0 \le T_B(u) + I_B(u) + F_B(u) \le 3. \tag{4}$$

For an SVNLS B in U, the SVNLN $\left\langle s_{\theta(u)}, (T_B(u), I_B(u), F_B(u)) \right\rangle$ is simply termed as $u = \left\langle s_{\theta(u)}, (T_u, I_u, F_u) \right\rangle$. The operational rules for SVNLNs $u_i = \left\langle s_{\theta(u_i)}, (T_{u_i}, I_{u_i}, F_{u_i}) \right\rangle (i = 1, 2)$ are defined as follows:

(1) $u_1 \oplus u_2 = \left\langle s_{\theta(u_1)+\theta(u_2)}, (T_{u_1} + T_{u_2} - T_{u_1} * T_{u_2}, I_{u_1} * T_{u_2}, F_{u_1} * F_{u_2}) \right\rangle$;
(2) $\lambda u_1 = \left\langle s_{\lambda\theta(u_1)}, (1-(1-T_{u_1})^\lambda, (I_{u_1})^\lambda, (F_{u_1})^\lambda) \right\rangle, \lambda > 0$;
(3) $u_1^\lambda = \left\langle s_{\theta^\lambda(u_1)}, ((T_{u_1})^\lambda, 1-(1-I_{u_1})^\lambda, 1-(1-F_{u_1})^\lambda) \right\rangle, \lambda > 0$.

Definition 3 *[7]. Given two SVNLNs $u_i = \left\langle s_{\theta(u_i)}, (T_{u_i}, I_{u_i}, F_{u_i}) \right\rangle (i = 1, 2)$, their distance measure is defined using the following formula:*

$$d(u_1, u_2) = \left[\left| \theta(u_1)T_{u_1} - \theta(u_2)T_{u_2} \right|^l + \left| \theta(u_1)I_{u_1} - \theta(u_2)I_{u_2} \right|^l + \left| \theta(u_1)F_{u_1} - \theta(u_2)F_{u_2} \right|^l \right]^{1/l}, \tag{5}$$

where $l \in (0, +\infty)$. If we consider different weights associated with individual distances of SVNLVs, then we can get the single-valued neutrosophic linguistic weighted distance (SVNLWD) measure [10].

Definition 4. *Let $u_j, u'_j (j = 1, 2, \ldots, n)$ be the two collections of SVNLNs, a single-valued neutrosophic linguistic weighted distance measure is defined as following formula:*

$$SVNLWD\big((u_1, u'_1), \ldots, (u_n, u'_n)\big) = \sum_{j=1}^{n} w_j d(u_j, u'_j), \tag{6}$$

where the associated weighting vector w_j satisfies $w_j \in [0, 1]$ and $\sum_{j=1}^{n} w_j = 1$.

2.4. The Single-Valued Neutrosophic Linguistic Set (SVNLS)

Motivated by the induced ordered weighted averaging (IOWA) operator [26], Merigó and Casanovas [19] developed the IOWAD operator. For two crisp sets $X = (x_1, \ldots, x_n)$ and $Y = (y_1, \ldots, y_n)$, the IOWAD operator be easily obtained as follows:

Definition 5. *An IOWAD operator is defined by a weight vector $W = (w_1, \ldots, w_n)^T$ with $0 \leq w_j \leq 1$ and $\sum_{j=1}^{n} w_j = 1$ and an order-inducing vector $T = (t_1, \ldots, t_n)$, such that:*

$$IOWAD(\langle t_1, x_1, y_1 \rangle, \ldots, \langle t_n, x_n, y_n \rangle) = \sum_{j=1}^{n} w_j D_j, \tag{7}$$

where $(D_1, \ldots, D_n)$ is recorded $(d_1, \ldots, d_n)$, induced by the decreasing order of $(t_1, \ldots, t_n)$, and $d_i = d(x_i, y_i) = |x_i - y_i|$ is the distance between x_i and y_i.

3. Single-Valued Neutrosophic Linguistic-Induced Aggregation Distance Measures

3.1. SVNLIOWAD Measure

Previous analysis has shown that the IOWAD is a very practical tool to measure deviation in many fields, such as clustering analysis and decision making. In this section, we explore the application of the IOWAD operator in an SVNL situation and develop the SVNLIOWAD operator.

Definition 6. *Let $u_j, u'_j (j = 1, 2, \ldots, n)$ be two sets of SVNLNs, then the SVNLIOWAD operator is defined by a weight vector $W = (w_1, \ldots, w_n)^T$ with $0 \leq w_j \leq 1$ and $\sum_{j=1}^{n} w_j = 1$ and an order-inducing vector $T = (t_1, \ldots, t_n)$, such that:*

$$SVNLIOWAD\big(\langle t_1, u_1, u'_1 \rangle, \ldots, \langle t_n, u_n, u'_n \rangle\big) = \sum_{j=1}^{n} w_j D_j, \tag{8}$$

where $(D_1, \ldots, D_n)$ is recorded $(d_1, \ldots, d_n)$, induced by the decreasing order of $(t_1, \ldots, t_n)$, $d_i = d(u_i, u'_i) = |u_i - u'_i|$ is the distance between SVNLNs, defined in Equation (5).

Using a similar analysis with the IOWAD operator [18,19,27,28], it is easy to derive the following useful properties for the SVNLIOWAD operator:

Theorem 1 *(Idempotency). If $d_i = d(u_i, u'_i) = |u_i - u'_i| = d$ for all i, then*

$$SVNLIOWAD\big(\langle t_1, u_1, u'_1 \rangle, \ldots, \langle t_n, u_n, u'_n \rangle\big) = d. \tag{9}$$

Theorem 2 *(Boundedness). Let* $\min_i(|u_i - u'_i|) = x$ *and* $\max_i(|u_i - u'_i|) = y$*, then*

$$x \leq SVNLIOWAD\left(\left\langle t_1, u_1, u'_1\right\rangle, \ldots, \left\langle t_n, u_n, u'_n\right\rangle\right) \leq y. \tag{10}$$

Theorem 3 *(Monotonicity). If* $|u_i - u'_i| \geq |v_i - v'_i|$ *for all i, then*

$$SVNLIOWAD\left(\left\langle t_1, u_1, u'_1\right\rangle, \ldots, \left\langle t_n, u_n, u'_n\right\rangle\right) \geq SVNLIOWAD\left(\left\langle t_1, v_1, v'_1\right\rangle, \ldots, \left\langle t_n, v_n, v'_n\right\rangle\right). \tag{11}$$

Theorem 4 *(Commutativity-IOWA operator aggregation). Let* $\left(\left\langle t_1, u_1, u'_1\right\rangle, \ldots, \left\langle t_n, u_n, u'_n\right\rangle\right)$ $(i = 1, 2, \ldots, n)$ *be any possible permutation of the argument vector* $\left(\left\langle t_1, v_1, v'_1\right\rangle, \ldots, \left\langle t_n, v_n, v'_n\right\rangle\right)$*, then*

$$SVNLIOWAD\left(\left\langle t_1, u_1, u'_1\right\rangle, \ldots, \left\langle t_n, u_n, u'_n\right\rangle\right) = SVNLIOWAD\left(\left\langle t_1, v_1, v'_1\right\rangle, \ldots, \left\langle t_n, v_n, v'_n\right\rangle\right). \tag{12}$$

We can also illustrate the property of commutativity by considering the distance measure:

$$SVNLIOWAD\left(\left\langle t_1, u_1, u'_1\right\rangle, \ldots, \left\langle t_n, u_n, u'_n\right\rangle\right) = SVNLIOWAD\left(\left\langle t_1, u'_1, u_1\right\rangle, \ldots, \left\langle t_n, u'_n, u_n\right\rangle\right). \tag{13}$$

By considering different cases of the weighted vector in the SVNLIOWAD operator, we can get several special distance measures. For example:

- If $w_1 = \cdots = w_n = \frac{1}{n}$, we obtain the SVNLWD;
- If the ordering of weight w_j is same as the order-inducing t_j for all j, then the SVNLIOWAD reduces to the SVNLOWAD measure [15];
- If $T = (t, 0, \cdots, 0)$, then

$$SVNLIOWAD\left(\left\langle t_1, u_1, u'_1\right\rangle, \ldots, \left\langle t_n, u_n, u'_n\right\rangle\right) = D_1. \tag{14}$$

Next, a numerical example is given to show the aggregation process of the SVNLIOWAD operator.

Example 1. *Assuming that:*

$$\begin{aligned} U &= (u_1, u_2, u_3, u_4, u_5) \\ &= (\langle s_2, (0.5, 0.3, 0.4)\rangle, \langle s_5, (0.3, 0.3, 0.6)\rangle, \langle s_5, (0.5, 0.2, 0.2)\rangle, \langle s_7, (0.5, 0.8, 0.2)\rangle, \langle s_2, (0.1, 0.4, 0.6)\rangle) \end{aligned}$$

and

$$\begin{aligned} V &= (v_1, v_2, v_3, v_4, v_5) \\ &= (\langle s_3, (0.7, 0.8, 0)\rangle, \langle s_5, (0.4, 0.4, 0.5)\rangle, \langle s_3, (0.5, 0.7, 0.2)\rangle, \langle s_3, (0.4, 0.2, 0.6)\rangle, \langle s_4, (0.5, 0.7, 0.2)\rangle), \end{aligned}$$

are two SVNLNs defined in linguist term set $S = \{s_1, s_2, s_3, s_4, s_5, s_6, s_7\}$ *and suppose* $w = (0.20, 0.30, 0.15, 0.10, 0.25)^T$ *and* $T = (5, 8, 4, 2, 7)$ *are the weight vector and order-inducing variable vector of the SVNLIOWAD operator, respectively. Then, the calculation steps of the SVNLIOWAD are displayed as follows:*

(1) Calculate the individual distances $d(u_i, v_i)$ $(i = 1, 2, \ldots, 5)$ *(let* $\lambda = 1$*) according to Equation (5):*

$$d(u_1, v_1) = |2 \times 0.5 - 3 \times 0.7| + |2 \times 0.3 - 3 \times 0.8| + |2 \times 0.4 - 3 \times 0| = 3.7.$$

Similarly, we get

$$d(u_2,v_2)=1.5,\ d(u_3,v_3)=2.4,\ d(u_4,v_4)=7.7,\ d(u_5,v_5)=3.2;$$

(2) *Sort the* $d(u_i,v_i)$ $(i=1,2,\ldots,5)$ *according to the decreasing order of the order-inducing variable:*

$$D_1=d(u_2,v_2)=1.5,\ D_2=d(u_5,v_5)=3.2,\ D_3=d(u_1,v_1)=3.7,$$
$$D_4=d(u_3,v_3)=2.4,\ d(u_4,v_4)=7.7;$$

(3) *Utilize the SVNLIOWAD operator defined in Equation (8) to perform the following aggregation:*

$$SVNLIOWAD(U,V)$$
$$=0.20\times1.5+0.30\times3.2+0.15\times3.7+0.10\times2.4+0.25\times7.7=3.71.$$

From the aggregation process of the SVNLIOWAD operator, as well as the existing other induced aggregation distances, we see that the order-inducing variables are not really infused in the aggregation results, which fail to express the variation caused by the change of order-inducing variables. Thus, we needed to develop a new induced aggregation distance operator for SVNLSs to overcome this defect.

3.2. SVNLWIOWAD Measure

The special feature of the SVNLWIOWAD operator is that its induced ordering-variables play a dual role in the aggregation process. One role is, as the previous SVNLIOWAD operator, to induce the order of the arguments and the other is to adjust the associated weights. Thus it can better reflect the influence of the induced variables on the ensemble results. The SVNLWIOWAD operator can be defined as follows.

Definition 7. *Let* $u_j,u'_j(j=1,2,\ldots,n)$ *be two sets of SVNLNs, the SVNLWIOWAD operator is defined by a weight vector* $W=(w_1,\ldots,w_n)^T$ *with* $0\le w_j\le1$ *and* $\sum_{j=1}^{n}w_j=1$; *and an order-inducing vector* $T=(t_1,\ldots,t_n)$, *such that:*

$$SVNLWIOWAD\big(\langle t_1,u_1,u'_1\rangle,\ldots,\langle t_n,u_n,u'_n\rangle\big)=\sum_{j=1}^{n}\varpi_jD_j, \tag{15}$$

where $(D_1,\ldots,D_n)$ *is recorded* $(d_1,\ldots,d_n)$ *induced by the decreasing order of* $(t_1,\ldots,t_n)$, $d_i=|u_i-u'_i|$ *is the distance between SVNLNs, defined in Equation (5).* $\varpi_j(j=1,2,\ldots,n)$ *is a moderated weight that is relatively determined by the weight* $w_j\in W$ *and order-inducing variable* $t_j\in T$:

$$\varpi_j=\frac{w_jt_{\sigma(j)}}{\sum_{j=1}^{n}w_jt_{\sigma(j)}}, \tag{16}$$

where $(\sigma(1),\ldots,\sigma(n))$ *is a permutation of* $(1,\ldots,n)$ *such that* $t_{\sigma(j-1)}\ge t_{\sigma(j)}$ *for all* $j>1$. *Example 2 illustrates the performance of the SVNLWIOWAD operator.*

Example 2 *(Example 1 continuation). To utilize the SVNLWIOWAD operator, we calculated the moderated weight ϖ_j defined in Equation (16):*

$$\varpi_1 = \frac{w_1 t_{\sigma(1)}}{\sum_{j=1}^{5} w_j t_{\sigma(j)}} = \frac{0.20 \times 8}{0.20 \times 8 + 0.30 \times 7 + 0.15 \times 5 + 0.10 \times 4 + 0.25 \times 4} = 0.274.$$

Similarly,

$$\varpi_2 = 0.359, \varpi_3 = 0.128, \varpi_4 = 0.068, \varpi_5 = 0.171.$$

Thus, based on the results of Example 1, we can get the aggregation result of the SVNLWIOWAD operator:

$$SVNLWIOWAD(U, V)$$
$$= 0.274 \times 1.5 + 0.359 \times 3.2 + 0.128 \times 3.7 + 0.068 \times 2.4 + 0.171 \times 7.7 = 3.462$$

Obviously, we got a different result compared with the SVNLIOWAD operator in Example 1. The main reason for the difference is that the order-inducing variables in the SVNLIOWAD operator (including the existing IOWAD and its numerous extensions) only act as inducers for the arguments, and do not participate in the actual calculation process. However, the SVNLWIOWAD's order-inducing variables can not only act as the inducer, but also participate in the actual calculation progress by adjusting the associated weights. Therefore, it can measure the effect of order-inducing variables on the aggregation results. Consequently, the SVNLWIOWAD can achieve a more reasonable and scientific measurement over the SVNLIOWAD operator.

The following theorems show some useful properties of the SVNLWIOWAD operator:

Theorem 5 *(Idempotency). Let Q be the SVNLWIOWAD operator, if all $d_i = |u_i - u_i'| = d$ for all i, then:*

$$Q\left(\left\langle t_1, u_1, u_1' \right\rangle, \ldots, \langle t_n, u_n, u_n' \rangle\right) = d. \tag{17}$$

Proof. Because $d_i = |u_i - u_i'| = d$, then $D_j = d$ for $j = 1, 2, \ldots, n$, and we have:

$$Q\left(\left\langle t_1, u_1, u_1' \right\rangle, \ldots, \langle t_n, u_n, u_n' \rangle\right) = \sum_{j=1}^{n} \varpi_j D_j = d \sum_{j=1}^{n} \varpi_j.$$

Note that $\sum_{j=1}^{n} \varpi_j = 1$, thus we obtain $Q\left(\left\langle t_1, u_1, u_1' \right\rangle, \ldots, \langle t_n, u_n, u_n' \rangle\right) = d \sum_{j=1}^{n} \varpi_j = d$. □

Theorem 6 *(Boundedness). Let $\min_i(|u_i - u_i'|) = x$ and $\max_i(|u_i - u_i'|) = y$, then:*

$$x \leq Q\left(\left\langle t_1, u_1, u_1' \right\rangle, \ldots, \langle t_n, u_n, u_n' \rangle\right) \leq y. \tag{18}$$

Proof. Because $\varpi_j \in [0, 1]$ and $\sum_{j=1}^{n} \varpi_j = 1$, then:

$$Q\left(\left\langle t_1, u_1, u_1' \right\rangle, \ldots, \langle t_n, u_n, u_n' \rangle\right) = \sum_{j=1}^{n} \varpi_j D_j \leq \sum_{j=1}^{n} \varpi_j y = y \sum_{j=1}^{n} \varpi_j = y.$$

Similarly,

$$Q\left(\left\langle t_1, u_1, u_1'\right\rangle, \ldots, \langle t_n, u_n, u_n'\rangle\right) = \sum_{j=1}^{n} \varpi_j D_j \geq \sum_{j=1}^{n} \varpi_j x = x \sum_{j=1}^{n} \varpi_j = x.$$

Thus, we get

$$x \leq Q\left(\left\langle t_1, u_1, u_1'\right\rangle, \ldots, \langle t_n, u_n, u_n'\rangle\right) \leq y$$

□

Theorem 7 *(Monotonicity). If* $\left|u_i - u_i'\right| \geq \left|v_i - v_i'\right|$ *for all i, then:*

$$Q\left(\left\langle t_1, u_1, u_1'\right\rangle, \ldots, \langle t_n, u_n, u_n'\rangle\right) \geq Q\left(\left\langle t_1, v_1, v_1'\right\rangle, \ldots, \langle t_n, v_n, v_n'\rangle\right). \tag{19}$$

Proof. Let

$$Q\left(\left\langle t_1, u_1, u_1'\right\rangle, \ldots, \langle t_n, u_n, u_n'\rangle\right) = \sum_{j=1}^{n} \varpi_j D_j,$$

$$Q\left(\left\langle t_1, v_1, v_1'\right\rangle, \ldots, \langle t_n, v_n, v_n'\rangle\right) = \sum_{j=1}^{n} \varpi_j D'_j.$$

As $\left|u_i - u_i'\right| \geq \left|v_i - v_i'\right|$ for all i, it follows $D_j \geq D_j'$ for all j, therefore

$$Q\left(\left\langle t_1, u_1, u_1'\right\rangle, \ldots, \langle t_n, u_n, u_n'\rangle\right) = \sum_{j=1}^{n} \varpi_j D_j \geq \sum_{j=1}^{n} \varpi_j D_j' = Q\left(\left\langle t_1, v_1, v_1'\right\rangle, \ldots, \langle t_n, v_n, v_n'\rangle\right)$$

□

Theorem 8 *(Commutativity-IOWA operator aggregation). Let* $\left(\left\langle t_1, u_1, u_1'\right\rangle, \ldots, \langle t_n, u_n, u_n'\rangle\right)$ $(i = 1, 2, \ldots, n)$ *be any possible permutation of the argument vector* $\left(\left\langle t_1, v_1, v_1'\right\rangle, \ldots, \langle t_n, v_n, v_n'\rangle\right)$*, then:*

$$Q\left(\left\langle t_1, u_1, u_1'\right\rangle, \ldots, \langle t_n, u_n, u_n'\rangle\right) = Q\left(\left\langle t_1, v_1, v_1'\right\rangle, \ldots, \langle t_n, v_n, v_n'\rangle\right). \tag{20}$$

Proof. The permutation between $\left(\left\langle t_1, u_1, u_1'\right\rangle, \ldots, \langle t_n, u_n, u_n'\rangle\right)$ and $\left(\left\langle t_1, v_1, v_1'\right\rangle, \ldots, \langle t_n, v_n, v_n'\rangle\right)$ $(i = 1, 2, \ldots, n)$ follows that the corresponding rearranged arguments $D_j = D_j'$ for all j, therefore

$$Q\left(\left\langle t_1, u_1, u_1'\right\rangle, \ldots, \langle t_n, u_n, u_n'\rangle\right) = \sum_{j=1}^{n} \varpi_j D_j = \sum_{j=1}^{n} \varpi_j D_j' = Q\left(\left\langle t_1, v_1, v_1'\right\rangle, \ldots, \langle t_n, v_n, v_n'\rangle\right)$$

We can also illustrate the property of commutativity by considering the distance measure:

$$Q\left(\left\langle t_1, u_1, u_1'\right\rangle, \ldots, \langle t_n, u_n, u_n'\rangle\right) = Q\left(\left\langle t_1, u_1', u_1\right\rangle, \ldots, \langle t_n, u_n', u_n\rangle\right). \tag{21}$$

Note that $\left|u_i - u_i'\right| = \left|u_i' - u_i\right|$ for all i, thus the Equation (20) is easy to prove. □

In light of the similar analysis methods in [29–34], some particular cases of the SVNLWIOWAD operator can be achieved by exploring the weight vector and order-inducing values.

4. A New MAGDM Approach Based on the SVNLWIOWAD Operator

4.1. Steps of the MAGDM Method Based on the SVNWIOWAD Operator

On the basis of the analysis reviewed in the Introduction, it is customary for decision makers to express their opinions on alternatives over attributes by SVNLNs because of their cognition with uncertainty and vagueness. Therefore, it is well worth investigating the application of the proposed SVNLWIOWAD under the SVNL framework. For an MAGDM problem with n alternatives $A = \{A_1, A_2, \ldots, A_n\}$ assessed by decision makers with respect to m schemes (attributes) $C = \{C_1, C_2, \ldots, C_m\}$, the decision steps based on the SVNLWIOWAD are listed as follows:

Step 1: Each expert $d_k(k = 1,2,\ldots,l)$ (whose weight is ε_k, meeting $\varepsilon_k \geq 0$ and $\sum_{k=1}^{l} \varepsilon_k = 1$) provides his or her performance of attributes by the SVNLNs. Afterwards, the individual decision matrix $U^k = \left(u_{ij}^{(k)}\right)_{m \times n}$ is obtained, where $u_{ij}^{(k)}$ is the k-th expert's evaluation of the alternative A_j with respect to the attribute C_i;

Step 2: Aggregate all performances of the individual experts into a collective one and then form the group decision matrix:

$$U = \left(u_{ij}\right)_{m \times n} = \begin{pmatrix} u_{11} & \cdots & u_{1n} \\ \vdots & \ddots & \vdots \\ u_{m1} & \cdots & u_{mn} \end{pmatrix}, \tag{22}$$

where $u_{ij} = \sum_{k=1}^{l} \varepsilon_k u_{ij}^{(k)}$;

Step 3: Find the ideal levels for each attribute to construct the ideal scheme, listed in the Table 1;

Table 1. Ideal scheme.

	C_1	C_2	$\cdots$	C_n
I	I_1	I_2	$\cdots$	I_n

Step 4: Utilize Equation (15) to calculate the distance $SVNLWIOWAD(A_i, I)$ between different alternatives $A_i(i = 1,2,\ldots,m)$ and the ideal scheme I;

Step 5: Rank the alternatives and identify the best one(s) according to $SVNLWIOWAD(A_i, I)$, where the smaller the value of $SVNLWIOWAD(A_i, I)$, the better the alternative $A_i(i = 1,2,\ldots,m)$.

4.2. An Illustrative Example: Investment Selection

We explored the application of the proposed approach in an investment selection problem where three decision makers were invited to assess a suitable strategy. There were four companies (alternatives) considered as potential investment options, chemical company (A_1), food company (A_2), car company (A_3) and furniture company (A_4), according to following possible situations (attributes) for the next year: C_1 was the risk, C_2 was the growth, C_3 was the environmental impact, and C_4 was other impacts. The evaluation presented by the decision makers with respect to the four attributes formed individual SVNL decision matrices under the linguistic term set $S = \{s_1$ = extremely poor, s_2 = very poor, s_3 = poor, s_4 = fair, s_5 = good, s_6 = very good, and s_7 = extremely good$\}$, as shown in Tables 2–4.

Table 2. Single-valued neutrosophic linguistic (SVNL) decision matrix U^1.

	C_1	C_2	C_3	C_4
A_1	$\left\langle s_4^{(1)}, (0.3, 0.2, 0.3)\right\rangle$	$\left\langle s_3^{(1)}, (0.5, 0.3, 0.1)\right\rangle$	$\left\langle s_4^{(1)}, (0.5, 0.2, 0.3)\right\rangle$	$\left\langle s_5^{(1)}, (0.3, 0.5, 0.2)\right\rangle$
A_2	$\left\langle s_6^{(1)}, (0.6, 0.1, 0.2)\right\rangle$	$\left\langle s_4^{(1)}, (0.5, 0.2, 0.2)\right\rangle$	$\left\langle s_5^{(1)}, (0.6, 0.1, 0.2)\right\rangle$	$\left\langle s_3^{(1)}, (0.6, 0.2, 0.4)\right\rangle$
A_3	$\left\langle s_5^{(1)}, (0.7, 0.0, 0.1)\right\rangle$	$\left\langle s_3^{(1)}, (0.3, 0.1, 0.2)\right\rangle$	$\left\langle s_4^{(1)}, (0.6, 0.1, 0.2)\right\rangle$	$\left\langle s_6^{(1)}, (0.6, 0.1, 0.2)\right\rangle$
A_4	$\left\langle s_5^{(1)}, (0.4, 0.2, 0.3)\right\rangle$	$\left\langle s_3^{(1)}, (0.3, 0.2, 0.5)\right\rangle$	$\left\langle s_5^{(1)}, (0.4, 0.2, 0.3)\right\rangle$	$\left\langle s_4^{(1)}, (0.5, 0.3, 0.3)\right\rangle$

Table 3. SVNL decision matrix U^2.

	C_1	C_2	C_3	C_4
A_1	$\left\langle s_6^{(2)}, (0.4, 0.2, 0.4)\right\rangle$	$\left\langle s_4^{(2)}, (0.6, 0.1, 0.3)\right\rangle$	$\left\langle s_6^{(2)}, (0.6, 0.3, 0.4)\right\rangle$	$\left\langle s_5^{(2)}, (0.4, 0.4, 0.1)\right\rangle$
A_2	$\left\langle s_6^{(2)}, (0.7, 0.2, 0.3)\right\rangle$	$\left\langle s_5^{(2)}, (0.6, 0.2, 0.2)\right\rangle$	$\left\langle s_6^{(2)}, (0.7, 0.2, 0.3)\right\rangle$	$\left\langle s_4^{(2)}, (0.5, 0.4, 0.2)\right\rangle$
A_3	$\left\langle s_4^{(2)}, (0.8, 0.1, 0.2)\right\rangle$	$\left\langle s_4^{(2)}, (0.4, 0.2, 0.2)\right\rangle$	$\left\langle s_5^{(2)}, (0.7, 0.2, 0.3)\right\rangle$	$\left\langle s_6^{(2)}, (0.6, 0.3, 0.3)\right\rangle$
A_4	$\left\langle s_5^{(2)}, (0.4, 0.3, 0.4)\right\rangle$	$\left\langle s_5^{(2)}, (0.3, 0.1, 0.6)\right\rangle$	$\left\langle s_6^{(2)}, (0.5, 0.1, 0.2)\right\rangle$	$\left\langle s_3^{(2)}, (0.7, 0.1, 0.1)\right\rangle$

Table 4. SVNL decision matrix U^3.

	C_1	C_2	C_3	C_4
A_1	$\left\langle s_6^{(3)}, (0.5, 0.1, 0.3)\right\rangle$	$\left\langle s_4^{(3)}, (0.6, 0.2, 0.1)\right\rangle$	$\left\langle s_5^{(3)}, (0.6, 0.1, 0.3)\right\rangle$	$\left\langle s_4^{(3)}, (0.3, 0.6, 0.2)\right\rangle$
A_2	$\left\langle s_5^{(3)}, (0.5, 0.2, 0.3)\right\rangle$	$\left\langle s_5^{(3)}, (0.7, 0.2, 0.1)\right\rangle$	$\left\langle s_4^{(3)}, (0.7, 0.2, 0.2)\right\rangle$	$\left\langle s_6^{(3)}, (0.4, 0.6, 0.2)\right\rangle$
A_3	$\left\langle s_4^{(3)}, (0.6, 0.1, 0.2)\right\rangle$	$\left\langle s_3^{(3)}, (0.4, 0.1, 0.1)\right\rangle$	$\left\langle s_4^{(3)}, (0.5, 0.2, 0.2)\right\rangle$	$\left\langle s_5^{(3)}, (0.7, 0.2, 0.1)\right\rangle$
A_4	$\left\langle s_6^{(3)}, (0.5, 0.2, 0.3)\right\rangle$	$\left\langle s_5^{(3)}, (0.2, 0.1, 0.6)\right\rangle$	$\left\langle s_6^{(3)}, (0.6, 0.2, 0.4)\right\rangle$	$\left\langle s_4^{(3)}, (0.5, 0.2, 0.3)\right\rangle$

Assuming that the weights of the experts were $\varepsilon_1 = 0.30$, $\varepsilon_2 = 0.37$, and $\varepsilon_3 = 0.33$, respectively, then the group SVNL decision matrix could be obtained through aggregating the three individual decision matrices. The results are listed in the Table 5.

Table 5. Group SVNL decision matrix U.

	C_1	C_2	C_3	C_4
A_1	$\langle s_{5.26}, (0.399, 0.163, 0.330)\rangle$	$\langle s_{3.37}, (0.566, 0.185, 0.144)\rangle$	$\langle s_{4.96}, (0.566, 0.186, 0.330)\rangle$	$\langle s_{4.70}, (0.335, 0.491, 0.159)\rangle$
A_2	$\langle s_{5.70}, (0.611, 0.155, 0.258)\rangle$	$\langle s_{2.37}, (0.602, 0.200, 0.162)\rangle$	$\langle s_{4.70}, (0.666, 0.155, 0.229)\rangle$	$\langle s_{4.23}, (0.514, 0.350, 0.258)\rangle$
A_3	$\langle s_{4.37}, (0.714, 0.000, 0.155)\rangle$	$\langle s_{3.67}, (0.365, 0.128, 0.163)\rangle$	$\langle s_{4.33}, (0.611, 0.155, 0.229)\rangle$	$\langle s_{5.70}, (0.633, 0.180, 0.186)\rangle$
A_4	$\langle s_{5.30}, (0.432, 0.229, 0.330)\rangle$	$\langle s_{2.37}, (0.271, 0.129, 0.561)\rangle$	$\langle s_{5.63}, (0.450, 0.159, 0.286)\rangle$	$\langle s_{3.67}, (0.578, 0.185, 0.209)\rangle$

.

The ideal scheme (Table 6) determined by experts represents the optimal results that a supplier should satisfy, which further serves as a reference point in the aggregation process.

Table 6. Ideal scheme.

	C_1	C_2	C_3	C_4
I	$\langle s_7, (0.9, 0, 0)\rangle$	$\langle s_7, (0.9, 0, 0.1)\rangle$	$\langle s_7, (1, 0, 0.1)\rangle$	$\langle s_6, (0.9, 0.1, 0)\rangle$

We assumed that the weight and the order-inducing vectors of the SVNLWIOWAD were $w = (0.2, 0.15, 0.3, 0.35)^T$ and $T = (5, 9, 7, 4)$, respectively. Based on the available information, we utilized the SVNLWIOWAD to calculate the distances between the alternative A_i and the ideal scheme I:

$$SVNLWIOWAD(A_1, I) = 6.440,\ SVNLWIOWAD(A_2, I) = 5.713,$$
$$SVNLWIOWAD(A_3, I) = 5.323,\ SVNLWIOWAD(A_4, I) = 6.810.$$

Therefore, the ordering of the alternatives through the values of $SVNLWIOWAD(A_i, I)(i = 1, 2, 3, 4)$ was $A_3 \succ A_2 \succ A_1 \succ A_4$, which implies that the optimal company A_3 is the best choice for investment.

To conduct a comparative analysis with the existing methods, in this example we utilized the SVNLWD, SVNLOWAD, and SVNLIOWAD to measure the relative performance of all alternatives to the ideal scheme, and the aggregation results are listed in the Table 7.

Table 7. Aggregation results.

	A_1	A_2	A_3	A_4	**Ranking**
$SVNLWD(A_i, I)$	6.828	5.836	5.048	6.444	$A_3 \succ A_2 \succ A_4 \succ A_1$
$SVNLOWAD(A_i, I)$	6.466	5.652	4.802	6.460	$A_3 \succ A_2 \succ A_4 \succ A_1$
$SVNLIOWAD(A_i, I)$	6.770	5.788	4.833	6.460	$A_3 \succ A_2 \succ A_1 \succ A_4$

From the Table 7, it is easy to see that the most desirable alternative was A_3 for the different distance measures used, which was the same as the result obtained from the SVNLWIOWAD operator. We also found that the ranking of alternatives may change for the different distance measures used because the different operators include different information. The SVNLWD uses the importance of attributes and the SVNLOWD focuses on the ordered location of the arguments. The SVNLIOWAD considers the attitudinal character of the decision-makers, while the SVNLIOWAD operator includes more information than the SVNLIOWAD as its design function of the order-induced variables. It is worth pointing out that the SVNLWIOWAD operator not only combines the advantages of the existing methods, but also overcomes some of their shortcomings, so that it can achieve a more scientific and reasonable result.

5. Conclusions

With the help of SVNLNs, decision makers may easily evaluate alternatives by linguistic terms as well as uncertainty degrees, which is very close to human cognition. In order to highlight the theory and application of SVNLS, in this paper, we explored some distance measures for SVNLSs from an induced aggregation point of view. Firstly, we put forward the SVNLIOWAD operator, which is a useful extension of the existing IOWAD operator. Then, a novel induced aggregation distance, namely the single valued neutrosophic linguistic weighted IOWAD (SVNLWIOWAD) operator, was developed to overcome the defects of the existing methods. The key feature of the SVNLWIOWAD is that it extends the functions of the order-inducing variables, which not only induce the order of arguments, but also moderate the associated weights. Compared with the existing methods, wherein the order-inducing variables just play the induced function, this dual role enables the SVNLWIOWAD operator to effectively measure the intrinsic variation of the induced variables on the integration results. Therefore, it can consider the complex attitudinal characteristics as well as reflect the influence of the induced variables on the aggregation results by moderating the associated weights. An MAGDM method, based on the SVNLWIOWAD operator, was further presented, which turned out to be a very powerful approach to handle decision making problems under SVNL situation. Finally, a numerical example on investment selection and comparative analysis were utilized to demonstrate the feasibility and effectiveness of the proposed method.

For future research, we will consider some methodological extensions and application of the proposed method with other decision making approaches, such as moving averaging and probability information.

Author Contributions: S.Z. and C.Z. drafted the initial manuscript and conceived the MADM framework. G.Y. provided the relevant literature review and the illustrated example. All authors have read and agree to the published version of the manuscript.

Funding: This paper was supported by China Postdoctoral Science Foundation (No. 2019M651403), Major Humanities and Social Sciences Research Projects in Zhejiang Universities (No. 2018QN058), Zhejiang Province Natural Science Foundation (No. LY18G010007; No. LQ20G010001), Ningbo Natural Science Foundation (No. 2019A610037) and First Class Discipline of Zhejiang - A (Zhejiang Gongshang University - Statistics).

Conflicts of Interest: The authors declare no conflict of interest.

References

1. Zadeh, L.A. Fuzzy sets. *Inf. Control.* **1965**, *18*, 338–353. [CrossRef]
2. Atanassov, K.T. Intuitionistic fuzzy sets. *Fuzzy Sets Syst.* **1986**, *20*, 87–96. [CrossRef]
3. Cuong, B.C. Picture fuzzy sets. *J. Comput. Sci. Cybern.* **2014**, *30*, 409–420.
4. Wei, G.W. Picture fuzzy aggregation operators and their application to multiple attribute decision making. *J. Intell. Fuzzy Syst.* **2017**, *33*, 713–724. [CrossRef]
5. Herrera, F.; Herrera-Viedma, E. Linguistic decision analysis: Steps for solving decision problems under linguistic information. *Fuzzy Sets Syst.* **2000**, *115*, 67–82. [CrossRef]
6. Smarandache, F. *A Unifying Field in Logics. Neutrosophy: Neutrosophic Probability, Set and Logic*; American Research Press: Rehoboth, DE, USA, 1999.
7. Ye, J. An extended TOPSIS method for multiple attribute group decision making based on single valued neutrosophic linguistic numbers. *J. Intell. Fuzzy Syst.* **2015**, *28*, 247–255. [CrossRef]
8. Wang, H.; Smarandache, F.; Zhang, Y.Q.; Sunderraman, R. Single valued neutrosophic sets. *Multispace Multistruct.* **2010**, *4*, 410–413.
9. Hwang, C.L.; Yoon, K. Multiple Attribute Decision Making: Methods and Applications. In *A State-of-the-Art Survey*; Springer: Berlin, Germany, 1981.
10. Wang, J.Q.; Yang, Y.; Li, L. Multi-criteria decision-making method based on single-valued neutrosophic linguistic Maclaurin symmetric mean operators. *Neural Comput. Appl.* **2018**, *30*, 1529–1547. [CrossRef]
11. Chen, J.; Zeng, S.Z.; Zhang, C.H. An OWA Distance-Based, Single-Valued Neutrosophic Linguistic TOPSIS Approach for Green Supplier Evaluation and Selection in Low-Carbon Supply Chains. *Int. J. Environ. Res. Public Health* **2018**, *15*, 1439. [CrossRef]
12. Wu, Q.; Wu, P.; Zhou, L. Some new Hamacher aggregation operators under single-valued neutrosophic 2-tuple linguistic environment and their applications to multi-attribute group decision making. *Comput. Ind. Eng.* **2018**, *116*, 144–162. [CrossRef]
13. Kazimieras, Z.E.; Bausys, R.; Lazauskas, M. Sustainable Assessment of Alternative Sites for the Construction of a Waste Incineration Plant by Applying WASPAS Method with Single-Valued Neutrosophic Set. *Sustainability* **2015**, *7*, 15923–15936. [CrossRef]
14. Garg, H.; Nancy. Linguistic single-valued neutrosophic prioritized aggregation operators and their applications to multiple-attribute group decision-making. *J. Ambient. Intell. Humaniz. Comput.* **2018**, *9*, 1975–1997. [CrossRef]
15. Cao, C.D.; Zeng, S.Z.; Luo, D.D. A Single-Valued Neutrosophic Linguistic Combined Weighted Distance Measure and Its Application in Multiple-Attribute Group Decision-Making. *Symmetry* **2019**, *11*, 275. [CrossRef]
16. Xu, Z.S.; Chen, J. Ordered weighted distance measure. *J. Syst. Sci. Syst. Eng.* **2008**, *16*, 529–555. [CrossRef]
17. Merigó, J.M.; Gil-Lafuente, A.M. New decision-making techniques and their application in the selection of financial products. *Inf. Sci.* **2010**, *180*, 2085–2094. [CrossRef]
18. Zeng, S.Z.; Xiao, Y. A method based on TOPSIS and distance measures for hesitant fuzzy multiple attribute decision making. *Technol. Econ. Dev. Econ.* **2018**, *24*, 969–983. [CrossRef]
19. Merigó, J.M.; Casanovas, M. Decision making with distance measures and induced aggregation operators. *Comput. Ind. Eng.* **2011**, *60*, 66–76. [CrossRef]

20. Zeng, Z.S.; Li, W.; Merigó, J.M. Extended induced ordered weighted averaging distance operators and their application to group decision-making. *Int. J. Inf. Technol. Decis. Mak.* **2013**, *12*, 1973–6845. [CrossRef]
21. Xian, S.D.; Sun, W.J. Fuzzy linguistic induced Euclidean OWA distance operator and its application in group linguistic decision making. *Int. J. Intell. Syst.* **2014**, *29*, 478–491. [CrossRef]
22. Zeng, S.Z.; Merigó, J.M.; Palacios-Marques, D.; Jin, H.H.; Gu, F.J. Intuitionistic fuzzy induced ordered weighted averaging distance operator and its application to decision making. *J. Intell. Fuzzy Syst.* **2017**, *32*, 11–22. [CrossRef]
23. Li, C.G.; Zeng, S.Z.; Pan, T.J.; Zheng, L.N. A method based on induced aggregation operators and distance measures to multiple attribute decision making under 2-tuple linguistic environment. *J. Comput. Syst. Sci.* **2014**, *80*, 1339–1349. [CrossRef]
24. Ye, J. Multicriteria decision-making method using the correlation coefficient under single-valued neutrosophic environment. *Int. J. Gen. Syst.* **2013**, *42*, 386–394. [CrossRef]
25. Xu, Z.S. A note on linguistic hybrid arithmetic averaging operator in multiple attribute group decision making with linguistic information. *Group Decis. Negot.* **2006**, *15*, 593–604. [CrossRef]
26. Yager, R.R.; Filev, D.P. Induced ordered weighted averaging operators. *IEEE Trans. Syst. Man Cybern. Part B* **1999**, *29*, 141–150. [CrossRef] [PubMed]
27. Yu, L.P.; Zeng, S.Z.; Merigo, J.M.; Zhang, C.H. A new distance measure based on the weighted induced method and its application to Pythagorean fuzzy multiple attribute group decision making. *Int. J. Intell. Syst.* **2019**, *34*, 1440–1454. [CrossRef]
28. Zhou, L.; Tao, Z.; Chen, H.; Liu, J. Generalized ordered weighted logarithmic harmonic averaging operators and their applications to group decision making. *Soft Comput.* **2014**, *19*, 715–730. [CrossRef]
29. Aggarwal, M. A new family of induced OWA operators. *Int. J. Intell. Syst.* **2015**, *30*, 170–205. [CrossRef]
30. Merigó, J.M.; Palacios-Marqués, D.; Soto-Acosta, P. Distance measures, weighted averages, OWA operators and Bonferroni means. *Appl. Soft Comput.* **2017**, *50*, 356–366. [CrossRef]
31. Balezentis, T.; Streimikiene, D.; Melnikienė, R.; Zeng, S.Z. Prospects of green growth in the electricity sector in Baltic States: Pinch analysis based on ecological footprint. *Resour. Conserv. Recycl.* **2019**, *142*, 37–48. [CrossRef]
32. Zeng, S.Z.; Mu, Z.M.; Balezentis, T. A novel aggregation method for Pythagorean fuzzy multiple attribute group decision making. *Int. J. Intell. Syst.* **2018**, *33*, 573–585. [CrossRef]
33. Zeng, S.Z.; Chen, S.M.; Kuo, L.W. Multiattribute decision making based on novel score function of intuitionistic fuzzy values and modified VIKOR method. *Inf. Sci.* **2019**, *488*, 76–92. [CrossRef]
34. Zeng, S.Z.; Peng, X.M.; Baležentis, T.; Streimikiene, D. Prioritization of low-carbon suppliers based on Pythagorean fuzzy group decision making with self-confidence level. *Econ. Res. Ekon. Istraž.* **2019**, *32*, 1073–1087. [CrossRef]

symmetry

MDPI

Article

A Hybrid Approach of Interval Neutrosophic Vague Sets and DEMATEL with New Linguistic Variable

Ashraf Al-Quran [1,*], Hazwani Hashim [2] and Lazim Abdullah [3]

[1] Preparatory Year Deanship, King Faisal University, Hofuf 31982, Al-Ahsa, Saudi Arabia
[2] Faculty of Computer and Mathematical Sciences, Universiti Teknologi Mara (UiTM), Campus Machang, Kelantan 18500, Malaysia; hazwanihashim@uitm.edu.my
[3] School of Informatics and Applied Mathematics, University Malaysia Terengganu, Terengganu, Kuala Nerus 21030, Malaysia; lazim_m@umt.edu.my
* Correspondence: aalquran@kfu.edu.sa

Received: 15 January 2020; Accepted: 8 February 2020; Published: 12 February 2020

check for updates

Abstract: Nowadays, real world problems are complicated because they deal with uncertainty and incomplete information. Obviously, such problems cannot be solved by a single technique because of the multiple perspectives that may arise. Currently, the combination of DEMATEL and the neutrosophic environment are still new and not fully explored. Previous studies of DEMATEL and this neutrosophic environment have been carried out based on numerical values to represent a new scale. Until now, little importance has been placed on the development of a linguistic variable for DEMATEL. It is important to develop a new linguistic variable to represent opinions based on human experience. Therefore, to fill this gap, the concept of Interval Neutrosophic Vague Sets (INVS) has been extended to the linguistic variable that can be used in the decision-making process. The INVS is useful tool to deal with uncertainty and incomplete information. Additionally, the advantages of the linguistic variable of INVS allows the greater range of value for membership functions. This study proposes a new framework for INVS and DEMATEL. In addition, a case study on the quality of hospital service has been evaluated to demonstrate the proposed approach. Finally, a comparative analysis to check the feasibility of the proposed method is presented. It demonstrates that different methods produce different relations and levels of importance. This is due to the inclusion of the INVS linguistic variable.

Keywords: INVS; DEMATEL; linguistic variable

---

1. Introduction

Multi Criteria Decision Making (MCDM) was introduced in the mid-1960s, and is still a hot topic in decision making. The application fields of the MCDM include in-system engineering [1], energy planning [2,3], supply chain-selection [4], risk management [5], water resources management [6], and so on. Besides that, Pamučar et al. [7] used the MCDM method to select of the optimal type of hotel for investment. MCDM can be defined as a systematic and standardized method of decision making to resolve complex problems [8]. This method requires decision makers to choose the best among a set of alternatives by comparing them according to the relevant criteria. Today, the Trial Evolution Laboratory (DEMATEL) method approach is one of the widely known MCDM methods. In the 1970s, the DEMATEL method was developed to solve complex problems in the identifying relationships between cause–effect [9]. In DEMATEL, there are formally four basic steps: the development of a direct influence matrix, establishing the direct influence matrix, constructing the total influence matrix and producing the influential relation map. DEMATEL's strengths are as a systematic tool for constructing and evaluating the structure of complex causal relationships between matrix or diagram variable.

Generally, crisp numbers are used to represent the existing scale in classical DEMATEL in order to reflect the ambiguity and vagueness that occur in the decision-making problems. However, several studies have criticized classical DEMATEL, which is insufficient to resolve ambiguity due to the input of linguistic experts into the information [10–13]. Thus, the DEMATEL method is extended by integrating with fuzzy set theory. The combination is called Fuzzy DEMATEL.

Zadeh [14] introduced Fuzzy Sets to overcome the confusion in decision making. Fuzzy DEMATEL has been successfully applied in various applications. Most of the linguistic variables in DEMATEL are constructed based on Fuzzy Set. This model has been applied in green supply chain management practices by [15–17]. Meanwhile Akyuz and Celik [18] used Fuzzy DEMATEL to evaluate critical operational hazards during the gas freeing process. Atanassov [19] extended the concept of the fuzzy set to the intuitionistic fuzzy set (IFS). An IFS consists of membership and non-membership to deal with uncertain information. A study by Govindan et al. [20] applied IFS with DEMATEL to handle the linguistic impression and ambiguity of human judgment. Another study by Li et al. [21] used IFS as a linguistic variable with a DEMATEL to identify critical success factors in emergency management. Hosseini et al. [22] proposed a fuzzy extension of the DAMATEL. In this study, the linguistic variable is in form of a type 2 fuzzy set to obtain the weight of criteria based on word. Later, research by Dalalah et al. [23] developed a modified fuzzy DEMATEL where the fuzzy distance measure is presented. The FPIS and FNIS are used to find similarities of the available alternatives. There was an attempt made by Abdullah and Zulkifli [24] to propose the integration of fuzzy AHP and interval type 2 fuzzy DEMATEL. The authors focus on linguistic variables in interval type-2 fuzzy sets (IT2FS) and the expected value for normalizing the upper and lower membership of IT2FS. Authors in [25] developed an interval type-2 fuzzy set based hierarchical MADM model by combining DEMATEL and TOPSIS. The inherent complexity that arises in the decision-making problem is solved using a hierarchical decomposition approach. The interval type-2 fuzzy DEMATEL is used to solve interdependencies among problem attributes. Baykasoğlu et al. [26] proposed fuzzy DEMATEL for the assessment of criteria, weight of criteria and the hierarchical fuzzy TOPSIS method for the assessment of alternatives by criteria.

Gray system theory is a good theory that combines with MCDM, and this set is being used with DEMATEL. Julong [27] implemented the gray system to solve uncertainties and incomplete information [28–31]. Besides that, the combination between gray–fuzzy and DEMATEL in expert judgment to evaluate interrelationship of service quality has been done by Tseng [32]. In short, several kinds of extensions of DEMATEL are used to model uncertainty inherent in the assessment. Nevertheless, some sources of uncertainty are partially or completely overlooked in the previous literature [28].

The neutrosophic set is a powerful tool for dealing with uncertainty-related issues, and consists of the level of truth, indeterminate and false degrees. In recent years, the theory extensions of neutrosophic have made rapid progress among scholars, such as [33–38]. A considerable amount of literature has been published on neutrosophic and MCDM, such as Dung et al. [39], who used interval neutrosophic set with TOPSIS to evaluate personnel selection. In addition, one work [40] suggested the TOPSIS method for MCDM under a single-valued neutrosophic set, and illustrated it by example. Abdel-Basset et al. [41] implemented the combination in the neutrosophic context of the Analytic Hierarchy Process (AHP) and Delphi Method. The authors have highlighted different techniques for monitoring consistency and evaluating the consensus level of expert opinions. Pamučar et al. [42] developed a new model which combines linguistic neutrosophic numbers (LNNs) and the weighted aggregated sum product assessment (WASPAS) for evaluating consultants' work in hazardous goods transport. In addition, Abdel-Basset et al. [43] developed a combination of the neutrosophic ANP and VIKOR method to achieve sustainable supplier choice. The triangular neutrosophic numbers (TriNs) are used in this study to represent a linguistic variable based on opinion experts and decision makers. However, a combination of neutrosophic, particularly with DEMATEL, has not yet been fully explored [44]. The literature published related to DEMATEL and the neutrosophic environment, such as Abdel-Basset et al. [45] simply represents numerical values without focusing on the linguistic variable.

Most experts cannot give accurate numerical values to represent opinions based on human experience and rather use linguistic assessments as opposed to numerical values to be more practical [10,46]. This method seems to lack information on the linguistic parameter, since the key shortcoming of DEMATEL is that it relies on the input of linguistic experts [12,47]. Hence to fill this gap, we develop a new linguistic variable under the neutrosophic environment. Our proposed method can be seen as a DEMATEL framework in which interval neutrosophic vague sets are used as the linguistic variable. The benefits of our new linguistic variable allow greater range of values for the membership functions, since a new parameter is added to the interval neutrosophic set. It considers more range of values while handling the uncertainty that arises in decision-making problems. The insertion of INVS in DEMATEL gives a new representation of the model.

The remainder of the paper is organized as follows: In Section 2, some fundamental concepts of interval neutrosophic vague sets is presented. Section 3 discusses the proposed method, and Section 4 introduces an implementation of the proposed method. Finally, Section 5 describes the findings and proposal for future study.

2. Preliminaries

This section introduces the basic definitions related to the interval neutrosophic vague set (INVS).

Definition 1. [48] *Let U be a universe discourse and the interval-valued neutrosophic set S is defined as follows:*

$$S = \left\{a, \left\langle \left[m_S^L(a), m_S^U(a)\right], \left[n_S^L(a), n_S^U(a)\right], \left[p_S^L(a), p_S^U(a)\right] \right\rangle \middle| a \in U \right\} \tag{1}$$

where $\left[m_S^L(a), m_S^U(a)\right] \in [0,1]$, $\left[n_S^L(a), n_S^U(a)\right] \in [0,1]$, $\left[p_S^L(a), p_S^U(a)\right] \in [0,1]$ *satisfies* $0 \le m_S(a) + n_S(a) + p_S(a) \le 3$. *When the upper and lower limits of* $m_S(a)$, $n_S(a)$, $p_S(a)$ *in INS are equal, the INS is reduced to SVNS. For notational convenience, we use* $S = \left\langle \left[m_S^L(a), m_S^U(a)\right], \left[n_S^L(a), n_S^U(a)\right], \left[p_S^L(a), p_S^U(a)\right] \right\rangle$ *to represent the element S in INS, while the element S refers to an interval-valued neutrosophic number (INN).*

Definition 2. [49] *Let S be a universe discourse U. Then an interval neutrosophic vague set denoted as* S_{INV} *is written as:*

$$S_{INV} = \left\{a, \left[\overline{m}_S^L(a), \overline{m}_S^U(a)\right], \left[\overline{n}_S^L(a), \overline{n}_S^U(a)\right], \left[\overline{p}_S^L(a), \overline{p}_S^U(a)\right] > \middle| a \in U \right\} \tag{2}$$

Whose truth membership, indeterminacy membership and falsity-membership functions are defined as:

$$\begin{aligned} &\overline{m}_S^L(a) = \left[m^{L^-}, m^{L+}\right], \overline{m}_S^U(a) = \left[m^{U-}, m^{U+}\right], \overline{n}_S^L(a) = \left[n^{L-}, n^{L+}\right], \overline{n}_S^U(a) = \left[n^{U-}, n^{U+}\right] \\ &\text{and } \overline{p}_S^L(e) = \left[p^{L-}, p^{L^+}\right], \overline{p}_S^U(e) = \left[p^{U-}, p^{U+}\right] \end{aligned} \tag{3}$$

where

$$\begin{aligned} m^{L+} &= 1 - p^{L-}, p^{L+} = 1 - m^{L-}, \\ m^{U+} &= 1 - p^{U-}, p^{U+} = 1 - m^{U-}, \\ {}^{-}0 \le m^{L-} &+ m^{U-} + n^{L-} + n^{U-} + p^{L-} + p^{U-} \le 4^{+}, \\ {}^{-}0 \le m^{L+} &+ m^{U+} + n^{L+} + n^{U+} + p^{L+} + p^{U+} \le 4^{+}. \end{aligned} \tag{4}$$

Definition 3. [49] *Let κ_{INV} be an INVS of the universe U where $\forall a_i \in U$,*

$$\overline{m}^L_{\kappa_{INV}}(a) = [1,1],\ \overline{m}^U_{\kappa_{INV}}(a) = [1,1],$$
$$\overline{n}^L_{\kappa_{INV}}(a) = [0,0],\ \overline{n}^U_{\kappa_{INV}}(a) = [0,0],$$
$$\overline{p}^L_{\kappa_{INV}}(a) = [0,0],\ \overline{p}^U_{\kappa_{INV}}(a) = [0,0].$$

Then, a unit INVS is denoted as κ_{INV} where $1 \le i \le n$.

Definition 4. [49] *Let η_{INV} be an INVS of the universe U where $\forall a_i \in U$,*

$$\overline{m}^L_{\eta_{INV}}(a) = [0,0],\ \overline{m}^U_{\eta_{INV}}(a) = [0,0],$$
$$\overline{n}^L_{\eta_{INV}}(a) = [1,1],\ \overline{n}^U_{\eta_{INV}}(a) = [1,1],$$
$$\overline{p}^L_{\eta_{INV}}(a) = [1,1],\ \overline{p}^U_{\eta_{INV}}(a) = [1,1].$$

Hence, a zero INVS is denoted as η_{INV} where $1 \le i \le n$.

3. Proposed Method

This section is presented mainly to discuss the development of the INVS-DEMATEL. In this study, a new linguistic variable for INVS DEMATEL is constructed, and some changes have been made to DEMATEL without the loss of originality of the DEMATEL method. Figure 1 demonstrates the overall structure of the proposed method.

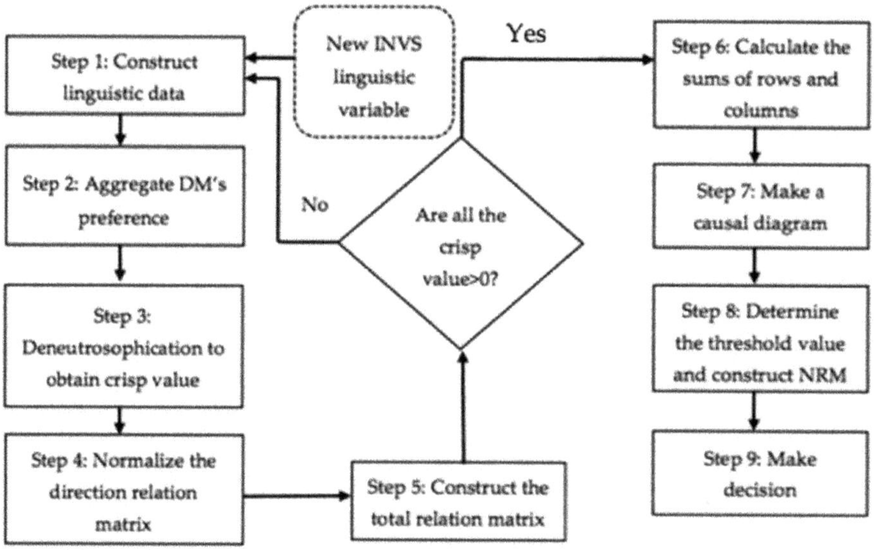

Figure 1. Algorithm of the proposed method.

The proposed method consists of nine steps, and is basically similar with the concept of DEMATEL. However, the difference in the proposed method is especially in the development of the linguistic variable. The proposed method INVS-DEMATEL uses the linguistic variable developed from the interval neutrosophic set. Definition of INS in [49] is extended to the new linguistic variable in the form of IVNS. The aggregation operator is used to aggregate all the experts' opinion. The important step is, the total relation matrix should be greater than zero before the casual diagram is obtained. The threshold value is setup and the degree of importance and net impact is obtained from the NRM.

3.1. *Construction of Linguistic Variable*

Gabus et al. [9] introduced a 4-degree scale in the classical DEMATEL. The most commonly used are: the original 4-degree scale and a 3-degree scale, but other scales such as a 5-degree scale or even an 8-degree scale are also available [50]. The linguistic variable under neutrosophic environment SVNS and INS have been developed by [39,40]. In this study, we have constructed a new linguistic variable for INVS based on Equation (1). The linguistic variable INVS consists of a 5-degree scale. Table 1 shows the linguistic variables for INS:

Table 1. Linguistic variable [39].

Linguistic Variable	Interval Neutrosophic Set
No Influence (NI)	$[0.1, 0.2], [0.5, 0.6], [0.7, 0.8]$
Very Low Influence (LI)	$[0.2, 0.4], [0.5, 0.6], [0.5, 0.6]$
Medium Influence (MI)	$[0.4, 0.6], [0.4, 0.5], [0.3, 0.4]$
High Influence (HI)	$[0.6, 0.8], [0.3, 0.4], [0.2, 0.4]$
Absolutely Influence (AI)	$[0.7, 0.9], [0.2, 0.3], [0.1, 0.2]$

In order to illustrate this conversion, the linguistic variable of "No Influence" from Table 1 is considered and calculated as follows:

Step 1: Convert linguistic variable of INS to INVS.

Using the definition of the interval neutrosophic set Equation (1), we have $S = \{\langle a, m_S(a), n_S(a), p_S(a)\rangle : a \in U\}$, where $m_S(a) = \left[m_S^L(a), m_S^U(a)\right] \subseteq [0,1]$, $n_S(a) = \left[n_S^L(a), n_S^U(a)\right] \subseteq [0,1]$ and $p_S(a) = \left[p_S^L(a), p_S^U(a)\right] \subseteq [0,1]$. Therefore, it is represented as $[0.1, 0.2], [0.5, 0.6], [0.7, 0.8]$.
Using definition of INVS Equation (2) $\left\langle\left\{\left[m^{L-}, m^{L+}\right], \left[m^{U-}, m^{U+}\right]\right\}, \left\{\left[n^{L-}, n^{L+}\right], \left[n^{U-}, n^{U+}\right]\right\}, \left\{\left[p^{L-}, p^{L+}\right], \left[p^{U-}, p^{U+}\right]\right\}\right\rangle$, therefore we obtain $\left\langle\left\{\left[0.1, m^{L+}\right], \left[m^{U-}, 0.2\right]\right\}, \left\{\left[0.5, n^{L+}\right], \left[n^{U-}, 0.6\right]\right\}, \left\{\left[0.7, p^{L+}\right], \left[p^{U-}, 0.8\right]\right\}\right\rangle$.

Step 2: Calculation of $m^{L+}, m^{U-}, n^{L+}, n^{U-}, p^{L+}, p^{L-}$ is obtained by condition of INVS.

Using Equations (3) and (4) and restated $m^{L+} = 1 - p^{L-} = 1 - 0.7 = 0.3$, $p^{L^+} = 1 - m^{L-} = 1 - 0.1 = 0.9$, $p^{U-} = 1 - m^{U+} = 1 - 0.2 = 0.8$ and $p^{U-} = 1 - m^{U+} = 1 - 0.2 = 0.8$. Therefore, we get:

$$\left\langle\{[0.1, 0.3], [0.2, 0.2]\}, \left\{\left[0.5, n^{L+}\right], \left[n^{U-}, 0.6\right]\right\}, \{[0.7, 0.9], [0.8, 0.8]\}\right\rangle$$

In the definition of INVS, the indeterminate value is free since vague set do not handle indeterminacy. Therefore, we can assign any value (if possible) for indeterminacy interval. Therefore, we reach:

$$\langle\{[0.1, 0.3], [0.2, 0.2]\}, \{[0.5, 0.65], [0.6, 0.6]\}, \{[0.7, 0.9], [0.8, 0.8]\}\rangle.$$

Step 3: verify the linguistic variable for INVS.

Using condition $^-0 \leq m^{L-} + m^{U-} + n^{L-} + n^{U-} + p^{L-} + p^{U-} \leq 4^+$, therefore, we have $0.1 + 0.2 + 0.5 + 0.6 + 0.7 + 0.8 = 2.9$ and $^-0 \leq m^{L+} + m^{U+} + n^{L+} + n^{U+} + p^{L+} + p^{U+} \leq 4^+$.

Therefore, we have $0.3 + 0.2 + 0.65 + 0.6 + 0.9 + 0.8 = 3.45$.

The rest of calculation for the linguistic variable INVS is calculated similarly. Finally, we propose the linguistic variables that are defined in INVS, as presented in Table 2.

Table 2. The new linguistic variable under the Interval Neutrosophic Vague Sets (INVS) concept.

Linguistic Variable	Interval Neutrosophic Vague Set
No Influence (NI)	$\langle\{[0.1, 0.3], [0.2, 0.2]\}, \{[0.5, 0.65], [0.6, 0.6]\}, \{[0.7, 0.9], [0.8, 0.8]\}\rangle$
Very Low Influence (LI)	$\langle\{[0.2, 0.5], [0.4, 0.4]\}, \{[0.5, 0.55], [0.5, 0.6]\}, \{[0.5, 0.8], [0.6, 0.6]\}\rangle$
Medium Influence (MI)	$\langle\{[0.4, 0.7], [0.6, 0.6]\}, \{[0.40.45], [0.4, 0.5]\}, \{[0.3, 0.6], [0.4, 0.4]\}\rangle$
High Influence (HI)	$\langle\{[0.6, 0.8], [0.6, 0.8]\}, \{[0.3, 0.35], [0.3, 0.4]\}, \{[0.2, 0.4], [0.2, 0.4]\}\rangle$
Absolutely Influence (AI)	$\langle\{[0.7, 0.9], [0.8, 0.9]\}, \{[0.2, 0.25], [0.2, 0.3]\}, \{[0.1, 0.3], [0.1, 0.2]\}\rangle$

3.2. The INVS DEMATEL Procedures

The procedures of INVS DEMATEL with the new linguistic variable are described as follows:

Step 1: Construct linguistic data using the new linguistic variable.

The decision makers (DMs) constructs a decision matrix based on the proposed INVS linguistic variable. DMs were asked to determine a score using five linguistic variables that ranged from no influence to absolute influence based on criteria. The *kth* DM gave the INVS score a^k_{ij} and the notation of a_{ij} shows the degree to which DM believes criteria i affects criteria j. The diagonal components are set to zero for decision making, where:

$$A^k = \begin{bmatrix} 0 & a^k_{12} & \cdots & a^k_{1n} \\ a^k_{21} & 0 & \cdots & a^k_{2n} \\ \vdots & \vdots & 0 & \vdots \\ a^k_{n1} & a^k_{n2} & \cdots & 0 \end{bmatrix} \tag{5}$$

The matrix contains INVSs in the form of

$$a^k_{ij} = \langle m_{ij}, n_{ij}, p_{ij}\rangle = \left\langle \left\{\left[m^{L-}_{11}, m^{L+}_{12}\right], \left[m^{U-}_{13}, m^{U+}_{14}\right]\right\}, \left\{\left[n^{L-}_{11}, n^{L+}_{12}\right], \left[n^{U-}_{13}, n^{U+}_{14}\right]\right\}, \left\{\left[p^{L-}_{11}, p^{L+}_{12}\right], \left[p^{U-}_{13}, p^{U+}_{14}\right]\right\}\right\rangle$$

Step 2: Aggregate DM's preferences using the mean operator of INVS.

The membership degrees obtained from the DMs are combined using mean operators of INVS as follows:

$$x_{ij} = \frac{1}{H}\sum_{k=1}^{H} a^k_{ij} \tag{6}$$

where H is the total number of DMs and $a^k_{ij} = \langle m_{ij}, n_{ij}, p_{ij}\rangle = \left\langle \left\{\left[m^{L-}_{11}, m^{L+}_{12}\right], \left[m^{U-}_{13}, m^{U+}_{14}\right]\right\}, \left\{\left[n^{L-}_{11}, n^{L+}_{12}\right], \left[n^{U-}_{13}, n^{U+}_{14}\right]\right\}, \left\{\left[p^{L-}_{11}, p^{L+}_{12}\right], \left[p^{U-}_{13}, p^{U+}_{14}\right]\right\}\right\rangle$.

Step 3: Deneutrosophication process to obtain crisp value.

Deneutrosophication is the method by which a crisp number is collected. The deneutrosophication formula is as follows:

Step 4: Normalizing the direct relation matrix.

$$\begin{aligned} B_{ij} = {} & \tfrac{m^{L-}+m^{U-}}{2} + \tfrac{m^{L+}+m^{U+}}{2} + \left(1 - \tfrac{n^{L-}+n^{U-}}{2}\right)l^{U-} + \left(1 - \tfrac{n^{L+}+n^{U+}}{2}\right)n^{U+} \\ & -\left(\tfrac{p^{L-}+p^{U-}}{2}\right)1 - p^{U-} - \left(\tfrac{p^{L+}+p^{U+}}{2}\right)1 - p^{U+} \end{aligned} \tag{7}$$

The initial direct-relation is normalized using $D = B \times S$ where

$$S = \frac{1}{\max\limits_{1 \le i \le n} \sum_{j=1}^{n} b_{ij}} \tag{8}$$

Step 5: Constructing the INVS total relation matrix.

In this step, from the normalized matrix D, the INVS total relation matrix is computed using Equation (9), where I denotes the identity matrix.

$$T = D \times (I - D)^{-1} \tag{9}$$

Step 6: Calculating the sum of the rows and columns.

The sum of rows denoted as R and the sum of columns denoted as C are both calculated as using Equations (10) and (11) as follows:

$$R = \left[\sum_{i=1}^{n} t_{ij} \right]_{n \times 1} \tag{10}$$

$$C = \left[\sum_{j=1}^{n} t_{ij} \right]_{1 \times n} \tag{11}$$

Step 7: Construct a causal diagram.

The graph is constructed by plotting the $(R + C, R - C)$ data set. The $R + C$ on the horizontal axis characterizes as "Prominence" and the vertical axis $R - C$ represents as "Relation". Generally, when $R - C$ is positive, the criterion belongs to the cause group. Otherwise, the criterion belongs to the effect group if $R - C$ is negative. This diagraph is very useful as a decision-making aid.

Step 8: Set up the threshold value and the network relationship map.

In this step, the threshold value referred as θ is calculated by measuring the average of the component in matrix T. Matrix T elements are considered to be zero if they are lower than θ, which means their effect is lower than other criteria. The network relationship map's advantages can reflect the MCDM flow. Each graph node represents the object examined, while the arc between two nodes shows the direction and strength of the influence relationship [50].

4. Illustrative Example: Hospital Service Quality

The proposed INVS DEMATEL with a new linguistic variable has been tested using a numerical example provided by [51].

Step 1: Construct the decision matrix with proposed INVS linguistic variable.

Three decision makers are selected to define key success factors for the performance of hospital service. There are seven criteria involved, which are: F_1: well-equipped medical facilities, F_2: service personnel with good communication skills, F_3: trusted medical staff with professional competence of health care, F_4: service personnel with immediate-solving abilities, F_5: detailed description of the patient's condition by the medical doctor, F_6: medical staff with professional skills and F_7: pharmacist's advice for taking medicine. Table 3 shows DMs analysis based on 7 criteria.

Table 3. Decision makers' (DMs) analysis of the criteria.

	F_1	F_2	F_3	F_4	F_5	F_6	F_7
F_1	0	HI, MI, MI	MI, MI, HI	LI, HI, MI	MI, LI, HI	MI, AI, HI	NI, HI, MI
F_2	MI, LI, HI	0	MI, AI, LI	HI, HI, NI	HI, HI, MI	LI, MI, HI	LI, HI, MI
F_3	M, VU, M	NI, MI, NI	0	HI, NI, LI	NI, MI, AI	NI, MI, HI	HI, MI, LI
F_4	HI, HI, MI	MI, AI, HI	LI, MI, MI	0	MI, HI, AI	NI, HI, MI	NI, AI, HI
F_5	HI, MI, MI	MI, NI, AI	MI, MI, HI	HI, MI, MI	0	NI, MI, HI	AI, HI, MI
F_6	MI, NI, MI	HI, MI, MI	MI, NI, AI	LI, HI, MI	MI, MI, MI	0	LI, MI, AI
F_7	HI, MI, LI	AI, HI, MI	LI, VI, LI	HI, HI, MI	AI, HI, NI	MI, HI, AI	0

Step 2: Aggregate DM's preferences using mean operator of INVS.

Equation (6) is used to aggregate the DM's opinion; for instance the element of a_{12} can be obtained as follows:

$$\begin{aligned} a_{12} = &\tfrac{1}{3}\{[0.4+0.2+0.6,\ 0.7+0.5+0.8], [0.6+0.4+0.6, 0.6+0.4+0.8]\}, \\ &\{[0.4+0.5+0.3, 0.45+0.55+0.35],\ [0.4+0.5+0.3, 0.5+0.6+0.4]\}, \\ &\{[0.3+0.5+0.2, 0.6+0.8+0.4],\ [0.4+0.6+0.2], [0.4+0.6+0.4]\} \\ = &\{[0.4, 0.67], [0.43, 0.6]\}, \{[0.4, 0.43],\ [0.4, 0.5]\}, \{[0.4, 0.46]\} \end{aligned}$$

The rest of the elements are calculated similarly.

Step 3: Deneutrosophication process to obtain crisp value.

Equation (7) is used to obtain crisp value and the result is presented in Table 4.

Table 4. The crisp values of matrix.

	F_1	F_2	F_3	F_4	F_5	F_6	F_7	
F_1	0.0000	1.2461	1.2461	1.1022	1.1022	1.4383	0.9911	7.1261
F_2	1.1022	0.0000	1.1789	1.0833	1.3519	1.1789	1.1022	6.9975
F_3	0.9011	0.7169	0.0000	0.8756	1.0589	0.7169	1.1022	5.3717
F_4	1.3519	1.4383	1.0044	0.0000	1.4383	0.9911	1.1528	7.3769
F_5	1.2461	1.0589	1.2461	1.2461	0.0000	0.9911	1.4383	7.2267
F_6	0.9011	1.2461	1.0589	1.1022	1.1425	0.0000	1.1789	6.6297
F_7	1.1022	1.4383	1.0408	1.3519	1.1528	1.4383	0.0000	7.5244

Step 4: Normalizing the INVS direct relation matrix.

Normalizing the direct relation matrix denoted as D can be achieved using Equation (8). The sum for each row is calculated, and the largest value is obtained by row 7 (see Table 4). Each element in Table 4 is divided by 7.5244. The result is shown in Table 5.

Table 5. The normalize direct relation matrix.

	F_1	F_2	F_3	F_4	F_5	F_6	F_7
F_1	0.0000	0.1656	0.1656	0.1465	0.1465	0.1912	0.1317
F_2	0.1465	0.0000	0.1567	0.1440	0.1797	0.1567	0.1465
F_3	0.1198	0.0953	0.0000	0.1164	0.1407	0.0953	0.1465
F_4	0.1797	0.1912	0.1335	0.0000	0.1912	0.1317	0.1532
F_5	0.1656	0.1407	0.1656	0.1656	0.0000	0.1317	0.1912
F_6	0.1198	0.1656	0.1407	0.1465	0.1518	0.0000	0.1567
F_7	0.1465	0.1912	0.1383	0.1797	0.1532	0.1912	0.0000

Step 5: Construct the INVS total relation matrix.

The total relation matrix T can be computed using Equation (9), where I is denoted as the identity matrix. Since we have 7 criteria, then identity matrix should be size of 7×7. In this step, Maple software is used to calculate total relation matrix. Table 6 shows total relation matrix.

Table 6. The Interval Neutrosophic Vague Sets (INVS) total relation matrix.

	F_1	F_2	F_3	F_4	F_5	F_6	F_7
F_1	1.562	1.8104	1.8104	1.7226	1.8214	1.7509	1.768
F_2	1.5936	1.5657	1.6528	1.6225	1.7405	1.627	1.6726
F_3	1.3341	1.3978	1.3108	1.3572	1.4518	1.336	1.4209
F_4	1.8383	1.9587	1.9587	1.7196	1.9871	1.83	1.9091
F_5	1.7028	1.7904	1.7904	1.7354	1.6903	1.7048	1.8053
F_6	1.6069	1.7427	1.7427	1.6584	1.757	1.5243	1.7163
F_7	1.8336	1.9825	1.9825	1.8937	1.9824	1.8957	1.7983

Step 6: Calculating the sum of the rows and columns.

The sums of rows are represented by R and sums of columns represented by C is calculated by Equations (10) and (11). The $R + C$ and $R - C$ values are calculated in which these values reflect the importance and relation values, respectively. Based on the information in Table 7, the importance degree $R + C$ of criteria towards hospital service quality is identified as $F_7 \phi F_4 \phi F_5 \phi F_2 \phi F_1 \phi F_6 \phi F_3$. The most important criteria that influence the hospital service quality are F_7 and F_4. Meanwhile, F_6 and F_3 are the least important. The details results are presented in Table 7.

Table 7. The total of rows and columns.

	R	C	R+C	**Rank of Importance**	R−C	**Rank of Effect**	**Cause/Effect**
F_1	12.2457	11.4713	23.717	5	0.7744	4	Cause
F_2	11.4747	12.2482	23.7229	4	−0.7735	6	Effect
F_3	9.6086	12.2483	21.8569	7	−2.6397	7	Effect
F_4	13.2015	11.7094	24.9109	2	1.4921	1	Cause
F_5	12.2194	12.4305	24.6499	3	−0.2111	5	Effect
F_6	11.7483	11.6687	23.417	6	0.0796	3	Cause
F_7	13.3687	12.0905	25.4592	1	1.2782	2	Cause

Step 7: Construct a causal diagram.

The complex causal relationships of criteria can be seen in the causal diagram illustrated in Figure 2. In addition, it provides valuable insight into solving problems. The horizontal in this diagram reflects the level of importance of each criterion, while the vertical axis classifies the criteria into the category of causes and effects.

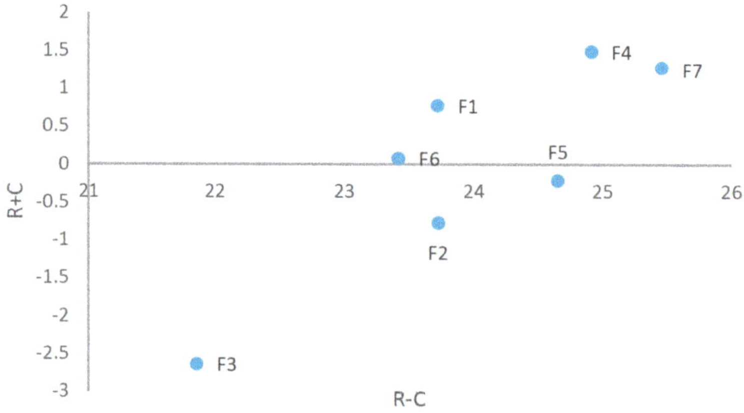

Figure 2. Causal diagram.

Figure 2 shows that the criteria with positive values of $R - C$ are F_1, F_4, F_6, F_7; these criteria are categorized into a cause group. On the other hand, F_6, F_3 and F_5 are categorized into the effect group.

Step 8: Setup a threshold value and construct the network relationship map.

The threshold value θ is obtained by taking the average of the INVS total relation matrix, $\theta = 1.7116$. The values below the θ are set by 0, and the values above the θ are set by 1. Table 8 shows the new total relation matrix denoted as T_θ. Figure 3 displays the graph of the network relationship map to visualize the existent of mutual influence among the criteria. This map is constructed based on new total influence in Table 8. It can be seen that F_1 (well-equipped medical equipment) has arrows pointing toward the other criteria, which indicates that it has an influence on them. On the other hand, there are arrows pointing toward F_1 (well-equipped medical equipment), which indicates that this criterion is affected by some other criteria.

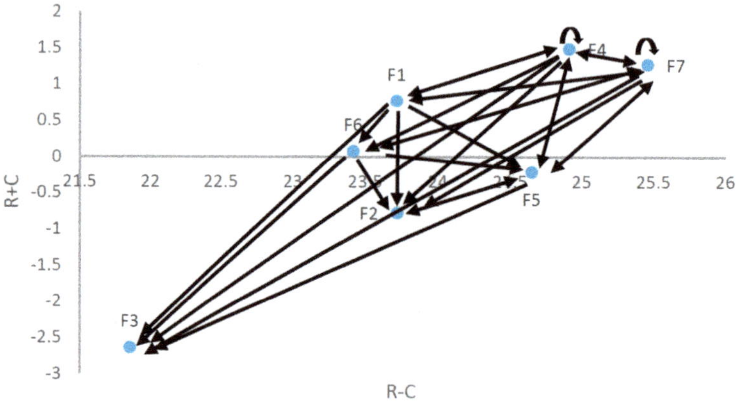

Figure 3. Network relationship map.

Table 8. New total influence.

	F_1	F_2	F_3	F_4	F_5	F_6	F_7
F_1	0	1	1	1	1	1	1
F_2	0	0	0	0	1	0	0
F_3	0	0	0	0	0	0	0
F_4	1	1	1	1	1	1	1
F_5	0	1	1	1	0	0	1
F_6	0	1	1	0	1	0	1
F_7	1	1	1	1	1	1	1

5. Comparison Analysis

In this section, comparison analysis is performed in order to validate the proposed method. For this purpose, firstly we present comparative analysis between DEMATEL, Neutrosophic DEMATEL and the proposed method INVS-DEMATEL. In second section, we compare the INVS-DEMATEL with two other existing models which are interval-valued hesitant fuzzy sets and Neutrosophic DEMATEL.

5.1. Comparative Analysis

Comparative analysis is performed in this study to observe the accuracy of the DEMATEL's modification. Table 9 represents comparative findings of the INVS-DEMATEL method against DEMATEL and Neutrosophic DEMATEL methods. It can be seen that INVS-DEMATEL reveals an obvious difference degree of importance and net impact. For example, by using the proposed method, F_7 is the most important criterion. The main reason is that the new linguistic variable is included. Even though a new linguistic variable is suggested, the proposed method was implemented without losing the originality of the DEMATEL method. In addition, INVS-DEMATEL introduces a new linguistic variable that consists of truth, falsity and indeterminacy degrees which are not limited to a single interval. This set provides interval-based membership when dealing with incomplete and inconsistent information. Meanwhile, classical DEMATEL uses crisp value to solve the uncertainty problems. The classical DEMATEL cannot represent the better decision under uncertainty. Besides that, Neutrosophic DEMATEL is characterized by truth membership, indeterminacy membership and falsity membership. In real life, some complicated problems cannot be solved by SVNS, but need to use several possible values. Therefore, INVS is blended with DEMATEL to accurately assess the relationship between the factors.

Table 9. Comparative results of the different models.

Type of Assessment	Degree of Importance & Net Impact
INVS-DEMATEL with new linguistic variable (proposed method)	$F_7 \succ F_4 \succ F_5 \succ F_2 \succ F_1 \succ F_6 \succ F_3$ Cause criterion: F_1, F_4, F_6, F_7 Effect criterion: F_2, F_3, F_5
Neutrosophic DEMATEL	$F_3 \succ F_2 \succ F_6 \succ F_1 \succ F_4 \succ F_5 \succ F_7$ Cause criterion: F_2, F_3, F_6 Effect criterion: F_1, F_4, F_5, F_7
DEMATEL	$F_4 \succ F_2 \succ F_6 \succ F_7 \succ F_5 \succ F_3 \succ F_1$ Cause criterion: F_1, F_4, F_7 Effect criterion: F_2, F_3, F_5, F_6

5.2. Comparison between INVS-DEMATEL and the Existing Models

In this study, we have used INVSs as a linguistic variable accompanied with DEMATEL. The development of a new linguistic variable is important to better represent the opinions of experts. The INVS-DEMATEL is applied in the case study of hospital service quality to represent the effectiveness of this model. The findings are in the form of a cause and effect group. In this section, we will compare our proposed INVS-DEMATEL with two other existing models which are interval-valued hesitant fuzzy sets DEMATEL [28] and Neutrosophic DEMATEL [45].

Umut et al. [28] integrate interval-valued hesitant fuzzy sets and DEMATEL to better represent the uncertainty and vagueness in the decision-making problem. The concept of interval-valued hesitant fuzzy set is a generalization of fuzzy set and hesitant fuzzy set [52]. It consists of the membership degrees of an element in the form of several possible interval values. However, the interval-valued hesitant fuzzy set is unable to solve problems that involved indeterminacy. Considering the complexity of the decision-making process, it is difficult to insert the indeterminacy degree during the process of collection data in the decision making. This is beyond the scope of interval-valued hesitant fuzzy sets. Therefore, this integration may lead to incomplete information and results.

Abdel-Baset et al. [45], developed a combination of Neutrosophic and DEMATEL. Neutrosophic sets involve an indeterminacy degree that helps experts to express their opinions more accurately. The authors examine the proposed model for selection of supplier. The Neutrosophic DEMATEL method represents a new scale from 0 to 1 and employs the maximum truth membership degree (α), the minimum indeterminacy membership degree (θ) and the minimum falsity membership degree (β) of a single value neutrosophic number. Nevertheless, in this analysis, authors focused on the numerical values in order to convey the opinion of the experts without emphasizing linguistic variable growth. The DEMATEL method requires qualitative evaluation by experts as data input. Therefore, it is better to represent the experts' opinion in the form of a linguistic variable.

INVS is able to overcome the classical challenges of DEMATEL methods. INVS is characterized by multiple intervals instead of a single interval. INVS allows a greater range of value when dealing with an uncertain and incomplete environment. The use of simple a linguistic variable is not suitable to express the real preferences of the expert. For this reason, our approach is more focused towards defining a new linguistic variable in the entire framework INVS-DEMATEL, and without losing the originality of DEMATEL.

6. Conclusions

A new INVS-DEMATEL has been successfully proposed. We have constructed a new linguistic variable for INVS based on the definition of INS from previous study. The proposed approach is used for identifying the key success factors of hospital service quality. The results show that F_7 is the most important criterion and the most influential criterion among these seven criteria, because it has the highest strength of relation to other criteria. The management should give attention on this criterion so that the hospital service quality is guaranteed. In addition, F_1,F_4, F_6 and F_7 are categorized as a cause criteria group. Meanwhile, the effect criterion group were F_2, F_3 and F_5. A comparative analysis between the proposed methods and the other existing method has been performed. The results show that different methods produce difference results. This research contributes to the literature by filling in the gap of linguistic variable in a neutrosophic environment.

In summary, this research's main results are as follows:

- In this research, the neutrosophic environment was used to establish the linguistic variable of the DEMATEL. The new INVS linguistic variable considers more range of value while handling uncertainty, since a new parameter is added to INS. This is accordance with recommendations by Rodríguez et al. [53]. It is useful to include the complex linguistic variable to capture information in different forms and to manage uncertainties of different types within a single framework.
- The combination of INVS and DEMATEL can manage the complex interactions between criteria.
- The insertion of a vague set with a neutrosophic set gives a new result on the degree of importance and net impact.

As an extension of this study, different types of threshold value should be explored [44]. Additionally, future studies can be extended to another types of aggregation operator. The sensitivity analysis is recommended for the future studies to show the robustness of INVS DEMATEL and its results by changing the criteria weights in different situations. Additionally, the weighted super-matrix should be computed to become a long-term stable super-matrix.

Author Contributions: Conceptualization, H.H. and A.A.-Q.; formal analysis, H.H., A.A.-Q. and L.A.; investigation, A.A.-Q.; methodology, A.A.-Q. and H.H.; writing-original draft, H.H.; writing-review and editing, A.A.-Q., supervision, L.A. and A.A.-Q.; funding acquisition, A.A.-Q. All authors have read and agreed to the published version of the manuscript.

Funding: This work was fully supported by the Deanship of Scientific Research, King Faisal University through the Nasher track under Grant 186311.

Acknowledgments: The authors acknowledge the Deanship of Scientific Research at King Faisal University for the financial support under Nasher Track (Grant No.186311).

Conflicts of Interest: The authors declare no conflict of interest.

References

1. Gou, X.; Xu, Z.; Liao, H. Hesitant fuzzy linguistic entropy and cross-entropy measures and alternative queuing method for multiple criteria decision making. *Inf. Sci.* **2017**, *388–389*, 225–246. [CrossRef]
2. Ekel, P.; Kokshenev, I.; Parreiras, R.; Pedrycz, W.; Pereira, J., Jr. Multiobjective and multiattribute decision making in a fuzzy environment and their power engineering applications. *Inf. Sci.* **2016**, *361–362*, 100–119. [CrossRef]
3. Wang, J.-J.; Jing, Y.-Y.; Zhang, C.-F.; Zhao, J.-H. Review on multi-criteria decision analysis aid in sustainable energy decision-making. *Renew. Sustain. Energy Rev.* **2009**, *13*, 2263–2278. [CrossRef]
4. Wan, S.; Xu, G.; Dong, J. Supplier selection using ANP and ELECTRE II in interval 2-tuple linguistic environment. *Inf. Sci.* **2017**, *385–386*, 19–38. [CrossRef]
5. Rodríguez, A.; Ortega, F.; Concepción, R. An intuitionistic method for the selection of a risk management approach to information technology projects. *Inf. Sci.* **2017**, *375*, 202–218. [CrossRef]
6. Kuang, H.; Kilgour, D.M.; Hipel, K.W. Grey-based PROMETHEE II with application to evaluation of source water protection strategies. *Inf. Sci.* **2015**, *294*, 376–389. [CrossRef]
7. Karabasevic, D.; Popovic, G.; Stanujkic, D.; Maksimovic, M.; Sava, C. An approach for hotel type selection based on the single-valued intuitionistic fuzzy numbers. *Int. Rev.* **2019**, *1–2*, 7–14. [CrossRef]
8. Smarandache, F. α-Discounting Method for Multi-Criteria Decision Making (α-D MCDM). In Proceedings of the 2010 13th Conference on Information Fusion (FUSION), Edinburgh, UK, 26–29 July 2010. [CrossRef]
9. Gabus, A.; Fontella, E. *Perceptions of the World Problematique*; Battelle Institute, Geneva Research Center: Geneva, Switzerland, 1975.
10. Lin, C.-J.; Wu, W.-W. A causal analytical method for group decision-making under fuzzy environment. *Expert Syst. Appl.* **2008**, *34*, 205–213. [CrossRef]
11. Bai, C.; Sarkis, J. A grey-based DEMATEL model for evaluating business process management critical success factors. *Int. J. Prod. Econ.* **2013**, *146*, 281–292. [CrossRef]
12. Haleem, A.; Khan, S.; Khan, M.I. Traceability implementation in food supply chain: A grey-DEMATEL approach. *Inf. Process. Agric.* **2019**, *6*, 335–348. [CrossRef]
13. Cui, L.; Chan, H.K.; Zhou, Y.; Dai, J.; Lim, J.J. Exploring critical factors of green business failure based on grey-decision making trial and evaluation laboratory (DEMATEL). *J. Bus. Res.* **2019**, *98*, 450–461. [CrossRef]
14. Zadeh, L.A. Fuzzy sets. *Inf. Control* **1965**, *8*, 338–353. [CrossRef]
15. Lin, R.-J. Using fuzzy DEMATEL to evaluate the green supply chain management practices. *J. Clean. Prod.* **2013**, *40*, 32–39. [CrossRef]
16. Malviya, R.K.; Kant, R. Identifying Critical Success Factors for Green Supply Chain Management Implementation Using Fuzzy DEMATEL Method. In Proceedings of the 2014 IEEE International Conference on Industrial Engineering and Engineering Management, Selangor, Malaysia, 9–12 December 2014; pp. 214–218. [CrossRef]
17. Lin, K.-P.; Hung, K.-C.; Lin, R.-J. Developing Tω Fuzzy DEMATEL Method for Evaluating Green Supply Chain Management Practices. In Proceedings of the 2014 IEEE International Conference on Fuzzy Systems (FUZZ-IEEE), Beijing, China, 6–11 July 2014; pp. 1422–1427. [CrossRef]
18. Akyuz, E.; Celik, E. A fuzzy DEMATEL method to evaluate critical operational hazards during gas freeing process in crude oil tankers. *J. Loss Prev. Process Ind.* **2015**, *38*, 243–253. [CrossRef]

19. Atanassov, K.T. Intuitionistic fuzzy sets. *Fuzzy Sets Syst.* **1986**, *20*, 87–96. [CrossRef]
20. Govindan, K.; Khodaverdi, R.; Vafadarnikjoo, A. Intuitionistic fuzzy based DEMATEL method for developing green practices and performances in a green supply chain. *Expert Syst. Appl.* **2015**, *42*, 7207–7220. [CrossRef]
21. Li, Y.; Hu, Y.; Zhang, X.; Deng, Y.; Mahadevan, S. An evidential DEMATEL method to identify critical success factors in emergency management. *Appl. Soft Comput.* **2014**, *22*, 504–510. [CrossRef]
22. Hosseini, M.B.; Tarokh, M.J. Type-2 fuzzy set extension of DEMATEL method combined with perceptual computing for decision making. *J. Ind. Eng. Int.* **2013**, *9*, 10. [CrossRef]
23. Dalalah, D.; Hayajneh, M.; Batieha, F. A fuzzy multi-criteria decision making model for supplier selection. *Expert Syst. Appl.* **2011**, *38*, 8384–8391. [CrossRef]
24. Abdullah, L.; Zulkifli, N. Integration of fuzzy AHP and interval Type-2 fuzzy DEMATEL: An application to human resource management. *Expert Syst. Appl.* **2015**, *42*, 4397–4409. [CrossRef]
25. Baykasoğlu, A.; Gölcük, İ. Development of an interval Type-2 fuzzy sets based hierarchical MADM model by combining DEMATEL and TOPSIS. *Expert Syst. Appl.* **2017**, *191*, 194–206. [CrossRef]
26. Baykasoğlu, A.; Kaplanoğlu, V.; Du, Z.D.; Şahin, C. Integrating fuzzy DEMATEL and fuzzy hierarchical TOPSIS methods for truck selection. *Expert Syst. Appl.* **2013**, *40*, 899–907. [CrossRef]
27. Julong Deynrt, D. Introduction to grey system theory. *Introd. Grey Syst. Theory* **1989**, *1*, 1–24.
28. Asan, U.; Kadaifci, C.; Bozdag, E.; Soyer, A.; Serdarasan, S. A new approach to DEMATEL based on interval-valued hesitant fuzzy sets. *Appl. Soft Comput.* **2018**, *66*, 34–49. [CrossRef]
29. Çelikbilek, Y.; Tüysüz, F. An integrated grey based multi-criteria decision making approach for the evaluation of renewable energy sources. *Energy* **2016**, *115*, 1246–1258. [CrossRef]
30. Moktadir, M.A.; Ali, S.M.; Rajesh, R.; Paul, S.K. Modeling the interrelationships among barriers to sustainable supply chain management in leather industry. *J. Clean. Prod.* **2018**, *181*, 631–651. [CrossRef]
31. Bouzon, M.; Govindan, K.; Taboada, C.M. Resources, conservation and recycling evaluating barriers for reverse logistics implementation under a multiple stakeholders' perspective analysis using grey decision making approach. *Resour. Conserv. Recycl.* **2018**, *128*, 315–335. [CrossRef]
32. Tseng, M.-L.; Lin, Y.H. Application of fuzzy DEMATEL to develop a cause and effect model of municipal solid waste management in metro manila. *Environ. Monit. Assess.* **2009**, *158*, 519–533. [CrossRef]
33. Ali, M.; Smarandache, F. Complex neutrosophic set. *Neural Comput. Appl.* **2017**, *28*, 1817–1834. [CrossRef]
34. Alias, S.; Mohamad, D.; Shuib, A. Rough neutrosophic multisets relation with application in marketing strategy. *Neutrosophic Sets Syst.* **2018**, *21*, 36–55.
35. Al-Quran, A.; Hassan, N. Neutrosophic vague soft set and its applications. *Malays. J. Math. Sci.* **2017**, *11*, 141–163.
36. Al-Quran, A.; Hassan, N. Neutrosophic vague soft multiset for decision under uncertainty. *Songklanakarin J. Sci. Technol.* **2018**, *40*, 290–305.
37. Alkhazaleh, S. Neutrosophic vague set theory. *Crit. Rev.* **2015**, *X*, 29–39.
38. Kumar Maji, P. Neutrosophic soft set. *Ann. Fuzzy Math. Inform.* **2013**, *5*, 157–168.
39. Dung, V.; Thu Thuy, L.; Quynh Mai, P.; Van Dan, N.; Thi Mai La, N. TOPSIS approach using interval neutrosophic sets for personnel selection. *Asian J. Sci. Res.* **2018**, *11*, 434–440. [CrossRef]
40. Biswas, P.; Pramanik, S.; Giri, B.C. TOPSIS method for multi-attribute group decision-making under single-valued neutrosophic environment. *Neural Comput. Appl.* **2016**, *27*, 727–737. [CrossRef]
41. Abdel-Basset, M.; Mohamed, M.; Sangaiah, A.K. Neutrosophic AHP-delphi group decision making model based on trapezoidal neutrosophic numbers. *J. Ambient Intell. Humaniz. Comput.* **2018**, *9*, 1427–1443. [CrossRef]
42. Pamučar, D.; Sremac, S.; Stević, Ž.; Ćirović, G.; Tomić, D. New multi-criteria LNN WASPAS model for evaluating the work of advisors in the transport of hazardous goods. *Neural Comput. Appl.* **2019**, *31*, 5045–5068. [CrossRef]
43. Abdel-Baset, M.; Chang, V.; Gamal, A.; Smarandache, F. An integrated neutrosophic ANP and VIKOR method for achieving sustainable supplier selection: A case study in importing field. *Comput. Ind.* **2019**, *106*, 94–110. [CrossRef]
44. Si, S.L.; You, X.Y.; Liu, H.C.; Zhang, P. DEMATEL technique: A systematic review of the state-of-the-art literature on methodologies and applications. *Math. Probl. Eng.* **2018**, *2018*, 1–33. [CrossRef]

45. Abdel-Basset, M.; Manogaran, G.; Gamal, A.; Smarandache, F. A hybrid approach of neutrosophic sets and DEMATEL method for developing supplier selection criteria. *Des. Autom. Embed. Syst.* **2018**, *22*, 257–278. [CrossRef]
46. Herrera, F.; Herrera-Viedma, E.; Martínez, L. A fusion approach for managing multi-granularity linguistic term sets in decision making. *Fuzzy Sets Syst.* **2000**, *114*, 43–58. [CrossRef]
47. Wei, D.; Liu, H.; Shi, K. What are the key barriers for the further development of shale gas in china? A grey-DEMATEL approach. *Energy Rep.* **2019**, *5*, 298–304. [CrossRef]
48. Wang, H.; Smarandache, F.; Zhang, Y.-Q.; Sunderraman, R. *Interval Neutrosophic Sets and Logic: Theory and Applications in Computing*; HEXIS: Phoenix, AZ, USA, 2005.
49. Hashim, H.; Abdullah, L.; Al-Quran, A. Interval neutrosophic vague sets. *Neutrosophic Sets Syst.* **2019**, *25*, 66–75.
50. Kobryń, A. DEMATEL as a weighting method in multi-criteria decision analysis. *Mult. Criteria Decis. Mak.* **2017**, *12*, 153–167. [CrossRef]
51. Shieh, J.-I.; Wu, H.-H.; Huang, K.-K. A DEMATEL method in identifying key success factors of hospital service quality. *Knowl. Based Syst.* **2010**, *23*, 277–282. [CrossRef]
52. Chen, N.; Xu, Z.; Xia, M. Interval-valued hesitant preference relations and their applications to group decision making. *Knowl. Based Syst.* **2013**, *37*, 528–540. [CrossRef]
53. Rodríguez, R.M.; Labella, Á.; Martínez, L. An overview on fuzzy modelling of complex linguistic preferences in decision making. *Int. J. Comput. Intell. Syst.* **2016**, *9*, 81–94. [CrossRef]

symmetry

MDPI

Article

Study of Imaginative Play in Children Using Single-Valued Refined Neutrosophic Sets

Vasantha W. B. [1], Ilanthenral Kandasamy [1,*], Florentin Smarandache [2], Vinayak Devvrat [1] and Shivam Ghildiyal [1]

1 School of Computer Science and Engineering, VIT, Vellore, Tamilnadu-632014, India; vasantha.wb@vit.ac.in (V.W.B.); vinayak.devvrat2015@vit.ac.in (V.D.); shivam.ghildiyal2015@vit.ac.in (S.G.)
2 Department of Mathematics, University of New Mexico, Albuquerque, NM 87301, USA; smarand@unm.edu
* Correspondence: ilanthenral.k@vit.ac.in

Received: 14 February 2020; Accepted: 2 March 2020; Published: 4 March 2020

check for updates

Abstract: This paper introduces Single Valued Refined Neutrosophic Set (SVRNS) which is a generalized version of the neutrosophic set. It consists of six membership functions based on imaginary and indeterminate aspect and hence, is more sensitive to real-world problems. Membership functions defined as complex (imaginary), a falsity tending towards complex and truth tending towards complex are used to handle the imaginary concept in addition to existing memberships in the Single Valued Neutrosophic Set (SVNS). Several properties of this set were also discussed. The study of imaginative pretend play of children in the age group from 1 to 10 years was taken for analysis using SVRNS, since it is a field which has an ample number of imaginary aspects involved. SVRNS will be more apt in representing these data when compared to other neutrosophic sets. Machine learning algorithms such as K-means, parallel axes coordinate, etc., were applied and visualized for a real-world application concerned with child psychology. The proposed algorithms help in analysing the mental abilities of a child on the basis of imaginative play. These algorithms aid in establishing a correlation between several determinants of imaginative play and a child's mental abilities, and thus help in drawing logical conclusions based on it. A brief comparison of the several algorithms used is also provided.

Keywords: neutrosophic sets; Single Valued Refined Neutrosophic Set; applications of Neutrosophic sets; k-means algorithm; clustering algorithms

---

1. Introduction

Neutrosophy is an emerging branch in modern mathematics. It is based on philosophy and was introduced by Smarandache and deals with the concept of indeterminacy [1]. Neutrosophic logic is a generalization of fuzzy logic proposed by Zadeh [2]. A proposition in Neutrosophic logic is either true (T), false (F) or indeterminate (I). This inclusion of indeterminacy makes the neutrosophic logic capable of analyzing uncertainty in datasets. Hence, it can be used to logically represent the uncertain and often inconsistent information in the real world problems. Single Valued Neutrosophic Sets (SVNS) [3] are an instance of a neutrosophic set which can be used in real scientific and engineering applications such as Decision-making problems [4–11], Image Processing [12–14], Social Network Analysis [15], Social problems [16,17] and psychology [18]. The distance and similarity measures have found practical applications in the fields of psychology for comparing different behavioural and cognitive patterns.

Imaginative or pretend play is one of the fascinating topics in child psychology. It begins around the age of 1 year or so. It is at its most prominent during the preschool years when children begin to interact with other children of their own age and begin to access more toys. It is crucial in child development as it helps in the development of language (sometimes the child language which cannot be deciphered by everyone) and also helps nurture the imagination of tiny-tots. However, the factors determining the level of imaginative play in children are varied and complicated and a study of them would help one to assess their mental development. It is here that fuzzy neutrosophic logic comes into play. In this paper, we propose a new notion of Single Valued Refined Neutrosophic Sets (SVRNS) which is a model structured on indeterminate and imaginary notions, coupled with machine learning techniques such as heat maps, clustering, parallel axes coordinate, etc., to study the factors that determine and influence imaginative play in children and how it differs in children with different abilities and skills.

Every child is born different. The personality and behaviour of children is an interplay of several different factors. Psychology is a complicated and varied science and open to subjective interpretations. The study of child psychology in an objective manner can help one uncover several aspects of child behaviour and also result in early detection of certain mental disorders. One of the key motivations of this research is to uncover the factors that determine the mental abilities of a child and the extent of their imagination which helps in predicting their academic and overall performance in later stages. Machine Learning is slowly but steadily becoming one of the hot topics of computer science. Amalgamation of machine learning algorithms and psychology on the basis of complex and neutrosophic logic is certainly exciting and will help to cover new bounds.

This study primarily focuses on the analysis of imaginative play in children on the basis of neutrosophic logic and draws conclusions on the same with the help of clustering algorithms. The approach is initialized by generating a finite number of complex and neutrosophic sets determined by several cognitive, psychological and biological factors that affect imaginative play in the mentioned age group. The primary advantage here is the ability of such sets to deal with the uncertainty, imagination and indeterminacy present in the study of pretend play in children in the age group from 1 to 10 years. With the help of this study, we aim to distinguish the contribution of several factors of imaginative play in children and conclude from the study whether the child has any mental disorders or not and about the general cognitive skills coupled with imagination. This model will also help in identifying factors which may contribute to potential psychological disorders in young children at an early stage and predict the academic performance of the child.

In this research, a new complex fuzzy neutrosophic set is defined which will be used as a model to study the imaginary and indeterminate behaviour in young children in the age group from 1 to 10 years by giving them suitable stimuli for imaginary play. The data were collected from different sources with the help of a questionnaire, observations, recorded sessions and interviews, and after transforming the data into the proposed new neutrosophic logic, they were fitted into the newly constructed model and conclusions were drawn from them using a child psychologist as an expert. This model attempts to discover the extent to which several factors contribute to imaginative play in children of the specified age group and to detect possibilities of mental disorders such as autism and hyperactivity in young children on the basis of the trained model.

The paper is organized into seven major sections which are further divided into a few subsections. Section one is introductory in nature. A detailed analysis of the works related to neutrosophy and its applications to a few relevant fields are presented in section two. It also provides the gaps that have been identified in those works. Section 3 introduces Single Valued Refined Neutrosophic Sets (SVRNS) along with their properties, such as distance measures and related algorithms. It also introduces and discusses several machine learning techniques used for assessment. The description of the dataset used for the application of algorithms such as K-means clustering, heat maps, parallel axes coordinate is given in section four. It also includes the approach involved in processing the data obtained appropriately into SVRNSs. Section 5 provides an illustrative example of the methods described in the preceding section. Section 6 details the results obtained from the application of the discussed algorithms and their respective visualizations. Section 7 discusses the conclusions based on our study and its future scope.

2. Related Works

Fink [19] explored the role of imaginative play in the attainment of conservation and perspectivism with the help of a training study paradigm. Kindergarten children were assigned to certain conditions such as free play in the presence of an experimenter and a control group. The method of their data collection was observation. The results indicate that imaginative play can result in new cognitive structures. The relationship between different types of play experiences and the construction of certain physical or social concepts were also discussed, along with educational implications.

Udwin [20] studied a group of children who had been removed from harmful family backgrounds and placed in institutional care. These children were exposed to imaginative play training sessions. These subjects showed an increase in imaginative behaviour. Age, non-verbal intelligence and fantasy predisposition were determinants of the subjects' response to the training programme, with younger, high-fantasy and high-IQ children being most susceptible to the influence of the training exercises.

Huston-Stein [21] attempted to establish a relationship between social structure and child psychology by employing methods of direct observations of field experiments. The behaviour was then categorised on the basis of a set of defined behavioural categories and evaluated on the basis of suitable metrics. The results focus on establishing correlations between these behavioural categories and classroom structure and draw conclusions on how such social structures impact imaginative play.

Bodrova [22] related another important parameter, namely academic performance, to imaginative play. They have established imaginative play as a necessary prerequisite and one of the major sources of child development. They deduced how imaginative play scenarios require a certain knowledge of environmental setting and how it affects the academic excellence of a child.

Seja [23] explored another important factor in child psychology—emotions. They attempted to determine how imaginative play helps to understand the emotional integration of children. The source of data collected in this study is elementary school children who were tested on verbal intelligence and by standard psychological tests. Conclusions were drawn on the basis of an extensive statistical analysis which also attempted to investigate gender differences.

Neutrosophy has given importance to the imprecision and complexity of data. This is an important reason behind using neutrosophic logic in real life applications. Dhingra et al. [24] attempted to classify a given leaf as diseased or healthy based on the membership functions of the neutrosophic sets. Image segmentation into true, false and indeterminate regions after preprocessing was used to extract features and several classifiers were used to arrive at a classification. A comparative analysis of these classifiers was also provided.

Several researchers [25–30] dealt with algebraic structures of neutrosophic duplets, which are a special case of neutrality. Single Valued Neutrosophic Sets (SVNS), which is particular cases of triplet following the fuzzy neutrosphic membership concepts in their mathematical properties and operations are dealt by Haibin [31].

Haibin [31] gave the notion of Single Valued Neutrosophic Sets (SVNS) along with their mathematical properties and set operations. Properties such as inclusion, complement and union were defined on SVNS. They also gave examples of how such sets can be used in practical engineering applications. SVNS has found a major application in medical diagnosis. Shehzadi [32] presented the use of Hamming distance and similarity measures of given SVNSs to diagnose a patient as having Diabetes, Dengue or Tuberculosis. The three membership functions (truth, falsity and indeterminacy) were assigned suitable values and distance and similarity measures were applied on them. These measures were then used to provide a medical diagnosis. Smarandache and Ali [33] provided the notion of complex neutrosophic sets (CNS). Membership values given to them were of the form a+bi. Several properties of these sets were defined. These sets find applications in electrical engineering and decision-making fields. Neutrosophic Refined Sets where defined in [34].

A more refined and precise view of indeterminacy is provided by Kandasamy [35]. The indeterminacy membership function was further categorized as indeterminacy tending towards truth and indeterminacy tending towards false. Hence, resulting in Double-Valued Neutrosophic

Set (DVNS). Their properties, such as complement, union and equality were also discussed and distance measures were also defined on them. On the basis of these properties, minimum spanning trees and clustering algorithms were described [36]. Dice measures on DVNS were proposed in [37]. The importance given to the indeterminacy of incomplete and imprecise data, as often found in the real world, is a major advantage of the DVNS and hence, is more apt for several engineering and medical applications.

The model of Triple Refined Indeterminate Neutrosophic Set (TRINS) was also introduced by Kandasamy and Smarandache [38]. It categorizes indeterminacy membership function as leaning towards truth and leaning towards false in addition to the traditional three membership functions of neutrosophic sets. After defining the several properties and distance measures, the TRINS was used for personality classification. The personality classification using TRINS has been found to be more accurate and realistic as compared to SVNS and DVNS. Indeterminate Likert scaling using five point scale was introduced in [39] and a sentiment analysis using Neutrosophic refined sets was conducted in [40,41].

To date, the study of imaginative play in children has not been analysed using neutrosophy coupled with an imaginary concept; thus, to cover this unexplored area, the new notion of Single Valued Refined Neutrosophic Sets (SVRNS) that represent imaginary and indeterminate memberships individually were defined. A study of imaginative play in children using Neutrosophic Cognitive Maps (NCM) model was carried out in [42].

3. Single Valued Refined Neutrosophic Set (SVRNS) and Its Properties

This section presents the definition of Single Valued Refined Neutrosophic Set (SVRNS). These sets are based on the essential concepts of real, complex and neutrosophic values which takes membership from the fuzzy interval [0,1]. In a way this can be realized as a mixture of refined neutrosophic sets coupled with real membership values for imaginary aspect. However SVRNS are different from traditional neutrosophic sets. The neutrosophic logic is powerful and can model concepts of arbitrary complexity covering incomplete and imprecise data. Children's behaviour is one such complicated and the imprecise branch that can be modelled as objectively as possible by coupling imaginary or complex nature of data with its indeterminacy.

The concept of SVRNS are defined, developed and described in the following.

3.1. Single Valued Refined Neutrosophic Set (SVRNS)

Definition 1. *Let X be a space of points (objects), with a generic element in X denoted by x. A neutrosophic set A in X is characterised by a truth membership function $T_A(x)$, a true tending towards complex membership function $TC_A(x)$, a complex membership function $C_A(x)$, a false tending towards complex membership function $FC_A(x)$, an indeterminacy membership function $I_A(x)$, and a falsity membership function $F_A(x)$. For each point x in X, there are $T_A(x)$, $TC_A(x)$, $C_A(x)$, $FC_A(x)$, $I_A(x)$, $F_A(x) \in [0, 1]$ and $0 \le T_A(x) + TC_A(x) + C_A(x) + FC_A(x) + I_A(x) + F_A(x) \le 6$. Therefore, a Single Valued Refined Neutrosophic Set (SVRNS) A can be represented by*

$$A = \{\langle T_A(x), TC_A(x), C_A(x), FC_A(x), I_A(x), F_A(x)\rangle | x \in X\}.$$

3.2. Distance Measures of SVRNS

The distance measures of SVRNSs are defined in this section and the related algorithm for determining the distance is given.

Definition 2. *Consider two SVRNSs A and B in a universe of discourse, $X = x_1, x_2, \ldots, x_n$, which are denoted by*

$$A = \{\langle T_A(x_i), TC_A(x_i), C_A(x_i), FC_A(x_i), I_A(x_i), F_A(x_i)\rangle | x_i \in X\},$$

and

$$B = \{\langle T_B(x_i), TC_B(x_i), C_B(x_i), FC_B(x_i), I_B(x_i), F_B(x_i)\rangle | x_i \in X\},$$

where $T_A(x_i)$, $TC_A(x_i)$, $C_A(x_i)$, $FC_A(x_i)$, $I_A(x_i)$, $F_A(x_i)$, $T_B(x_i)$, $TC_B(x_i)$, $C_B(x_i)$, $FC_B(x_i)$, $I_B(x_i)$, $F_B(x_i) \in [0, 1]$ *for every* $x_i \in X$. *Let* w_i $(i = 1, 2, \ldots, n)$ *be the weight of an element* x_i $(i = 1, 2, \ldots, n)$, *with* $w_i \geq 0$ $(i = 1, 2, \ldots, n)$ *and* $\sum_{i=1}^{n} w_i = 1$. *Then, the generalised SVRNS weighted distance is defined as follows:*

$$d_\lambda(A,B) = \{\frac{1}{6}\sum_{i=1}^{n} w_i[|T_A(x_i) - T_B(x_i)|^\lambda + |TC_A(x_i) - TC_B(x_i)|^\lambda + |C_A(x_i) - C_B(x_i)|^\lambda +$$

$$|FC_A(x_i) - FC_B(x_i)|^\lambda + |I_A(x_i) - I_B(x_i)|^\lambda + |F_A(x_i) - F_B(x_i)|^\lambda]\}^{\frac{1}{\lambda}}$$

where $\lambda > 0$.
The above equation reduces to the SVRNS weighted Hamming distance and the SVRNS weighted Euclidean distance, when $\lambda = 1, 2$, *respectively. The SVRNS weighted Hamming distance is given as*

$$d_\lambda(A,B) = \{\frac{1}{6}\sum_{i=1}^{n} w_i[|T_A(x_i) - T_B(x_i)| + |TC_A(x_i) - TC_B(x_i)| + |C_A(x_i) - C_B(x_i)| +$$

$$|FC_A(x_i) - FC_B(x_i)| + |I_A(x_i) - I_B(x_i)| + |F_A(x_i) - F_B(x_i)|]\}$$

where $\lambda = 1$.

The SVRNS weighted Euclidean distance is given as

$$d_\lambda(A,B) = \{\frac{1}{6}\sum_{i=1}^{n} w_i[|T_A(x_i) - T_B(x_i)|^2 + |TC_A(x_i) - TC_B(x_i)|^2 + |C_A(x_i) - C_B(x_i)|^2 +$$

$$|FC_A(x_i) - FC_B(x_i)|^2 + |I_A(x_i) - I_B(x_i)|^2 + |F_A(x_i) - F_B(x_i)|^2]\}^{\frac{1}{2}}$$

where $\lambda = 2$.

The algorithm to obtain the generalized SVRNS weighted distance $d_\lambda(A,B)$ between two SVRNS A and B is given in Algorithm 1.

Algorithm 1: Generalized SVRNS weighted distance $d_\lambda(A,B)$

Input: $X = x_1, x_2, \ldots, x_n$, SVRNS A, B where
$A = \{\langle T_A(x_i), TC_A(x_i), C_A(x_i), FC_A(x_i), I_A(x_i), F_A(x_i)\rangle | x_i \in X\}$,
$B = \{\langle T_B(x_i), TC_B(x_i), C_B(x_i), FC_B(x_i), I_B(x_i), F_B(x_i)\rangle | x_i \in X\}$, $w_i (i = 1, 2, \ldots, n)$
Output: $d_\lambda(A,B)$
begin
$d_\lambda \leftarrow 0$
for $i = 1$ *to* n **do**

$$d_\lambda \leftarrow d_\lambda + \sum_{i=1}^{n} w_i[|T_A(x_i) - T_B(x_i)|^\lambda + |TC_A(x_i) - TC_B(x_i)|^\lambda +$$
$$|C_A(x_i) - C_B(x_i)|^\lambda + |FC_A(x_i) - FC_B(x_i)|^\lambda +$$
$$|I_A(x_i) - I_B(x_i)|^\lambda + |F_A(x_i) - F_B(x_i)|^\lambda]$$

end
$d_\lambda \leftarrow d_\lambda / 6$
$d_\lambda \leftarrow d_\lambda^{\{\frac{1}{\lambda}\}}$
end

The related flowchart is given in Figure 1.

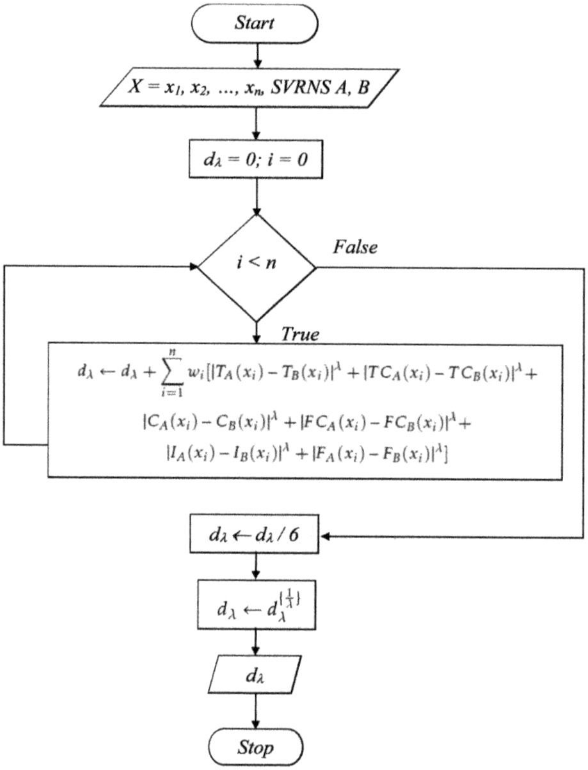

Figure 1. Flow Chart for Generalized SVRNS weighted distance $d_{(\lambda)}$.

The generalised SVRNS weighted distance $d_\lambda\,(A, B)$ for $\lambda > 0$ satisfies the following properties:

1. $d_\lambda\,(A, B) \geq 0$
2. $d_\lambda\,(A, B) = 0$ if and only if $A = B$
3. $d_\lambda\,(A, B) = d_\lambda\,(B, A)$
4. If $A \subseteq B \subseteq C$, C is a SVRNS in X, then $d_\lambda\,(A, C) \geq d_\lambda\,(A, B)$ and $d_\lambda\,(A, C) \geq d_\lambda\,(B, C)$

3.3. *K-Means Algorithm*

The K-means algorithm for SVRNS is given in Algorithm 2.

Algorithm 2: K-means algorithm for clustering SVRNS values

Input: $A_1, A_2, \ldots, A_n$ SVRNS, K—Number of Clusters
Output: K Clusters
begin
Step 1: Choose K different SVRNS A_j as the initial centroids, denoted as α_j, $j = 1, \ldots, K$
Step 2: Initialize $\beta_j \leftarrow 0$, $j = 1, \ldots, K$; `// 0 is a vector with all 0's`
Step 3: Initialize $n_j \leftarrow 0$, $j = 1, \ldots, K$; `// n_j is the number of points in cluster j`
Step 4: Creation of Clusters **repeat**
for $i = 1$ *to* n **do**

$$j \leftarrow \underset{j \in \{1,\ldots,K\}}{\operatorname{argmin}} d_\lambda\left(A_i, \alpha_j\right)$$

`// From Algorithm 1`
assign A_i *to cluster* j
$\beta_j \leftarrow \beta_j + a$
$n_j \leftarrow n_j + 1$
end

$$\alpha_j \leftarrow \frac{\beta_j}{n_j}, \ j = 1, \ldots, K$$

until *Clusters do not change*
end

The related flowchart is given in Figure 2.

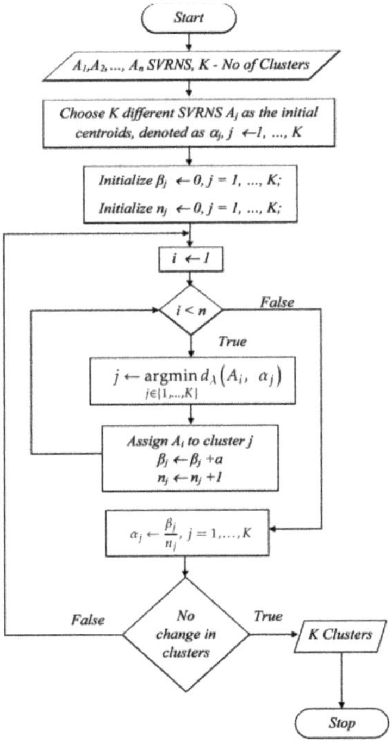

Figure 2. Flow Chart for k means clustering of SVRNS values.

We used the following machine learning techniques in this paper after obtaining and processing the data.

3.4. Other Machine Learning Techniques

The Elbow method is a technique used to find the value of appropriate value of K(Number of clusters) in K-means clustering. It makes the cluster analysis design consistent. A heat map is a data visualization technique used to show correlation between two attributes in the form of a matrix where each value is represented as colours. The Principal Component Analysis (PCA) makes use of orthogonal transformation to convert a set of observations of variables which might be possibly correlated, into a set of values of linearly uncorrelated variables called principal components. It is a widely used statistical technique. Parallel coordinates (also known as Parallel Axes Chart (PAC)) are highly used for the visualization of multi-dimensional geometry and analysis of multivariate data. Easy visualization of multiple dimensions is an innate feature of PAC plot, making it simple to analyse attributes which are associated with other attributes in a similar manner.

4. Dataset Description

Imaginative play is defined as "a form of symbolic play where children use objects, actions or ideas to represent other objects, actions, or ideas using their imaginations to assign roles to inanimate objects or people". During the early stage, "toddlers begin to develop their imaginations, with sticks becoming boats and brooms becoming horses. Their play is mostly solitary, assigning roles to inanimate objects like their dolls and teddy bears". It has proven to be highly beneficial as it results in early use of language and proper use of tenses and adjectives. It gives the children a sense of freedom and allows them to be creative in their own space. It helps children make sense of the physical world and also their inner selves. It can develop with the help of the most basic tools such as a toy mobile or a cardboard tube.

The data regarding imaginative play in children were collected from the local school and an orphanage in Vellore, India.

A child psychologist was present throughout the sessions, analyzed and suggested the various parameters and recorded the observations about each session. The session at each of these places began with the expert talking to the child about general things and everyday life as an ice-breaker exercise. This included talking about his/her favourite subjects, parents and treating him/her with biscuits or chocolates. The surroundings were made as comfortable as possible. The child was then asked to conduct an imaginary phone call in whichever way he/she liked. The imaginary conversation was then recorded as video on a phone. The expert made observations that were recorded on paper in a running hand description. This signified the end of the session.

Overall, 10 such sessions were conducted at the school and 2 were conducted at the orphanage. The children belonged to the age group of 6 to 8 years. Additionally, in order to make the dataset diverse as suggested by the expert, 7 videos were taken from the internet in which children conducted imaginary conversations over the phone. The running hand description thus collected was used by this expert to assign values to the six membership functions based on which the SVRNS is constructed.

Table 1 provides the parameters which have been used to study imaginative play along with their description. The parameters 1 to 11 are available in [19] and the other 4 parameters from 12 to 15 are introduced by us.

Table 1. Parameter Description.

S.No	Parameter Name	Description
1	Imaginative Theme (IT)	The theme of the imaginative play is assumed by the child and can be based on a real or imaginative situation and/or setting.
2	Physical Movements (PM)	The movements a child may make while s/he conducts the imaginative play are also an important determinant of the child's cognitive patterns. They are the ways in which the child uses his/her body during the play.
3	Gestures (G)	They are the ways in which the child moves a part of the body in order to express an idea or some meaning. They are the non-verbal means of communication using hands, head, etc.
4	Facial Expressions (FE)	The movement of facial muscles for non-verbal communication and also convey the emotions experienced by the child.
5	Nature and Length of Social Interaction (NoI/LoI)	The time duration during which the child engages in the imaginative play activity can determine the extent of his/her imagination. The nature of any form of interaction which may take place during the imaginative play be day-to-day, meaningful in some way, etc. and even the combination of the two.
6	Play Materials Used (PMU)	They are the objects provided to the child to conduct an imaginative play activity. The play material used here was a play mobile phone to conduct an imaginary talk.
7	Way Play Materials were Used (WPMwu)	The child's approach to using the play material provided can give an insight into his/her imaginative capabilities.
8	Verbalisation (V)	It is the way in which the child is expressing his/her feelings or emotions during the imaginative play activity.
9	Tone of Voice (ToI)	It is an important aspect that child's mood and state of mind as in if the child is happy, sad or nervous. For example, a high pitched voice may indicate happiness or excitement.
10	Role Identification (RI)	It is the role a child assumes during the imaginative play and the role s/he assigns to other people.
11	Engagement Level (EL)	It is the extent to which the child involves in the activity of imaginative play.
12	Eye Reaction (ER)	It refers to the movement of the eyes during the imaginative play activity. It can give insight into the child's emotions during the play.
13	Cognitive Response (CR)	It is the mental process by which the child forms association between things.
14	Grammar and Linguistics (GaL)	It refers to the ability of a child to make grammatically correct sentences with proper sentence structure and syntax.
15	Coherence (C)	Whether the child is making sense of the talks, i.e., if the sentences formed are related to one another is called coherence.

Method of Evaluation

The running hand description of the above-mentioned parameters was transformed into a complex fuzzy neutrosophic sets by the expert/child psychologist, for applying machine learning algorithms discussed in the earlier section. The methods of evaluation for each parameter as suggested by the expert are discussed below.

1. *Imaginative Theme:* An imaginative theme that is based on the real situation will result in the increase in the truth membership function and otherwise if the theme is entirely imaginative. However, since there is always a degree of complex and indeterminacy in this parameter, the complex and indeterminate membership functions was also assigned certain values from [0,1].
2. *Physical Movements:* If physical movements are made, the value of truth membership function will increase else the falsity membership function will increase. Complex and indeterminacy values from [0,1] shall be assigned values if movements are difficult to interpret properly or happened to be imaginary.
3. *Gestures:* Similar to physical movements, any gestures made in accordance with the imaginative activity will result in an increase in the truth membership value and in falsity value otherwise. Any indeterminate or complex feature will result in values being assigned to indeterminate and complex respectively from [0,1].
4. *Facial Expressions:* Any facial expressions made in accordance with the imaginative activity conducted will lead to an increase in the truth membership and in falsity membership function otherwise. Complex and indeterminacy membership functions shall be assigned values if facial expressions are difficult to interpret properly.

5. *Nature and Length of Social Interaction:* Any interaction that is made in accordance with the play activity will result in an increase in truth membership functions and in falsity membership functions otherwise. Indeterminate and complex membership functions shall be assigned values if the interactions are difficult to interpret properly.
6. *Play Materials Used:* These are nouns and need not be translated to SVRNS.
7. *Way Play Materials were Used:* Any usage of play materials in a realistic manner will lead to an increase in the truth membership and in falsity membership function otherwise. Complex and indeterminacy membership functions shall be assigned values if usage is difficult to interpret properly.
8. *Verbalisation:* Any verbalisation that is made in accordance with the play activity will result in an increase in truth membership functions and in falsity membership functions otherwise. Complex and indeterminacy membership functions shall be assigned values if the verbalization is difficult to interpret properly.
9. *Tone of Voice:* If the tone of voice is in accordance with the situation of play activity and high, it will result in an increase in truth membership functions and in falsity membership functions otherwise. Complex and indeterminacy membership functions shall be assigned values if the interactions are difficult to interpret properly.
10. *Role Identification:* Any role identification that is realistic will lead to an increase in the truth membership and in falsity membership function otherwise. Complex and indeterminacy membership functions shall be assigned values if role identification is difficult to interpret properly.
11. *Engagement Level:* If the engagement level is high but the theme and role identification are realistic, truth membership function value increases. If the engagement level is high but the theme and role identification are imaginative, falsity membership function value increases. Other combinations of engagement level, theme and role identification will result in assigning values to the other membership functions.
12. *Eye Reaction:* Any eye reaction that is made in accordance with the play activity will result in an increase in truth membership functions and in falsity membership functions otherwise. Complex and indeterminacy membership functions shall be assigned values if the eye reaction is difficult to interpret properly.
13. *Cognitive Response:* Any cognitive response that is made in accordance with the play activity will result in an increase in truth membership functions and in falsity membership functions otherwise. Complex and indeterminacy membership functions shall be assigned values if the cognitive is difficult to interpret properly.
14. *Grammar and Linguistics:* If the grammar, sentence structure and syntax are correct, the value of truth membership function will increase. Any error in grammar, syntax or sentence structure will lead to an increase in the value of falsity membership function. If, however, the linguistics are difficult to comprehend, indeterminate and complex membership functions' value will increase.
15. *Coherence:* If the sentences made are related to one another, the value of truth membership function will increase. Any incoherence, i.e., making sentences are not related to one another will lead to an increase in the value of falsity membership function. If, however, the coherence of sentences is difficult to comprehend, indeterminate and complex membership functions' value will increase.

5. Illustrative Example

This section provides an example on processing of the data obtained as a running hand description. On the basis of this description, the expert estimated and evaluated the child. The following example is based on a video of 3-year-old child and the following observations given by the expert are made in form of running hand descriptions of the 15 parameters given in Table 2.

Table 2. Parameter Description for Example.

S.No	Parameter Name	Description
1	Imaginative Theme	The child talks to Mickey Mouse over the phone. The child attempts to discuss something she describes "gross".
2	Physical Movements	The child does not use a lot of her body during the conversation.
3	Gestures	The child does not use any significant gestures during the conversation.
4	Facial Expressions	The child is cheerful, serious and astonished when she initiates the conversation, asks something to the receiver and when comes to know about something "gross" respectively.
5	Nature and Length of Social Interaction	The child engages in the conversation for about a minute. The interaction is mostly day-to-day and the child is rather expressive of her emotions.
6	Play Materials Used	The child uses a toy mobile to conduct an imaginative conversation between herself and Mickey Mouse.
7	Way Play Materials were Used	The child uses the mobile in a very realistic way.
8	Verbalisation	The child makes sound and noises in accordance with the mood of the conversation.
9	Tone of Voice	The tone of the child's voice is high-pitched. She is very expressive.
10	Role Identification	The child does not assume any role other than herself. However, she does imagine herself to be a friend of Mickey Mouse.
11	Engagement Level	The child's engagement level is high and she is attentive throughout the play activity.
12	Eye Reaction	The child's eyes widen and narrow during different points of the play activity.
13	Cognitive Response	The cognitive response is direct, quick and coherent.
14	Grammar and Linguistics	The child makes grammatically correct sentences except she does skip supportive verbs like "will".
15	Coherence	The sentences made are coherent and in sync with the imaginative conversation.

Table 2 depicts a running hand description of the discussed parameters. These parameters are then assigned real values by the expert. These values are discussed in Table 3.

Table 3. SVRNS for Example.

S.No	Parameter	Description	SVRNS
1	IT	Entirely imaginative theme though the conversation was realistic	⟨0.75, 0, 0, 0, 0.25, 0⟩
2	PM	Not a lot	⟨0, 0, 0, 0, 0.25, 0.75⟩
3	G	Not a lot	⟨0, 0, 0, 0, 0.25, 0.75⟩
4	FE	Cheerful, confident, serious	⟨0, 0.75, 0.25, 0, 0, 0⟩
5	NoI/LoI	1 minute; day-to-day, verbal	⟨0.5, 0.25, 0.25, 0, 0, 0⟩
6	PMU	Mobile	NA
7	WPMwu	Realistic	⟨0.75,0,0,0,0.25⟩
8	V	In accordance with imaginative play	⟨0.5, 0.25, 0.25, 0, 0, 0⟩
9	ToI	In accordance with imaginative play; high pitched	⟨0.5, 0.25, 0.25, 0, 0, 0⟩
10	RI	Self	⟨0.5, 0, 0.25, 0, 0.25, 0⟩
11	EL	High	⟨0.5, 0.25, 0.25, 0, 0, 0⟩
12	ER	Widening, narrowing; In accordance with imaginative play	⟨0, 0, 0.5, 0, 0.5, 0⟩
13	CR	Direct; In accordance with imaginative play	⟨0.75, 0, 0, 0, 0.25, 0⟩
14	GaL	Partially correct; In accordance with imaginative play	⟨0.75, 0, 0.25, 0, 0, 0⟩
15	C	In accordance with imaginative play	⟨0.75, 0, 0, 0.25, 0, 0⟩

Likewise the SVRNS tuples for the other data sets was done with the help of the expert. Then these SVRNS sets are used for analysis using machine learning algorithms.

6. Results and Discussions

Several libraries such as pandas, numpy, matplotlib, sklearn, seaborn and pylab associated with Python were used for data visualization. Programming was carried out using python for the visualization of the previous discussed algorithms, based on the result of elbow curve, K-means clustering was done. Logical conclusions have been drawn from these visualizations and the role several determinants play in determining the imaginative capabilities of the child has also been highlighted.

Heat map, which strongly demonstrates the factors of correlation and associativity, has a colour scale in which lighter shades signify positive correlation and darker shades signify a negative correlation. Correlation between any two parameters signifies their associated relation. Positive correlation happens when an increase in one attribute shows an increase in another attribute as well. Negative correlation happens when an increase in one attribute shows a decrease in another attribute. The heat map, which strongly demonstrates the factors of correlation and associativity, has a colour scale in which lighter shades signify positive correlation and darker shades signify a negative correlation. For example, in Figure 3, which is a heat map for feature T, Grammar and Coherence show extremely positive correlation whereas Eye Reaction and Role Identification show a negative correlation.

The results from the Figure 3 shows the heatmap for feature T (Truth membership).

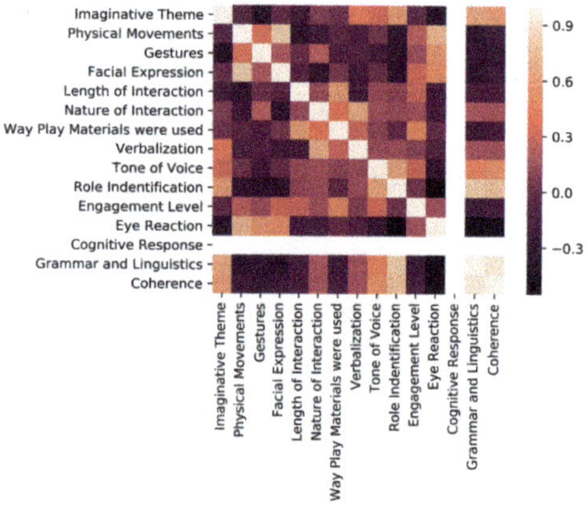

Figure 3. Heat map for feature T.

An elbow curve was plotted to determine the optimal number of clusters for K-means and PCA K-means clustering. Figure 4 shows our elbow curve for feature T where we can see that the sharp bend comes at k = 4, thus, 4 clusters are optimal.

In Figure 5, while testing K-means on feature T for the parameters 'Facial Expression' on the y-axis against 'Imaginative Theme' on the x-axis, it was found that higher concentration of points lies near x = 0.5 and y = 0.2.

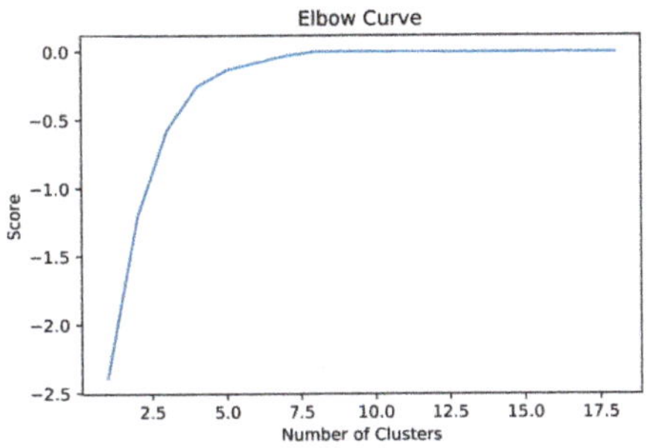

Figure 4. Elbow curve for feature T.

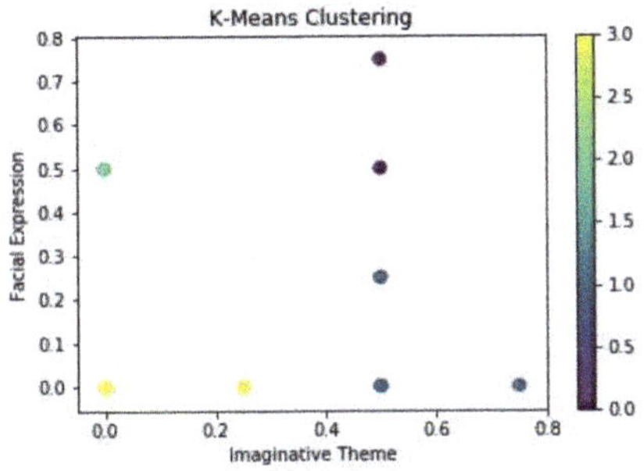

Figure 5. K-means for feature T.

Then, the data was resolved along its principal components, thus giving a new spatial arrangement of the feature, which was then clustered again using K-Means. Figure 6 shows the output for PCA K-Means Clustering for T. A significant deviation of the spatial arrangement of data points is seen in the figure. Now, the higher concentration of points shift to x = 0.2, y = 0.08. 'Tone of Voice' and 'Engagement Level' are similarly associated with 'Role Identification' as the co-ordinate axis is symmetrical about it, as shown in Figure 7.

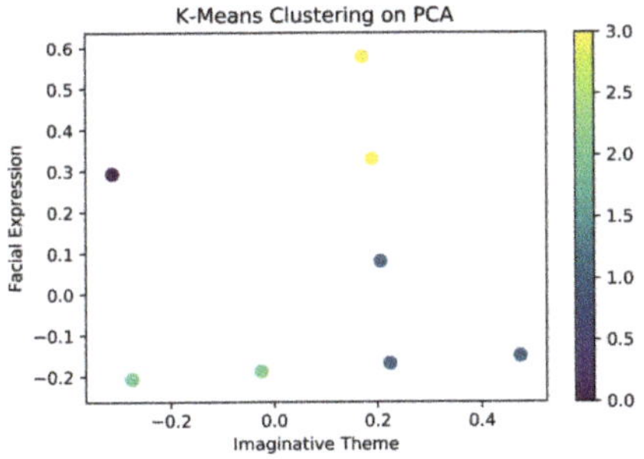

Figure 6. PCA K-means for feature T.

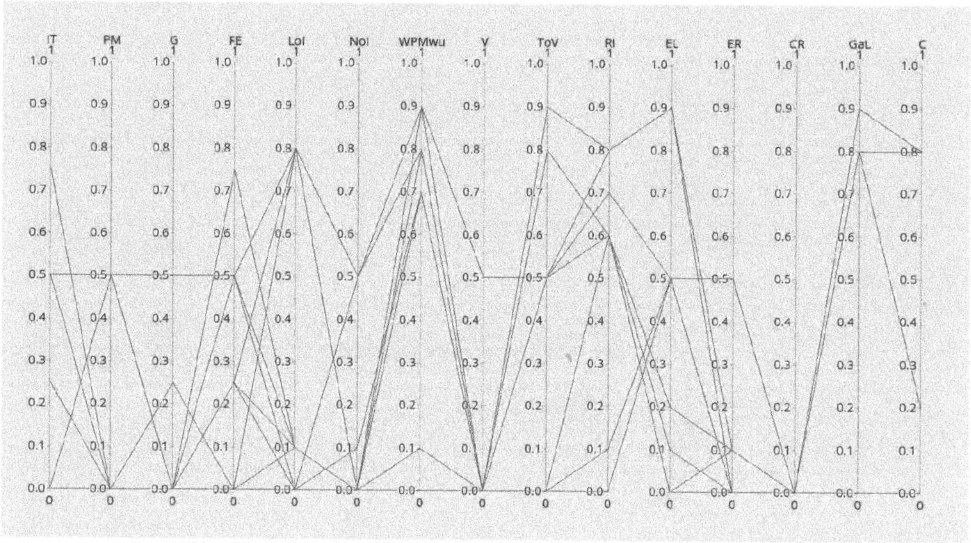

Figure 7. PAC for feature T.

The comparative analysis in Table 4 focuses on five common factors between the four algorithms. The correlation between any two parameters signifies their associated relation. A positive correlation happens when an increase in one attribute shows an increase in another attribute as well. A negative correlation happens when an increase in one attribute shows a decrease in another attribute. The heat map, which strongly demonstrates the factors of correlation and associativity, has a colour scale in which lighter shades signify positive correlation and darker shades signify a negative correlation. For example, in Figure 3, which is a heat map for feature T, Grammar and Coherence show extremely positive correlation whereas Eye Reaction and Role Identification show a negative correlation. The visibility of data points is best observed in the PAC graph while the least was observed in the Heat Map, which focused more on their associativity. Associativity, the reverse of this happened in PAC Graphs and Heat Maps where associativity in the former decreased due to conflict of interest in the arrangement of axes. The dynamicity of PAC, unlike for all other graphs, is the highest because the axes can be rearranged to see which arrangement gives us the best results. However, in K-Means, PCA K-Means and Heat map, the axes are static and rearranging them does not show any significant change. Scalability is a measure of how many data points can be represented in the same graph without the loss of visibility. This was found to be strongest in K-Means and PCA K-means as each point could be seen uniquely on a 2D Cartesian space.

Table 4. Comparative Analysis.

Factors	Heat Map	K-Means	PCA K-Means	PAC Graph
Correlation	Strong	Weak	Weak	Weak
Visibility	Weak	Medium	Medium	Strong
Associativity	Strong	Strong	Strong	Medium
Dynamicity	Medium	Strong	Strong	Very Strong
Scalability	Medium	Strong	Strong	Medium

7. Conclusions and Future Work

The authors have defined the new concept of Single Valued Refined Neutrosophic Sets (SVRNS) which is a generalized version of neutrosophic sets which functions using six memberships values. Furthermore, these SVRNS make use of imaginary values for the memberships. This newly defined concept of SVRNS was used to study the imaginative play in children. The model proposed also consists of distance measures such as Hamming distance and Euclidean distance for two given SVRNSs.

On the basis of expert opinion, the data was successfully transformed into SVRNS. These sets were helpful in drawing clusters, heat maps, parallel axes coordinate and so on. The pictorial representation of the results of these algorithms has helped to gain useful insight into the data collected. We were able to objectively interpret, for instance, the role of factors such as grammar in imaginative play in children.

On the basis of the data collected and processed to form SVRNSs, we will be able to successfully develop an artificial neural network (ANN), decision trees and other supervised learning algorithms in this domain for future research and they will be useful for drawing insights into the role of these parameters by varying the values of the parameters. Other quality measures such as p-value, confusion matrix and accuracy can also be drawn from it. Since the data under consideration were small, we were not able to construct ANN.

For future work, we will study the mentally retarded children in this age group and perform a comparative analysis with the normal children in this age group.

The model will help us in identifying children with autism and attention deficit hyperactivity disorder (ADHD) and other psychological disorders. The detection of such disorders if any at an early stage with the help of our model will help parents and doctors to use the necessary measures to treat and control them quickly.

The model can be further used for other psychological studies like for modeling destructive behaviours of alcoholics and bulimic children and/or adults.

With this given dataset, cross culture validation was not done. For future research, we shall consider the study of cross culture among children and try to generate a variation from cross culture and its effect or influence on the cognitive and language abilities of children.

Author Contributions: Conceptualization, methodology, V.W.B., F.S., and I.K.; software, and validation, V.D. and S.G.; formal analysis, investigation, I.K., V.D. and S.G.; resources, and data curation, I.K., V.D. and S.G.; writing—original draft preparation, I.K., V.D. and S.G.; writing—review and editing, V.W.B. and F.S.; visualization, V.D. and S.G.; supervision, V.W.B. All authors have read and agreed to the published version of the manuscript.

Funding: This research received no external funding.

Conflicts of Interest: The authors declare no conflict of interest.

Abbreviations

The following abbreviations are used in this manuscript:

SVNS	Single Valued Neutrosophic Sets
SVRNS	Single Valued Refined Neutrosophic Sets
CNS	Complex Neutrosophic Sets
DVNS	Double Valued Neutrosophic Sets
TRINS	Triple Refined Indeterminate Neutrosophic Sets
NCM	Neutrosophic Cognitive Maps
PCA	Principal Component Analysis
PAC	Parallel Axes Chart
ANN	Artificial Neural Networks
ADHD	Attention deficit hyperactivity disorder

References

1. Smarandache, F. *A Unifying Field in Logics: Neutrosophic Logic. Neutrosophy, Neutrosophic Set, Probability, and Statistics*; American Research Press: Rehoboth, DE, USA, 2000.
2. Zadeh, L.A. Fuzzy sets. *Inf. Control.* **1965**, *8*, 338–353. [CrossRef]
3. Wang, H.; Smarandache, F.; Zhang, Y.; Sunderraman, R. *Single Valued Neutrosophic Sets*; Infinite Study: Phoenix, AZ, USA, 2010; p. 10.

4. Liu, P.; Wang, Y. Multiple attribute decision-making method based on single-valued neutrosophic normalized weighted Bonferroni mean. *Neural Comput. Appl.* **2014**, *25*, 2001–2010. [CrossRef]
5. Liu, P.; Shi, L. The generalized hybrid weighted average operator based on interval neutrosophic hesitant set and its application to multiple attribute decision making. *Neural Comput. Appl.* **2015**, *26*, 457–471. [CrossRef]
6. Liu, P.; Teng, F. Multiple attribute group decision making methods based on some normal neutrosophic number Heronian Mean operators. *J. Intell. Fuzzy Syst.* **2017**, *32*, 2375–2391. [CrossRef]
7. Liu, P.; Li, H. Multiple attribute decision-making method based on some normal neutrosophic Bonferroni mean operators. *Neural Comput. Appl.* **2017**, *28*, 179–194. [CrossRef]
8. Ye, J. Multicriteria decision-making method using the correlation coefficient under single-valued neutrosophic environment. *Int. J. Gen. Syst.* **2013**, *42*, 386–394. [CrossRef]
9. Ye, J. A multicriteria decision-making method using aggregation operators for simplified neutrosophic sets. *J. Intell. Fuzzy Syst.* **2014**, *26*, 2459–2466. [CrossRef]
10. Ye, J. Single valued neutrosophic cross-entropy for multicriteria decision making problems. *Appl. Math. Model.* **2014**, *38*, 1170–1175. [CrossRef]
11. Ye, J. Similarity measures between interval neutrosophic sets and their applications in multicriteria decision-making. *J. Intell. Fuzzy Syst.* **2014**, *26*, 165–172. [CrossRef]
12. Cheng, H.D.; Guo, Y. A new neutrosophic approach to image thresholding. *New Math. Nat. Comput.* **2008**, *4*, 291–308. [CrossRef]
13. Sengur, A.; Guo, Y. Color texture image segmentation based on neutrosophic set and wavelet transformation. *Comput. Vis. Image Underst.* **2011**, *115*, 1134–1144. [CrossRef]
14. Zhang, M.; Zhang, L.; Cheng, H. A neutrosophic approach to image segmentation based on watershed method. *Signal Process.* **2010**, *90*, 1510–1517. [CrossRef]
15. Salama, A.; Haitham, A.; Manie, A.; Lotfy, M. Utilizing Neutrosophic Set in Social Network Analysis e-Learning Systems. *Int. J. Inf. Sci. Intell. Syst.* **2014**, *3*, 61–72.
16. Vasantha, W.; Smarandache, F. *Fuzzy Cognitive Maps and Neutrosophic Cognitive Maps*; Infinite Study: Phoenix, AZ, USA, 2003
17. Vasantha, W.; Smarandache, F. *Analysis of Social Aspects of Migrant Labourers Living With HIV/AIDS Using Fuzzy Theory and Neutrosophic Cognitive Maps: With Special Reference to Rural Tamil Nadu in India*; Infinite Study: Phoenix, AZ, USA, 2004.
18. Smarandache, F. Neutropsychic personality. In *A Mathematical Approach to Psychology*; Pons: Brussels, Belgium, 2018
19. Fink, R.S. Role of imaginative play in cognitive development. *Psychol. Rep.* **1976**, *39*, 895–906. [CrossRef]
20. Udwin, O. Imaginative play training as an intervention method with institutionalised preschool children. *Br. J. Educ. Psychol.* **1983**, *53*, 32–39. [CrossRef]
21. Huston-Stein, A.; Friedrich-Cofer, L.; Susman, E.J. The relation of classroom structure to social behavior, imaginative play, and self-regulation of economically disadvantaged children. *Child Dev.* **1977**, *48*, 908–916. [CrossRef]
22. Bodrova, E. Make-believe play versus academic skills: A Vygotskian approach to today's dilemma of early childhood education. *Eur. Early Child. Educ. Res. J.* **2008**, *16*, 357–369. [CrossRef]
23. Seja, A.L.; Russ, S.W. Children's fantasy play and emotional understanding. *J. Clin. Child Psychol.* **1999**, *28*, 269–277. [CrossRef]
24. Dhingra, G.; Kumar, V.; Joshi, H.D. A novel computer vision based neutrosophic approach for leaf disease identification and classification. *Measurement* **2019**, *135*, 782–794. [CrossRef]
25. Vasantha, W.B.; Kandasamy, I.; Smarandache, F. Neutrosophic Duplets of Zpn,× and Zpq,× and Their Properties. *Symmetry* **2018**, *10*, 345. [CrossRef]
26. Vasantha, W.; Kandasamy, I.; Smarandache, F. Algebraic Structure of Neutrosophic Duplets in Neutrosophic Rings. *Neutrsophic Sets Syst.* **2018**, *23*, 85–95.
27. Vasantha, W.B.; Kandasamy, I.; Smarandache, F. A Classical Group of Neutrosophic Triplet Groups Using Z2p, ×. *Symmetry* **2018**, *10*, 194. [CrossRef]
28. Kandasamy W.B.; Kandasamy, I.; Smarandache, F. Semi-Idempotents in Neutrosophic Rings. *Mathematics* **2019**, *7*, 507. [CrossRef]
29. Kandasamy, W.B.; Kandasamy, I.; Smarandache, F. Neutrosophic Triplets in Neutrosophic Rings. *Mathematics* **2019**, *7*, 563. [CrossRef]

30. Kandasamy, W.V.; Kandasamy, I.; Smarandache, F. Neutrosophic Quadruple Vector Spaces and Their Properties. *Mathematics* **2019**, *7*, 758.
31. Haibin, W.; Smarandache, F.; Zhang, Y.; Sunderraman, R. *Single Valued Neutrosophic Sets*; Infinite Study: Tamil Nadu, India, 2010; Volume 1, pp. 10–14.
32. Shahzadi, G.; Akram, M.; Saeid, A.B. An application of single-valued neutrosophic sets in medical diagnosis. *Neutrosophic Sets Syst.* **2018**, *18*, 80–88.
33. Ali, M.; Smarandache, F. Complex neutrosophic set. *Neural Comput. Appl.* **2017**, *28*, 1817–1834. [CrossRef]
34. Smarandache, F. n-Valued Refined Neutrosophic Logic and Its Applications to Physics. *Prog. Phys.* **2013**, *4*, 143–146.
35. Kandasamy, I. Double-valued neutrosophic sets, their minimum spanning trees, and clustering algorithm. *J. Intell. Syst.* **2018**, *27*, 163–182. [CrossRef]
36. Kandasamy, I.; Smarandache, F. Multicriteria Decision Making Using Double Refined Indeterminacy Neutrosophic Cross Entropy and Indeterminacy Based Cross Entropy. *Appl. Mech. Mater.* **2016**, *859*, 129–143. [CrossRef]
37. Khan, Q.; Liu, P.; Mahmood, T. Some Generalized Dice Measures for Double-Valued Neutrosophic Sets and Their Applications. *Mathematics* **2018**, *6*, 121. [CrossRef]
38. Kandasamy, I.; Smarandache, F. Triple Refined Indeterminate Neutrosophic Sets for Personality Classification. In Proceedings of the 2016 IEEE Symposium Series on Computational Intelligence (SSCI), Athens, Greece, 6–9 December 2016; IEEE: Piscataway, NJ, USA, 2016; pp. 1–8. [CrossRef]
39. Kandasamy, I.; Vasantha, W.B.; Obbineni, J.; Smarandache, F. Indeterminate Likert Scaling. *Soft Comput.* **2019**, 1–10. [CrossRef]
40. Kandasamy, I.; Vasantha, W.; Mathur, N.; Bisht, M.; Smarandache, F. Chapter 6 Sentiment analysis of the MeToo movement using neutrosophy: Application of single-valued neutrosophic sets. In *Optimization Theory Based on Neutrosophic and Plithogenic Sets*; Elsevier: Amsterdam, The Netherlands, 2020. [CrossRef]
41. Kandasamy, I.; Vasantha, W.; Obbineni, J.M.; Smarandache, F. Sentiment analysis of tweets using refined neutrosophic sets. *Comput. Ind.* **2020**, *115*, 103180. [CrossRef]
42. Kandasamy W.B., V.; Kandasamy, I.; Smarandache, F.; Devvrat, V.; Ghildiyal, S. Study of Imaginative Play in Children using Neutrosophic Cognitive Maps Model. *Neutrosophic Sets Syst.* **2019**, *30*, 241–252.

symmetry

MDPI

Article

The Generalized Neutrosophic Cubic Aggregation Operators and Their Application to Multi-Expert Decision-Making Method

Majid Khan [1], **Muhammad Gulistan** [1,*], **Mumtaz Ali** [2] and **Wathek Chammam** [3,*]

1 Department of Mathematics and Statistics, Hazara University, Mansehra 21310, Pakistan; majid_swati@yahoo.com
2 Deakin-SWU Joint Research Centre on Big Data, School of Information Technology Deakin University, Melbourne, VIC 3125, Australia; mumtaz.ali@deakin.edu.au
3 Department of Mathematics, College of Science Al-Zulfi, Majmaah University, Al-Zufi, P.O. Box 66, Majmaah 11952, Saudi Arabia
* Correspondence: gulistanmath@hu.edu.pk (M.G.); w.chammam@mu.edu.sa (W.C.)

Received: 24 February 2020; Accepted: 12 March 2020; Published: 27 March 2020

check for updates

Abstract: In the modern world, the computation of vague data is a challenging job. Different theories are presented to deal with such situations. Amongst them, fuzzy set theory and its extensions produced remarkable results. Samrandache extended the theory to a new horizon with the neutrosophic set (NS), which was further extended to interval neutrosophic set (INS). Neutrosophic cubic set (NCS) is the generalized version of NS and INS. This characteristic makes it an exceptional choice to deal with vague and imprecise data. Aggregation operators are key features of decision-making theory. In recent times several aggregation operators were defined in NCS. The intent of this paper is to generalize these aggregation operators by presenting neutrosophic cubic generalized unified aggregation (NCGUA) and neutrosophic cubic quasi-generalized unified aggregation (NCQGUA) operators. The accuracy and precision are a vital tool to minimize the potential threat in decision making. Generally, in decision making methods, alternatives and criteria are considered to evaluate the better outcome. However, sometimes the decision making environment has more components to express the problem completely. These components are named as the state of nature corresponding to each criterion. This complex frame of work is dealt with by presenting the multi-expert decision-making method (MEDMM).

Keywords: neutrosophic cubic set (NCS); neutrosophic cubic generalized unified aggregation (NCGUA); neutrosophic cubic quasi-generalized unified aggregation (NCQGUA); multi-expert decision-making method (MEDMM)

1. Introduction

In real-life problems complex phenomena occur. One of the complex phenomena is to deal with the vagueness and uncertainty in data. Because uncertainty is inevitable in problems in different areas of life, conventional methods have failed to cope with such problems. The big task was to deal with uncertain information for many years. Many models have been introduced to incorporate uncertainty into the description of the system. *Zadeh presented their theory of* Fuzzy sets [1]. The possibilistic nature of fuzzy set theory attracted researchers to apply it in different fields of sciences like artificial intelligence, decision making theory, information sciences, medical sciences and more. Due to its applicability in sciences and daily life problems, fuzzy set has been extended to interval valued fuzzy sets (IVFS) [2,3], intuitionistic fuzzy sets (IFS) [4], interval valued intuitionistic fuzzy sets (IVIFS) [5], cubic sets [6], etc. Over the last decades, researchers used it for decision making problems [7–12].

Smrandache presented the idea of neutrosophic sets (NS) [13]. NS provide the more general plate form to extend the ideas of classic theory and fuzzy set theory. The NS consists of three components: truth, indeterminacy and falsehood; all three components are independent, and this makes NS more general than IFS. In fact, NS is a generalization of IFS [14]. For sciences and engineering problems, Wang et al. [15] presented single value neutrosophic set (SVNS), which is a class of NS. Wang et al. [16] further characterized neutrosophic sets to interval neutrosophic set (INS). INS range to an interval value within [0, 1], which comfort the selectors to make an appropriate choice. Jun et al. [17] combined INS and NS to form neutrosophic cubic set (NCS). NCS enables us to choose both interval value and single value membership and indeterminacy and falsehood components. The NCS is the generalization of all abovementioned predeccessors. For example, if the interval part is not taken into account, the set becomes NS, and if the second part is not considered, the NS is dealt with. Since the NS is a generalization of IFS, it is concluded that NCS is the generalization of NS, INS, CS, IFS and fuzzy set. Due to this nature of NCS, it provides more general plate form for uncertain and vague data.

The aggregation operators are an important component of decision making. The insufficient and vague data make it challenging for a decision maker to compute the exact decision. This situation can be minimized by the vague nature of NS and its extensions. The vague nature of NS attracted researchers to implement it in the different fields of science, engineering and decision-making theory. The researchers proposed different aggregation operators and multi-criteria decision-making methods in NS and INS [18–27]. The NCS is the more general form of both NS and INS. This nature attracted researchers to apply it in different fields like science, engineering and decision-making theory. Khan et al. [19] presented neutrosophic cubic Einstein geometric aggregation operators. Zhan et al. [28] worked on multi-criteria decision making on neutrosophic cubic sets. Banerjee et al. [29] used grey rational analysis (GRA) techniques to neutrosophic cubic sets. Lu and Ye [30] defined cosine measure to neutrosophic cubic set. Pramanik et al. [31] used similarity measure to neutrosophic cubic set. Shi and Ji [32] defined Dombi aggregation operators on neutrosophic cubic sets. Ye [33] defined aggregation operators over the neutrosophic cubic numbers. Alhazaymeh et al. [34] presented hybrid geometric aggregation operator with application to multi-attribute decision-making method on neutrosophic cubic sets.

Contribution. The methodologies to measure the generalized aggregations of neutrosophic cubic values.

- First neutrosophic cubic generalized unified aggregation operators are proposed.
- Second neutrosophic cubic quasi-generalized unified aggregation operators are proposed.
- The multi-expert decision-making method is proposed.
- The method is furnished upon numeric data of EMU European Monitory union as an application.
- Comparison is given between some aggregation operators.

Organization. The remaining manuscript is structured as follows. In Section 2, the preliminary work is reviewed. In Section 3, the NCGUA operators are defined. In Section 4, the NCQGU operators are defined. In Section 5, MEDMM is proposed and applied to a numeric data of EMU as an application. Lastly, the comparison of some aggregation operators is provided.

2. Preliminaries

This section consists of some work that provides the foundation for our work.

Definition 1. [13] *A structure* $N = \{(T_N(u), I_N(u), F_N(u)) | u \in U\}$ *is NS, where* $\{T_N(u), I_N(u), F_N(u) \in [0^-, 1^+]\}$ *and* $T_N(u), I_N(u), F_N(u)$ *are truth, indeterminacy and falsehood respectively.*

Definition 2. [15] *A structure* $N = \{(T_N(u), I_N(u), F_N(u)) | u \in U\}$ *is SVNS, where* $\{T_N(u), I_N(u), F_N(u) \in [0,1]\}$ *respectively called truth, indeterminancy and falsehood are simply denoted by* $N = (T_N, I_N, F_N)$.

Definition 3. [16] *An INS in U is a structure* $N = \{(\widetilde{T}_N(u), \widetilde{I}_N(u), \widetilde{F}_N(u)) | u \in U\}$ *where* $\{\widetilde{T}_N(u), \widetilde{I}_N(u), \widetilde{F}_N(u) \in D[0,1]\}$ *are respectively called truth, indeterminacy and falsehood in U, is simply denoted by* $N = (\widetilde{T}_N, \widetilde{I}_N, \widetilde{F}_N)$. *For convenience denoted by* $N = (\widetilde{T}_N, \widetilde{I}_N, \widetilde{F}_N)$ *by* $N = (\widetilde{T}_N = [T_N^L, T_N^U], \widetilde{I}_N = [I_N^L, I_N^U], \widetilde{F}_N = [F_N^L, F_N^U])$.

Definition 4. [16] *A structure* $N = \{(u, \widetilde{T}_N(u), \widetilde{I}_N(u), \widetilde{F}_N(u), T_N(u), I_N(u), F_N(u)) | u \in U\}$ *is NCS in U, in which* $(\widetilde{T}_N = [T_N^L, T_N^U], \widetilde{I}_N = [I_N^L, I_N^U], \widetilde{F}_N = [F_N^L, F_N^U])$ *is an INS and* (T_N, I_N, F_N) *is NS in U, is simply denoted by* $N = (\widetilde{T}_N, \widetilde{I}_N, \widetilde{F}_N, T_N, I_N, F_N)$, $[0,0] \leq \widetilde{T}_N + \widetilde{I}_N + \widetilde{F}_N \leq [3,3]$, *and* $0 \leq T_N + I_N + F_N \leq 3.N^U$ *denotes the collection of NCS in U, which is simply denoted by* $N = (\widetilde{T}_N, \widetilde{I}_N, \widetilde{F}_N, T_N, I_N, F_N)$.

Definition 5. [28] *The sum of two NCS,* $A = (\widetilde{T}_A, \widetilde{I}_A, \widetilde{F}_A, T_A, I_A, F_A)$, *where* $\widetilde{T}_A = [T_A^L, T_A^U], \widetilde{I}_A = [I_A^L, I_A^U], \widetilde{F}_A = [F_A^L, F_A^U]$ *and* $B = (\widetilde{T}_B, \widetilde{I}_B, \widetilde{F}_B, T_B, I_B, F_B)$, *where* $\widetilde{T}_B = [T_B^L, T_B^U], \widetilde{I}_B = [I_B^L, I_B^U], \widetilde{F}_B = [F_B^L, F_B^U]$ *is defined as*

$$A \oplus B = \left([T_A^L + T_B^L - T_A^L T_B^L, T_A^U + T_B^U - T_A^U T_B^U], [I_A^L + I_B^L - I_A^L I_B^L, I_A^U + I_B^U - I_A^U I_B^U], [F_A^L F_B^L, F_A^U F_B^U], T_A T_B, I_A I_B, F_A + F_B - F_A F_B\right)$$

Definition 6. [28] *The product of two NCS,* $A = (\widetilde{T}_A, \widetilde{I}_A, \widetilde{F}_A, T_A, I_A, F_A)$, *where* $\widetilde{T}_A = [T_A^L, T_A^U], \widetilde{I}_A = [I_A^L, I_A^U], \widetilde{F}_A = [F_A^L, F_A^U]$ *and* $B = (\widetilde{T}_B, \widetilde{I}_B, \widetilde{F}_B, T_B, I_B, F_B)$, *where* $\widetilde{T}_B = [T_B^L, T_B^U], \widetilde{I}_B = [I_B^L, I_B^U], \widetilde{F}_B = [F_B^L, F_B^U]$ *is defined as*

$$A \otimes B = \left([T_A^L T_B^L, T_A^U T_B^U], [I_A^L I_B^L, I_A^U I_B^U], [F_A^L + F_B^L - F_A^L F_B^L, F_A^U + F_B^U - F_A^U F_B^U], T_A + T_B - T_A T_B, I_A + I_B - I_A I_B, F_A F_B\right)$$

Definition 7. [28] *The scalar multiplication on a NCS,* $A = (\widetilde{T}_A, \widetilde{I}_A, \widetilde{F}_A, T_A, I_A, F_A)$ *where* $\widetilde{T}_A = [T_A^L, T_A^U], \widetilde{I}_A = [I_A^L, I_A^U], \widetilde{F}_A = [F_A^L, F_A^U]$ *and a scalar k is defined*

$$kA = \left(\left[1 - (1 - T_A^L)^k, 1 - (1 - T_A^U)^k\right], \left[1 - (1 - I_A^L)^k, 1 - (1 - I_A^U)^k\right], \left[(F_A^L)^k, (F_A^U)^k\right], (T_A)^k, (I_A)^k, 1 - (1 - F_A)^k\right)$$

Definition 8. [28] *Let* $A = ([T_A^L, T_A^U], [I_A^L, I_A^U], [F_A^L, F_A^U], T_A, I_A, F_A)$, *be an NCS and* $A^* = ([1, 1], [1, 1], [0, 0], 0, 0, 1)$ *be maximum NCS, then the cosine measure* (C_m) *is defined as*

$$C_m(A) = \left\{\frac{\pi}{18}\left(1 - T_A^L + 1 - T_A^U + 1 - I_A^L + 1 - I_A^U + F_A^L + F_A^U + T_A + I_A + 1 - F_A\right)\right\}, \; C_m(A) \in [0, 1]$$

For comparison of two NCS cosine measure is used.

3. The Neutrosophic Cubic Generalized Unified Aggregation Operator

The NCGUA operators are the generalization of many aggregation operators. The NCGUA operators unify several aggregations operators consequent upon their importance to analyze the imprecise data according to their importance. Moreover, it allows to use arithmetic, quadratic and geometric aggregation operators. By including a wide range of systems, it can adopt different scenario without losing any information.

Definition 9. *The NCGUA is defined as,* $NCGUA(A_1, A_2, \ldots, A_n) = \sum_{j=1}^{m} C_j\left(\sum_{i=1}^{n} w_i^j A_i^{\lambda_j}\right)^{1/\lambda_j}$ *where* C_j *is the relevance that each sub-aggregation has in the system with* $C_j \in [0, 1]$ *and* $\sum_{j=1}^{m} C_j = 1$;w_i^j *is the ith weight of the jth weighing vector W with* $w_i^j \in [0, 1]$,$_j = 1$,λ_h *is the parameter such that* $\lambda_j \in R$ *and* A_i *is the argument value of neutrosophic cubic value.*

Definition 10. *The further generalization can be expressed as,*

$$NCGUA(A_1, A_2, \ldots, A_n) = \left(\left(\sum_{j=1}^{m} C_j\left(\sum_{i=1}^{n} w_i^j A_i^{\lambda_j}\right)^{1/\lambda_j}\right)^{\delta}\right)^{1/\delta}$$

where $\delta \in (-\infty, \infty)$. *Usually* λ_j *remain same but for complex type of aggregation different values can be assumed.*

The operation used on NC is defined by [28].

NCGUA operators accomplish properties like monotonicity, boundness and idempotency.

Families of NCGUA Operators

The main aspect of NCGUA operator is that it characterizes a variety of aggregation operators. The aim of this section is to analyze these sub aggregations operators. First generalized NCOWA, NCWA and NCPOWA operators are analyzed.

$$NCGUA(A_1, A_2, \ldots, A_n) = C_1 \sum_{i=1}^{n} w_i^1 A_i + C_2 \sum_{i=1}^{n} w_i^2 A_i + C_3 \sum_{i=1}^{n} w_i^3 A_i \quad (1)$$

where A_i is the *ith* largest of A_n, and w^1, w^2, w^3 represent the weights corresponding to the NCOWA, NCWA and NCPOWA operators, $\lambda_j = \delta = 1$.

Note that in NCOWA operator, an additional order is made of $(A_1, A_2, \ldots, A_n)$, then it is weighted.

$$C_1 = 1,\ C_2 = C_3 = 0 \Rightarrow NCOWA \quad (2)$$

$$C_2 = 1,\ C_1 = C_3 = 0 \Rightarrow NCWA \quad (3)$$

$$C_3 = 1,\ C_1 = C_2 = 0 \Rightarrow NCPOWA \quad (4)$$

Some other family of aggregation operators can be analyzed by assigning values to λ_j and δ, these values depend on the type of problem under discussion.

The averaging aggregation operators have a most practical operator among their competitors, but in some situations, other operators like geometric, quadratic, cubic operators are in a much better position to evaluate the values.

- If $\lambda = \delta = 1$ and for all j, the aggregation operator is deduced to NCUA.

$$NCUA(A_1, A_2, \ldots, A_n) = \sum_{j=1}^{m} C_j \sum_{i=1}^{n} w_i^j A_i \quad (5)$$

- If $\lambda \to 1$, $\delta = 1$ and for all j, the aggregation operator is deduced to NCUG.

$$NCUG(A_1, A_2, \ldots, A_n) = \sum_{j=1}^{m} C_j \prod_{i=1}^{n} A_i^{w_j} \quad (6)$$

- If $\lambda = 2,\ \delta = 1$ and for all j, the aggregation operator is deduced to NCUQA.

$$NCUQA(A_1,\ A_2, \ldots,\ A_n) = \sum_{j=1}^{m} C_j \left(\sum_{i=1}^{n} w_i^j A_i^2 \right)^{1/2} \tag{7}$$

- If $\lambda \to 0,\ \delta \to 0$ and for all j, the aggregation operator is deduced to NCGUG.

$$NCGUG(A_1,\ A_2, \ldots,\ A_n) = \prod_{j=1}^{m} C_j \prod_{i=1}^{n} A_i^{w_j} \tag{8}$$

- If $\lambda = 2,\ \delta = 2$ and for all j, the aggregation operator is deduced to NCQUQA.

$$NCQUQA(A_1,\ A_2, \ldots,\ A_n) = \left(\sum_{j=1}^{m} C_j^2 \left(\sum_{i=1}^{n} w_i^j A_i^2 \right)^{1/2} \right)^{1/2} \tag{9}$$

The particular cases of NCUA operators can be analyzed by different values of λ and δ as shown in Table 1.

Table 1. Types of family of neutrosophic cubic generalized unified aggregation (NCGUA) operators.

	$\lambda = -\infty$	$\lambda \to 0$	$\lambda = 2$	$\lambda = \infty$
$\delta = -\infty$	Min	Min	Min	Max
$\delta \to 0$	Min	NCUGA		Max
$\delta = 1$	Min	NCUGA		max
$\delta = 2$	Min	NCUGA		Max
$\delta = \infty$	Min	NCUGA		Max

λ and δ have different values in the sub-aggregation operators. Thus, the aggregation operators have listed aggregation operators above. For further analysis, assume that aggregation operators follow the weighted averaging aggregation approach. Observe that a different scenario may be constructed by assigning different values to λ and δ. Today's world has much more complex phenomena; to deal with such situation, complex aggregation operators become a vital tool. Note that the complex and simple aggregation operators can be studied by assigning different values not only to λ and δ but to the weight as well.

4. Neutrosophic Cubic Quasi-Generalized Unified Aggregation Operators

The NCGUA operator can further be generalized by quasi- arithmetic means by neutrosophic cubic quasi-generalized unified aggregation (NCQGUA) operator. The characteristic of NCQGUA operators is that they are a generalization of not only NCGUA operators but of some other aggregation operators by the function introduced in NCQGUA.

Definition 11. *The NCQGUA operator can be defined as*

$$NCQGUA(A_1,\ A_2, \ldots,\ A_n) = f_j^{-1} \left(\sum_{j=1}^{m} mC_j f_j \left(g_j^{-1} \left(\sum_{i=1}^{n} w_i^j g_j(A_i) \right) \right)^{1/2} \right)^{1/2}$$

where f_j and g_j are strictly continuous monotone functions.

The NCQGUA operator can be deduced to a wide range of aggregation operators. The Table 2 illustrates the range generated by NCQGUA operators.

Table 2. Types of family neutrosophic cubic quasi-generalized unified aggregation (NCQGUA) operators.

g_j=a	f_j=C_j	**NCAO Operator**
$g_j = a^2$	$f_j = C_j$	NCQA operator
$g_j = a^{-1}$	$f_j = C_j$	NCHA operator
$g_j = a^3$	$f_j = C_j$	NCCA operator
$g_j = a^0$	$f_j = C_j$	NCGA operator
$g_j = a^2$	$f_j = C_j^2$	NCAUAQA operator
$g_j = a^{-1}$	$f_j = C_j^-$	NCHUHA operator
$g_j = a^3$	$f_j = C_j^3$	NCUCA operator
$g_j = a^0$	$f_j = C_j^0$	NCUGA operator

In Table 1 it is assumed that $g_1 = g_2 = \ldots = g_m$ for all j.

5. The Application of NCQGUA and NCGUA Operators to Multi-Expert Decision-Making Method

The NCQGUA and NCGUA aggregation operators are generalized forms of most of the aggregation operators that can easily be deduced under some special conditions. This characteristic offers a great advantage when applying them to different MCDM. For this purpose, the multi-expert decision-making method MEDMM is proposed. This method is specially designed for complex situations where need of more than one criterion has a further classification, which is the state of nature. In the modern world, the areal study becomes a vital tool to set foreign policy or investments to a country or region. The choice of country or region to invest is a risk-taking job. The countries and multinational companies hire experts for these regions to come up with profitable decisions. For such situation MEDMM is developed under a NC environment that is a neutrosophic cubic multi-expert decision-making method (NCMEDMM).

5.1. Algorithm

This decision-making technique, which is specially designed for the problem, has different criteria, and each criterion has different classifications named as state of nature. The choice of alternatives in such a situation becomes different from the group decision making studied until now. Due to this phenomenon the following method is proposed.

The problem consist of n alternatives $A = \{A_1, A_2, A_3, \ldots, A_n\}$ corresponding to k $C = \{C_1, C_2, C_3, \ldots, C_k\}$ criteria and has $S = \{S_1, S_2, S_3, \ldots, S_m\}$ m state of nature corresponding to each criterion.

- ➢ Construction of expert's criteria matrices for each criterion corresponding to the given alternatives and finite state of nature.
- ➢ Transformation of expert criteria matrices to general group expert's matrix by aggregation operator.
- ➢ Transformation of all the general group experts' matrices to a single matrix by aggregation operator.
- ➢ Ranking of alternatives.

5.2. Model Formulation

European Monetary Union EMU is an organization working for the benefit of the EU. The EMU formation has a lot of micro- and macro-finance decisions to build a productive state. All the areas like economics, finance, politics, marketing, management and EU laws are taken into account for making any decision.

Consider an illustrative example in multi-person decision making in the EMU [35–38]. This type of problem in macroeconomics usually deals with huge amounts of capital or other variables. This makes it critical to find the accurate decision; otherwise, a small deviation may cause huge economic difference in the region. The two different criteria are considered to analyze MPDMM. It is very common that a decision of EMU is usually influenced by several experts and criteria.

The model consists of the following data.

Criteria

$Cr1$: Internal economic condition.
$Cr2$: Global economic condition.

Alternative

A_1 : Increase the rates 1%.
A_2 : Increase the rates 0.5%.
A_3 : No change in rates.
A_4 : Decrease the rates 0.5%.
A_5 : Decrease the rates 1%.

State of nature

S_1 : Negative growth.
S_2 : Growth close to 0.
S_3 : Positive growth.

The company collected the data regarding these alternatives. The assumption is that this decision has potential states of nature benefit corresponding to the two criteria.
$Cr1$: Internal economic condition
S_1 : Negative growth.
S_2 : Growth close to 0.
S_3 : Positive growth.
$Cr2$: Global economic condition.
S_1 : Negative growth.
S_2 : Growth close to 0.
S_3 : Positive growth.

The EMU nominates individuals responsible for this decision which is divided into three groups; each group provides their opinion regarding the outcome and the possible strategy. The data of expert 1 subject to criterion 1 is shown in Table 3.

Table 3. Expert 1–Criterion 1.

	S_1	S_2	S_3
A_1	$\begin{pmatrix} [0.25,0.35],[0.35,0.45], \\ [0.45,0.55],0.30,0.40,0.50 \end{pmatrix}$	$\begin{pmatrix} [0.55,0.65],[0.45,0.55], \\ [0.75,0.85],0.60,0.70,0.80 \end{pmatrix}$	$\begin{pmatrix} [0.75,0.85],[0.87,0.97], \\ [0.75,0.85],0.30,0.50,0.60 \end{pmatrix}$
A_2	$\begin{pmatrix} [0.45,0.60],[0.35,0.50], \\ [0.73,0.89],0.30,0.45,0.80 \end{pmatrix}$	$\begin{pmatrix} [0.85,0.95],[0.55,0.65], \\ [0.30,0.40],0.75,0.40,0.59 \end{pmatrix}$	$\begin{pmatrix} [0.45,0.55],[0.35,0.48], \\ [0.61,0.81],0.40,0.60,0.70 \end{pmatrix}$
A_3	$\begin{pmatrix} [0.45,0.55],[0.65,0.75], \\ [0.25,0.45],0.50,0.30,0.60 \end{pmatrix}$	$\begin{pmatrix} [0.40,0.55],[0.43,0.53], \\ [0.45,0.55],0.35,0.50,0.40 \end{pmatrix}$	$\begin{pmatrix} [0.60,0.75],[0.25,0.35], \\ [0.40,0.55],0.70,0.20,0.30 \end{pmatrix}$
A_4	$\begin{pmatrix} [0.28,0.37],[0.41,0.53], \\ [0.40,0.50],0.33,0.52,0.50 \end{pmatrix}$	$\begin{pmatrix} [0.40,0.50],[0.25,0.35], \\ [0.65,0.78],0.30,0.50,0.80 \end{pmatrix}$	$\begin{pmatrix} [0.28,0.40],[0.35,0.43], \\ [0.25,0.45],0.30,0.20,0.50 \end{pmatrix}$
A_5	$\begin{pmatrix} [0.35,0.55],[0.30,0.50], \\ [0.50,0.60],0.30,0.40,0.40 \end{pmatrix}$	$\begin{pmatrix} [0.60,0.70],[0.48,0.60], \\ [0.70,0.80],0.60,0.50,0.80 \end{pmatrix}$	$\begin{pmatrix} [0.45,0.55],[0.53,0.65], \\ [0.80,0.85],0.50,0.58,0.70 \end{pmatrix}$

The data of expert 1 subject to criterion 2 is shown in Table 4.

Table 4. Expert 1–Criterion 2.

	S_1	S_2	S_3
A_1	$\begin{pmatrix} [0.50, 0.60], [0.25, 0.35], \\ [0.45, 0.55], 0.55.40, 0.50 \end{pmatrix}$	$\begin{pmatrix} [0.25, 0.35], [0.45, 0.55], \\ [0.50, 0.60], 0.30, 0.45, 0.50 \end{pmatrix}$	$\begin{pmatrix} [0.35, 0.45], [0.45, 0.55], \\ [0.50, 0.55], 0.40, 0.50, 0.60 \end{pmatrix}$
A_2	$\begin{pmatrix} [0.65, 0.75], [0.50, 0.55], \\ [0.75, 0.85], 0.70, 0.80, 0.90 \end{pmatrix}$	$\begin{pmatrix} [0.40, 0.55], [0.35, 0.45], \\ [0.50, 0.60], 0.40, 0.50, 0.50 \end{pmatrix}$	$\begin{pmatrix} [0.50, 0.60], [0.55, 0.65], \\ [0.65, 0.75], 0.40, 0.50, 0.60 \end{pmatrix}$
A_3	$\begin{pmatrix} [0.25, 0.35], [0.45, 0.55], \\ [0.75, 0.85], 0.30, 0.50, 0.60 \end{pmatrix}$	$\begin{pmatrix} [0.65, 0.75], [0.75, 0.85], \\ [0.85, 0.95], 0.70, 0.80, 0.90 \end{pmatrix}$	$\begin{pmatrix} [0.20, 0.35], [0.35, 0.45], \\ [0.45, 0.55], 0.30, 0.40, 0.50 \end{pmatrix}$
A_4	$\begin{pmatrix} [0.55, 0.65], [0.45, 0.55], \\ [0.75, 0.85], 0.50, 0.70, 0.80 \end{pmatrix}$	$\begin{pmatrix} [0.45, 0.55], [0.55, 0.60], \\ [0.70, 0.80], 0.50, 0.50, 0.70 \end{pmatrix}$	$\begin{pmatrix} [0.45, 0.55], [0.50, 0.55], \\ [0.80, 0.90], 0.50, 0.60, 0.70 \end{pmatrix}$
A_5	$\begin{pmatrix} [0.20, 0.30], [0.30, 0.40], \\ [0.50, 0.60], 0.30, 0.40, 0.60 \end{pmatrix}$	$\begin{pmatrix} [0.60, 0.70], [0.55, 0.65], \\ [0.6, 0.75], 0.70, 0.50, 0.70 \end{pmatrix}$	$\begin{pmatrix} [0.35, 0.45], [0.35, 0.45], \\ [0.75, 0.85], 0.70, 0.50, 0.70 \end{pmatrix}$

Subject to this information, the first criterion is weighted as 0.70 and the second criterion as 0.30. The NCGUA operators are applied to form the general matrix, which represents the matrix of the group of experts illustrated in Table 5.

Table 5. Expert 1–General result.

	S_1	S_2	S_3
A_1	$\begin{pmatrix} [0.4967, 0.5981], \\ [0.3214, 0.4217], \\ [0.4500, 0.5500], \\ 0.3598, 0.4000, 0.5000 \end{pmatrix}$	$\begin{pmatrix} [0.4754, 0.5785], \\ [0.4500, 0.5500], \\ [0.6641, 0.7656], \\ 0.4873, 0.6131, 0.7367 \end{pmatrix}$	$\begin{pmatrix} [0.6670, 0.7785], \\ [0.7996, 0.9324], \\ [0.6641, 0.7459], \\ 0.3270, 0.5000, 0.6000 \end{pmatrix}$
A_2	$\begin{pmatrix} [0.5167, 0.6526], \\ [0.3992, 0.5155], \\ [0.6314, 0.8618], \\ 0.3868, 0.5348, 0.8375 \end{pmatrix}$	$\begin{pmatrix} [0.7726, 0.9033], \\ [0.4975, 0.5992], \\ [0.3497, 0.4517], \\ 0.6211, 0.4277, 0.5648 \end{pmatrix}$	$\begin{pmatrix} [0.4655, 0.5656], \\ [0.4179, 0.5382], \\ [0.6217, 0.7915], \\ 0.4000, 0.56810, 0.7121 \end{pmatrix}$
A_3	$\begin{pmatrix} [0.3964, 0.4975], \\ [0.5991, 0.7017], \\ [0.3476, 0.5445], \\ 0.4289, 0.3497, 0.8375 \end{pmatrix}$	$\begin{pmatrix} [0.4896, 0.6227], \\ [0.5549, 0.6663], \\ [0.7044, 0.6036], \\ 0.4309, 0.5757, 0.8375 \end{pmatrix}$	$\begin{pmatrix} [0.3502, 0.6670], \\ [0.3500, 0.4361], \\ [0.2982, 0.4779], \\ 0.6131, 0.2662, 0.3868 \end{pmatrix}$
A_4	$\begin{pmatrix} [0.3746, 0.4718], \\ [0.4223, 0.5361], \\ [0.4830, 0.5862], \\ 0.3738, 0.5139, 0.6202 \end{pmatrix}$	$\begin{pmatrix} [0.4155, 0.5155], \\ [0.3565, 0.4381], \\ [0.6646, 0.7859], \\ 0.3496, 0.5000, 0.7741 \end{pmatrix}$	$\begin{pmatrix} [0.3358, 0.4496], \\ [0.3500, 0.4360], \\ [0.3544, 0.5540], \\ 0.3496, 0.2780, 0.5710 \end{pmatrix}$
A_5	$\begin{pmatrix} [0.3353, 0.4862], \\ [0.3000, 0.4718], \\ [0.5000, 0.6000], \\ 0.3000, 0.6000, 0.4687 \end{pmatrix}$	$\begin{pmatrix} [0.6000, 0.7000], \\ [0.5020, 0.6157], \\ [0.6683, 0.7846], \\ 0.6283, 0.5000, 0.607741 \end{pmatrix}$	$\begin{pmatrix} [0.3817, 0.5228], \\ [0.4819, 0.5992], \\ [0.7646, 0.8500], \\ 0.6328, 0.5547, 0.7000 \end{pmatrix}$

The data of expert 2 subject to criterion 1 is shown in Table 6.

Table 6. Expert 2–Criterion 1.

	S_1	S_2	S_3
A_1	$\begin{pmatrix} [0.22,0.35],[0.40,0.50], \\ [0.45,0.55],0.30,0.45,0.55 \end{pmatrix}$	$\begin{pmatrix} [0.70,0.75],[0.75,0.85], \\ [0.70,0.80],0.60,0.70,0.85 \end{pmatrix}$	$\begin{pmatrix} [0.70,0.85],[0.15,0.30], \\ [0.25,0.35],0.80,0.50,0.20 \end{pmatrix}$
A_2	$\begin{pmatrix} [0.20,0.30],[0.30,0.40], \\ [0.60,0.70],0.25,0.45,0.70 \end{pmatrix}$	$\begin{pmatrix} [0.35,0.45],[0.45,0.60], \\ [0.30,0.40],0.15,0.40,0.10 \end{pmatrix}$	$\begin{pmatrix} [0.50,0.60],[0.45,0.55], \\ [0.40,0.53],0.70,0.30,0.60 \end{pmatrix}$
A_3	$\begin{pmatrix} [0.10,0.25],[0.25,0.75], \\ [0.25,0.45],0.30,0.20,0.50 \end{pmatrix}$	$\begin{pmatrix} [0.25,0.35],[0.40,0.55], \\ [0.40,0.50],0.35,0.50,0.40 \end{pmatrix}$	$\begin{pmatrix} [0.60,0.75],[0.20,0.35], \\ [0.45,0.55],0.70,0.20,0.40 \end{pmatrix}$
A_4	$\begin{pmatrix} [0.55,0.65],[0.40,0.50], \\ [0.45,0.50],0.70,0.50,0.40 \end{pmatrix}$	$\begin{pmatrix} [0.60,0.70],[0.25,0.35], \\ [0.45,0.55],0.80,0.50,0.30 \end{pmatrix}$	$\begin{pmatrix} [0.25,0.40],[0.35,0.43], \\ [0.75,0.85],0.30,0.40,0.85 \end{pmatrix}$
A_5	$\begin{pmatrix} [0.70,0.85],[0.20,0.40], \\ [0.40,0.50],0.90,0.20,0.40 \end{pmatrix}$	$\begin{pmatrix} [0.70,0.90],[0.40,0.50], \\ [0.50,0.60],0.90,0.50,0.40 \end{pmatrix}$	$\begin{pmatrix} [0.15,0.25],[0.25,0.35], \\ [0.80,0.85],0.30,0.30,0.70 \end{pmatrix}$

The data of expert 2 subject to criterion 2 is shown in Table 7.

Table 7. Expert 2–Criterion 2.

	S_1	S_2	S_3
A_1	$\begin{pmatrix} [0.40,0.50],[0.50,0.60], \\ [0.55,0.65],0.55.20,0.60 \end{pmatrix}$	$\begin{pmatrix} [0.55,0.65],[0.40,0.50], \\ [0.50,0.65],0.60,0.45,0.60 \end{pmatrix}$	$\begin{pmatrix} [0.15,0.25],[0.35,0.55], \\ [0.55,0.65],0.40,0.50,0.70 \end{pmatrix}$
A_2	$\begin{pmatrix} [0.30,0.45],[0.50,0.55], \\ [0.70,0.80],0.20,0.60,0.70 \end{pmatrix}$	$\begin{pmatrix} [0.10,0.20],[0.30,0.40], \\ [0.40,0.50],0.10,0.40,0.45 \end{pmatrix}$	$\begin{pmatrix} [0.40,0.50],[0.25,0.35], \\ [0.65,0.75],0.40,0.30,0.60 \end{pmatrix}$
A_3	$\begin{pmatrix} [0.70,0.85],[0.45,0.55], \\ [0.25,0.35],0.80,0.50,0.40 \end{pmatrix}$	$\begin{pmatrix} [0.15,0.25],[0.70,0.80], \\ [0.75,0.85],0.20,0.80,0.90 \end{pmatrix}$	$\begin{pmatrix} [0.30,0.45],[0.35,0.45], \\ [0.55,0.65],0.30,0.40,0.65 \end{pmatrix}$
A_4	$\begin{pmatrix} [0.55,0.65],[0.25,0.35], \\ [0.85,0.95],0.50,0.30,0.80 \end{pmatrix}$	$\begin{pmatrix} [0.35,0.45],[0.55,0.60], \\ [0.20,0.30],0.50,0.50,0.30 \end{pmatrix}$	$\begin{pmatrix} [0.50,0.50],[0.50,0.55], \\ [0.40,0.50],0.50,0.50,0.45 \end{pmatrix}$
A_5	$\begin{pmatrix} [0.70,0.80],[0.30,0.40], \\ [0.40,0.60],0.85,0.40,0.35 \end{pmatrix}$	$\begin{pmatrix} [0.30,0.40],[0.55,0.65], \\ [0.40,0.50],0.40,0.50,0.50 \end{pmatrix}$	$\begin{pmatrix} [0.35,0.45],[0.35,0.45], \\ [0.70,0.80],0.40,0.40,0.70 \end{pmatrix}$

Subject to this information, the first criterion is weighted as 0.70 and the second criterion as 0.30. The NCGUA operators are applied to form the general matrix, which represents the matrix of the group of experts illustrated in Table 8.

Table 8. Expert 2–General result.

	S_1	S_2	S_3
A_1	$\begin{pmatrix} [0.2832,0.4235], \\ [0.4489,0.5625], \\ [0.5423,0.6235], \\ 0.4623,0.3233,0.5436 \end{pmatrix}$	$\begin{pmatrix} [0.6235,0.7135], \\ [0.6145,0.7023], \\ [0.6235,0.7235], \\ 0.6000,0.6123,0.7127 \end{pmatrix}$	$\begin{pmatrix} [0.3523,0.5515], \\ [0.2523,0.3124], \\ [0.3931,0.4956], \\ 0.6123,0.5000,0.4456 \end{pmatrix}$
A_2	$\begin{pmatrix} [0.2645,0.3845], \\ [0.4124,0.5142], \\ [0.6496,0.7485], \\ 0.2124,0.5217,0.7000 \end{pmatrix}$	$\begin{pmatrix} [0.2345,0.3325], \\ [0.3854,0.5123], \\ [0.3485,0.4489], \\ 0.1223,0.4000,0.6625 \end{pmatrix}$	$\begin{pmatrix} [0.4659,0.7152], \\ [0.3645,0.4153], \\ [0.5189,0.6478], \\ 0.6478,0.2485,0.5123 \end{pmatrix}$
A_3	$\begin{pmatrix} [0.4125,0.5689], \\ [0.4351,0.6628], \\ [0.2500,0.3485], \\ 0.5489,0.5000,0.4658 \end{pmatrix}$	$\begin{pmatrix} [0.2134,0.3125], \\ [0.5681,0.7125], \\ [0.5678,0.5000], \\ 0.2189,0.6456,0.6478 \end{pmatrix}$	$\begin{pmatrix} [0.4691,0.6123], \\ [0.3500,0.4485], \\ [0.5986,0.7458], \\ 0.3000,0.4489,0.7674 \end{pmatrix}$
A_4	$\begin{pmatrix} [0.5500,0.6500], \\ [0.3456,0.4657], \\ [0.6647,0.7456], \\ 0.5987,0.3985,0.6123 \end{pmatrix}$	$\begin{pmatrix} [0.4875,0.5825], \\ [0.3645,0.5621], \\ [0.3245,0.6987], \\ 0.4989,0.6487,0.3000 \end{pmatrix}$	$\begin{pmatrix} [0.3825,0.4658], \\ [0.4360,0.5032], \\ [0.5687,0.6689], \\ 0.3958,0.4489,0.6658 \end{pmatrix}$
A_5	$\begin{pmatrix} [0.7000,0.8354], \\ [0.2658,0.4680], \\ [0.4000,0.5489], \\ 0.887200,0.3152,0.3254 \end{pmatrix}$	$\begin{pmatrix} [0.5125,0.6652], \\ [0.4658,0.5684], \\ [0.4458,0.5482], \\ 0.6685,0.5000,0.4458 \end{pmatrix}$	$\begin{pmatrix} [0.2145,0.3641], \\ [0.3125,0.4128], \\ [0.7458,0.8285], \\ 0.3456,0.3456,0.7000 \end{pmatrix}$

The data of expert 3 subject to criterion 1 is shown in Table 9.

Table 9. Expert 3–Criterion 1.

	S_1	S_2	S_3
A_1	$\left(\begin{array}{c}[0.20,0.40],[0.30,0.40],\\ [0.55,0.65],0.30,0.40,0.60\end{array}\right)$	$\left(\begin{array}{c}[0.55,0.65],[0.60,0.75],\\ [0.70,0.80],0.60,0.70,0.75\end{array}\right)$	$\left(\begin{array}{c}[0.30,0.40],[0.40,0.50],\\ [0.60,0.70],0.30,0.45,0.65\end{array}\right)$
A_2	$\left(\begin{array}{c}[0.35,0.45],[0.35,0.45],\\ [0.60,0.75],0.30,0.45,0.70\end{array}\right)$	$\left(\begin{array}{c}[0.50,0.60],[0.55,0.65],\\ [0.20,0.30],0.50,0.55,0.20\end{array}\right)$	$\left(\begin{array}{c}[0.50,0.60],[0.40,0.50],\\ [0.60,0.75],0.50,0.50,0.70\end{array}\right)$
A_3	$\left(\begin{array}{c}[0.55,0.65],[0.60,0.75],\\ [0.70,0.80],0.60,0.70,0.75\end{array}\right)$	$\left(\begin{array}{c}[0.55,0.65],[0.60,0.75],\\ [0.70,0.80],0.60,0.70,0.75\end{array}\right)$	$\left(\begin{array}{c}[0.60,0.75],[0.25,0.35],\\ [0.50,0.60],0.70,0.20,0.50\end{array}\right)$
A_4	$\left(\begin{array}{c}[0.60,0.75],[0.25,0.35],\\ [0.50,0.60],0.70,0.20,0.50\end{array}\right)$	$\left(\begin{array}{c}[0.55,0.65],[0.60,0.75],\\ [0.70,0.80],0.60,0.70,0.75\end{array}\right)$	$\left(\begin{array}{c}[0.60,0.75],[0.25,0.35],\\ [0.50,0.60],0.70,0.20,0.50\end{array}\right)$
A_5	$\left(\begin{array}{c}[0.30,0.40],[0.30,0.40],\\ [0.55,0.65],0.30,0.40,0.60\end{array}\right)$	$\left(\begin{array}{c}[0.60,0.70],[0.45,0.60],\\ [0.65,0.80],0.60,0.50,0.70\end{array}\right)$	$\left(\begin{array}{c}[0.20,0.40],[0.30,0.40],\\ [0.55,0.65],0.30,0.40,0.60\end{array}\right)$

The data of expert 3 subject to criterion 2 is shown in Table 10.

Table 10. Expert 3–Criterion 2.

	S_1	S_2	S_3
A_1	$\left(\begin{array}{c}[0.35,0.45],[0.35,0.45],\\ [0.60,0.75],0.30,0.45,0.70\end{array}\right)$	$\left(\begin{array}{c}[0.55,0.65],[0.60,0.75],\\ [0.70,0.80],0.60,0.70,0.75\end{array}\right)$	$\left(\begin{array}{c}[0.20,0.40],[0.30,0.40],\\ [0.55,0.65],0.30,0.40,0.60\end{array}\right)$
A_2	$\left(\begin{array}{c}[0.55,0.65],[0.60,0.75],\\ [0.70,0.80],0.60,0.70,0.75\end{array}\right)$	$\left(\begin{array}{c}[0.40,0.55],[0.35,0.45],\\ [0.50,0.60],0.40,0.50,0.50\end{array}\right)$	$\left(\begin{array}{c}[0.35,0.45],[0.35,0.45],\\ [0.60,0.75],0.30,0.45,0.70\end{array}\right)$
A_3	$\left(\begin{array}{c}[0.45,0.55],[0.20,0.30],\\ [0.60,0.75],0.50,0.20,0.60\end{array}\right)$	$\left(\begin{array}{c}[0.45,0.55],[0.20,0.30],\\ [0.60,0.75],0.50,0.20,0.60\end{array}\right)$	$\left(\begin{array}{c}[0.60,0.70],[0.55,0.65],\\ [0.45,0.55],0.70,0.40,0.60\end{array}\right)$
A_4	$\left(\begin{array}{c}[0.55,0.65],[0.45,0.55],\\ [0.75,0.85],0.50,0.70,0.80\end{array}\right)$	$\left(\begin{array}{c}[0.45,0.55],[0.55,0.60],\\ [0.70,0.80],0.50,0.50,0.70\end{array}\right)$	$\left(\begin{array}{c}[0.35,0.45],[0.35,0.45],\\ [0.60,0.75],0.30,0.45,0.70\end{array}\right)$
A_5	$\left(\begin{array}{c}[0.35,0.45],[0.35,0.45],\\ [0.60,0.75],0.30,0.45,0.70\end{array}\right)$	$\left(\begin{array}{c}[0.45,0.55],[0.20,0.30],\\ [0.60,0.75],0.50,0.20,0.60\end{array}\right)$	$\left(\begin{array}{c}[0.35,0.45],[0.30,0.40],\\ [0.75,0.85],0.70,0.50,0.70\end{array}\right)$

Subject to this information, the first criterion is weighted as 0.70 and the second criterion as 0.30. The NCGUA operators are applied to form the general matrix, which represents the matrix of the group of experts illustrated in Table 11.

Table 11. Expert 3–General result.

	S_1	S_2	S_3
A_1	$\left(\begin{array}{c}[0.2325,0.4103],\\ [0.3102,0.4103],\\ [0.5645,0.6625],\\ 0.3000,0.4102,0.6354\end{array}\right)$	$\left(\begin{array}{c}[0.5500,0.6500],\\ [0.6000,0.7500],\\ [0.7000,0.8000],\\ 0.6000,0.7000,0.7500\end{array}\right)$	$\left(\begin{array}{c}[0.2145,0.4000],\\ [0.3185,0.4152],\\ [0.5624,0.6625],\\ 0.3000,0.4123,0.6124\end{array}\right)$
A_2	$\left(\begin{array}{c}[0.3960,0.4952],\\ [0.4101,0.4963],\\ [0.6000,0.7652],\\ 0.3446,0.4915,0.7124\end{array}\right)$	$\left(\begin{array}{c}[0.4215,0.5685],\\ [0.3925,0.4952],\\ [0.2402,0.3446],\\ 0.4781,0.5396,0.2235\end{array}\right)$	$\left(\begin{array}{c}[0.4730,0.5736],\\ [0.3912,0.4925],\\ [0.6000,0.7500],\\ 0.4623,0.4921,0.7000\end{array}\right)$
A_3	$\left(\begin{array}{c}[0.5315,0.6319],\\ [0.5405,0.6928],\\ [0.6725,0.7821],\\ 0.5748,0.6289,0.7253\end{array}\right)$	$\left(\begin{array}{c}[0.5315,0.6319],\\ [0.5405,0.6928],\\ [0.6721,0.7808],\\ 0.5742,0.6235,0.7253\end{array}\right)$	$\left(\begin{array}{c}[0.6000,0.7407],\\ [0.3228,0.5344],\\ [0.4852,0.5842],\\ 0.7000,0.2347,0.5218\end{array}\right)$
A_4	$\left(\begin{array}{c}[0.5904,0.6500],\\ [0.2951,0.3960],\\ [0.5547,0.6672],\\ 0.6581,0.3324,0.5837\end{array}\right)$	$\left(\begin{array}{c}[0.5315,0.6319],\\ [0.5904,0.7253],\\ [0.7000,0.8000],\\ 0.5714,0.6582,0.7407\end{array}\right)$	$\left(\begin{array}{c}[0.5592,0.7072],\\ [0.2711,0.3713],\\ [0.5182,0.6235],\\ 0.6446,0.2577,0.5485\end{array}\right)$
A_5	$\left(\begin{array}{c}[0.3713,0.4103],\\ [0.3102,0.4103],\\ [0.5886,0.6758],\\ 0.3000,0.4101,0.6822\end{array}\right)$	$\left(\begin{array}{c}[0.5736,0.6746],\\ [0.4072,0.5526],\\ [0.4458,0.5482],\\ 0.5972,0.4475,0.6822\end{array}\right)$	$\left(\begin{array}{c}[0.2325,0.4103],\\ [0.3000,0.4000],\\ [0.5823,0.6952],\\ 0.6321,0.4103,0.6223\end{array}\right)$

Now applying NCGUA operator subject to the weight $(0.45, 0.35, 0.20)$, the following matrix is obtained illustrated in Table 12.

Table 12. Collective results of experts.

	S_1	S_2	S_3
A_1	$\left(\begin{array}{c}[0.3802, 0.5076],\\ [0.3670, 0.5291],\\ [0.5026, 0.5964],\\ 0.3787, 0.3731, 0.5453\end{array}\right)$	$\left(\begin{array}{c}[0.5470, 0.6452],\\ [0.5443, 0.6537],\\ [0.6565, 0.7572],\\ 0.6000, 0.6123, 0.6293\end{array}\right)$	$\left(\begin{array}{c}[0.5009, 0.6539],\\ [0.5941, 0.7656],\\ [0.5347, 0.6313],\\ 0.4003, 0.4811, 0.5544\end{array}\right)$
A_2	$\left(\begin{array}{c}[0.4145, 0.5426],\\ [0.4060, 0.5112],\\ [0.6232, 0.8101],\\ 0.3064, 0.5213, 0.7742\end{array}\right)$	$\left(\begin{array}{c}[0.5808, 0.7435],\\ [0.4399, 0.5504],\\ [0.3240, 0.3805],\\ 0.5053, 0.4376, 0.5529\end{array}\right)$	$\left(\begin{array}{c}[0.4771, 0.6266],\\ [0.3943, 0.4888],\\ [0.5794, 0.6548],\\ 0.4874, 0.4579, 0.6167\end{array}\right)$
A_3	$\left(\begin{array}{c}[0.4316, 0.6806],\\ [0.5354, 0.6867],\\ [0.3534, 0.5007],\\ 0.4957, 0.4457, 0.7262\end{array}\right)$	$\left(\begin{array}{c}[0.4178, 0.5368],\\ [0.5523, 0.6884],\\ [0.5812, 0.5949],\\ 0.3600, 0.6088, 0.7634\end{array}\right)$	$\left(\begin{array}{c}[0.4869, 0.6263],\\ [0.3764, 0.4808],\\ [0.4194, 0.4701],\\ 0.4902, 0.3178, 0.5282\end{array}\right)$
A_4	$\left(\begin{array}{c}[0.4879, 0.5788],\\ [0.3720, 0.4862],\\ [0.5552, 0.6544],\\ 0.4935, 0.4309, 0.6104\end{array}\right)$	$\left(\begin{array}{c}[0.4659, 0.5647],\\ [0.4492, 0.5537],\\ [0.5225, 0.7569],\\ 0.4368, 0.5786, 0.6550\end{array}\right)$	$\left(\begin{array}{c}[0.4512, 0.5314],\\ [0.3764, 0.5032],\\ [0.4512, 0.6059],\\ 0.4126, 0.3178, 0.6028\end{array}\right)$
A_5	$\left(\begin{array}{c}[0.5024, 0.6455],\\ [0.2903, 0.4587],\\ [0.4777, 0.5955],\\ 0.4384, 0.4438, 0.4788\end{array}\right)$	$\left(\begin{array}{c}[0.5658, 0.6821],\\ [0.4715, 0.5874],\\ [0.5348, 0.6441],\\ 0.6356, 0.4890, 0.5755\end{array}\right)$	$\left(\begin{array}{c}[0.2979, 0.4495],\\ [0.3925, 0.5034],\\ [0.7177, 0.7881],\\ 0.5119, 0.4425, 0.6860\end{array}\right)$

Now applying NCGUA operator subject to the weight $(0.33, 0.33, 0.34)$ the following matrix is obtained illustrated in Table 13.

Table 13. Aggregated result of experts.

A_1	$\left(\begin{array}{c}[0.4808, 0.6079],\\ [0.5117, 0.6644],\\ [0.5606, 0.6578],\\ 0.4491, 0.4790, 0.5778\end{array}\right)$
A_2	$\left(\begin{array}{c}[0.5358, 0.5426],\\ [0.4135, 0.5172],\\ [0.4899, 0.5872],\\ 0.4232, 0.4708, 0.6613\end{array}\right)$
A_3	$\left(\begin{array}{c}[0.4467, 0.6191],\\ [0.4927, 0.6286],\\ [0.4414, 0.5187],\\ 0.4443, 0.4403, 0.6860\end{array}\right)$
A_4	$\left(\begin{array}{c}[0.4684, 0.5585],\\ [0.4000, 0.5151],\\ [0.5071, 0.6688],\\ 0.4460, 0.4282, 0.6214\end{array}\right)$
A_5	$\left(\begin{array}{c}[0.4670, 0.6028],\\ [0.3892, 0.5193],\\ [0.5694, 0.6722],\\ 0.5224, 0.4577, 0.5900\end{array}\right)$

For comparison of NC values, the cosine measure is computed of the alternatives, so that the values are ranked.

$$C(A_1) = 0.04055, C(A_2) = 0.4058,\ C(A_3) = 0.04394,\ C(A_4) = 0.03890,\ C(A_5) = 0.03750.$$

From this data the ranking is $A_3 > A_2 > A_1 > A_4 > A_5$.

For comparison purposes, some other aggregation operators are considered, and their results are computed.

The aggregation operators like neutrosophic cubic weighted geometric (NCWG), neutrosophic cubic probabilistic (NCP), neutrosophic cubic maximum (NCmax) and neutrosophic cubic minimum (NCmin) operators are also applied to the data. The following results are obtained.

The graphical comparison of these operators is shown in the following figure.

Figure 1: Graphical representation of Operators.

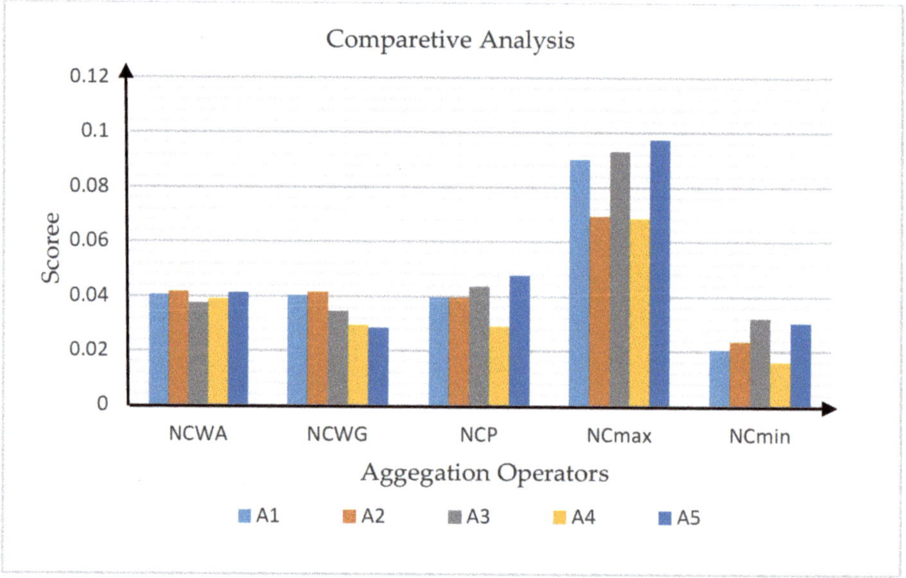

Figure 1. Graphical Comparison of different methods.

The following table also illustrates the comparison of these aggregation operators under the MPDMM.

From the above table with the different alternatives of each decision maker, the alternative in closest accordance to his interests will be selected; this data indicates that the optimal decision is A_5. Thus, the optimal alternative is A_5.

6. Discussion

Jose et al. [39] analyze the data on fuzzy sets and conclude with the following results.

Comparison of Tables 14 and 15. It is observed that in both tables, the optimal alternative is A_5, although individually, some aggregation operators may have different results. The reason may be that NC has more components compared to the FS, like indeterminate and falsehood. Overall comparison has the same result, which ensures the validity of this method. But the advantage of NCS over table FS is that NCS is a generalized version of FS. That is, it enables the expert to deal with inconsistent and indeterminate data more efficiently than the FS.

Table 14. Ranking using different operators.

Operators	Ranking
NCWA	$A_5 > A_2 > A_1 > A_4 > A_3$
NCWG	$A_2 > A_1 > A_3 > A_4 > A_5$
NCP	$A_5 > A_3 > A_1 > A_2 > A_4$
NCmax	$A_5 > A_3 > A_1 > A_2 > A_4$
NCmin	$A_3 > A_5 > A_2 > A_1 > A_4$

Table 15. Ranking using different operators in fuzzy data.

Operators	Ranking
FWA	$A_5 > A_3 > A_1 > A_2 > A_4$
FOWA	$A_3 > A_5 > A_2 > A_1 > A_4$
FPA	$A_5 > A_3 > A_2 > A_1 > A_4$
FMax	$A_5 > A_3 > A_1 > A_2 > A_4$
FMin	$A_2 > A_3 > A_5 > A_1 > A_4$

7. Conclusions

This paper generalized aggregation operators, such as NCGUA operators, which provide a single plate form for researchers and experts to deal with different types of aggregation operators. This generalization helps to work in a complex frame of work where more than one operator is required. The data involves inconsistent and indeterminate factors and is furnished upon a daily life problem. The NCGUA operator is further generalized to NCQGUA aggregation operators. Moreover, it can include complex aggregation operators that deal with complex environments such as problems with a wide range of sub-aggregations. It is shown that some particular aggregation operators like NCWA, NCOWA, NCGA, NCOGA, NCmax, NCmin and other aggregation operators can easily be deduced from NCQGUA or NCGUA operator. A new decision-making method is formed to deal with the problem in which each criterion is further classified into nature of stats. A numeric example involving the EMU is provided as an application. It is concluded that the proposed method provides the best results in comparison to the method previously proposed due to the indeterminate factors involved in data.

Author Contributions: The author contributed as follow. "Conceptualization, M.K. and M.G.; methodology, M.K. and M.G.; software, M.K.; validation, M.A., W.C. And M.G.; formal analysis, M.K. and M.G.; investigation, M.A. and W.C.; resources, M.G.; data curation, M.A.; writing—original draft preparation, M.K.; writing—review and editing, M.K. and M. Gulistan.; visualization, M.K.; supervision, M. Gulistan and M.A.; project administration, W.C.; funding acquisition, W.C.". All authors have read and agreed to the published version of the manuscript.

Funding: The authors extend their appreciation to the Deanship of Scientific Research at Majmaah University for funding this work under Project Number No. RGP-2019- 5.

Conflicts of Interest: The authors declare no conflicts of interest.

References

1. Zadeh, L.A. Fuzzy sets. *Inform. Control* **1965**, *8*, 338–353. [CrossRef]
2. Zadeh, L.A. Outline of a new approach to the analysis of complex system and decision processes interval-valued fuzzy sets. *IEEE Trans.* **1973**, *1*, 28–44.
3. Turksen, I.B. Interval-valued strict preference with Zadeh triples. *Fuzzy Sets Syst.* **1996**, *78*, 183–195. [CrossRef]
4. Atanassov, K.; Gargov, G. Interval-valued intuitionistic fuzzy sets. *Fuzzy Sets Syst.* **1989**, *31*, 343–349. [CrossRef]
5. Atanassov, K. Intuitionistic fuzzy sets. *Fuzzy Sets Syst.* **1986**, *20*, 87–96. [CrossRef]
6. Jun, Y.B.; Kim, C.S.; Yang, K.O. Cubic sets. *Ann. Fuzzy Math. Inform.* **2012**, *1*, 83–98.

7. Garg, H.; Singh, S. A novel triangular interval type-2 intuitionistic fuzzy sets and their aggregation operators. *Iran. J. Fuzzy Syst.* **2018**, *15*, 69–93.
8. Kumar, K.; Garg, H. TOPSIS method based on the connection number of set pair analysis under interval-valued intuitionistic fuzzy set environment. *Comput. Appl. Math.* **2018**, *37*, 1319–1329. [CrossRef]
9. Kaur, G.; Garg, H. Cubic intuitionistic fuzzy aggregation operators. *Int. J. Uncertain. Quantif.* **2018**, *8*, 405–427. [CrossRef]
10. Rani, D.; Garg, H. Distance measures between the complex intuitionistic fuzzy sets and its applications to the decision—Making process. *Int. J. Uncertain. Quantif.* **2007**, *7*, 423–439. [CrossRef]
11. Xu, Z.S. Intuitionistic fuzzy aggregation operators. *IEEE Trans. Fuzzy Syst.* **2007**, *15*, 1179–1187.
12. Xu, Z.S.; Yager, R.R. Some geometric aggregation operators based on intuitionistic fuzzy sets. *Int. J. Gen. Syst.* **2006**, *35*, 417–433. [CrossRef]
13. Smarandache, F. *A Unifying Field in Logics: Neutrosophic Logic. Neutrosophy, Neutrosophic Set, Neutrosophic Probability*; American Research Press: Rehoboth, NM, USA, 1999.
14. Smarandache, F. Neutrosophic Set—A Generalization of the Intuitionistic Fuzzy Set. In Proceedings of the IEEE International Conference on Granular Computing, Atlanta, GA, USA, 10–12 May 2006.
15. Wang, H.; Smarandache, F.; Zhang, Y.Q.; Sunderraman, R. Single valued neutrosophic sets. *Multispace Multistruct.* **2010**, *4*, 410–413.
16. Wang, H.; Smarandache, F.; Zhang, Y.Q. *Interval Neutrosophic Sets and Logic: Theory Applications Computing*; Hexis: Phoenix, AZ, USA, 2005.
17. Jun, Y.B.; Smarandache, F.; Kim, C.S. Neutrosophic cubic sets. *New Math. Nat. Comput.* **2015**, *8*, 41. [CrossRef]
18. Liu, P.; Wang, Y. Multiple attribute decision-making method based on single-valued neutrosophic normalized weighted bonferroni mean. *Neural Comput. Appl.* **2014**, *25*, 2001–2010. [CrossRef]
19. Abdel-Basset, M.; Zhou, Y.; Mohamed, M.; Chang, V. A group decision making framework based on neutrosophic vikor approach for e-government website evaluation. *J. Intell. Fuzzy Syst.* **2018**, *34*, 4213–4224. [CrossRef]
20. Nancy, G.H.; Garg, H. Novel single-valued neutrosophic decision making operators under frank norm operations and its application. *Int. J. Uncertain. Quantif.* **2014**, *6*, 361–375. [CrossRef]
21. Peng, X.D.; Dai, J.G. A bibliometric analysis of neutrosophic set: Two decades review from 1998–2017. *Artif. Intell. Rev.* **2018**. [CrossRef]
22. Li, B.; Wang, J.; Yang, L.; Li, X. A novel generalized simplified neutrosophic number Einstein aggregation operator. *Int. J. Appl. Math.* **2018**, *48*, 67–72.
23. Biswas, P.; Pramanik, S.; Giri, B.C. Topsis method for multi-attribute group decision-making under single-valued neutrosophic environment. *Neural Comput. Appl.* **2016**, *27*, 727–737. [CrossRef]
24. Ye, J. Exponential operations and aggregation operators of interval neutrosophic sets and their decision making methods. *SpringerPlus* **2016**, *5*, 1488. [CrossRef]
25. Garg, H.; Nancy, G. Multi-criteria decision-making method based on prioritized muirhead mean aggregation operator under neutrosophic set environment. *Symmetry* **2018**, *10*, 280. [CrossRef]
26. Garg, H.; Nancy, B. Some hybrid weighted aggregation operators under neutrosophic set environment and their applications to multicriteria decision-making. *Appl. Intell.* **2018**, 1–18. [CrossRef]
27. Jha, S.; Kumar, R.; Chatterjee, J.M.; Khari, M.; Yadav, N.; Smarandache, F. Neutrosophic soft set decision making for stock trending analysis. *Evol. Syst.* **2019**, 1–7. [CrossRef]
28. Khan, M.; Gulistan, M.; Khan, M.; Smaradache, F. Neutrosophic cubic Einstein geometric aggregatio operatorswith application to multi-creiteria decision making theory method. *Symmetry* **2019**, *11*, 247.
29. Zhan, J.; Khan, M.; Gulistan, M.; Ali, A. Applications of neutrosophic cubic sets in multi-criteria decision making. *Int. J. Uncertain. Quantif.* **2017**, *7*, 377–394. [CrossRef]
30. Banerjee, D.; Giri, B.C.; Pramanik, S.; Smarandache, F. GRA for multi attribute decision making in neutrosophic cubic set environment. *Neutrosophic Sets Syst.* **2017**, *15*, 64–73.
31. Lu, Z.; Ye, J. Cosine measure for neutrosophic cubic sets for multiple attribte decision making. *Symmetry* **2017**, *10*, 121–131.
32. Pramanik, S.; Dalapati, S.; Alam, S.; Roy, S.; Smarandache, F. Neutrosophic cubic MCGDM method based on similarity measure. *Neutrosophic Sets Syst.* **2017**, *16*, 44–56.
33. Shi, L.; Ye, J. Dombi aggregation operators of neutrosophic cubic set for multiple attribute deicision making. *Algorithms* **2018**, *11*, 29. [CrossRef]

34. Ye, J. Operations and aggregation methods of neutrosophic cubic numbers for multiple attribute decision-making. *Soft Comput.* **2018**, *22*, 7435–7444. [CrossRef]
35. Alhazaymeh, K.; Gulistan, M.; Khan, M.; Kadry, S. Neutrosophic cubic Einstein hybrid geometric aggregation operators with application in prioritization using multiple attribute decision-making method. *Mathematics* **2019**, *7*, 346.
36. Figueira, J.; Greco, S.; Ehrgott, M. *Multiple Criteria Decision Analysis: State of the Art Surveys*; Springer: Boston, MA, USA, 2005.
37. Zavadskas, E.K.; Turskis, Z. Multiple criteria decision making (MCDM) methods: An overview. *Technol. Econ. Dev. Econ.* **2011**, *17*, 397–427. [CrossRef]
38. Zhou, L.G.; Wu, J.X.; Chen, H.Y. Linguistic continuous ordered weighted distance measure and its application to multiple attributes group decision making. *Appl. Soft Comput.* **2014**, *25*, 266–276. [CrossRef]
39. Merigó, J.M.; Gil-Lafuente, A.M.; Yu, D.; Llopis-Albert, C. Fuzzy decision making in complex frameworks with generalized aggregation operators. *Appl. Soft Comput.* **2018**. [CrossRef]

symmetry

MDPI

Article

Triangular Single Valued Neutrosophic Data Envelopment Analysis: Application to Hospital Performance Measurement

Wei Yang [1], Lulu Cai [2], Seyed Ahmad Edalatpanah [3,*] and Florentin Smarandache [4]

1 STATE GRID Quzhou Power Supply Company, Quzhou University, Quzhou 324000, China; easteryang@163.com
2 Department of Electrical Automation, Quzhou University, Quzhou 324000, China; ysu-fbg@163.com
3 Department of Applied Mathematics, Quzhou University, Quzhou 324000, China
4 Department of Mathematics, University of New Mexico, Gallup, NM 87301, USA; smarand@unm.edu
* Correspondence: saedalat@yahoo.com

Received: 22 February 2020; Accepted: 18 March 2020; Published: 8 April 2020

Abstract: The foremost broadly utilized strategy for the valuation of the overall performance of a set of identical decision-making units (DMUs) that use analogous sources to yield related outputs is data envelopment analysis (DEA). However, the witnessed values of the symmetry or asymmetry of different types of information in real-world applications are sometimes inaccurate, ambiguous, inadequate, and inconsistent, so overlooking these conditions may lead to erroneous decision-making. Neutrosophic set theory can handle these occasions of data and makes an imitation of the decision-making procedure with the aid of thinking about all perspectives of the decision. In this paper, we introduce a model of DEA in the context of neutrosophic sets and sketch an innovative process to solve it. Furthermore, we deal with the problem of healthcare system evaluation with inconsistent, indeterminate, and incomplete information using the new model. The triangular single-valued neutrosophic numbers are also employed to deal with the mentioned data, and the proposed method is utilized in the assessment of 13 hospitals of Tehran University of Medical Sciences of Iran. The results exhibit the usefulness of the suggested approach and point out that the model has practical outcomes for decision-makers.

Keywords: single-valued neutrosophic set; triangular neutrosophic number; data envelopment analysis; healthcare systems; performance evaluation

1. Introduction

As a strong analytical tool for benchmarking and efficiency evaluation, DEA (data envelopment analysis) is a technique for evaluating the relation efficiency of decision-making units (DMUs), developed initially by Charens et al. [1] on a printed paper named the Charnes, Cooper, and Rhodes (CCR) model. They extended the nonparametric method introduced by Farrell [2] to gauge DMUs with multiple inputs and outputs. The Banker, Charnes, and Cooper (BCC) model is an extension of the previous model under the assumption of variable returns-to-scale (VRS) [3]. With this technique, managers can obtain the relative efficiency of a set of DMUs. In time, many theoretical and empirical studies have applied DEA to several fields of science and engineering, such as healthcare, agriculture, banking supply chains, and financial services, among others. For more details, the reader is referred to the studies of [4–14].

Conventional DEA models require crisp information that may not be permanently accessible in real-world applications. Nevertheless, in numerous cases, data are unstable, uncertain, and complicated; therefore, they cannot be accurately measured. Zadeh [15] first proposed the theory of fuzzy sets (FSs) against certain logic. After this work, many researchers studied this topic; details of some approaches can be observed in [16–20]. Several researchers also proposed some models of DEA under a fuzzy environment [21–25].

However, Zadeh's fuzzy sets consider only the membership function and cannot deal with other parameters of vagueness. To overcome this lack of information, Atanassov [26] introduced an extension of FSs called intuitionistic fuzzy sets (IFSs). There are also several models of DEA with intuitionistic fuzzy data: see [27–30].

Although the theory of IFSs can handle incomplete information for various real-world issues, it cannot address all types of uncertainty such as inconsistent and indeterminate evidence. Therefore, Smarandache [31,32] established the neutrosophic set (NS) as a robust overall framework that generalizes classical and all kinds of fuzzy sets (FSs and IFSs).

NSs can accommodate indeterminate, ambiguous, and conflicting information where the indeterminacy is clearly quantified, and define three kinds of membership function independently.

In the past years, some versions of NSs such as interval neutrosophic sets [33,34], bipolar neutrosophic sets [35,36], single-valued neutrosophic sets [37–39], and neutrosophic linguistic sets [40] have been presented. In addition, in the field of neutrosophic sets, logic, measure, probability, statistics, pre-calculus and calculus, and their applications in multiple areas have been extended: see [41–44].

In real circumstances, some data in DEA may be uncertain, indeterminate, and inconsistent, and considering truth, falsity, and indeterminacy membership functions for each input/output of DMUs in the neutrosophic sets help decision-makers to obtain a better interpretation of information. In addition, by using the NS in DEA, analysts can better set their acceptance, indeterminacy, and rejection degrees regarding each datum. Moreover, with NSs, we can obtain a better depiction of reality through seeing all features of the decision-making procedure. Therefore, the NS can embrace imprecise, vague, incomplete, and inconsistent evidence powerfully and efficiently. Although there are several approaches to solve various problems under neutrosophic environments, there are not many studies that have dealt with DEA under NSs.

The utilization of neutrosophic logic in DEA can be traced to Edalatpanah [45]. Kahraman et al. [46] proposed a hybrid algorithm based on a neutrosophic analytic hierarchy process (AHP) and DEA for bringing a solution to the efficiency of private universities. Edalatpanah and Smarandache [47], based on some operators and natural logarithms, proposed an input-oriented DEA model with simplified neutrosophic numbers. Abdelfattah [48], by converting a neutrosophic DEA into an interval DEA, developed a new DEA model under neutrosophic numbers. Although these approaches are interesting, some restrictions exist. One of them is that these methods have high running times, mainly when we have many inputs and outputs. Furthermore, the main flaw of [48] is the existence of several production frontiers in the steps of efficiency measure, and this leads to the lack of comparability between efficiencies.

Therefore, in this paper, we design an innovative simple model of DEA in which all inputs and outputs are triangular single-valued neutrosophic numbers (TSVNNs), and establish a new efficient strategy to solve it. Furthermore, we use the suggested technique for the performance assessment of 13 hospitals of Tehran University of Medical Sciences (TUMS) of Iran.

The paper unfolds as follows: some basic knowledge, concepts, and arithmetic operations on NSs and TSVNNs are discussed in Section 2. In Section 3, some concepts of DEA and the CCR model are reviewed. In Section 4, we establish the mentioned model of DEA under the neutrosophic environment and propose a method to solve it. In Section 5, the suggested model is utilized for a case study of TUMS. Lastly, conclusions and future directions are presented in Section 6.

2. Preliminaries

In this section, we discuss some basic definitions related to neutrosophic sets and single-valued neutrosophic numbers, respectively.

Smarandache put forward an indeterminacy degree of membership as an independent component in his papers [31,32], and since the principle of excluded middle cannot be applied to new logic, he combines non-standard analysis with three-valued logic, set theory, probability theory, and philosophy. As a result, neutrosophic means "neutral thinking knowledge." Given this meaning and the use of the term neutral, along with the components of truth (membership) and falsity (non-membership), its distinction is marked by fuzzy sets and intuitionistic fuzzy sets. Here, it is appropriate to give a brief explanation of the non-standard analysis.

In the early 1960s, Robinson developed non-standard analysis as a form of analysis and a branch of logic in which infinitesimals are precisely defined [49]. Formally, x is called an infinitesimal number if and only if for any non-null positive integer n we have $|x| \leq \frac{1}{n}$. Let $\varepsilon > 0$ be an infinitesimal number; then, the extended real number set is an extension of the set of real numbers that contains the classes of infinite numbers and the infinitesimal numbers. If we consider non-standard finite numbers $1^{+} = 1 + \varepsilon$ and $-0 = 0 - \varepsilon$, where 0 and 1 are the standard parts and ε is the non-standard part, then $]^{-}0, 1^{+}[$ is a non-standard unit interval. It is clear that 0, 1, as well as the non-standard infinitesimal numbers that are less than zero and infinitesimal numbers that are more than one belong to this non-standard unit interval. Now, let us define a neutrosophic set:

Definition 1 ([31,32,41]) (neutrosophic set). *A neutrosophic set in universal U is defined by three membership functions for the truth, indeterminacy, and falsity of x in the real non-standard* $]^{-}0, 1^{+}[$*, where the summation of them belongs to [0, 3].*

Definition 2 ([34]). *If the three membership functions of a NS are singleton in the real standard [0, 1], then a single-valued neutrosophic set (SVNS)* ψ *is denoted by:*

$$\psi = \left\{\left(x,\ \tau_{\psi}(x),\ \iota_{\psi}(x),\ \nu_{\psi}(x)\right)\middle| x \in U\right\},$$

which satisfies the following condition:

$$0 \leq \tau_{\psi}(x) + \iota_{\psi}(x) + \nu_{\psi}(x) \leq 3.$$

Definition 3 ([38]). *A TSVNN* $A^{\aleph} = \left\langle \left(a^{l}, a^{m}, a^{u}\right), \left(b^{l}, b^{m}, b^{u}\right), \left(c^{l}, c^{m}, c^{u}\right) \right\rangle$ *is a particular single-valued neutrosophic number (SVNN) whose* $\tau_{A^{\aleph}}(x)$, $\iota_{A^{\aleph}}(x)$, *and* $\nu_{A^{\aleph}}(x)$ *are presented as follows:*

$$\tau_{A^{\aleph}}(x) = \begin{cases} \frac{(x-a^{l})}{(a^{m}-a^{l})} & a^{l} \leq x < a^{m}, \\ 1 & x = a^{m}, \\ \frac{(a^{u}-x)}{(a^{u}-a^{m})} & a^{m} < x \leq a^{u}, \\ 0 & otherwise. \end{cases},$$

$$\iota_{A^{\aleph}}(x) = \begin{cases} \frac{(b^{m}-x)}{(b^{m}-b^{l})} & b^{l} \leq x < b^{m}, \\ 0 & x = b^{m}, \\ \frac{(x-b^{m})}{(b^{u}-b^{m})} & b^{m} < x \leq b^{u}, \\ 1 & otherwise. \end{cases},$$

$$\nu_{A^{\aleph}}(x) = \begin{cases} \frac{(c^{m}-x)}{(c^{m}-c^{l})} & c^{l} \leq x < c^{m}, \\ 0 & c = c^{m}, \\ \frac{(x-c^{m})}{(c^{u}-c^{m})} & c^{m} < x \leq c^{u}, \\ 1 & otherwise. \end{cases},$$

Definition 4 ([38]). *Let* $A^{\aleph} = \langle (a^l, a^m, a^u), (b^l, b^m, b^u), (c^l, c^m, c^u) \rangle$ *and* $B^{\aleph} = \langle (d^l, d^m, d^u), (e^l, e^m, e^u), (f^l, f^m, f^u) \rangle$ *be two TSVNNs, where their elements are in* $[L_1, U_1]$. *Then, Equations (1) to (3) are true:*

$$
\begin{aligned}
(i) A^{\aleph} \oplus B^{\aleph} = & \langle (min(a^l + d^l, U_1), min(a^m + d^m, U_1), min(a^u + d^u, U_1); \\
& (min(b^l + e^l, U_1), min(b^m + e^m, U_1), min(b^u + e^u, U_1); \\
& (min(c^l + f^l, U_1), min(c^m + f^m, U_1), min(c^u + f^u, U_1) \rangle,
\end{aligned}
\tag{1}
$$

$$(ii) - A^{\aleph} = \langle (-a^u, -a^m, -a^l), (-b^u, -b^m, -b^l), (-c^u, -c^m, -c^l) \rangle, \tag{2}$$

$$(iii) \lambda A^{\aleph} = \langle (\lambda a^l, \lambda a^m, \lambda a^u), (\lambda b^l, \lambda b^m, \lambda b^u), (\lambda c^l, \lambda c^m, \lambda c^u) \rangle, \quad \lambda > 0. \tag{3}$$

Definition 5 ([38]). *Consider* $A^{\aleph} = \langle (a^l, a^m, a^u), (b^l, b^m, b^u), (c^l, c^m, c^u) \rangle$ *as a TSVNN. Then, the ranking function of* $A^{\aleph}$ *can be defined with Equation (4):*

$$\xi(A^{\aleph}) = \frac{(a^l + b^l + c^l) + 2(a^m + b^m + c^m) + (a^u + b^u + c^u)}{12} \tag{4}$$

Definition 6 ([20]). *Suppose* $P^{\aleph}$ *and* $Q^{\aleph}$ *are two TSVNNs, then:*

(i) $P^{\aleph} \leq Q^{\aleph}$ *if and only if* $\xi(P^{\aleph}) \leq \xi(Q^{\aleph})$,

(ii) $P^{\aleph} < Q^{\aleph}$ *if and only if* $\xi(P^{\aleph}) < \xi(Q^{\aleph})$.

3. Data Envelopment Analysis

Let a set of n DMUs, with each DMUj $(j = 1, 2, \ldots, n)$ using m inputs p_{ij} $(i = 1, 2, \ldots, m)$ produce s outputs $q_{rj} (r = 1, 2, \ldots, s)$. If DMU_o is under consideration, then the *input-oriented CCR multiplier* model for the relative efficiency is computed on the basis of Equation (5) [1]:

$$\theta_o^* = max \frac{\sum_{r=1}^{s} v_r q_{ro}}{\sum_{i=1}^{m} u_i p_{io}} \tag{5}$$

s.t:

$$
\begin{aligned}
& \frac{\sum_{r=1}^{s} v_r q_{rj}}{\sum_{i=1}^{m} u_i p_{ij}} \leq 1, \qquad j = 1, 2, \ldots, n \\
& v_r, u_i \geq 0 r = 1, \ldots, s, i = 1, \ldots, m.
\end{aligned}
$$

where v_r and u_i are the related weights. The above nonlinear programming may be converted as Equation (6) to simplify the computation:

$$\theta_o^* = max \sum_{r=1}^{s} v_r q_{ro} \tag{6}$$

s.t:

$$
\begin{aligned}
& \sum_{i=1}^{m} u_i p_{io} = 1 \\
& \sum_{r=1}^{s} v_r q_{rj} - \sum_{i=1}^{m} u_i p_{ij} \leq 0, \qquad j = 1, 2, \ldots, n \\
& v_r, u_i \geq 0 r = 1, \ldots, s, i = 1, \ldots, m.
\end{aligned}
$$

The DMU_o is efficient if $\theta_o^* = 1$; otherwise, it is inefficient.

4. Neutrosophic Data Envelopment Analysis

Like every other model, DEA has been the subject of evolution. One of the critical improvements in this field is related to circumstances where the information of DMUs is characterized and measured beneath conditions of uncertainty and indeterminacy. Indeed, one of the traditional DEA models' assumptions is their crispness of inputs and outputs.

However, it seems questionable to assume the data and observations are crisp in situations where uncertainty and indeterminacy are inevitable features of a real environment. In addition, most management decisions are not made based on known calculations, and there is a lot of uncertainty, indeterminacy, and ambiguity in decision-making problems. The DEA under a neutrosophic environment is more advantageous than a crisp DEA because a decision-maker, in the preparation of the problem, is not obliged to make a subtle formulation. Furthermore, because of a lack of comprehensive knowledge and evidence, precise mathematics are not sufficient to model a complex system. Therefore, the approach based on neutrosophic logic seems fit for such problems [31,32]. In this section, we establish DEA under a neutrosophic environment.

Consider the input and output for the jth DMU as follows:

$$\dddot{p}_{ij} = \left\langle \overset{a_i}{p}_{ij}, \overset{b_i}{p}_{ij}, \overset{c_i}{p}_{ij} \right\rangle = \left\langle [\overset{a_1}{p}_{ij}, \overset{a_2}{p}_{ij}, \overset{a_3}{p}_{ij}], [\overset{b_1}{p}_{ij}, \overset{b_2}{p}_{ij}, \overset{b_3}{p}_{ij}], [\overset{c_1}{p}_{ij}, \overset{c_2}{p}_{ij}, \overset{c_3}{p}_{ij}] \right\rangle,$$

$$\dddot{q}_{rj} = \left\langle \overset{a_i}{q}_{rj}, \overset{b_i}{q}_{rj}, \overset{c_i}{q}_{rj} \right\rangle = \left\langle [\overset{a_1}{q}_{rj}, \overset{a_2}{q}_{rj}, \overset{a_3}{q}_{rj}], [\overset{b_1}{q}_{rj}, \overset{b_2}{q}_{rj}, \overset{b_3}{q}_{rj}], [\overset{c_1}{q}_{rj}, \overset{c_2}{q}_{rj}, \overset{c_3}{q}_{rj}] \right\rangle,$$

which are TSVNNs. Then, the triangular single-valued neutrosophic CCR model called TSVNN-CCR is defined as follows:

$$\theta_o^{\aleph^*} = max \sum_{r=1}^{s} v_r \dddot{q}_{ro} \tag{7}$$

s.t:

$$\sum_{i=1}^{m} u_i \dddot{p}_{io} = 1$$
$$\sum_{r=1}^{s} v_r \dddot{q}_{rj} - \sum_{i=1}^{m} u_i \dddot{p}_{ij} \leq 0, \qquad j = 1, 2, \ldots, n$$
$$v_r, u_i \geq 0 r = 1, \ldots, s, i = 1, \ldots, m.$$

Next, to solve Model (7), we propose the following algorithm:

Algorithm 1. The solution of TSVNN-CCR Model

Step 1. Construct the problem based on Model (8).

Step 2. Using Definition 3 (ii, iii), transform the TSVNN-CCR model of Step 1 into Model (8):

$$\theta_o^{\aleph} = max \sum_{r=1}^{s} \left\langle [v_r \overset{a_1}{q}_{ro}, v_r \overset{a_2}{q}_{ro}, v_r \overset{a_3}{q}_{ro}], [v_r \overset{b_1}{q}_{ro}, v_r \overset{b_2}{q}_{ro}, v_r \overset{b_3}{q}_{ro}], [v_r \overset{c_1}{q}_{ro}, v_r \overset{c_2}{q}_{ro}, v_r \overset{c_3}{q}_{ro}] \right\rangle \quad (8)$$

s.t:

$$\sum_{i=1}^{m} \left\langle [u_i \overset{a_1}{p}_{io}, u_i \overset{a_2}{p}_{io}, u_i \overset{a_3}{p}_{io}], [u_i \overset{b_1}{p}_{io}, u_i \overset{b_2}{p}_{io}, u_i \overset{b_3}{p}_{io}], [u_i \overset{c_1}{p}_{io}, u_i \overset{c_2}{p}_{io}, u_i \overset{c_3}{p}_{io}] \right\rangle = 1$$

$$\sum_{r=1}^{s} \left\langle [v_r \overset{a_1}{q}_{rj}, v_r \overset{a_2}{q}_{rj}, v_r \overset{a_3}{q}_{rj}], [v_r \overset{b_1}{q}_{rj}, v_r \overset{b_2}{q}_{rj}, v_r \overset{b_3}{q}_{rj}], [v_r \overset{c_1}{q}_{rj}, v_r \overset{c_2}{q}_{rj}, v_r \overset{c_3}{q}_{rj}] \right\rangle \oplus$$

$$\sum_{i=1}^{m} \left\langle [-u_i \overset{a_3}{p}_{ij}, -u_i \overset{a_2}{p}_{ij}, -u_i \overset{a_1}{p}_{ij}], [-u_i \overset{b_3}{p}_{ij}, -u_i \overset{b_2}{p}_{ij}, -u_i \overset{b_1}{p}_{ij}], -[u_i \overset{c_3}{p}_{ij}, -u_i \overset{c_2}{p}_{ij}, -u_i \overset{c_1}{p}_{ij}] \right\rangle \leq 0,$$

$$v_r, u_i \geq 0 r = 1, \ldots, s, i = 1, \ldots, m.$$

Step 3. Transform Model (8) into the following model:

$$\theta_o^{\aleph} = max \left\langle \left(\sum_{r=1}^{s} v_r \overset{a_1}{q}_{ro}, \sum_{r=1}^{s} v_r \overset{a_2}{q}_{ro}, \sum_{r=1}^{s} v_r \overset{a_3}{q}_{ro} \right), \left(\sum_{r=1}^{s} v_r \overset{b_1}{q}_{ro}, \sum_{r=1}^{s} v_r \overset{b_2}{q}_{ro}, \sum_{r=1}^{s} v_r \overset{b_3}{q}_{ro} \right), \left(\sum_{r=1}^{s} v_r \overset{c_1}{q}_{ro}, \sum_{r=1}^{s} v_r \overset{c_2}{q}_{ro}, \sum_{r=1}^{s} v_r \overset{c_3}{q}_{ro} \right) \right\rangle \quad (9)$$

s.t:

$$\left\langle \left(\sum_{i=1}^{m} u_i \overset{a_1}{p}_{io}, \sum_{i=1}^{m} u_i \overset{a_2}{p}_{io}, \sum_{i=1}^{m} u_i \overset{a_3}{p}_{io} \right), \left(\sum_{i=1}^{m} u_i \overset{b_1}{p}_{io}, \sum_{i=1}^{m} u_i \overset{b_2}{p}_{io}, \sum_{i=1}^{m} u_i \overset{b_3}{p}_{io} \right), \left(\sum_{i=1}^{m} u_i \overset{c_1}{p}_{io}, \sum_{i=1}^{m} u_i \overset{c_2}{p}_{io}, \sum_{i=1}^{m} u_i \overset{c_3}{p}_{io} \right) \right\rangle = 1$$

$$\left\langle \left(\sum_{r=1}^{s} v_r \overset{a_1}{q}_{rj} \oplus \sum_{i=1}^{m} -u_i \overset{a_3}{p}_{ij}, \sum_{r=1}^{s} v_r \overset{a_2}{q}_{rj} \oplus \sum_{i=1}^{m} -u_i \overset{a_2}{p}_{ij}, \sum_{r=1}^{s} v_r \overset{a_3}{q}_{rj} \oplus \sum_{i=1}^{m} -u_i \overset{a_1}{p}_{ij} \right), \right.$$

$$\left(\sum_{r=1}^{s} v_r \overset{b_1}{q}_{rj} \oplus \sum_{i=1}^{m} -u_i \overset{b_3}{p}_{ij}, \sum_{r=1}^{s} v_r \overset{b_2}{q}_{rj} \oplus \sum_{i=1}^{m} -u_i \overset{b_2}{p}_{ij}, \sum_{r=1}^{s} v_r \overset{b_3}{q}_{rj} \oplus \sum_{i=1}^{m} -u_i \overset{b_1}{p}_{ij} \right),$$

$$\left. \left(\sum_{r=1}^{s} v_r \overset{c_1}{q}_{rj} \oplus \sum_{i=1}^{m} -u_i \overset{c_3}{p}_{ij}, \sum_{r=1}^{s} v_r \overset{c_2}{q}_{rj} \oplus \sum_{i=1}^{m} -u_i \overset{c_2}{p}_{ij}, \sum_{r=1}^{s} v_r \overset{c_3}{q}_{rj} \oplus \sum_{i=1}^{m} -u_i \overset{c_1}{p}_{ij} \right) \right\rangle \leq 0,$$

$$v_r, u_i \geq 0 r = 1, \ldots, s, i = 1, \ldots, m.$$

Step 4. Based on Definitions 4–5, convert TSVNN-CCR Model (9) into crisp Model (10):

$$\theta_o{}^* \approx \xi\left(\theta_o^{\aleph}\right) = \quad (10)$$

$$max \sum_{r=1}^{s} \xi\left(\left\langle [v_r \overset{a_1}{q}_{ro}, v_r \overset{a_2}{q}_{ro}, v_r \overset{a_3}{q}_{ro}], [v_r \overset{b_1}{q}_{ro}, v_r \overset{b_2}{q}_{ro}, v_r \overset{b_3}{q}_{ro}], [v_r \overset{c_1}{q}_{ro}, v_r \overset{c_2}{q}_{ro}, v_r \overset{c_3}{q}_{ro}] \right\rangle \right)$$

s.t:

$$\sum_{i=1}^{m} \xi\left(\left\langle [u_i \overset{a_1}{p}_{io}, u_i \overset{a_2}{p}_{io}, u_i \overset{a_3}{p}_{io}], [u_i \overset{b_1}{p}_{io}, u_i \overset{b_2}{p}_{io}, u_i \overset{b_3}{p}_{io}], [u_i \overset{c_1}{p}_{io}, u_i \overset{c_2}{p}_{io}, u_i \overset{c_3}{p}_{io}] \right\rangle \right) = 1$$

$$\sum_{r=1}^{s} \xi\left(\left\langle [v_r \overset{a_1}{q}_{rj}, v_r \overset{a_2}{q}_{rj}, v_r \overset{a_3}{q}_{rj}], [v_r \overset{b_1}{q}_{rj}, v_r \overset{b_2}{q}_{rj}, v_r \overset{b_3}{q}_{rj}], [v_r \overset{c_1}{q}_{rj}, v_r \overset{c_2}{q}_{rj}, v_r \overset{c_3}{q}_{rj}] \right\rangle \right) \oplus$$

$$\sum_{i=1}^{m} \xi\left(\left\langle \left[-u_i \overset{a_3}{p}_{ii}, -u_i \overset{a_2}{p}_{ij}, -u_i \overset{a_1}{p}_{ij} \right], \left[-u_i \overset{b_3}{p}_{ij}, -u_i \overset{b_2}{p}_{ij}, -u_i \overset{b_1}{p}_{ij} \right], \left[-u_i \overset{c_3}{p}_{ij}, -u_i \overset{c_2}{p}_{ij}, -u_i \overset{c_1}{p}_{ij} \right] \right\rangle \right) \leq 0,$$

$$v_r, u_i \geq 0 r = 1, \ldots, s, i = 1, \ldots, m.$$

Step 5. Run Model (10) and get the optimal efficiency of each DMU.

5. Numerical Experiment

In this section, a case study of a DEA problem under a neutrosophic environment is used to reveal the validity and usefulness of the proposed model.

Case Study: The Efficiency of the Hospitals of TUMS

Performance assessments in healthcare frameworks are a noteworthy worry of policymakers so that reforms to improve performance in the health sector are on the policy agenda of numerous national governments and worldwide agencies. In the related literature, various methods such as least squares and simple ratio analysis have been applied to assess the performance of healthcare systems (see for instance: [50–52]). Nonetheless, due to the applicability of DEA in the solution of problems with multiple inputs and outputs, it is most commonly used in healthcare systems [53]. The utilizations of DEA in the healthcare sector can be found in several works of literature, including for crisp data [54–56], fuzzy data [57,58], and intuitionistic fuzzy data [59]. To the best of our knowledge, none of these current works assessed the efficiency of healthcare organizations with neutrosophic sets. Therefore, to assess the efficiency of the mentioned systems under a neutrosophic environment, we used the proposed model to evaluate 13 hospitals of TUMS. It is worth emphasizing that due to privacy policies, the names of these hospitals are not shared. Furthermore, for the selection of the most suitable and acceptable items of the healthcare system, which are commonly used for measuring efficiency in the literature, we considered two inputs, namely the number of doctors and number of beds, and three outputs, namely the total yearly days of hospitalization of all patients, number of outpatient department visits, and overall patient satisfaction.

For each hospital, we gathered the related data from the medical records unit of the hospitals, Center of Statistics of the University of Medical Sciences, the reliable library, online resources, and the judgments of some experts. After collecting data, we found that the information was sometimes inconsistent, indeterminate, and incomplete. The investigation revealed that several reforms by the mentioned hospitals and other issues have led to considerable uncertainty and indeterminacy about the data. As a result, we identified them as triangular single-valued neutrosophic numbers (TSVNNs). For example, for "Patient Satisfaction," we collected data in terms of "satisfaction," "dissatisfaction," and "abstention," and for each term, the related data was expressed by a triangular fuzzy number. In addition, each triangular fuzzy number was constructed based on min, average, and max. All data were expressed by using TSVNNs, and can be found in Tables 1 and 2.

Table 1. Input information of the nominee hospitals.

DMU	Inputs 1 Number of Doctors	Inputs 2 Number of Beds
1	$\langle[404, 540, 674], [350, 440, 560], [420, 645, 700]\rangle$	$\langle[520, 530, 535], [520, 525, 530], [532, 534, 540]\rangle$
2	$\langle[119, 136, 182], [122, 125, 137], [125, 178, 200]\rangle$	$\langle[177, 180, 188], [173, 175, 179], [185, 189, 195]\rangle$
3	$\langle[139, 145, 158], [139, 140, 147], [146, 155, 167]\rangle$	$\langle[208, 214, 218], [195, 209, 215], [210, 217, 230]\rangle$
4	$\langle[86, 93, 151], [83, 85, 87], [89, 138, 160]\rangle$	$\langle[114, 116, 118], [114, 115, 117], [116, 118, 125]\rangle$
5	$\langle[84, 93, 143], [84, 89, 120], [90, 140, 155]\rangle$	$\langle[110, 117, 121], [105, 112, 120], [113, 119, 128]\rangle$
6	$\langle[101, 113, 170], [110, 112, 115], [112, 120, 177]\rangle$	$\langle[101, 107, 111], [95, 100, 104], [108, 112, 115]\rangle$
7	$\langle[561, 694, 864], [510, 640, 750], [582, 857, 930]\rangle$	$\langle[492, 495, 508], [492, 494, 500], [493, 506, 520]\rangle$
8	$\langle[123, 179, 199], [122, 125, 130], [195, 200, 205]\rangle$	$\langle[66, 68, 73], [63, 67, 69], [68, 70, 78]\rangle$
9	$\langle[101, 153, 155], [140, 145, 150], [145, 149, 167]\rangle$	$\langle[192, 195, 198], [185, 193, 197], [194, 196, 205]\rangle$
10	$\langle[147, 164, 170], [147, 160, 167], [165, 169, 180]\rangle$	$\langle[333, 340, 357], [335, 338, 350], [338, 347, 364]\rangle$
11	$\langle[130, 158, 192], [110, 144, 173], [146, 177, 205]\rangle$	$\langle[96, 100, 114], [97, 99, 103], [99, 110, 129]\rangle$
12	$\langle[128, 137, 187], [128, 133, 164], [134, 184, 199]\rangle$	$\langle[213, 220, 224], [208, 215, 223], [216, 222, 231]\rangle$
13	$\langle[151, 160, 210], [151, 156, 187], [157, 207, 222]\rangle$	$\langle[320, 327, 331], [315, 322, 330], [323, 329, 338]\rangle$

Next, we used Algorithm 1 to solve the performance valuation problem. For example, Algorithm 1 for DMU_1 can be used as follows:

First, we construct a DEA model with the mentioned TSVNNs:

$$\begin{aligned} max\ \widetilde{\theta}_1 \approx\ &\langle[121.13,\ 139.24,\ 140.04], [138.64,\ 139.14,\ 139.81], [139.14,\ 140.02,\ 141.17]\rangle v_1 \oplus \\ &\langle[38, 41,\ 45], [38,\ 40,\ 43], [41,\ 44,\ 49]\rangle v_2 \oplus \\ &\langle[104.23,\ 114.04,\ 278.51], [102.37,\ 109.15,\ 235.72], [104.81,\ 275.25,\ 279.88]\rangle v_3 \end{aligned}$$

s.t:

$$\begin{aligned} &\langle[404,\ 540,\ 674], [350,\ 440,\ 560], [420,\ 645, 700]\rangle u_1 \\ &\oplus\langle[520,\ 530,\ 535], [520,\ 525,\ 530], [532,\ 534,\ 540]\rangle u_2 = 1, \end{aligned}$$

$$\begin{aligned} &(\langle[121.13,\ 139.24,\ 140.04], [138.64,\ 139.14,\ 139.81], [139.14,\ 140.02,\ 141.17]\rangle v_1 \oplus \langle[38, 41,\ 45], [38,\ 40,\ 43], \\ &[41,\ 44,\ 49]\rangle v_2 \oplus \langle[104.23,\ 114.04,\ 278.51], [102.37, 109.15,\ 235.72], [104.81, 275.25,\ 279.88]\rangle v_3) - \\ &(\langle[404,\ 540,\ 674], [350,\ 440,\ 560], [420,\ 645, 700]\rangle u_1 \oplus \langle[520,\ 530,\ 535], [520,\ 525,\ 530], [532,\ 534,\ 540]\rangle u_2) \le 0, \end{aligned}$$

$$\begin{aligned} &(\langle[31.54,\ 34.93,\ 38.89], [31.54,\ 34.15,\ 38.27], [34.86,\ 38.15,\ 39.83]\rangle v_1 \oplus \langle[40,\ 44,\ 47],\ [35,\ 52,\ 45], \\ &[41,\ 46,\ 50]\rangle v_2 \oplus \langle[34.54,\ 36.98,\ 54.82\ ], [36.45,\ 36.80,\ 41.57], [47.61,\ 54.25,\ 55.35] > v_3\rangle) - \\ &(\langle < [109,\ 126,\ 172]\ ,\ [112,\ 115,\ 127],\ [115,\ 168, 190]\rangle u_1 \oplus [\langle 177, 180,\ 188],\ [173, 175, 179],\ [185, 189, 195]\rangle u_2) \le 0, \end{aligned}$$

$$\begin{gathered}
(\langle[81.62, 82.07, 85.51], [81.41, 81.94, 83.35], [81.78, 85.49, 88.16]\rangle v_1 \oplus \langle[18, 20, 29], [19, 21, 23], \\
[28, 30, 35]\rangle v_2 \oplus [\langle 157.75, 177.57, 264.52], [157.75, 176.68, 250.75], [180.29, 263.98, 272.16]\rangle v_3) - \\
(\langle[139, 145, 158], [139, 140, 147], [146, 155, 167]\rangle u_1 \oplus \langle[208, 214, 218], [195, 209, 215], [210, 217, 230]\rangle u_2) \le 0,
\end{gathered}$$

$$\begin{gathered}
(\langle[19.54, 20.41, 20.59], [20.15, 20.25, 20.32], [20.54, 20.58, 20.70]\rangle v_1 \oplus \langle[18, 21, 25], [15, 19, 23], \\
[20, 24, 30]\rangle v_2 \oplus \langle[32.89, 35.56, 87.74], [35.25, 35.50, 35.61], [87.50, 87.94, 88.30]\rangle v_3) - \\
(\langle[86, 93, 151], [83, 85, 87], [89, 138, 160]\rangle u_1 \oplus \langle[114, 116, 118], [114, 115, 117], [116, 118, 125]\rangle u_2) \le 0,
\end{gathered}$$

$$\begin{gathered}
(\langle[23.89, 24.60, 26.09], [23.56, 23.60, 23.68], [25.97, 26.35, 26.72]\rangle v_1 \oplus \langle[30, 36, 41], [34, 35, 37], \\
[35, 40, 57]\rangle v_2 \oplus \langle[63.23, 69.58, 120.73], [63, 65.17, 94.93], [64.47, 118.75, 124.75]\rangle v_3) - \\
(\langle[84, 93, 143], [84, 89, 120], [90, 140, 155]\rangle u_1 \oplus \langle[110, 117, 121], [105, 112, 120], [113, 119, 128]\rangle u_2) \le 0,
\end{gathered}$$

$$\begin{gathered}
(\langle[21.33, 21.49, 23.31], [20.94, 24.25, 22.68], [21.38, 23.14, 23.94]\rangle v_1 \oplus \langle[50, 55, 60], [50, 53, 57], \\
[56, 59, 70]\rangle v_2 \oplus \langle[72.84, 82.84, 94.18], [82.15, 82.68, 84.89], [85.75, 93.50, 97.18]\rangle v_3) - \\
(\langle[101, 113, 170], [110, 112, 115], [112, 120, 177]\rangle u_1 \oplus \langle[101, 107, 111], [95, 100, 104], [108, 112, 115]\rangle u_2) \le 0,
\end{gathered}$$

$$\begin{gathered}
(\langle[145.77, 148.28, 169.01], [145.77, 147.16, 168.31], [150.69, 168.95, 175.18]\rangle v_1 \oplus \langle[40, 44, 46], [42, 43, 45], \\
[43, 44, 55]\rangle v_2 \oplus \langle[147.59, 150.37, 227.12], [147.30, 147.45, 148.25], [218.24, 224.61, 229.63]\rangle v_3) - \\
(\langle[561, 694, 864], [510, 640, 750], [582, 857, 930]\rangle u_1 \oplus \langle[492, 495, 508], [492, 494, 500], [493, 506, 520]\rangle u_2) \le 0,
\end{gathered}$$

$$\begin{gathered}
(\langle[11.56, 11.74, 12.96], [11.42, 11.61, 11.98], [11.58, 12.64, 13.16]\rangle v_1 \oplus \langle[60, 75, 80], [55, 60, 62], \\
[78, 83, 85]\rangle v_2 \oplus \langle[189.37, 202.08, 284.99], [189.37, 200.52, 281.63], [270.16, 284.55, 289.12]\rangle v_3) - \\
(\langle[123, 179, 199], [122, 125, 130], [195, 200, 205]\rangle u_1 \oplus \langle[66, 68, 73], [63, 67, 69], [68, 70, 78]\rangle u_2) \le 0,
\end{gathered}$$

$$\begin{gathered}
(\langle[57.55, 62.67, 63.03], [62.15, 62.50, 62.93], [62.50, 62.97, 63.61]\rangle v_1 \oplus \langle[32, 35, 38], [32, 33, 35], \\
[34, 36, 45]\rangle v_2 \oplus \langle[14.63, 14.85, 29.40], [14.70, 14.75, 15.25], [24.75, 28.36, 32.64]\rangle v_3) - \\
(\langle[101, 153, 155], [140, 145, 150], [145, 149, 167]\rangle u_1 \oplus \langle[192, 195, 198], [185, 193, 197], [194, 196, 205]\rangle u_2) \le 0,
\end{gathered}$$

$$\begin{gathered}
(\langle[73.21, 76.03, 81.90], [75.76, 76.05, 76.25], [81.67, 82.27, 82.64]\rangle v_1 \oplus \langle[22, 25, 40], [20, 24, 27], \\
[23, 25, 29]\rangle v_2 \oplus \langle[96.77, 97.27, 110.39], [96.77, 96.89, 105.14], [99.76, 108.62, 115.27]\rangle v_3) - \\
(\langle[147, 164, 170], [147, 160, 167], [165, 169, 180]\rangle u_1 \oplus \langle[333, 340, 357], [335, 338, 350], [338, 347, 364]\rangle u_2) \le 0,
\end{gathered}$$

$$\begin{gathered}
(\langle[22.90, 27.71, 35.56], [22.90, 26.45, 31.28], [27.92, 34.62, 39.41]\rangle v_1 \oplus \langle[20, 23, 26], [21, 22, 24], \\
[22, 25, 30]\rangle v_2 \oplus \langle[171.53, 182.46, 384.99], [171.12, 178.65, 210.34], [175.59, 270.65, 400.12]\rangle v_3) - \\
(\langle[130, 158, 192], [110, 144, 173], [146, 177, 205]\rangle u_1 \oplus \langle[96, 100, 114], [97, 99, 103], [99, 110, 129]\rangle u_2) \le 0,
\end{gathered}$$

$$\begin{gathered}
(\langle[58.41, 59.12, 60.61], [58.08, 58.12, 58.20], \langle[60.49, 60.87, 61.24]\rangle v_1 \oplus \langle[25, 31, 37], [29, 30, 32], \\
\langle[30, 35, 52]\rangle v_2 \oplus \Big\langle [59.87, 66.22, 117.37], [59.64, 61.81, 91.57], [61.11, 115.39, 121.39]\rangle v_3) - \\
(\langle[128, 137, 187], [128, 133, 164], [134, 184, 199]\rangle u_1 \oplus \langle[213, 220, 224], [208, 215, 223], [216, 222, 231]\rangle u_2) \le 0,
\end{gathered}$$

$$\begin{gathered}
(\langle[66.97, 67.68, 69.17], [66.64, 66.68, 66.76], [69.05, 69.43, 69.80]\rangle v_1 \oplus \langle[20, 27, 31], [23, 26, 28], \\
[24, 30, 46]\rangle v_2 \oplus \langle[96.97, 103.32, 154.47], [96.74, 98.91, 128.67], [98.21, 152.49, 158.50]\rangle v_3) - \\
(\langle[151, 160, 210], [151, 156, 187], [157, 207, 222]\rangle u_1 \oplus \langle[320, 327, 331], [315, 322, 330], [323, 329, 338]\rangle u_2) \le 0,
\end{gathered}$$

$$v_r, u_i \ge 0, r = 1, 2, 3, i = 1, 2.$$

Table 2. Output information of the nominee hospitals.

DMU	Outputs 1 Days of Hospitalization (in Thousands)	Outputs 2 Patient Satisfaction (%)	Outputs 3 Number of Outpatients (in Thousands)
1	⟨[121.13, 139.24, 140.04], [138.64, 139.14, 139.81], [139.14, 140.02, 141.17]⟩	⟨[38,41, 45], [38, 40, 43], [41, 44, 49]⟩	⟨[104.23, 114.04, 278.51], [102.37, 109.15, 235.72], [104.81, 275.25, 279.88]⟩
2	⟨[31.54, 34.93, 38.89], ⟨[31.54, 34.15, 38.27], ⟨[34.86, 38.15, 39.83]⟩	⟨[40, 44, 47], [35, 42, 45], [41, 46, 50]⟩	⟨[34.54, 36.98, 54.82], [36.45, 36.80, 41.57], [47.61, 54.25, 55.35]⟩
3	⟨[81.62, 82.07, 85.51], [81.41, 81.94, 83.35], [81.78, 85.49, 88.16]⟩	⟨[18,20,29], [19,21,23], [28,30,35]⟩	⟨[157.75, 177.57, 264.52], [157.75,176.68, 250.75], [180.29, 263.98, 272.16]⟩
4	⟨[19.54, 20.41, 20.59], [20.15, 20.25, 20.32], [20.54, 20.58, 20.70]⟩	⟨[18, 21, 25], [15, 19, 23], [20, 24, 30]⟩	⟨[32.89, 35.56, 87.74], [35.25, 35.50, 35.61], [87.50, 87.94, 88.30]⟩
5	⟨[23.89, 24.60, 26.09], [23.56, 23.60, 23.68], [25.97, 26.35, 26.72]⟩	⟨[30, 36,41], [34,35, 37], [35, 40, 57]⟩	⟨[63.23, 69.58, 120.73], [63, 65.17, 94.93], [64.47, 118.75, 124.75]⟩
6	⟨[21.33, 21.49, 23.31], [20.94, 24.25, 22.68], [21.38, 23.14, 23.94]⟩	⟨[50, 55, 60], [50, 53, 57], [56, 59, 70]⟩	⟨[72.84, 82.84, 94.18], [82.15, 82.68, 84.89], [85.75, 93.50, 97.18]⟩
7	⟨[145.77, 148.28, 169.01], [145.77, 147.16, 168.31], [150.69, 168.95, 175.18]⟩	⟨[40, 44, 46], [42, 43,45], [43, 44,55]⟩	⟨[147.59, 150.37, 227.12], [147.30, 147.45, 148.25], [218.24, 224.61, 229.63]⟩
8	⟨[11.56, 11.74, 12.96], [11.42, 11.61, 11.98], [11.58, 12.64, 13.16]⟩	⟨[60, 75,80], [55,60,62], [78,83,85]⟩	⟨[189.37, 202.08, 284.99], [189.37, 200.52, 281.63], [270.16, 284.55, 289.12]⟩
9	⟨[57.55, 62.67, 63.03], [62.15, 62.50, 62.93], [62.50, 62.97, 63.61]⟩	⟨[32, 35,38], [32, 33, 35], [34, 36, 45]⟩	⟨[14.63, 14.85, 29.40], [14.70, 14.75, 15.25], [24.75, 28.36, 32.64]⟩
10	⟨[73.21, 76.03, 81.90], [75.76, 76.05, 76.25], [81.67, 82.27, 82.64]⟩	⟨[22, 25, 40], [20, 24, 27], [23, 25, 29]⟩	⟨[96.77, 97.27, 110.39], [96.77, 96.89, 105.14], [99.76, 108.62, 115.27]⟩
11	⟨[22.90, 27.71, 35.56], [22.90, 26.45, 31.28], [27.92, 34.62, 39.41]⟩	⟨[20, 23, 26], [21,22, 24], [22, 25,30]⟩	⟨[171.53, 182.46, 384.99], [171.12, 178.65, 210.34], [175.59, 270.65, 400.12]⟩
12	⟨[58.41, 59.12, 60.61], [58.08, 58.12, 58.20], [60.49, 60.87, 61.24]⟩	⟨[25, 31,37], [29, 30, 32], [30, 35, 52]⟩	⟨[59.87, 66.22, 117.37], [59.64, 61.81, 91.57], [61.11, 115.39, 121.39]⟩
13	⟨[66.97, 67.68, 69.17], [66.64, 66.68, 66.76], [69.05, 69.43, 69.80]⟩	⟨[20, 27, 31], [23, 26, 28], [24, 30, 46]⟩	[96.97, 103.32, 154.47], [96.74, 98.91, 128.67], ⟨[98.21, 152.49, 158.50]⟩

Finally, based on Definition 4, we convert the above model to the following model:

$$max\ \widetilde{\theta}_1 \approx 138.0608v_1 + 42v_2 + 175.2v_3$$

s.t:

$$
\begin{gathered}
529.8333u_1 + 529.5833u_2 = 1, \\
138.0608v_1 + 42v_2 + 175.2v_3 - 529.8333u_1 - 529.5833u_2 \leq 0, \\
35.7792v_1 + 43.5v_2 + 43.8667v_3 - 146.9167u_1 - 182.0833u_2 \leq 0, \\
83.4025v_1 + 24.5v_2 + 209.9733v_3 - 148u_1 - 213u_2 \leq 0, \\
20.36v_1 + 21.5833v_2 + 57.1075v_3 - 104.3333u_1 - 116.8333u_2 \leq 0, \\
24.9175v_1 + 38v_2 + 86.5092v_3 - 110u_1 - 116.0833u_2 \leq 0, \\
22.6117v_1 + 56.4167v_2 + 86.2525v_3 - 122.9167u_1 - 106u_2 \leq 0, \\
156.9592v_1 + 44.4167v_2 + 180.2492v_3 - 714.9167u_1 - 499.5833u_2 \leq 0, \\
12.0533v_1 + 71.3333v_2 + 239.9117v_3 - 165.1667u_1 - 68.9167u_2 \leq 0, \\
62.3375v_1 + 35.3333v_2 + 20.6075v_3 - 146u_1 - 194.9167u_2 \leq 0, \\
78.3442v_1 + 25.75v_2 + 102.4717v_3 - 163.5u_1 - 343.9167u_2 \leq 0, \\
29.7942v_1 + 23.5833v_2 + 231.4342v_3 - 159.5u_1 - 104.6667u_2 \leq 0, \\
59.4375v_1 + 33.0833v_2 + 83.1492v_3 - 154u_1 - 219.0833u_2 \leq 0, \\
67.9975v_1 + 28.1667v_2 + 120.25v_3 - 177u_1 - 326.0833u_2 \leq 0, \\
v_r, u_i \geq 0, r = 1,2,3, i = 1,2.
\end{gathered}
$$

After computations with Lingo, we obtained $\theta_1^* = 0.6673$ for DMU_1. Similarly, for the other DMUs, we reported the results in Table 3. From these results, we can see that DMUs 3, 6, 8, and 11 are efficient and others are inefficient.

Table 3. The efficiencies of the decision-making units (DMUs) by the triangular single-valued neutrosophic number-Charnes, Cooper, and Rhodes (TSVNN-CCR) model.

DMUs	Efficiency	Ranking
1	0.6673	9
2	0.8057	6
3	1.00	1
4	0.5950	10
5	0.8754	4
6	1.00	1
7	0.7024	7
8	1.00	1
9	0.9116	2
10	0.8751	3
11	1.00	1
12	0.8536	5
13	0.7587	8

To authenticate the suggested efficiencies, these efficiencies were compared with the efficiencies obtained by the crisp CCR (Model (6)), and are given in Figure 1. In this figure, the efficiencies of DMUs are found to be smaller for TSVNN-CCR compared to crisp CCR.

It is interesting that DMU 12 is efficient in crisp DEA, but it is inefficient with an efficiency score of 0.8536 using TSVNN-CCR. Therefore, TSVNN-CCR is more realistic than crisp CCR. In addition, crisp CCR and TSVNN-CCR may give the same efficiencies for certain data. However, the crisp CCR model does not deal with the uncertain, indeterminate, and incongruous information. Therefore, TSVNN-CCR is more realistic than crisp CCR.

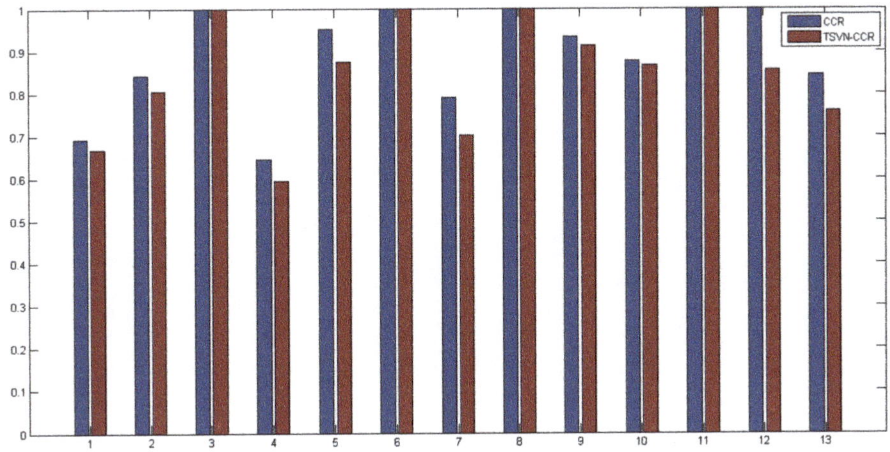

Figure 1. Comparison of suggested and crisp models.

6. Conclusions and Future Work

In this paper, a new approach for data envelopment analysis was proposed in that indeterminacy, uncertainty, vagueness, inconsistent, and incompleteness of data were shown by neutrosophic sets. Furthermore, the sorting of DMUs in DEA has been presented, and using a de-neutrosophication technique, a ranking order has been extracted. The efficiency scores of the proposed model have a similar meaning and interpretation with the conventional CCR model. Finally, the application of the proposed model was examined in a real-world case study of 13 hospitals of TUMS. The new model is appropriate in situations where some inputs or outputs do not have an exact quantitative value, and the proposed approach has produced promising results from computing efficiency and performance aspects.

The proposed study had some barriers: first, the indeterminacy, uncertainty, and ambiguity in the present report was limited to triangular single-valued neutrosophic numbers, but the other forms of NSs such as bipolar NSs and interval-valued neutrosophic numbers can also be used to indicate variables characterizing the neutrosophic core in global problems. Second, the presented model was investigated under a constant returns-to-scale (CRS), but the suggested method can also be extended under a VRS assumption, so we plan to extend this model to the VRS. Moreover, although the arithmetic operations, model, and results presented here demonstrate the effectiveness of our methodology, it could also be considered in other types of DEA models such as network DEA and its applications to banks, supplier selection, tax offices, police stations, schools, and universities. While developing data envelopment analysis, models based on bipolar and interval-valued neutrosophic data is another area for further studies. As for future research, we intend to study these problems.

Author Contributions: The authors contributed equally to writing this article. All authors have read and agreed to the published version of the manuscript.

Funding: This work was supported by Quzhou University.

Conflicts of Interest: The authors declare no conflicts of interest.

Abbreviations: List of Acronyms

DEA	Data Envelopment Analysis
DMU	Decision-Making Units
CCR model	Charnes, Cooper, Rhodes model
BCC model	Banker, Charnes, Cooper model
CRS	Constant Returns-to-Scale
VRS	Variable Returns-to-Scale
AHP	Analytic Hierarchy Process
TUMS	Tehran University of Medical Sciences
FS	Fuzzy Set
IFS	Intuitionistic Fuzzy Set
NS	Neutrosophic Set
SVNS	Single-Valued Neutrosophic Set
TSVNN	Triangular Single-Valued Neutrosophic number

References

1. Charnes, A.; Cooper, W.W.; Rhodes, E. Measuring the efficiency of decision making units. *Eur. J. Oper. Res.* **1978**, *2*, 429–444. [CrossRef]
2. Farrell, M.J. The measurement of productive efficiency. *J. R. Stat. Soc.* **1957**, *120*, 253–290. [CrossRef]
3. Banker, R.D.; Charnes, A.; Cooper, W.W. Some models for estimating technical and scale inefficiencies in data envelopment analysis. *Manag. Sci.* **1984**, *30*, 1078–1092. [CrossRef]
4. Sahoo, B.K.; Tone, K. Decomposing capacity utilization in data envelopment analysis: An application to banks in India. *Eur. J. Oper. Res.* **2009**, *195*, 575–594. [CrossRef]
5. Lee, Y.J.; Joo, S.J.; Park, H.G. An application of data envelopment analysis for Korean banks with negative data. *Benchmarking Int. J.* **2017**, *24*, 1052–1067. [CrossRef]
6. Jiang, H.; He, Y. Applying Data Envelopment Analysis in Measuring the Efficiency of Chinese Listed Banks in the Context of Macroprudential Framework. *Mathematics* **2018**, *6*, 184. [CrossRef]
7. Karasakal, E.; Aker, P. A multicriteria sorting approach based on data envelopment analysis for R&D project selection problem. *Omega* **2017**, *73*, 79–92.
8. Lacko, R.; Hajduová, Z.; Gábor, V. Data Envelopment Analysis of Selected Specialized Health Centres and Possibilities of its Application in the Terms of Slovak Republic Health Care System. *J. Health Manag.* **2017**, *19*, 144–158. [CrossRef]
9. Ertay, T.; Ruan, D.; Tuzkaya, U.R. Integrating data envelopment analysis and analytic hierarchy for the facility layout design in manufacturing systems. *Inf. Sci.* **2006**, *176*, 237–262. [CrossRef]
10. Düzakın, E.; Düzakın, H. Measuring the performance of manufacturing firms with super slacks based model of data envelopment analysis: An application of 500 major industrial enterprises in Turkey. *Eur. J. Oper. Res.* **2007**, *182*, 1412–1432. [CrossRef]
11. Jahanshahloo, G.R.; Lotfi, F.H.; Valami, H.B. Malmquist productivity index with interval and fuzzy data, an application of Data envelopment analysis. *Int. Math. Forum* **2006**, *1*, 1607–1623. [CrossRef]
12. Shafiee, M.; Lotfi, F.H.; Saleh, H. Supply chain performance evaluation with data envelopment analysis and balanced scorecard approach. *Appl. Math. Model.* **2014**, *38*, 5092–5112. [CrossRef]
13. Soheilirad, S.; Govindan, K.; Mardani, A.; Zavadskas, E.K.; Nilashi, M.; Zakuan, N. Application of data envelopment analysis models in supply chain management: A systematic review and meta-analysis. *Ann. Oper. Res.* **2018**, *271*, 915–969. [CrossRef]
14. Krmac, E.; Djordjević, B. A New DEA Model for Evaluation of Supply Chains: A Case of Selection and Evaluation of Environmental Efficiency of Suppliers. *Symmetry* **2019**, *11*, 565. [CrossRef]
15. Zadeh, L.A. Fuzzy sets. *Inf. Control* **1965**, *8*, 338–353. [CrossRef]
16. Hsu, T.; Tsai, Y.; Wu, H. The preference analysis for tourist choice of destination: A case study of Taiwan. *Tour. Manag.* **2009**, *30*, 288–297. [CrossRef]
17. Finol, J.; Guo, Y.K.; Jing, X.D. A rule based fuzzy model for the prediction of petrophysical rock parameters. *J. Pet. Sci. Eng.* **2001**, *29*, 97–113. [CrossRef]

18. Najafi, H.S.; Edalatpanah, S.A. An improved model for iterative algorithms in fuzzy linear systems. *Comput. Math. Modeling* **2013**, *24*, 443–451. [CrossRef]
19. Hosseinzadeh, A.; Edalatpanah, S.A. A new approach for solving fully fuzzy linear programming by using the lexicography method. *Adv. Fuzzy Syst.* **2016**. [CrossRef]
20. Das, S.K.; Edalatpanah, S.A.; Mandal, T. A proposed model for solving fuzzy linear fractional programming problem: Numerical Point of View. *J. Comput. Sci.* **2018**, *25*, 367–375. [CrossRef]
21. Hatami-Marbini, A.; Emrouznejad, A.; Tavana, M. A taxonomy and review of the fuzzy data envelopment analysis literature: Two decades in the making. *Eur. J. Oper. Res.* **2011**, *214*, 457–472. [CrossRef]
22. Emrouznejad, A.; Tavana, M.; Hatami-Marbini, A. The state of the art in fuzzy data envelopment analysis. In *Performance Measurement with Fuzzy Data Envelopment Analysis*; Springer: Berlin/Heidelberg, Germany, 2014; pp. 1–45.
23. Emrouznejad, A.; Yang, G.L. A survey and analysis of the first 40 years of scholarly literature in DEA: 1978–2016. *Socio-Econ. Plan. Sci.* **2018**, *61*, 4–8. [CrossRef]
24. Yen, B.T.; Chiou, Y.C. Dynamic fuzzy data envelopment analysis models: Case of bus transport performance assessment. *RAIRO-Oper. Res.* **2019**, *53*, 991–1005. [CrossRef]
25. Lotfi, F.H.; Ebrahimnejad, A.; Vaez-Ghasemi, M.; Moghaddas, Z. *Data Envelopment Analysis with R*; Springer: Cham, Switzerland, 2020.
26. Atanassov, K.T. Intuitionistic fuzzy sets. *Fuzzy Sets Syst.* **1986**, *20*, 87–96. [CrossRef]
27. Rouyendegh, B.D. The DEA and intuitionistic fuzzy TOPSIS approach to departments' performances: A pilot study. *J. Appl. Math.* **2011**, 1–16. [CrossRef]
28. Puri, J.; Yadav, S.P. Intuitionistic fuzzy data envelopment analysis: An application to the banking sector in India. *Expert Syst. Appl.* **2015**, *42*, 4982–4998. [CrossRef]
29. Edalatpanah, S.A. A data envelopment analysis model with triangular intuitionistic fuzzy numbers. *Int. J. Data Envel. Anal.* **2019**, *7*, 47–58.
30. Arya, A.; Yadav, S.P. Development of intuitionistic fuzzy data envelopment analysis models and intuitionistic fuzzy input–output targets. *Soft Comput.* **2019**, *23*, 8975–8993. [CrossRef]
31. Smarandache, F. *A Unifying Field in Logics. Neutrosophy: Neutrosophic Probability, Set and Logic*; American Research Press: Rehoboth, MA, USA, 1999.
32. Smarandache, F. A unifying field in logics: Neutrosophic logic. In *Neutrosophy, Neutrosophic Set, Neutrosophic Probability and Statistics*, 3rd ed.; American Research Press: Rehoboth, MA, USA, 2003.
33. Broumi, S.; Smarandache, F. Correlation coefficient of interval neutrosophic set. *Appl. Mech. Mater.* **2013**, *436*, 511–517. [CrossRef]
34. Ye, J. Similarity measures between interval neutrosophic sets and their applications in multicriteria decision-making. *J. Intell. Fuzzy Syst.* **2014**, *26*, 165–172. [CrossRef]
35. Broumi, S.; Smarandache, F.; Talea, M.; Bakali, A. An introduction to bipolar single valued neutrosophic graph theory. *Appl. Mech. Mater.* **2016**, *841*, 184–191. [CrossRef]
36. Wang, L.; Zhang, H.Y.; Wang, J.Q. Frank Choquet Bonferroni mean operators of bipolar neutrosophic sets and their application to multi-criteria decision-making problems. *Int. J. Fuzzy Syst.* **2018**, *20*, 13–28. [CrossRef]
37. Ye, J. Multicriteria decision-making method using the correlation coefficient under single-valued neutrosophic environment. *Int. J. Gen. Syst.* **2013**, *42*, 386–394. [CrossRef]
38. Chakraborty, A.; Mondal, S.P.; Ahmadian, A.; Senu, N.; Alam, S.; Salahshour, S. Different Forms of Triangular Neutrosophic Numbers, De-Neutrosophication Techniques, and their Applications. *Symmetry* **2018**, *10*, 327. [CrossRef]
39. Garg, H. New Logarithmic operational laws and their applications to multiattribute decision making for single-valued neutrosophic numbers. *Cogn. Syst. Res.* **2018**, *52*, 931–946. [CrossRef]
40. Garg, H. Linguistic single-valued neutrosophic prioritized aggregation operators and their applications to multiple-attribute group decision-making. *J. Ambient Intell. Hum. Comput.* **2018**, *9*, 1975–1997. [CrossRef]
41. Smarandache, F. *About Nonstandard Neutrosophic Logic: Answers to Imamura's "Note on the Definition of Neutrosophic Logic"*; Infinite Study: Coimbatore, India, 2019.
42. Garg, H. Algorithms for possibility linguistic single-valued neutrosophic decision-making based on COPRAS and aggregation operators with new information measures. *Measurement* **2019**, *138*, 278–290. [CrossRef]
43. Kumar, R.; Edalatpanah, S.A.; Jha, S.; Broumi, S.; Dey, A. Neutrosophic shortest path problem. *Neutrosophic Sets Syst.* **2018**, *23*, 5–15.

44. Edalatpanah, S.A. Nonlinear approach for neutrosophic linear programming. *J. Appl. Res. Ind. Eng.* **2019**, *6*, 367–373.
45. Edalatpanah, S.A. Neutrosophic perspective on DEA. *J. Appl. Res. Ind. Eng.* **2018**, *5*, 339–345.
46. Kahraman, C.; Otay, I.; Öztayşi, B.; Onar, S.C. An Integrated AHP & DEA Methodology with Neutrosophic Sets. In *Fuzzy Multi-Criteria Decision-Making Using Neutrosophic Sets*; Springer: Cham, Switzerland, 2019; pp. 623–645.
47. Edalatpanah, S.A.; Smarandache, F. Data Envelopment Analysis for Simplified Neutrosophic Sets. *Neutrosophic Sets Syst.* **2019**, *29*, 215–226.
48. Abdelfattah, W. Data envelopment analysis with neutrosophic inputs and outputs. *Expert Syst.* **2019**, *36*, e12453. [CrossRef]
49. Robinson, A. *Non-Standard Analysis*; Princeton University Press: Princeton, NJ, USA, 2016.
50. Aristovnik, A. Measuring relative efficiency in health and education sector: The case of East European countries. *Actual Probl. Econ.* **2012**, *136*, 305–314.
51. Barros, C.P.; de Menezes, A.G.; Vieira, J.C. Measurement of hospital efficiency, using a latent class stochastic frontier model. *Appl. Econ.* **2013**, *45*, 47–54. [CrossRef]
52. Colombi, R.; Martini, G.; Vittadini, G. Determinants of transient and persistent hospital efficiency: The case of Italy. *Health Econ.* **2017**, *26*, 5–22. [CrossRef] [PubMed]
53. Bryce, C.L.; Engberg, J.B.; Wholey, D.R. Comparing the agreement among alternative models in evaluating HMO efficiency. *Health Serv. Res.* **2000**, *35*, 509. [PubMed]
54. Kalhor, R.; Amini, S.; Sokhanvar, M.; Lotfi, F.; Sharifi, M.; Kakemam, E. Factors affecting the technical efficiency of general hospitals in Iran: Data envelopment analysis. *J. Egypt. Public Health Assoc.* **2016**, *91*, 20–25. [CrossRef]
55. Chen, H.; Liu, J.; Li, Y.; Chiu, Y.-H.; Lin, T.Y. A Two-stage Dynamic Undesirable Data Envelopment Analysis Model Focused on Media Reports and the Impact on Energy and Health Efficiency. *Int. J. Environ. Res. Public Health* **2019**, *16*, 1535. [CrossRef]
56. Kohl, S.; Schoenfelder, J.; Fügener, A.; Brunner, J.O. The use of Data Envelopment Analysis (DEA) in healthcare with a focus on hospitals. *Health Care Manag. Sci.* **2019**, *22*, 245–286. [CrossRef]
57. Ji, A.B.; Qiao, Y.; Liu, C. Fuzzy DEA-based classifier and its applications in healthcare management. *Health Care Manag. Sci.* **2019**, *22*, 560–568. [CrossRef]
58. Dotoli, M.; Epicoco, N.; Falagario, M.; Sciancalepore, F. A cross-efficiency fuzzy data envelopment analysis technique for performance evaluation of decision making units under uncertainty. *Comput. Ind. Eng.* **2015**, *79*, 103–114. [CrossRef]
59. Otay, İ.; Oztaysi, B.; Onar, S.C.; Kahraman, C. Multi-expert performance evaluation of healthcare institutions using an integrated intuitionistic fuzzy AHP&DEA methodology. *Knowl. Based Syst.* **2017**, *133*, 90–106.

symmetry

MDPI

Article

A Novel Dynamic Multi-Criteria Decision Making Method Based on Generalized Dynamic Interval-Valued Neutrosophic Set

Nguyen Tho Thong [1,2], Florentin Smarandache [3], Nguyen Dinh Hoa [1], Le Hoang Son [1], Luong Thi Hong Lan [4,5], Cu Nguyen Giap [6], Dao The Son [7] and Hoang Viet Long [8,9,*]

1 VNU Information Technology Institute, Vietnam National University, Hanoi 010000, Vietnam; nguyenthothongtt89@gmail.com or 17028003@vnu.edu.vn (N.T.T.); hoand@vnu.edu.vn (N.D.H.); sonlh@vnu.edu.vn (L.H.S.)

2 Center for High-Performance Computing, VNU University of Engineering and Technology, Vietnam National University, Hanoi 010000, Vietnam

3 Department of Mathematics, University of New Mexico, 705 Gurley Ave, Gallup, NM 87301, USA; fsmarandache@gmail.com

4 Faculty of Computer Science and Engineering, Thuyloi University, 175 Tay Son, Dong Da, Hanoi 010000, Vietnam; lanlhbk@tlu.edu.vn

5 Institute of Information Technology, Vietnam Academy of Science and Technology, Hanoi 010000, Vietnam

6 Faculty of Management Information System & E-commerce, Thuongmai University, 79 Ho Tung Mau, Nam Tu Liem, Hanoi 010000, Vietnam; cunguyengiap@tmu.edu.vn

7 Department of Economics, Thuongmai University, 79 Ho Tung Mau, Nam Tu Liem, Hanoi 010000, Vietnam; daotheson@tmu.edu.vn

8 Division of Computational Mathematics and Engineering, Institute for Computational Science, Ton Duc Thang University, Ho Chi Minh City 700000, Vietnam

9 Faculty of Mathematics and Statistics, Ton Duc Thang University, Ho Chi Minh City 700000, Vietnam

* Correspondence: hoangvietlong@tdtu.edu.vn

Received: 23 February 2020; Accepted: 25 March 2020; Published: 14 April 2020

check for updates

Abstract: Dynamic multi-criteria decision-making (DMCDM) models have many meaningful applications in real life in which solving indeterminacy of information in DMCDMs strengthens the potential application of DMCDM. This study introduces an extension of dynamic internal-valued neutrosophic sets namely generalized dynamic internal-valued neutrosophic sets. Based on this extension, we develop some operators and a TOPSIS method to deal with the change of both criteria, alternatives, and decision-makers by time. In addition, this study also applies the proposal model to a real application that facilitates ranking students according to attitude-skill-knowledge evaluation model. This application not only illustrates the correctness of the proposed model but also introduces its high potential appliance in the education domain.

Keywords: generalized dynamic interval-valued neutrosophic set; hesitant fuzzy set; dynamic neutrosophic environment; dynamic TOPSIS method; neutrosophic data analytics

---

1. Introduction

Multi-criteria decision-making (MCDM) in real world is often dynamic [1]. In the dynamic MCDM (DMCDM) model, neither alternatives nor criteria are constant throughout the whole problem and do not change over time. Besides, the DMCDM model has to cope with both dynamic and indeterminate problems of data. For example, when ranking tertiary students during learning time in a university by the set of criteria based on attitudes-skills-knowledge model (ASK), the criteria, students and lecturers are changing during semesters. The lecturers' evaluations using scores, or other ordered scales, are also subject to indeterminacy because of lecturers' personal experiences and biases. Therefore, a ranking model that can handle these issues is necessary.

In [2], Smarandache introduced neutrosophic set including truth-membership, an indeterminacy-membership and a falsity-membership to well treat the problem of information indeterminacy. Since then, variant forms of MCDM and DMCDM models have been proposed as in [3–15]. In order to consider the time dimension, Wang [16] proposed the interval neutrosophic set and its mathematical operators. Ye [9] proposed MCDM in interval-valued neutrosophic set. Dynamic MCDM for dynamic interval-valued neutrosophic set (DIVNS) was proposed in [14]. The authors have developed mathematical operators for TOPSIS method in DIVNSs.

In some cases, criteria, alternatives and decision-makers are changing by time. This fact requires a new method for DMCDM using TOPSIS method in the interval-valued neutrosophic set [17] with diversion of history data. The TOPSIS method for DIVNS in [14] did not solve the problem with the changing criteria, alternatives, and decision-makers. Liu et al. [13] combined the theory of both interval-valued neutrosophic set and hesitant fuzzy set to solve the MCDM problem. However, this study did not use TOPSIS method, and it did not consider the change of criteria also. In order to take the history data into account, Je [10] proposed two hesitant interval neutrosophic linguistic weighted operators to ranking alternatives in dynamic environment. In short, the DMCDM model in DIVNS based on TOPSIS method has not been addressed before.

The purpose of this paper is to deal with the change of criteria, alternatives, and decision-makers during time. We define generalized dynamic interval-valued neutrosophic set (GDIVNS) and some operators. Based on mathematical operators in GDIVNS (distance and weighted aggregation operators), a framework of dynamic TOPSIS is introduced. The proposed method is applied for ranking students of Thuongmai University, Vietnam on attributes of ASK model. ASK model is applicable for evaluation of tertiary students' performance, and it gives more information that support employers besides a set of university exit benchmark. It also facilitates students to make proper self-adjustments and help them pursue appropriate professional orientation for their future career [18–21]. This application proves the suitability of the proposed model for real ranking problems.

This paper is structured as follows: The Section 1 is an introduction, and the Section 2 provides the brief preliminaries for DMCDM model in both legacy environment and interval-valued neutrosophic set. The Section 3 presents the definition of GDIVNS and some mathematical operators on this set. The Section 4 introduces the framework of dynamic TOPSIS method in GDIVNSs environment. The Section 5 presents the application of dynamic TOPSIS method in the problem of ranking students based on attributes of ASK model. The Section 6 compares the result of proposed model with previous TOPSIS model in DIVNS. The last section mentions the brief summary of this study and intended future works.

2. Preliminary

2.1. Multi-Criteria Decision-Making Model Based on History

A dynamic multi-criteria decision-making model introduced by Campanella and Ribeiro [1] is a DMCDM in which all alternatives and criteria are subject to change. The model gives decisions at all periods or just at the last one. The final rating of alternatives is calculated as:

$$E_t(a) = \begin{cases} R_t(a), & a \in A_t \backslash H_{t-1}^A \\ D_E(E_{t-1}(a), R_t(a)), & a \in A_t \cap H_{t-1}^A \\ E_{t-1}(a), & a \in H_{t-1}^A \backslash A_t \end{cases} \quad (1)$$

where A_t is a set of alternatives at period t, H_{t-1}^A is a historical set of alternatives at period $t-1$ ($H_0^A = \emptyset$), $R_t(a)$ is rating of alternative a at period t, and D_E is an aggregation operator.

2.2. Dynamic Interval-Valued Neutrosophic Set and Hesitant Fuzzy Set

Thong et al. [14] introduced the concept of dynamic interval-valued neutrosophic set (DIVNS).

Definition 1. [14] *Let U be a universe of discourse, and A be a dynamic interval-valued neutrosophic Set (DIVNS) expressed by,*

$$A = \left\{x, \left\langle \left[T_x^L(\tau), T_x^U(\tau)\right], \left[I_x^L(\tau), I_x^U(\tau)\right], \left[F_x^L(\tau), F_x^U(\tau)\right] \right\rangle \middle| x \in U\right\} \tag{2}$$

where T_x, I_x, F_x are the truth-membership, indeterminacy-membership, falsity-membership respectively, $\tau = \{\tau_1, \tau_2, \ldots, \tau_k\}$ is set of time sequence and

$$\left[T_x^L(\tau), T_x^U(\tau)\right] \subseteq [0,1]; \left[I_x^L(\tau), I_x^U(\tau)\right] \subseteq [0,1]; \left[F_x^L(\tau), F_x^U(\tau)\right] \subseteq [0,1]$$

Example 1. *A DIVNS in time sequence $\tau = \{\tau_1, \tau_2\}$ and universal $U = \{x_1, x_2, x_3\}$ is:*

$$A = \left\{ \begin{array}{l} x_1, \langle([0.5,0.6],[0.1,0.3],[0.2,0.4]),([0.4,0.55],[0.25,0.3],[0.3,0.42])\rangle \\ x_2, \langle([0.7,0.81],[0.1,0.2],[0.1,0.2]),([0.72,0.8],[0.11,0.25],[0.2,0.4])\rangle \\ x_3, \langle([0.3,0.5],[0.4,0.5],[0.6,0.7]),([0.4,0.5],[0.5,0.6],[0.66,0.73])\rangle \end{array} \right\}$$

Hesitant fuzzy set (HFS) first introduced by Torra and Narukawa [19] and Torra [20] is defined as follows.

Definition 2. *[20] A hesitant fuzzy set E on U is defined by the function $h_E(x)$. When $h_E(x)$ is applied to U, it returns a finite subset of [0, 1], which can be represented as*

$$E = \{\langle x, h_E(x)\rangle | x \in U\} \tag{3}$$

where $h_E(x)$ is a set of some values in $[0,1]$.

Example 2. *Let $X = \{x_1, x_2, x_3\}$ be the discourse set, and $h_E(x_1) = \{0.1, 0.2\}$, $h_E(x_2) = \{0.3\}$ and $h_E(x_3) = \{0.2, 0.3, 0.5\}$. Then, E can be considered as a HFS:*

$$E = \{\langle x_1, \{0.1, 0.2\}\rangle, \langle x_2, \{0.3\}\rangle, \langle x_3, \{0.2, 0.3, 0.5\}\rangle\}$$

3. Generalized Dynamic Interval-Valued Neutrosophic Set

Extending DIVNS by the concept of HFS is considered how to express the criteria, alternatives, and DMs that are changing during time criteria, alternatives and decision-makers are changing by time.

In this section, we propose the concepts of generalized dynamic interval-valued neutrosophic set (GDIVNS) and generalized dynamic interval-valued neutrosophic element (GDIVNE) including fundamental elements, operational laws as well as the score functions. Then, GDIVNS's theory is applied for the decision-making model in Section 4.

Definition 3. *Let U be a universe of discourse. A generalized dynamic interval-valued neutrosophic set (GDIVNS) in U can be expressed as,*

$$\widetilde{E} = \left\{\left\langle x, \widetilde{h_{\widetilde{E}}}(x(t_r))\right\rangle | x \in U; \forall t_r \in t;\right\} \tag{4}$$

where $\widetilde{h_{\widetilde{E}}}(x(t_r))$ is expressed for importing HFS into DIVNS. $\widetilde{h_{\widetilde{E}}}(x(t_r))$ is a set of DIVNSs at period t_r and $t = \{t_1, t_2, t_3, \ldots, t_s\}$, which denotes the possible DIVNSs of the element $x \in X$ to the set $\widetilde{E}$, $\widetilde{h_{\widetilde{E}}}(x(t_r))$ can be represented by a generalized dynamic interval-valued neutrosophic element (GDIVNE). When $s = 1$ and

$\left|\widetilde{h}_{\widetilde{E}}(x(t_r))\right| = 1$, *GDIVNS simplifies to DIVNS [14]. For convenience, we denote* $\widetilde{h} = \widetilde{h}_{\widetilde{E}}(x(t)) = \{\gamma | \gamma \in \widetilde{h}\}$, *where*

$$\gamma = \left(\left[T^L(x(\tau)), T^U(x(\tau))\right], \left[I^L(x(\tau)), I^U(x(\tau))\right], \left[F^L(x(\tau)), F^U(x(\tau))\right]\right)$$

is a dynamic interval-valued neutrosophic number.

Example 3. *Let* $t = \{t_1, t_2\}$; $\tau = \{\tau_1, \tau_2\}$ *and an universal* $X = \{x_1, x_2, x_3\}$. *A GDIVNS in X is given as:*

$$\widetilde{E} = \left\{ \begin{array}{l} \left\langle x_1, \left\{ \begin{array}{l} \langle([0.2, 0.33], [0.4, 0.5], [0.6, 0.7]), ([0.24, 0.39], [0.38, 0.47], [0.56, 0.7])\rangle, \\ \langle([0.29, 0.37], [0.3, 0.5], [0.4, 0.58]), ([0.4, 0.5], [0.2, 0.3], [0.35, 0.42])\rangle \end{array} \right\} \right\rangle, \\ \left\langle x_2, \left\{ \begin{array}{l} \langle([0.8, 0.9], [0.1, 0.2], [0.1, 0.2]), ([0.72, 0.8], [0.11, 0.25], [0.23, 0.45])\rangle, \\ \langle([0.4, 0.6], [0.2, 0.4], [0.3, 0.4]), ([0.41, 0.5], [0.26, 0.39], [0.2, 0.3])\rangle \end{array} \right\} \right\rangle, \\ \left\langle x_3, \left\{ \begin{array}{l} \langle([0.6, 0.7], [0.2, 0.3], [0.4, 0.5]), ([0.52, 0.66], [0.34, 0.4], [0.6, 0.77])\rangle, \\ \langle([0.54, 0.62], [0.15, 0.3], [0.2, 0.4]), ([0.4, 0.5], [0.25, 0.32], [0.39, 0.43])\rangle \end{array} \right\} \right\rangle \end{array} \right\}$$

Definition 4. *Let* $\widetilde{h}$, $\widetilde{h}_1$ *and* $\widetilde{h}_2$ *be three GDIVNEs. When* $\lambda > 0$, *the operations of GDIVNEs are defined as follows:*

(i) *Addition*

$$\begin{array}{l} \widetilde{h}_1 \oplus \widetilde{h}_2 = \cup_{\forall \gamma_1 \in \widetilde{h}_1; \forall \gamma_2 \in \widetilde{h}_2} \{\gamma_1 \oplus \gamma_2\} \\ = \left\{ \left\langle \begin{array}{l} \left[T^L_{\gamma_1}(x(\tau)) + T^L_{\gamma_2}(x(\tau)) - T^L_{\gamma_1}(x(\tau)) \times T^L_{\gamma_2}(x(\tau)), T^U_{\gamma_1}(x(\tau)) + T^U_{\gamma_2}(x(\tau)) - T^U_{\gamma_1}(x(\tau)) \times T^U_{\gamma_2}(x(\tau))\right], \\ \left[I^L_{\gamma_1}(x(\tau)) \times I^L_{\gamma_2}(x(\tau)), I^U_{\gamma_1}(x(\tau)) \times I^U_{\gamma_2}(x(\tau))\right], \left[F^L_{\gamma_1}(x(\tau)) \times F^L_{\gamma_2}(x(\tau)), F^U_{\gamma_1}(x(\tau)) \times F^U_{\gamma_2}(x(\tau))\right] \end{array} \right\rangle \right\} \end{array}$$

(ii) *Multiplication*

$$\begin{array}{l} \widetilde{h}_1 \otimes \widetilde{h}_2 = \cup_{\forall \gamma_1 \in \widetilde{h}_1; \forall \gamma_2 \in \widetilde{h}_2} \{\gamma_1 \otimes \gamma_2\} \\ = \left\{ \left\langle \begin{array}{l} \left[T^L_{\gamma_1}(x(\tau)) \times T^L_{\gamma_2}(x(\tau)), T^U_{\gamma_1}(x(\tau)) \times T^U_{\gamma_2}(x(\tau))\right], \\ \left[I^L_{\gamma_1}(x(\tau)) + I^L_{\gamma_2}(x(\tau)) - I^L_{\gamma_1}(x(\tau)) \times I^L_{\gamma_2}(x(\tau)), I^U_{\gamma_1}(x(\tau)) + I^U_{\gamma_2}(x(\tau)) - I^U_{\gamma_1}(x(\tau)) \times I^U_{\gamma_2}(x(\tau))\right], \\ \left[F^L_{\gamma_1}(x(\tau)) + F^L_{\gamma_2}(x(\tau)) - F^L_{\gamma_1}(x(\tau)) \times F^L_{\gamma_2}(x(\tau)), F^U_{\gamma_1}(x(\tau)) + F^U_{\gamma_2}(x(\tau)) - F^U_{\gamma_1}(x(\tau)) \times F^U_{\gamma_2}(x(\tau))\right] \end{array} \right\rangle \right\} \end{array}$$

(iii) *Scalar Multiplication*

$$\begin{array}{l} \lambda\widetilde{h} = \cup_{\forall \gamma \in \widetilde{h}} \{\lambda\gamma\} \\ = \cup_{\forall \gamma \in \widetilde{h}} \left\{ \left\langle \begin{array}{l} \left[1 - \left(1 - T^L(x(\tau))\right)^\lambda, 1 - \left(1 - T^U(x(\tau))\right)^\lambda\right], \\ \left[\left(I^L(x(\tau))\right)^\lambda, \left(I^U(x(\tau))\right)^\lambda\right], \left[\left(F^L(x(\tau))\right)^\lambda, \left(F^U(x(\tau))\right)^\lambda\right] \end{array} \right\rangle \right\} \end{array}$$

(iv) *Power*

$$\begin{array}{l} \widetilde{h}^\lambda = \cup_{\forall \gamma \in \widetilde{h}} \{\gamma^\lambda\} \\ = \cup_{\forall \gamma \in \widetilde{h}} \left\{ \left\langle \begin{array}{l} \left[\left(T^L(x(\tau))\right)^\lambda, \left(T^U(x(\tau))\right)^\lambda\right], \left[1 - \left(1 - I^L(x(\tau))\right)^\lambda, 1 - \left(1 - I^U(x(\tau))\right)^\lambda\right], \\ \left[1 - \left(1 - F^L(x(\tau))\right)^\lambda, 1 - \left(1 - F^U(x(\tau))\right)^\lambda\right] \end{array} \right\rangle \right\} \end{array}$$

Definition 5. *Let* $\widetilde{h}$ *be a GDIVNE. Then, the score functions of the GDIVNE* $\widetilde{h}$ *are defined by,*

$$S(\widetilde{h}) = \frac{1}{\#\widetilde{h}} \times \frac{1}{k} \sum_{\forall \gamma \in \widetilde{h}} \sum_{l=1}^{k} \left(\left(\frac{T^L(\tau_l) + T^U(\tau_l)}{2} + \left(1 - \frac{I^L(\tau_l) + I^U(\tau_l)}{2}\right) + \left(1 - \frac{F^L(\tau_l) + F^U(\tau_l)}{2}\right) \right) / 3 \right) \quad (5)$$

where $\tau = \{\tau_1, \tau_2, \ldots, \tau_k\}$, *and* $\#\widetilde{h}$ *is number of elements in* $\widetilde{h}$. *Obviously,* $S(\widetilde{h}) \in [0,1]$. *If* $S(\widetilde{h}_1) \geq S(\widetilde{h}_2)$, *then* $\widetilde{h}_1 \geq \widetilde{h}_2$.

Example 4. *Let three GDIVNEs:*

$$\begin{aligned}
\widetilde{h}_1 &= \{\langle([1,1],[0,0],[0,0]),([1,1],[0,0],[0,0])\rangle,\langle([1,1],[0,0],[0,0]),([1,1],[0,0],[0,0])\rangle\} \\
\widetilde{h}_2 &= \{\langle([0,0],[1,1],[0,0]),([0,0],[1,1],[0,0])\rangle,\langle([0,0],[1,1],[0,0]),([0,0],[1,1],[0,0])\rangle\} \\
\widetilde{h}_3 &= \{\langle([0,0],[1,1],[1,1]),([0,0],[1,1],[1,1])\rangle,\langle([0,0],[1,1],[1,1]),([0,0],[1,1],[1,1])\rangle\}
\end{aligned}$$

According to Equation (5), we have $S(\widetilde{h}_1) = 1$; $S(\widetilde{h}_2) = \frac{1}{3}$; $S(\widetilde{h}_3) = 0$. Thus, $\widetilde{h}_1 > \widetilde{h}_2 > \widetilde{h}_3$.

Definition 6. *Let* $\widetilde{h}_j (j = 1, 2, \ldots, n)$ *be a collection of GDIVNEs. Generalized dynamic interval-valued neutrosophic weighted average (GDIVNWA) operator is defined as*

$$\begin{aligned}
GDIVNWA(\widetilde{h}_1, \widetilde{h}_2, \ldots, \widetilde{h}_n) &= \sum_{j=1}^{n} w_j \widetilde{h}_j \\
&= \bigcup_{\gamma_1 \in \widetilde{h}_1, \gamma_2 \in \widetilde{h}_2, \ldots, \gamma_n \in \widetilde{h}_n} \left\{ \left(\left(\begin{array}{c} \left[1 - \prod_{j=1}^{n} \left(1 - T^L_{\gamma_j}(\tau)\right)^{w_j}, 1 - \prod_{j=1}^{n} \left(1 - T^U_{\gamma_j}(\tau)\right)^{w_j} \right], \\ \left[\prod_{j=1}^{n} \left(I^L_{\gamma_j}(\tau)\right)^{w_j}, \prod_{j=1}^{n} \left(I^U_{\gamma_j}(\tau)\right)^{w_j} \right], \left[\prod_{j=1}^{n} \left(F^L_{\gamma_j}(\tau)\right)^{w_j}, \prod_{j=1}^{n} \left(F^U_{\gamma_j}(\tau)\right)^{w_j} \right] \end{array} \right) \right) \right\}
\end{aligned} \tag{6}$$

Theorem 1. *Let* $\widetilde{h}_j (j = 1, 2, \ldots, n)$ *be the collection of GDIVNEs. The result aggregated from GDIVNWA operator is still a GDIVNE.*

Proof. The Equation (6) is proved by mathematical inductive reasoning method. □

When $n = 1$, Equation (6) holds because it simplifies to the trivial outcome, which is obviously GDIVNE as,

$$GDIVNWA(\widetilde{h}_1) = \left(\begin{array}{c} \left[1 - \left(1 - T^L_{\gamma_1}(\tau)\right)^{w_1}, 1 - \left(1 - T^U_{\gamma_1}(\tau)\right)^{w_1} \right], \\ \left[\left(I^L_{\gamma_1}(\tau)\right)^{w_1}, \left(I^U_{\gamma_1}(\tau)\right)^{w_1} \right], \left[\left(F^L_{\gamma_1}(\tau)\right)^{w_1}, \left(F^U_{\gamma_1}(\tau)\right)^{w_1} \right] \end{array} \right) \tag{7}$$

Let us assume that (6) is true for $n = z$,

$$\sum_{j=1}^{z} w_j \widetilde{h}_j = \bigcup_{\gamma_1 \in \widetilde{h}_1, \gamma_2 \in \widetilde{h}_2, \ldots, \gamma_z \in \widetilde{h}_z} \left\{ \left(\left(\begin{array}{c} \left[1 - \prod_{j=1}^{z} \left(1 - T^L_{\gamma_j}(\tau)\right)^{w_j}, 1 - \prod_{j=1}^{z} \left(1 - T^U_{\gamma_j}(\tau)\right)^{w_j} \right], \\ \left[\prod_{j=1}^{z} \left(I^L_{\gamma_j}(\tau)\right)^{w_j}, \prod_{j=1}^{z} \left(I^U_{\gamma_j}(\tau)\right)^{w_j} \right], \left[\prod_{j=1}^{z} \left(F^L_{\gamma_j}(\tau)\right)^{w_j}, \prod_{j=1}^{z} \left(F^U_{\gamma_j}(\tau)\right)^{w_j} \right] \end{array} \right) \right) \right\} \tag{8}$$

When $n = z + 1$

$$\sum_{j=1}^{z+1} w_j\widetilde{h}_j = \sum_{j=1}^{z} w_j\widetilde{h}_j \oplus w_{z+1}\widetilde{h}_{z+1}$$
$$= \bigcup_{\gamma_1\in\widetilde{h}_1,\gamma_2\in\widetilde{h}_2,\ldots,\gamma_z\in\widetilde{h}_z}\left\{\left(\begin{array}{l}\left[1-\prod_{j=1}^{z}\left(1-T^L_{\gamma_j}(\tau)\right)^{w_j}, 1-\prod_{j=1}^{z}\left(1-T^U_{\gamma_j}(\tau)\right)^{w_j}\right],\\ \left[\prod_{j=1}^{z}\left(I^L_{\gamma_j}(\tau)\right)^{w_j}, \prod_{j=1}^{z}\left(I^U_{\gamma_j}(\tau)\right)^{w_j}\right],\left[\prod_{j=1}^{z}\left(F^L_{\gamma_j}(\tau)\right)^{w_j}, \prod_{j=1}^{z}\left(F^U_{\gamma_j}(\tau)\right)^{w_j}\right]\end{array}\right)\right\}$$
$$\oplus\left(\begin{array}{l}\left[1-\left(1-T^L_{\gamma_{k+1}}(\tau)\right)^{w_{z+1}}, 1-\left(1-T^U_{\gamma_{z+1}}(\tau)\right)^{w_{z+1}}\right],\\ \left[\left(I^L_{\gamma_{z+1}}(\tau)\right)^{w_{z+1}}, \left(I^U_{\gamma_{z+1}}(\tau)\right)^{w_{z+1}}\right],\left[\left(F^L_{\gamma_{z+1}}(\tau)\right)^{w_{z+1}}, \left(F^U_{\gamma_{z+1}}(\tau)\right)^{w_{z+1}}\right]\end{array}\right) \tag{9}$$
$$= \bigcup_{\gamma_1\in\widetilde{h}_1,\gamma_2\in\widetilde{h}_2,\ldots,\gamma_{z+1}\in\widetilde{h}_{z+1}}\left\{\left(\begin{array}{l}\left[1-\prod_{j=1}^{z+1}\left(1-T^L_{\gamma_j}(\tau)\right)^{w_j}, 1-\prod_{j=1}^{z+1}\left(1-T^U_{\gamma_j}(\tau)\right)^{w_j}\right],\\ \left[\prod_{j=1}^{z+1}\left(I^L_{\gamma_j}(\tau)\right)^{w_j}, \prod_{j=1}^{z+1}\left(I^U_{\gamma_j}(\tau)\right)^{w_j}\right],\left[\prod_{j=1}^{z+1}\left(F^L_{\gamma_j}(\tau)\right)^{w_j}, \prod_{j=1}^{z+1}\left(F^U_{\gamma_j}(\tau)\right)^{w_j}\right]\end{array}\right)\right\}$$

It follows that if (6) holds for $n = z$, then it holds for $n = z + 1$. Because it is also true for $n = 1$, according to the method of mathematical inductive reasoning, Equation (6) holds for natural numbers N and Theorem 1 is proven.

Definition 7. *Let $\widetilde{h}_j (j = 1, 2, \ldots, n)$ be a collection of GDIVNEs. Generalized dynamic interval-valued neutrosophic weighted geometric (GDIVNWG) operator is defined as*

$$GDIVNWG\left(\widetilde{h}_1, \widetilde{h}_2, \ldots, \widetilde{h}_n\right) = \prod_{j=1}^{n} \widetilde{h}_j^{w_j}$$
$$= \bigcup_{\gamma_1\in\widetilde{h}_1,\gamma_2\in\widetilde{h}_2,\ldots,\gamma_n\in\widetilde{h}_n}\left\{\left(\begin{array}{l}\left[\prod_{j=1}^{n}\left(T^L_{\gamma_j}(\tau)\right)^{w_j}, \prod_{j=1}^{n}\left(T^U_{\gamma_j}(\tau)\right)^{w_j}\right],\left[1-\prod_{j=1}^{n}\left(1-I^L_{\gamma_j}(\tau)\right)^{w_j}, 1-\prod_{j=1}^{n}\left(1-I^U_{\gamma_j}(\tau)\right)^{w_j}\right],\\ \left[1-\prod_{j=1}^{n}\left(1-F^L_{\gamma_j}(\tau)\right)^{w_j}, 1-\prod_{j=1}^{n}\left(1-F^U_{\gamma_j}(\tau)\right)^{w_j}\right]\end{array}\right)\right\} \tag{10}$$

Theorem 2. *Let $\widetilde{h}_j (j = 1, 2, \ldots, n)$ be the collection of GDIVNEs. The result aggregated from GDIVNWG operator is still a GDIVNE.*

Proof. The Equation (10) is proved by mathematical inductive reasoning method. □

When $n = 1$, Equation (10) is true because it simplifies to the trivial outcome, which is obviously GDIVNE,

$$GDIVNWG\left(\widetilde{h}_1\right) = \left(\begin{array}{l}\left[\left(T^L_{\gamma_1}(\tau)\right)^{w_1}, \left(T^U_{\gamma_1}(\tau)\right)^{w_1}\right],\left[1-\left(1-I^L_{\gamma_1}(\tau)\right)^{w_1}, 1-\left(1-I^U_{\gamma_1}(\tau)\right)^{w_1}\right],\\ \left[1-\left(1-F^L_{\gamma_1}(\tau)\right)^{w_1}, 1-\left(1-F^U_{\gamma_1}(\tau)\right)^{w_1}\right]\end{array}\right) \tag{11}$$

Let us assume that (10) is true for $n = z$.

$$\prod_{j=1}^{z} \widetilde{h}_j^{w_j} = \bigcup_{\gamma_1\in\widetilde{h}_1,\gamma_2\in\widetilde{h}_2,\ldots,\gamma_z\in\widetilde{h}_z}\left\{\left(\begin{array}{l}\left[\prod_{j=1}^{z}\left(T^L_{\gamma_j}(\tau)\right)^{w_j}, \prod_{j=1}^{z}\left(T^U_{\gamma_j}(\tau)\right)^{w_j}\right],\left[1-\prod_{j=1}^{z}\left(1-I^L_{\gamma_j}(\tau)\right)^{w_j}, 1-\prod_{j=1}^{z}\left(1-I^U_{\gamma_j}(\tau)\right)^{w_j}\right],\\ \left[1-\prod_{j=1}^{z}\left(1-F^L_{\gamma_j}(\tau)\right)^{w_j}, 1-\prod_{j=1}^{z}\left(1-F^U_{\gamma_j}(\tau)\right)^{w_j}\right]\end{array}\right)\right\} \tag{12}$$

When $n = z + 1$

$$
\begin{aligned}
&\prod_{j=1}^{z+1} \widetilde{h}_j^{w_j} = \prod_{j=1}^{z} \widetilde{h}_j^{w_j} \otimes \widetilde{h}_{z+1}^{w_{z+1}} \\
&= \bigcup_{\gamma_1 \in \widetilde{h}_1, \gamma_2 \in \widetilde{h}_2, \ldots, \gamma_z \in \widetilde{h}_z} \left\{ \left(\begin{array}{l} \left[\prod_{j=1}^{k} \left(T^L_{\gamma_j}(\tau)\right)^{w_j}, \prod_{j=1}^{k} \left(T^U_{\gamma_j}(\tau)\right)^{w_j} \right], \left[1 - \prod_{j=1}^{k} \left(1 - I^L_{\gamma_j}(\tau)\right)^{w_j}, 1 - \prod_{j=1}^{k} \left(1 - I^U_{\gamma_j}(\tau)\right)^{w_j} \right], \\ \left[1 - \prod_{j=1}^{k} \left(1 - F^L_{\gamma_j}(\tau)\right)^{w_j}, 1 - \prod_{j=1}^{k} \left(1 - F^U_{\gamma_j}(\tau)\right)^{w_j} \right] \end{array} \right) \right\} \\
&\otimes \left(\begin{array}{l} \left[\left(T^L_{\gamma_j}(\tau)\right)^{w_{z+1}}, \left(T^U_{\gamma_{z+1}}(\tau)\right)^{w_{z+1}} \right], \left[1 - \left(1 - I^L_{\gamma_{z+1}}(\tau)\right)^{w_{z+1}}, 1 - \left(1 - I^U_{\gamma_{z+1}}(\tau)\right)^{w_{z+1}} \right], \\ \left[1 - \left(1 - F^L_{\gamma_{z+1}}(\tau)\right)^{w_{z+1}}, 1 - \left(1 - F^U_{\gamma_{z+1}}(\tau)\right)^{w_{z+1}} \right] \end{array} \right) \\
&= \bigcup_{\gamma_1 \in \widetilde{h}_1, \gamma_2 \in \widetilde{h}_2, \ldots, \gamma_{z+1} \in \widetilde{h}_{z+1}} \left\{ \left(\begin{array}{l} \left[\prod_{j=1}^{z+1} \left(T^L_{\gamma_j}(\tau)\right)^{w_j}, \prod_{j=1}^{z+1} \left(T^U_{\gamma_j}(\tau)\right)^{w_j} \right], \left[1 - \prod_{j=1}^{z+1} \left(1 - I^L_{\gamma_j}(\tau)\right)^{w_j}, 1 - \prod_{j=1}^{z+1} \left(1 - I^U_{\gamma_j}(\tau)\right)^{w_j} \right], \\ \left[1 - \prod_{j=1}^{z+1} \left(1 - F^L_{\gamma_j}(\tau)\right)^{w_j}, 1 - \prod_{j=1}^{z+1} \left(1 - F^U_{\gamma_j}(\tau)\right)^{w_j} \right] \end{array} \right) \right\}
\end{aligned}
\tag{13}
$$

It follows that if (10) holds for $n = z$, then it holds for $n = z + 1$. Because it is also true for $n = 1$, according to the method of mathematical inductive reasoning, Equation (10) holds for all natural numbers N and Theorem 2 is proven.

Herein, we define the generalized dynamic interval-valued neutrosophic hybrid weighted averaging (GDIVNHWA) operator to combine the effects of attribute weight vector and the positional weight vector, which are mentioned in Definitions 6 and 7.

Definition 8. *Let $\lambda > 0$ and $\widetilde{h}_j (j = 1, 2, \ldots, n)$ be a collection of GDIVNEs. Generalized dynamic interval-valued neutrosophic hybrid weighted averaging (GDIVNHWA) operator is defined as,*

$$
\begin{aligned}
&DIVHNWG\left(\widetilde{h}_1, \widetilde{h}_2, \ldots, \widetilde{h}_n\right) = \left(\sum_{j=1}^{n} w_j \widetilde{h}_j^{\lambda} \right)^{\frac{1}{\lambda}} \\
&= \bigcup_{\gamma_1 \in \widetilde{h}_1, \gamma_2 \in \widetilde{h}_2, \ldots, \gamma_n \in \widetilde{h}_n} \left\{ \left(\begin{array}{l} \left[\left(1 - \prod_{j=1}^{n} \left(1 - \left(T^L_{\gamma_j}(\tau)\right)^{\lambda}\right)^{w_j}\right)^{\frac{1}{\lambda}}, \left(1 - \prod_{j=1}^{n} \left(1 - \left(T^U_{\gamma_j}(\tau)\right)^{\lambda}\right)^{w_j}\right)^{\frac{1}{\lambda}} \right], \\ \left[1 - \left(1 - \prod_{j=1}^{n} \left(1 - \left(1 - I^L_{\gamma_j}(\tau)\right)^{\lambda}\right)^{w_j}\right)^{\frac{1}{\lambda}}, 1 - \left(1 - \prod_{j=1}^{n} \left(1 - \left(1 - I^U_{\gamma_j}(\tau)\right)^{\lambda}\right)^{w_j}\right)^{\frac{1}{\lambda}} \right], \\ \left[1 - \left(1 - \prod_{j=1}^{n} \left(1 - \left(1 - F^L_{\gamma_j}(\tau)\right)^{\lambda}\right)^{w_j}\right)^{\frac{1}{\lambda}}, 1 - \left(1 - \prod_{j=1}^{n} \left(1 - \left(1 - F^U_{\gamma_j}(\tau)\right)^{\lambda}\right)^{w_j}\right)^{\frac{1}{\lambda}} \right] \end{array} \right) \right\}
\end{aligned}
\tag{14}
$$

Theorem 3. *Let $\widetilde{h}_j (j = 1, 2, \ldots, n)$ be the collection of GDIVNEs. The result aggregated from GDIVNHWA operator is still a GDIVNE.*

Proof. The Equation (14) can be proved by mathematical inductive reasoning method. □

We first prove that (15) is a collection of GDIVNEs,

$$
\sum_{j=1}^{n} w_j \widetilde{h}_j^{\lambda} = \bigcup_{\gamma_1 \in \widetilde{h}_1, \gamma_2 \in \widetilde{h}_2, \ldots, \gamma_n \in \widetilde{h}_n} \left\{ \left(\begin{array}{l} \left[1 - \prod_{j=1}^{n} \left(1 - \left(T^L_{\gamma_j}(\tau)\right)^{\lambda}\right)^{w_j}, 1 - \prod_{j=1}^{n} \left(1 - \left(T^U_{\gamma_j}(\tau)\right)^{\lambda}\right)^{w_j} \right], \\ \left[1 - \prod_{j=1}^{n} \left(1 - \left(1 - I^L_{\gamma_j}(\tau)\right)^{\lambda}\right)^{w_j}, 1 - \prod_{j=1}^{n} \left(1 - \left(1 - I^U_{\gamma_j}(\tau)\right)^{\lambda}\right)^{w_j} \right], \\ \left[1 - \prod_{j=1}^{n} \left(1 - \left(1 - F^L_{\gamma_j}(\tau)\right)^{\lambda}\right)^{w_j}, 1 - \prod_{j=1}^{n} \left(1 - \left(1 - F^U_{\gamma_j}(\tau)\right)^{\lambda}\right)^{w_j} \right] \end{array} \right) \right\}
\tag{15}
$$

When $n = 1$, Equation (15) is true because it simplifies to the trivial outcome, which is obviously GDIVNE,

$$w_1\widetilde{h}_1^{\lambda} = \left(\begin{array}{l}\left[1-\left(1-\left(T_{\gamma_1}^L(\tau)\right)^{\lambda}\right)^{w_1}, 1-\left(1-\left(T_{\gamma_1}^U(\tau)\right)^{\lambda}\right)^{w_1}\right],\\ \left[1-\left(1-\left(1-I_{\gamma_1}^L(\tau)\right)^{\lambda}\right)^{w_1}, 1-\left(1-\left(1-I_{\gamma_1}^U(\tau)\right)^{\lambda}\right)^{w_1}\right],\\ \left[1-\left(1-\left(1-F_{\gamma_1}^L(\tau)\right)^{\lambda}\right)^{w_1}, 1-\left(1-\left(1-F_{\gamma_1}^U(\tau)\right)^{\lambda}\right)^{w_1}\right]\end{array}\right) \tag{16}$$

Let us assume that (15) is true for $n = z$,

$$\sum_{j=1}^{z} w_j\widetilde{h}_j^{\lambda} = \bigcup_{\gamma_1\in\widetilde{h}_1,\gamma_2\in\widetilde{h}_2,\ldots,\gamma_z\in\widetilde{h}_z}\left\{\left(\begin{array}{l}\left[1-\prod_{j=1}^{z}\left(1-\left(T_{\gamma_j}^L(\tau)\right)^{\lambda}\right)^{w_j}, 1-\prod_{j=1}^{z}\left(1-\left(T_{\gamma_j}^U(\tau)\right)^{\lambda}\right)^{w_j}\right],\\ \left[1-\prod_{j=1}^{z}\left(1-\left(1-I_{\gamma_j}^L(\tau)\right)^{\lambda}\right)^{w_j}, 1-\prod_{j=1}^{z}\left(1-\left(1-I_{\gamma_j}^U(\tau)\right)^{\lambda}\right)^{w_j}\right],\\ \left[1-\prod_{j=1}^{z}\left(1-\left(1-F_{\gamma_j}^L(\tau)\right)^{\lambda}\right)^{w_j}, 1-\prod_{j=1}^{z}\left(1-\left(1-F_{\gamma_j}^U(\tau)\right)^{\lambda}\right)^{w_j}\right]\end{array}\right)\right\} \tag{17}$$

When $n = z + 1$,

$$\begin{array}{l}\sum_{j=1}^{z+1} w_j\widetilde{h}_j^{\lambda} = \sum_{j=1}^{z} w_j\widetilde{h}_j^{\lambda} \oplus w_{z+1}\widetilde{h}_{z+1}^{\lambda}\\ = \bigcup_{\gamma_1\in\widetilde{h}_1,\gamma_2\in\widetilde{h}_2,\ldots,\gamma_z\in\widetilde{h}_z}\left\{\left(\begin{array}{l}\left[1-\prod_{j=1}^{z}\left(1-\left(T_{\gamma_j}^L(\tau)\right)^{\lambda}\right)^{w_j}, 1-\prod_{j=1}^{z}\left(1-\left(T_{\gamma_j}^U(\tau)\right)^{\lambda}\right)^{w_j}\right],\\ \left[1-\prod_{j=1}^{z}\left(1-\left(1-I_{\gamma_j}^L(\tau)\right)^{\lambda}\right)^{w_j}, 1-\prod_{j=1}^{z}\left(1-\left(1-I_{\gamma_j}^U(\tau)\right)^{\lambda}\right)^{w_j}\right],\\ \left[1-\prod_{j=1}^{z}\left(1-\left(1-F_{\gamma_j}^L(\tau)\right)^{\lambda}\right)^{w_j}, 1-\prod_{j=1}^{z}\left(1-\left(1-F_{\gamma_j}^U(\tau)\right)^{\lambda}\right)^{w_j}\right]\end{array}\right)\right\}\\ \oplus\left\{\left(\begin{array}{l}\left[1-\left(1-\left(T_{\gamma_{z+1}}^L(\tau)\right)^{\lambda}\right)^{w_{z+1}}, 1-\left(1-\left(T_{\gamma_{z+1}}^U(\tau)\right)^{\lambda}\right)^{w_{z+1}}\right],\\ \left[1-\left(1-\left(1-I_{\gamma_{z+1}}^L(\tau)\right)^{\lambda}\right)^{w_{z+1}}, 1-\left(1-\left(1-I_{\gamma_{z+1}}^U(\tau)\right)^{\lambda}\right)^{w_{z+1}}\right],\\ \left[1-\left(1-\left(1-F_{\gamma_{z+1}}^L(\tau)\right)^{\lambda}\right)^{w_{z+1}}, 1-\left(1-\left(1-F_{\gamma_{z+1}}^U(\tau)\right)^{\lambda}\right)^{w_{z+1}}\right]\end{array}\right)\right\}\\ = \bigcup_{\gamma_1\in\widetilde{h}_1,\gamma_2\in\widetilde{h}_2,\ldots,\gamma_{k+1}\in\widetilde{h}_{k+!}}\left\{\left(\begin{array}{l}\left[1-\prod_{j=1}^{k+1}\left(1-\left(T_{\gamma_j}^L(\tau)\right)^{\lambda}\right)^{w_j}, 1-\prod_{j=1}^{k+!}\left(1-\left(T_{\gamma_j}^U(\tau)\right)^{\lambda}\right)^{w_j}\right],\\ \left[1-\prod_{j=1}^{k+1}\left(1-\left(1-I_{\gamma_j}^L(\tau)\right)^{\lambda}\right)^{w_j}, 1-\prod_{j=1}^{k+1}\left(1-\left(1-I_{\gamma_j}^U(\tau)\right)^{\lambda}\right)^{w_j}\right],\\ \left[1-\prod_{j=1}^{k+1}\left(1-\left(1-F_{\gamma_j}^L(\tau)\right)^{\lambda}\right)^{w_j}, 1-\prod_{j=1}^{k+1}\left(1-\left(1-F_{\gamma_j}^U(\tau)\right)^{\lambda}\right)^{w_j}\right]\end{array}\right)\right\}\end{array} \tag{18}$$

It follows that if (15) holds for $n = z$, then it holds for $n = z + 1$. Because it is also true for $n = 1$, according to the method of mathematical inductive reasoning, Equation (15) holds for natural numbers N. According to Equation (15) and Definition 4, we have,

$$DIVHNWG\left(\widetilde{h}_1,\widetilde{h}_2,\ldots,\widetilde{h}_n\right)=\left(\sum_{j=1}^{n}w_j\widetilde{h}_j^{\lambda}\right)^{\frac{1}{\lambda}}$$

$$=\bigcup_{\gamma_1\in\widetilde{h}_1,\gamma_2\in\widetilde{h}_2,\ldots,\gamma_n\in\widetilde{h}_n}\left\{\left(\left(\begin{array}{l}\left[\left(1-\prod_{j=1}^{n}\left(1-\left(T_{\gamma_j}^{L}(\tau)\right)^{\lambda}\right)^{w_j}\right)^{\frac{1}{\lambda}},\left(1-\prod_{j=1}^{n}\left(1-\left(T_{\gamma_j}^{U}(\tau)\right)^{\lambda}\right)^{w_j}\right)^{\frac{1}{\lambda}}\right],\\ \left[1-\left(1-\prod_{j=1}^{n}\left(1-\left(1-I_{\gamma_j}^{L}(\tau)\right)^{\lambda}\right)^{w_j}\right)^{\frac{1}{\lambda}},1-\left(1-\prod_{j=1}^{n}\left(1-\left(1-I_{\gamma_j}^{U}(\tau)\right)^{\lambda}\right)^{w_j}\right)^{\frac{1}{\lambda}}\right],\\ \left[1-\left(1-\prod_{j=1}^{n}\left(1-\left(1-F_{\gamma_j}^{L}(\tau)\right)^{\lambda}\right)^{w_j}\right)^{\frac{1}{\lambda}},1-\left(1-\prod_{j=1}^{n}\left(1-\left(1-F_{\gamma_j}^{U}(\tau)\right)^{\lambda}\right)^{w_j}\right)^{\frac{1}{\lambda}}\right]\end{array}\right)\right)\right\}$$

Thus, Theorem 3 is proven.

4. Dynamic TOPSIS Method

Based on the theory of GDVINS, the dynamic decision-making model is proposed to deal with the change of criteria, alternatives, and decision-makers during time.

For each period $t=\{t_1,t_2,\ldots,t_s\}$, assume $\widetilde{A}(t_r)=\{A_1,A_2,\ldots,A_{v_r}\}$ and $\widetilde{C}(t_r)=\left\{C_1,C_2,\ldots,C_{n_r}\right\}$ and $\widetilde{D}(t_r)=\left\{D_1,D_2,\ldots,D_{h_r}\right\}$ being the sets of alternatives, criteria, and decision-makers at period r^{th}, $r=\{1,2,\ldots,s\}$. For a decision-maker $D_q; q=1,\ldots,h_r$, the evaluation of an alternative $A_m; m=1,\ldots,v_r$, on a criteria $C_p; p=1,\ldots,n_r$, in time sequence $\tau=\left\{\tau_1,\tau_2,\ldots,\tau_{k_r}\right\}$ is represented by the Neutrosophic decision matrix $\Re^q(t_r)=\left(\xi_{mp}^q(\tau)\right)_{v_r\times n_r}; l=1,2,\ldots,k_r$. where

$$\xi_{mp}^q(\tau)=\left\langle x_{d_{mp}}^q(\tau),\left(T^q\left(d_{mp},\tau\right),I^q\left(d_{mp},\tau\right),F^q\left(d_{mp},\tau\right)\right)\right\rangle;$$

taken by GDIVNSs evaluated by decision maker D_q.

Step 1. Calculate aggregate ratings at period r^{th}.

Let $x_{mpq}(\tau_l)=\left\{\left[T_{mpq}^L\left(x_{\tau_l}\right),T_{mpq}^U\left(x_{\tau_l}\right)\right],\left[I_{mpq}^L\left(x_{\tau_l}\right),I_{mpq}^U\left(x_{\tau_l}\right)\right],\left[F_{mpq}^L\left(x_{\tau_l}\right),F_{mpq}^U\left(x_{\tau_l}\right)\right]\right\}$ be the appropriateness rating of alternative A_m for criterion C_p by decision-maker D_q in time sequence τ_l, where: $m=1,\ldots,v_r; p=1,\ldots,n_r; q=1,\ldots,h_r; l=1,\ldots,k_r$. The averaged appropriateness rating $\overline{x_{mp}}=\left\{\left[\overline{T_{mp}^L(x),T_{mp}^U(x)}\right],\left[\overline{I_{mp}^L(x),I_{mp}^U(x)}\right],\left[\overline{F_{mp}^L(x),F_{mp}^U(x)}\right]\right\}$ can be evaluated as:

$$\overline{x_{mp}}=\frac{1}{h_r\times k_r}\times\left\langle\begin{array}{l}\left[1-\left\{1-\left(1-\sum_{q=1}^{h_r}T_{pmq}^L\left(x_{\tau_l}\right)\right)^{\frac{1}{h_r}}\right\}^{\frac{1}{k_r}},1-\left\{1-\left(1-\sum_{q=1}^{h}T_{pmq}^U\left(x_{\tau_l}\right)\right)^{\frac{1}{h_r}}\right\}^{\frac{1}{k_r}}\right],\\ \left[\left(\sum_{q=1}^{h}I_{pmq}^L\left(x_{\tau_l}\right)\right)^{\frac{1}{h_r\times k_r}},\left(\sum_{q=1}^{h}I_{pmq}^U\left(x_{\tau_l}\right)\right)^{\frac{1}{h_r\times k_r}}\right],\\ \left[\left(\sum_{q=1}^{h}F_{pmq}^L\left(x_{\tau_l}\right)\right)^{\frac{1}{h_r\times k_r}},\left(\sum_{q=1}^{h}F_{pmq}^U\left(x_{\tau_l}\right)\right)^{\frac{1}{h_r\times k_r}}\right]\end{array}\right\rangle \quad (19)$$

Step 2. Calculate importance weight aggregation at period r^{th}.

Let $y_{pq}(\tau_l)=\left\{\left[T_{pq}^L\left(y_{\tau_l}\right),T_{pq}^U\left(y_{\tau_l}\right)\right],\left[I_{pq}^L\left(y_{\tau_l}\right),I_{pq}^U\left(y_{\tau_l}\right)\right],\left[F_{pq}^L\left(y_{\tau_l}\right),F_{pq}^U\left(y_{\tau_l}\right)\right]\right\}$ be the weight of D_q to criterion C_p in time sequence τ_l, where: $p=1,\ldots,n_r; q=1,\ldots,h_r; l=1,\ldots,k$. The average weight $\overline{w_p}=\left\{\left[\overline{T_p^L(y),T_p^U(y)}\right],\left[\overline{I_p^L(y),I_p^U(y)}\right],\left[\overline{F_p^L(y),F_p^U(y)}\right]\right\}$ can be evaluated as:

$$\overline{w_p} = \frac{1}{h_r \times k_r} \times \left\langle \begin{array}{l} \left[1-\left\{1-\left(1-\sum_{q=1}^{h_r} T_{pq}^{L}\left(y_{\tau_l}\right)\right)^{\frac{1}{h_r}}\right\}^{\frac{1}{k_r}}, 1-\left\{1-\left(1-\sum_{q=1}^{h} T_{pq}^{U}\left(y_{\tau_l}\right)\right)^{\frac{1}{h_r}}\right\}^{\frac{1}{k_r}}\right], \\ \left[\left(\sum_{q=1}^{h_r} I_{pq}^{L}\left(y_{\tau_l}\right)\right)^{\frac{1}{h_r \times k_r}}, \left(\sum_{q=1}^{h_r} I_{pq}^{U}\left(y_{\tau_l}\right)\right)^{\frac{1}{h_r \times k_r}}\right], \\ \left[\left(\sum_{q=1}^{h_r} F_{pq}^{L}\left(y_{\tau_l}\right)\right)^{\frac{1}{h_r \times k_r}}, \left(\sum_{q=1}^{h_r} F_{pq}^{U}\left(y_{\tau_l}\right)\right)^{\frac{1}{h_r \times k_r}}\right] \end{array} \right\rangle, \quad (20)$$

Step 3. Evaluation for aggregate ratings of alternatives with history data.
Using Equation (21), evaluate aggregate ratings and importance weight aggregation.

$$\widetilde{A}(t_r^*) = \{A_1, A_2, \ldots, A_{v_r}\} \cup \widetilde{A}(t_{r-1})$$

$$\overline{x_{mp}^*} = \begin{cases} \overline{x_{mp}^r} \quad if \left(\begin{array}{l} A_m \in \widetilde{A}(t_r) \backslash \widetilde{A}(t_{r-1}) \& C_p \in \widetilde{C}(t_r) \backslash \widetilde{C}(t_{r-1}) \\ or \quad A_m \in \widetilde{A}(t_{r-1}) \backslash \widetilde{A}(t_r) \& C_p \in \widetilde{C}(t_r) \backslash \widetilde{C}(t_{r-1}) \\ or \quad A_m \in \widetilde{A}(t_r) \backslash \widetilde{A}(t_{r-1}) \& C_p \in \widetilde{C}(t_{r-1}) \backslash \widetilde{C}(t_r) \end{array} \right) \\ \overline{x_{mp}^r} \oplus \overline{x_{mp}^{r-1}} \quad if \quad A_m \in \widetilde{A}(t_r) \cap \widetilde{A}(t_{r-1}) \& C_p \in \widetilde{C}(t_r) \cap \widetilde{C}(t_{r-1}) \\ \overline{x_{mp}^{r-1}} \quad if \quad A_m \in \widetilde{A}(t_{r-1}) \backslash \widetilde{A}(t_r) \& C_p \in \widetilde{C}(t_{r-1}) \backslash \widetilde{C}(t_r) \end{cases} \quad (21)$$

Step 4. Evaluation for importance weight aggregation of criteria with history data.
Using Equation (22), evaluate aggregate ratings and importance weight aggregation.

$$\widetilde{C}(t_r^*) = \left\{C_1, C_2, \ldots, C_{n_r}\right\} \cup \widetilde{C}(t_{r-1})$$

$$\overline{w_p^*} = \begin{cases} \overline{w_p^r} & if \quad C_p \in \widetilde{C}(t_r) \backslash \widetilde{C}(t_{r-1}) \\ \overline{w_p^r} \oplus \overline{w_p^{r-1}} & if \quad C_p \in \widetilde{C}(t_r) \cap \widetilde{C}(t_{r-1}) \\ \overline{w_p^{r-1}} & if \quad C_p \in \widetilde{C}(t_{r-1}) \backslash \widetilde{C}(t_r) \end{cases} \quad (22)$$

Step 5. Calculate the average weighted ratings at period r^{th}.
The average weighted ratings of alternatives at period t_r, can be calculated by:

$$\Theta_m = \frac{1}{n_r^*} \sum_{p=1}^{n_r^*} \overline{x_{mp}^*} * \overline{w_p^*}; m = 1, \ldots, v_r^*; p = 1, \ldots, n_r^*; \quad (23)$$

Step 6. Determination of A_r^+, A_r^- and d_r^+, d_r^- at period r^{th}.

Interval-valued neutrosophic positive ideal solution (PIS, A_r^+) and interval-valued neutrosophic negative ideal solution (NIS, A_r^-) are:

$$A_r^+ = \left\{x, \left\{([1,1],[0,0],[0,0])_1, ([1,1],[0,0],[0,0])_2, \ldots, ([1,1],[0,0],[0,0])_{n_r^*}\right\}\right\} \quad (24)$$

$$A_r^- = \left\{x, \left\{([0,0],[1,1],[1,1])_1, ([0,0],[1,1],[1,1])_2, \ldots, ([0,0],[1,1],[1,1])_{n_r^*}\right\}\right\} \quad (25)$$

The distances of each alternative A_m, $m = 1, 2, \ldots, n^*$ from A_r^+ and A_r^- at period t_r, are calculated as:

$$d_m^+ = \sqrt{\left(\Theta_m - A_r^+\right)^2} \quad (26)$$

$$d_m^- = \sqrt{(\Theta_m - A_r^-)^2} \tag{27}$$

where d_m^+ and d_m^- respectively represent the shortest and farthest distances of A_m.

Step 7. Determination the closeness coefficient.

The closeness coefficient at period t_r, is calculated in Equation (28), where an alternative that is close to interval-valued neutrosophic PIS and far from interval-valued neutrosophic NIS, has high value:

$$BC_m = \frac{d_m^-}{d_m^+ + d_m^-} \tag{28}$$

Step 8. Rank the alternatives.

The alternatives are ranked by their closeness coefficient values. See Figure 1 for illustration.

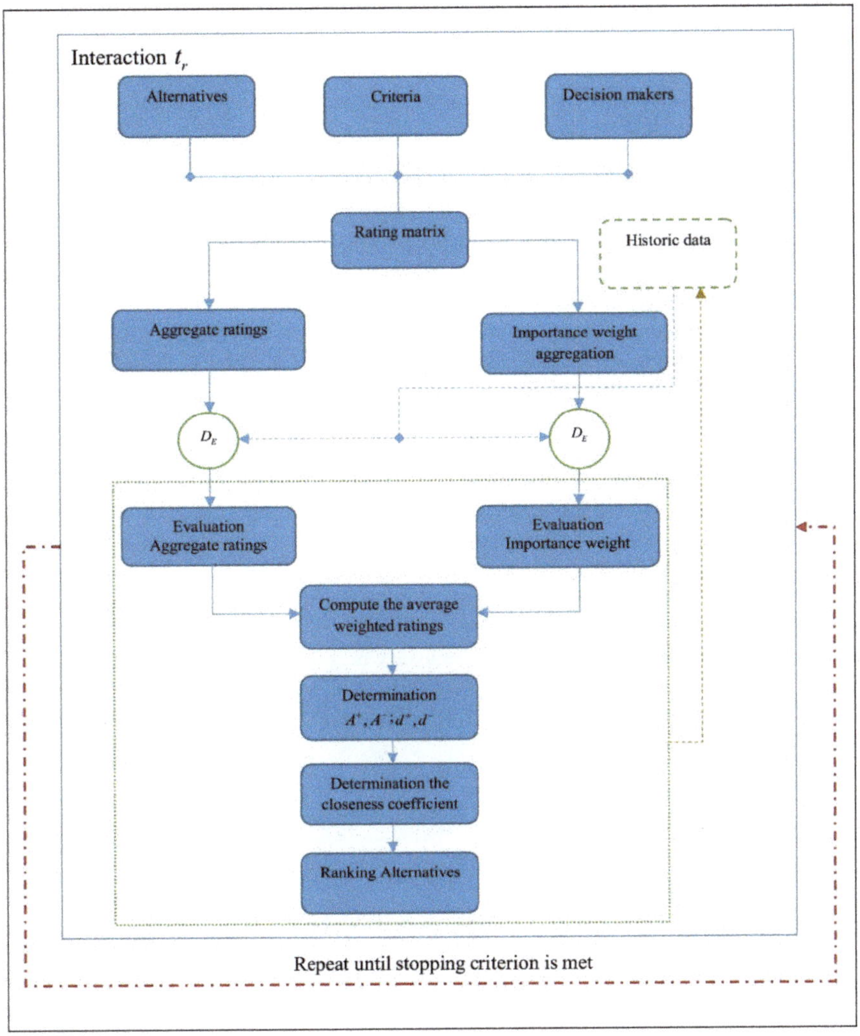

Figure 1. Dynamic TOPSIS method.

5. Applications

5.1. ASK Model for Ranking Students

Human resources recruitment plays a pivotal role in any enterprise as it exerts tremendous impact on its sustainable development. Thus, the selection of competent and job-relevant staff will lay the solid foundation for the successful performance of an enterprise. Notably, every year most of the businesses invest a large sum of money for job vacancy advertisements (on newspapers, websites, and in job fairs) and recruitment activities including application screening and interview. However, to recruit new graduated student the organizations are likely to encounter high potential risks as there are definitely inevitable employee turnovers or the selected candidates fall short of employers' expectation [22]. Mis-assessment of candidate's competence might be rooted from assessors' criteria and model for new graduated student evaluation.

The above problems underline the need for making the right assessment of potential employees. Currently, ASK model (attitude, skill and knowledge) has been widely used by many organizations because of its comprehensive assessment. This model was initially proposed by Bloom [11] with three factors including knowledge which is acquired through education, comprehension, analysis, and application skills which are the ability to process the knowledge to perform activities or tasks, and attitude which is concerned with feeling, emotions, or motivation toward employment. These elements are given divergent weights in the assessment model according to positions and requirements of the job. ASK is applicable to evaluate tertiary students' performance to give more information that support employers besides a set of university exit benchmark. It also facilitates students to make proper self-adjustments and pursue appropriate professional orientation for their future career [23,24]. Ranking students based on attributes of ASK model requires a dynamic multi-criteria decision-making model that is able to combine the estimations of different lecturers in different periods. The proposed DTOPSIS completely fit to this complex task, and the application model is depicted bellow.

5.2. Application Model

As mentioned above, the proposed method is applied to rank students of Thuongmai University, Hanoi, Vietnam. In this research, the datasets were surveyed through three consecutive semesters under three criteria (attitudes-skills-knowledge). Each student will be surveyed at the beginning of semester and by the end of semester. With the model assessing student competence, it will be conducted over semesters and over school years. This is the way of setting the time period in the decision-making model of this research.

Figure 2 shows the ASK model for ranking students where three lectures i.e., D_1, D_2, D_3 are chosen. According to the language labels shown in Tables 1 and 2, rating of five students and criteria' weights are done by the lectures based on fourteenth criteria in three groups: attitude, skill, and knowledge. The attitude group includes five criteria [25], the skill group includes six criteria [26], and the knowledge group includes three criteria [23].

The criteria used for ranking Thuongmai university's students contain 14 criteria divided into three groups (attitudes-skills-knowledge) in the ASK model. In the early stage of each semester, the knowledge criteria will not cause many impacts on student competency assessment so that we only pay attention to 11 criteria in the two remaining groups: attitudes and skills. In the following semesters, the knowledge criterion shall be supplemented that why all 14 criteria in three group shall be conducted.

(1) Period t_1 (the first semester): the decision-maker provides assessments of three students A_1, A_2, A_3 according to 11 criteria in two groups: attitude, skill. Tables 3 and 4 show the steps of the model at time t_1 and Table 5 shows the ranking order as $A_1 > A_2 > A_3$. Thus, the best student is A_1.

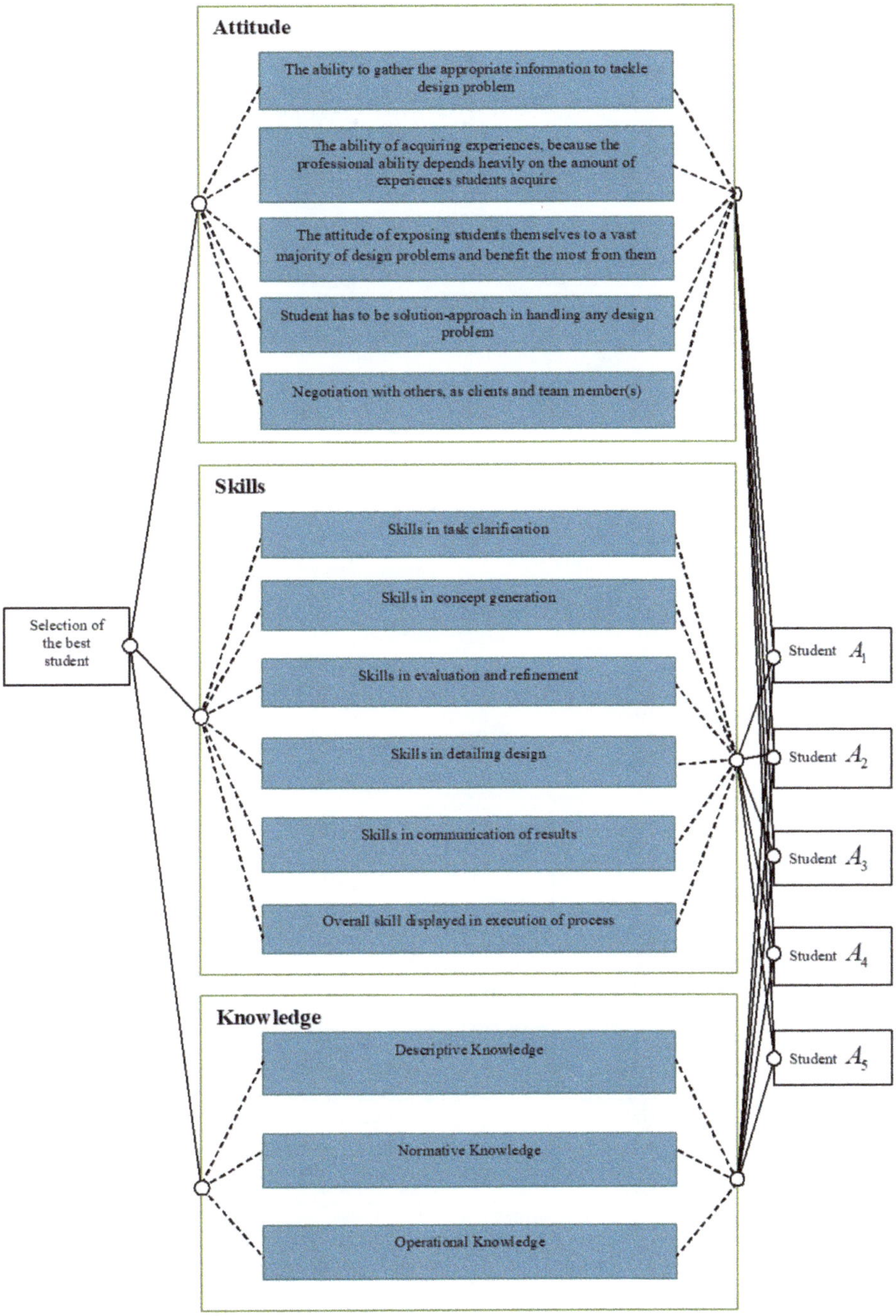

Figure 2. Attitudes-skills-knowledge (ASK) model for recruitment of ternary students.

Table 1. Appropriateness ratings.

Language Labels	Values
Very Poor	([0.1, 0.26], [0.4, 0.5], [0.63, 0.76])
Poor	([0.26, 0.38], [0.47, 0.6], [0.51, 0.6])
Medium	([0.38, 0.5], [0.4, 0.61], [0.44, 0.55])
Good	([0.5, 0.65], [0.36, 0.5], [0.31, 0.48])
Very Good	([0.65, 0.8], [0.1, 0.2], [0.12, 0.2])

Table 2. The importance of criteria.

Language Labels	Values
Unimportant	([0.1, 0.19], [0.32, 0.47], [0.64, 0.8])
Slightly Important	([0.2, 0.38], [0.46, 0.62], [0.36, 0.55])
Important	([0.45, 0.63], [0.41, 0.53], [0.2, 0.42])
Very Important	([0.66, 0.8], [0.3, 0.39], [0.22, 0.32])
Absolutely Important	([0.8, 0.94], [0.18, 0.29], [0.1, 0.2])

Table 3. Aggregated ratings at period t_1.

Criteria	Students		
	A_1	A_2	A_3
C_1	([0.463, 0.606], [0.018, 0.054], [0.014, 0.044])	([0.38, 0.5], [0.022, 0.082], [0.029, 0.059])	([0.43, 0.577], [0.021, 0.053], [0.017, 0.048])
C_2	([0.488, 0.632], [0.005, 0.025], [0.008, 0.021])	([0.419,0.578], [0.011,0.037], [0.011,0.026])	([0.419,0.578], [0.011,0.037], [0.011,0.026])
C_3	([0.463, 0.606], [0.018, 0.054], [0.014, 0.044])	([0.463, 0.606], [0.018, 0.054], [0.014, 0.044])	([0.423, 0.556], [0.02, 0.066], [0.02, 0.051])
C_4	([0.423, 0.556], [0.02, 0.066], [0.02, 0.051])	([0.463, 0.606], [0.018, 0.054], [0.014, 0.044])	([0.388, 0.523], [0.023, 0.065], [0.024, 0.056])
C_5	([0.523, 0.673], [0.005, 0.021], [0.005, 0.018])	([0.423, 0.556], [0.02, 0.066], [0.02, 0.051])	([0.43, 0.577], [0.021, 0.053], [0.017, 0.048])
C_6	([0.43, 0.577], [0.021, 0.053], [0.017, 0.048])	([0.38, 0.5], [0.022, 0.082], [0.029, 0.059])	([0.342, 0.463], [0.026, 0.081], [0.034, 0.065])
C_7	([0.38, 0.5], [0.022, 0.082], [0.029, 0.059])	([0.388, 0.523], [0.023, 0.065], [0.024, 0.056])	([0.342, 0.463], [0.026, 0.081], [0.034, 0.065])
C_8	([0.26, 0.38], [0.036, 0.078], [0.046, 0.078])	([0.38, 0.5], [0.022, 0.082], [0.029, 0.059])	([0.38, 0.5], [0.022, 0.082], [0.029, 0.059])
C_9	([0.463, 0.606], [0.018, 0.054], [0.014, 0.044])	([0.523, 0.673], [0.005, 0.021], [0.005, 0.018])	([0.463, 0.606], [0.018, 0.054], [0.014, 0.044])
C_{10}	([0.5, 0.65], [0.016, 0.044], [0.01, 0.038])	([0.38, 0.5], [0.022, 0.082], [0.029, 0.059])	([0.43, 0.577], [0.021, 0.053], [0.017, 0.048])
C_{11}	([0.463, 0.606], [0.018, 0.054], [0.014, 0.044])	([0.302, 0.423], [0.03, 0.079], [0.04, 0.071])	([0.38, 0.5], [0.022, 0.082], [0.029, 0.059])

Table 4. Aggregated weights at period t_1.

Criteria	Importance Aggregated Weights
C_1	([0.963, 0.996], [0.022, 0.06], [0.004, 0.027])
C_2	([0.908, 0.968], [0.041, 0.094], [0.017, 0.056])
C_3	([0.758, 0.89], [0.077, 0.174], [0.014, 0.097])
C_4	([0.648, 0.816], [0.087, 0.204], [0.026, 0.127])
C_5	([0.604, 0.794], [0.06, 0.154], [0.046, 0.185])
C_6	([0.963, 0.992], [0.022, 0.06], [0.004, 0.027])
C_7	([0.834, 0.925], [0.069, 0.149], [0.008, 0.074])
C_8	([0.758, 0.89], [0.077, 0.174], [0.014, 0.097])
C_9	([0.758, 0.89], [0.077, 0.174], [0.014, 0.097])
C_{10}	([0.936, 0.975], [0.037, 0.081], [0.01, 0.043])
C_{11}	([0.897, 0.959], [0.05, 0.11], [0.009, 0.056])

Table 5. Weighted ratings at period t_1.

Students	Weighted Ratings
A_1	([0.368, 0.409], [0.069, 0.168], [0.03, 0.114])
A_2	([0.34, 0.382], [0.071, 0.181], [0.035, 0.12])
A_3	([0.338, 0.377], [0.072, 0.178], [0.035, 0.121])

(2) Period t_2 (the second semester): At this stage, a new student A_4 is added with new criteria in knowledge group. The steps are shown in Tables 6–12. Finally, the ranking is obtained: $A_1 \succ A_2 \succ A_3 \succ A_4$. Thus, the best student is A_1.

Table 6. The distance of each student from $A_{t_1}^+$ and $A_{t_1}^-$ at period t_1.

Students	$d_{t_1}^+$	$d_{t_1}^-$
A_1	0.364193	0.773329
A_2	0.380989	0.763987
A_3	0.382736	0.763579

Table 7. Closeness coefficient at period t_1.

Students	Closeness Coefficients	Ranking Order
A_1	0.679837	1
A_2	0.667251	2
A_3	0.666116	3

Table 8. Aggregated ratings at period t_2.

Criteria	Students			
	A_1	A_2	A_3	A_4
C_1	([0.699, 0.83], [0.001, 0.005], [0, 0.002])	([0.566, 0.75], [0.001, 0.009], [0.001, 0.003])	([0.637, 0.759], [0.001, 0.007], [0.001, 0.003])	([0.5, 0.6], [0.022, 0.046], [0.009, 0.022])
C_2	([0.707, 0.852], [0.001, 0.007], [0, 0.002])	([0.686, 0.834], [0.001, 0.008], [0, 0.003])	([0.72, 0.862], [0.001, 0.006], [0, 0.002])	([0.498, 0.6], [0.023, 0.049], [0.009, 0.023])
C_3	([0.709, 0.848], [0.003, 0.016], [0, 0.005])	([0.643, 0.783], [0.003, 0.018], [0, 0.006])	([0.603, 0.767], [0.003, 0.019], [0.001, 0.006])	([0.56, 0.669], [0.008, 0.029], [0.004, 0.016])
C_4	([0.598, 0.766], [0.004, 0.022], [0.001, 0.007])	([0.639, 0.782], [0.004, 0.021], [0.001, 0.007])	([0.634, 0.793], [0.004, 0.02], [0.001, 0.008])	([0.506, 0.643], [0.009, 0.034], [0.006, 0.021])
C_5	([0.721, 0.866], [0.002, 0.012], [0.001, 0.015])	([0.651, 0.823], [0.002, 0.013], [0.002, 0.016])	([0.616, 0.765], [0.002, 0.014], [0.002, 0.017])	([0.461, 0.604], [0.013, 0.042], [0.009, 0.035])
C_6	([0.685, 0.81], [0.001, 0.005], [0, 0.002])	([0.623, 0.803], [0.001, 0.007], [0, 0.002])	([0.546, 0.72], [0.001, 0.009], [0.001, 0.004])	([0.3, 0.5], [0.022, 0.08], [0.022, 0.044])
C_7	([0.62, 0.802], [0.002, 0.013], [0, 0.004])	([0.618, 0.769], [0.002, 0.013], [0.001, 0.005])	([0.543, 0.72], [0.002, 0.015], [0.001, 0.006])	([0.438, 0.569], [0.024, 0.061], [0.012, 0.03])
C_8	([0.491, 0.648], [0.005, 0.025], [0.002, 0.013])	([0.686, 0.862], [0.004, 0.02], [0, 0.006])	([0.499, 0.709], [0.005, 0.025], [0.001, 0.009])	([0.43, 0.567], [0.026, 0.071], [0.012, 0.033])
C_9	([0.702, 0.847], [0.004, 0.021], [0, 0.007])	([0.761, 0.891], [0.004, 0.019], [0, 0.006])	([0.682, 0.828], [0.004, 0.022], [0, 0.007])	([0.488, 0.598], [0.026, 0.062], [0.009, 0.027])
C_{10}	([0.687, 0.8], [0.002, 0.01], [0, 0.003])	([0.663, 0.836], [0.001, 0.008], [0, 0.003])	([0.718, 0.842], [0.001, 0.008], [0, 0.003])	([0.534, 0.636], [0.012, 0.032], [0.006, 0.018])
C_{11}	([0.608, 0.751], [0.001, 0.009], [0.001, 0.003])	([0.557, 0.722], [0.001, 0.01], [0.001, 0.006])	([0.565, 0.75], [0.001, 0.011], [0.001, 0.004])	([0.499, 0.6], [0.023, 0.048], [0.009, 0.023])
C_{12}	([0.36, 0.533], [0.043, 0.12], [0.021, 0.06])	([0.4, 0.516], [0.049, 0.11], [0.023, 0.065])	([0.463, 0.606], [0.033, 0.089], [0.012, 0.047])	([0.258, 0.439], [0.049, 0.133], [0.037, 0.087])
C_{13}	([0.229, 0.373], [0.05, 0.119], [0.055, 0.108])	([0.229, 0.373], [0.05, 0.119], [0.055, 0.108])	([0.43, 0.568], [0.038, 0.095], [0.017, 0.047])	([0.43, 0.568], [0.038, 0.095], [0.017, 0.047])
C_{14}	([0.284, 0.408], [0.083, 0.167], [0.046, 0.123])	([0.284, 0.408], [0.083, 0.167], [0.046, 0.123])	([0.269, 0.486], [0.071, 0.179], [0.03, 0.098])	([0.431, 0.592], [0.061, 0.137], [0.017, 0.076])

Table 9. Aggregated weights at period t_2.

Criteria	Importance Aggregated Weights
C_1	([0.999, 1], [0, 0.003], [0, 0.001])
C_2	([0.997, 1], [0.001, 0.006], [0, 0.002])
C_3	([0.985, 0.998], [0.003, 0.014], [0, 0.004])
C_4	([0.978, 0.997], [0.003, 0.016], [0, 0.005])
C_5	([0.959, 0.993], [0.002, 0.011], [0.001, 0.015])
C_6	([0.999, 1], [0, 0.003], [0, 0.001])
C_7	([0.993, 0.999], [0.002, 0.009], [0, 0.002])
C_8	([0.975, 0.997], [0.004, 0.019], [0, 0.005])
C_9	([0.975, 0.997], [0.004, 0.019], [0, 0.005])
C_{10}	([0.996, 1], [0.001, 0.006], [0, 0.002])
C_{11}	([0.998, 1], [0.001, 0.005], [0, 0.001])
C_{12}	([0.963, 0.996], [0.022, 0.06], [0.004, 0.027])
C_{13}	([0.977, 0.998], [0.016, 0.044], [0.005, 0.02])
C_{14}	([0.897, 0.973], [0.05, 0.11], [0.009, 0.056])

Table 10. Weighted ratings at period t_2.

Students	Weighted Ratings
A_1	([0.605, 0.76], [0.004, 0.02], [0.001, 0.009])
A_2	([0.594, 0.761], [0.004, 0.02], [0.001, 0.009])
A_3	([0.581, 0.744], [0.004, 0.021], [0.001, 0.009])
A_4	([0.458, 0.588], [0.022, 0.058], [0.011, 0.031])

Table 11. The distance of each student from $A_{t_2}^+$ and $A_{t_2}^-$ at period t_2.

Students	$d_{t_2}^+$	$d_{t_2}^-$
A_1	0.188874	0.901553
A_2	0.192392	0.900405
A_3	0.200641	0.896588
A_4	0.279475	0.848118

Table 12. Closeness coefficient at period t_2.

Students	Closeness Coefficients	Ranking Order
A_1	0.826789	1
A_2	0.823945	2
A_3	0.817138	3
A_4	0.752149	4

(3) Period t_3 (the third semester): At this stage, a new student A_5 is considered and an existing student A_2 is discarded. The criteria remain the same as in the previous period t_2. Tables 13–17 show the steps of this stage. Finally, the ranking is obtained: $A_5 \succ A_4 \succ A_2 \succ A_1 \succ A_3$. Thus, the best student is A_5.

5.3. Comparison with the Related Methods

The proposed dynamic TOPSIS method has superior features compared to the method in [14]. In Table 18, the ranking order of five students in three periods are presented. We can observe that at period t_1, the results of the both methods are the same i.e., $A_1 \succ A_2 \succ A_3$.

Table 13. Aggregated ratings at period t_3.

Criteria	Students				
	A_1	A_2	A_3	A_4	A_5
C_1	([0.794, 0.9], [0, 0], [0, 0])	([0.51, 0.75], [0, 0.006], [0, 0.002])	([0.764, 0.893], [0, 0], [0, 0])	([0.711, 0.822], [0, 0.003], [0, 0.001])	([0.441, 0.569], [0.022, 0.053], [0.012, 0.027])
C_2	([0.871, 0.951], [0, 0], [0, 0])	([0.675, 0.818], [0, 0.002], [0, 0.001])	([0.881, 0.96], [0, 0], [0, 0])	([0.788, 0.891], [0, 0.001], [0, 0])	([0.441, 0.569], [0.022, 0.053], [0.012, 0.027])
C_3	([0.829, 0.918], [0, 0.001], [0, 0])	([0.608, 0.785], [0.001, 0.005], [0, 0.001])	([0.728, 0.884], [0, 0.001], [0, 0])	([0.817, 0.91], [0, 0.001], [0, 0])	([0.569, 0.67], [0.005, 0.016], [0.004, 0.012])
C_4	([0.711, 0.875], [0, 0.001], [0, 0])	([0.608, 0.785], [0.001, 0.005], [0, 0.001])	([0.816, 0.922], [0, 0.001], [0, 0])	([0.663, 0.795], [0, 0.002], [0, 0.001])	([0.569, 0.67], [0.005, 0.016], [0.004, 0.012])
C_5	([0.81, 0.912], [0, 0.001], [0, 0.001])	([0.635, 0.804], [0, 0.003], [0, 0.001])	([0.777, 0.889], [0, 0.001], [0, 0.001])	([0.751, 0.872], [0, 0.001], [0, 0.001])	([0.48, 0.608], [0.011, 0.032], [0.008, 0.021])
C_6	([0.832, 0.923], [0, 0], [0, 0])	([0.608, 0.785], [0, 0.004], [0, 0.001])	([0.744, 0.902], [0, 0], [0, 0])	([0.482, 0.69], [0.001, 0.006], [0.001, 0.003])	([0.536, 0.637], [0.011, 0.026], [0.006, 0.016])
C_7	([0.689, 0.86], [0, 0.001], [0, 0])	([0.591, 0.759], [0.001, 0.004], [0, 0.002])	([0.682, 0.86], [0, 0.001], [0, 0])	([0.586, 0.733], [0.001, 0.004], [0.001, 0.002])	([0.441, 0.569], [0.022, 0.053], [0.012, 0.027])
C_8	([0.751, 0.898], [0, 0.001], [0, 0])	([0.662, 0.822], [0, 0.003], [0, 0.001])	([0.732, 0.89], [0, 0.001], [0, 0])	([0.699, 0.83], [0, 0.004], [0, 0.001])	([0.268, 0.441], [0.027, 0.079], [0.033, 0.062])
C_9	([0.874, 0.95], [0, 0.002], [0, 0])	([0.749, 0.861], [0, 0.002], [0, 0.001])	([0.889, 0.963], [0, 0.002], [0, 0])	([0.743, 0.853], [0, 0.003], [0, 0.001])	([0.418, 0.578], [0.011, 0.039], [0.011, 0.026])
C_{10}	([0.757, 0.891], [0, 0], [0, 0])	([0.636, 0.804], [0, 0.003], [0, 0.001])	([0.837, 0.926], [0, 0], [0, 0])	([0.712, 0.818], [0, 0.002], [0, 0.001])	([0.5, 0.6], [0.022, 0.044], [0.009, 0.022])
C_{11}	([0.753, 0.88], [0, 0], [0, 0])	([0.521, 0.71], [0.001, 0.005], [0.001, 0.003])	([0.696, 0.875], [0, 0.001], [0, 0])	([0.651, 0.769], [0.001, 0.004], [0, 0.002])	([0.569, 0.67], [0.005, 0.015], [0.004, 0.012])
C_{12}	([0.753, 0.884], [0.001, 0.007], [0, 0.002])	([0.53, 0.662], [0.002, 0.011], [0.001, 0.007])	([0.778, 0.903], [0.001, 0.007], [0, 0.002])	([0.544, 0.72], [0.002, 0.013], [0.001, 0.005])	([0.534, 0.636], [0.012, 0.032], [0.006, 0.018])
C_{13}	([0.677, 0.845], [0, 0.002], [0, 0.001])	([0.338, 0.521], [0.001, 0.006], [0.003, 0.009])	([0.759, 0.881], [0, 0.002], [0, 0])	([0.699, 0.83], [0.001, 0.004], [0, 0.001])	([0.374, 0.536], [0.022, 0.065], [0.016, 0.035])
C_{14}	([0.688, 0.837], [0.001, 0.005], [0, 0.001])	([0.407, 0.555], [0.002, 0.008], [0.002, 0.008])	([0.777, 0.916], [0.001, 0.005], [0, 0.001])	([0.699, 0.826], [0.001, 0.007], [0, 0.002])	([0.44, 0.569], [0.023, 0.057], [0.012, 0.029])

Table 14. Aggregated weights at period t_3.

Criteria	Importance Aggregated Weights
C_1	([0.99999, 1], [0, 0.00009], [0, 0.00001])
C_2	([0.99995, 1], [0.00001, 0.00019], [0, 0.00002])
C_3	([0.99964, 1], [0.00005, 0.00062], [0, 0.00009])
C_4	([0.99912, 0.99998], [0.00009, 0.00097], [0, 0.00018])
C_5	([0.99776, 0.99995], [0.00004, 0.00077], [0.00001, 0.00053])
C_6	([0.99999, 1], [0, 0.00006], [0, 0])
C_7	([0.99985, 1], [0.00003, 0.00039], [0, 0.00005])
C_8	([0.99907, 0.99999], [0.00009, 0.00114], [0, 0.00015])
C_9	([0.99842, 0.99996], [0.00014, 0.00154], [0, 0.00024])
C_{10}	([0.99991, 1], [0.00002, 0.00029], [0, 0.00004])
C_{11}	([0.99997, 1], [0.00001, 0.00016], [0, 0.00001])
C_{12}	([0.99615, 0.99988], [0.00112, 0.00657], [0.00004, 0.00152])
C_{13}	([0.99969, 1], [0.00016, 0.00145], [0.00001, 0.00026])
C_{14}	([0.99762, 0.99993], [0.00082, 0.00483], [0.00004, 0.00116])

Table 15. Weighted ratings at period t_3.

Students	Weighted Ratings
A_1	([0.78, 0.901], [0, 0.001], [0, 0])
A_2	([0.589, 0.759], [0.001, 0.004], [0, 0.002])
A_3	([0.785, 0.91], [0, 0.001], [0, 0])
A_4	([0.693, 0.822], [0, 0.003], [0, 0.001])
A_5	([0.476, 0.599], [0.014, 0.037], [0.009, 0.022])

Table 16. The distance of each student from $A^+_{t_3}$ and $A^-_{t_3}$ at period t_3.

Students	$d^+_{t_3}$	$d^-_{t_3}$
A_1	0.37844	0.776416
A_2	0.352522	0.752181
A_3	0.381797	0.777005
A_4	0.358066	0.764391
A_5	0.325366	0.738391

Table 17. Closeness coefficient at period t_3.

Students	Closeness Coefficients	Ranking Order
A_1	0.672305	4
A_2	0.680890	3
A_3	0.670525	5
A_4	0.680998	2
A_5	0.694135	1

Table 18. The dynamic rankings obtained at periods.

Time Period	The Method in [14]	The Proposed Method
t_1	$A_1 \succ A_2 \succ A_3$	$A_1 \succ A_2 \succ A_3$
t_2	$A_4 \succ A_2 \succ A_3 \succ A_1$	$A_1 \succ A_2 \succ A_3 \succ A_4$
t_3	$A_5 \succ A_3 \succ A_4 \succ A_1$	$A_5 \succ A_4 \succ A_2 \succ A_1 \succ A_3$

At period t_2, the method in [14] and the proposed method show difference in ranking order of A_1 and A_4. In this period, $A_2 \succ A_3$ according to both methods, and the method in [14] ranks A_4 at the top, meanwhile, A_1 is ranked at the top by the proposed method. The reason is that A_4 is evaluated at the first time and it has not appeared while A_1 has historical data, particularly A_1 were ranked at the top in the previous period. In this circumstance, the proposed model better utilizes the effect of historical data on the alternatives A_1 and A_4. The result of the dynamic TOPSIS model is time-dependent and combines the effect of current and historical data.

At period t_3, the result shows difference in the number of ranked alternatives and in their preferential order. In this period, the alternative A_2 is not evaluated by decision-makers and it has only historical data. The method in [14] could not process alternative A_2, meanwhile the proposed model could. Moreover, the alternative A_5 is highly ranked by both methods. However, there is a change in the relative order of A_3 and A_4. The method in [14] ranks $A_3 \succ A_4$, but the proposed method ranks $A_4 \succ A_3$.

The comparison between the methods again illustrates the effect of historical data over the output of the proposed decision-making model. If an alternative is considered and it has good evaluation in the previous periods, this alternative will have high potential to reach high order. From that point of view, the proposed model presents good compliance with the perceived dynamic rules. It illustrates the advantage and applicability of the model.

6. Conclusions

The proposed dynamic TOPSIS (DTOPSIS) model in dynamic interval-valued neutrosophic sets presents its advantages to cope with dynamic and indeterminate information in decision-making model. DTOPSIS model handles historical data including the change of criteria, alternatives, and decision-makers during periods. The concepts of generalized dynamic interval-valued neutrosophic set, GDIVNS, and their mathematical operators on GDIVNSs have been proposed. Distance and weighted aggregation operators are used to construct a framework of DTOPSIS in DIVNS environment. The proposed DTOPSIS fulfills the requirement of an issue that is evaluates tertiary students' performance based on the attributes of ASK model. Data of Thuongmai University students were used to illustrate the proper of DTOPSIS model which opened the potential application in larger scale also. For the future works, we will extend generalized dynamic interval-valued neutrosophic sets for some other real-world applications [27–35]. Furthermore, we hope to apply GDIVNS for dealing with the unlimited time problems in decision-making model in dynamic neutrosophic environment based on the idea in [36,37].

Author Contributions: Conceptualization, N.T.T. and N.D.H.; methodology, N.T.T. and L.T.H.L.; software, C.N.G. and D.T.S.; validation, L.T.H.L. and H.V.L.; formal analysis, L.H.S. and F.S.; investigation, L.T.H.L.; data curation, D.T.S.; writing—original draft preparation, N.T.T.; writing—review and editing, L.H.S., H.V.L., and F.S.; visualization, C.N.G. and D.T.S.; supervision, L.T.S. and H.V.L. All authors have read and agreed to the published version of the manuscript.

Funding: This research is funded by the Ministry of Education and Training and Thuongmai University under grant number B2019-TMA-02.

Acknowledgments: The authors would like to thank the Editor-in-Chief and the anonymous reviewers for their valuable comments and suggestions. The authors are grateful for the support from Thuongmai University to have the necessary data to implement the proposed approach.

Conflicts of Interest: The authors declare no conflict of interest.

References

1. Campanella, G.; Ribeiro, R.A. A framework for dynamic multiple-criteria decision making. *Decis. Support Syst.* **2011**, *52*, 52–60. [CrossRef]
2. Smarandache, F. *Neutrosophy: Neutrosophic Probability, Set, and Logic: Analytic Synthesis & Synthetic Analysis*; American Research Press: Santa Fe, NM, USA, 1998.
3. Abdel-Basset, M.; Manogaran, G.; Gamal, A.; Smarandache, F. A hybrid approach of neutrosophic sets and DEMATEL method for developing supplier selection criteria. *Des. Autom. Embed. Syst.* **2018**, *22*, 257–278. [CrossRef]
4. Deli, I.; Broumi, S.; Smarandache, F. On neutrosophic refined sets and their applications in medical diagnosis. *J. New Theory* **2015**, *6*, 88–98.
5. Peng, J.-J.; Wang, J.-Q.; Wang, J.; Zhang, H.-Y.; Chen, X.-H. Simplified neutrosophic sets and their applications in multi-criteria group decision-making problems. *Int. J. Syst. Sci.* **2015**, *47*, 1–17. [CrossRef]
6. Şahin, R.; Liu, P. Possibility-induced simplified neutrosophic aggregation operators and their application to multi-criteria group decision-making. *J. Exp. Theor. Artif. Intell.* **2016**, *29*, 1–17. [CrossRef]
7. Ye, J. Multicriteria decision-making method using the correlation coefficient under single-valued neutrosophic environment. *Int. J. Gen. Syst.* **2013**, *42*, 386–394. [CrossRef]
8. Ye, J. Single-Valued Neutrosophic Minimum Spanning Tree and Its Clustering Method. *J. Intell. Syst.* **2014**, *23*, 311–324. [CrossRef]
9. Ye, J. Similarity measures between interval neutrosophic sets and their applications in multicriteria decision-making. *J. Intell. Fuzzy Syst.* **2014**, *26*, 165–172. [CrossRef]
10. Ye, J. Hesitant interval neutrosophic linguistic set and its application in multiple attribute decision making. *Int. J. Mach. Learn. Cybern.* **2017**, *10*, 667–678. [CrossRef]
11. Bloom, B.S.; Davis, A.; Hess, R.O.B.E.R.T. *Com-Pensatory Education for Cultural Depri-Vation*; Holt, Rinehart, &Winston: New York, NY, USA, 1965.
12. Huang, Y.-H.; Wei, G.; Wei, C. VIKOR Method for Interval Neutrosophic Multiple Attribute Group Decision-Making. *Information* **2017**, *8*, 144. [CrossRef]
13. Liu, P.; Shi, L. The generalized hybrid weighted average operator based on interval neutrosophic hesitant set and its application to multiple attribute decision making. *Neural Comput. Appl.* **2014**, *26*, 457–471. [CrossRef]
14. Thong, N.T.; Dat, L.Q.; Son, L.H.; Hoa, N.D.; Ali, M.; Smarandache, F. Dynamic interval valued neutrosophic set: Modeling decision making in dynamic environments. *Comput. Ind.* **2019**, *108*, 45–52. [CrossRef]
15. Wang, L.; Zhang, H.-Y.; Wang, J.-Q. Frank Choquet Bonferroni Mean Operators of Bipolar Neutrosophic Sets and Their Application to Multi-criteria Decision-Making Problems. *Int. J. Fuzzy Syst.* **2017**, *20*, 13–28. [CrossRef]

16. Wang, H.; Smarandache, F.; Sunderraman, R.; Zhang, Y.Q. *Interval Neutrosophic Sets and Logic: Theory and Applications in Computing: Theory and Applications in Computing*; Infinite Study; Hexis: Phoenix, AZ, USA, 2005; Volume 5.
17. Lupiáñez, F.G. Interval neutrosophic sets and topology. *Kybernetes* **2009**, *38*, 621–624. [CrossRef]
18. Bera, T.; Mahapatra, N.K. Distance Measure Based MADM Strategy with Interval Trapezoidal Neutrosophic Numbers. *Neutrosophic Sets Syst.* **2018**, *20*, 7.
19. Torra, V.; Narukawa, Y. On hesitant fuzzy sets and decision. In Proceedings of the 2009 IEEE International Conference on Fuzzy Systems, Jeju Island, Korea, 20–24 August 2009; Institute of Electrical and Electronics Engineers (IEEE); pp. 1378–1382.
20. Torra, V. Hesitant fuzzy sets. *Int. J. Intell. Syst.* **2010**, *25*, 529–539. [CrossRef]
21. Rubb, S. Overeducation, undereducation and asymmetric information in occupational mobility. *Appl. Econ.* **2013**, *45*, 741–751. [CrossRef]
22. Bakarman, A.A. Attitude, Skill, and Knowledge: (ASK) a New Model for Design Education. In Proceedings of the Canadian Engineering Education Association (CEEA), Queen's University Library, Kingston, ON, Canada, 6–8 June 2011.
23. Bell, R. Unpacking the link between entrepreneurialism and employability. *Educ. Train.* **2016**, *58*, 2–17. [CrossRef]
24. Cross, N. Expertise in design: An overview. *Des. Stud.* **2004**, *25*, 427–441. [CrossRef]
25. Lewis, W.; Bonollo, E. An analysis of professional skills in design: Implications for education and research. *Des. Stud.* **2002**, *23*, 385–406. [CrossRef]
26. Son, N.T.K.; Dong, N.P.; Long, H.V.; Son, L.H.; Khastan, A. Linear quadratic regulator problem governed by granular neutrosophic fractional differential equations. *ISA Trans.* **2020**, *97*, 296–316. [CrossRef] [PubMed]
27. Thong, N.T.; Giap, C.N.; Tuan, T.M.; Chuan, P.M.; Hoang, P.M. Modeling multi-criteria decision-making in dynamic neutrosophic environments bases on Choquet integral. *J. Comput. Sci. Cybern.* **2020**, *36*, 33–47. [CrossRef]
28. Son, N.T.K.; Dong, N.P.; Son, L.H.; Long, H.V. Towards granular calculus of single-valued neutrosophic functions under granular computing. *Multimedia Tools Appl.* **2019**, *78*, 1–37. [CrossRef]
29. Son, N.T.K.; Dong, N.P.; Son, L.H.; Abdel-Basset, M.; Manogaran, G.; Long, H.V. On the Stabilizability for a Class of Linear Time-Invariant Systems under Uncertainty. *Circuits, Syst. Signal Process.* **2019**, *39*, 919–960. [CrossRef]
30. Long, H.V.; Ali, M.; Son, L.H.; Khan, M.; Tu, D.N. A novel approach for fuzzy clustering based on neutrosophic association matrix. *Comput. Ind. Eng.* **2019**, *127*, 687–697. [CrossRef]
31. Jha, S.; Kumar, R.; Son, L.H.; Chatterjee, J.M.; Khari, M.; Yadav, N.; Smarandache, F. Neutrosophic soft set decision making for stock trending analysis. *Evol. Syst.* **2018**, *10*, 621–627. [CrossRef]
32. Ngan, R.T.; Son, L.H.; Cuong, B.C.; Ali, M. H-max distance measure of intuitionistic fuzzy sets in decision making. *Appl. Soft Comput.* **2018**, *69*, 393–425. [CrossRef]
33. Ali, M.; Dat, L.Q.; Son, L.H.; Smarandache, F. Interval Complex Neutrosophic Set: Formulation and Applications in Decision-Making. *Int. J. Fuzzy Syst.* **2017**, *20*, 986–999. [CrossRef]
34. Nguyen, G.N.; Son, L.H.; Ashour, A.S.; Dey, N. A survey of the state-of-the-arts on neutrosophic sets in biomedical diagnoses. *Int. J. Mach. Learn. Cybern.* **2017**, *10*, 1–13. [CrossRef]
35. Thong, N.T.; Lan, L.T.H.; Chou, S.-Y.; Son, L.H.; Dong, D.D.; Ngan, T.T. An Extended TOPSIS Method with Unknown Weight Information in Dynamic Neutrosophic Environment. *Mathematics* **2020**, *8*, 401. [CrossRef]
36. Ali, M.; Son, L.H.; Deli, I.; Tien, N.D. Bipolar neutrosophic soft sets and applications in decision making. *J. Intell. Fuzzy Syst.* **2017**, *33*, 4077–4087. [CrossRef]
37. Alcantud, J.C.R.; Torrecillas, M.M. Intertemporal Choice of Fuzzy Soft Sets. *Symmetry* **2018**, *10*, 371. [CrossRef]

symmetry

MDPI

Article

A Kind of Variation Symmetry: Tarski Associative Groupoids (TA-Groupoids) and Tarski Associative Neutrosophic Extended Triplet Groupoids (TA-NET-Groupoids)

Xiaohong Zhang [1,*], Wangtao Yuan [1], Mingming Chen [1] and Florentin Smarandache [2]

[1] Department of Mathematics, Shaanxi University of Science & Technology, Xi'an 710021, China; 1809007@sust.edu.cn (W.Y.); chenmingming@sust.edu.cn (M.C.)

[2] Department of Mathematics, University of New Mexico, Gallup, NM 87301, USA; smarand@unm.edu

* Correspondence: zhangxiaohong@sust.edu.cn

Received: 21 February 2020; Accepted: 9 April 2020; Published: 2 May 2020

check for updates

Abstract: The associative law reflects symmetry of operation, and other various variation associative laws reflect some generalized symmetries. In this paper, based on numerous literature and related topics such as function equation, non-associative groupoid and non-associative ring, we have introduced a new concept of Tarski associative groupoid (or transposition associative groupoid (TA-groupoid)), presented extensive examples, obtained basic properties and structural characteristics, and discussed the relationships among few non-associative groupoids. Moreover, we proposed a new concept of Tarski associative neutrosophic extended triplet groupoid (TA-NET-groupoid) and analyzed related properties. Finally, the following important result is proved: every TA-NET-groupoid is a disjoint union of some groups which are its subgroups.

Keywords: Tarski associative groupoid (TA-groupoid); TA-NET-groupoid; semigroup; subgroup

---

1. Introduction

Generally, group and semigroup [1–5] are two basic mathematical concepts which describe symmetry. As far as we know the term semigroup was firstly introduced in 1904 in a French book (see book review [1]). A semigroup is called right commutative if it satisfies the identity $a*(x*y) = a*(y*x)$ [4]. When we combine right commutative with associative law, we can get the identity:

$$(x * y) * z = x * (z * y) \text{ (Tarski associative law).}$$

In this study we focused on the non-associative groupoid satisfying Tarski associative law (it is also called transposition associative law), and this kind of groupoid is called Tarski associative groupoid (TA-groupoid). From a purely algebraic point of view, these structures are interesting. They produce innovative ideas and methods that help solve some old algebraic problems.

In order to express the general symmetry and algebraic operation laws which are similar with the associative law, scholars have studied various generalized associative laws. As early as in 1924, Suschkewitsch [6] studied the following generalized associative law (originally called "Postulate A"):

$$(x * a) * b = x * c,$$

where the element c depended upon the element a and b only, and not upon x. Apparently, the associative law is a special case of this Postulate A when $c = a * b$, and Tarski associative law explained above is also a special case of this Postulate A when $c = b * a$. This fact shows that Tarski associative groupoid (TA-groupoid) studied in our research is a natural generalization of the semigroup. At the same time, Hosszu studied the function equations satisfying Tarski associative law in 1954 (see [7–9]); Thedy [10] studied rings satisfying $x(yz) = (yx)z$, and it is symmetric to Tarski associative groupoid, since defining $x*y = yx$, $x(yz) = (yx)z$ is changed to $(z*y)*x = z*(x*y)$; Phillips (see the Table 12 in [11]) and Pushkashu [12] also referred to Tarski associative law. These facts show that the systematic study of Tarski associative groupoid (TA-groupoid) is helpful to promote the study of non-associative rings and other non-associative algebraic systems.

In recent years, a variety of non-associative groupoids have been studied in depth (it should be noted that the term "groupoid" has many different meanings, such as the concept in category theory and algebraic topology, see [13]). An algebraic structure midway between a groupoid and a commutative semigroup appeared in 1972, Kazim and Naseeruddin [14] introduced the concept of left almost semigroup (LA-semigroup) as a generalization of commutative semigroup and it is also called Abel-Grassmann's groupoid (or simply AG-groupoid). Many different aspects of AG-groupoids have been studied in [15–22]. Moreover, Mushtaq and Kamran [19] in 1989 introduced the notion of AG*-groupoids: one AG-groupoid $(S, *)$ is called AG*-groupoid if it satisfies

$$(x * y) * z = y * (x * z), \text{ for any } x, y, z \in S.$$

Obviously, when we reverse the above equation, we can get $(z*x)*y = z*(y*x)$, which is the Tarski associative law (transposition associative law). In [23], a new kind of non-associative groupoid (cyclic associative groupoid, shortly, CA-groupoid) is proposed, and some interesting results are presented.

Moreover, this paper also involves with the algebraic system "neutrosophic extended triplet group", which has been widely studied in recent years. The concept of neutrosophic extended triplet group (NETG) is presented in [24], and the close relationship between NETGs and regular semigroups has been established [25]. Many other significant results on NETGs and related algebraic systems can be found, see [25,26]. In this study, combining neutrosophic extended triplet groups (NETGs) and Tarski associative groupoids (TA-groupoids), we proposed the concept of Tarski associative neutrosophic extended triplet groupoid (TA-NET-groupoid).

This paper has been arranged as follows. In Section 2, we give some definitions and properties on groupoid, CA-groupoid, AG-groupoid and NETG. In Section 3, we propose the notion of Tarski associative groupoid (TA-groupoid), and show some examples. In Section 4, we study its basic properties, and, moreover, analyze the relationships among some related algebraic systems. In Section 5, we introduce the new concept of Tarski associative NET-groupoid (TA-NET-groupoid) and weak commutative TA-NET-groupoid (WC-TA-NET-Groupoid), investigate basic properties of TA-NET-groupoids and weak commutative TA-NET-groupoids (WC-TA-NET-Groupoids). In Section 6, we prove a decomposition theorem of TA-NET-groupoid. Finally, Section 7 presents the summary and plans for future work.

2. Preliminaries

In this section, some notions and results about groupoids, AG-groupoids, CA-groupoids and neutrosophic triplet groups are given. A groupoid is a pair $(S, *)$ where S is a non-empty set with a binary operation $*$. Traditionally, when the $*$ operator is omitted, it will not be confused. Suppose $(S, *)$ is a groupoid, we define some concepts as follows:

(1) $\forall a, b, c \in S$, $a*(b*c) = a*(c*b)$, S is called right commutative; if $(a*b)*c = (b*a)*c$, S is called left commutative. When S is right and left commutative, then it is called bi-commutative groupoid.
(2) If $a^2 = a$ $(a \in S)$, the element a is called idempotent.
(3) If for all $x, y \in S$, $a*x = a*y \Rightarrow x = y$ $(x*a = y*a \Rightarrow x = y)$, the element $a \in S$ is left cancellative (respectively right cancellative). If an element is a left and right cancellative, the element is

cancellative. If ($\forall a \in S$) a is left (right) cancellative or cancellative, then S is left (right) cancellative or cancellative.

(4) If $\forall a, b, c \in S$, $a*(b*c) = (a*b)*c$, S is called semigroup. If $\forall a, b \in S$, $a * b = b * a$, then a semigroup $(S, *)$ is commutative.

(5) If $\forall a \in S$, $a^2 = a$, a semigroup $(S, *)$ is called a band.

Definition 1. *([14,15]) Assume that $(S, *)$ is a groupoid. If S satisfying the left invertive law: $\forall$ $a, b, c \in S$, $(a*b)*c = (c*b)*a$. S is called an Abel-Grassmann's groupoid (or simply AG-groupoid).*

Definition 2. *([21,22]) Let $(S, *)$ be an AG-groupoid, for all $a, b, c \in S$.*

(1) If $(a*b)*c = b*(a*c)$, then S is called an AG-groupoid.*
*(2) If $a*(b*c) = b*(a*c)$, then S is called an AG**-groupoid.*
(3) If $a*(b*c) = c*(a*b)$, then S is called a cyclic associative AG-groupoid (or CA-AG-groupoid).

Definition 3. *[23] Let $(S, *)$ be a groupoid. S is called a cyclic associative groupoid (shortly, CA-groupoid), if S satisfying the cyclic associative law:$\forall a, b, c \in S$, $a*(b*c) = c*(a*b)$.*

Proposition 1. *[23] Let $(S, *)$ be a CA-groupoid, then:*

(1) For any $a, b, c, d, x, y \in S$, $(a * b) * (c * d) = (d * a) * (c * b)$;
(2) For any $a, b, c, d, x, y \in S$, $(a * b) * ((c * d) * (x * y)) = (d * a) * ((c * b) * (x * y))$.

Definition 4. *([24,26]) Suppose S be a non-empty set with the binary operation *. If for any $a \in S$, there is a neutral "a" (denote by neut(a)), and the opposite of "a" (denote by anti(a)), such that $neut(a) \in S$, $anti(a) \in S$, and: $a * neut(a) = neut(a) * a = a$; $a * anti(a) = anti(a) * a = neut(a)$. Then, S is called a neutrosophic extended triplet set.*

Note: For any $a \in S$, *neut(a)* and *anti(a)* may not be unique for the neutrosophic extended triplet set $(S, *)$. To avoid ambiguity, we use the symbols $\{neut(a)\}$ and $\{anti(a)\}$ to represent the sets of *neut(a)* and *anti(a)*, respectively.

Definition 5. *([24,26]) Let $(S, *)$ be a neutrosophic extended triplet set. Then, S is called a neutrosophic extended triplet group (NETG), if the following conditions are satisfied:*

(1) $(S, *)$ is well-defined, that is, for any $a, b \in S$, $a * b \in S$.
(2) $(S, *)$ is associative, that is, for any $a, b, c \in S$, $(a * b) * c = a * (b * c)$.

A NETG S is called a commutative NETG if $a * b = b * a$, $\forall a, b \in S$.

Proposition 2. *([25]) Let $(S, *)$ be a NETG. Then ($\forall a \in S$) neut(a) is unique.*

Proposition 3. *([25]) Let $(S, *)$ be a groupoid. Then S is a NETG if and only if it is a completely regular semigroup.*

3. Tarski Associative Groupoids (TA-Groupoids)

Definition 6. *Let $(S, *)$ be a groupoid. S is called a Tarski associative groupoid (shortly, TA-groupoid), if S satisfying the Tarski associative law (it is also called transposition associative law): $(a * b) * c = a * (c * b)$, $\forall$ $a, b, c \in S$.*

The following examples depict the wide existence of TA-groupoids.

Example 1. *For the regular hexagon as shown in Figure 1, denote* $S = \{\theta, G, G^2, G^3, G^4, G^5\}$*, where* G, G^2, G^3, G^4, G^5 *and* θ *represent rotation 60, 120, 180, 240, 300 and 360 degrees clockwise around the center, respectively.*

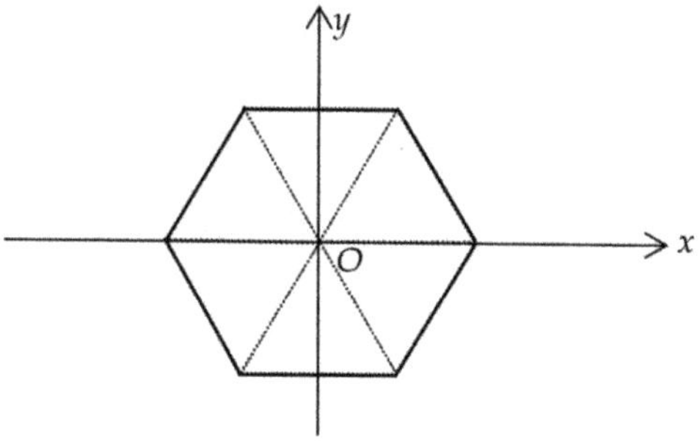

Figure 1. Regular hexagon.

Define the binary operation ∘ as a composition of functions in S, that is, $\forall\, U, V \in S$, $U \circ V$ is that the first transforming V and then transforming U. Then $(S, \circ)$ is a TA-groupoid (see Table 1).

Table 1. Cayley table on $S = \{\theta, G, G^2, G^3, G^4, G^5\}$.

$\circ$	θ	G	G^2	G^3	G^4	G^5
θ	θ	G	G^2	G^3	G^4	G^5
G	G	G^2	G^3	G^4	G^5	θ
G^2	G^2	G^3	G^4	G^5	θ	G
G^3	G^3	G^4	G^5	θ	G	G^2
G^4	G^4	G^5	θ	G	G^2	G^3
G^5	G^5	θ	G	G^2	G^3	G^4

Example 2. *Let* $S = [n, 2n]$ *(real number interval, n is a natural number),* $\forall\, x, y \in S$. *Define the multiplication* $*$ *by*

$$x * y = \begin{cases} x + y - n, & if\ x + y \le 3n \\ x + y - 2n, & if\ x + y > 3n \end{cases}$$

Then $(S, *)$ *is a TA-groupoid, since it satisfies* $(x * y) * z = x * (z * y)$, $\forall\, x, y, z \in S$, *the proof is as follows:*

Case 1: $x + y + z - n \le 3n$. *It follows that* $y + z \le x + y + z - n \le 3n$ *and* $x + y \le x + y + z - n \le 3n$. *Then* $(x * y)*z = (x + y - n) *z = x + y + z - 2n = x * (z + y - n) = x * (z * y)$.

Case 2: $x + y + z - n > 3n$, $y + z \le 3n$ *and* $x + y \le 3n$. *Then*$(x * y)*z = (x + y - n)*c = x + y + z - 3n = x * (z + y - n) = x * (z * y)$.

Case 3: $x + y + z - n > 3n$, $y + z \le 3n$ *and* $x + y > 3n$. *It follows that* $x + y + z - 2n \le x + 3n - 2n = x + n \le 3n$. *Then* $(x * y) *z = (x + y - 2n) *c = x + y + z - 3n = x * (z + y - n) = x * (z * y)$.

Case 4: $x + y + z - n > 3n$, $y + z > 3n$ *and* $x + y \le 3n$. *It follows that* $x + y + z - 2n \le 3n + c - 2n = z + n \le 3n$. *Then* $(x * y) *z = (x + y - n) *z = x + y + z - 3n = x * (z + y - 2n) = x * (z * y)$.

Case 5: $x + y + z - n > 3n$, $y + z > 3n$ *and* $x + y > 3n$. *When* $x + y + c - 2n \le 3n$, $(x * y) *z = (x + y - 2n) *z = x + y + z - 3n = x * (z + y - 2n) = x * (z * y)$; *When* $x + y + z - 2n > 3n$, $(x * y) *z = (x + y - 2n)*z = x + y + z - 4n = x * (z + y - 2n) = x * (z * y)$.

Example 3. *Let*

$$S=\left\{\begin{pmatrix} x & 0 \\ 0 & 0 \end{pmatrix}: x \text{ is a integralnumber}\right\}\cup\left\{\begin{pmatrix} 1 & 0 \\ 0 & 1 \end{pmatrix}, \begin{pmatrix} 1 & 0 \\ 0 & -1 \end{pmatrix}\right\}.$$

Denote $S_1=\left\{\begin{pmatrix} a & 0 \\ 0 & 0 \end{pmatrix}: a \text{ isaintegralnumber}\right\}$, $S_2=\left\{\begin{pmatrix} 1 & 0 \\ 0 & 1 \end{pmatrix}, \begin{pmatrix} 1 & 0 \\ 0 & -1 \end{pmatrix}\right\}$. *Define the operation* * *on* S: $\forall$ $x, y \in S$, *(1) if* $x \in S_1$ *or* $y \in S_1$, x^*y *is common matrix multiplication; (2) if* $x \in S_2$ *and* $y \in S_2$, $x^*y = \begin{pmatrix} 1 & 0 \\ 0 & 1 \end{pmatrix}$. *Then* $(S, {}^*)$ *is a TA-groupoid. In fact, we can verify that* $(x^*y)^*z = x^*(z^*y)$ $\forall$ $x, y, z \in S$, *since*

(i) *if* $x, y, z \in S_1$, *by the definition of operation* * *we can get* $(x^*y)^*z = x^*(y^*z) = x^*(z^*y)$;

(ii) *if* $x, y, z \in S_2$, *then* $(x^*y)^*z = \begin{pmatrix} 1 & 0 \\ 0 & 1 \end{pmatrix} = x^*(z^*y)$, *by (2) in the definition of operation* *;

(iii) *if* $x \in S_2$, $y, z \in S_1$, *then* $(x^*y)^*z = y^*z = z^*y = x^*(z^*y)$, *by (1) in the definition of operation* *;

(iv) *if* $x \in S_2$, $y \in S_2$, $z \in S_1$, *then* $(x^*y)^*z = \begin{pmatrix} 1 & 0 \\ 0 & 1 \end{pmatrix}^*z = z = z^*y = x^*(z^*y)$, *by the definition of operation* *;

(v) *if* $x \in S_2$, $y \in S_1$, $z \in S_2$, *then* $(x^*y)^*z = y^*z = y = z^*y = x^*(z^*y)$, *by the definition of operation* *;

(vi) *if* $x \in S_1$, $y \in S_2$, $z \in S_1$, *then* $(x^*y)^*z = x^*z = x^*(z^*y)$, *by (1) in the definition of operation* *;

(vii) *if* $x \in S_1$, $y \in S_1$, $z \in S_2$, *then* $(x^*y)^*z = x^*y = x^*(z^*y)$, *by (1) in the definition of operation* *;

(vii) *if* $x \in S_1$, $y \in S_2$, $z \in S_2$, *then* $(x^*y)^*z = x^*z = x = x^*\begin{pmatrix} 1 & 0 \\ 0 & 1 \end{pmatrix} = x^*(z^*y)$, *by (1) and (2) in the definition of operation* *.

Example 4. *Table 2 shows the non-commutative TA-groupoid of order 5. Since* $(b * a) * b \neq b * (a * b)$, $(a * b) * b \neq (b * b) * a$, *so* $(S, *)$ *is not a semigroup, and it is not an AG-groupoid.*

Table 2. Cayley table on $S = \{a, b, c, d, e\}$.

*	a	b	c	d	e
a	a	a	a	a	a
b	d	d	c	c	b
c	d	c	c	c	c
d	d	d	c	c	c
e	d	c	c	c	e

From the following example, we know that there exists TA-groupoid which is a non- commutative semigroup, moreover, we can generate some semirings from a TA-groupoid.

Example 5. *As shown in Table 3, put* $S = \{s, t, u, v, w\}$, *and define the operations* * *on S. Then we can verify through MATLAB that* $(S, *)$ *is a TA-groupoid, and* $(S, *)$ *is a semigroup.*

Table 3. Cayley table on $S = \{s, t, u, v, w\}$.

*	s	t	u	v	w
s	s	s	s	s	s
t	t	t	t	t	t
u	s	s	u	u	s
v	s	s	u	v	s
w	t	t	w	w	t

*Now, define the operation + on S as Table 4 (or Table 5), then (∀m, n, p∈S) (m + n) * p = m * p + n * p and (S; +, *) is a semiring (see* [27]*).*

Table 4. A Commutative semigroup (*S*, +).

+	*s*	*t*	*u*	*v*	*w*
s	*s*	*t*	*u*	*u*	*w*
t	*t*	*s*	*w*	*w*	*u*
u	*u*	*w*	*u*	*u*	*w*
v	*u*	*w*	*u*	*u*	*w*
w	*w*	*u*	*w*	*w*	*u*

Table 5. Another commutative semigroup (*S*, +) with unit *s*.

+	*s*	*t*	*u*	*v*	*w*
s	*s*	*t*	*u*	*v*	*w*
t	*t*	*t*	*w*	*w*	*w*
u	*u*	*w*	*u*	*u*	*w*
v	*v*	*w*	*u*	*u*	*w*
w	*w*	*w*	*w*	*w*	*w*

Proposition 4. *(1) If (S, *) is a commutative semigroup, then (S, *) is a TA-groupoid. (2) Let (S, *) be a commutative TA-groupoid. Then (S, *) is a commutative semigroup.*

Proof. It is easy to verify from the definitions. □

4. Various Properties of Tarski Associative Groupoids (TA-Groupoids)

In this section, we discussed the basic properties of TA-groupoids, gave some typical examples, and established its relationships with CA-AG-groupoids and semigroups (see Figure 2). Furthermore, we discussed the cancellative and direct product properties that are important for exploring the structure of TA-groupoids.

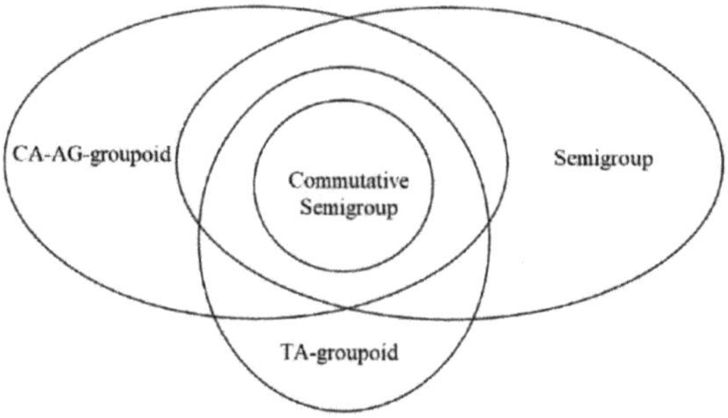

Figure 2. The relationships among some algebraic systems.

Proposition 5. *Let (S, *) be a TA-groupoid. Then ∀ m, n, p, r, s, t ∈S:*

(1) *(m*n) *(p*r) = (m*r)*(p*n);*

(2) *((m*n)*(p*r))*(s*t) = (m*r)*((s*t)*(p*n)).*

Proof. (1) Assume that $(S, *)$ is a TA-groupoid, then for any $m, n, p, r \in S$, by Definition 6, we have

$$(m * n) * (p * r) = m * ((p * r) * n) = m * (p * (n * r)) = (m * (n * r)) * p = ((m * r) * n) * p = (m * r) * (p * n).$$

(2) For any $m, n, p, r, s, t \in S$, by Definition 6, we have

$$((m * n) * (p * r)) * (s * t) = (m * n) * ((s * t) * (p * r)) = (m * n) * ((s * r) * (p * t)) = ((m * n) * (p * t)) * (s * r)$$
$$= ((m*n)*r)*(s* (p*t)) = ((m*n) *r)*((s*t) *p) = ((m*n) *p)*((s*t) *r)$$
$$= (m*(p*n))*((s*t)*r) = (m*r)*((s*t)*(p*n)).$$ □

Theorem 1. *Assume that $(S, *)$ is a TA-groupoid.*

(1) *If $\exists e \in S$ such that $(\forall a \in S)$ $e*a=a$, then $(S, *)$ is a commutative semigroup.*
(2) *If $e \in S$ is a left identity element in S, then e is an identity element in S.*
(3) *If S is a right commutative CA-groupoid, then S is an AG-groupoid.*
(4) *If S is a right commutative CA-groupoid, then S is a left commutative CA-groupoid.*
(5) *If S is a left commutative CA-groupoid, then S is a right commutative CA-groupoid.*
(6) *If S is a left commutative CA-groupoid, then S is an AG-groupoid.*
(7) *If S is a left commutative semigroup, then S is a CA-groupoid.*

Proof. It is easy to verify from the definitions, and the proof is omitted. □

From the following example, we know that a right identity element in S may be not an identity element in S.

Example 6. *TA-groupoid of order 6 is given in Table 6, and e_6 is a right identity element in S, but e_6 is not a left identity element in S.*

Table 6. Cayley table on $S = \{e_1, e_2, e_3, e_4, e_5, e_6\}$.

*	e_1	e_2	e_3	e_4	e_5	e_6
e_1	e_1	e_1	e_1	e_1	e_1	e_1
e_2	e_2	e_2	e_2	e_2	e_2	e_2
e_3	e_1	e_1	e_4	e_6	e_1	e_3
e_4	e_1	e_1	e_6	e_3	e_1	e_4
e_5	e_2	e_2	e_5	e_5	e_2	e_5
e_6	e_1	e_1	e_3	e_4	e_1	e_6

By Theorem 1 (1) and (2) we know that the left identity element in a TA-groupoid is unique. But the following example shows that the right identity element in a TA-groupoid may be not unique.

Example 7. *The following non-commutative TA-groupoid of order 5 given in Table 7. Moreover, x_1 and x_2 are right identity elements in S.*

Table 7. Cayley table on $S = \{x_1, x_2, x_3, x_4, x_5\}$.

*	x_1	x_2	x_3	x_4	x_5
x_1	x_1	x_1	x_3	x_3	x_5
x_2	x_2	x_2	x_4	x_4	x_5
x_3	x_3	x_3	x_1	x_1	x_5
x_4	x_4	x_4	x_2	x_2	x_5
x_5	x_5	x_5	x_5	x_5	x_5

Theorem 2. *Let (S, *) be a TA-groupoid.*

(1) *If S is a left commutative AG-groupoid, then S is a CA-groupoid.*
(2) *If S is a left commutative AG-groupoid, then S is a right commutative TA-groupoid.*
(3) *If S is a right commutative AG-groupoid, then S is a left commutative TA-groupoid*
(4) *If S is a right commutative AG-groupoid, then S is a CA-groupoid.*
(5) *If S is a left commutative semigroup, then S is an AG-groupoid.*

Proof. It is easy to verify from the definitions, and the proof is omitted. □

Theorem 3. *Let (S, *) be a groupoid.*

(1) *If S is a CA-AG-groupoid and a semigroup, then S is a TA-groupoid.*
(2) *If S is a CA-AG-groupoid and a TA-groupoid, then S is a semigroup.*
(3) *If S is a semigroup, TA-groupoid and CA-groupoid, then S is an AG-groupoid.*
(4) *If S is a semigroup, TA-groupoid and AG-groupoid, S is a CA-groupoid.*

Proof. (1) If $(S, *)$ is a CA-AG-groupoid and a semigroup, then by Definition 2, $\forall\, a, b, c \in S$:

$$b * (c * a) = c * (a * b) = (c * a) * b = (b * a) * c.$$

It follows that $(S, *)$ is a TA-groupoid by Definition 6.
(2) Assume that $(S, *)$ is a CA-AG-groupoid and a TA-groupoid, by Definition 2, $\forall\, a, b, c \in S$:

$$a * (b * c) = c * (a * b) = (c * b) * a = (a * b) * c.$$

This means that $(S, *)$ is a semigroup.
(3) Assume that $(S, *)$ is a semigroup, TA-groupoid and CA-groupoid. Then, we have ($\forall\, a, b, c \in S$):

$$(a * b) * c = a * (b * c) = c * (a * b) = (c * b) * a.$$

Thus, $(S, *)$ is an AG-groupoid.
(4) Suppose that $(S, *)$ is a semigroup, TA-groupoid and AG-groupoid. $\forall\, a, b, c \in S$:

$$c * (b * a) = (c * b) * a = (a * b) * c = a * (c * b).$$

That is, $(S, *)$ is a CA-groupoid by Definition 3. □

Example 8. *Put $S = \{e, f, g, h, i\}$. The operation * is defined on S in Table 8. We can get that $(S, *)$ is a CA-AG-groupoid. But $(S, *)$ is not a TA-groupoid, due to the fact that $(i * h) * i \neq i * (i * h)$. Moreover, $(S, *)$ is not a semigroup, because $(i * i) * i \neq i * (i * i)$.*

Table 8. Cayley table on $S = \{e, f, g, h, i\}$.

*	e	f	g	h	i
e	e	e	e	e	e
f	e	e	e	e	e
g	e	e	e	e	f
h	e	e	e	e	f
i	e	e	e	g	h

From Proposition 4, Theorems 1–3, Examples 4–5 and Example 8, we get the relationships among TA-groupoids and its closely linked algebraic systems, as shown in Figure 2.

Theorem 4. *Let (S, *) be a TA-groupoid.*

(1) *Every left cancellative element in S is right cancellative element;*
(2) *if* $x, y \in S$ *and they are left cancellative elements, then* x^*y *is a left cancellative element;*
(3) *if x is left cancellative and y is right cancellative, then* x^*y *is left cancellative;*
(4) *if* x^*y *is right cancellative, then y is right cancellative;*
(5) *If for all* $a \in S$, $a^2 = a$, *then it is associative. That is, S is a band.*

Proof. (1) Suppose that x is a left cancellative element in S. If $(\forall p, q \in S)\ p^*x = q^*x$, then:

$$\begin{aligned} x^*(x^*(x^*p)) &= (x^*(x^*p))^*x = ((x^*p)^*x)^*x = (x^*p)^*(x^*x) \\ &= x^*((x^*x)^*p) = x^*(x^*(p^*x)) = x^*(x^*(q^*x)) \\ &= x^*((x^*x)^*q) = (x^*q)^*(x^*x) = ((x^*q)^*x)^*x \\ &= (x^*(x^*q))^*x = x^*(x^*(x^*q)). \end{aligned}$$

From this, applying left cancellability, $x^*(x^*p) = x^*(x^*q)$. From this, applying left cancellability two times, we get that $p= q$. Therefore, x is right cancellative.

(2) If x and y are left cancellative, and $(\forall p, q \in S)\ (x^*y)^*p = (x^*y)^*q$, there are:

$$\begin{aligned} x^*(x^*(y^*p)) &= x^*((x^*p)^*y) = (x^*y)^*(x^*p) \\ &= (x^*p)^*(x^*y) \text{ (by Proposition 5 (1))} \\ &= x((x^*y)^*p) = x((xy)^*p) = x((xy)^*q) = x((x^*y)^*q) \\ &= (x^*q)^*(x^*y) = (x^*y)^*(x^*q) = x^*((x^*q)^*y) \\ &= x\ ^*(x^*(y^*q)). \end{aligned}$$

Applying the left cancellation property of x, we have $y^*p = y^*q$. Moreover, since y is left cancellative, we can get that $p = q$. Therefore, x^*y is left cancellative.

(3) Suppose that x is left cancellative and y is right cancellative. If $(\forall p, q \in S)\ (x^*y)^*p = (x^*y)^*q$, there are:

$$x^*(p^*y) = (x^*y)^*p = (x^*y)^*q = x^*(q^*y).$$

Applying the left cancellation property of x, we have $p^*y = q^*y$. Moreover, since y is right cancellative, we can get that $p = q$. Therefore, x^*y is left cancellative.

(4) If x^*y is right cancellative, and $p^*y = q^*y$, $p, q \in S$, there are:

$$p^*(x^*y) = (p^*y)^*x = (q^*y)^*x = q^*(x^*y).$$

Applying the right cancellation property of x^*y, we have $p = q$. Hence, we get that y is right cancellative. □

(5) Assume that for all $a \in S$, $a^2=a$. Then, $\forall\ r, s, t \in S$,

$$\begin{aligned} r^*(s^*t) &= (r^*(s^*t))^*(r^*(s^*t)) = r^*((r^*(s^*t))^*(s^*t)) \\ &= r^*(r^*((s^*t)^*(s^*t))) = r^*(r^*(s^*t)). \end{aligned} \quad (1)$$

Similarly, according to (1) we can get $r^*(t^*s) = r^*(r^*(t^*s))$. And, by Proposition 5 (1), we have

$$\begin{aligned} r^*(r^*(s^*t)) &= r^*((r^*t)^*s) = (r^*s)^*(r^*t) = (r^*t)^*(r^*s) \\ &= r^*((r^*s)^*t) = r^*(r^*(t^*s)). \end{aligned}$$

Combining the results above, we get that $r^*(s^*t) = r^*(r^*(s^*t)) = r^*(r^*(t^*s)) = r^*(t^*s)$. Moreover, by Definition 6, $(r^*s)^*t = r^*(t^*s)$. Thus

$$(r^*s)^*t = r^*(t^*s) = r^*(s^*t).$$

This means that S is a semigroup, and for all $a \in S$, $a^2 = a$.

Therefore, we get that S is a band. □

Example 9. *TA-groupoid of order 4, given in Table 9. It is easy to verify that (S, *) is a band, due to the fact that* $x * x = x$, $y * y = y$, $z * z = z$, $u * u = u$.

Table 9. Cayley table on $S = \{x, y, z, u\}$.

*	x	y	z	u
x	x	x	x	x
y	y	y	z	y
z	u	u	z	u
u	u	u	u	u

Definition 7. *Assume that* $(S_1, *_1)$ *and* $(S_2, *_2)$ *are TA-groupoids,* $S_1 \times S_2 = \{(a, b) | a \in S_1, b \in S_2\}$. *Define the operation * on* $S_1 \times S_2$ *as follows:*

$$(a_1, a_2) * (b_1, b_2) = (a_1 *_1 b_1, a_2 *_2 b_2), \text{ for any } (a_1, a_2), (b_1, b_2) \in S_1 \times S_2.$$

Then $(S_1 \times S_2, *)$ *is called the direct product of* $(S_1, *_1)$ *and* $(S_2, *_2)$.

Theorem 5. *If* $(S_1, *_1)$ *and* $(S_2, *_2)$ *are TA-groupoids, then their direct product* $(S_1 \times S_2, *)$ *is a TA-groupoid.*

Proof. Assume that $(a_1, a_2), (b_1, b_2), (c_1, c_2) \in S_1 \times S_2$. Since

$$\begin{aligned} ((a_1, a_2) * ((b_1, b_2)) * (c_1, c_2) &= (a_1 *_1 b_1, a_2 *_2 b_2) * (c_1, c_2) \\ = ((a_1 *_1 b_1) *_1 c_1, (a_2 *_2 b_2) *_2 c_2) &= (a_1 *_1 (c_1 *_1 b_1), a_2 *_2 (c_2 *_2 b_2)) \\ = (a_1, a_2) * (c_1 *_1 b_1, c_2 *_2 b_2) &= (a_1, a_2) * ((c_1, c_2) * (b_1, b_2)). \end{aligned}$$

Hence, $(S_1 \times S_2, *)$ is a TA-groupoid. □

Theorem 6. *Let* $(S_1, *_1)$ *and* $(S_2, *_2)$ *be two TA-groupoids, if x and y are cancellative* $(x \in S_1, y \in S_2)$, *then* $(x, y) \in S_1 \times S_2$ *is cancellative.*

Proof. Using Theorem 5, we can get that $S_1 \times S_2$ is a TA-groupoid. Moreover, for any $(s_1, s_2), (t_1, t_2) \in S_1 \times S_2$, if $(x, y) * (s_1, s_2) = (x, y) * (t_1, t_2)$, there are:

$$(xs_1, ys_2) = (xt_1, yt_2)$$
$$xs_1 = xt_1, ys_2 = yt_2.$$

Since x and y are cancellative, so $s_1 = t_1$, $s_2 = t_2$, and $(s_1, s_2) = (t_1, t_2)$.

Therefore, (x, y) is cancellative. □

5. Tarski Associative Neutrosophic Extended Triplet Groupoids (TA-NET-Groupoids) and Weak Commutative TA-NET-Groupoids (WC-TA-NET-Groupoids)

In this section, we first propose a new concept of TA-NET-groupoids and discuss its basic properties. Next, this section will discuss an important kind of TA-NET-groupoids, called weak commutative TA-NET-groupoids (WC-TA-NET-groupoids). In particular, we proved some well-known properties of WC-TA-NET-groupoids.

Definition 8. *Let (S, *) be a neutrosophic extended triplet set. If*

(1) *(S, *) is well-defined, that is, (∀x, y∈S) x*y ∈ S;*

(2) *(S, *) is Tarski associative, that is, for any x, y, z ∈S (x*y)*z = x*(z*y).*

*Then (S, *) is called a Tarski associative neutrosophic extended triplet groupoid (or TA-NET-groupoid). A TA-NET-groupoid (S, *) is called to be commutative, if (∀ x, y ∈ S) x*y = y*x.*

According to the definition of the TA-NET-groupoid, element *a* may have multiple neutral elements *neut(a)*. We tried using the MATLAB math tools to find an example showing that an element's neutral element is not unique. Unfortunately, we did not find this example. This leads us to consider another possibility: every element in a TA-NET-groupoid has a unique neutral element? Fortunately, we successfully proved that this conjecture is correct.

Theorem 7. *Let (S, *) be a TA-NET-groupoid. Then the local unit element neut(a) is unique in S.*

Proof. For any $a \in S$, if there exists $s, t \in \{neut(a)\}$, then $\exists\, m, n \in S$ there are:

$$a * s = s * a = a \text{ and } a * m = m * a = s; a * t = t * a = a \text{ and } a * n = n * a = t.$$

(1) $s = t * s$. Since

$$s = a * m = (t * a) * m = t * (m * a) = t * s.$$

(2) $t = t * s$. Since

$$t = n * a = n * (s * a) = (n * a) * s = t * s.$$

Hence $s = t$ and $neut(a)$ is unique for any $a \in S$. □

Remark 1. *For element a in TA-NET-groupoid (S, *), although neut(a) is unique, we know from Example 10 that anti(a) may be not unique.*

Example 10. *TA-NET-groupoid of order 6, given in Table 10. While neut(Δ) = Δ, {anti(Δ)} = {Δ, Γ, I, ϑ, K}.*

Table 10. Cayley table on $S = \{\Delta, \Gamma, I, \vartheta, K, \Lambda\}$.

*	Δ	Γ	I	ϑ	K	Λ
Δ	Δ	Δ	Δ	Δ	Δ	Δ
Γ	Δ	Γ	I	ϑ	K	Δ
I	Δ	I	K	Γ	ϑ	Δ
ϑ	Δ	ϑ	Γ	K	I	Δ
K	Δ	K	ϑ	I	Γ	Δ
Λ	Λ	Λ	Λ	Λ	Λ	Λ

Theorem 8. *Let (S, *) be a TA-NET-groupoid. Then ∀ x∈S:*

(1) $neut(x) * neut(x) = neut(x)$;

(2) $neut(neut(x)) = neut(x)$;

(3) $anti(neut(x)) \in \{anti(neut(x))\}$, $x = anti(neut(x)) * x$.

Proof. (1) For any $x \in S$, according to $x * anti(x) = anti(x) * x = neut(x)$, we have

$$neut(x)*neut(x) = neut(x)*((anti(x)*x) = (neut(x)*x)*anti(x) = x*(anti(x)) = neut(x).$$

(2) $\forall x \in S$, by the definition of $neut(neut(x))$, there are:

$$neut(neut(x))*neut(x) = neut(x)*neut(neut(x)) = neut(x).$$

Thus,

$$neut(neut(x))*x = neut(neut(x))*(x*neut(x)) = (neut(neut(x))*neut(x))*x = neut(x)*x = x;$$
$$x*neut(neut(x)) = (x*neut(x))*neut(neut(x)) = x*(neut(neut(x))*neut(x)) = x*neut(x) = x.$$

Moreover, we can get:

$$anti(neut(x))*neut(x) = neut(x)*anti(neut(x)) = neut(neut(x)).$$

Then,

$$(anti(neut(x))*anti(x))*x = anti(neut(x))*(x*anti(x)) = anti(neut(x))*neut(x) = neut(neut(x));$$
$$x*(anti(neut(x))*anti(x)) = (x*anti(x))*anti(neut(x)) = neut(x)*anti(neut(x)) = neut(neut(x)).$$

Combining the results above, we get

$$neut(neut(x))*x = x*neut(neut(x)) = x;$$
$$(anti(neut(x))*anti(x))*x = x*(anti(neut(x))*anti(x)) = neut(neut(x)).$$

This means that $neut(neut(x))$ is a neutral element of x (see Definition 4). Applying Theorem 6, we get that $neut(neut(x)) = neut(x)$.

(3) For all $x \in S$, using Definition 8 and above (2),

$$anti(neut(x))*x = anti(neut(x))*(x*(neut(x))) = (anti(neut(x))*neut(x))*x$$
$$= neut(neut(x))*x = neut(x)*x = x.$$

Thus, $anti(neut(x))*x = x$. □

Example 11. *TA-NET-groupoid of order 4, given in Table 11. And* $neut(\alpha) = \alpha$, $neut(\beta) = \beta$, $neut(\delta) = \delta$, $\{anti(\alpha)\} = \{\alpha, \delta, \varepsilon\}$. *While* $anti(\alpha) = \delta$, $neut(anti(\alpha)) = neut(\delta) = \delta \neq \alpha = neut(\alpha)$.

Table 11. Cayley table on $S = \{\alpha, \beta, \delta, \varepsilon\}$.

*	α	β	δ	ε
α	α	α	α	α
β	β	β	β	β
δ	α	α	δ	δ
ε	α	α	δ	ε

Theorem 9. *Let* $(S, *)$ *be a TA-NET-groupoid. Then* $\forall x \in S$, $\forall m, n \in \{anti(a)\}$, $\forall anti(a) \in \{anti(a)\}$:

(1) $m*(neut(x)) = neut(x)*n$;

(2) $anti(neut(x))*anti(x) \in \{anti(x)\}$;

(3) $neut(x)*anti(n) = x*neut(n)$;

(4) $neut(m)*neut(x) = neut(x)*neut(m) = neut(x)$;

(5) $(n*(neut(x))*x = x*(neut(x)*n) = neut(x)$;

(6) $neut(n)*x = x$.

Proof. (1) By the definition of neutral and opposite element (see Definition 4), applying Theorem 6, there are:

(2) By Theorem 7(2), there are:

$$m^*x = x^*m = neut(x),\ n^*x = x^*n = neut(x).$$
$$m^*(neut(x)) = m^*(n^*x) = (m^*x)^*n = neut(x)^*n.$$
$$x^*[anti(neut(x))^*anti(x)] = [x^*(anti(x))]^*anti(neut(x)) = neut(x)^*anti(neut(x)) = neut(neut(x)) = neut(x).$$
$$[anti(neut(x))^*anti(x)]^*x = anti(neut(x))^*[x^*(anti(x)] = anti(neut(x))^*neut(x) = neut(neut(x)) = neut(x).$$

Thus, anti(neut(x))*anti(x)∈{anti(x)}.

(3) For any $x \in S$, $n \in \{anti(a)\}$, by $x^*n = n^*x = neut(x)$ and $n^*anti(n) = anti(n)^*n = neut(n)$, we get

$$x^*neut(n) = x^*[anti(n)^*n] = (x^*n)^*anti(n) = neut(x)^*anti(n).$$

This shows that $neut(x)^*anti(n) = x^*neut(n)$.

(4) For any $x \in S$, $m \in \{anti(x)\}$, by $x^*m = m^*x = neut(x)$ and $anti(m)^*m = m^*anti(m) = neut(m)$, there are:

$$neut(m)^*neut(x) = neut(m)^*(x^*m) = (neut(m)^*m)^*x = m^*x = neut(x).$$
$$neut(x)^*neut(m) = neut(x)^*[m^*(anti(m))] = [neut(x)^*anti(m)]^*m.$$

Applying (3), there are:

$$neut(x)^*neut(m) = [neut(x)^*anti(m)]^*m = [x^*(neut(m))]^*m = x^*(m^*(neut(m)) = x^*m = neut(x).$$

That is,

$$neut(m)^*neut(x) = neut(x)^*neut(m) = neut(x).$$

(5) By $x^*n = n^*x = neut(x)$, there are:

$$[n^*(neut(x))]^*x = n^*(x^*(neut(x))) = n^*x = neut(x).$$
$$x^*[neut(x)^*n] = (x^*n)^*(neut(x)) = neut(x)^*neut(x) = neut(x).$$

Thus, $[n^*(neut(x))]^*x = x^*[neut(x)^*n] = neut(x)$.

(6) For any $x \in S$, $n \in \{anti(x)\}$, by $x^*n = n^*x = neut(x)$,

$$neut(n)^*x = neut(n)^*[x^*(neut(x))] = [neut(n)^*neut(x)]^*x.$$

From this, applying (4), there are:

$$neut(n)^*x = [neut(n)^*neut(x)]^*x = neut(x)^*x = x.$$

Hence, $neut(n)^*x = x$. □

Proposition 6. *Let (S, *) be a TA-NET-groupoid. Then* $\forall\, x, y, z \in S$:

(1) $y^*x = z^*x$, *implies* $neut(x)^*y = neut(x)^*z$;
(2) $y^*x = z^*x$, *if and only if* $y^*neut(x) = z^*neut(x)$.

Proof. (1) For any $x, y \in S$, if $y^*x = z^*x$, then $anti(x)^*(y^*x) = anti(x)^*(z^*x)$. By Definition 6 and Definition 8 there are:

$$anti(x)^*(y^*x) = (anti(x)^*x)^*y = neut(x)^*y;$$
$$anti(x)^*(z^*x) = (anti(x)^*x)^*z = neut(x)^*z.$$

Thus $neut(x)^*y = anti(x)^*(y^*x) = anti(x)^*(z^*x) = neut(x)^*z$.

(2) For any $x, y \in S$, if $y^*x = z^*x$, then $(y^*x)^*anti(x) = (z^*x)^*anti(x)$. Since

$$(y^*x)^*anti(x) = y^*(anti(x)^*x) = y^*neut(x);\ (z^*x)^*anti(x) = z^*(anti(x)^*x) = z^*neut(x).$$

It follows that $y^*neut(x) = z^*neut(x)$. This means that $y^*x = z^*x$ implies $y^*neut(x) = z^*neut(x)$. Conversely, if $y^*neut(x) = z^*neut(x)$, then $(y^*neut(x))^*x = (z^*neut(x))^*x$. Since

$$(y^*neut(x))^*x = y^*(x^*neut(x)) = y^*x;\ (z^*neut(x))^*x = z^*(x^*neut(x)) = z^*x.$$

Thus, $y^*x = z^*x$. Hence, $y^*neut(x) = z^*neut(x)$ implies $y^*x = z^*x$. □

Proposition 7. *Suppose that (S, *) is a commutative TA-NET-groupoid.* $\forall x, y \in S$:

(1) $neut(x) * neut(y) = neut(x * y)$;
(2) $anti(x) * anti(y) \in \{anti(x * y)\}$.

Proof. (1) For any $x, y \in S$, since S is commutative, so $x * y = y * x$. From this, by Proposition 5(1), we have

$$(x^*y)^*(neut(x)^*neut(y)) = (y^*x)^*(neut(x)^*neut(y)) = (y^*neut(y))^*(neut(x)^*x) = y^*x = x^*y;$$
$$(neut(x)^*neut(y))^*(x^*y) = (neut(x)^*neut(y))^*(y^*x) = (neut(x)^*x)^*(y^*(neut(y)) = x^*y.$$

Moreover, using Proposition 5(1),

$$(anti(x)^*anti(y))^*(x^*y) = (anti(x)^*anti(y))^*(y^*x) = (anti(x)^*x)^*(y^*anti(y)) = neut(x)^*neut(y);$$
$$(x^*y)^*(anti(x)^*anti(y)) = (x^*y)^*(anti(y)^*anti(x)) = (x^*anti(x))^*(anti(y)^*y) = neut(x)^*neut(y).$$

This means that $neut(x)^*neut(y)$ is a neutral element of x^*y (see Definition 4). Applying Theorem 6, we get that $neut(x)^*neut(y) = neut(x^*y)$.

(2) For any $anti(x) \in \{anti(x)\}$, $anti(y) \in \{anti(y)\}$, by the proof of (1) above,

$$(anti(x)^*anti(y))^*(x^*y) = (x^*y)^*(anti(x)^*anti(y)) = neut(x)^*neut(y).$$

From this and applying (1), there are:

$$(anti(x)^*anti(y))^*(x^*y) = (x^*y)^*(anti(x)^*anti(y)) = neut(x^*y).$$

Hence, $anti(x)^*anti(y) \in \{anti(x^*y)\}$. □

Definition 9. *Let (S, *) be a TA-NET-groupoid. If* $(\forall x, y \in S)\ x * neut(y) = neut(y) * x$, *then we said that S is a weak commutative TA-NET-groupoid (or WC-TA-NET-groupoid).*

Proposition 8. *Let (S, *) be a TA-NET-groupoid. Then (S, *) is weak commutative* ⇔ *S satisfies the following conditions* $(\forall x, y \in S)$:

(1) $neut(x)^*neut(y) = neut(y)^*neut(x)$.
(2) $neut(x)^*(neut(y)^*x) = neut(x)^*(x^*neut(y))$.

Proof. Assume that $(S, *)$ is a weak commutative TA-NET-groupoid, using Definition 9, there are $(\forall x, y \in S)$:

$$neut(x)*neut(y) = neut(y)*neut(x),$$
$$neut(x)*(neut(y)*x) = neut(x)*(x*neut(y)).$$

In contrast, suppose that S satisfies the above conditions (1) and (2). there are$(\forall x, y \in S)$:

$$x*neut(y) = (neut(x)*x)*neut(y) = neut(x)*(neut(y)*x) = neut(x)*(x*neut(y))= (neut(x)*neut(y))*x =(neut(y)*neut(x))*x = neut(y)*(x*neut(x)) = neut(y)*x.$$

From Definition 9 and this we can get that $(S, *)$ is a weak commutative TA-NET-groupoid. □

Theorem 10. *Assume that $(S, *)$ is a weak commutative TA-NET-groupoid. Then $\forall$ $x, y \in S$:*

(1) $neut(x)*neut(y) = neut(y*x)$;
(2) $anti(x)*anti(y) \in \{anti(y*x)\}$;
(3) *(S is commutative)* $\Leftrightarrow$ *(S is weak commutative).*

Proof. (1) By Proposition 5 (1)), there are:

$$[neut(x)*neut(y)]*(y*x) = [neut(x)*x]*[y*neut(y)] = [neut(x)*x]*[neut(y)*y] = [neut(x)*y]*[neut(y)*x] = [y*neut(x)]*[x*neut(y)] = [y*neut(y)]*[x*neut(x)] = y*x.$$

And, $(y*x)*[neut(x)*neut(y)] = [y*neut(y)]*[neut(x)*x] = y*x$. That is,

$$[neut(x)*neut(y)]*(y*x) = (y*x)*[neut(x)*neut(y)] = y*x.$$

And that, there are:

$$[anti(x)*anti(y)]*(y*x) = [anti(x)*x]*[y*anti(y)] = neut(x)*neut(y);$$
$$(y*x)*[anti(x)*anti(y)] = [y*anti(y)] * [anti(x)*x] = neut(y)*neut(x) = neut(x)*neut(y).$$

That is,

$$[anti(x)*anti(y)]*(y*x) = (y*x)*[anti(x)*anti(y)] = neut(x)*neut(y).$$

Thus, combining the results above, we know that $neut(x)*neut(y)$ is a neutral element of $y*x$. Applying Theorem 6, we get $neut(x)*neut(y) = neut(y*x)$.

(2) Using (1) and the following result (see the proof of (1))

$$[anti(x)*anti(y)]*(y*x) = (y*x)*[anti(x)*anti(y)] = neut(x)*neut(y)$$

we can get that $anti(x)*anti(y) \in \{anti(y*x)\}$.

(3) If S is commutative, then S is weak commutative.

On the other hand, suppose that S is a TA-NET-groupoid and S is weak commutative. By Proposition 5 (1) and Definition 9, there are:

$$x*y = (x*neut(x))*(y*neut(y)) = (x*neut(y))*(y*neut(x)) = (neut(y)*x)*(neut(x)*y) = (neut(y)*y)*(neut(x)*x) = y*x.$$

Therefore, S is a commutative TA-NET-groupoid. □

6. Decomposition Theorem of TA-NET-Groupoids

This section generalizes the well-known Clifford's theorem in semigroup to TA-NET-groupoid, which is very exciting.

Theorem 11. *Let (S, *) be a TA-NET-groupoid. Then for any* $x \in S$, *and all* $m \in \{anti(a)\}$:

(1) $neut(x)*m \in \{anti(x)\}$;

(2) $m*neut(x) = (neut(x)*m)*neut(x)$;

(3) $neut(x)*m = (neut(x)*m)*neut(x)$;

(4) $m*neut(x) = neut(x)*m$;

(5) $neut(m*(neut(x))) = neut(x)$.

Proof. (1) For any $x \in S$, $m \in \{anti(x)\}$, we have $m*x = x*m = neut(x)$. Then, by Definition 6, Theorem 7 (1) and Proposition 5 (1), there are:

$$x*[neut(x)*m] = (x*m)*neut(x) = neut(x)*neut(x) = neut(x);$$
$$[neut(x)*m]*x = [neut(x)*m]*[x*neut(x)] = [neut(x)*neut(x)]*(x*m) = [neut(x)*neut(x)]*neut(x) = neut(x).$$

This means that $neut(x)*m \in \{anti(x)\}$.

(2) If $x \in S$, $m \in \{anti(x)\}$, then $m*x = x*m = neut(x)$. Applying (1) and Theorem 8 (1),

$$m*neut(x) = neut(x)*[neut(x)*m].$$

On the other hand, using Theorem 7 (1) and Proposition 5 (1), there are:

$$neut(x)*[neut(x)*m] = (neut(x)*neut(x))*[neut(x)*m] = [neut(x)*m]*[neut(x)*neut(x)] = [neut(x)*m]*neut(x).$$

Combining two equations above, we get $m*neut(x) = (neut(x)*m)*neut(x)$.

(3) Assume that $m \in \{anti(x)\}$, then $x*m = m*x = neut(x)$ and $m*neut(m) = neut(m)*m = m$. By Theorem 7 (1), Proposition 5 (1) and Theorem 8 (4), there are:

$$neut(x)*m = [neut(x)*neut(x)]*(neut(m)*m) = (neut(x)*m)[neut(m)*neut(x)] = (neut(x)*m)*neut(x).$$

That is, $neut(x)*m = (neut(x)*m)*neut(x)$.

(4) It follows from (2) and (3).

(5) Assume $m \in \{anti(x)\}$, then $x*m = m*x = neut(x)$. Denote $t = m*neut(x)$. We prove the following equations,

$$t*neut(x) = neut(x)*t = t;\ t*x = x*t = neut(x).$$

By (3) and (4), there are:

$$t*neut(x) = (m*neut(x))*neut(x) = (neut(x)*m)*neut(x) = neut(x)*m = m*neut(x) = t.$$

Using Definition 6, Theorem 7 (1) and Theorem 8 (1), there are:

$$neut(x)*t = neut(x)*[m*(neut(x))] = (neut(x)*neut(x))*m = neut(x)*m = m*neut(x) = t.$$

Moreover, applying Proposition 5 (1), Theorem 7 (1) and Definition 6, there are:

$$t^*x = [m^*(neut(x))]^*x = [m^*neut(x)] * (neut(x)^*x) = (m^*x)^*[neut(x)^*neut(x)]$$
$$= neut(x)^*[neut(x)^*neut(x)] = neut(x).$$
$$x^*t = x^*[m^*(neut(x))] = [x^*neut(x)]^*m = x^*m = neut(x).$$

Thus,

$$t^*neut(x) = neut(x)^*t = t;\ t^*x = x^*t = neut(x).$$

By the definition of neutral element and Theorem 6, we get that $neut(x)$ is the neutral element of $t = m^*neut(x)$. This means that $neut(m^*(neut(x))) = neut(x)$. □

Theorem 12. *Let (S, *) be a TA-NET-groupoid. Then the product of idempotents is still idempotent. That is for any* $y_1, y_2 \in S$, $(y_1 * y_2) * (y_1 * y_2) = y_1 * y_2$.

Proof. Assume that $y_1, y_2 \in S$ and ($y_1{}^*y_1 = y_1$, $y_2{}^*y_2 = y_2$), then:

$$(y_1{}^*y_2)^*(y_1{}^*y_2) = y_1{}^*[(y_1{}^*y_2)^*y_2] = y_1{}^*[y_1{}^*(y_2{}^*y_2)] = y_1{}^*(y_1{}^*y_2).$$

From this, applying Definition 4 and Definition 6,

$$y_1{}^*y_2 = [neut(y_1{}^*y_2)]^*(y_1{}^*y_2) = [anti(y_1{}^*y_2)^*(y_1{}^*y_2)]^*(y_1{}^*y_2) = anti(y_1{}^*y_2)^*[(y_1{}^*y_2)^*(y_1{}^*y_2)]$$
$$= anti(y_1{}^*y_2)^*[y_1{}^*(y_1{}^*y_2)] \text{ (By } (y_1{}^*y_2)^*(y_1{}^*y_2) = y_1{}^*(y_1{}^*y_2))$$
$$= [anti(y_1{}^*y_2)^*(y_1{}^*y_2)]^*y_1 = neut(y_1{}^*y_2)^*y_1.$$

Thus,

$$(y_1{}^*y_2)^*(y_1{}^*y_2) = y_1{}^*(y_1{}^*y_2) = (y_1{}^*y_2)^*y_1$$
$$= [neut(y_1{}^*y_2)^*y_1]^*y_1 \text{ (By } y_1{}^*y_2 = [neut(y_1{}^*y_2)]^*y_1)$$
$$= neut(y_1{}^*y_2)^*(y_1{}^*y_1) = neut(y_1{}^*y_2)^*y_1 = y_1{}^*y_2.$$

This means that the product of idempotents is still idempotent. □

Example 12. TA-NET-groupoid of order 4, given in Table 12, and the product of any two idempotent elements is still idempotent, due to the fact that,

$$(z_1{}^*\, z_2)^*(z_1{}^*\, z_2) = z_1{}^*\, z_2,\ (z_1{}^*\, z_3)^*(z_1{}^*\, z_3) = z_1{}^*\, z_3,\ (z_1{}^*\, z_4) * (z_1{}^*\, z_4) = z_1{}^*\, z_4,$$
$$(z_2{}^*\, z_3)^*(z_2{}^*\, z_3) = z_2{}^*\, z_3,\ (z_2{}^*\, z_4)^*(z_2{}^*\, z_4) = z_2{}^*\, z_4,\ (z_3{}^*\, z_4)^*(z_3{}^*\, z_4) = z_3{}^*\, z_4.$$

Table 12. Cayley table on $S = \{z_1, z_2, z_3, z_4\}$.

*	z_1	z_2	z_3	z_4
z_1	z_1	z_1	z_1	z_4
z_2	z_2	z_2	z_2	z_4
z_3	z_1	z_1	z_3	z_4
z_4	z_4	z_4	z_4	z_4

Theorem 13. *Let (S, *) be a TA-NET-groupoid. Denote E(S) be the set of all different neutral element in S,* $S(e) = \{a \in S \mid neut(a) = e\}$ $(\forall\ e \in E(S))$. *Then:*

(1) S(e) is a subgroup of S.
(2) for any $e_1, e_2 \in E(S)$, $e_1 \neq e_2 \Rightarrow S(e_1) \cap S(e_2) = \emptyset$.
(3) $S = \cup_{e \in E(S)} S(e)$.

Proof. (1) For any $m \in S(e)$, $neut(m) = e$. That is, e is an identity element in $S(e)$. And, using Theorem 7 (1), we get $e * e = e$.

Assume that $m, n \in S(e)$, then $neut(m) = neut(n) = e$. We're going to prove that $neut(m*n) = e$. Applying Definition 6, Proposition 5 (1),

$$(m*n)*e = m*(e*n) = m*n;$$
$$e*(m*n) = (e*e)*(m*n) = (e*n)*(m*e) = (e*n)*m$$
$$= (e*n)*(e*m) = (e*m)*(e*n) = m*n.$$

On the other hand, for any $anti(m) \in \{anti(m)\}$, $anti(n) \in \{anti(n)\}$, by Proposition 5 (1), we have

$$(m*n)*[anti(m)*anti(n)] = (m*anti(n))*(anti(m)*n) = [(m*anti(n))*n]*anti(m)$$
$$= [m*(n*anti(n))]*anti(m) = (m*neut(n))*anti(m) = (m*e)*anti(m)$$
$$= m*anti(m) = neut(m) = e.$$
$$[anti(m)*anti(n)]*(m*n) = [anti(m)*n]*[m*anti(n)] = anti(m)*[(m*anti(n))*n]$$
$$= anti(m)* [m*(n*anti(n))] = anti(m)*(m*neut(n)) = anti(m)*(m*e)$$
$$= anti(m)*m = neut(m) = e.$$

From this, using Theorem 6 and Definition 4, we know that $neut(m*n) = e$. Therefore, $m*n \in S(e)$, i.e., $(S(e), *)$ is a sub groupoid.

Moreover, $\forall\ m \in S(e)$, $\exists q \in S$ such that $q \in \{anti(m)\}$. Applying Theorem 10 (1)(2)(3), $q*neut(m) \in \{anti(m)\}$; and applying Theorem 10 (5), $neut(q*neut(m)) = neut(m)$.

Put $t = q*neut(m)$, we get

$$t = q*neut(m) \in \{anti(m)\},$$
$$neut(t) = neut(q*neut(m)) = neut(m) = e.$$

Thus $t \in \{anti(m)\}$, $neut(t) = e$, i.e., $t \in S(e)$ and t is the inverse element of m in $S(e)$.
Hence, $(S(e), *)$ is a subgroup of S.

(2) Let $x \in S(e_1) \cap S(e_2)$ and $e_1, e_2 \in E(S)$. We have $neut(x) = e_1$, $neut(x) = e_2$. Using Theorem 6, $e_1 = e_2$. Therefore, $e_1 \neq e_2 \Rightarrow S(e_1) \cap S(e_2) = \emptyset$.

(3) For any $x \in S$, there exists $neut(x) \in S$. Denote $e = neut(x)$, then $e \in E(S)$ and $x \in S(e)$.
This means that $S = \cup_{e \in E(S)} S(e)$. □

Example 13. *Table 13 represents a TA-NET-groupoid of order 5. And,*

$$neut(m_1) = m_4,\ anti(m_1) = m_1;\ neut(m_2) = m_3,\ anti(m_2) = m_2;$$
$$neut(m_3) = m_3,\ anti(m_3) = \{m_3, m_5\};\ neut(m_4) = m_4,\ anti(m_4) = m_4;\ neut(m_5) = m_5,\ anti(m_5) = m_5.$$

Table 13. Cayley table on $S = \{m_1, m_2, m_3, m_4, m_5\}$.

*	m_1	m_2	m_3	m_4	m_5
m_1	m_4	m_4	m_1	m_1	m_1
m_2	m_3	m_3	m_2	m_2	m_2
m_3	m_2	m_2	m_3	m_3	m_3
m_4	m_1	m_1	m_4	m_4	m_4
m_5	m_2	m_2	m_3	m_3	m_5

Denote $S_1 = \{m_1, m_4\}$, $S_2 = \{m_2, m_3\}$, $S_3 = \{m_5\}$, then S_1, S_2 and S_3 are subgroup of S, and $S = S_1 \cup S_2 \cup S_3$, $S_1 \cap S_2 = \emptyset$, $S_1 \cap S_3 = \emptyset$, $S_2 \cap S_3 = \emptyset$.

Example 14. *Table 14 represents a TA-NET-groupoid of order 5. And,*

$$neut(x) = x,\ anti(x) = x;\ neut(y) = y,\ \{anti(y)\} = \{y, v\};$$
$$neut(z) = y,\ \{anti(z)\} = \{z, v\};\ neut(u) = u,\ \{anti(u)\} = \{y, z, u, v\};\ neut(v) = v,\ anti(v) = v.$$

Denote $S_1 = \{x\}, S_2 = \{y, z\}, S_3 = \{u\}, S_4 = \{v\}$, *then* S_1, S_2, S_3 *and* S_4 *are subgroup of* S, *and* $S = S_1 \cup S_2 \cup S_3 \cup S_4$, $S_1 \cap S_2 = \emptyset$, $S_1 \cap S_3 = \emptyset$, $S_1 \cap S_4 = \emptyset$, $S_2 \cap S_3 = \emptyset$, $S_2 \cap S_4 = \emptyset$, $S_3 \cap S_4 = \emptyset$.

Table 14. Cayley table on $S = \{x, y, z, u, v\}$.

*	x	y	z	u	v
x	x	x	x	x	x
y	u	y	z	u	y
z	u	z	y	u	z
u	u	u	u	u	u
v	u	y	z	u	v

Open Problem. *Are there some TA-NET-groupoids which are not semigroups?*

7. Conclusions

In this study, we introduce the new notions of TA-groupoid, TA-NET-groupoid, discuss some fundamental characteristics of TA-groupoids and established their relations with some related algebraic systems (see Figure 2), and prove a decomposition theorem of TA-NET-groupoid (see Theorem 13). Studies have shown that TA-groupoids have important research value, provide methods for studying other non-associated algebraic structures, and provide new ideas for solving algebraic problems. This study obtains some important results:

(1) The concepts of commutative semigroup and commutative TA-groupoid are equivalent.
(2) Every TA-groupoid with left identity element is a monoid.
(3) A TA-groupoid is a band if each element is idempotent (see Theorem 4 and Example 9).
(4) In a Tarski associative neutrosophic extended triplet groupoid (TA-NET-groupoid), the local unit element *neut(a)* is unique (see Theorem 7).
(5) The concepts of commutative TA-groupoid and WC-TA-groupoid are equivalent.
(6) In a TA-NET-groupoid, the product of two idempotent elements is still idempotent (see Theorem 12 and Example 12).
(7) Every TA-NET-groupoid is factorable (see Theorem 13 and Example 13–14).

Those results are of great significance to study the structural characteristics of TA-groupoids and TA-NET-groupoids. As the next research topic, we will study the Green relations on TA-groupoids and some relationships among related algebraic systems (see [23,25,28]).

Author Contributions: X.Z., W.Y. and M.C. initiated the research and wrote the paper, F.S. gave related guidance. All authors have read and agreed to the published version of the manuscript.

Funding: This research was funded by National Natural Science Foundation of China grant number 61976130.

Conflicts of Interest: The authors declare no conflict of interest.

References

1. Dickson, L.E. Book Review: Éléments de la Théorie des Groupes Abstraits. *Bull. Amer. Math. Soc.* **1904**, *11*, 159–162. [CrossRef]
2. Clifford, A.H.; Preston, G.B. *The Algebraic Theory of Semigroups*; American Mathematical Society: Providence, RI, USA, 1961.
3. Holgate, P. Groupoids satisfying a simple invertive law. *Math. Stud.* **1992**, *61*, 101–106.
4. Howie, J.M. *Fundamentals of Semigroup Theory*; Oxford University Press: Oxford, UK, 1995.
5. Akinmoyewa, J.T. A study of some properties of generalized groups. *Octogon* **2009**, *17*, 599–626.
6. Suschkewitsch, A. On a generalization of the associative law. *Transactions of the American Mathematical Society.* **1929**, *31*, 204–214. [CrossRef]
7. Hosszu, M. Some functional equations related with the associative law. *Publ. Math. Debrecen* **1954**, *3*, 205–214.
8. Maksa, G. CM solutions of some functional equations of associative type. *Ann. Univ. Sci. Budapest. Sect. Comput.* **2004**, *24*, 125–132.
9. Schölzel, K.; Tomaschek, J. Power series solutions of Tarski's associativity law and of the cyclic associativity law. *Aequat. Math.* **2016**, *90*, 411–425. [CrossRef]
10. Thedy, A. Ringe mit x(yz) = (yx)z. *Mathematische Zeitschriften.* **1967**, *99*, 400–404. [CrossRef]
11. Phillips, J.D.; Vojtechovský, P. Linear groupoids and the associated wreath products. *J. Symb. Comput.* **2005**, *40*, 1106–1125. [CrossRef]
12. Pushkashu, D.I. Para-associative groupoids. *Quasigroups Relat. Syst.* **2010**, *18*, 187–194.
13. Weinstein, A. Groupoids: Unifying internal and external symmetry. A tour through some examples. *Not. AMS* **1996**, *43*, 744–752.
14. Kazim, M.A.; Naseeruddin, M. On almost semigroups. *Alig. Bull. Math.* **1972**, *2*, 1–7.
15. Shah, M.; Shah, T.; Ali, A. On the cancellativity of AG-groupoids. *Int. Math. Forum* **2011**, *6*, 2187–2194.
16. Stevanovic, N.; Protic, P.V. Some decomposition on Abel-Grassmann's groupoids. *PU. M. A.* **1997**, *8*, 355–366.
17. Shah, M.; Ali, A.; Ahmad, I. On introduction of new classes of AG-groupoids. *Res. J. Recent Sci.* **2013**, *2*, 67–70.
18. Mushtaq, Q.; Khan, M. Direct product of Abel-Grassmann's groupoids. *J. Interdiscip. Math.* **2008**, *11*, 461–467. [CrossRef]
19. Mushtaq, Q.; Kamran, M.S. On LA-semigroups with weak associative law. *Sci. Khyber.* **1989**, *1*, 69–71.
20. Ahmad, I.; Rashad, M.; Shah, M. Some properties of AG*-groupoid. *Res. J. Recent Sci.* **2013**, *2*, 91–93.
21. Iqbal, M.; Ahmad, I.; Shah, M.; Ali, M.I. On cyclic associative Abel-Grassman groupoids. *Br. J. Math. Comput. Sci.* **2016**, *12*, 1–16. [CrossRef]
22. Iqbal, M.; Ahmad, I. On further study of CA-AG-groupoids. *Proc. Pakistan Acad. Sci. A. Phys. Comput. Sci.* **2016**, *53*, 325–337.
23. Zhang, X.H.; Ma, Z.R.; Yuan, W.T. Cyclic associative groupoids (CA-groupoids) and cyclic associative neutrosophic extended triplet groupoids (CA-NET-groupoids). *Neutrosophic Sets Syst.* **2019**, *29*, 19–29.
24. Smarandache, F. *Neutrosophic Perspectives: Triplets, Duplets, Multisets, Hybrid Operators, Modal Logic, Hedge Algebras and Applications*; Pons Publishing House: Brussels, Belgium, 2017.
25. Zhang, X.H.; Wu, X.Y.; Mao, X.Y.; Smarandache, F.; Park, C. On neutrosophic extended triplet groups (loops) and Abel-Grassmann's groupoids (AG-groupoids). *J. Intell. Fuzzy Syst.* **2019**, *37*, 5743–5753. [CrossRef]
26. Smarandache, F.; Ali, M. Neutrosophic triplet group. *Neural Comput. Appl.* **2018**, *29*, 595–601.
27. Golan, J.S. *Semirings and Their Applications*; Springer: Dordrecht, The Netherlands, 1999.
28. Zhang, X.H.; Borzooei, R.A.; Jun, Y.B. Q-filters of quantum B-algebras and basic implication algebras. *Symmetry* **2018**, *10*, 573. [CrossRef]

symmetry

MDPI

Article

Neutrosophic Components Semigroups and Multiset Neutrosophic Components Semigroups

Vasantha W.B. [1], **Ilanthenral Kandasamy** [1,*], **Florentin Smarandache** [2]

[1] School of Computer Science and Engineering, VIT, Vellore, Tamilnadu 632014, India; vasantha.wb@vit.ac.in

[2] Department of Mathematics, University of New Mexico, Gallup Campus, NM 87131, USA; smarand@unm.edu

* Correspondence: ilanthenral.k@vit.ac.in

Received: 15 April 2020; Accepted: 11 May 2020; Published: 16 May 2020

check for updates

Abstract: Neutrosophic components (NC) under addition and product form different algebraic structures over different intervals. In this paper authors for the first time define the usual product and sum operations on NC. Here four different NC are defined using the four different intervals: (0, 1), [0, 1), (0, 1] and [0, 1]. In the neutrosophic components we assume the truth value or the false value or the indeterminate value to be from the intervals (0, 1) or [0, 1) or (0, 1] or [0, 1]. All the operations defined on these neutrosophic components on the four intervals are symmetric. In all the four cases the NC collection happens to be a semigroup under product. All of them are torsion free semigroups or weakly torsion free semigroups. The NC defined on the interval [0, 1) happens to be a group under addition modulo 1. Further it is proved the NC defined on the interval [0, 1) is an infinite commutative ring under addition modulo 1 and usual product with infinite number of zero divisors and the ring has no unit element. We define multiset NC semigroup using the four intervals. Finally, we define n-multiplicity multiset NC semigroup for finite n and these two structures are semigroups under + modulo 1 and $\{M(S), +, \times\}$ and $\{n\text{-}M(S), +, \times\}$ are NC multiset semirings. Several interesting properties are discussed about these structures.

Keywords: neutrosophic components (NC); NC semigroup; multiset NC; n-multiplicity; multiset NC semigroup; special zero divisors; torsion free semigroup; weakly torsion free semigroup; infinite commutative ring; group under addition modulo 1; infinite neutrosophic communicative ring; multiset NC semirings

---

1. Introduction

Semigroups play a vital role in algebraic structures [1–5] and they are applied in several fields and it is a generalization of groups, as all groups are semigroups and not vice versa. Neutrosophic sets proposed by Smarandache in [6] has become an interesting area of major research in recent days both in the area of algebraic structures [7–11] as well as in applications ranging from medical diagnosis to sentiment analysis [12,13]. The study of neutrosophic triplets happens to be a special form of neutrosophic sets. Extensive study in this direction have been carried out by several researchers in [8,14–17]. Here we are interested in the study of neutrosophic components (NC) over the intervals (0, 1), (0, 1], [0, 1) and [0, 1]. So far researchers have studied and applied NC only on the interval [0, 1] though they were basically defined by Smarandache [18] on all intervals. Further they have not studied them under the usual operation + and $\times$. Here we venture to study NC on all the four intervals and obtain several interesting algebraic properties about them.

Smarandache multiset semigroup studied in [19] is different from these semigroups. Further these multiset NC semigroups are also different from multi semigroups in [20] which deals with multi structures on semigroups.

Any algebraic structure becomes more efficient for application only when it enjoys some strong properties. In fact a set endowed with closed associative binary operation happens to be a semigroup. This semigroup structure does not yield many applications like algebraic codes or commutative rings or commutative semirings. Basically to have a vector space one needs at least the basic algebraic structure to be a group under addition. The same is true in case of algebraic codes. However none of the intervals [0, 1] or (0, 1) or (0, 1] can afford to have a group structure under +. One can not imagine of a group structure under product for no inverse element can be got for any element in these intervals. But when we consider the interval [0, 1) we see it is a group under addition modulo 1.

In fact for any collection of NC which are triplets to have a stronger structure than a semigroup we need to have a strong structure on the interval over which it is built. That is why this paper studies the NC on the interval [0, 1). These commutative rings in [0, 1) can be used to built both algebraic codes on the NC for which we basically need these NC to be at least a commutative ring. With this motivation, we have developed this paper.

This paper further proves that multiset NC built on the interval [0, 1) happens to be a commutative semiring paving way to build multiset NC algebraic codes and multiset neutrosophic algebraic codes which can be applied to cryptography with indeterminacy.

The paper is organized as follows. Section one is introductory in nature. Section 2 recalls the basic concepts of partial order, torsion free semigroup and neutrosophic set. Section 3 introduces NC on the four intervals [0, 1], (0,1), [0, 1) and (0, 1] and mainly prove they are infinite NC semigroups which are torsion free. The new notion of weakly torsion free elements in a semigroup is introduced in this paper and it is proved that NC semigroups built on intervals [0, 1] and [0, 1) are weakly torsion free under usual product $\times$. We further prove the NC built using the interval [0, 1) happens to be an infinite order commutative ring with infinite number of zero divisors and it has no unit. In Section 4 we prove multiset NC built using these four intervals are multiset neutrosophic semigroups under usual product $\times$. We prove only in case of [0, 1) the multiset NC is a ring with infinite number of zero divisors and in all the other interval, $M(S)$ is a torsion free or weakly torsion free semigroup under $\times$. Only in case of the interval [0, 1), $M(S)$ is semigroup under modulo addition 1. In Section 5 we define n-multiplicity multiset NC on all the intervals and obtain several interesting properties. Discussions about this study are given in Section 6 and the final section gives conclusions and future research based on their structures.

2. Basic Concepts

In this section we introduce the basic concepts needed to make this paper a self contained one. We first recall the definition of partially ordered set.

Definition 1. *There exist some distinct elements $a, b \in S$ such that $a < b$ or $a > b$, and other distinct elements $b, c \in S$ such that neither $b < c$ nor $b > c$, then we say $(S, <)$ is a partially ordered set. We say $(S, \leq)$ is a totally ordered set if for every pair $a, b, \in S$ we have $a \leq b$ or $b \geq a$.*

The set of integers is a totally ordered set and the power set of a set X; P(X) is only a partially ordered set.

Next we proceed on to define torsion free semigroup.

Definition 2. *A semigroup $\{S, \times\}$ is said to be a torsion free semigroup if for $a, b \in S, a \neq b, a^n \neq b^n$ for any $1 \leq n < \infty$.*

We recall the definition of semiring in the following from [21].

Definition 3. *For a non empty set $S, \{S, +, \times\}$ is defined as a semiring if the following conditions are true*

1. *$\{S, +\}$ is a commutative semigroup with 0 as its additive identity.*
2. *$\{S, \times\}$ is a semigroup.*

3. $a \times (b+c) = a \times b + a \times c$ *for all* $a, b, c, \in S$ *follows distribution law.*

If $\{S, \times\}$ *is a commutative semigroup we call* $\{S, +, \times\}$ *as a commutative semiring.*

For more, see [21].

For example, set of integers under product is a torsion free semigroup. Finally we give the basic definition of neutrosophic set.

Definition 4. *The Neutrosophic components (NC) is a triplet (a, b, c) where a is the truth membership function from the unit interval [0, 1], b is the indeterminacy membership function and c is the falsity membership function all of them are from the unit interval [0, 1].*

For more about Neutrosophic components (NC), sets and their properties please refer [6].

Next we proceed onto define the notion of multiset.

Definition 5. *A neutrosophic multiset is a neutrosophic set where one or more elements are repeated with same neutrosophic components or with different neutrosophic components.*

Example 1. $M = \{a(0.3, 0.4, 0.5), a(0.3, 0.4, 0.5), b(1, 0, 0.2), b(1, 0, 0.2), c(0.7, 1, 0)\}$ *is a neutrosophic multiset. For more refer [18]. However we in this paper use the term multiset NC to denote elements of the form* $\{5(0.3, 0.4, 1), 3(0.6, 0, 1), (0, 0.7, 0.5)\}$ *so 5 is the multiplicity of the NC (0.3, 0.4, 1) and 3 is the multiplicity of the NC (0.6, 0, 1) and 1 is the multiplicity of the NC (0, 0.7, 0.5).*

For more about multisets and multiset graphs [18,22].

3. Neutrosophic Components (NC) Semigroups under Usual Product and Sum

Throughout this section $\{x, y, z\}$ will denote the truth value, indeterminate value, false value where x, y, z belongs to $[0, 1]$, the neutrosophic set. However we define special NC on the intervals (0, 1), (0, 1] and [0, 1). We first prove $S_1 = \{(x, y, z)/x, y, z \in (0, 1)\}$ is a semigroup under product and obtain several interesting properties about NC semigroups using the four intervals (0, 1), (0, 1], [0, 1) and [0, 1].

Example 2. *Let* $a = (0.3, 0.8, 0.5)$ *and* $b = (0.9, 0.2, 0.7)$ *be any two NC in* S_1*. We define product* $a \times b =$ *(0.3, 0.8, 0.5)* $\times$ *(0.9, 0.2, 0.7) = (0.3* $\times$ *0.9, 0.8* $\times$ *0.2, 0.5* $\times$ *0.7) = (0.27, 0.16, 0.35). It is again a neutrosophic set in* S_1.

Definition 6. *The four NC* $S_1 = \{(x, y, z)/x, y, z \in (0, 1)\}$, $S_2 = \{(x, y, z)/x, y, z \in [0, 1)\}$, $S_3 = \{(x, y, z)/x, y, z \in (0, 1]\}$ *and* $S_4 = \{(x, y, z)/x, y, z \in [0, 1]\}$ *are all only partially ordered sets for if a = (x, y, z) and b = (s, r, t) are in* S_i *then* $a < b$ *if and only if* $x < s$, $y < r$, $z < t$*; but not all elements are ordered in* S_i*, that is why we say* S_i *are only partially ordered sets, and denote it by* $(S_i, \leq)$*;where* $\leq$ *denotes the classical order relation over reals;* $1 \leq i \leq 4$.

For instance if $a = (0.3, 0.7, 0.5)$ and $b = (0.5, 0.2, 0.3)$ are in S_i then a and b cannot be compared. If $d = (0.8, 0.5, 0.7)$ and $c = (0.6, 0.2, 0.5)$, then $d > c$ or $c < d$.

In view of this we have the following theorem.

Theorem 1. *Let* $S_1 = \{(x, y, z)/x, y, z \in (0, 1)\}$ *be the collection of all NC which are such that the elements x, y and z do not take any extreme values.*

1. $\{S_1, \times\}$ *is an infinite order commutative semigroup which is not a monoid and has no zero divisors.*
2. *Every* $a = (x, y, z)$ *in* S_1 *will generate an infinite cyclic subsemigroup under product of* S_1 *denoted by* $(P, \times)$.

3. *The elements of P forms a totally ordered set, (for if $a = (x, y, z) \in P$ we see $a^2 = a \times a < a$).*
4. *$\{S_1, \times\}$ has no idempotents and $\{S_1, \times\}$ is a torsion free semigroup.*

Proof. Proof of 1: Clearly if $a = (x, y, z)$ and $b = (r, s, t)$ are in S_1, then $a \times b = (x \times r, y \times s, z \times t)$ is in S_1; as $x \times r, y \times s$ and $z \times t \in (0, 1)$. Hence, $\{S_1, \times\}$ is a semigroup under product. Further as number of elements in (0, 1) is infinite so is S_1. Finally as the product in (0, 1) is commutative so is the product in S_1. Hence the claim. (1, 1, 1) is not in S_1 as we have used only the open interval (0, 1), we see $\{S_1, \times\}$ is not a monoid. S_1 has no zero divisors as the elements are from the open interval which does not include 0, hence the claim.

Proof of 2: Let $a = (x, y, z)$ be in S, we see $a \times a = (x \times x, y \times y, z \times z) = a^2$, and so on $a \times a \times \ldots \times a = a^n = (x^n, y^n, z^n)$ and n can take values from $(0, \infty)$. Thus a in S generates a cyclic subsemigroup of infinite order, hence the claim.

Proof of 3: Let $P = \langle a \rangle$, a generates the semigroup under product, it is of infinite order and from the property of elements in (0, 1); $a > a^2 > a^3 >$ and so on $> a^n$. Hence the claim.

Proof of 4: If any $a = (x, y, z) \in S_1$ as $x, y, z \in (0, 1)$, and x, y and z are torsion free so is a. We see $a^2 \neq a$ for any $a \in S_1$. Further if $a \neq b$ for no $n \in (0, \infty)$; $a^n = b^n$. Hence the claim.

□

Definition 7. *The four NC S_1, S_2, S_3 and S_4 mentioned in definition 6 under the usual product $\times$ forms a commutative semigroup of infinite order defined as the NC semigroups.*

Theorem 2. *Let $S_2 = \{(x, y, z)/x, y, z \in [0, 1)\}$ be the collection of NC. $\{S_2, \times\}$ is only a semigroup and not a monoid and has infinite number of zero divisors. Further all other results mentioned in Theorem 1 are true with an additional property if $a \neq b$; $(a, b \in S_2)$ we have*

$$\lim_{n \to \infty} a^n = \lim_{n \to \infty} b^n = (0, 0, 0)$$

as (0, 0, 0) $\in S_2$.

Proof as in case of Theorem 1.

In view of this we define an infinite torsion free semigroup to be weakly torsion free if $a \neq b$; but

$$\lim_{n \to \infty} a^n = \lim_{n \to \infty} b^n$$

Thus S_2 is only a weakly torsion free semigroup.

It is interesting to note S_1 is contained in S_2 and in fact S_1 is a subsemigroup of S_2.The differences between S_1 and S_2 is that S_2 has infinite number of zero divisors and the $\lim_{n \to \infty} a^n = (0,0,0)$ exists in S_2 and S_1 is torsion free but S_2 is weakly torsion free.

Theorem 3. *Let $S_3 = \{(x, y, z)/x, y, z \in (0, 1]\}$ be the collection of NC. $\{S_3, \times\}$ is a monoid and has no zero divisors.*

Results 2 to 4 of Theorem 1 are true. Finally S_1 is a subset of S_3, in fact S_1 is a subsemigroup of S_3. The main difference between S_1 and S_3 is that S_3 is a monoid and S_1 is not a monoid. The difference between S_2 and S_3 is that S_3 has no zero divisors but S_2 has zero divisors and S_3 is a monoid.

Next we prove a theorem for S_4.

Theorem 4. *Let $S_4 = \{(x, y, z)/x, y, z \in [0, 1]\}$. $\{S_4, \times\}$ is a semigroup and is a monoid and has zero divisors. Other three conditions of Theorem 1 is true, but S_4 like S_2 is only a weakly torsion free semigroup.*

Proof as in case of Theorem 1. We have S_1 contained in S_2 and S_2 is contained in S_4 and S_1 contained in S_3 and S_3 is contained in S_4.

However, it is interesting to note S_2 and S_3 are not related in spite of the above relations.

Now we analyse all these four neutrosophic semigroups to find out, on which of them we can define addition modulo 1. S_1 does not include the element (0, 0, 0) as 0 is not in (0, 1), so S_1 is not even closed under addition modulo 1. So S_1 in not a semigroup or a group under plus modulo 1. Since S_3 and S_4 contains (1, 1, 1) we cannot define addition modulo 1; hence, they can not have any algebraic structure under addition modulo 1. Now consider $\{S_2, +\}$, clearly $\{S_2, +\}$ is a group under addition modulo 1.

In view of all these we have the following theorem.

Definition 8. *The NC $\{S_2, +\}$ under usual addition modulo 1 is a group defined as the NC group denoted by $\{S_2, +\}$*

Theorem 5. *$\{S_2, +\}$ is a group under addition modulo 1.*

Proof. For any $y, x \in S_2$, x + y (mod 1) $\in S_2$. $(0, 0, 0) \in S_2$ acts as additive identity. Further for every x there is a unique $y \in S_2$ with $x + y = (0, 0, 0)$. Hence the theorem. □

Definition 9. *The NC S_2 under the operations of the usual addition + modulo 1 and usual product × forms a commutative ring of infinite order defined as the NC commutative ring denoted by $\{S_2, +, \times\}$*

Theorem 6. *$\{S_2, +, \times\}$ is a commutative ring with infinite number of zero divisors and has no multiplicative identity (1, 1, 1).*

Proof. Follows from the Theorem 1 and the fact S_2 is closed under + modulo 1 by Theorem 5. The distributive property is inherited from the number theoretic properties of modulo integers. As 1 is not in [0, 1); (1, 1, 1) is not in S_2, hence the result. □

Next we proceed on to define multiset NC semigroups in the following section.

4. Multiset NC Semigroups

In this section we proceed on to define multiset NC semigroups using S_1, S_2, S_3 and S_4. We see $M(S_1) = \{$Collection of all multiset NC using elements of $S_1\}$. On similar lines we define $M(S_2), M(S_3)$ and $M(S_4)$ using S_2, S_3 and S_4 respectively. We prove $\{M(S_2), +, \times\}$ is a multiset neutrosophic semiring of infinite order.

Recall [18], A is a multi neutrosophic set, then $A = \{5(0.3, 0.7, 0.9), 12(0.6.0.2, 0.7), 8(0.1, 0.5, 0.1), (0.6, 0.7, 0.5)\}$; that is in the multiset neutrosophic set A; (0.3, 0.7, 0.9) has occurred 5 times; (0.6, 0.2, 0.7) has occurred 12 times or its multiplicity is 12 in A and so on.

Let $M(S_1) = \{$Collection of all multisets using the elements from $S_1\}$, $M(S_1)$ is an infinite collection. We just show how the classical product is defined on $M(S_1)$.

Let $A = \{9(0.3, 0.2, 0.4), 2(0.6, 0.7, 0.1), (0.1, 0.3, 0.2)\}$ and $B = \{5(0.1, 0.2, 0.5), 10(0.8, 0.4, 0.5)\}$ in $M(S_1)$ be any two multisets. We define the classical product × of A and B as follows;

$$\begin{aligned} A \times B = \{&9(0.3,0.2,0.4) \times 5(0.1,0.2,0.5), 9(0.3,0.2,0.4) \times 10(0.8,0.4,0.5), \\ &2(0.6,0.7,0.1) \times 5(0.1,0.2,0.5), 2(0.6,0.7,0.1) \times 10(0.8,0.4,0.5), \\ &(0.1,0.3,0.2) \times 5(0.1,0.2,0.5), (0.1,0.2,0.5) \times 10(0.8,0.4,0.5)\} \\ = \{&45(0.03,0.04,0.2), 90(0.24,0.08,0.2), 10(0.06,0.14,0.05), \\ &20(0.48,0.28,0.05), 5(0.01,0.06,0.1), 10(0.08,0.08,0.25)\}; \end{aligned}$$

$A \times B$ is in $M(S_1)$, thus $\{M(S_1), \times\}$ is a commutative semigroup of infinite order defined as the multiset NC semigroup.

Definition 10. *Let $M(S_i)$ be the multi NC using elements of $S_i(i = 1,2,3,4)$, $\{M(S_i), \times\}$ on the usual product $\times$ is defined as the multiset neutrosophic semigroup for $i = 1,2,3$ and 4.*

Definition 11. *Let $\{S_2, \times\}$ be the multiset NC semigroup under $\times$, elements of the form $(a,0,0), (0,b,c)$ and so on which are infinite in number with $a,b,c \in S_2$ contribute to zero divisors. Hence multisets using these types of elements contribute to zeros of the form $n(0,0,0)$; $1 < n < \infty$. As the zeros are of varying multiplicity we call these zero divisors as special type of zero divisors.*

We will provide examples of them.

Example 3. *Let $R = \{(S_2), \times\}$ be the multiset NC semigroup under product. Let A = (0.6, 0, 0) and B = (0, 0.4, 0.5) be in R, A × B = (0, 0, 0). Take $D = \{9(0.6,0.9,0)\}$ and E = 9(0, 0, 0.4) in R; we get $D \times E = \{81(0,0,0)\}$. Take $W = \{7(0,0.5,0), 4(0,0.6,0)\}$ and $V = \{(0.7,0,0.4), 20(0.8,0,0)\}$ be two multisets in R; $W \times V = \{7 \times 44(0,0,0) + 7 \times 20(0,0,0) + 4 \times 44(0,0,0) + 4 \times 20(0,0,0)\} = \{704(0,0,0)\}$ is a special type of zero divisor of R.*

Thus $M(S_2)$ is closed under the binary operation $\times$.

Theorem 7. *The neutrosophic multiset semigroups $\{M(S_i), \times\}$ for i = 1, 2, 3, 4 are commutative and of infinite order satisfying, the following properties for each $M(S_i)$; i = 1, 2, 3, 4.*

1. *$\{M(S_1), \times\}$ has no trivial or non-trivial special type of zero divisors and no trivial or non-trivial idempotents.*
2. *$\{M(S_2), \times\}$ has infinite number of special type of zero divisors and no non-trivial idempotents.*
3. *$\{M(S_3), \times\}$ has no trivial or non-trivial special zero divisors but has (1, 1, 1) as identity and has no non trivial idempotents.*
4. *$\{M(S_4), \times\}$ has non-trivial special type of zero divisors and has (1, 1, 1) as its identity and has idempotents of the form $\{(0,1,0), (1,1,0), (0,0,1), (1,0,1)$ and so on $\}$.*

Proof. 1. Follows from the fact that S_1 has no zero divisors and idempotents as it is built on the interval (0, 1).
2. Evident from the fact S_2 is built on [0, 1) so has special type of zero divisors by definition but no idempotent.
3. True from the fact S_3 is built on (0, 1], so (1, 1, 1) ∈ $M(S_3)$.
4. S_4 which is built on [0, 1] has infinite special type of zero divisors as (0, 0, 0) ∈ S_4 by Definition 11 and (1, 1, 1) ∈ $M(S_4)$ and has idempotents of the form $\{(0,1,0), (1,1,0), (0,0,1), (1,0,1)$ and so on $\}$.

Hence the claims of the theorem. □

Now we proceed onto define usual addition on $M(S_1)$

$S_1 = \{(x,y,z)/x,y,z \in (0,1)\}$ in not even closed under addition. For there are $x, y \in (0,1)$ such that $x + y$ is 1 or greater than 1, so these elements are not in (0 , 1), hence our claim.

Recall $S_2 = \{(x,y,z)/x,y,z \in [0,1)\}$. We can define addition modulo 1 and product under that addition both S_2 and $[0,1)$ are closed.

Let $a = (0.7,0.6,0.9)$ and $b = (0.5,0.9,0.4)$ be in S_2, we find $a + b$ mod 1.

$a + b = (0.7,0.6,0.9) + (0.5,0.9,0.4) = (0.7 + 0.5(mod\ 1), 0.6 + 0.9(mod\ 1), 0.9 + 0.4(mod\ 1)) = (0.2,0.5,0.3)$ is in S_2. (0, 0, 0) in S_2 acts as the additive identity.

For every $a \in S_2$ there is a unique $b \in S_2$ such that $a + b = (0,0,0) mod\ 1$. Thus $(S_2, +)$ is a NC group of infinite under addition modulo 1. Further $(S_2, \times)$ is a semigroup under product of infinite order which is commutative and not a monoid as (1, 1, 1) is not in S_2.

Now we illustrate how addition is performed on any two neutrosophic multisets in $M(S_2)$.

Let $A = \{7(0.3,0.8,0.45), 9(0.02,0.41,0.9), (0.6,0.3,0.2)\}$ and $B = \{5(0.1,0,0.9), 2(0.6,0.5,0)\}$ be any two multisets of $M(S_2)$. To find the sum of A with B under addition modulo 1.

$A + B$ = { 35[(0.3, 0.8, 0.45) + (0.1, 0, 0.9)]mod 1, 45[(0.02, 0.41, 0.9) + (0.1, 0, 0.9)]mod 1, 5[(0.6, 0.3, 0.2) + (0.1,0, 0.9)]mod 1, 14[(0.3, 0.8, 0.45) + (0.6, 0.5,0)]mod 1, 18[(0.02, 0.41, 0.9) + (0.6, 0.5, 0)]mod 1, 2[(0.6, 0.3, 0.2) + (0.6, 0.5, 0)]mod 1} $=$ {35(0.4, 0.8, 0.35), 45(0.12, 0.41, 0.8), 5(0.7, 0.3, 0.1), 14(0.9, 0.3, 0.45), 18(0.62, 0.91, 0.9), 2(0.2, 0.8, 0.2)}

is in $M(S_2)$. This is the way addition modulo 1 operation is performed. For $M(S_3)$ and $M(S_4)$ we can not define usual addition modulo 1 as (1, 1, 1) $\in M(S_3)$ and $M(S_4)$.

Next we proceed on to describe the product of any two elements in $M(S_2)$. We take the above A and B and find $A \times B$. $A \times B$ = {35[(0.3,0.8,0.45) $\times$ (0.1, 0, 0.9)], 45[(0.02, 0.41, 0.9) $\times$ (0.1, 0, 0.9)], 5[(0.6, 0.3, 02)$\times$ (0.1, 0, 0.9)], 14[(0.3, 0.8, 0.45)$\times$(0.6, 0.5 0)], 18[(0.02, 0.41. 0.9)$\times$ (0.0.6, 0.5, 0)], 2[(0.6, 0.3, 0.2) $\times$ (0.6, 0.5, 0)]} = {35(0.03, 0, 0.405), 45(0.002,0, 0.81), 5(0.06, 0, 0.18), 14(0.18, 0.4, 0), 18(0.012, 0.205, 0), 2(0.36, 0.15, 0)}, is in $M(S_2)$.

Theorem 8. $\{M(S_2), +\}$ *is a multiset NC semigroup under addition modulo 1.*

Proof. $M(S_2)$ is closed under the binary operation addition modulo 1. Thus $M(S_2)$ is the neutrosophic multiset semigroup under + modulo 1. □

Now we proceed on to define a special type of zero divisors. In view of this we have the following theorem.

Theorem 9. $R = \{M(S_2), \times\}$ *is an infinite commutative multiset NC semigroup, which is not a monoid and has special type of zero divisors.*

Proof. We see $M(S_2)$ under the binary operation product is closed and is associative as the base set S_2 is associative and commutative and is closed under the binary operation product. Thus $\{(S_2), \times\}$ is commutative semigroup of infinite order. Further $M(S_2)$ does not contain (1, 1, 1) so $\{M(S_2), \times\}$ is not a monoid.

From the above definition and description of special zero divisors R has infinite number of them. □

We have the following theorem.

Theorem 10. $\{M(S_2), +, \times\}$ *is a NC multiset commutative semiring of infinite order which has infinite numbers of special type of zero divisors.*

Proof. Follows from Theorem 8 and Theorem 9. □

Next we proceed on to define n- multiplicity neutrosophic multisets and derive some properties related with them. $M(S_3)$ and $M(S_4)$ are just multiset NC semigroups under product and in fact they are monoids. Further $M(S_4)$ has infinite number of special zero divisors.

5. n-Multiplicity Neutrosophic Set Semigroups Using S_1, S_2, S_3 and S_4

In this section we define the new notion of n-multiplicity NC using S_1, S_2, S_3 and S_4. We prove these n-multiplicity NC are of infinite order but what is restricted is the multiplicity n, that is any element cannot exceed multiplicity n; it can maximum be n, where n is a positive finite integer. Finally we prove $\{M(S_2), +, \times\}$ where $S_2 = [0, 1)$ is a NC n-multiset commutative semiring of infinite order.

We will first illustrate this situation by some examples before we make an abstract definition of them.

Example 4. *Let* 4-$M(S_1)$ $=$ {*collection all multisets with entries from* $S_1 = \{(x, y, z)/x, y, z \in (0, 1)\}$, *such that any element in* S_1 *can maximum repeat itself only four times*}. *Here* $n = 4$, $A = \{4(0.5, 0.7, 0.4), 3(0.1, 0.9, 0.7), 4(0.1, 0.2, 0.3), 4(0.7, 0.8, 0.4), 4(0.8, 0.8, 0.8), 2(0.9, 0.9, 0.9),$

$3(0.7,0.9,0.6),(0.6,0.1,0.1)\}$ be a 4-multiplicity multiset from 4-$M(S_1)$. We see the NC $(0.5,0.7,0.4),(0.1,0.2,0.3),(0.7,0.8,0.4)$ and (0.8, 0.8,0.8) have multiplicity four which is the highest multiplicity an element of 4-$M(S_1)$ can have. The NC (0.1, 0.9, 0.7) and (0.7,0.9,0.6) have multiplicity 3. The multiplicity of (0.9, 0.9, 0.9) is two and that of (0.6, 0.1, 0.1) is one. Clearly S_1 does not contain the extreme values 0 and 1 as S_1 is built using the open interval (0, 1). However on $M(S_1)$ we can not define addition.

Thus 4-$M(S_1)$ can not have the operation of addition defined on it. Now we show how the operation $\times$ is defined on 4-$M(S_1)$ for the some $A, B \in$ 4-$M(S_1)$. Now

$$\begin{aligned} A \times B &= \{3(0.3,0.7,0.8),2(0.5,0.9,0.6),4(0.2,0.3,0.4)\} \times \{(0.1,0.3,0.7),2(0.5,0.7,0.1)\} \\ &= \{3(0.03,0.21,0.56),2(0.05,0.27,0.42),4(0.02,0.09,0.28), \\ &\quad 6(0.15,0.49,0.08),4(0.25,0.63,0.06),8(0.1,0.21,0.04)\} \end{aligned}$$

we now use the fact we can have maximum only 4 multiplicity of an element so we replace 6(0.15, 0.49, 0.08) by 4(0.15, 0.49, 0.08) and 8(0.1, 0.21, 0.04) by 4(0.1, 0.21, 0.04). Now the thresholded product is $\{$(3(0.03, 0.21, 0.56), 2(0.05, 0.27, 0.42), 4(0.02, 0.09, 0.28), 4(0.15, 0.49, 0.08), 4(0.25, 0.63, 0.06), 4(0.1, 0.21, 0.04))$\} \in$ 4-$M(S_1)$.

$\{$4-$M(S_1), \times\}$ is a commutative neutrosophic multiset semigroup of infinite order and the multiplicity of any element cannot exceed 4.

This semigroup is not a monoid and it has no special zero divisors or zero divisors or units.

Definition 12. *12 Let n-$M(S_i)$ =$\{$ collection of all multisets with entries from S_i of at-most multiplicity $n; 2 \leq n < \infty\}(1 \leq i < 4)$. n-$M(S_i)$ under usual product, $\times$ is defined as the n-multiplicity NC semigroup, $1 \leq i \leq 4$.*

In view of this we have the following theorem.

Theorem 11. *Let n-$M(S_i) = \{t(x,y,z) | x,y,z \in S_i; 1 \leq t \leq n\}$ be the n-multiplicity neutrosophic multisets $(1 \leq i \leq 4)$.*

1. *n-$M(S_i)$ is not closed under the binary operation '+' under usual addition, for $i = 1, 3$ and 4.*
2. *n-$M(S_i)$ is a (n-multiplicity neutrosophic multiset) semigroup under the usual product for i = 1, 2, 3 and 4.*
3. *$\{$n-$M(S_i), \times\}$ is a monoid for i = 3 and 4. .*
4. *$\{$n-$M(S_i), \times\}$ has no special zero divisors if $S_i = S_1$ and S_3 but they have no non trivial idempotents. S_2 and special zero divisors and no non trivial idempotents, but S_4 has both non trivial special zero divisors and non trivial idempotents.*

Proof. Proof of 1: If $A = \{(0.3,0.8,0.9)\}$ and $B = \{(0.4,0.3,0.1)\} \in$ n-$M(S_i)$. $A + B = \{(0.7, 1.1, 1.0)\} \notin$ n-$M(S_i)$ as S_i when built using S_3 and S_4 and by example 4 n-$M(S_1)$. Only $M(S_2)$ is closed under addition.

Proof of 2: Since $(S_i, \times)$ is closed under product so is n-$M(S_i)$ with replacing the numbers greater than n by n in the resultant product; $i = 1,2,3$ and 4 are semigroups, hence the claim.

Proof of 3: As $(1,1,1) \in S_3$ and S_4 so is in n-$M(S_3)$ and n-$M(S_4)$ respectively so they are monoids.

Proof of 4: n-$M(S_i)$ has no special zero divisors in case of S_1 and S_3. Finally $S_i = \{(x,y,z) | x,y,z \in S_i\}$, has zero divisors and special zero divisors in case of S_2 and S_4 for $i = 2$ and 4, and non trivial idempotents contributed by 0's and 1's only in case of S_4. Hence the theorem. □

Example 5. *Let 5-$M(S_2)$ = $\{$Collection of all neutrosophic multisets which can occur at most 5-times that is the multiplicity is 5 with elements from $S_2 = \{(x,y,z) | x,y,z \in [0,1)\}\}$ Let $A =$*

$4(0.2, 0.5, 0.7), 3(0.1, 0.2, 0.3), 5(0.3, 0.1, 0.2), (0.1, 0.2, 0.8) \in$ *5-*$M(S_2)$ *We see the multiplicity of (0.3, 0.1, 0.2) is 5 others are less than 5.*

Let $A = \{3(0.3, 0.2, 0), 4(0.5, 0.6, 0.9), 5(0.1, 0.2, 0.7)\}$ *and* $B = \{4(0.8, 0.1, 0.9), 2(0.6, 0.6, 0.6)\} \in$ *5-*$M(S_2)$. *Now we first find* $A \times B = \{5(0.24, 0.02, 0), 5(0.4, 0.06, 0.81), 5(0.08, 0.02, 0.63), 5(0.06, 0.12, 0.42)\} \in 5(M(S_2)$.

$A + B = \{5(0.1, 0.3, 0.9), 5(0.9, 0.8, 0.6), 5(0.3, 0.7, 0.8), 5(0.9, 0.3, 0.6), 5(0.1, 0.2, 0.5), 5(0.7, 0.8, 0.3)\} \in$ *5-*$M(S_2)$. *Addition is done modulo 1. However we have closure axiom to be true under + for elements in* S_2 *and in case of* S_1; $0 \notin S_1 = (0, 1))$. *This closure axiom is flouted.*

If addition modulo 1 is done we have to see that 1 is not included in the interval and 0 is included in that interval so we need to have only closed open interval [0, 1). Under these two constraints only we can make S_2 *as well as* $M(S_2)$ *and n-*$M(S_2)$ *as semigroups under addition modulo 1.*

We can built strong structure only using the [0, 1).

Theorem 12. *Let n-*$M(S_2)$ *= Collection of all multisets of S built using* $S_2 = \{(x, y, z) | x, y, z \in [0, 1)\}$ *with multiplicity less than or equal to n;* $2 \leq n \leq \infty$

$\{$ *n-*$M(S_2), \times\}$ *is a commutative neutrosophic multiset semigroup of infinite order and is not a monoid, n-*$M(S_2)$ *has infinite number of zero divisors.*

Proof. If A and $B \in$ n-$M(S_2)$ we find $A \times B$ and update the multiplicities in $A \times B$ to be less than or equal to n so that $A \times B \in$ n-$M(S_2)$. by Theorem 11(2).

Clearly $(1, 1, 1) \notin$ n-$M(S_2)$ so is not a monoid.

□

Theorem 13. $B = \{$*n-*$M(S_2), +, \times\}$, *the n-multiplicity multiset NC is a commutative semiring of infinite order and has no unit, where* $S_2 = [0, 1)$.

Proof. Follows from the fact { n-$M(S_2)$, +} is a commutative semigroup under addition modulo 1, Theorem 11(1) and Theorem 12 and {n-$M(S_2)$, $\times$} is a commutative semigroup under $\times$. Hence the claim.

□

6. Discussions

The main motive of this paper is to construct strong algebraic structures with two binary operations on the NC. Here we are able to get a NC commutative ring structure using the base interval as [0, 1). This will lead to future research of constructing Smarandache neutrosophic vector spaces and Smarandache neutrosophic algebraic codes using the same interval [0, 1). Now using the same interval [0, 1), we construct multiset NC and n-multiset NC $2 \leq n < \infty$. On these we were able to built only neutrosophic multiset(n-multiplication set) commutative semiring structure. Now using these we can construct Smarandache multiset neutrosophic semi vector spaces which will be taken as future research. So this is significant first step to develop other strong structures and apply them to NC codes and NC cryptography.

7. Conclusions

In this paper, authors have made a study of NC on the 4-intervals (0, 1) (0, 1], [0, 1] and [0, 1). We define usual + and × on these intervals which is very different from the study taken so far. The main properties enjoyed by these NC semigroups are developed. Further of these intervals only the interval [0, 1) gives a nice algebraic structure viz an abelian group under usual addition modulo 1, which in turn helps in constructing NC commutative ring under usual addition modulo 1 and product, the ring has infinite number of zero divisors, whereas all the other intervals are semigroups/monoids which are torsion free or weakly torsion free of infinite order under $\times$. Further in this paper we introduce the notion of multiset NC semigroups using these four intervals under product. Furthermore, the multiset NC forms a commutative semiring with zero divisors only when the interval [0, 1) is used. Finally we introduce n-multiplicity multiset using these NC. They are also semigroups which is torsion free or weakly torsion free under product.

For future research we will be using the product and addition modulo 1 in the place of min and max in Single Valued Neutrosophic Set (SVNS) and would compare the results with the existing ones when applied as SVNS models to real world problems.

Apart from all these we can use these NC, multiset NC and n-multiplicity multiset NC to built NC codes which is one of the applications to neutrosophic cryptography which will be taken up by the authors for future research.

Author Contributions: Conceptualization, V.W.B.; writing–original draft preparation, I.K.; writing–review and editing, F.S. All authors have read and agreed to the published version of the manuscript.

Funding: This research received no external funding.

Acknowledgments: We would like thanks the reviewers for their valuable suggestions.

Conflicts of Interest: The authors declare no conflict of interest.

Abbreviations

The following abbreviations are used in this manuscript:

SVNS	Single Valued Neutrosophic Set

References

1. Herstein, I.N. *Topics in Algebra*; John Wiley & Sons: Hoboken, NJ, USA, 2006.
2. Hall, M. *The Theory of Groups*; Courier Dover Publications: Mineola, NY, USA, 2018.
3. Howie, J.M. *Fundamentals of Semigroup Theory*; Clarendon Oxford: Oxford, UK, 1995.
4. Godin, T.; Klimann, I.; Picantin, M. On torsion-free semigroups generated by invertible reversible Mealy automata. In *International Conference on Language and Automata Theory and Applications*; Springer: Berlin, Germany, 2015; pp. 328–339.
5. East, J.; Egri-Nagy, A.; Mitchell, J.D.; Peresse, Y. Computing finite semigroups. *J. Symb. Comput.* **2019**, *92*, 110–155.
6. Smarandache, F. *A Unifying Field in Logics: Neutrosophic Logic. Neutrosophy, Neutrosophic Set, Probability, and Statistics*; American Research Press: Rehoboth, DE, USA, 2000.
7. Smarandache, F.; Ali, M. Neutrosophic triplet group. *Neural Comput Applic* **2018**, *29*, 595–601.
8. Kandasamy W.B., V.; Kandasamy, I.; Smarandache, F. Semi-Idempotents in Neutrosophic Rings. *Mathematics* **2019**, *7*, 507.
9. Kandasamy W. B., V.; Kandasamy, I.; Smarandache, F. Neutrosophic Triplets in Neutrosophic Rings. *Mathematics* **2019**, *7*, 563.
10. Kandasamy, W.V.; Kandasamy, I.; Smarandache, F. Neutrosophic Quadruple Vector Spaces and Their Properties. *Mathematics* **2019**, *7*, 758.
11. Saha, A.; Broumi, S. New Operators on Interval Valued Neutrosophic Sets. *Neutrosophic Sets Syst.* **2019**, *28*, 10.
12. Sahin, R.; Karabacak, M. A novel similarity measure for single-valued neutrosophic sets and their applications in medical diagnosis, taxonomy, and clustering analysis. In *Optimization Theory Based on Neutrosophic and Plithogenic Sets*; Elsevier: Amsterdam, Netherlands, 2020; pp. 315–341.
13. Jain, A.; Nandi, B.P.; Gupta, C.; Tayal, D.K. Senti-NSetPSO: Large-sized document-level sentiment analysis using Neutrosophic Set and particle swarm optimization. *Soft Comput.* **2020**, *24*, 3–15.
14. Wu, X.; Zhang, X. The Decomposition Theorems of AG-Neutrosophic Extended Triplet Loops and Strong AG-(l, l)-Loops. *Mathematics* **2019**, *7*, 268.

15. Ma, Y.; Zhang, X.; Yang, X.; Zhou, X. Generalized Neutrosophic Extended Triplet Group. *Symmetry* **2019**, *11*, 327.
16. Li, Q.; Ma, Y.; Zhang, X.; Zhang, J. Neutrosophic Extended Triplet Group Based on Neutrosophic Quadruple Numbers. *Symmetry* **2019**, *11*, 696.
17. Ali, M.; Smarandache, F.; Khan, M. Study on the Development of Neutrosophic Triplet Ring and Neutrosophic Triplet Field. *Mathematics* **2018**, *6*, 46.
18. Smarandache, F. *Neutrosophic Perspectives: Triplets, Duplets, Multisets, Hybrid Operators, Modal Logic, Hedge Algebras. And Applications.*; EuropaNova: Bruxelles, Belgium, 2017.
19. Kandasamy, W.V.; Ilanthenral, K. *Smarandashe Special Elements in Multiset Semigroups*; EuropaNova ASBL: Brussels, Belgium, 2018.
20. Forsberg, L. Multisemigroups with multiplicities and complete ordered semi-rings. *Beitr Algebra Geom* **2017**, *58*, 405–426.
21. Kandasamy, W.V. Smarandache Semirings, Semifields, And Semivector Spaces. *Smarandache Notions J.* **2002**, *13*, 88.
22. Blizard, W.D.; others. The development of multiset theory. *Mod. Log.* **1991**, *1*, 319–352.

symmetry

MDPI

Article

Solution and Interpretation of Neutrosophic Homogeneous Difference Equation

Abdul Alamin [1], Sankar Prasad Mondal [1], Shariful Alam [2], Ali Ahmadian [3,*], Soheil Salahshour [4] and Mehdi Salimi [5]

1 Department of Applied Science, Maulana Abul Kalam Azad University of Technology, West Bengal, Haringhata 741249, Nadia, West Bengal, India; avishek.chakraborty@nit.ac.in (A.A.); sankar.prasad.mondal@midnaporecollege.ac.in (S.P.M.)
2 Department of Mathematics, Indian Institute of Engineering Science and Technology, Shibpur, Howrah 711103, West Bengal, India; salam@math.iiests.ac.in
3 Institute of Industry Revolution 4.0, The National University of Malaysia, Bangi 43600, Selangor, Malaysia
4 Faculty of Engineering and Natural Sciences, Bahcesehir University, Istanbul, Turkey; soheil.salahshour@eng.bau.edu.tr
5 Center for Dynamics, Department of Mathematics, Technische Universität Dresden, 01062 Dresden, Germany; mehdi.salimi@medalics.org
* Correspondence: ali.ahmadian@ukm.edu.my; Tel.: +60-12-668-7968

Received: 23 May 2020; Accepted: 19 June 2020; Published: 1 July 2020

check for updates

Abstract: In this manuscript, we focus on the brief study of finding the solution to and analyzingthe homogeneous linear difference equation in a neutrosophic environment, i.e., we interpreted the solution of the homogeneous difference equation with initial information, coefficient and both as a neutrosophic number. The idea for solving and analyzing the above using the characterization theorem is demonstrated. The whole theoretical work is followed by numerical examples and an application in actuarial science, which shows the great impact of neutrosophic set theory in mathematical modeling in a discrete system for better understanding the behavior of the system in an elegant manner. It is worthy to mention that symmetry measure of the systems is employed here, which shows important results in neutrosophic arena application in a discrete system.

Keywords: fuzzy set theory; difference equation; neutrosophic number; simplified neutrosophic symmetry measure

1. Introduction

1.1. Uncertainty Theory and Neutrosophic Sets

The uncertainty theory becomes a very helpful tool for real life modeling in discrete and continuous systems. The different theories of the fuzzy uncertainty theory have been given a new direction since the setting of the fuzzy set, invented by Professor Zadeh [1]. This is generalized representation of [1] is established as an intuitionistic fuzzy set theory by Atanassov [2]. Atarasov gave a novel designusing the intuitionistic fuzzy theory, where he demonstrated the idea of a membership function and non-membership function by which degree of belongingness and non-belongingness, respectively, can be measured in a set. Liu and Yuan [3] ignited the perception of a triangular intuionistic fuzzy set, which is the affable blend of a triangular fuzzy number and a intuionistic fuzzy set theory. Ye [4] set up the idea for a trapezoidal intuionistic fuzzy set. Smarandache [5] found his more generalized idea as a neutrosophic set, considering terms of the truth membership function, the indeterminacy membership function, and the falsity membership function. This theory become more beneficial and germane, rather than the common fuzzy and intuitionistic fuzzy theory settings.

Several researchershave already worked in the neutrosophic field, some of which have developed the theory [6,7], while some have applied the related theories in an applied field [8,9]. Various kinds of forms and extensions of the Neutrosophic set, such as the triangular neutrosophic set [10], the bipolar neutrosophic sets [11–14], and the multi-valued neutrosophic sets [15], were also found.

1.2. Difference Equation in an Uncertain Environment

There exist some works associated with difference equation and uncertainty. Mostly, researchers have worked on the difference equation allied with fuzzy and intuitionistic fuzzy environments. We are now giving details descriptions of some related published work. In the literature [16], Deebaet al. found a strategy for solving the fuzzy difference equation with an interesting application. The model involving CO2 levels in blood streamflow is thinkingin the view ofthe fuzzy difference equation by Deebaet al. [17]. Lakshmikantham and Vatsala [18] talk abou tdifferent basic theories and properties of fuzzy difference equations. Papaschinopoulos et al. [19,20] and PapaschinopoulosandSchinas [21] discuss more findings in a similar context. Papaschinopoulos and Stefanidou [22] provide an explanation on boundedness with asymptotic behavior of a fuzzy difference equation. Umekkanet al. [23] give a finance application based on discrete system modeling in a fuzzy environment. Stefanidou et al. [24] treat the exponential-type fuzzy difference equation. The asymptotic behavior of a second order fuzzy difference equation is considered by Din [25]. The fuzzy non-linear difference equation is considered by Zhang et al. [26], where Memarbashi and Ghasemabadi [27] corporate with a volterra type rational form by Stefanidou and Papaschinopoulos [28]. The economics application is considered by Konstantinos et al. [29]. Mondal et al. [30] solve the second-order intuitionistic difference equation. Non-linear interval-valued fuzzy numbers and their relevance to difference equations are shown in [31]. National income determination models with fuzzy stability analysis in a discrete system are elaborately discussed by Sarkar et al. [32]. The fuzzy discrete logistic equation is taken and stability situations are found in the literature [33]. Zhang et al. [34] show the asymptotic performance of a discrete time fuzzy single species population model. On discrete time, a Beverton–Holt population replica with fuzzy environment is illustrated in [35]. Additionally, a different view of the fuzzy discrete logistic equation is taken under uncertainty in [36]. The existence and stability situation of the difference equation with a fuzzy setting is found by Mondal et al. [37]. Important results are also found for fuzzy difference equations by Khastan and Alijani [38] and Khastan [39].

1.3. Novelties of the Work

In this connection of the above idea, few advances can still be prepared, which include:

(1) The homogeneous difference equation, solved and analyzed with a neutrosophic initial condition, neutrosophic coefficient, and neutrosophic coefficient and initial together as a different section, which was not done earlier.
(2) Establishment of the corresponding characterization theorem for the neutrosophic set with a difference equation.
(3) Different theorems, lemmas, and corollary drawn for the purpose of the study.
(4) Numerical examples of the difference equation with a neutrosophic number, solved and illustrated for better understanding of our observations.
(5) An application in actuarial science, illustrated in a neutrosophic environment for better understanding of the practical application of the proposed theoretical results.

1.4. Structure of the Paper

In Section 1, we recall the related work and write the novelties of our study. The preliminary concepts are addressed in Section 2. The difference equation with a neutrosophic variable is defined and corresponds with a necessary theory, for which a lemma is prepared for the study in Section 3. Section 4 shows the solution of the neutrosophic homogeneous difference equation. Two numerical examples are shown in Section 5. In Section 6, we take an appliance of an actuarial science problem in the neutrosophic data and solve it. The conclusion and future research scope are written in Section 7.

2. PreliminaryIdea

Definition 1. ***Neutrosophicset:*** *[6] Let X be a universe set. A single-valued neutrosophic set A on X is distinct as* $A = \{(T_A(x),\ I_A(x),\ F_A(x)) : x \in X\}$, *where* $T_A(x),\ I_A(x),\ F_A(x) : X \rightarrow [0,1]$ *is the degree of membership, degree of indeterministic, and degree of non-membership, respectively, of the element* $x \in X$, *such that* $0 \le T_A(x) + I_A(x) + F_A(x) \le 3$.

Definition 2. ***Neutrosophicfunction:*** *If we take the set of all real numbers as notation* $\mathcal{R}$ *and real valued fuzzy numbers as notation* $\mathcal{R}_{\mathcal{F}}$, *then the function* $W : \mathcal{R} \rightarrow [0,1]$ *is called a fuzzy number valued function if w satisfies the subsequent properties.*

(1) W is the upper semi continuous.
(2) W is the fuzzy convex, i.e., $W(\lambda s_1 + (1-\lambda)s_2) \ge min\{W(s_1), W(s_2)\}$ for all $s_1,\ s_2 \in \mathcal{R}$ and $\lambda \in [0,1]$.
(3) W is normal, i.e., $\exists$ a $s_0 \in \mathcal{R}$, such that $W(s_0) = 1$
(4) Closure of $supp(W)$ is compact, where $supp(W) = \{s \in \mathcal{R} |\ W(s) > 0\}$.

Definition 3. *Triangular neutrosophic number: [40] If we consider the measure of the truth, for which indeterminacy and falsity are not dependent, then a Triangular Neutrosophic number is taken as* $\widetilde{N} = (r_0, r_1, r_2; s_0, s_1, s_2; w_0, w_1, w_2)$, *where the truth membership, falsity, and indeterminacy membership function is treated as follows:*

$$T_{\widetilde{N}}(y) = \begin{cases} \frac{y-r_0}{r_1-r_0} & \text{when } r_0 \le y < r_1 \\ 1 & \text{when } y = r_1 \\ \frac{r_2-y}{r_2-r_1} & \text{when } r_1 < y \le r_2 \\ 0 & \text{otherwise} \end{cases}$$

and

$$F_{\widetilde{N}}(y) = \begin{cases} \frac{s_1-y}{s_1-s_0} & \text{when } s_0 \le y < s_1 \\ 0 & \text{when } y = s_1 \\ \frac{y-s_1}{s_2-s_2} & \text{when } s_1 < y \le s_2 \\ 1 & \text{otherwise} \end{cases}$$

$$I_{\widetilde{N}}(y) = \begin{cases} \frac{w_1-y}{w_1-w_0} & \text{when } w_0 \le y < w_1 \\ 0 & \text{when } y = w_1 \\ \frac{y-w_1}{w_2-w_1} & \text{when } w_1 < y \le w_2 \\ 1 & \text{otherwise} \end{cases}$$

where $0 \le T_{\widetilde{N}}(y) + F_{\widetilde{N}}(y) + I_{\widetilde{N}}(y) \le 1,\ y \in \widetilde{N}$.

The parametric setting of the above number is $(\widetilde{N})_{\alpha,\beta,\gamma} = [T_{Neu1}(\alpha), T_{Neu2}(\alpha); I_{Neu1}(\beta), I_{Neu2}(\beta); F_{Neu1}(\gamma), F_{Neu2}(\gamma)]$,

where

$$N_L^1(\alpha) = r_0 + \alpha(r_1 - r_0)$$

$$N_R^1(\alpha) = r_2 - \alpha(r_2 - r_1)$$

$$N_L^2(\beta) = s_1 - \beta(s_1 - s_0)$$

$$N_R^2(\beta) = s_1 + \beta(s_2 - s_1)$$

$$N_L^3(\gamma) = w_1 - \gamma(w_1 - w_0)$$
$$N_R^3(\gamma) = w_1 + \gamma(w_2 - w_1)$$

Here, $0 < \alpha, \beta, \gamma \leq 1$ and $0 < \alpha + \beta + \gamma \leq 3$

The verbal phrase with the number can be written as in Table 1:

Table 1. The verbal phrase of different uncertain settings and neutrosophic numbers.

Type of Uncertain Parameter	Verbal Phrase	Used Functionsand Their Roles
Triangular Fuzzy Number	[Low, Medium, High]	Membership function for measuring degree of belongingness
Triangular Intuitionistic Fuzzy Number	[Low, Medium, High; Very Low, Medium, Very High]	Membership and non-membership function for measuring degree of belongingness and non-belongingness
Triangular NeutrosophicNumber	[Low, Medium, High; Very Low, Medium, Very High; Between low and very low; Medium; Between high and very high]	Truthiness, falsity, and indeterminacy function for measuring the degree of truth belongingness, strictly non-belongingness and indeterminacy

Definition 4. ***Hukuhara difference on neutrosophic function:*** *Let E^* be the set of all neutrosophicfunctions, $\widetilde{s}, \widetilde{t} \in E^*$. If $\exists$ is a neutrosophic number, $\widetilde{w} \in E^*$ and $\widetilde{w}$ suit the relation $\widetilde{s} = \widetilde{w} + \widetilde{t}$, then $\widetilde{w}$ is assumed to be the Hukuhara difference of $\widetilde{s}$ and $\widetilde{t}$, denoted by $\widetilde{w} = \widetilde{s} \ominus \widetilde{t}$.*

3. Difference Equation with a Neutrosophic Variable

Definition 5. *A difference equation (sometime named as a recurrence relation) is an equation that relates the consecutive terms of a sequence of numbers.*

A qth order difference equation in the linear form can be articulated:

$$x_{n+q} = d_1 x_{n+q-1} + d_2 x_{n+q-2} + \cdots + d_q x_n + b_n \tag{1}$$

where $d_1, d_2, \ldots, d_q$ and b_n are constants, which are known.

If $b_n = 0$ for all n, then Equation (1) is the homogeneous difference equation. On the other hand, it will be the non-homogeneous difference equation if $b_n \neq 0$, where b_n is treated as the forcing factor.

We consider an autonomous linear homogeneous difference equation of the form:

$$x_{n+1} = \sigma x_n \, , (\sigma \neq 0) \tag{2}$$

with the initial condition $x_{n=0} = x_0$. The solution of Equation (2) can then be written as:

$$x_n = \sigma^n x_0 \tag{3}$$

Theorem 1. *[41] Let $m \in \mathbb{N}$, $m \geq 2$. A linear homogeneous system of them first order difference equation is given in matrix form as:*

$$X_{n+1} = AX_n \tag{4}$$

where, $X_n = \left(X_n^1, X_n^2, \ldots, X_n^m\right)^T$ and $A = \left(a_{ij}\right)_{m \times m}$, $i, j = 1, 2, \ldots, m$

The solution of Equation (3) can then be written as:

$$X_n = A^n X_0, n \in \mathbb{N} \tag{5}$$

The difference Equation (1) is considered as the neutrosophic difference equation if any one of the following conditions are added:

(i) The initial condition or conditions are the neutrosophic number (Type I);
(ii) The coefficient or coefficients are the neutrosophic number (Type II);
(iii) The initial conditions and coefficient or the coefficients are both neutrosophic numbers (Type III).

Theorem 2. ***Characterization theorem:*** *Let us consider the neutrosophic difference equation problem:*

$$\widetilde{x}_{n+1} = \widetilde{f}(x_n, n), \tag{6}$$

with initial value $\widetilde{x}_{n=0} = \widetilde{x}_0$ *as a neutrospohic number, where* $f : E^* \times \mathbb{Z}_{\geq 0} \to E^*$, *such that*

(1) *The parametric form of the function is:*

$$\left[\widetilde{f}((x_n, n))\right]_{(\alpha,\,\beta,\gamma)} = \begin{bmatrix} f^1_{L,n}\left(x^1_{L,n}(\alpha),\ x^1_{R,n}(\alpha), n, \alpha\right),\ f^1_{R,n}\left(x^1_{L,n}(\alpha),\ x^1_{R,n}(\alpha), n, \alpha\right); \\ f^2_{L,n}\left(x^2_{L,n}(\beta),\ x^2_{R,n}(\beta), n, \beta\right),\ f^2_{R,n}\left(x^2_{L,n}(\beta),\ x^2_{R,n}(\beta), n, \beta\right); \\ f^3_{L,n}\left(x^3_{L,n}(\gamma),\ x^3_{R,n}(\gamma), n, \gamma\right),\ f^3_{R,n}\left(x^3_{L,n}(\gamma),\ x^3_{R,n}(\gamma), n, \gamma\right) \end{bmatrix}$$

(2) *The functions* $f^1_{L,n}\left(x^1_{L,n}(\alpha),\ x^1_{R,n}(\alpha), n, \alpha\right)$, $f^1_{R,n}\left(x^1_{L,n}(\alpha),\ x^1_{R,n}(\alpha), n, \alpha\right)$, $f^2_{L,n}\left(x^2_{L,n}(\beta),\ x^2_{R,n}(\beta), n, \beta\right)$, $f^2_{R,n}\left(x^2_{L,n}(\beta),\ x^2_{R,n}(\beta), n, \beta\right)$, $f^3_{L,n}\left(x^3_{L,n}(\gamma),\ x^3_{R,n}(\gamma), n, \gamma\right)$ *and* $f^3_{R,n}\left(x^3_{L,n}(\gamma),\ x^3_{R,n}(\gamma), n, \gamma\right)$ *are taken as continuous functions, i.e., for any* $\epsilon_1 > 0\ \exists$ *a* $\delta_1 > 0$, *such that:*

$$\left|f^1_{L,n}\left(x^1_{L,n}(\alpha),\ x^1_{R,n}(\alpha), n, \alpha\right) - f^1_{L,n_1}\left(x^1_{L,n_1}(\alpha),\ x^1_{R,n_1}(\alpha), n_1, \alpha\right)\right| < \epsilon_1$$

for all $\alpha \in [0,1]$ *with* $\left\|\left(x^1_{L,n}(\alpha),\ x^1_{R,n}(\alpha), n, \alpha\right) - \left(x^1_{L,n_1}(\alpha),\ x^1_{R,n_1}(\alpha), n_1, \alpha\right)\right\| < \delta_1$ *and for any* $\epsilon_2 > 0\ \exists$ *an* $\delta_2 > 0$, *such that:*

$$\left|f^1_{R,n}\left(x^1_{L,n}(\alpha),\ x^1_{R,n}(\alpha), n, \alpha\right) - f^1_{R,n_2}\left(x^1_{L,n_2}(\alpha),\ x^1_{R,n_2}(\alpha), n_2, \alpha\right)\right| < \epsilon_2 \text{ for all } \alpha \in [0,1]$$

with $\left\|\left(x^1_{L,n}(\alpha),\ x^1_{R,n}(\alpha), n, \alpha\right) - \left(x^1_{L,n_2}(\alpha),\ x^1_{R,n_2}(\alpha), n_2, \alpha\right)\right\| < \delta_2$, *where* n, n_1 *and* $n_2 \in \mathbb{Z}_{\geq 0}$.

In a similar way, the continuity of the remaining four functions, $f^2_{L,n}\left(x^2_{L,n}(\beta),\ x^2_{R,n}(\beta), n, \beta\right)$, $f^2_{R,n}\left(x^2_{L,n}(\beta),\ x^2_{R,n}(\beta), n, \beta\right)$, $f^3_{L,n}\left(x^3_{L,n}(\gamma),\ x^3_{R,n}(\gamma), n, \gamma\right)$ and $f^3_{R,n}\left(x^3_{L,n}(\gamma),\ x^3_{R,n}(\gamma), n, \gamma\right)$, can be defined.

The difference Equation (6) then reduces to the system of six difference equations, as follows:

$$x^1_{L,n+1}(\alpha) = f^1_{L,n}\left(x^1_{L,n}(\alpha),\ x^1_{R,n}(\alpha), n, \alpha\right)$$

$$x^1_{R,n+1}(\alpha) = f^1_{R,n}\left(x^1_{L,n}(\alpha),\ x^1_{R,n}(\alpha), n, \alpha\right)$$

$$x^2_{L,n+1}(\beta) = f^2_{L,n}\left(x^2_{L,n}(\beta),\ x^2_{R,n}(\beta), n, \beta\right)$$

$$x^2_{R,n+1}(\beta) = f^2_{R,n}\left(x^2_{L,n}(\beta),\ x^2_{R,n}(\beta), n, \beta\right)$$

$$x^3_{L,n+1}(\gamma) = f^3_{L,n}\left(x^3_{L,n}(\gamma),\ x^3_{R,n}(\gamma), n, \gamma\right)$$

$$x^3_{R,n+1}(\gamma) = f^3_{R,n}\left(x^3_{L,n}(\gamma),\ x^3_{R,n}(\gamma), n, \gamma\right)$$

with the initial conditions:

$$x^1_{L,n=0}(\alpha) = x^1_{L,0}(\alpha)$$

$$x^1_{R,n=0}(\alpha) = x^1_{R,0}(\alpha)$$

$$x^1_{L,n=0}(\beta) = x^1_{L,0}(\beta)$$

$$x^1_{R,n=0}(\beta) = x^1_{R,0}(\beta)$$

$$x^1_{L,n=0}(\gamma) = x^1_{L,0}(\gamma)$$
$$x^1_{R,n=0}(\gamma) = x^1_{R,0}(\gamma)$$

Note 1. *By the characterization theorem, we can see that a neutrosopic difference equation is transformed into a system of six difference equations in crisp form. In this article, we have taken only a single neutrosophic difference equation in a neutrosophic environment. Hence, the difference equation converted into six crisp difference equations.*

Definition 6. ***Strong and weak solutions of a neutrosophic difference equation:*** *The solutions of difference Equation (6), with initial condition (3.7) to be regarded as:*

(1) A strong solution if

$$x^1_{L,n}(\alpha) \le x^1_{R,n}(\alpha)$$
$$x^1_{L,n}(\beta) \le x^1_{R,n}(\beta)$$
$$x^1_{L,n}(\gamma) \le x^1_{R,n}(\gamma)$$

and

$$\frac{\partial}{\partial\alpha}\left[x^1_{L,n}(\alpha)\right] > 0,\ \frac{\partial}{\partial\alpha}\left[x^1_{R,n}(\alpha)\right] < 0$$
$$\frac{\partial}{\partial\beta}\left[x^1_{L,n}(\beta)\right] < 0,\ \frac{\partial}{\partial\beta}\left[x^1_{R,n}(\beta)\right] > 0$$
$$\frac{\partial}{\partial\gamma}\left[x^1_{L,n}(\gamma)\right] < 0,\ \frac{\partial}{\partial\gamma}\left[x^1_{R,n}(\gamma)\right] > 0$$

for every $\alpha,\ \beta, \gamma \in [0,1]$.

(2) A weak solution if

$$x^1_{L,n}(\alpha) \ge x^1_{R,n}(\alpha)$$
$$x^1_{L,n}(\beta) \ge x^1_{R,n}(\beta)$$
$$x^1_{L,n}(\gamma) \ge x^1_{R,n}(\gamma)$$

and

$$\frac{\partial}{\partial\alpha}\left[x^1_{L,n}(\alpha)\right] < 0,\ \frac{\partial}{\partial\alpha}\left[x^1_{R,n}(\alpha)\right] > 0$$
$$\frac{\partial}{\partial\beta}\left[x^1_{L,n}(\beta)\right] > 0,\ \frac{\partial}{\partial\beta}\left[x^1_{R,n}(\beta)\right] < 0$$
$$\frac{\partial}{\partial\gamma}\left[x^1_{L,n}(\gamma)\right] > 0,\ \frac{\partial}{\partial\gamma}\left[x^1_{R,n}(\gamma)\right] < 0$$

for every $\alpha,\ \beta, \gamma \in [0,1]$.

Definition 7. *Let p and q be neutrosophic numbers, where* $[\widetilde{p}]_{(\alpha,\beta,\gamma)} = \left[p^1_L(\alpha), p^1_R(\alpha);\ p^2_L(\beta), p^2_R(\beta); p^3_L(\gamma), p^3_R(\gamma)\right]$, $[\widetilde{q}]_{(\alpha,\beta,\gamma)} = \left[q^1_L(\alpha), q^1_R(\alpha);\ q^2_L(\beta), q^2_R(\beta); q^3_L(\gamma), q^3_R(\gamma)\right]$, *for all* $\alpha, \beta, \gamma \in [0,1]$. *The metric on the neutrosophic number space is then defined as:*

$$d(p,q) = sup \max_{\alpha,\beta,\gamma\in[0,1]}\left\{\left|p^1_L(\alpha) - q^1_L(\alpha)\right|, \left|p^1_R(\alpha) - q^1_R(\alpha)\right|, \left|p^2_L(\beta) - q^2_L(\beta)\right|, \left|p^2_R(\beta) - q^2_R(\beta)\right|, \left|p^3_L(\gamma) - q^3_L(\gamma)\right|, \left|p^3_R(\gamma) - q^3_R(\gamma)\right|\right\}.$$

Note 2. *For some cases, the solution may not become strictly strong or weak solution type. In this scenario, a specific time interval or specific interval of* α, β, *or* γ *becomes the strong or weak solution. The main objective is to find the strong solutions. For scenariosin which neitherthe strong nor weak solutions occur, we call them non-recommended neutrosophic solutions. We strongly recommended taking strong solutions.*

4. Solution of Neutrosophic Homogeneous Difference Equation

Considering linear homogeneous difference equations:

$$u_{n+1} = au_n \tag{7}$$

In a neutrosophic sense, another inequivalent form of is (7) taken as:

$$u_{n+1} - au_n = 0 \tag{8}$$

Remarks 1. *Equations (7) and (8) are equivalent in a crisp sense, but in fuzzy sense they are not equivalent.*

Proof 1. If we take the fuzzy difference Equation (7), it becomes Theorem 1.

$$[u_{n+1}]_{(\alpha,\beta,\gamma)} = [au_n]_{(\alpha,\beta,\gamma)}$$

or

$$\begin{aligned}&\left[u^1_{L,n+1}(\alpha), u^1_{R,n+1}(\alpha); u^2_{L,n+1}(\beta), u^2_{R,n+1}(\beta); u^3_{L,n+1}(\gamma), u^3_{R,n+1}(\gamma)\right]\\ &= a\left[\left[u^1_{L,n}(\alpha), u^1_{R,n}(\alpha); u^2_{L,n}(\beta), u^2_{R,n}(\beta); u^3_{L,n}(\gamma), u^3_{R,n}(\gamma)\right]\right],\end{aligned}$$

i.e.,

$$\begin{cases} u^1_{L,n+1}(\alpha) = au^1_{L,n}(\alpha) \\ u^1_{R,n+1}(\alpha) = au^1_{R,n}(\alpha) \\ u^2_{L,n+1}(\beta) = au^2_{L,n}(\beta) \\ u^2_{R,n+1}(\beta) = au^2_{R,n}(\beta) \\ u^3_{L,n+1}(\gamma) = au^3_{L,n}(\gamma) \\ u^3_{R,n+1}(\gamma) = au^3_{R,n}(\gamma) \end{cases} \tag{9}$$

but when we take (8), it becomes Theorem 1.

$$[u_{n+1}]_{(\alpha,\beta,\gamma)} - [au_n]_{(\alpha,\beta,\gamma)} = 0$$

or

$$\begin{aligned}&\left[u^1_{L,n+1}(\alpha), u^1_{R,n+1}(\alpha); u^2_{L,n+1}(\beta), u^2_{R,n+1}(\beta); u^3_{L,n+1}(\gamma), u^3_{R,n+1}(\gamma)\right]\\ &-a\left[\left[u^1_{L,n}(\alpha), u^1_{R,n}(\alpha); u^2_{L,n}(\beta), u^2_{R,n}(\beta); u^3_{L,n}(\gamma), u^3_{R,n}(\gamma)\right]\right] = 0,\end{aligned}$$

i.e.,

$$\begin{cases} u^1_{L,n+1}(\alpha) - au^1_{R,n}(\alpha) = 0 \\ u^1_{R,n+1}(\alpha) - au^1_{L,n}(\alpha) = 0 \\ u^2_{L,n+1}(\beta) - au^2_{R,n}(\beta) = 0 \\ u^2_{R,n+1}(\beta) - au^2_{L,n}(\beta) = 0 \\ u^3_{L,n+1}(\gamma) - au^3_{R,n}(\gamma) = 0 \\ u^3_{R,n+1}(\gamma) - au^3_{L,n}(\gamma) = 0 \end{cases}$$

or

$$\begin{cases} u^1_{L,n+1}(\alpha) = au^1_{R,n}(\alpha) \\ u^1_{R,n+1}(\alpha) = au^1_{L,n}(\alpha) \\ u^2_{L,n+1}(\beta) = au^2_{R,n}(\beta) \\ u^2_{R,n+1}(\beta) = au^2_{L,n}(\beta) \\ u^3_{L,n+1}(\gamma) = au^3_{R,n}(\gamma) \\ u^3_{R,n+1}(\gamma) = au^3_{L,n}(\gamma) \end{cases} \tag{10}$$

Clearly, from (9) and (10), we conclude that they are different.

Therefore, in a crisp sense, (7) and (8) are the same, but not in a neutrosophic sense. □

Theorem 3. *Supposea and u_0 are positive neutrosophic numbers, then $\exists$ is a unique positive solution for Equation (7).*

Proof 2. Let the (α, β, γ)-cut of the positive neutrosophic number $\widetilde{u}_0$ be defined as $[\widetilde{u}_0]_{(\alpha,\beta,\gamma)} = \left[u^1_{L,0}(\alpha), u^1_{R,0}(\alpha); u^2_{L,0}(\beta), u^2_{R,0}(\beta); u^3_{L,0}(\gamma), u^3_{R,0}(\gamma)\right]$ and $[\widetilde{a}]_{(\alpha,\beta,\gamma)} = \left[a^1_L(\alpha), a^1_R(\alpha); a^2_L(\beta), a^2_R(\beta); a^3_L(\gamma), a^3_R(\gamma)\right], \forall \alpha, \beta, \gamma \in [0,1]$, and $0 \le \alpha+\beta+\gamma \le 1$, and if $\widetilde{u}_0 = [\xi_1, \xi_2, \xi_3; \eta_1, \eta_2, \eta_3; \zeta_1, \zeta_2, \zeta_3]$ then,

$$\begin{cases} u^1_{L,0}(\alpha) = \xi_1 + \alpha(\xi_2 - \xi_1) \\ u^1_{R,0}(\alpha) = \xi_3 - \alpha(\xi_3 - \xi_2) \\ u^2_{L,0}(\beta) = \eta_2 - \beta(\eta_2 - \eta_1) \\ u^2_{L,0}(\beta) = \eta_2 + \beta(\eta_3 - \eta_2) \\ u^3_{L,0}(\gamma) = \zeta_2 - \gamma(\zeta_2 - \zeta_1) \\ u^3_{L,0}(\gamma) = \zeta_2 + \gamma(\zeta_3 - \zeta_2) \end{cases}$$

Suppose there exists a sequence of netrosophic numbers u_n of Equation (7), with the positive netrosophic number u_0. Taking the (α, β, γ)-cut of Equation (7), we have:

$$[u_{n+1}]_{(\alpha,\beta,\gamma)} = [au_n]_{(\alpha,\beta,\gamma)} = [a]_{(\alpha,\beta,\gamma)}[u_n]_{(\alpha,\beta,\gamma)}$$

or

$$\begin{aligned} &\left[u^1_{L,n+1}(\alpha), u^1_{R,n+1}(\alpha); u^2_{L,n+1}(\beta), u^2_{R,n+1}(\beta); u^3_{L,n+1}(\gamma), u^3_{R,n+1}(\gamma)\right] \\ &= \left[a^1_L(\alpha), a^1_R(\alpha); a^2_L(\beta), a^2_R(\beta); a^3_L(\gamma), a^3_R(\gamma)\right]\left[u^1_{L,0}(\alpha), u^1_{R,0}(\alpha); u^2_{L,0}(\beta), u^2_{R,0}(\beta); u^3_{L,0}(\gamma), u^3_{R,0}(\gamma)\right] \end{aligned} \quad (11)$$

Equation (11) then forwards the following system of the crisp homogeneous linear difference equation for all α, β, and $\gamma \in [0,1]$, as follows:

$$\begin{cases} u^1_{L,n+1}(\alpha) = a^1_L(\alpha)u^1_{L,n}(\alpha) \\ u^1_{R,n+1}(\alpha) = a^1_R(\alpha)u^1_{R,n}(\alpha) \\ u^2_{L,n+1}(\beta) = a^2_L(\beta)u^2_{L,n}(\beta) \\ u^2_{R,n+1}(\beta) = a^2_R(\beta)u^2_{R,n}(\beta) \\ u^3_{L,n+1}(\gamma) = a^3_L(\gamma)u^3_{L,n}(\gamma) \\ u^3_{R,n+1}(\gamma) = a^3_R(\gamma)u^3_{R,n}(\gamma) \end{cases} \quad (12)$$

and Equation (12) has unique solutions $\left[u^1_{L,n}(\alpha), u^1_{R,n}(\alpha); u^2_{L,n+1}(\beta), u^2_{R,n}(\beta); u^3_{L,n}(\gamma), u^3_{R,n}(\gamma)\right]$ with an initial condition $\left[u^1_{L,0}(\alpha), u^1_{R,0}(\alpha); u^2_{L,0}(\beta), u^2_{R,0}(\beta); u^3_{L,0}(\gamma), u^3_{R,0}(\gamma)\right]$.

(The unique solution concept of a difference equation is taken from [42])

Therefore, using Equation (3), solutions are as follows:

$$\begin{cases} u^1_{L,n}(\alpha) = \left(a^1_L(\alpha)\right)^n u^1_{L,0}(\alpha) \\ u^1_{R,n}(\alpha) = \left(a^1_R(\alpha)\right)^n u^1_{R,0}(\alpha) \\ u^2_{L,n}(\beta) = \left(a^2_L(\beta)\right)^n u^2_{L,0}(\beta) \\ u^2_{R,n}(\beta) = \left(a^2_R(\beta)\right)^n u^2_{R,0}(\beta) \\ u^3_{L,n}(\gamma) = \left(a^3_L(\gamma)\right)^n u^3_{L,0}(\gamma) \\ u^3_{R,n}(\gamma) = \left(a^3_R(\gamma)\right)^n u^3_{R,0}(\gamma) \end{cases} \quad (13)$$

We show that $\left[u^1_{L,n}(\alpha),\ u^1_{R,n}(\alpha);\ u^2_{L,n}(\beta),\ u^2_{R,n}(\beta);\ u^3_{L,n}(\gamma),\ u^3_{R,n}(\gamma)\right]$, where each components are given (by 4.5) with the initial condition $\left[u^1_{L,0}(\alpha),\ u^1_{R,0}(\alpha);\ u^2_{L,0}(\beta),\ u^2_{R,0}(\beta);\ u^3_{L,0}(\gamma),\ u^3_{R,0}(\gamma)\right]$, which indicates the $(\alpha,\ \beta,\ \gamma)$-cut of solution $\widetilde{u}_n$ of (7) with initial condition $\widetilde{u}_0$, so that:

$$[u_n]_{(\alpha,\beta,\gamma)} = \left[u^1_{L,n}(\alpha),\ u^1_{R,n}(\alpha);\ u^2_{L,n}(\beta),\ u^2_{R,n}(\beta);\ u^3_{L,n}(\gamma),\ u^3_{R,n}(\gamma)\right] \tag{14}$$

Now,

$$\begin{aligned}&\left[u^1_{L,n}(\alpha),\ u^1_{R,n}(\alpha);\ u^2_{L,n}(\beta),\ u^2_{R,n}(\beta);\ u^3_{L,n}(\gamma),\ u^3_{R,n}(\gamma)\right]\\ &= \begin{bmatrix} \left(a^1_L(\alpha)\right)^n u^1_{L,0}(\alpha), \left(a^1_R(\alpha)\right)^n u^1_{R,0}(\alpha); \\ \left(a^2_L(\beta)\right)^n u^2_{L,0}(\beta), \left(a^2_R(\beta)\right)^n u^2_{R,0}(\beta); \\ \left(a^3_L(\gamma)\right)^n u^3_{L,0}(\gamma), \left(a^3_R(\gamma)\right)^n u^3_{R,0}(\gamma) \end{bmatrix}\\ &= [au_n]_{(\alpha,\beta,\gamma)}\end{aligned}$$

Therefore, $\left[u^1_{L,n}(\alpha),\ u^1_{R,n}(\alpha);\ u^2_{L,n}(\beta),\ u^2_{R,n}(\beta);\ u^3_{L,n}(\gamma),\ u^3_{R,n}(\gamma)\right]$ represents a positive neutrosophic number, such that $u_n = a^n u_0$ is the solution of (7).

To prove the uniqueness of the solution, let us assume that there exists an alternative solution $\hat{u}_n$ for Equation (4.1). Proceeding in asimilar way, we then have:

$$[\hat{u}_n]_{(\alpha,\beta,\gamma)} = \left[u^1_{L,n}(\alpha),\ u^1_{R,n}(\alpha);\ u^2_{L,n}(\beta),\ u^2_{R,n}(\beta);\ u^3_{L,n}(\gamma),\ u^3_{R,n}(\gamma)\right] \text{ for all } (\alpha,\beta,\gamma) \in [0,1]. \tag{15}$$

Therefore, from Equations (14) and (15), we obtain $[\hat{u}_n]_{(\alpha,\beta,\gamma)} = [u_n]_{(\alpha,\beta,\gamma)}$ for all $(\alpha,\beta,\gamma) \in [0,1]$, i.e., $\hat{u}_n = u_n$. Thus, the theorem is proved. □

Theorem 4. *Let a and u_0 are positive neutrosophic numbers. There also exists a unique positive solution for Equation (8).*

Proof 3. The proof of this theorem is almost similar to Theorem (3). □

Theorem 5. *Let a and u_0b epositive neutrosophic numbers, and $\max\left\{a^1_L(\alpha), a^1_R(\alpha); a^2_L(\beta), a^2_R(\beta);\ a^3_L(\gamma), a^3_R(\gamma)\right\} < 1,\ \forall\ \alpha\ ,\beta,\gamma \in [0,1]$ and $supp(u_0) \subset [M_1, N_1]$, where M_1, N_1 are finite positive real numbers. All the sequences of positive neutrosophic solution of Equation (7) are then bounded and persist.*

Proof 4. Let u_n be a sequence of positive neutrosophic solutions of Equation (7). Since $\max\left\{a^1_L(\alpha), a^1_R(\alpha); a^2_L(\beta), a^2_R(\beta);\ a^3_L(\gamma), a^3_R(\gamma)\right\} < 1, \forall \alpha,\beta,\gamma \in [0,1]$ and $supp(u_0) \subset [M_1, N_1]$, where M_1, N_1 are finite positive real numbers, it is evident from Equation (9) that all the component solutions of neutrosophic positive solution u_n converge to 0 as $n \to \infty$ i.e., $u_n \to 0_{netro}$ as $n \to \infty$, where $(0_{neutro})_{(\alpha,\beta,\gamma)} = [0,0;0,0;0,0]$. Since every convergent sequence is bounded, the sequence of positive neutrosophic solutions u_n of Equation (7) is bounded. □

Theorem 6. *Let a and u_0 bepositive neutrosophic numbers and $\max\left\{a^1_L(\alpha), a^1_R(\alpha); a^2_L(\beta), a^2_R(\beta);\ a^3_L(\gamma), a^3_R(\gamma)\right\} < 1,\ \forall\ \alpha\ ,\beta,\gamma \in [0,1]$ and $supp(u_0) \subset [M_1, N_1]$, where M_1, N_1 are finite positive real numbers. All the sequences of positive neutrosophic solutions of Equation (8) are then bounded and persist.*

4.1. Solution of Homogeneous Difference Equation of Type I

Consider Equation (4.1) with the fuzzy initial condition $\widetilde{u}_{n=0} = \widetilde{u}_0$ as a neutrosophic number.

Let $[\widetilde{u}_0]_{(\alpha,\beta,\gamma)} = \left[u^1_{L,0}(\alpha), u^1_{R,0}(\alpha); u^2_{L,0}(\beta), u^2_{R,0}(\beta); u^3_{L,0}(\gamma), u^3_{R,0}(\gamma)\right]$, $\forall\, \alpha,\beta,\gamma \in [0,1]$, and $0 < \alpha+\beta+\gamma < 3$, where, $[\widetilde{u}_0]_{(\alpha,\beta,\gamma)}$ is the (α,β,γ)-cut of $\widetilde{u}_0$ and, if $\widetilde{u}_0 = [\xi_1,\xi_2,\xi_3;\eta_1,\eta_2,\eta_3;\zeta_1,\zeta_2,\zeta_3]$, then

$$\begin{cases} u^1_{L,0}(\alpha) = \xi_1 + \alpha(\xi_2 - \xi_1) \\ u^1_{R,0}(\alpha) = \xi_3 - \alpha(\xi_3 - \xi_2) \\ u^2_{L,0}(\beta) = \eta_2 - \beta(\eta_2 - \eta_1) \\ u^2_{L,0}(\beta) = \eta_2 + \beta(\eta_3 - \eta_2) \\ u^3_{L,0}(\gamma) = \zeta_2 - \gamma(\zeta_2 - \zeta_1) \\ u^3_{L,0}(\gamma) = \zeta_2 + \gamma(\zeta_3 - \zeta_2) \end{cases} \tag{16}$$

4.1.1. The Solution When a > 0 Is a Crisp Number and u_0 Is a Neutrosophic Number

Taking the (α,β,γ)-cut of Equation (7), we have the following equations:

$$\begin{cases} u^1_{L,n+1}(\alpha) = au^1_{L,n}(\alpha) \\ u^1_{R,n+1}(\alpha) = au^1_{R,n}(\alpha) \\ u^2_{L,n+1}(\beta) = au^2_{L,n}(\beta) \\ u^2_{R,n+1}(\beta) = au^2_{R,n}(\beta) \\ u^3_{L,n+1}(\gamma) = au^3_{L,n}(\gamma) \\ u^3_{R,n+1}(\gamma) = au^3_{R,n}(\gamma) \end{cases} \tag{17}$$

Solutions of the above equations are:

$$\begin{cases} u^1_{L,n}(\alpha) = a^n u^1_{L,0}(\alpha) \\ u^1_{R,n}(\alpha) = a^n u^1_{R,0}(\alpha) \\ u^2_{L,n}(\beta) = a^n u^2_{L,0}(\beta) \\ u^2_{R,n}(\beta) = a^n u^2_{R,0}(\beta) \\ u^3_{L,n}(\gamma) = a^n u^3_{L,0}(\gamma) \\ u^3_{R,n}(\gamma) = a^n u^3_{R,0}(\gamma) \end{cases} \tag{18}$$

4.1.2. The Solution When a = 1 and the Initial Value u_0 is a Neutrosophic Number

In this case, a sequence of solutions is given by

$$\begin{cases} u^1_{L,n}(\alpha) = u^1_{L,0}(\alpha) \\ u^1_{R,n}(\alpha) = u^1_{R,0}(\alpha) \\ u^2_{L,n}(\beta) = u^2_{L,0}(\beta) \\ u^2_{R,n}(\beta) = u^2_{R,0}(\beta) \\ u^3_{L,n}(\gamma) = u^3_{L,0}(\gamma) \\ u^3_{R,n}(\gamma) = u^3_{R,0}(\gamma) \end{cases} \tag{19}$$

which lead to convergent solutions.

4.1.3. The Solution When $a < 0$ and the Initial Value u_0 Is a Neutrosophic Number

Let $a = -\mu,\ \mu > 0$, the real valued number.
From Equation (7), we then have

$$\left[\underline{u}_{n+1}(\alpha), \overline{u}_{n+1}(\alpha)\right] = -\mu\left[\underline{u}_n(\alpha), \overline{u}_n(\alpha)\right] \tag{20}$$

Therefore, we obtain the following:

$$\begin{cases} u^1_{L,n+1}(\alpha) = -\mu u^1_{R,n}(\alpha) \\ u^1_{R,n+1}(\alpha) = -\mu u^1_{L,n}(\alpha) \\ u^2_{L,n+1}(\beta) = -\mu u^2_{R,n}(\beta) \\ u^2_{R,n+1}(\beta) = -\mu u^2_{L,n}(\beta) \\ u^3_{L,n+1}(\gamma) = -\mu u^3_{R,n}(\gamma) \\ u^3_{R,n+1}(\gamma) = -\mu u^3_{L,n}(\gamma) \end{cases} \tag{21}$$

The first pairs of equations can be written in the matrix form as:

$$\begin{pmatrix} u^1_{L,n+1}(\alpha) \\ u^1_{R,n+1}(\alpha) \end{pmatrix} = \begin{pmatrix} 0 & -\mu \\ -\mu & 0 \end{pmatrix} \begin{pmatrix} u^1_{L,n}(\alpha) \\ u^1_{R,n}(\alpha) \end{pmatrix} \tag{22}$$

From Equation (22), let the co-efficient matrix be $A_1 = \begin{pmatrix} 0 & -\mu \\ -\mu & 0 \end{pmatrix}$

Therefore,

$$A_1{}^n = \begin{cases} \begin{pmatrix} \mu^n & 0 \\ 0 & \mu^n \end{pmatrix} \textit{when n is an even natural number} \\ \begin{pmatrix} 0 & -\mu^n \\ -\mu^n & 0 \end{pmatrix} \textit{when n is an odd natural number} \end{cases}$$

Therefore, the solution of (4.1.6), using Theorem (3.1), is given by:

$$\begin{pmatrix} u^1_{L,n}(\alpha) \\ u^1_{R,n}(\alpha) \end{pmatrix} = A_1^n \begin{pmatrix} u^1_{L,0}(\alpha) \\ u^1_{R,0}(\alpha) \end{pmatrix} \tag{23}$$

When n is an even natural number, the general solutions are:

$$\begin{cases} u^1_{L,n}(\alpha) = \mu^n u^1_{L,0}(\alpha) \\ u^1_{R,n}(\alpha) = \mu^n u^1_{R,0}(\alpha) \\ u^2_{L,n}(\beta) = \mu^n u^2_{L,0}(\beta) \\ u^2_{R,n}(\beta) = \mu^n u^2_{R,0}(\beta) \\ u^3_{L,n}(\gamma) = \mu^n u^3_{L,0}(\gamma) \\ u^3_{R,n}(\gamma) = \mu^n u^3_{R,0}(\gamma) \end{cases} \tag{24}$$

When n is odd natural number, the general solutions are:

$$\begin{cases} u^1_{L,n}(\alpha) = -\mu^n u^1_{R,0}(\alpha) \\ u^1_{R,n}(\alpha) = -\mu^n u^1_{L,0}(\alpha) \\ u^2_{L,n}(\beta) = -\mu^n u^2_{R,0}(\beta) \\ u^2_{R,n}(\beta) = -\mu^n u^2_{L,0}(\beta) \\ u^3_{L,n}(\gamma) = -\mu^n u^3_{R,0}(\gamma) \\ u^3_{R,n}(\gamma) = -\mu^n u^3_{L,0}(\gamma) \end{cases} \tag{25}$$

4.1.4. The Solution When a > 0 Is Aneutrosophic Number and the Initial Value u_0 Is a Crisp Number

Let $[\overline{a}]_{(\alpha,\beta,\gamma)} = \left[a^1_L(\alpha), a^1_R(\alpha);\ a^2_L(\beta), a^2_R(\beta);\ a^3_L(\gamma), a^3_R(\gamma)\right]$, $\forall\ \alpha, \beta, \gamma \in [0,1]$, and $0 \le \alpha + \beta + \gamma \le 3$.

Taking the (α, β, γ)-cut of Equation (7), we have the following equation:

$$\begin{cases} u^1_{L,n+1}(\alpha) = a^1_L(\alpha)u^1_{L,n}(\alpha) \\ u^1_{R,n+1}(\alpha) = a^1_R(\alpha)u^1_{R,n}(\alpha) \\ u^2_{L,n+1}(\beta) = a^2_L(\beta)u^2_{L,n}(\beta) \\ u^2_{R,n+1}(\beta) = a^2_R(\beta)u^2_{R,n}(\beta) \\ u^3_{L,n+1}(\gamma) = a^3_L(\gamma)u^3_{L,n}(\gamma) \\ u^3_{R,n+1}(\gamma) = a^3_R(\gamma)u^3_{R,n}(\gamma) \end{cases} \tag{26}$$

where u_0 is the initial value. The solutions are as follows:

$$\begin{cases} u^1_{L,n}(\alpha) = \left(a^1_L(\alpha)\right)^n u_0 \\ u^1_{R,n}(\alpha) = \left(a^1_R(\alpha)\right)^n u_0 \\ u^2_{L,n}(\beta) = \left(a^2_L(\beta)\right)^n u_0 \\ u^2_{R,n}(\beta) = \left(a^2_R(\beta)\right)^n u_0 \\ u^3_{L,n}(\gamma) = \left(a^3_L(\gamma)\right)^n u_0 \\ u^3_{R,n}(\gamma) = \left(a^3_R(\gamma)\right)^n u_0 \end{cases} \tag{27}$$

4.1.5. The Solution When $a < 0$ Is a Neutrosophic Number and the Initial Value u_0 Is a Crisp Number

Let $a = -\mu$, where μ is a positive fuzzy number. $[\widetilde{\mu}]_{(\alpha,\,\beta,\gamma)} = \left[\mu^1_L(\alpha), \mu^1_R(\alpha);\ \mu^2_L(\beta), \mu^2_R(\beta);\ \mu^3_L(\gamma), \mu^3_R(\gamma)\right]$, $\forall\ \alpha, \beta, \gamma \in [0,1]$, and $0 \le \alpha + \beta + \gamma \le 3$.

Equation (7) then splits into the following equations:

$$\begin{cases} u^1_{L,n+1}(\alpha) = -\mu^1_R(\alpha)u^1_{R,n}(\alpha) \\ u^1_{R,n+1}(\alpha) = -\mu^1_L(\alpha)u^1_{L,n}(\alpha) \\ u^2_{L,n+1}(\beta) = -\mu^2_R(\beta)u^2_{R,n}(\beta) \\ u^2_{R,n+1}(\beta) = -\mu^2_L(\beta)u^2_{L,n}(\beta) \\ u^3_{L,n+1}(\gamma) = -\mu^3_R(\gamma)u^3_{R,n}(\gamma) \\ u^3_{R,n+1}(\gamma) = -\mu^3_L(\gamma)u^3_{L,n}(\gamma) \end{cases} \tag{28}$$

In the matrix form, the first pairs of equations of Equation (28) can be written as:

$$\begin{pmatrix} u^1_{L,n+1}(\alpha) \\ u^1_{R,n+1}(\alpha) \end{pmatrix} = \begin{pmatrix} 0 & -\mu^1_R(\alpha) \\ -\mu^1_L(\alpha) & 0 \end{pmatrix} \begin{pmatrix} u^1_{L,n}(\alpha) \\ u^1_{R,n}(\alpha) \end{pmatrix} \tag{29}$$

The solution of (29) is given by:

$$\begin{pmatrix} u^1_{L,n}(\alpha) \\ u^1_{R,n}(\alpha) \end{pmatrix} = A_2^n \begin{pmatrix} u_0 \\ u_0 \end{pmatrix} \tag{30}$$

where,

$$A_2 = \begin{pmatrix} 0 & -\mu^1_R(\alpha) \\ -\mu^1_L(\alpha) & 0 \end{pmatrix}$$

and

$$A_2^n = \begin{cases} \begin{pmatrix} \left(\mu^1_L(\alpha)\mu^1_R(\alpha)\right)^{\frac{n}{2}} & 0 \\ 0 & \left(\mu^1_L(\alpha)\mu^1_R(\alpha)\right)^{\frac{n}{2}} \end{pmatrix} \text{when } n \text{ is even} \\ \begin{pmatrix} 0 & -\left(\mu^1_L(\alpha)\right)^{\frac{n-1}{2}} \left(\mu^1_R(\alpha)\right)^{\frac{n+1}{2}} \\ -\left(\mu^1_L(\alpha)\right)^{\frac{n+1}{2}} \left(\mu^1_R(\alpha)\right)^{\frac{n-1}{2}} & 0 \end{pmatrix} \text{when } n \text{ is odd} \end{cases}$$

The solution of Equation (30) when n is even is:

$$\begin{cases} u^1_{L,n}(\alpha) = \left(\mu^1_L(\alpha)\ \mu^1_R(\alpha)\right)^{\frac{n}{2}} u_0 \\ u^1_{R,n}(\alpha) = \left(\mu^1_L(\alpha)\ \left(\mu^1_R\alpha\right)\right)^{\frac{n}{2}} u_0 \\ u^2_{L,n}(\beta) = \left(\mu^2_L(\beta)\ \mu^2_R(\beta)\right)^{\frac{n}{2}} u_0 \\ u^2_{R,n}(\beta) = \left(\mu^2_L(\beta)\ \mu^2_R(\beta)\right)^{\frac{n}{2}} u_0 \\ u^3_{L,n}(\gamma) = \left(\mu^3_L(\gamma)\ \mu^3_R(\gamma)\right)^{\frac{n}{2}} u_0 \\ u^3_{R,n}(\gamma) = \left(\mu^3_L(\gamma)\ \mu^3_R(\gamma)\right)^{\frac{n}{2}} u_0 \end{cases} \tag{31}$$

In this case, solutions become crisp numbers, i.e., $u_n(\alpha) = \left(\underline{\mu}(\alpha)\ \overline{\mu}(\alpha)\right)^{\frac{n}{2}} u_0$.
The solution of Equation (30) when n is odd:

$$\begin{cases} u^1_{L,n}(\alpha) = -\left(\mu^1_L(\alpha)\right)^{\frac{n-1}{2}} \left(\mu^1_R(\alpha)\right)^{\frac{n+1}{2}} u_0 \\ u^1_{R,n}(\alpha) = -\left(\mu^1_L(\alpha)\ )\right)^{\frac{n+1}{2}} \left(\mu^1_R(\alpha)\right)^{\frac{n-1}{2}} u_0 \\ u^2_{L,n}(\beta) = -\left(\mu^2_L(\beta)\right)^{\frac{n-1}{2}} \left(\mu^2_R(\beta)\right)^{\frac{n+1}{2}} u_0 \\ u^2_{R,n}(\beta) = -\left(\mu^2_L(\beta)\ )\right)^{\frac{n+1}{2}} \left(\mu^2_R(\beta)\right)^{\frac{n-1}{2}} u_0 \\ u^3_{L,n}(\gamma) = -\left(\mu^3_L(\gamma)\right)^{\frac{n-1}{2}} \left(\mu^3_R(\gamma)\right)^{\frac{n+1}{2}} u_0 \\ u^3_{R,n}(\gamma) = -\left(\mu^3_L(\gamma)\ )\right)^{\frac{n+1}{2}} \left(\mu^3_R(\gamma)\right)^{\frac{n-1}{2}} u_0 \end{cases} \tag{32}$$

4.1.6. The Solution When a > 0 and u_0 Are Bothneutrosophic Numbers

Let

$$[\widetilde{a}]_{(\alpha,\ \beta,\ \gamma)} = \left[a^1_L(\alpha), a^1_R(\alpha);\ a^2_L(\beta), a^2_R(\beta);\ a^3_L(\gamma), a^3_R(\gamma)\right]$$

$$[\widetilde{u}_0]_{(\alpha,\ \beta,\ \gamma)} = \left[u^1_{L,0}(\alpha),\ u^1_{R,0}(\alpha);\ u^2_{L,0}(\beta),\ u^2_{R,0}(\beta);\ u^3_{L,0}(\gamma),\ u^3_{R,0}(\gamma)\right]$$

$\forall\ \alpha, \beta, \gamma \in [0,1]$ and $0 \le \alpha + \beta + \gamma \le 3$.
The solution of Equation (16), which follows from Equation (26), is thengiven by:

$$\begin{cases} u^1_{L,n}(\alpha) = \left(a^1_L(\alpha)\right)^n u^1_{L,0}(\alpha) \\ u^1_{R,n}(\alpha) = \left(a^1_R(\alpha)\right)^n u^1_{R,0}(\alpha) \\ u^2_{L,n}(\beta) = \left(a^2_L(\beta)\right)^n u^2_{L,0}(\beta) \\ u^2_{R,n}(\beta) = \left(a^2_R(\beta)\right)^n u^2_{R,0}(\beta) \\ u^3_{L,n}(\gamma) = \left(a^3_L(\gamma)\right)^n u^3_{L,0}(\gamma) \\ u^3_{R,n}(\gamma) = \left(a^3_R(\gamma)\right)^n u^3_{R,0}(\gamma) \end{cases} \tag{33}$$

4.1.7. The Solution When a < 0 and u_0 Are Both Neutrosophic Numbers

Let $a = -\mu$, $\mu > 0$. Let $[\widetilde{\mu}]_{(\alpha,\ \beta,\ \gamma)} = \left[\mu^1_L(\alpha), \mu^1_R(\alpha);\ \mu^2_L(\beta), \mu^2_R(\beta);\ \mu^3_L(\gamma), \mu^3_R(\gamma)\right]$ and $[\widetilde{u}_0]_{(\alpha,\ \beta,\ \gamma)} = \left[u^1_{L,0}(\alpha),\ u^1_{R,0}(\alpha);\ u^2_{L,0}(\beta),\ u^2_{R,0}(\beta);\ u^3_{L,0}(\gamma),\ u^3_{R,0}(\gamma)\right]$
$\forall\ \alpha, \beta, \gamma \in [0,1]$ and $0 \le \alpha + \beta + \gamma \le 3$.

The solution of Equation (16), which is follows from Equation (31), is then given by:

$$\begin{cases} u^1_{L,n}(\alpha) = \left(\mu^1_L(\alpha)\,\mu^1_R(\alpha)\right)^{\frac{n}{2}} u^1_{L,0}(\alpha) \\ u^1_{R,n}(\alpha) = \left(\mu^1_L(\alpha)\,\mu^1_R(\alpha)\right)^{\frac{n}{2}} u^1_{R,0}(\alpha) \\ u^2_{L,n}(\beta) = \left(\mu^2_L(\beta)\,\mu^2_R(\beta)\right)^{\frac{n}{2}} u^2_{L,0}(\beta) \\ u^2_{R,n}(\beta) = \left(\mu^2_L(\beta)\,\mu^2_R(\beta)\right)^{\frac{n}{2}} u^2_{R,0}(\beta) \\ u^3_{L,n}(\gamma) = \left(\mu^3_L(\gamma)\,\mu^3_R(\gamma)\right)^{\frac{n}{2}} u^3_{L,0}(\gamma) \\ u^3_{R,n}(\gamma) = \left(\mu^3_L(\gamma)\,\mu^3_R(\gamma)\right)^{\frac{n}{2}} u^3_{R,0}(\gamma) \end{cases} \tag{34}$$

The above equations show that the solution for n is even only. When n is odd, the solutions, which follow from Equation (32), are as follows:

$$\begin{cases} u^1_{L,n}(\alpha) = -\left(\mu^1_L(\alpha)\right)^{\frac{n-1}{2}}\left(\mu^1_R(\alpha)\right)^{\frac{n+1}{2}} u^1_{L,0}(\alpha) \\ u^1_{R,n}(\alpha) = -\left(\mu^1_L(\alpha)\right)^{\frac{n+1}{2}}\left(\mu^1_R(\alpha)\right)^{\frac{n-1}{2}} u^1_{R,0}(\alpha) \\ u^2_{L,n}(\beta) = -\left(\mu^2_L(\beta)\right)^{\frac{n-1}{2}}\left(\mu^2_R(\beta)\right)^{\frac{n+1}{2}} u^2_{L,0}(\beta) \\ u^2_{R,n}(\beta) = -\left(\mu^2_L(\beta)\right)^{\frac{n+1}{2}}\left(\mu^2_R(\beta)\right)^{\frac{n-1}{2}} u^2_{R,0}(\beta) \\ u^3_{L,n}(\gamma) = -\left(\mu^3_L(\gamma)\right)^{\frac{n-1}{2}}\left(\mu^3_R(\gamma)\right)^{\frac{n+1}{2}} u^3_{L,0}(\gamma) \\ u^3_{R,n}(\gamma) = -\left(\mu^3_L(\gamma)\right)^{\frac{n+1}{2}}\left(\mu^3_R(\gamma)\right)^{\frac{n-1}{2}} u^3_{R,0}(\gamma) \end{cases} \tag{35}$$

4.2. *Solution of Homogeneous Difference Equation of Type II*

4.2.1. The Solution When $a = 1$ and the Initial Condition u_0 Is a Neutrosophic Number

Taking the $(\alpha,\ \beta,\ \gamma)$-cut of Equation (8), we have the following:

$$\begin{cases} u^1_{L,n+1}(\alpha) = u^1_{R,n}(\alpha) \\ u^1_{R,n+1}(\alpha) = u^1_{L,n}(\alpha) \\ u^2_{L,n+1}(\beta) = u^2_{R,n}(\beta) \\ u^2_{R,n+1}(\beta) = u^2_{L,n}(\beta) \\ u^3_{L,n+1}(\alpha) = u^3_{R,n}(\alpha) \\ u^3_{R,n+1}(\alpha) = u^3_{L,n}(\alpha) \end{cases} \tag{36}$$

In the matrix form, the first pairs of Equation (36) can be written as:

$$\begin{pmatrix} u^1_{L,n+1}(\alpha) \\ u^1_{R,n+1}(\alpha) \end{pmatrix} = \begin{pmatrix} 0 & 1 \\ 1 & 0 \end{pmatrix} \begin{pmatrix} u^1_{L,n}(\alpha) \\ u^1_{R,n}(\alpha) \end{pmatrix} \tag{37}$$

The solution of Equation (37) is, when n is even:

$$\begin{cases} u^1_{L,n}(\alpha) = u^1_{L,0}(\alpha) \\ u^1_{R,n}(\alpha) = u^1_{R,0}(\alpha) \end{cases} \tag{38}$$

When n is odd, the solutions are:

$$\begin{cases} u^1_{L,n}(\alpha) = u^1_{R,0}(\alpha) \\ u^1_{R,n}(\alpha) = u^1_{L,0}(\alpha) \end{cases} \tag{39}$$

For both cases, when either n is even or odd, $u^1_{L,n}(\alpha)$ and $u^1_{R,n}(\alpha)$ leads to a convergent solution.

In a similar way, solutions of remaining equations are as follows:
when n is even:

$$\begin{cases} u^1_{L,n}(\beta) = u^1_{L,0}(\beta) \\ u^1_{R,n}(\beta) = u^1_{R,0}(\beta) \\ u^1_{L,n}(\gamma) = u^1_{L,0}(\gamma) \\ u^1_{R,n}(\gamma) = u^1_{R,0}(\gamma) \end{cases} \tag{40}$$

When n is odd:

$$\begin{cases} u^1_{L,n}(\beta) = u^1_{R,0}(\beta) \\ u^1_{R,n}(\beta) = u^1_{L,0}(\beta) \\ u^1_{L,n}(\gamma) = u^1_{R,0}(\gamma) \\ u^1_{R,n}(\gamma) = u^1_{L,0}(\gamma) \end{cases} \tag{41}$$

4.2.2. The Solution When $a > 0$, a Real Valued Number, and the Initial Condition u_0 Is a Neutrosophic Number

Taking the (α, β, γ)-cut of (8), we get the following equations:

$$\begin{cases} u^1_{L,n+1}(\alpha) - au^1_{R,n}(\alpha) = 0 \\ u^1_{R,n+1}(\alpha) - au^1_{L,n}(\alpha) = 0 \\ u^1_{L,n+1}(\beta) - au^1_{R,n}(\beta) = 0 \\ u^1_{R,n+1}(\beta) - au^1_{L,n}(\beta) = 0 \\ u^1_{L,n+1}(\gamma) - au^1_{R,n}(\gamma) = 0 \\ u^1_{R,n+1}(\gamma) - au^1_{L,n}(\gamma) = 0 \end{cases} \tag{42}$$

In the matrix form, the first pair of Equation (42) can be written as:

$$\begin{pmatrix} u^1_{L,n+1}(\alpha) \\ u^1_{R,n+1}(\alpha) \end{pmatrix} = \begin{pmatrix} 0 & a \\ a & 0 \end{pmatrix} \begin{pmatrix} u^1_{L,n}(\alpha) \\ u^1_{R,n}(\alpha) \end{pmatrix} \tag{43}$$

The solutions of (43) are, when n is even:

$$\begin{cases} u^1_{L,n}(\alpha) = a^n u^1_{L,0}(\alpha) \\ u^1_{R,n}(\alpha) = a^n u^1_{R,0}(\alpha) \end{cases} \tag{44}$$

The solutions of (44) and (45) are, when n is odd:

$$\begin{cases} u^1_{L,n}(\alpha) = a^n u^1_{R,0}(\alpha) \\ u^1_{R,n}(\alpha) = a^n u^1_{L,0}(\alpha) \end{cases} \tag{45}$$

In a similar way, the solutions of the remaining Equation (42) are as follows:
When n is even:

$$\begin{cases} u^1_{L,n}(\beta) = a^n u^1_{L,0}(\beta) \\ u^1_{R,n}(\beta) = a^n u^1_{R,0}(\beta) \\ u^1_{L,n}(\gamma) = a^n u^1_{L,0}(\gamma) \\ u^1_{R,n}(\gamma) = a^n u^1_{R,0}(\gamma) \end{cases} \tag{46}$$

When n is odd:

$$\begin{cases} u^1_{L,n}(\beta) = a^n u^1_{R,0}(\beta) \\ u^1_{R,n}(\beta) = a^n u^1_{L,0}(\beta) \\ u^1_{L,n}(\gamma) = a^n u^1_{R,0}(\gamma) \\ u^1_{R,n}(\gamma) = a^n u^1_{L,0}(\gamma) \end{cases} \tag{47}$$

4.2.3. The Solution When $a < 0$ and When the Initial Condition u_0 Is a Neutrosophic Number

Let $a = -m$, $m > 0$, a real valued number.

From Equation (8), after taking the $(\alpha,\ \beta,\ \gamma)$-cut, we have the following sets of equations:

$$\begin{cases} u^1_{L,n+1}(\alpha) + mu^1_{L,n}(\alpha) = 0 \\ u^1_{R,n+1}(\alpha) + mu^1_{R,n}(\alpha) = 0 \\ u^1_{L,n+1}(\beta) + mu^1_{L,n}(\beta) = 0 \\ u^1_{R,n+1}(\beta) + mu^1_{R,n}(\beta) = 0 \\ u^1_{L,n+1}(\gamma) + mu^1_{L,n}(\gamma) = 0 \\ u^1_{R,n+1}(\gamma) + mu^1_{R,n}(\gamma) = 0 \end{cases} \tag{48}$$

Solving the above equations, we get:

$$\begin{cases} u^1_{L,n}(\alpha) = (-m)^n u^1_{L,0}(\alpha) \\ u^1_{R,n}(\alpha) = (-m)^n u^1_{R,0}(\alpha) \\ u^2_{L,n}(\beta) = (-m)^n u^2_{L,0}(\beta) \\ u^2_{R,n}(\beta) = (-m)^n u^2_{R,0}(\beta) \\ u^3_{L,n}(\gamma) = (-m)^n u^3_{L,0}(\gamma) \\ u^3_{R,n}(\gamma) = (-m)^n u^3_{R,0}(\gamma) \end{cases} \tag{49}$$

4.2.4. The Solution When $a > 0$ Is a Positive Neutrosophic Number and the Initial Condition u_0 Is Nota Neutrosophic Number

Let $[\overline{a}]_{(\alpha,\ \beta,\ \gamma)} = \left[a^1_L(\alpha), a^1_R(\alpha);\ a^2_L(\beta), a^2_R(\beta);\ a^3_L(\gamma), a^3_R(\gamma)\right], \forall\ \alpha\ , \beta, \gamma \in [0,1]$, and $0 \le \alpha + \beta + \gamma \le 3$.

Taking the $(\alpha,\ \beta,\ \gamma)$-cut of Equation (8), we have the following equation:

$$\begin{cases} u^1_{L,n+1}(\alpha) = a^1_L(\alpha)u^1_{R,n}(\alpha) \\ u^1_{R,n+1}(\alpha) = a^1_R(\alpha)u^1_{L,n}(\alpha) \\ u^2_{L,n+1}(\beta) = a^2_L(\beta)u^2_{R,n}(\beta) \\ u^2_{R,n+1}(\beta) = a^2_R(\beta)u^2_{L,n}(\beta) \\ u^3_{L,n+1}(\gamma) = a^3_L(\gamma)u^3_{R,n}(\gamma) \\ u^3_{R,n+1}(\gamma) = a^3_R(\gamma)u^3_{L,n}(\gamma) \end{cases} \tag{50}$$

In the matrix form, among the above equations, the first pair of Equation (50) can be written as:

$$\begin{pmatrix} u^1_{L,n+1}(\alpha) \\ u^1_{R,n+1}(\alpha) \end{pmatrix} = \begin{pmatrix} 0 & a^1_R(\alpha) \\ a^1_L(\alpha) & 0 \end{pmatrix} \begin{pmatrix} u^1_{L,n}(\alpha) \\ u^1_{R,n}(\alpha) \end{pmatrix} \tag{51}$$

The solution of Equation (51), when n is even:

$$\begin{cases} u^1_{L,n}(\alpha) = \left(a^1_L(\alpha)\, a^1_R(\alpha)\right)^{\frac{n}{2}} u_0 \\ u^1_{R,n}(\alpha) = \left(a^1_L(\alpha)\, a^1_R(\alpha)\right)^{\frac{n}{2}} u_0 \\ u^2_{L,n}(\beta) = \left(a^2_L(\beta)\, a^2_R(\beta)\right)^{\frac{n}{2}} u_0 \\ u^2_{R,n}(\beta) = \left(a^2_L(\beta)\, a^2_R(\beta)\right)^{\frac{n}{2}} u_0 \\ u^3_{L,n}(\gamma) = \left(a^3_L(\gamma)\, a^3_R(\gamma)\right)^{\frac{n}{2}} u_0 \\ u^3_{R,n}(\gamma) = \left(a^3_L(\gamma)\, a^3_R(\gamma)\right)^{\frac{n}{2}} u_0 \end{cases} \tag{52}$$

When n is odd:

$$\begin{cases} u^1_{L,n}(\alpha) = \left(a^1_L(\alpha)\right)^{\frac{n-1}{2}}\left(a^1_R(\alpha)\right)^{\frac{n+1}{2}} u_0 \\ u^1_{R,n}(\alpha) = \left(a^1_L(\alpha)\right)^{\frac{n+1}{2}}\left(a^1_R(\alpha)\right)^{\frac{n-1}{2}} u_0 \\ u^2_{L,n}(\beta) = \left(a^2_L(\beta)\right)^{\frac{n-1}{2}}\left(a^2_R(\beta)\right)^{\frac{n+1}{2}} u_0 \\ u^2_{R,n}(\beta) = \left(a^2_L(\beta)\right)^{\frac{n+1}{2}}\left(a^2_R(\beta)\right)^{\frac{n-1}{2}} u_0 \\ u^3_{L,n}(\gamma) = \left(a^3_L(\gamma)\right)^{\frac{n-1}{2}}\left(a^3_R(\gamma)\right)^{\frac{n+1}{2}} u_0 \\ u^3_{R,n}(\gamma) = \left(a^3_L(\gamma)\right)^{\frac{n+1}{2}}\left(a^3_R(\gamma)\right)^{\frac{n-1}{2}} u_0 \end{cases} \quad (53)$$

4.2.5. The Solution When a < 0 Is a Neutrosophic Number and when the Initial Condition u_0 Is a Crisp Number

Let $a = -m$, $m > 0$. Let $[\widetilde{m}]_{(\alpha,\,\beta,\,\gamma)} = \left[m^1_L(\alpha), m^1_R(\alpha);\ m^2_L(\beta), m^2_R(\beta);\ m^3_L(\gamma), m^3_R(\gamma)\right]$ and $[\widetilde{u}_0]_{(\alpha,\,\beta,\,\gamma)} = \left[u^1_{L,0}(\alpha),\ u^1_{R,0}(\alpha);\ u^2_{L,0}(\beta),\ u^2_{R,0}(\beta);\ u^3_{L,0}(\gamma),\ u^3_{R,0}(\gamma)\right]$
$\forall\ \alpha, \beta, \gamma \in [0,1]$ and $0 \le \alpha + \beta + \gamma \le 3$.
Taking the $(\alpha,\ \beta,\ \gamma)$-cut of Equation (8), we have the following equations:

$$\begin{cases} u^1_{L,n+1}(\alpha) = -m^1_L(\alpha)u^1_{L,n}(\alpha) \\ u^1_{R,n+1}(\alpha) = -m^1_{R,n}(\alpha)u^1_{R,n}(\alpha) \\ u^2_{L,n+1}(\beta) = -m^2_L(\beta)u^2_{L,n}(\beta) \\ u^2_{R,n+1}(\beta) = -m^2_{R,n}(\beta)u^2_{R,n}(\beta) \\ u^3_{L,n+1}(\gamma) = -m^3_L(\gamma)u^3_{L,n}(\gamma) \\ u^3_{R,n+1}(\gamma) = -m^3_{R,n}(\gamma)u^3_{R,n}(\gamma) \end{cases} \quad (54)$$

The general solutions of the above equations are as follows:

$$\begin{cases} u_{L,n}(\alpha) = (-m_L(\alpha))^n u_0 \\ u_{R,n}(\alpha) = (-m_R(\alpha))^n u_0 \\ u_{L,n}(\beta) = (-m_L(\beta))^n u_0 \\ u_{R,n}(\beta) = (-m_R(\beta))^n u_0 \\ u_{L,n}(\gamma) = (-m_L(\gamma))^n u_0 \\ u_{R,n}(\gamma) = (-m_R(\gamma))^n u_0 \end{cases} \quad (55)$$

4.2.6. The Solution When the Initial Condition u_0 and a > 0 Are Both Neutrosophic Numbers

Let $[\widetilde{a}]_{(\alpha,\,\beta,\,\gamma)} = \left[a^1_L(\alpha), a^1_R(\alpha);\ a^2_L(\beta), a^2_R(\beta);\ a^3_L(\gamma), a^3_R(\gamma)\right]$ and $[\widetilde{u}_0]_{(\alpha,\,\beta,\,\gamma)} = \left[u^1_{L,0}(\alpha),\ u^1_{R,0}(\alpha);\ u^2_{L,0}(\beta),\ u^2_{R,0}(\beta);\ u^3_{L,0}(\gamma),\ u^3_{R,0}(\gamma)\right]$
$\forall\ \alpha, \beta, \gamma \in [0,1]$ and $0 \le \alpha + \beta + \gamma \le 3$.
In this case, the solutions are given, following from Equation (50):
when n is even:

$$\begin{cases} u^1_{L,n}(\alpha) = \left(a^1_L(\alpha)\,a^1_R(\alpha)\right)^{\frac{n}{2}} u^1_{L,0}(\alpha) \\ u^1_{R,n}(\alpha) = \left(a^1_L(\alpha)\,a^1_R(\alpha)\right)^{\frac{n}{2}} u^1_{R,0}(\alpha) \\ u^2_{L,n}(\beta) = \left(a^2_L(\beta)\,a^2_R(\beta)\right)^{\frac{n}{2}} u^2_{L,0}(\beta) \\ u^2_{R,n}(\beta) = \left(a^2_L(\beta)\,a^2_R(\beta)\right)^{\frac{n}{2}} u^2_{R,0}(\beta) \\ u^3_{L,n}(\gamma) = \left(a^3_L(\gamma)\,a^3_R(\gamma)\right)^{\frac{n}{2}} u^3_{L,0}(\gamma) \\ u^3_{R,n}(\gamma) = \left(a^3_L(\gamma)\,a^3_R(\gamma)\right)^{\frac{n}{2}} u^3_{R,0}(\gamma) \end{cases} \quad (56)$$

when n is odd:

$$\begin{cases} u^1_{L,n}(\alpha) = \left(a^1_L(\alpha)\right)^{\frac{n-1}{2}}\left(a^1_R(\alpha)\right)^{\frac{n+1}{2}} u^1_{R,0}(\alpha) \\ u^1_{R,n}(\alpha) = \left(a^1_L(\alpha)\right)^{\frac{n+1}{2}}\left(a^1_R(\alpha)\right)^{\frac{n-1}{2}} u^1_{L,0}(\alpha) \\ u^2_{L,n}(\beta) = \left(a^2_L(\beta)\right)^{\frac{n-1}{2}}\left(a^2_R(\beta)\right)^{\frac{n+1}{2}} u^2_{R,0}(\beta) \\ u^2_{R,n}(\beta) = \left(a^2_L(\beta)\right)^{\frac{n+1}{2}}\left(a^2_R(\beta)\right)^{\frac{n-1}{2}} u^2_{L,0}(\beta) \\ u^3_{L,n}(\gamma) = \left(a^3_L(\gamma)\right)^{\frac{n-1}{2}}\left(a^3_R(\gamma)\right)^{\frac{n+1}{2}} u^3_{R,0}(\gamma) \\ u^3_{R,n}(\gamma) = \left(a^3_L(\gamma)\right)^{\frac{n+1}{2}}\left(a^3_R(\gamma)\right)^{\frac{n-1}{2}} u^3_{L,0}(\gamma) \end{cases} \tag{57}$$

4.2.7. The Solution When the Initial Condition u_0 and $a < 0$ Are Both Neutrosophic Numbers

Let $a = -m$, $m > 0$. Let $[\widetilde{m}]_{(\alpha,\,\beta,\,\gamma)} = \left[m^1_L(\alpha), m^1_R(\alpha);\ m^2_L(\beta), m^2_R(\beta);\ m^3_L(\gamma), m^3_R(\gamma)\right]$ and $[\widetilde{u}_0]_{(\alpha,\,\beta,\,\gamma)} = \left[u^1_{L,0}(\alpha),\ u^1_{R,0}(\alpha);\ u^2_{L,0}(\beta),\ u^2_{R,0}(\beta);\ u^3_{L,0}(\gamma),\ u^3_{R,0}(\gamma)\right]$
$\forall\ \alpha, \beta, \gamma \in [0,1]$ and $0 \le \alpha + \beta + \gamma \le 3$.

In a similar way, as seen in Equation (54), we have the following solutions.

The general solutions of the above equations are as follows:

$$\begin{cases} u^1_{L,n}(\alpha) = (-m_L(\alpha))^n u^1_{L,0}(\alpha) \\ u^1_{R,n}(\alpha) = (-m_R(\alpha))^n u^1_{R,0}(\alpha) \\ u^2_{L,n}(\beta) = (-m_L(\beta))^n u^2_{L,0}(\beta) \\ u^2_{R,n}(\beta) = (-m_R(\beta))^n u^2_{R,0}(\beta) \\ u^3_{L,n}(\gamma) = (-m_L(\gamma))^n u^3_{L,0}(\gamma) \\ u^3_{R,n}(\gamma) = (-m_R(\gamma))^n u^3_{R,0}(\gamma) \end{cases} \tag{58}$$

5. Numerical Example

Example 1. *Solve the difference equation:*

$$u_{n+1} = (2,4,6;1,4,5;2,4,5)u_n \tag{59}$$

with the initial condition $\widetilde{u}_{n=0} = (50,60,70;55,60,75;50,60,80)$

Solution 1. *If the* $[\widetilde{u}_n]_{(\alpha,\,\beta,\,\gamma)}$ *is the* $(\alpha,\ \beta,\ \gamma)$*-cut of a sequence of neutrosophic numbers, then its components are as follows:*

$$\begin{cases} u^1_{L,n}(\alpha) = (2+2\alpha)^n(50+10\alpha) \\ u^1_{R,n}(\alpha) = (6-2\alpha)^n(70-10\alpha) \\ u^2_{L,n}(\beta) = (4-3\beta)^n(60-5\beta) \\ u^2_{R,n}(\beta) = (4+\beta)^n(60+15\beta) \\ u^3_{L,n}(\gamma) = (4-2\gamma)^n(60-10\gamma) \\ u^3_{R,n}(\gamma) = (4+\gamma)^n(60+20\gamma) \end{cases} \tag{60}$$

Remarks 2. *We plot the solution for* $n = 2$*. From the above Table 2 and Figure 1, we see that* $u^1_{L,n}(\alpha)$ *is an increasing function and* $u^1_{R,n}(\alpha)$ *is a decreasing function, with respect to* α*. On the other hand,* $u^2_{L,n}(\beta)$ *is a decreasing function and* $u^2_{R,n}(\beta)$ *is an increasing function, with respect to* β*. Additionally,* $u^3_{L,n}(\gamma)$ *is a decreasing function and* $u^3_{R,n}(\gamma)$ *is an increasing function, with respect to* γ*. Therefore, using the concept of Definition 3.2, we call the solution a strong solution.*

Table 2. Solution for $n = 2$.

α,β,γ	$u^1_{L,n}(\alpha)$	$u^1_{R,n}(\alpha)$	$u^2_{L,n}(\beta)$	$u^2_{R,n}(\beta)$	$u^3_{L,n}(\gamma)$	$u^3_{R,n}(\gamma)$
0	200.00	2520.00	960.00	960.00	960.00	960.00
0.1	246.84	2321.16	814.55	1033.81	851.96	1042.22
0.2	299.52	2132.48	682.04	1111.32	751.68	1128.96
0.3	358.28	1953.72	562.18	1192.60	658.92	1220.34
0.4	423.36	1784.64	454.72	1277.76	573.44	1316.48
0.5	495.00	1625.00	359.37	1366.87	495.00	1417.50
0.6	573.44	1474.56	275.88	1460.04	423.36	1523.52
0.7	658.92	1333.08	203.96	1557.34	358.28	1634.66
0.8	751.68	1200.32	143.36	1658.88	299.52	1751.04
0.9	851.96	1076.04	93.79	1764.73	246.84	1872.78
1	960.00	960.00	55.00	1875.00	200.00	2000.00

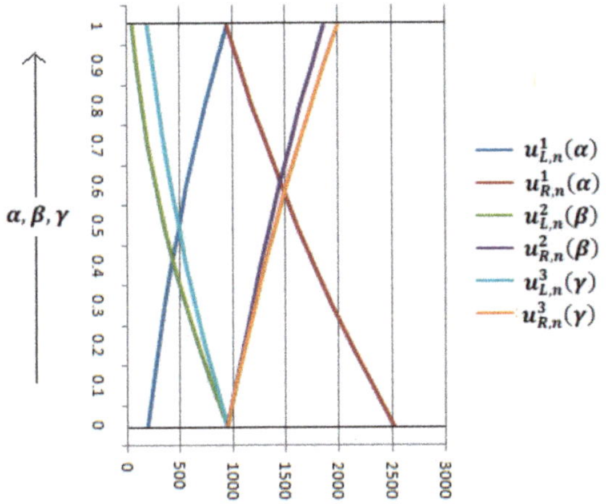

Figure 1. Graph for $n = 2$.

Remarks 3. *We plotted the solution for $n = 5$. From the above Table 3 and Figure 2, we see that $u^1_{L,n}(\alpha)$ is an increasing function and $u^1_{R,n}(\alpha)$ is a decreasing function, with respect to α. On the other hand, $u^2_{L,n}(\beta)$ is a decreasing function and $u^2_{R,n}(\beta)$ is an increasing function, with respect to β. Additionally, $u^3_{L,n}(\gamma)$ is a decreasing function and $u^3_{R,n}(\gamma)$ is an increasing function, with respect to γ. Therefore, using the concept of Definition 6, we call the solution a strong solution.*

Table 3. Solution for $n = 5$.

α,β,γ	$u^1_{L,n}(\alpha)$	$u^1_{R,n}(\alpha)$	$u^2_{L,n}(\beta)$	$u^2_{R,n}(\beta)$	$u^3_{L,n}(\gamma)$	$u^3_{R,n}(\gamma)$
0	1600.00	544,320.00	61,440.00	61,440.00	61,440.00	61,440.00
0.1	2628.35	452,886.16	41,259.65	71,251.56	46,748.74	71,830.84
0.2	4140.56	374,497.60	26,806.90	82,335.47	35,070.38	83,642.38
0.3	6297.12	307,640.56	16,748.05	94,820.44	25,898.19	97,025.57
0.4	9293.59	250,934.66	9982.01	108,844.70	18,790.48	112,143.03
0.5	13,365.00	203,125.00	5615.23	124,556.48	13,365.00	129,169.68
0.6	18,790.48	163,074.53	2937.57	142,114.45	9293.59	148,293.34
0.7	25,898.19	129,756.67	1398.99	161,688.22	6297.12	169,715.30
0.8	35,070.38	102,248.05	587.20	183,458.85	4140.56	193,651.01
0.9	46,748.74	79,721.65	206.06	207,619.30	2628.35	220,330.69
1	61,440.00	61,440.00	55.00	234,375.00	1600.00	250,000.00

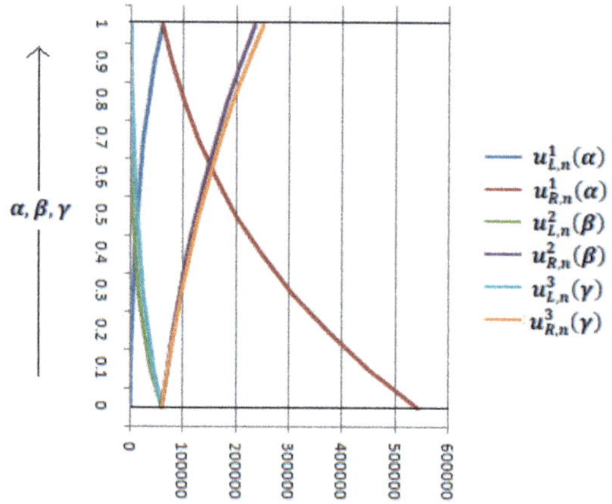

Figure 2. Graph for $n = 5$.

We interpret the solution for fixed $\alpha, \beta, \gamma = 0.4$ and different n in Table 4 and Figure 3.

Table 4. Solution for $\alpha, \beta, \gamma = 0.4$ and different n.

n	$u^1_{L,n}(\alpha)$	$u^1_{R,n}(\alpha)$	$u^2_{L,n}(\beta)$	$u^2_{R,n}(\beta)$	$u^3_{L,n}(\gamma)$	$u^3_{R,n}(\gamma)$
1	151.20	343.20	162.40	290.40	179.20	299.20
2	381.93	1513.38	410.23	1101.80	510.47	1135.19
3	964.79	6673.49	1036.26	4180.33	1454.14	4307.01
4	2437.12	29,427.69	2617.65	15,860.57	4142.31	16,341.19
5	6156.31	129,765.52	6612.33	60,176.40	11,799.88	61,999.93
6	15,551.16	572,219.06	16,703.10	228,314.58	33,613.43	235,233.20
7	39,283.04	2,523,279.35	42,192.90	866,245.64	95,752.00	892,495.51
8	99,231.00	11,126,750.35	106,581.44	3,286,612.28	272,761.37	3,386,206.59
9	250,662.64	49,064,949.17	269,230.24	12,469,696.48	776,994.32	12,847,566.07
10	633,186.78	216,358,699.65	680,089.51	47,311,126.78	2,213,363.93	48,744,797.29

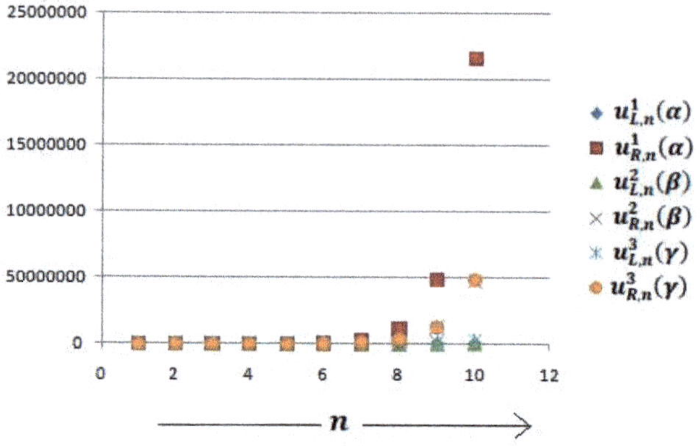

Figure 3. Graph for $\alpha, \beta, \gamma = 0.4$ and different n.

Example 2. *Solve the difference equation:*

$$u_{n+1} - 4u_n = 0 \tag{61}$$

with initial condition $\widetilde{u}_{n=0} = (50, 60, 70; 55, 60, 75; 50, 60, 80)$

Solution 2. *If* $[\widetilde{u}_n]_{(\alpha,\ \beta,\ \gamma)}$ *is the* $(\alpha,\ \beta,\ \gamma)$*-cut of a sequence of neutrosophic numbers, then it's components are as follows:*

when n is even:

$$\begin{cases} u^1_{L,n}(\alpha) = 4^n(50 + 10\alpha) \\ u^1_{R,n}(\alpha) = 4^n(70 - 10\alpha) \\ u^1_{L,n}(\beta) = 4^n(60 - 5\beta) \\ u^1_{R,n}(\beta) = 4^n(60 + 15\beta) \\ u^1_{L,n}(\gamma) = 4^n(60 - 10\gamma) \\ u^1_{R,n}(\gamma) = 4^n(60 + 20\gamma) \end{cases} \quad (62)$$

when n is odd:

$$\begin{cases} u^1_{L,n}(\alpha) = 4^n(70 - 10\alpha) \\ u^1_{R,n}(\alpha) = 4^n(50 + 10\alpha) \\ u^1_{L,n}(\beta) = 4^n(60 + 15\beta) \\ u^1_{R,n}(\beta) = 4^n(60 - 5\beta) \\ u^1_{L,n}(\gamma) = 4^n(60 + 20\gamma) \\ u^1_{R,n}(\gamma) = 4^n(60 - 10\gamma) \end{cases} \quad (63)$$

As previous examples, we easily interpret the solutions in a different manner.

6. Application of the Method in Actuarial Science

Let us consider that a sum S_0 is invested at a compound interest of i per unit amount and per unit of time and S_t is the amount at the end of time t. We then get the difference equation associated with the problem, which is:

$$S_{t+1} = S_t + iS_t = (1 + i)S_t \quad (64)$$

If, for some reason, i may vary, we are interested to find the possible amount after a certain time interval.

For this problem, let us consider hypothetical data and solve it. Suppose a person has initially invested $S_{t=0} = 10000\$$ in a firm, where they get about 4% interest (which may be considered a neutrosophic value).

As per Table 1, if we take the verbal phrase for a triangular neutrosophic number, we then set the interest rate as follows:

For the truth part: low as 3%, medium as 4%, high as 5%;
For the falsity portion: very low as 2%, medium as 4%, very high as 6%;
For the indeterminacy part: between low and very low 2.5%, medium 4%, between high and very high 5.5%,

i.e., we can take $\widetilde{i} = (3, 4, 5; 2, 4, 6; 2.5, 4, 5.5)\%$ per annum rate. We wish to predict the amount of money after 10 years.

Therefore, we get the fuzzy difference equation

$$S_{t+1} = S_t + iS_t = \left(1 + \widetilde{i}\right)S_t \tag{65}$$

With the initial conditions $S_{t=0} = 10000\$$ and $\widetilde{i} = (3,4,5;2,4,6;2.5,4,5.5)\%$.

Solution 3. *Equation (65) is equivalent to*

$$S_{t+1} = S_t + iS_t = (1 + (0.03, 0.04, 0.05; 0.02, 0.04, 0.06; 0.025, 0.04, 0.055))S_t$$

or

$$S_{t+1} = (1.03, 1.04, 1.05; 1.02, 1.04, 1.06; 1.025, 1.04, 1.055)S_t \tag{66}$$

with the initial condition $S_{t=0} = 10000\$$.

The solution of (66) can be written using the concept of (19), as follows:

$$\begin{cases} S^1_{L,t}(\alpha) = 10000(1.03 + 0.01\alpha)^t \\ S^1_{R,t}(\alpha) = 10000(1.05 - 0.01\alpha)^t \\ S^2_{L,t}(\beta) = 10000(1.04 - 0.02\beta)^t \\ S^2_{R,t}(\beta) = 10000(1.04 + 0.02\beta)^t \\ S^3_{L,t}(\gamma) = 10000(1.04 - 0.015\gamma)^t \\ S^3_{R,t}(\gamma) = 10000(1.04 + 0.015\gamma)^t \end{cases} \tag{67}$$

Remarks 4. *(1) We plot the solution for* $t = 10$. *From the above Table 5 and Figure 4, we see that* $S^1_{L,n}(\alpha)$ *is an increasing function and* $S^1_{R,n}(\alpha)$ *is a decreasing function, with respect to* α. *On the other hand,* $S^2_{L,n}(\beta)$ *is a decreasing function and* $S^2_{R,n}(\beta)$ *is an increasing function, with respect to* β. *Additionally,* $S^3_{L,n}(\gamma)$ *is a decreasing function and* $S^3_{R,n}(\gamma)$ *is an increasing function, with respect to* γ. *Therefore, using the concept of Definition 3.2, we call the solution a strong solution. (2) From Table 5, we can see that we find the crisp solution at* $\alpha = 1$, $\beta, \gamma = 0$ *(since, at* $\alpha = 1$, $\beta, \gamma = 0$, *the neutrosophic number becomes a crisp number) and for* $t = 10$ *is equal to 14802.4428. Therefore, we can say that after 10 years, the most probable chance to get the money is 14802.4428\$.(3) If we consider* $\alpha = 0$ *and* $\beta, \gamma = 1$, *i.e., in the case that we get the most uncertain solution interval, we observe that the truthiness of the solution belongs to the interval* [13439.1638, 16288.9463], *the falsity belongs to the interval* [12189.9442, 17908.4770], *and the indeterminacy belongs to the interval* [12800.8454, 17081.4446].

Table 5. Solution for $t = 10$.

α,β,γ	$S^1_{L,t}(\alpha)$	$S^1_{R,t}(\alpha)$	$S^2_{L,t}(\beta)$	$S^2_{R,t}(\beta)$	$S^3_{L,t}(\gamma)$	$S^3_{R,t}(\gamma)$
0	13,439.1638	16,288.9463	14,802.4428	14,802.4428	14,802.4428	14,802.4428
0.1	13,570.2126	16,134.4766	14,520.2313	15,089.5813	14,590.3264	15,017.3306
0.2	13,702.4105	15,981.3266	14,242.8714	15,381.7230	14,380.9496	15,235.0219
0.3	13,835.7662	15,829.4861	13,970.2889	15,678.9453	14,174.2808	15,455.5492
0.4	13,970.2889	15,678.9453	13,702.4105	15,981.3266	13,970.2889	15,678.9453
0.5	14,105.9876	15,529.6942	13,439.1638	16,288.9463	13,768.9430	15,905.2433
0.6	14,242.8714	15,381.7230	13,180.4776	16,601.8849	13,570.2126	16,134.4766
0.7	14,380.9496	15,235.0219	12,926.2814	16,920.2240	13,374.0675	16,366.6791
0.8	14,520.2313	15,089.5813	12,676.5060	17,244.0464	13,180.4776	16,601.8849
0.9	14,660.7259	14,945.3915	12,431.0828	17,573.4357	12,989.4133	16,840.1284
1	14,802.4428	14,802.4428	12,189.9442	17,908.4770	12,800.8454	17,081.4446

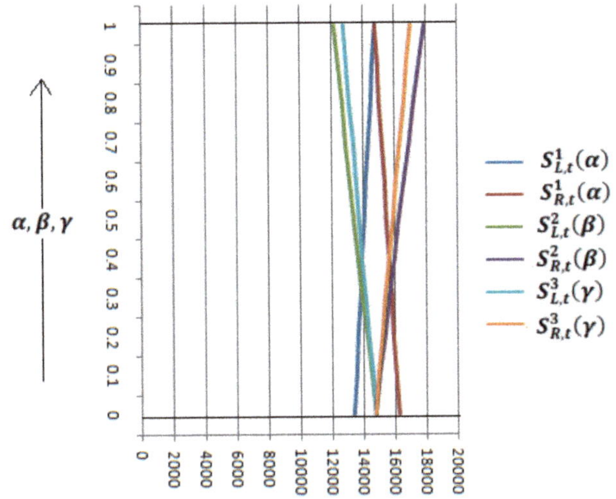

Figure 4. Graph for $t = 10$.

7. Conclusions and Future Research Scope

In this paper, we find the solution strategy for solving and analyzing homogeneous linear difference equations with neutrosophic numbers, i.e., we found the solutions of the homogeneous difference equations with initial conditions and coefficients, both as neutrosophic numbers. We demonstrate the solution of different cases using the neutrosophic characterization theorem, which is established in this paper. The strong and weak solution concepts are also applied to different results.

Moreover, the outcomes of the study are as follows:

(1) The difference type of the homogeneous difference equation is solved in a neutrosophic environment and the symmetric behavior between them is discussed.
(2) The characterization theorem for the neutrosophic difference equations are established.
(3) The strong and weak solution concept is applied for the neutrosophic difference equation.
(4) Different examples and real-life applications in actuarial science are illustrated for better understanding of neutrosophic difference equations.

For some limitations, we did not study the different perspectives of related research in the theory of difference equations with uncertainty in this present work. From this work, anyone can take motivation and find a new theory and results in the following field, as follows:

(1) The solution of difference equation can be found with different types of uncertainty, such as Type 2 fuzzy, interval valued fuzzy, hesitant fuzzy, rough fuzzy environment.
(2) Finding several methods (analytical and numerical) for solving non-linear first and higher order difference equations or system of difference equations with uncertainty.
(3) Solving the real-life model associated with the discrete system modeling with uncertain data.

As a final argument, we can surely say that this research is very helpful to the research community who deals with discrete system modeling with uncertainty.

Author Contributions: Conceptualization, S.P.M.; Data curation, A.A. (Abdul Alamin); Formal analysis, S.A.; Resources, S.P.M. Software; Supervision, A.A. (Ali Ahmadian); Visualization, A.A. (Ali Ahmadian); Writing—original draft, A.A. (Abdul Alamin) and S.A.; Writing—review & editing, A.A. (Ali Ahmadian), S.S. and M.S. All authors have read and agreed to the published version of the manuscript.

Funding: This research received no external funding.

Conflicts of Interest: The authors declare no conflict of interest.

References

1. Zadeh, L. Fuzzy sets. *Inf. Control.* **1965**, *8*, 338–353. [CrossRef]
2. Atanassov, K.T. Intuitionistic fuzzy sets. *Fuzzy Sets Syst.* **1986**, *20*, 87–96. [CrossRef]
3. Li, J.; Niu, Q.; Dong, X.-C. Similarity Measure and Fuzzy Entropy of Fuzzy Number Intuitionistic Fuzzy Sets. *Adv. Intell. Soft Comput.* **2009**, *54*, 373–379. [CrossRef]
4. Ye, J. Prioritized aggregation operators of trapezoidal intuitionistic fuzzy sets and their application to multicriteria decision-making. *Neural Comput. Appl.* **2014**, *25*, 1447–1454. [CrossRef]
5. Smarandache, F. *A Unifying Field in Logics Neutrosophy: Neutrosophic Probability, Set and Logic*; American Research Press: Rehoboth, DE, USA, 1998.
6. Wang, H.; Smarandache, F.; Zhang, Y.; Sunderraman, R. Single Valued Neutrosophic Sets. *Multisspace Multistructure* **2010**, *4*, 410–413.
7. Ye, J. Single-Valued Neutrosophic Minimum Spanning Tree and Its Clustering Method. *J. Intell. Syst.* **2014**, *23*, 311–324. [CrossRef]
8. Peng, J.-J.; Wang, J.-Q.; Wang, J.; Zhang, H.-Y.; Chen, X.-H. Simplified neutrosophic sets and their applications in multi-criteria group decision-making problems. *Int. J. Syst. Sci.* **2015**, *47*, 1–17. [CrossRef]
9. Peng, X.; Smarandache, F. New multiparametric similarity measure for neutrosophic set with big data industry evaluation. *Artif. Intell. Rev.* **2019**, *53*, 3089–3125. [CrossRef]
10. Chakraborty, A.; Mondal, S.P.; Ahmadian, A.; Senu, N.; Salahshour, S.; Alam, S. Different Forms of Triangular Neutrosophic Numbers, De-Neutrosophication Techniques, and their Applications. *Symmetry* **2018**, *10*, 327. [CrossRef]
11. Chakraborty, A.; Mondal, S.P.; Alam, S.; Ahmadian, A.; Senu, N.; De, D.; Salahshour, S. Disjunctive Representation of Triangular Bipolar Neutrosophic Numbers, De-Bipolarization Technique and Application in Multi-Criteria Decision-Making Problems. *Symmetry* **2019**, *11*, 932. [CrossRef]
12. Deli, I.; Ali, M.; Smarandache, F. Bipolar Neutrosophic Sets and their Application Based on Multi-Criteria Decision Making Problems. In Proceedings of the 2015 International Conference on Advanced Mechatronic Systems (ICAMechS), Beijing, China, 22–24 August 2015; Institute of Electrical and Electronics Engineers (IEEE): Piscataway, NJ, USA, 2015; pp. 249–254.
13. Lee, K.M. Bipolar-Valued Fuzzy Sets and their Operations. In Proceedings of the International Conference on Intelligent Technologies, Bangkok, Thailand, 12–14 December 2000; pp. 307–312.
14. Kang, M.K.; Kang, J.G. Bipolar fuzzy set theory applied to sub-semigroups with operators in semi groups. *J. Korean Soc. Math. Educ. Ser. B Pure Appl. Math.* **2012**, *19*, 23–35.
15. Smarandache, F. *Neutrosophic Perspectives: Triplets, Duplets, Multisets, Hybrid Operators, Modal Logic, Hedge Algebras and Applications*; Degree of Dependence and Independence of the (Sub) Components of Fuzzy Set and Neutrosophic Set. Neutrosophic Sets and Systems; Pons Publishing House: Stuttgart, Germany, 2016; Volume 11, pp. 95–97.
16. Deeba, E.Y.; De Korvin, A.; Koh, E.L. A fuzzy difference equation with an application. *J. Differ. Equ. Appl.* **1996**, *2*, 365–374. [CrossRef]
17. Deeba, E.; De Korvin, A. Analysis by fuzzy difference equations of a model of CO2 level in the blood. *Appl. Math. Lett.* **1999**, *12*, 33–40. [CrossRef]
18. Lakshmikantham, V.; Vatsala, A. Basic Theory of Fuzzy Difference Equations. *J. Differ. Equ. Appl.* **2002**, *8*, 957–968. [CrossRef]
19. Papaschinopoulos, G.; Papadopoulos, B.K. On the fuzzy difference equation x_(n+1)=A+B⁄x_n. *Soft Comput.* **2002**, *6*, 456–461. [CrossRef]
20. Papaschinopoulos, G.; Papadopoulos, B.K. On the fuzzy difference equation x_(n+1)=A+x_n⁄x_(n-m). *Fuzzy Sets Syst.* **2002**, *129*, 73–81. [CrossRef]

21. Papaschinopoulos, G.; Schinas, C.J. On the fuzzy difference equation $x_{n+1} \sum_{k=0}^{k=1} \frac{Ai}{x_{n-i}^{pi}} + \frac{1}{x_{n-k}^{pk}}$. *J. Differ. Equ. Appl.* **2000**, *6*, 85–89. [CrossRef]
22. Papaschinopoulos, G.; Stefanidou, G. Boundedness and asymptotic behavior of the solutions of a fuzzy difference equation. *Fuzzy Sets Syst.* **2003**, *140*, 523–539. [CrossRef]
23. Umekkan, S.A.; Can, E.; Bayrak, M.A. Fuzzy difference equation in finance. *IJSIMR* **2014**, *2*, 729–735.
24. Stefanidou, G.; Papaschinopoulosand, G.; Schinas, C.J. On an exponential–Type fuzzy Difference equation. *Advanced in difference equations. Adv. Differ. Equ.* **2010**, *2010*, 1–19. [CrossRef]
25. Din, Q. Asymptotic Behavior of a Second-Order Fuzzy Rational Difference Equation. *J. Discret. Math.* **2015**, *2015*, 1–7. [CrossRef]
26. Zhang, Q.H.; Yang, L.H.; Liao, D.X. Behaviour of solutions of to a fuzzy nonlinear difference equation. *Iran. J. Fuzzy Syst.* **2012**, *9*, 1–12.
27. Memarbashi, R.; Ghasemabadi, A. Fuzzy difference equations of volterra type. *Int. J. Nonlinear Anal. Appl.* **2013**, *4*, 74–78.
28. Stefanidou, G.; Papaschinopoulos, G. A fuzzy difference equation of a rational form. *J. Nonlinear Math. Phys.* **2005**, *12*, 300–315. [CrossRef]
29. Chrysafis, K.A.; Papadopoulos, B.K.; Papaschinopoulos, G. On the fuzzy difference equations of finance. *Fuzzy Sets Syst.* **2008**, *159*, 3259–3270. [CrossRef]
30. Mondal, S.P.; Vishwakarma, D.K.; Saha, A.K. Solution of second order linear fuzzy difference equation by Lagrange's multiplier method. *J. Soft Comput. Appl.* **2016**, *2016*, 11–27. [CrossRef]
31. Mondal, S.P.; Mandal, M.; Bhattacharya, D. Non-linear interval-valued fuzzy numbers and their application in difference equations. *Granul. Comput.* **2017**, *3*, 177–189. [CrossRef]
32. Sarkar, B.; Mondal, S.P.; Hur, S.; Ahmadian, A.; Salahshour, S.; Guchhait, R.; Iqbal, M.W. An optimization technique for national income determination model with stability analysis of differential equation in discrete and continuous process under the uncertain environment. *Rairo Oper. Res.* **2019**, *53*, 1649–1674. [CrossRef]
33. Zhang, Q.; Lin, F. On Dynamical Behavior of Discrete Time Fuzzy Logistic Equation. *Discret. Dyn. Nat. Soc.* **2018**, *2018*, 1–8. [CrossRef]
34. Zhang, Q.; Lin, F.; Zhong, X. Asymptotic Behavior of Discrete Time Fuzzy Single Species Model. *Discret. Dyn. Nat. Soc.* **2019**, *2019*, 1–9. [CrossRef]
35. Zhang, Q.; Lin, F.B.; Zhong, X.Y. On discrete time Beverton-Holt population model with fuzzy environment. *Math. Biosci. Eng.* **2019**, *16*, 1471–1488. [CrossRef] [PubMed]
36. Khastan, A. Fuzzy logistic difference equation. *Iran. J. Fuzzy Syst.* **2018**, *15*, 55–66.
37. Mondal, S.P.; Alam Khan, N.; Vishwakarma, D.; Saha, A.K. Existence and Stability of Difference Equation in Imprecise Environment. *Nonlinear Eng.* **2018**, *7*, 263–271. [CrossRef]
38. Khastan, A.; Alijani, Z. On the new solutions to the fuzzy difference equation xn+1=A+Bxn. *Fuzzy Sets Syst.* **2019**, *358*, 64–83. [CrossRef]
39. Khastan, A. New solutions for first order linear fuzzy difference equations. *J. Comput. Appl. Math.* **2017**, *312*, 156–166. [CrossRef]
40. Abdel-Basset, M.; Mohamed, M.; Hussien, A.N.; Sangaiah, A.K. A novel group decision-making model based on triangular neutrosophic numbers. *Soft Comput.* **2018**, *22*, 6629–6643. [CrossRef]
41. Jensen, A. *Lecture Notes on Difference Equation*; Department of mathematical Science, Aalborg University: Aalborg, Denmark, 18 July 2011; (It is lecture notes of the courses "Introduction to Mathematical Methods" and "Introduction to Mathematical Methods in Economics").
42. Elaydi, S.N. *An Introduction to Difference Equations*; Springer: Berlin/Heidelberg, Germany, 1995.

symmetry

MDPI

Article

Neutrosophic Modeling of Talcott Parsons's Action and Decision-Making Applications for It

Cahit Aslan [1], Abdullah Kargın [2,*] and Memet Şahin [2]

[1] Department of Sociology, Psychology and Philosophy Teaching, Education Faculty, Çukurova University, Balcali, 01330 Saricam/Adana, Turkey; aslanc@cu.edu.tr

[2] Department of Mathematics, Gaziantep University, 27310 Gaziantep, Turkey; mesahin@gantep.edu.tr

* Correspondence: ak23977@mail2.gantep.edu.tr; Tel.: +90-055-4270-6621

Received: 15 May 2020; Accepted: 3 July 2020; Published: 13 July 2020

Abstract: The grand theory of action of Parsons has an important place in social theories. Furthermore, there are many uncertainties in the theory of Parsons. Classical math logic is often insufficient to explain these uncertainties. In this study, we explain the grand theory of action of Parsons in neutrosociology for the first time. Thus, we achieve a more effective way of dealing with the uncertainties in the theory of Parsons as in all social theories. We obtain a similarity measure for single-valued neutrosophic numbers. In addition, we show that this measure of similarity satisfies the similarity measure conditions. By making use of this similarity measure, we obtain applications that allow finding the ideal society in the theory of Parsons within the theory of neutrosociology. In addition, we compare the results we obtained with the data in this study with the results of the similarity measures previously defined. Thus, we have checked the appropriateness of the decision-making application that we obtained.

Keywords: neutrosociology; modeling of grand theory of action of Talcott Parsons; single-valued neutrosophic number; measure of similarity; decision-making applications

---

1. Introduction

There are many uncertainties in the world. Classical math logic is usually insufficient to explain uncertainties. Thus, we are not always able to say for a situation or an event whether it is true or wrong in an absolute manner. For example, we cannot always say the weather is hot or cold. While the weather is hot according to some, it may be cold for others. Therefore, Smarandache obtained the neutrosophic logic and neutrosophic set to deal with uncertainties more objectively in 1998 [1]. 'T' is the membership degree, 'I' is the uncertainty degree and 'F' is the non-membership degree in the neutrosophic logic and neutrosophic sets. "T, I, F" are defined independently. In addition, a neutrosophic number has the form (T, I, F). Furthermore, neutrosophic logic is a generalization of fuzzy logic [2] and intuitionistic fuzzy logic [3] since fuzzy and intuitionistic fuzzy logic's membership, non–membership degrees are defined dependently. Thus, many researchers have obtained new structures and new applications on neutrosophic logic and sets [4–15].

In Section 2 of this study, we provide a literature review. In Section 3, we give related works. In Section 4, we include the definitions of neutrosophic sets [1], single-valued sets [6], similarity measures in [7] and [16], the theory of social action of Parsons [17], Hausdorff measure [18] and Hamming measure [18]. In Section 5, we re-model the social action theory of Parsons which is modeled in neutrosociology. In Section 6, we obtain a similarity measure for single-valued neutrosophic sets and prove that this measure meets the requirements of the similarity measure. In Section 7, we create the decision-making algorithm that we can choose the ideal society among the societies for the social action theory of Parsons in neutrosociology with the help of similarity measure in Section 6. In Section 8, we give sensitivity analysis for numeric example in Section 7; in Section 9, we give comparison methods. We compare the results we obtained with the data in this study with the results of the similarity measures previously defined. Thus, we have checked the appropriateness of the decision-making application we obtained; in Section 10, we discuss what we obtained in this study and make suggestions for studies that can be obtained by making use of this study; in Section 11, we give conclusions.

2. Literature Review

Similarity measure and decision-making practices emerge as an important application theory, especially after the definition of the fuzzy sets and neutrosophic sets. Many researchers have tried to deal with uncertainties by making new applications on neutrosophic sets, using similarity measures, TOPSIS method, VIKOR method, multicriteria method, Maximizing deviation method, decision tree methods, gray relational analysis method, etc. Recently, Şahin et al. studied combined classic neutrosophic sets and double neutrosophic sets [19]; Şahin et al. obtained decision-making applications for professional proficiencies in neutrosophic theory [16]; Uluçay et al. introduced decision-making applications for neutrosophic soft expert graphs [20]; Olgun et al. studied neutrosophic logic on the decision tree [21]; Wang et al. studied an extended VIKOR method with triangular fuzzy neutrosophic numbers [22]; Biswas et al. introduced TOPSIS method for decision–making applications [23]; Şahin et al. obtained a maximizing deviation method in neutrosophic theory [24]; Biswas et al. studied gray relational analysis method for decision-making applications [25].

3. Related Works

Smarandache claims that sociopolitical events can be studied mathematically [4]. In addition, he claims that it is possible to design a tool to describe an equation, an operator, a mathematical structure or a social phoneme. Studying the past gives us an idea about the future, at least partially. For this reason, we need to construct neutrosophic theories that may describe the new possible types of social structures with a neutrosophic number form. Since the social word contains a high degree of subjectivity that causes a low level of unanimity, these theories necessarily address uncertainty. Most of the data we come across in the field of sociology may be vague, incomplete, contradictory, biased, hybrid, ignorant, redundant, etc. Therefore, they are neutrosophic in nature and neutrosophic sciences dealing with indeterminacy should be involved in the study of sociology [4].

For the very same reasons, Smarandache proposed a model to be used in neutrosophic studies. He states that a neutrosophic extension of an element x with a neutrosophic number form.

Parsons, who built his theory on methodological and meta-theoretical debates in the field of social science, also paid special attention to hermeneutic to explain the extent of the individual's voluntary involvement in action [26]. He made structural and functional explanations to maintain social balance and harmony [21]. While Parsons saw culture as values and norms that guide the actions of individuals in social life, he conceptualized the structure as a system of intertwined and independent parts [27]. According to Parsons, cultural objects are autonomous. He did this by distinguishing between the cultural and social systems. He also viewed society as a general system of action. In addition, many researchers have studied Parsons's social action theory [26–36].

In this study, Parsons's social action theory was aimed to re-model neutrosociology. As in all social theories, the social action theory of Parsons could not escape uncertainty [21]. Hence, the handling of it in neutrosociology theory would make this theory more useful. Therefore, we have obtained a similarity measure with single-valued neutrosophic numbers and included applications where this measure can be used as the neutrosophic equivalent of the ideal society in this theory.

4. Preliminaries

This section includes the definitions of neutrosophic sets [1], single-valued neutrosophic sets [6], similarity measures [7,16] and theory of social action of Parsons [17], Hausdorff measure [18] and Hamming measure [18].

Definition 1 ([17]). *Parsons, who built his theory on methodological and meta-theoretical debates in the field of social science, also paid special attention to hermeneutic to explain the extent of the individual's voluntary involvement in action (so, which is neutrosophic). He made structural and functional explanations to maintain social balance and harmony. While Parsons saw culture as values and norms that guide the actions of individuals in social life, he conceptualized the structure as a system of intertwined and independent parts. According to Parsons, cultural objects are autonomous. He did this by distinguishing between the cultural and social systems. He also viewed society as a general system of action.*

Definition 2 ([1]). *Let X be a universal set. Neutrosophic set S; is identified as* $S = \{(x{:}T_{A(x)}, I_{A(x)}, F_{A(x)}>, x \in X\}$*. Where; on the condition that* $0^- \leq T_{A(x)} + I_{A(x)} + F_{A(x)} \leq 3^+$*; the functions* $T{:}U \to]0^-,1^+[$ *is truth function,* $I{:}U \to]0^-,1^+[$ *is uncertain function and* $F{:}U \to]0^-,1^+[$ *is falsity function.*

Definition 3 ([6]). *Let X be a universal set. Single-valued neutrosophic number set S; is identified as* $S = \{(x{:}T_{A(x)}, I_{A(x)}, F_{A(x)}>, x \in X\}$*. Where; on condition that* $0 \leq T_{A(x)} + I_{A(x)} + F_{A(x)} \leq 3$*; the functions* $T{:}X \to [0,1]$ *is truth function,* $I{:}X \to [0,1]$ *is uncertainly function and* $F{:}X \to [0,1]$ *is falsity function.*

Definition 4 ([6]). *Let* $A = \{(x{:}<T_{A(x)}, I_{A(x)}, F_{A(x)}>\}$ *and* $B = \{(x{:}<T_{B(x)}, I_{B(x)}, F_{B(x)}>\}$ *are single-valued neutrosophic numbers. If A = B; then* $T_{A(x)} = T_{B(x)}, I_{A(x)} = I_{B(x)}$ *and* $F_{A(x)} = F_{B(x)}$.

Definition 5 ([6]). *Let* $A = \{(x{:}<T_{A(x)}, I_{A(x)}, F_{A(x)}>\}$ *and* $B = \{(x{:}<T_{B(x)}, I_{B(x)}, F_{B(x)}>\}$ *are single-valued neutrosophic sets for* $x \in U$*. If A < B; then for* $\forall\, x \in U$*;* $T_{A(x)} < T_{B(x)}, I_{A(x)} < I_{B(x)}$ *and* $F_{A(x)} < F_{B(x)}$.

Properties 1 ([7]). *Let* A_1*,* A_2 *and* A_3 *are three single-valued neutrosophic numbers and S be a similarity measure. S provides the following conditions.*

i. $0 \leq S(A_1, A_2) \leq 1$
ii. $S(A_1, A_2) = S(A_2, A_1)$
iii. $S(A_1, A_2) = 1 \Leftrightarrow A_1 = A_2$.
iv. *If* $A_1 \leq A_2 \leq A_3$ *then,* $S(A_1, A_3) \leq S(A_1, A_2)$.

Definition 6 ([16]). *Let* $A_1 = <T_1, I_1, F_1>$ *and* $A_2 = <T_2, I_2, F_2>$ *be two single-valued neutrosophic numbers.*

$$S_N(A_1, A_2) =$$

$$1-(2/3)\left[\frac{min\{|3(T_1-T_2)-2(F_1-F_2)|,|F_1-F_2|\}}{\{max\{|3(T_1-T_2)-2(F_1-F_2)|,|F_1-F_2|\}/5\}+1}+\frac{min\{|4(T_1-T_2)-3(I_1-I_2)|,|I_1-I_2|\}}{\{max\{|4(T_1-T_2)-3(I_1-I_2)|,|I_1-I_2|\}/7\}+1}+\frac{min\{|5(T_1-T_2)-2(F_1-F_2)-3(I_1-I_2)|,|T_1-T_2|\}}{\{max\{|5(T_1-T_2)-2(F_1-F_2)-3(I_1-I_2)|,|T_1-T_2|\}/10\}+1}\right]$$

is a similarity measure.

Definition 7 ([18]). *Let* $A_1 = <T_1, I_1, F_1>$ *and* $A_2 = <T_2, I_2, F_2>$ *be two single-valued neutrosophic numbers.*

$$S_h(A_1,A_2) = 1 - max\{|T_1 - T_2|, |I_1 - I_2|, |F_1 - F_2|\}$$

is a Hausdorff similarity measure.

Definition 8 ([18]). *Let* $A_1 = <T_1, I_1, F_1>$ *and* $A_2 = <T_2, I_2, F_2>$ *be two single-valued neutrosophic numbers.*

$$S_H(A_1,A_2) = 1 - (|T_1 - T_2| + |I_1 - I_2| + |F_1 - F_2|)/3$$

is a Hamming similarity measure.

5. Neutrosophic Modeling of Parson's Theory of Action

According to the perfection of action categories of Parsons, it is inevitable to have deep doubts in every society and between layers of a particular society. However, "there is no ideal society in the sense that Marx defines, within each society the definition of ideal changes according to the place of a person within the society. By those who are at the top layer, the society is defined as ideal, by those at the lowest layer, it is far from being ideal, and by those in the mid-layer, who can sometimes be completely ignorant of what is an ideal society, it can be described as a fluctuating phenomenon depending on circumstances. Therefore, we always have a neutrosophic ideal society with an opposite and neutral triad. Naturally, this is valid for all societies since there are always people with more privileges than the others. Even in any a democratic society, some people have more privileges although they may form a small minority" [4].

Parsons developed a theory of action to explain how the macro and micro aspects of a particular social order show structural integrity together with the participation of its members. He took into account the voluntary participation of the individual in the social life on one hand, and structural continuity on the other. Here, it is assumed that the individual acts under the motivation of the social structure while taking action. According to him, social sciences should consider a trio considering the purposes, ends and ideals when examining actions.

Grand Theory of Action

The basic paradigm of Parson viewed society as a general system of action is based on the understanding of 'rational social action' of Weber [28]. However, according to Weber, sociology is a science that tries its interpretive understanding of social action to achieve a causal explanation of its course and its effects [36].

This interpretation is enriched from the perspective of the sociologist. Thus, social actions become neutrosophic. Others may agree, partially agree or disagree (1, 0, 0). Likewise, in the theory of Parson, the possibility of all members of society to participate in social values and norms that regulate, and guide human relations rather than individual activities is questionable, uncertain. Here we must see neutrosophic triplets.

According to Parson's theory, all social actions are based on five pattern variables. These:

1. Affectivity versus affective neutrality;
2. Self-orientation versus collective orientation;
3. Universalism versus particularism;
4. Quality versus performance;
5. Specificity versus diffuseness.

Parsons believes that these variables classify expectations and the structure of relationships, making the intangible action theory more understandable. However, according to Parsons, pattern variables are twofold, and each pattern variable indicates a problem or riddle that must be solved by the actor before the action can be performed. At the same time, there is a wide variety between the traditional society and the modern society. However, these can be seen as binary for neutrosophic sociological analysis (1, 0), it is very difficult to determine which of the individual's behaviors are modern or traditional. Therefore, each of them should be considered as triple neutrosophic (1, 0, 0). The feminists' response to Parsons' family view can be given as an example. According to Parsons, the instrumental leadership role in the family structure in modern societies should be given to the spouse–father, on which the family's reputation and income are based [32]. However, according to feminists, this statement by Parsons is nothing more than the continuation of the status quo [35]. In addition, these pattern variables (stereotypes) do not say how people will behave when faced with role conflict, and we will once again encounter uncertainty. This uncertainty can only be answered by neutrosociology.

The society model that Parsons has compared to the biologic model of an organism is based on the understanding of "living systems" that continues in a balanced way. According to him, a change in any part of the social system leads to adaptive changes in other parts [33]. There are four main problems an all-action system must solve. These are adaptation, goal-attainment, integration and latent pattern maintenance (AGIL). In short, these are referred to as AGIL in Table 1.

Table 1. Pure adaptation, goal-attainment, integration and latent pattern maintenance (AGIL) model for all living systems [33].

A	Instrumental	Consummatory	G
External	Adaptation	Goal-attainment	
Internal	Latent pattern maintenance	Integration	
L			I

"Adaptation" (A) is concerned with meeting the needs of the system from its environment and how resources are distributed within the system. Here, the system should provide sufficient resources from the environment and distribute it within itself. Social institutions are related to interrelated social rules and roles system that will meet social needs or functions and help solve social system problems. For example, economy, political order, law, religion, education and family are basic institutions for these. If a social system will continue to live, it needs structures and organizations that will function to adapt to its environment. The most dominant of these institutions is the economy. In "achieving the goal" (G), it is determined that the system reaches the specific target and which of these targets has priority. In other words, it should mobilize the resources and energies of the system and determine the priorities among them. "Integration" (I) refers to the coordination and harmony of parts of the system so that the system functions as a whole. To keep the system running, it must coordinate, correct, and regulate the relationships between the various actors or units in the system. "Latent pattern maintenance" (L) shows how to ensure the continuity of the action within the system according to a certain order or norm. The system should protect its values from deterioration and ensure the transfer of social values. Thus, it ensures the compliance of the members of the system. Especially family, religion, media and education have basic functions. Thanks to these, individuals gain a moral commitment to values shared socially [30].

The General Action Level is as follows in Table 2:

Table 2. General Action Level [30].

A		G	
	The behavioral organism	The personality system	
	The cultural system	The social system	
L			I

Ultimately we get this series: The social the system, the fiduciary the cognitive.

Let us rebuild this series neutrosociology: (1, 0, 0) (1, 0, 0) (1, 0, 0).

If we go back to the beginning, "Behavioral organic, Personality system, Cultural system and Social system" must work continuously to ensure social balance. This will be through "socialization" and "social control". If socialization "works", all members of the society will adhere to shared values, make appropriate choices between pattern variables, and do what is expected of them in harmony, integration and other issues. For example, people will marry and socialize their children (L), and the father in the family will gain bread as it should be (A) [35].

6. A New Measurement of Similarity for Single-Valued Neutrosophic Numbers

Definition 9. *Let $A_1 = <T_1, I_1, F_1>$, $A_2 = <T_2, I_2, F_2>$ be two single-valued neutrosophic numbers. We define measure of similarity between A_1 and A_2 as follows*

$$S_N(A_1, A_2) = 1 - (2/3)[\frac{min\{\sqrt{3(T_1-T_2)^2+(I_1-I_2)^2}, |2(T_1-T_2)-(I_1-I_2)|/3\}}{\{max\{\sqrt{3(T_1-T_2)^2+(I_1-I_2)^2}, |2(T_1-T_2)-(I_1-I_2)|/3\}/2\}+1}$$
$$+\frac{min\{\sqrt{3(T_1-T_2)^2+(F_1-F_2)^2}, |2(T_1-T_2)-(F_1-F_2)|/3\}}{\{max\{\sqrt{3(T_1-T_2)^2+(F_1-F_2)^2}, |2(T_1-T_2)-(F_1-F_2)|/3\}/2\}+1}$$
$$+\frac{min\{\sqrt{2(T_1-T_2)^2+(I_1-I_2)^2+(F_1-F_2)^2}, |3(T_1-T_2)-(I_1-I_2)-(F_1-F_2)|/5\}}{\{max\{\sqrt{2(T_1-T_2)^2+(I_1-I_2)^2+(F_1-F_2)^2}, |3(T_1-T_2)-(I_1-I_2)-(F_1-F_2)|/5\}/2\}+1}]$$

We show that the measure of similarity in Definition 9 meets the requirements in Properties 1.

Theorem 1. *Let S_N be the measure of similarity in Definition 9. S_N provides the following features.*

i. $0 \leq S_N(A_1, A_2) \leq 1$
ii. $S_N(A_1, A_2) = S_N(A_2, A_1)$
iii. $S_N(A_1, A_2) = 1$ *if and only if* $A_1 = A_2$.
iv. *If* $A_1 \leq A_2 \leq A_3$, *then* $S_N(A_1, A_3) \leq S_N(A_1, A_2)$.

Proof: (i) Since A_1 and A_2 are single-valued neutrosophic numbers, we have

$$max\{\frac{min\{\sqrt{3(T_1-T_2)^2+(I_1-I_2)^2}, |2(T_1-T_2)-(I_1-I_2)|/3\}}{\{max\{\sqrt{3(T_1-T_2)^2+(I_1-I_2)^2}, |2(T_1-T_2)-(I_1-I_2)|/3\}/2\}+1}\} = 1/2,$$
$$min\{\frac{min\{\sqrt{3(T_1-T_2)^2+(I_1-I_2)^2}, |2(T_1-T_2)-(I_1-I_2)|/3\}}{\{max\{\sqrt{3(T_1-T_2)^2+(I_1-I_2)^2}, |2(T_1-T_2)-(I_1-I_2)|/3\}/2\}+1}\} = 0,$$
$$max\{\frac{min\{\sqrt{3(T_1-T_2)^2+(F_1-F_2)^2}, |2(T_1-T_2)-(F_1-F_2)|/3\}}{\{max\{\sqrt{3(T_1-T_2)^2+(F_1-F_2)^2}, |2(T_1-T_2)-(F_1-F_2)|/3\}/2\}+1}\} = 1/2,$$
$$min\{\frac{min\{\sqrt{3(T_1-T_2)^2+(F_1-F_2)^2}, |2(T_1-T_2)-(F_1-F_2)|/3\}}{\{max\{\sqrt{3(T_1-T_2)^2+(F_1-F_2)^2}, |2(T_1-T_2)-(F_1-F_2)|/3\}/2\}+1}\} = 0,$$
$$max\{\frac{min\{\sqrt{2(T_1-T_2)^2+(I_1-I_2)^2+(F_1-F_2)^2}, |3(T_1-T_2)-(I_1-I_2)-(F_1-F_2)|/5\}}{\{max\{\sqrt{2(T_1-T_2)^2+(I_1-I_2)^2+(F_1-F_2)^2}, |3(T_1-T_2)-(I_1-I_2)-(F_1-F_2)|/5\}/2\}+1}\} = 1/2,$$
$$min\{\frac{min\{\sqrt{2(T_1-T_2)^2+(I_1-I_2)^2+(F_1-F_2)^2}, |3(T_1-T_2)-(I_1-I_2)-(F_1-F_2)|/5\}}{\{max\{\sqrt{2(T_1-T_2)^2+(I_1-I_2)^2+(F_1-F_2)^2}, |3(T_1-T_2)-(I_1-I_2)-(F_1-F_2)|/5\}/2\}+1}\} = 0.$$

Therefore,

$$min\{S_N(A_1, A_2)\} = 1-2/3(1/2+1/2+1/2) = 1-1 = 0,$$
$$max\{S_N(A_1, A_2)\} = 1-2/3(0 + 0 + 0) = 1-0 = 1.$$

Hence, $0 \leq S_N(A_1, A_2) \leq 1$.

(ii)

$$\begin{aligned}
S_N(A_1,\, A_2) \;=\; & 1-(2/3)[\frac{min\{\sqrt{3(T_1-T_2)^2+(I_1-I_2)^2},\ |2(T_1-T_2)-(I_1-I_2)|/3\}}{\{max\{\sqrt{3(T_1-T_2)^2+(I_1-I_2)^2},\ |2(T_1-T_2)-(I_1-I_2)|/3\}/2\}+1} \\
& +\frac{min\{\sqrt{3(T_1-T_2)^2+(F_1-F_2)^2},\ |2(T_1-T_2)-(F_1-F_2)|/3\}}{\{max\{\sqrt{3(T_1-T_2)^2+(F_1-F_2)^2},\ |2(T_1-T_2)-(F_1-F_2)|/3\}/2\}+1} \\
& +\frac{min\{\sqrt{2(T_1-T_2)^2+(I_1-I_2)^2+(F_1-F_2)^2},\ |3(T_1-T_2)-(I_1-I_2)-(F_1-F_2)|/5\}}{\{max\{\sqrt{2(T_1-T_2)^2+(I_1-I_2)^2+(F_1-F_2)^2},\ |3(T_1-T_2)-(I_1-I_2)-(F_1-F_2)|/5\}/2\}+1}] \\
= & \,1-2/3.\,\{\frac{min\{\sqrt{3(T_2-T_1)^2+(I_2-I_1)^2},\ |2(T_2-T_1)-(I_2-I_1)|/3\}}{\{max\{\sqrt{3(T_2-T_1)^2+(I_2-I_1)^2},\ |2(T_2-T_1)-(I_2-I_1)|/3\}/2\}+1} \\
& +\frac{min\{\sqrt{3(T_2-T_{12})^2+(F_2-F_1)^2},\ |2(T_2-T_1)-(F_2-F_1)|/3\}}{\{max\{\sqrt{3(T_2-T_1)^2+(F_2-F_1)^2},\ |2(T_2-T_1)-(F_2-F_1)|/3\}/2\}+1} \\
& +\frac{min\{\sqrt{2(T_2-T_1)^2+(I_2-I_1)^2+(F_2-F_1)^2},\ |3(T_2-T_1)-(I_2-I_1)-(F_2-F_1)|/5\}}{\{max\{\sqrt{2(T_2-T_1)^2+(I_2-I_1)^2+(F_2-F_1)^2},\ |3(T_2-T_1)-(I_2-I_1)-(F_2-F_1)|/5\}/2\}+1}\} \\
= & \,S_N(A_2\,,\, A_1).
\end{aligned}$$

(iii) We assume that

$$\begin{aligned}
S_N(A_1,\, A_2) \;=\; & 1-(2/3)[\frac{min\{\sqrt{3(T_1-T_2)^2+(I_1-I_2)^2},\ |2(T_1-T_2)-(I_1-I_2)|/3\}}{\{max\{\sqrt{3(T_1-T_2)^2+(I_1-I_2)^2},\ |2(T_1-T_2)-(I_1-I_2)|/3\}/2\}+1} \\
& +\frac{min\{\sqrt{3(T_1-T_2)^2+(F_1-F_2)^2},\ |2(T_1-T_2)-(F_1-F_2)|/3\}}{\{max\{\sqrt{3(T_1-T_2)^2+(F_1-F_2)^2},\ |2(T_1-T_2)-(F_1-F_2)|/3\}/2\}+1} \\
& +\frac{min\{\sqrt{2(T_1-T_2)^2+(I_1-I_2)^2+(F_1-F_2)^2},\ |3(T_1-T_2)-(I_1-I_2)-(F_1-F_2)|/5\}}{\{max\{\sqrt{2(T_1-T_2)^2+(I_1-I_2)^2+(F_1-F_2)^2},\ |3(T_1-T_2)-(I_1-I_2)-(F_1-F_2)|/5\}/2\}+1}] = 1
\end{aligned}$$

Therefore,

$$\begin{aligned}
& -(2/3)[\frac{min\{\sqrt{3(T_1-T_2)^2+(I_1-I_2)^2},\ |2(T_1-T_2)-(I_1-I_2)|/3\}}{\{max\{\sqrt{3(T_1-T_2)^2+(I_1-I_2)^2},\ |2(T_1-T_2)-(I_1-I_2)|/3\}/2\}+1} \\
& +\frac{min\{\sqrt{3(T_1-T_2)^2+(F_1-F_2)^2},\ |2(T_1-T_2)-(F_1-F_2)|/3\}}{\{max\{\sqrt{3(T_1-T_2)^2+(F_1-F_2)^2},\ |2(T_1-T_2)-(F_1-F_2)|/3\}/2\}+1} \\
& +\frac{min\{\sqrt{2(T_1-T_2)^2+(I_1-I_2)^2+(F_1-F_2)^2},\ |3(T_1-T_2)-(I_1-I_2)-(F_1-F_2)|/5\}}{\{max\{\sqrt{2(T_1-T_2)^2+(I_1-I_2)^2+(F_1-F_2)^2},\ |3(T_1-T_2)-(I_1-I_2)-(F_1-F_2)|/5\}/2\}+1}] = 0
\end{aligned}$$

So,

$$\begin{aligned}
& \frac{min\{\sqrt{3(T_1-T_2)^2+(I_1-I_2)^2},\ |2(T_1-T_2)-(I_1-I_2)|/3\}}{\{max\{\sqrt{3(T_1-T_2)^2+(I_1-I_2)^2},\ |2(T_1-T_2)-(I_1-I_2)|/3\}/2\}+1} = 0 \text{ and} \\
& \frac{min\{\sqrt{3(T_1-T_2)^2+(F_1-F_2)^2},\ |2(T_1-T_2)-(F_1-F_2)|/3\}}{\{max\{\sqrt{3(T_1-T_2)^2+(F_1-F_2)^2},\ |2(T_1-T_2)-(F_1-F_2)|/3\}/2\}+1} = 0, \\
& \frac{min\{\sqrt{2(T_1-T_2)^2+(I_1-I_2)^2+(F_1-F_2)^2},\ |3(T_1-T_2)-(I_1-I_2)-(F_1-F_2)|/5\}}{\{max\{\sqrt{2(T_1-T_2)^2+(I_1-I_2)^2+(F_1-F_2)^2},\ |3(T_1-T_2)-(I_1-I_2)-(F_1-F_2)|/5\}/2\}+1} = 0.
\end{aligned}$$

Therefore,

$$\begin{aligned}
min\{\sqrt{3(T_1-T_2)^2+(I_1-I_2)^2},\ |2(T_1-T_2)-(I_1-I_2)|/3\} &= 0, \\
min\{\sqrt{3(T_1-T_2)^2+(F_1-F_2)^2},\ |2(T_1-T_2)-(F_1-F_2)|/3\} &= 0, \\
min\{\sqrt{2(T_1-T_2)^2+(I_1-I_2)^2+(F_1-F_2)^2},\ |3(T_1-T_2)-(I_1-I_2)-(F_1-F_2)|/5\} &= 0.
\end{aligned}$$

Now, we write all the cases that can make these statements 0 one-by-one.

(a) We assume that

$$\sqrt{2(T_1-T_2)^2+(I_1-I_2)^2+(F_1-F_2)^2} = 0. \tag{1}$$

Therefore, it is

$$2(T_1-T_2)^2+(I_1-I_2)^2+(F_1-F_2)^2 = 0.$$

Here, it is obtained that $T_1 - T_2 = 0$, $I_1 - I_2 = 0$ and $F_1 - F_2 = 0$. Hence, we get $T_1 = T_2$, $I_1 = I_2$ and $F_1 = F_2$. By Definition 4, $A_1 = A_2$.

(b) Let

$$\sqrt{3(T_1-T_2)^2+(I_1-I_2)^2} = 0, \tag{2}$$

$$\sqrt{3(T_1-T_2)^2+(F_1-F_2)^2} = 0. \tag{3}$$

By (2) and (3), we obtain $3(T_1-T_2)^2+(I_1-I_2)^2 = 0$ and $3(T_1-T_2)^2+(F_1-F_2)^2 = 0$.

Therefore, we obtain $T_1 - T_2 = 0$, $I_1 - I_2 = 0$ and $F_1 - F_2 = 0$. Hence, we obtain $T_1 = T_2$, $I_1 = I_2$ and $F_1 = F_2$. By Definition 4, $A_1 = A_2$.

(c) We assume that

$$\sqrt{3(T_1-T_2)^2+(I_1-I_2)^2} = 0, \tag{4}$$

$$|2(T_1-T_2)-(F_1-F_2)| = 0, \tag{5}$$

By (4), we have

$$T_1 - T_2 = 0 \text{ and } I_1 - I_2 = 0. \tag{6}$$

Hence, we obtain that $F_1 - F_2 = 0$ by (5) and (6).

Hence, $T_1 = T_2$, $I_1 = I_2$ and $F_1 = F_2$. By Definition 4, we get $A_1 = A_2$.

(d) We assume that

$$|2(T_1-T_2)-(I_1-I_2)|/3 = 0, \tag{7}$$

$$|2(T_1-T_2)-(F_1-F_2)|/3 = 0, \tag{8}$$

$$|3(T_1-T_2)-(I_1-I_2)-(F_1-F_2)|/5 = 0. \tag{9}$$

By (7) and (8), we obtain

$$T_1 - T_2 = I_1 - I_2 = F_1 - F_2. \tag{10}$$

Hence, $T_1 - T_2 = 0$ by (9) and (10).

Hence, $T_1 = T_2$, $I_1 = I_2$ and $F_1 = F_2$. By Definition 4, $A_1 = A_2$.

We assume that $A_1 = A_2$. Therefore, by Definition 4, it is $T_1 = T_2$, $I_1 = I_2$, $F_1 = F_2$. Because of this, we have

$$\begin{aligned} S_N(A_1, A_2) &= 1-(2/3)\Big[\frac{min\{\sqrt{3(T_1-T_2)^2+(I_1-I_2)^2}, |2(T_1-T_2)-(I_1-I_2)|/3\}}{\{max\{\sqrt{3(T_1-T_2)^2+(I_1-I_2)^2}, |2(T_1-T_2)-(I_1-I_2)|/3\}/2\}+1} \\ &+ \frac{min\{\sqrt{3(T_1-T_2)^2+(F_1-F_2)^2}, |2(T_1-T_2)-(F_1-F_2)|/3\}}{\{max\{\sqrt{3(T_1-T_2)^2+(F_1-F_2)^2}, |2(T_1-T_2)-(F_1-F_2)|/3\}/2\}+1} \\ &+ \frac{min\{\sqrt{2(T_1-T_2)^2+(I_1-I_2)^2+(F_1-F_2)^2}, |3(T_1-T_2)-(I_1-I_2)-(I_1-I_2)-(F_1-F_2)|/5\}}{\{max\{\sqrt{2(T_1-T_2)^2+(I_1-I_2)^2+(F_1-F_2)^2}, |3(T_1-T_2)-(I_1-I_2)-(F_1-F_2)|/5\}/2\}+1}\Big] = 0 \end{aligned}$$

(iv) We assume that $A_1 \le A_2 \le A_3$. By Definition 5, it is $T_1 \le T_2 \le T_3$, $I_1 \ge I_2 \ge I_3$, $F_1 \ge F_2 \ge F_3$. Hence, we obtain that

$$
\begin{aligned}
&min\{\sqrt{3(T_1-T_2)^2+(I_1-I_2)^2},\ |2(T_1-T_2)-(I_1-I_2)|/3\} \le 1,\\
&max\{\sqrt{3(T_1-T_2)^2+(I_1-I_2)^2},\ |2(T_1-T_2)-(I_1-I_2)|/3\}/2\} \le 1,\\
&min\{\sqrt{3(T_1-T_3)^2+(I_1-I_3)^2},\ |2(T_1-T_3)-(I_1-I_3)|/3\} \le 1,\\
&max\{\sqrt{3(T_1-T_3)^2+(I_1-I_3)^2},\ |2(T_1-T_3)-(I_1-I_3)|/3\}/2\} \le 1.
\end{aligned}
$$

Therefore, we have

$$
\begin{aligned}
&min\{\sqrt{3(T_1-T_2)^2+(I_1-I_2)^2},\ |2(T_1-T_2)-(I_1-I_2)|/3\} \le\\
&min\{\sqrt{3(T_1-T_3)^2+(I_1-I_3)^2},\ |2(T_1-T_3)-(I_1-I_3)|/3\},\\
&max\{\sqrt{3(T_1-T_2)^2+(I_1-I_2)^2},\ |2(T_1-T_2)-(I_1-I_2)|/3\}/2\} \le\\
&max\{\sqrt{3(T_1-T_3)^2+(I_1-I_3)^2},\ |2(T_1-T_3)-(I_1-I_3)|/3\}/2\}.
\end{aligned}
$$

Hence,

$$
\frac{min\{\sqrt{3(T_1-T_2)^2+(I_1-I_2)^2},\ |2(T_1-T_2)-(I_1-I_2)|/3\}}{\{max\{\sqrt{3(T_1-T_2)^2+(I_1-I_2)^2},\ |2(T_1-T_2)-(I_1-I_2)|/3\}/2\}+1} \le
$$

$$
\frac{min\{\sqrt{3(T_1-T_2)^2+(I_1-I_2)^2},\ |2(T_1-T_2)-(I_1-I_2)|/3\}}{\{max\{\sqrt{3(T_1-T_2)^2+(I_1-I_2)^2},\ |2(T_1-T_2)-(I_1-I_2)|/3\}/2\}+1}. \tag{11}
$$

In addition,

$$
\begin{aligned}
&min\{\sqrt{3(T_1-T_2)^2+(F_1-F_2)^2},\ |2(T_1-T_2)-(F_1-F_2)|/3\} \le 1,\\
&max\{\sqrt{3(T_1-T_2)^2+(F_1-F_2)^2},\ |2(T_1-T_2)-(F_1-F_2)|/3\}/2\} \le 1\\
&min\{\sqrt{3(T_1-T_3)^2+(F_1-F_3)^2},\ |2(T_1-T_3)-(F_1-F_3)|/3\} \le 1,\\
&max\{\sqrt{3(T_1-T_3)^2+(F_1-F_3)^2},\ |2(T_1-T_3)-(F_1-F_3)|/3\}/2\} \le 1
\end{aligned}
$$

Therefore, we obtain that

$$
\begin{aligned}
&min\{\sqrt{3(T_1-T_2)^2+(F_1-F_2)^2},\ |2(T_1-T_2)-(F_1-F_2)|/3\} \le\\
&min\{\sqrt{3(T_1-T_3)^2+(F_1-F_3)^2},\ |2(T_1-T_3)-(F_1-F_3)|/3\},\\
&max\{\sqrt{3(T_1-T_2)^2+(F_1-F_2)^2},\ |2(T_1-T_2)-(F_1-F_2)|/3\}/2\} \le\\
&max\{\sqrt{3(T_1-T_3)^2+(F_1-F_3)^2},\ |2(T_1-T_3)-(F_1-F_3)|/3\}/2\}.
\end{aligned}
$$

Hence,

$$
\frac{min\{\sqrt{3(T_1-T_2)^2+(F_1-F_2)^2},\ |2(T_1-T_2)-(F_1-F_2)|/3\}}{\{max\{\sqrt{3(T_1-T_2)^2+(F_1-F_2)^2},\ |2(T_1-T_2)-(F_1-F_2)|/3\}/2\}+1} \le
$$

$$\frac{min\{\sqrt{3(T_1-T_2)^2+(F_1-F_2)^2},\ \left|2(T_1-T_2)-(F_1-F_2)\right|/3\}}{\{max\{\sqrt{3(T_1-T_2)^2+(F_1-F_2)^2},\ \left|2(T_1-T_2)-(F_1-F_2)\right|/3\}/2\}+1}. \tag{12}$$

In addition,

$$\begin{gathered}
min\{\sqrt{2(T_1-T_2)^2+(I_1-I_2)^2+(F_1-F_2)^2},\ \left|3(T_1-T_2)-(I_1-I_2)-(F_1-F_2)\right|/5\}\leq 1\\
max\{\sqrt{2(T_1-T_2)^2+(I_1-I_2)^2+(F_1-F_2)^2},\ \left|3(T_1-T_2)-(I_1-I_2)-(F_1-F_2)\right|/5\}/3\leq 1,\\
min\{\sqrt{2(T_1-T_3)^2+(I_1-I_3)^2+(F_1-F_3)^2},\ \left|3(T_1-T_3)-(I_1-I_3)-(F_1-F_3)\right|/5\}\leq 1\\
max\{\sqrt{2(T_1-T_3)^2+(I_1-I_3)^2+(F_1-F_3)^2},\ \left|3(T_1-T_3)-(I_1-I_3)-(F_1-F_3)\right|/5\}/3\leq 1,
\end{gathered}$$

Hence, we have

$$\begin{gathered}
min\{\sqrt{2(T_1-T_2)^2+(I_1-I_2)^2+(F_1-F_2)^2},\ \left|3(T_1-T_2)-(I_1-I_2)-(F_1-F_2)\right|/5\}\leq\\
min\{\sqrt{2(T_1-T_3)^2+(I_1-I_3)^2+(F_1-F_3)^2},\ \left|3(T_1-T_3)-(I_1-I_3)-(F_1-F_3)\right|/5\},\\
max\{\sqrt{2(T_1-T_2)^2+(I_1-I_2)^2+(F_1-F_2)^2},\ \left|3(T_1-T_2)-(I_1-I_2)-(F_1-F_2)\right|/5\}/2\leq 1,\\
max\{\sqrt{2(T_1-T_3)^2+(I_1-I_3)^2+(F_1-F_3)^2},\ \left|3(T_1-T_3)-(I_1-I_3)-(F_1-F_3)\right|/5\}/2.
\end{gathered}$$

Hence,

$$\begin{aligned}
&\frac{min\{\sqrt{2(T_1-T_2)^2+(I_1-I_2)^2+(F_1-F_2)^2},\ \left|3(T_1-T_2)-(I_1-I_2)-(F_1-F_2)\right|/5\}}{\{max\{\sqrt{2(T_1-T_2)^2+(I_1-I_2)^2+(F_1-F_2)^2},\ \left|3(T_1-T_2)-(I_1-I_2)-(F_1-F_2)\right|/5\}/2\}+1}\leq\\
&+\frac{min\{\sqrt{2(T_1-T_3)^2+(I_1-I_3)^2+(F_1-F_3)^2},\ \left|3(T_1-T_3)-(I_1-I_3)-(F_1-F_3)\right|/5\}}{\{max\{\sqrt{2(T_1-T_3)^2+(I_1-I_3)^2+(F_1-F_3)^2},\ \left|3(T_1-T_3)-(I_1-I_3)-(F_1-F_3)\right|/5\}/2\}+1}.
\end{aligned} \tag{13}$$

By (11), (12) and (13), we have

$$\begin{aligned}
&1-(2/3)[\frac{min\{\sqrt{3(T_1-T_3)^2+(I_1-I_3)^2},\ |2(T_1-T_3)-(I_1-I_3)|/3\}}{\{max\{\sqrt{3(T_1-T_3)^2+(I_1-I_3)^2},\ |2(T_1-T_3)-(I_1-I_3)|/3\}/2\}+1}\\
&+\frac{min\{\sqrt{3(T_1-T_3)^2+(F_1-F_3)^2},\ |2(T_1-T_3)-(F_1-F_3)|/3\}}{\{max\{\sqrt{3(T_1-T_3)^2+(F_1-F_3)^2},\ |2(T_1-T_3)-(F_1-F_3)|/3\}/2\}+1}\\
&+\frac{min\{\sqrt{2(T_1-T_3)^2+(I_1-I_3)^2+(F_1-F_3)^2},\ |3(T_1-T_3)-(I_1-I_3)-(I_1-I_3)-(F_1-F_3)|/5\}}{\{max\{\sqrt{2(T_1-T_3)^2+(I_1-I_3)^2+(F_1-F_3)^2},\ |3(T_1-T_3)-(I_1-I_3)-(F_1-F_3)|/5\}/2\}+1}]\leq\\
&1-(2/3)[\frac{min\{\sqrt{3(T_1-T_2)^2+(I_1-I_2)^2},\ |2(T_1-T_2)-(I_1-I_2)|/3\}}{\{max\{\sqrt{3(T_1-T_2)^2+(I_1-I_2)^2},\ |2(T_1-T_2)-(I_1-I_2)|/3\}/2\}+1}\\
&+\frac{min\{\sqrt{3(T_1-T_2)^2+(F_1-F_2)^2},\ |2(T_1-T_2)-(F_1-F_2)|/3\}}{\{max\{\sqrt{3(T_1-T_2)^2+(F_1-F_2)^2},\ |2(T_1-T_2)-(F_1-F_2)|/3\}/2\}+1}\\
&+\frac{min\{\sqrt{2(T_1-T_2)^2+(I_1-I_2)^2+(F_1-F_2)^2},\ |3(T_1-T_2)-(I_1-I_2)-(F_1-F_3)|/5\}}{\{max\left\{\sqrt{2(T_1-T_2)^2+(I_1-I_2)^2+(F_1-F_2)^2},\ |3(T_1-T_2)-(I_1-I_2)-(F_1-F_2)|/5\right\}/2\}+1}].
\end{aligned}$$

Hence, we get $S_N(A_1,\ A_3)\leq S_N(A_1,\ A_2)$ as desired. □

7. Decision-Making Applications for Neutrosophic Modeling of Talcott Parsons's Action

In this section, we give an algorithm for applications that allow us to find the ideal society in the grand theory of action of Parsons by taking advantage of the similarity measure in Definition 9. In addition, we give a numeric example to this algorithm.

7.1. Algorithm

1. Step: To find out which societies are closer to the ideal society, the criteria to be considered are determined. The criteria of the ideal society in the grand theory of action of Parsons [17] are taken as below:

c_1 = affectivity versus affective neutrality
c_2 = self-orientation versus collective orientation
c_3 = universalism versus particularism
c_4 = quality versus performance
c_5 = specificity versus diffuseness

Let the set of these criteria be $C = \{c_1, c_2, \ldots, c_5\}$.

2. Step: Let the set of weighted values of the criteria be $W = \{w_1, w_2, \ldots, w_m\}$ and let the weighted values be taken as below:

the weighted value of the criterion c_1 is w_1,
the weighted value of the criterion c_2 is w_2,
the weighted value of the criterion c_3 is w_3,
the weighted value of the criterion c_4 is w_4 and
the weighted value of the criterion c_5 is w_5.

In addition, it must be $\sum_{i=1}^{m} w_i = 1$ and $w_1, w_2, \ldots, w_m \in [0,1]$.

In this study, we will take the weighted value of each criterion as equal. If necessary, different weighted values can be selected for each criterion.

3. Step: Each society that will be taken into ideal society assessment should be evaluated by sociologists determined as a single-valued neutrosophic number. Let $T = \{t_1, t_2, \ldots, t_n\}$ be set of societies. Symbolic representation of societies as single-valued neutrosophic sets are denoted as:

$$t_1 = \{c_1{:}< T_{t_1(c_1)}, I_{t_1(c_1)}, F_{t_1(c_1)}>, c_2{:}< T_{t_1(c_2)}, I_{t_1(c_2)}, F_{t_1(c_2)}>, \ldots, c_5{:}< T_{t_1(c_5)}, I_{t_1(c_5)}, F_{t_1(c_5)}>; c_i \in C\ (i = 1, 2, \ldots, 5)\},$$
$$t_2 = \{c_1{:}< T_{t_2(c_1)}, I_{t_2(c_1)}, F_{t_2(c_1)}>, c_2{:}< T_{t_2(c_2)}, I_{t_2(c_2)}, F_{t_2(c_2)}>, \ldots, c_5{:}< T_{t_2(c_5)}, I_{t_2(c_5)}, F_{t_2(c_5)}>; c_i \in C\ (i = 1, 2, \ldots, 5)\},$$
$$t_3 = \{c_1{:}< T_{t_3(c_1)}, I_{t_3(c_1)}, F_{t_3(c_1)}>, c_2{:}< T_{t_3(c_2)}, I_{t_3(c_2)}, F_{t_3(c_2)}>, \ldots, c_5{:}< T_{t_3(c_5)}, I_{t_3(c_5)}, F_{t_3(c_5)}>; c_i \in C\ (i = 1, 2, \ldots, 5)\},$$
$$t_n = \{c_1{:}< T_{t_n(c_1)}, I_{t_n(c_1)}, F_{t_n(c_1)}>, c_2{:}< T_{t_n(c_2)}, I_{t_n(c_2)}, F_{t_n(c_2)}>, \ldots, c_5{:}< T_{t_n(c_5)}, I_{t_n(c_5)}, F_{t_n(c_5)}>; c_i \in C\ (i = 1, 2, \ldots, 5)\}.$$

Here, $c_1, c_2, \ldots, c_5$ are the criteria in Step 1. Thus, each society will be obtained as a single-valued neutrosophic number according to the given criteria.

4. Step: To compare how close the societies are to ideal society in the theory of Parsons, an imaginary perfect society is determined. Perfect society under the similarity measure we have obtained should be as

$$I = \{c_1 :< 1, 0, 0 >, x_2 :< 1, 0, 0 >, \ldots, c_5 : < 1, 0, 0 >; c_i \in C\ (i = 1, 2, \ldots, 5)\}.$$

Hence, we will accept the existence of an imaginary society that includes 100% truth, 0% uncertainty and 0% falsity according to each criterion.

5. Step: We express the societies given as a single-valued neutrosophic set in step 3 in a table according to criteria. Thus, we will obtain Table 3.

Table 3. Criteria table of societies.

	c_1	c_2	c_3	c_4	c_5
t_1	$<T_{t_1(c_1)}, I_{t_1(c_1)}, F_{t_1(c_1)}>$	$\ldots$	$<T_{t_1(c_3)}, I_{t_1(c_3)}, F_{t_1(c_3)}>$	$\ldots$	$<T_{t_1(c_5)}, I_{t_1(c_5)}, F_{t_1(c_5)}>$
t_2	$<T_{t_2(c_1)}, I_{t_2(c_1)}, F_{t_2(c_1)}>$	$\ldots$	$<T_{t_2(c_3)}, I_{t_2(c_3)}, F_{t_2(c_3)}>$	$\ldots$	$<T_{t_2(c_5)}, I_{t_2(c_5)}, F_{t_2(c_5)}>$
$\vdots$	$\vdots$	$\ldots$	$\vdots$	$\ldots$	$\vdots$
t_n	$<T_{t_n(c_1)}, I_{t_n(c_1)}, F_{t_n(c_1)}>$	$\ldots$	$<T_{t_n(c_3)}, I_{t_n(c_3)}, F_{t_n(c_3)}>$	$\ldots$	$<T_{t_n(c_5)}, I_{t_n(c_5)}, F_{t_n(c_5)}>$

6. Step: We will process each criterion values given for each society separately and each criterion values of the perfect society I in Step 4 separately with similarity measure. Hence, we will obtain Table 4.

Table 4. Similarity table for each social criteria to perfect society criteria.

	c_1	c_2	c_3	c_4	c_5
t_1	$S_N(I_{c_1}, t_{1_{c_1}})$	$\ldots$	$S_N(I_{c_3}, t_{1_{c_3}})$	$\ldots$	$S_N(I_{c_5}, t_{1_{c_5}})$
t_2	$S_N(I_{c_1}, t_{2_{c_1}})$	$\ldots$	$S_N(I_{c_3}, t_{2_{c_3}})$	$\ldots$	$S_N(I_{c_5}, t_{2_{c_5}})$
$\vdots$	$\vdots$	$\ldots$	$\vdots$	$\ldots$	$\vdots$
t_n	$S_N(I_{c_1}, t_{n_{c_1}})$		$S_N(I_{c_3}, t_{n_{c_3}})$	$\ldots$	$S_N(I_{c_5}, t_{n_{c_5}})$

7. Step: In this step, we will obtain a weighted similarity table (Table 5).

Table 5. Weighted similarity table for each social criteria to perfect society criteria.

	w_1c_1	w_2c_2	w_3c_3	w_4c_4	w_5c_5
t_1	$w_1S_N(I_{c_1}, t_{1_{c_1}})$	$\ldots$	$w_3S_N(I_{c_3}, t_{1_{c_3}})$	$\ldots$	$w_5S_N(I_{c_5}, t_{1_{c_5}})$
t_2	$w_1S_N(I_{c_1}, t_{2_{c_1}})$	$\ldots$	$w_3S_N(I_{c_3}, t_{2_{c_3}})$	$\ldots$	$w_5S_N(I_{c_5}, t_{2_{c_5}})$
$\vdots$	$\vdots$	$\ldots$	$\vdots$	$\ldots$	$\vdots$
t_n	$w_1S_N(I_{c_1}, t_{n_{c_1}})$		$w_3S_N(I_{c_3}, t_{n_{c_3}})$	$\ldots$	$w_5S_N(I_{c_5}, t_{n_{c_5}})$

In this study, this step is not needed since we take the same weighted value of each criterion. More precisely, Tables 4 and 5 will be the same since the weighted values are equal. This step can be used if necessary.

8. Step:

In this last step, we will obtain a similarity value table (Table 6) by applying $S_{Nk}(t_k, \mathrm{I}) = \sum_{i=1}^{n} w_i.S_N(I_{c_i}, t_{k_{c_i}})$.

Table 6. Similarity value table of societies to the perfect society.

	The Similarity Value
t_1	$S_{N1}(t_1, \mathrm{I})$
t_2	$S_{N2}(t_2, \mathrm{I})$
$\vdots$	$\vdots$
t_n	$S_{Nn}(t_n, \mathrm{I})$

See Figure 1.

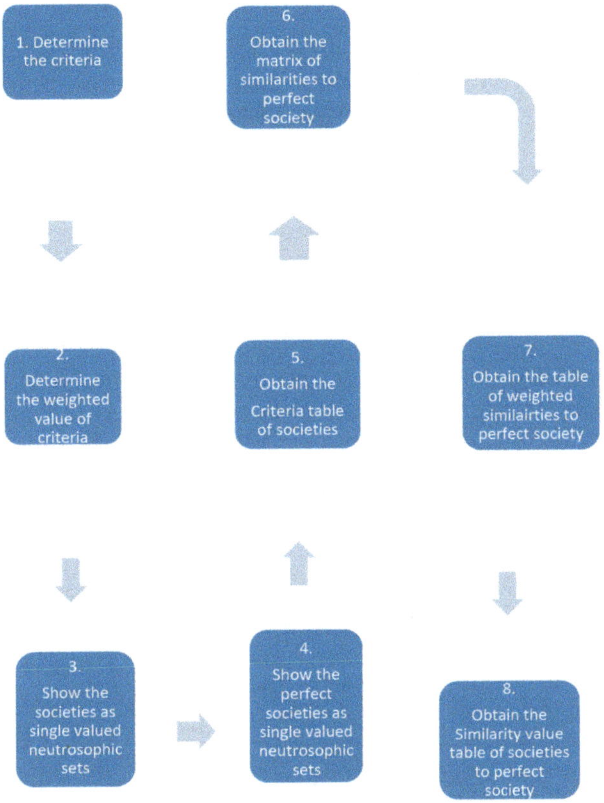

Figure 1. Diagram of the algorithm.

7.2. Numeric Example

Using the steps in Section 7.1, we show how close the 4 societies are to the ideal society.

1. Step: Let the criteria of an ideal society in the theory of Parsons be as it is in Step 1 of Section 7.1;

c_1 = affectivity versus affective neutrality
c_2 = self-orientation versus collective orientation
c_3 = universalism versus particularism
c_4 = quality versus performance
c_5 = specificity versus diffuseness

Let C = $\{c_1, c_2, \ldots, c_5\}$ be the set of criteria.

2. Step: In this example, we will take the weight values of each criterion equal so that $w_1 = w_2 = \ldots = w_5 = 0.2$.

3. Step: Let the set of societies be T = $\{t_1, t_2, t_3, t_4\}$. We assume that the single-valued neutrosophic set with evaluation of societies by sociologists according to the criteria in Step 1 will be as below:

$t_1 = \{c_1{:}< 0.6, 0.2, 0.1 >, c_2{:}< 0.7, 0.2, 0.1 >, c_3{:}< 0.4, 0.1, 0.2 >, c_4{:}< 0.8, 0.1, 0 >, c_5{:}< 0.5, 0.1, 0.2 >\}$
$t_2 = \{c_1{:}< 0.5, 0.2, 0.3 >, c_2{:}< 0.6, 0.1, 0.3 >, c_3{:}< 0.8, 0.1, 0.2 >, c_4{:}< 0.4, 0.1, 0.4 >, c_5{:}< 0.9, 0, 0.1 >\}$
$t_3 = \{c_1{:}< 0.5, 0.2, 0.1 >, c_2{:}< 0.8, 0.1, 0.1 >, c_3{:}< 0.8, 0.1, 0 >, c_4{:}< 0.7, 0.2, 0.1 >, c_5{:}< 0.7, 0.2, 0.3 >\}$
$t_4 = \{c_1{:}< 0.7, 0.2, 0.1 >, c_2{:}< 0.6, 0.2, 0.2 >, c_3{:}< 0.7, 0.2, 0.1 >, c_4{:}< 0.7, 0.1, 0.2 >, c_5{:}< 0.8, 0.1, 0.1 >\}$

4. Step: Let the dream perfect society that we compare societies be

$$\text{I} = \{c_1 :< 1, 0, 0 >, c_2 :< 1, 0, 0 >, c_3 :< 1, 0, 0 >, c_3 :< 1, 0, 0 >, c_4 :< 1, 0, 0 >, c_5 :< 1, 0, 0 >\}.$$

5. Step: Let us express the societies as a single-valued neutrosophic set in Step 3 in Table 7.

Table 7. The criteria table of societies.

	c_1	c_2	c_3	c_4	c_5
t_1	<0.6, 0.2, 0.1>	<0.7, 0.2, 0.1>	<0.4, 0.1, 0.2>	<0.8, 0.1, 0>	<0.5, 0.1, 0.2>
t_2	<0.5, 0.2, 0.3>	<0.6, 0.1, 0.3>	<0.8, 0.1, 0.2>	<0.4, 0.1, 0.4>	<0.9, 0, 0.1>
t_3	<0.5, 0.2, 0.1>	<0.8, 0.1, 0.1>	<0.8, 0.1, 0>	<0.7, 0.2, 0.1>	<0.7, 0.2, 0.3>
t_4	<0.7, 0.2, 0.1>	<0.6, 0.2, 0.2>	<0.7, 0.2, 0.1>	<0.7, 0.1, 0.2>	<0.8, 0.1, 0.1>

6. Step: Using the similarity measure, we obtain the similarity table (Table 8) which is the similarity of the criteria of societies to the criteria of the perfect society.

Table 8. The similarity table of the criteria of societies to the criteria of the perfect society.

	c_1	c_2	c_3	c_4	c_5
t_1	0.5351	0.6088	0.4121	0.7489	0.4700
t_2	0.4263	0.5132	0.6930	0.3734	0.8494
t_3	0.4700	0.7196	0.7489	0.6088	0.5610
t_4	0.6088	0.5112	0.6088	0.6088	0.7196

7. Step: In this example, there is no need to make any changes in Table 8 since we take the weighted value of each criterion as equal.

8. Step: In this step, we obtain similarity values of the societies in Table 8 to the perfect society.

Now, we obtain the similarity values of the societies in Table 9 and we obtain Table 10 by dividing the values in Table 9 by 5, taking the weighted values as equal for each society on 5 criteria and, hence, getting the results in the range [0,1].

Table 9. The similarity value table of the societies to the perfect society.

	The Similarity Value
t_1	$S_{N1}(t_1, \text{I}) = 2.7749$
t_2	$S_{N2}(t_2, \text{I}) = 2.8553$
t_3	$S_{N3}(t_3, \text{I}) = 3.1083$
t_4	$S_{N4}(t_4, \text{I}) = 3.0572$

Table 10. The similarity rate of the societies to the perfect society.

	The Similarity Value
t_1	$S_{N1}(t_1, \text{I}) = 0.5549$
t_2	$S_{N2}(t_2, \text{I}) = 0.5710$
t_3	$S_{N3}(t_3, \text{I}) = 0.6216$
t_4	$S_{N4}(t_4, \text{I}) = 0.6114$

In addition, the similarity value of each society to the perfect society in Table 10 is obtained. The result of the evaluation is given. Thus, societies closest to the perfect society are obtained as $\boldsymbol{t_3}$, $\boldsymbol{t_4}$, $\boldsymbol{t_2}$ and $\boldsymbol{t_1}$ respectively.

8. Sensitivity Analysis

In 7.1 Numeric example, we take the weighted values of criteria $W = \{w_1, w_2, \ldots, w_m\}$ equal such that

the weighted value of c_1 criteria $w_1 = 0.2$
the weighted value of c_2 criteria $w_2 = 0.2$
the weighted value of c_3 criteria $w_3 = 0.2$
the weighted value of c_4 criteria $w_4 = 0.2$
the weighted value of c_5 criteria $w_5 = 0.2$

Thus, societies closest to the perfect society are obtained as $\boldsymbol{t_3}, \boldsymbol{t_4}, \boldsymbol{t_2}, \boldsymbol{t_1}$ respectively.

(a) If we take the $W = \{w_1 = 0.1, w_2 = 0.3, w_3 = 0.2, w_4 = 0.2, w_5 = 0.2\}$, then we obtain that societies closest to the perfect society are obtained as t_3, t_4, t_2 and t_1 respectively (Table 11).

Table 11. The similarity rate of the societies to the perfect society for $W = \{w_1 = 0.1, w_2 = 0.3, w_3 = 0.2, w_4 = 0.2, w_5 = 0.2\}$.

	The Similarity Value
t_1	$S_{N1}(t_1, I) = 0.55235$
t_2	$S_{N2}(t_2, I) = 0.57975$
t_3	$S_{N3}(t_3, I) = 0.64662$
t_4	$S_{N4}(t_4, I) = 0.60168$

Thus, we obtain the same result with the Numeric Example 7.1.

(b) If we take the $W = \{w_1 = 0.2, w_2 = 0.2, w_3 = 0.3, w_4 = 0.1, w_5 = 0.2\}$, then we obtain that societies closest to the perfect society are obtained as t_3, t_4, t_2 and t_1 respectively respectively (Table 12).

Table 12. The similarity rate of the societies to the perfect society for $W = \{w_1 = 0.2, w_2 = 0.2, w_3 = 0.3, w_4 = 0.1, w_5 = 0.2\}$.

	The Similarity Value
t_1	$S_{N1}(t_1, I) = 0.5213$
t_2	$S_{N2}(t_2, I) = 0.60302$
t_3	$S_{N3}(t_3, I) = 0.63567$
t_4	$S_{N4}(t_4, I) = 0.61144$

Thus, we obtain same result with Numeric Example 7.1.

(c) If we take the $W = \{w_1 = 0.3, w_2 = 0.1, w_3 = 0.2, w_4 = 0.2, w_5 = 0.2\}$, then we obtain that societies closest to the perfect society are obtained as t_4, t_3, t_2 and t_1 respectively respectively (Table 13).

Table 13. The similarity rate of the societies to the perfect society for $W = \{w_1 = 0.3, w_2 = 0.1, w_3 = 0.2, w_4 = 0.2, w_5 = 0.2\}$.

	The Similarity Value
t_1	$S_{N1}(t_1, I) = 0.54761$
t_2	$S_{N2}(t_2, I) = 0.56237$
t_3	$S_{N3}(t_3, I) = 0.5967$
t_4	$S_{N4}(t_4, I) = 0.6212$

Thus, we obtain a different result from Numeric Example 7.1.

(d) If we take the $W = \{w_1 = 0.2, w_2 = 0.2, w_3 = 0.1, w_4 = 0.3, w_5 = 0.2\}$, then we obtain that societies closest to the perfect society are obtained as t_4, t_1, t_2 and t_3 respectively respectively (Table 14).

Table 14. The similarity rate of the societies to the perfect society for W = $\{w_1 = 0.2, w_2 = 0.2, w_3 = 0.1, w_4 = 0.3, w_5 = 0.2\}$.

	The Similarity Value
t_1	$S_{N1}(t_1, I) = 0.58866$
t_2	$S_{N2}(t_2, I) = 0.5391$
t_3	$S_{N3}(t_3, I) = 0.36379$
t_4	$S_{N4}(t_4, I) = 0.61144$

Thus, we obtain a different result with Numeric Example 7.1.

(e) If we take the W = $\{w_1 = 0.2, w_2 = 0.2, w_3 = 0.2, w_4 = 0.3, w_5 = 0.1\}$, then we obtain that societies closest to the perfect society are obtained as t_3, t_4, t_1, and t_2 respectively respectively (Table 15).

Table 15. The similarity rate of the societies to the perfect society for W = $\{w_1 = 0.2, w_2 = 0.2, w_3 = 0.2, w_4 = 0.3, w_5 = 0.1\}$.

	The Similarity Value
t_1	$S_{N1}(t_1, I) = 0.58287$
t_2	$S_{N2}(t_2, I) = 0.52346$
t_3	$S_{N3}(t_3, I) = 0.62644$
t_4	$S_{N4}(t_4, I) = 0.60036$

Thus, we obtain a different result from Numeric Example 7.1.

(f) If we take the W = $\{w_1 = 0.2, w_2 = 0.2, w_3 = 0.2, w_4 = 0.1, w_5 = 0.3\}$, then we obtain that societies closest to the perfect society are obtained as t_4, t_2, t_3, and t_1 respectively respectively (Table 16).

Table 16. The similarity rate of the societies to the perfect society for W = $\{w_1 = 0.2, w_2 = 0.2, w_3 = 0.2, w_4 = 0.1, w_5 = 0.3\}$.

	The Similarity Value
t_1	$S_{N1}(t_1, I) = 0.52709$
t_2	$S_{N2}(t_2, I) = 0.61866$
t_3	$S_{N3}(t_3, I) = 0.61688$
t_4	$S_{N4}(t_4, I) = 0.62252$

Thus, we obtain a different result from Numeric Example 7.1.

Now, we give results in (a), (b), (c), (d), (e) and (f) in Table 17.

Table 17. Ideal societies according to weighted values.

	Ideal Societies Respectively
W = $\{w_1 = 0.2, w_2 = 0.2, w_3 = 0.2, w_4 = 0.1, w_5 = 0.3\}$	t_3, t_4, t_1, t_2
W = $\{w_1 = 0.2, w_2 = 0.2, w_3 = 0.2, w_4 = 0.3, w_5 = 0.1\}$	t_4, t_3, t_2, t_1
W = $\{w_1 = 0.2, w_2 = 0.1, w_3 = 0.3, w_4 = 0.2, w_5 = 0.2\}$	t_3, t_4, t_2, t_1
W = $\{w_1 = 0.2, w_2 = 0.3, w_3 = 0.1, w_4 = 0.2, w_5 = 0.2\}$	t_4, t_1, t_2, t_3
W = $\{w_1 = 0.3, w_2 = 0.1, w_3 = 0.2, w_4 = 0.2, w_5 = 0.2\}$	t_4, t_3, t_2, t_1
W = $\{w_1 = 0.1, w_2 = 0.3, w_3 = 0.2, w_4 = 0.2, w_5 = 0.2\}$	t_3, t_4, t_2, t_1

As seen in Table 17, if we take W = $\{w_1 = 0.2, w_2 = 0.2, w_3 = 0.2, w_4 = 0.1, w_5 = 0.3\}$ or W = $\{w_1 = 0.1, w_2 = 0.3, w_3 = 0.2, w_4 = 0.2, w_5 = 0.2\}$, then we obtain same result with Numeric Example 7.1. In other cases, we obtain different results from Numeric Example 7.1.

9. Study Comparison Methods

In this section, we have compared the obtained results of the data in our Example 1 with the results of the similarity measures, Hausdorff measure [18], Hamming measure [18] and the previously defined similarity measure [16].

1. If we use the similarity measure in Definition 6 [16] for Example 1, we obtain Table 18 as a result.

Table 18. The similarity rate according to similarity measure, in Definition 6 [16], of the societies to the perfect society.

	The Similarity Value
t_1	$S_{N1}(t_1, \text{I}) = 0.661445$
t_2	$S_{N2}(t_2, \text{I}) = 0.639916$
t_3	$S_{N3}(t_3, \text{I}) = 0.691014$
t_4	$S_{N4}(t_4, \text{I}) = 0.678023$

Thus, societies closest to the perfect society are obtained as t_3, t_4, t_1 and t_2 respectively according to similarity measure in Definition 6 [16].

2. If we use the Hausdorff measure [18] for Example 1, we obtain Table 19 as a result.

Table 19. The similarity rate according to Hausdorff measure, in Definition 7 [18], of the societies to the perfect society.

	The Similarity Value
t_1	$S_h(t_1, \text{I}) = 0.6$
t_2	$S_h(t_2, \text{I}) = 0.64$
t_3	$S_h(t_3, \text{I}) = 0.7$
t_4	$S_h(t_4, \text{I}) = 0.7$

Thus, societies closest to the perfect society are obtained as $t_3 = t_4$, t_2 and t_1 respectively according to Hausdorff similarity measure in Definition 7 [18].

3. If we use the Hamming measure [18] for Example 1, we obtain Table 20 as a result.

Table 20. The similarity rate according to Hamming similarity measure, in Definition 8 [18], of the societies to the perfect society.

	The Similarity Value
t_1	$S_H(t_1, \text{I}) = 0.78$
t_2	$S_H(t_2, \text{I}) = 0.76$
t_3	$S_H(t_3, \text{I}) = 0.806667$
t_4	$S_H(t_4, \text{I}) = 0.8$

Thus, societies closest to the perfect society are obtained as t_3, t_4, t_1 and t_2 respectively according to Hamming similarity measure, in Definition 8 [18].

As a result,

according to our similarity measure, the perfect society is obtained as t_3, t_4, t_2, t_1 respectively;
according to similarity measure [16], the perfect society is obtained as t_3, t_4, t_1, t_2 respectively;
according to Hausdorff measure [18], the perfect society is obtained as $t_3 = t_4$, t_2, t_1 respectively;
according to Hamming measure [18], the perfect society is obtained as t_3, t_4, t_1, t_2 respectively.

10. Discussions

In this study, we explained the grand theory of action of Parsons, which has an important place in social theories, for the first time in neutrosociology. Thus, like all social theories, we have achieved a more effective way of dealing with uncertainties in the theory of Parsons. In addition, we have obtained a similarity measure for single-valued neutrosophic numbers. By making use of this similarity measure, we have obtained applications that allow finding the ideal society in the theory of Parsons within the theory of neutrosociology. Hence, we have added a new structure to neutrosophic theory, neutrosociology theory. In addition, by utilizing this study, other social theories can be explained in neutrosociology. Thus, the uncertainties encountered can be dealt with more easily. In addition, by using neutrosophic numbers and sets related to other social theories, new similarity measures can be obtained, and the consistency of these measures can be checked.

11. Conclusions

In Section 9, we obtained different results for the similarity measure [16]; Hausdorff measure [18]; and Hamming measure [18]. In addition, we give a comparison in Table 21.

Table 21. Comparison methods.

	Ideal Societies, Respectively
Similarity measure in definition 9	t_3, t_4, t_1, t_2
Similarity measure in definition 6 [16]	t_3, t_4, t_1, t_2
Hausdorff measure in definition 7 [18]	$t_3 = t_4, t_2, t_1$
Hamming measure in definition 8 [18]	t_3, t_4, t_1, t_2

Author Contributions: In this article, each author contributed equally. C.A. obtained Neutrosophic Modeling of Grand Theory of Action of Talcott Parsons; A.K. introduced similarity measure and algorithm; M.Ş. obtained examples and organized the study. All authors have read and agreed to the published version of the manuscript.

Funding: This research received no external funding.

Conflicts of Interest: The authors declare no conflicts of interest.

References

1. Smarandache, F. *A Unifying Field in Logics. Neutrosophy: Neutrosophic Probability, Set and Logic*; American Research Press: Rehoboth, DE, USA, 1999.
2. Zadeh, L.A. Fuzzy sets. *Inf. Control* **1965**, *8*, 338–353. [CrossRef]
3. Atanassov, T.K. Intuitionistic fuzzy sets. *Fuzzy Sets Syst.* **1986**, *20*, 87–96. [CrossRef]
4. Smarandache, F. *Introduction to Neutrosophic Sociology (Neutrosociology)*; Pons Publishing House/Pons asblQuai du Batelage: Bruxelles, Belgium, 2019.
5. Broumi, S.; Topal, S.; Bakali, A.; Talea, M.; Smarandache, F. A novel python toolbox for single and interval-valued neutrosophic matrices. In *Neutrosophic Sets in Decision Analysis and Operations Research*; IGI Global: Hershey, PA, USA, 2020; pp. 281–330.
6. Wang, H.; Smarandache, F.; Zhang, Y.; Sunderraman, R. Single valued neutrosophic sets. *Multispace Multistruct.* **2010**, *4*, 410–413.
7. Ye, J. Similarity measures between interval neutrosophic sets and their applications in multicriteria decision-making. *J. Intell. Fuzzy Syst.* **2014**, *26*, 165–172. [CrossRef]
8. Broumi, S.; Bakali, A.; Talea, M.; Smarandache, F.; Singh, P.K.; Uluçay, V.; Khan, M. Bipolar complex neutrosophic sets and its application in decision making problem. In *Fuzzy Multi-Criteria Decision-Making Using Neutrosophic Sets*; Springer: Cham, Switzerland, 2019; pp. 677–710.

9. Şahin, M.; Olgun, N.; Uluçay, V.; Kargın, A.; Smarandache, F. A new similarity measure on falsity value between single valued neutrosophic sets based on the centroid points of transformed single valued neutrosophic numbers with applications to pattern recognition. *Neutrosophic Sets Syst.* **2017**, *15*, 31–48. [CrossRef]
10. Şahin, M.; Ecemiş, O.; Uluçay, V.; Kargın, A. Some new generalized aggregation operators based on centroid single valued triangular neutrosophic numbers and their applications in multi-attribute decision making. *Asian J. Math. Comput. Res.* **2017**, *16*, 63–84.
11. Şahin, M.; Kargın, A. Neutrosophic triplet groups based on set valued neutrosophic quadruple numbers. *Neutrosophic Set Syst.* **2019**, *30*, 122–131.
12. Uluçay, V.; Şahin, M. Neutrosophic multigroups and applications. *Mathematics* **2019**, *7*, 95. [CrossRef]
13. Dey, A.; Broumi, S.; Bakali, A.; Talea, M.; Smarandache, F. A new algorithm for finding minimum spanning trees with undirected neutrosophic graphs. *Granul. Comput.* **2019**, *4*, 63–69. [CrossRef]
14. Khalid, H.E. Neutrosophic Geometric Programming (NGP) with (max-product) operator, an innovative model. *Neutrosophic Sets Syst.* **2020**, *32*, 16.
15. Bakbak, D.; Uluçay, V.; Şahin, M. Neutrosophic soft expert multiset and their application to multiple criteria decision making. *Mathematics* **2019**, *7*, 50. [CrossRef]
16. Şahin, M.; Kargın, A. New similarity measure between single-valued neutrosophic sets and decision-making applications in professional proficiencies. In *Neutrosophic Sets in Decision Analysis and Operations Research*; IGI Global: Hershey, PA, USA, 2020; pp. 129–149.
17. Parsons, T. *Social Systems and The Evolution of Action Theory*; The Free Press: New York, NY, USA, 1975.
18. Mukherjee, A.; Sarkar, S. Several Similarity Measures of Neutrosophic Soft Sets and its Application in Real Life Problems. *Ann. Pure Appl. Math.* **2014**, *7*, 1–6.
19. Şahin, M.; Kargın, A.; Smarandache, F. Combined classical-neutrosophic sets and numbers, Double Neutrosophic sets and numbers. In *Quadruple Neutrosophic Theory and Applications*; Pons Editions: Brussels, Belgium, 2020; Volume 18, pp. 254–265.
20. Uluçay, V.; Şahin, M. Decision-making method based on neutrosophic soft expert graphs. In *Neutrosophic Graph Theory and Algorithms*; IGI Global: Hershey, PA, USA, 2020; pp. 33–76.
21. Olgun, N.; Hatip, A. The effect of the neutrosophic logic on the decision tree. In *Quadruple Neutrosophic Theory and Applications*; Pons Editions: Brussels, Belgium, 2020; Volume 7, pp. 238–253.
22. Wang, J.; Wei, G.; Lu, M. An extended VIKOR method for multiple criteria group decision making with triangular fuzzy neutrosophic numbers. *Symmetry* **2018**, *10*, 497. [CrossRef]
23. Biswas, P.; Pramanik, S.; Giri, B.C. TOPSIS Method for Multi-Attribute Group Decision-Making Under Single-Valued Neutrosophic Environment. *Neural Comput. Appl.* **2016**, *27*, 727–737. [CrossRef]
24. Şahin, R.; Liu, P. Maximizing deviation method for neutrosophic multiple attribute decision making with incomplete weight information. *Neural Comput. Appl.* **2016**, *27*, 2017–2029. [CrossRef]
25. Biswas, P.; Pramanik, S.; Giri, B.C. Entropy based grey relational analysis method for multi-attribute decision making under single valued neutrosophic assessments. *Neutrosophic Sets Syst.* **2014**, *2*, 102–110.
26. Bourricaud, F. *The Sociology of Talcott Parsons*; University of Chicago Press: Chicago, IL, USA, 1981.
27. Polama, M.M. *Contemporary Sociological Theory*; Macmillan: New York, NY, USA, 1979.
28. Dillon, M. Talcott Parsons and Robert Merton, Functionalism and Modernization. In *Introduction to Sociological Theory: Theorists, Concepts, and their Applicability to the Twenty-First Century*; Wiley: Hoboken, NJ, USA, 2013; pp. 156–157.
29. Parsons, T. *Action Theory and the Human Condition*; Free Press: New York, NY, USA, 1978.
30. Parsons, T. *The System of Modern Societies*; Prentice-Hall: Englewood Cliffs, NJ, USA, 1971.
31. Parsons, T. *Societies: Evolutionary and Comparative Perspectives*; Prentice-Hall: Englewood Cliffs, NJ, USA, 1966.
32. Parsons, T. *Essays in Sociological Theory, Revised Edition*; The Free Press: New York, NY, USA, 1954.
33. Parsons, T.; Shils, E. *Toward a General Theory of Action*; Harvard University Press: Cambridge, MA, USA, 1951.
34. Parsons, T. *The Structure of Social Action: A Study in Social Theory with Special Reference to a Group of Recent European Writers*; Free Press: New York, NY, USA, 1968.

35. Wallace, R.A.; Wolf, A. *Contemporary Sociological Theory: Expanding the Classical Tradition*, Subsequent ed.; Prentice Hall, Pearson Education: New York, NY, USA, 1995.
36. Weber, M. *The Theory of Social and Economic Organization*; The Free Press: New York, NY, USA, 1947.

symmetry

MDPI

Article

Combination of the Single-Valued Neutrosophic Fuzzy Set and the Soft Set with Applications in Decision-Making

Ahmed Mostafa Khalil [1], Dunqian Cao [2,*], Abdelfatah Azzam [3,4] and Florentin Smarandache [5] and Wedad R. Alharbi [6]

1 Department of Mathematics, Faculty of Science, Al-Azhar University, Assiut 71524, Egypt; a.khalil@azhar.edu.eg
2 School of Mathematics and Physics, Guangxi University of Nationalities, Nanning 530006, China
3 Department of Mathematics, Faculty of Science and Humanities, Prince Sattam Bin Abdulaziz University, Alkharj 11942, Saudi Arabia; aa.azzam@psau.edu.sa
4 Department of Mathematics, Faculty of Science, New Valley University, Elkharga 72511, Egypt
5 Department of Mathematics, University of New Mexico, 705 Gurley Ave., Gallup, NM 87301, USA; smarand@unm.edu
6 Physics Department, Faculty of Science, University of Jeddah, Jeddah 23890, Saudi Arabia; 04103076@uj.edu.sa
* Correspondence: caodunqian@gxun.edu.cn; Tel.: +86-150-7715-5355

Received: 16 July 2020; Accepted: 8 August 2020; Published: 14 August 2020

check for updates

Abstract: In this article, we propose a novel concept of the single-valued neutrosophic fuzzy soft set by combining the single-valued neutrosophic fuzzy set and the soft set. For possible applications, five kinds of operations (e.g., subset, equal, union, intersection, and complement) on single-valued neutrosophic fuzzy soft sets are presented. Then, several theoretical operations of single-valued neutrosophic fuzzy soft sets are given. In addition, the first type for the fuzzy decision-making based on single-valued neutrosophic fuzzy soft set matrix is constructed. Finally, we present the second type by using the AND operation of the single-valued neutrosophic fuzzy soft set for fuzzy decision-making and clarify its applicability with a numerical example.

Keywords: single-valued neutrosophic fuzzy set; soft set; Algorithm 1; Algorithm 2; decision-making

1. Introduction

Many areas (e.g., physics, social sciences, computer sciences, and medicine) work with vague data that require fuzzy sets [1], intuitionistic fuzzy sets [2], picture fuzzy sets [3], and other mathematical tools. Molodtsov [4] presented a novel approach termed "soft set theory", which plays a very significant role in different fields. Therefore, several researchers have developed some methods and operations of soft set theory. For instance, Maji et al. [5] introduced some notions of and operations on soft sets. In addition, Maji et al. [6] gave an application of soft sets to solve fuzzy decision-making. Maji et al. [7] proposed the notion of fuzzy soft sets, followed by studies on inverse fuzzy soft sets [8], belief interval-valued soft sets [9], interval-valued intuitionistic fuzzy soft sets [10], interval-valued picture fuzzy soft sets [11], interval-valued neutrosophic soft sets [12], and generalized picture fuzzy soft sets [13]. Furthermore, several expansion models of soft sets have been developed very quickly, such as possibility Pythagorean fuzzy soft sets [14], possibility m-polar fuzzy soft sets [15], possibility neutrosophic soft sets [16], and possibility multi-fuzzy soft sets [17]. Karaaslan and Hunu [18] defined the notion of type-2 single-valued neutrosophic sets and gave several distance measure methods: Hausdorff, Hamming, and Euclidean distances for Type-2 single-valued neutrosophic sets. Al-Quran et al. [19] presented the notion of fuzzy parameterized complex neutrosophic soft expert sets and gave a novel approach by transforming from the complex case to the real case for decision-making. Qamar and Hassan [20] proposed a novel approach to Q-neutrosophic soft sets and studied several operations of Q-neutrosophic soft sets. Further, they generalized Q-neutrosophic soft expert sets based on uncertainty for decision-making [21]. On the other hand, Uluçay et al. [22] presented the concept of generalized neutrosophic soft expert sets and applied a novel algorithm for multiple-criteria decision-making. Zhang et al. [23] gave novel algebraic operations of totally dependent neutrosophic sets and totally dependent neutrosophic soft sets. In 2018, Smarandache [24] generalized the soft set to the hypersoft set by transforming the function F into a multi-argument function.

Fuzzy sets are used to tackle uncertainty using the membership grade, whereas neutrosophic sets are used to tackle uncertainty using the truth, indeterminacy, and falsity membership grades, which are considered as independent. As the motivation of this article, we present a novel notion of the single-valued neutrosophic fuzzy soft set, which can be seen as a novel single-valued neutrosophic fuzzy soft set model, which gives rise to some new concepts. Since neutrosophic fuzzy soft sets have some difficulties in dealing with some real-life problems due to the nonstandard interval of neutrosophic components, we introduce the single-valued neutrosophic fuzzy soft set (i.e., the single-valued neutrosophic set has a symmetric form, since the membership (T) and nonmembership (F) are symmetric with each other, while indeterminacy (I) is in the middle), which is considered as an instance of neutrosophic fuzzy soft sets. The structural operations (e.g., subset, equal, union, intersection, and complement) on single-valued neutrosophic fuzzy soft sets, and several fundamental properties of the five operations above are introduced. Lastly, two novel approaches (i.e., Algorithms 1 and 2) to fuzzy decision-making depending on single-valued neutrosophic fuzzy soft sets are discussed, in addition to a numerical example to show the two approaches we have developed.

The rest of this article is arranged as follows. Section 2 briefly introduces several notions related to fuzzy sets, neutrosophic sets, single-valued neutrosophic sets, neutrosophic fuzzy sets, single-valued neutrosophic fuzzy sets, soft sets, fuzzy soft sets, and neutrosophic soft sets. Section 3 discusses single-valued neutrosophic fuzzy soft sets (along with their basic operations and structural properties). Section 4 gives two algorithms for single-valued neutrosophic fuzzy soft sets for decision-making. Lastly, the conclusions are given in Section 5.

2. Preliminaries

In the following, we present a short survey of seven definitions which are necessary to this paper.

2.1. Fuzzy Set

Definition 1 (cf. [1]). *Assume that X (i.e., $X = \{x_1, x_2, ..., x_p\}$) is a set of elements and $\mu(x_p)$ is a membership function of element $x_p \in X$. Then*

(1) The following mapping (called fuzzy set), is given by

$$\mu : X \longrightarrow [0,1]$$

and $[0,1]^X$ is a set of whole fuzzy subset over X.

(2) Let

$$\mu = \left\{ \frac{\mu(x_1)}{x_1}, \frac{\mu(x_2)}{x_2}, \cdots, \frac{\mu(x_p)}{x_p} \middle| \, x_p \in X \right\} \in [0,1]^X$$

and

$$\nu = \left\{ \frac{\nu(x_1)}{x_1}, \frac{\nu(x_2)}{x_2}, \cdots, \frac{\nu(x_p)}{x_p} \middle| \, x_p \in X \right\} \in [0,1]^X.$$

Then

(1) *The union* $\mu \cup v$, *is defined as*

$$\mu \cup v = \left\{ \frac{\mu(x_1) \vee v(x_1)}{x_1}, \frac{\mu(x_2) \vee v(x_2)}{x_2}, \ldots, \frac{\mu(x_p) \vee v(x_p)}{x_p} \middle| x_p \in X \right\}.$$

(2) *The intersection* $\mu \cap v$, *is defined as*

$$\mu \cap v = \left\{ \frac{\mu(x_1) \wedge v(x_1)}{x_1}, \frac{\mu(x_2) \wedge v(x_2)}{x_2}, \ldots, \frac{\mu(x_p) \wedge v(x_p)}{x_p} \middle| x_p \in X \right\}.$$

2.2. Neutrosophic Set and Single-Valued Neutrosophic Set

Definition 2 (cf. [25,26]). *Assume that X (i.e.,* $X = \{x_1, x_2, ..., x_p\}$*) is a set of elements and*

$$\Phi = \left\{ \frac{(T_\Phi(x_p), I_\Phi(x_p), F_\Phi(x_p))}{x_p} \middle| x_p \in X,\ 0 \leq T_{\hat{\Phi}}(x_p) + I_{\hat{\Phi}}(x_p) + F_{\hat{\Phi}}(x_p) \leq 3 \right\}.$$

(1) *If* $T_\Phi(x_p)\ \in]0^-, 1^+[$ *(i.e., the degree of truth membership),* $I_\Phi(x_p)\ \in]0^-, 1^+[$ *(i.e., the degree of indeterminacy membership), and* $F_\Phi(x_p)$ *(i.e., the degree of falsity membership), then* Φ *is called a neutrosophic set on X, denoted by* $(\mathbb{NS})^X$.

(2) *If* $T_\Phi(x_p) \in [0,1]$ *(i.e., the degree of truth membership),* $I_\Phi(x_p) \in [0,1]$ *(i.e., the degree of indeterminacy membership), and* $F_\Phi(x_p) \in [0,1]$ *(i.e., the degree of falsity membership), then* Φ *is called a single-valued neutrosophic set on X, denoted by* $(\mathbb{SVNS})^X$.

2.3. Neutrosophic Fuzzy Set and Single-Valued Neutrosophic Fuzzy Set

Definition 3 (cf. [27]). *Assume that X (i.e.,* $X = \{x_1, x_2, ..., x_p\}$*) is a set of elements and*

$$\hat{\Phi} = \left\{ \frac{(T_{\hat{\Phi}}(x_p), I_{\hat{\Phi}}(x_p), F_{\hat{\Phi}}(x_p), \mu(x_p))}{x_p} \middle| x_p \in X,\ 0 \leq T_{\hat{\Phi}}(x_p) + I_{\hat{\Phi}}(x_p) + F_{\hat{\Phi}}(x_p) \leq 3 \right\}.$$

(1) *If* $T_{\hat{\Phi}}(x_p)\ \in]0^-, 1^+[$ *(i.e., the degree of truth membership),* $I_{\hat{\Phi}}(x_p)\ \in]0^-, 1^+[$ *(i.e., the degree of indeterminacy membership), and* $F_{\hat{\Phi}}(x_p)$ *(i.e., the degree of falsity membership), then* $\hat{\Phi}$ *is called a neutrosophic fuzzy set on X, denoted by* $(\mathbb{NFS})^X$.

(2) *If* $T_{\hat{\Phi}}(x_p) \in [0,1]$ *(i.e., the degree of truth membership),* $I_{\hat{\Phi}}(x_p) \in [0,1]$ *(i.e., the degree of indeterminacy membership), and* $F_{\hat{\Phi}}(x_p) \in [0,1]$ *(i.e., the degree of falsity membership), then* $\hat{\Phi}$ *is called a single-valued neutrosophic fuzzy set on X, denoted by* $(\mathbb{SVNFS})^X$.

Definition 4 (cf. [27]). *Let* $\hat{\Phi}, \hat{\Psi} \in (\mathbb{SVNFS})^X$, *where*

$$\hat{\Phi} = \left\{ \frac{(T_{\hat{\Phi}}(x_p), I_{\hat{\Phi}}(x_p), F_{\hat{\Phi}}(x_p), \mu(x_p))}{x_p} \middle| x_p \in X,\ 0 \leq T_{\hat{\Phi}}(x_p) + I_{\hat{\Phi}}(x_p) + F_{\hat{\Phi}}(x_p) \leq 3 \right\}$$

and

$$\hat{\Psi} = \left\{ \frac{(T'_{\hat{\Psi}}(x_p), I'_{\hat{\Psi}}(x_p), F'_{\hat{\Psi}}(x_p), \mu'(x_p))}{x_p} \middle| x_p \in X,\ 0 \leq T_{\hat{\Psi}}(x_p) + I_{\hat{\Psi}}(x_p) + F_{\hat{\Psi}}(x_p) \leq 3 \right\}.$$

The following operations (i.e., complement, inclusion, equal, union, and intersection) are defined by

(1) $$\hat{\Phi}^c = \left\{ \frac{(F_{\hat{\Phi}}(x_p), 1 - I_{\hat{\Phi}}(x_p), T_{\hat{\Phi}}(x_p), 1 - \mu(x_p))}{x_p} \middle| x_p \in X \right\}.$$

(2) $\hat{\Phi} \subseteq \hat{\Psi} \Longleftrightarrow T_{\hat{\Phi}}(x_p) \leq T'_{\hat{\Psi}}(x_p), I_{\hat{\Phi}}(x_p) \geq I'_{\hat{\Psi}}(x_p), F_{\hat{\Phi}}(x_p) \geq F'_{\hat{\Psi}}(x_p)$ *and* $\mu(x_p) \leq \mu'(x_p)$ $(\forall x_p \in X)$.

(3) $\hat{\Phi} = \hat{\Psi} \Longleftrightarrow \hat{\Phi} \subseteq \hat{\Psi}$ *and* $\hat{\Psi} \subseteq \hat{\Phi}$.

(4) $\hat{\Phi} \cup \hat{\Psi} = \left\{ \frac{(F_{\hat{\Phi}}(x_p) \vee F'_{\hat{\Psi}}(x_p), I_{\hat{\Phi}}(x_p) \wedge I'_{\hat{\Psi}}(x_p), T_{\hat{\Phi}}(x_p) \wedge T'_{\hat{\Psi}}(x_p), \mu(x_p) \vee \mu'(x_p))}{x_p} \middle| x_p \in X \right\}.$

(5) $\hat{\Phi} \cap \hat{\Psi} = \left\{ \frac{(F_{\hat{\Phi}}(x_p) \wedge F'_{\hat{\Psi}}(x_p), I_{\hat{\Phi}}(x_p) \vee I'_{\hat{\Psi}}(x_p), T_{\hat{\Phi}}(x_p) \vee T'_{\hat{\Psi}}(x_p), \mu(x_p) \wedge \mu'(x_p))}{x_p} \middle| x_p \in X \right\}.$

2.4. Soft Set, Fuzzy Soft Set, and Neutrosophic Soft Set

Definition 5 (cf. [4,7,28]). *Assume that* X *(i.e.,* $X = \{x_1, x_2, ..., x_p\}$*) is a set of elements and* I *(i.e.,* $I = \{i_1, i_2, ..., i_q\}$*) is a set of parameters, where* $(p, q \in N, N$ *are natural numbers). Then*

(1) The following mapping (called a soft set), is given by

$$S : I \to P(X),$$

where $P(X)$ *is a set of all subsets over* X.

(2) The following mapping (called a fuzzy soft set), is given by

$$\widetilde{S} : I \to [0,1]^X,$$

where $[0,1]^X$ *is a set of whole fuzzy subset over* X.

(3) The following mapping (called a neutrosophic soft set), is given by

$$\widetilde{\hat{S}} : I \to (\mathbb{NS})^X,$$

where $(\mathbb{NS})^X$ *is a set of whole neutrosophic subset over* X.

Example 1. *Assume that the two brothers Mr. Z and Mr. M plan to go the car dealership office to purchase a new car. Suppose that the car dealership office contains types of new cars* $X = \{x_1, x_2, x_3, x_4\}$ *and* $I = \{i_1, i_2, i_3\}$ *characterize three parameters, where* i_1 *is "cheap",* i_2 *is "expensive", and* i_3 *is "beautiful". Then*

(1) *By Definition 5(1) we can describe the soft sets as* $S_{(i_1)} = \{x_1, x_3\}, S_{(i_2)} = \{x_3, x_4\}$, *and* $S_{(i_3)} = \{x_2\}$. *Therefore,*

$$\mathcal{S} = \left\{ \frac{\{x_1, x_3\}}{i_1}, \frac{\{x_3, x_4\}}{i_2}, \frac{\{x_2\}}{i_3} \right\}.$$

(2) *It is obvious to replace the crisp number 0 or 1 by a membership of fuzzy information. Therefore, by Definition 5(2) we can describe the fuzzy soft sets by* $\widetilde{S}_{(i_1)} = \left\{\frac{0.3}{x_1}, \frac{0.4}{x_2}, \frac{0.6}{x_3}, \frac{0.5}{x_4}\right\}$, $\widetilde{S}_{(i_2)} = \left\{\frac{0.6}{x_1}, \frac{0.9}{x_2}, \frac{0.1}{x_3}, \frac{0.2}{x_4}\right\}$, $\widetilde{S}_{(i_3)} = \left\{\frac{0.7}{x_1}, \frac{0.5}{x_2}, \frac{0.2}{x_3}, \frac{0.9}{x_4}\right\}$. *Then,*

$$\widetilde{S} = \left\{ \frac{\left\{\frac{0.3}{x_1}, \frac{0.4}{x_2}, \frac{0.6}{x_3}, \frac{0.5}{x_4}\right\}}{i_1}, \frac{\left\{\frac{0.6}{x_1}, \frac{0.9}{x_2}, \frac{0.1}{x_3}, \frac{0.2}{x_4}\right\}}{i_2}, \frac{\left\{\frac{0.7}{x_1}, \frac{0.5}{x_2}, \frac{0.2}{x_3}, \frac{0.9}{x_4}\right\}}{i_3} \right\}.$$

(3) *By Definition 5(3) we can describe the neutrosophic soft sets as*

$$\widetilde{\hat{S}}_{(i_1)} = \left\{ \frac{(0.3, 0.7, 0.5)}{x_1}, \frac{(0.1, 0.8, 0.5)}{x_2}, \frac{(0.2, 0.6, 0.8)}{x_3}, \frac{(0.4, 0.7, 0.6)}{x_4} \right\},$$

$$\widetilde{\hat{S}}_{(i_2)} = \left\{ \frac{(0.3, 0.7, 0.5)}{x_1}, \frac{(0.1, 0.8, 0.5)}{x_2}, \frac{(0.2, 0.6, 0.8)}{x_3}, \frac{(0.5, 0.8, 0.3)}{x_4} \right\},$$

and

$$\tilde{\hat{S}}_{(i_3)} = \left\{ \frac{(0.3, 0.7, 0.5)}{x_1}, \frac{(0.1, 0.8, 0.5)}{x_2}, \frac{(0.2, 0.6, 0.8)}{x_3}, \frac{(0.8, 0.9, 0.2)}{x_4} \right\}.$$

3. Single-Valued Neutrosophic Fuzzy Soft Set

In the following, we propose the concept of a single-valued neutrosophic fuzzy soft set and study some definitions, propositions, and examples.

Definition 6. *Assume that X (i.e., $X = \{x_1, x_2, ..., x_p\}$) is a set of elements, I (i.e., $I = \{i_1, i_2, ..., i_q\}$) is a set of parameters, and $\mathbb{S}^{XI}$ is called a soft universe. A single-valued neutrosophic fuzzy soft set $\hat{\Phi}_{(i_q)}$ over X, denoted by $(\mathbb{SVNFS})^{XI}$, is defined by*

$$\hat{\Phi}_{(i_q)} = \left\{ \frac{(T_{\hat{\Phi}_{(i_q)}}(x_p), I_{\hat{\Phi}_{(i_q)}}(x_p), F_{\hat{\Phi}_{(i_q)}}(x_p), \mu(x_p))}{x_p} \,\middle|\, i_q \in I,\ x_p \in X,\ 0 \le T_{\hat{\Phi}_{(i_q)}}(x_p) + I_{\hat{\Phi}_{(i_q)}}(x_p) + F_{\hat{\Phi}_{(i_q)}}(x_p) \le 3 \right\},$$

where $p, q \in N$ (N are natural numbers) and $\mu(x_p) \in [0,1]$. For each parameter $i_q \in I$ and for each $x_p \in X$, $T_{\hat{\Phi}_{(i_q)}}(x_p) \in [0,1]$ (i.e., the degree of truth membership), $I_{\hat{\Phi}_{(i_q)}}(x_p) \in [0,1]$ (i.e., the degree of indeterminacy membership), and $F_{\hat{\Phi}_{(i_q)}}(x_p) \in [0,1]$ (i.e., the degree of falsity membership).

Example 2. *Assume that $X = \{x_1, x_2, x_3\}$ are three kinds of novel cars and $I = \{i_1, i_2, i_3\}$ are three parameters, where i_1 is "cheap", i_2 is "expensive", and i_3 is "beautiful". Let $\mu \in [0,1]^X$ and $\hat{\Phi}_{(i_q)} \in (\mathbb{SVNFS})^{XI}$ are defined as follows ($q = 1, 2, 3$):*

$$\hat{\Phi}_{(i_1)} = \left\{ \frac{(0.3, 0.7, 0.5, 0.2)}{x_1}, \frac{(0.1, 0.8, 0.5, 0.5)}{x_2}, \frac{(0.2, 0.6, 0.8, 0.7)}{x_3} \right\},$$

$$\hat{\Phi}_{(i_2)} = \left\{ \frac{(0.9, 0.4, 0.5, 0.7)}{x_1}, \frac{(0.3, 0.7, 0.5, 0.4)}{x_2}, \frac{(0.8, 0.2, 0.6, 0.8)}{x_3} \right\},$$

$$\hat{\Phi}_{(i_3)} = \left\{ \frac{(0.6, 0.3, 0.5, 0.6)}{x_1}, \frac{(0.3, 0.5, 0.6, 0.4)}{x_2}, \frac{(0.7, 0.1, 0.6, 0.3)}{x_3} \right\}.$$

Additionally, we can write by matrix form as

$$\hat{\Phi} = \left(\begin{array}{c|ccc} I & x_1 & x_2 & x_3 \\ \hline i_1 & (0.3, 0.7, 0.5, 0.2) & (0.1, 0.8, 0.5, 0.5) & (0.2, 0.6, 0.8, 0.7) \\ i_2 & (0.9, 0.4, 0.5, 0.7) & (0.3, 0.7, 0.5, 0.4) & (0.8, 0.2, 0.6, 0.8) \\ i_3 & (0.6, 0.3, 0.5, 0.6) & (0.3, 0.5, 0.6, 0.4) & (0.7, 0.1, 0.6, 0.3) \end{array} \right).$$

Definition 7. *Let $\hat{\Phi}_{(i_q)}, \hat{\Psi}_{(i_q)} \in (\mathbb{SVNFS})^{XI}$ over $\mathbb{S}^{XI}$ and $\mu, \mu' \in [0,1]^X$, where*

$$\hat{\Phi}_{(i_q)} = \left\{ \frac{(T_{\hat{\Phi}_{(i_q)}}(x_p), I_{\hat{\Phi}_{(i_q)}}(x_p), F_{\hat{\Phi}_{(i_q)}}(x_p), \mu(x_p))}{x_p} \,\middle|\, i_q \in I,\ x_p \in X,\ 0 \le T_{\hat{\Phi}_{(i_q)}}(x_p) + I_{\hat{\Phi}_{(i_q)}}(x_p) + F_{\hat{\Phi}_{(i_q)}}(x_p) \le 3 \right\}$$

and

$$\hat{\Psi}_{(i_q)} = \left\{ \frac{(T'_{\hat{\Psi}_{(i_q)}}(x_p), I'_{\hat{\Psi}_{(i_q)}}(x_p), F'_{\hat{\Psi}_{(i_q)}}(x_p), \mu'(x_p))}{x_p} \,\middle|\, i_q \in I,\ x_p \in X,\ 0 \le T'_{\hat{\Psi}_{(i_q)}}(x_p) + I'_{\hat{\Psi}_{(i_q)}}(x_p) + F'_{\hat{\Psi}_{(i_q)}}(x_p) \le 3 \right\}.$$

Then, $\hat{\Phi}_{(i_q)} \Subset \hat{\Psi}_{(i_q)}$ (i.e., $\hat{\Phi}_{(i_q)}$ is a single-valued neutrosophic fuzzy soft subset of $\hat{\Psi}_{(i_q)}$) if

(1) $\mu(x_p) \le \mu'(x_p)\ \forall x_p \in X$;
(2) For all $i_q \in I, x_p \in X, T_{\hat{\Phi}_{(i_q)}}(x_p) \le T'_{\hat{\Psi}_{(i_q)}}(x_p), I_{\hat{\Phi}_{(i_q)}}(x_p) \ge I'_{\hat{\Psi}_{(i_q)}}(x_p), F_{\hat{\Phi}_{(i_q)}}(x_p) \ge F'_{\hat{\Psi}_{(i_q)}}(x_p)$.

Example 3. *(Continued from Example 2). Let $\hat{\Psi}_{(i_q)} \in (\mathbb{SVNFS})^{XI}$ be defined as follows ($q = 1, 2, 3$):*

$$\hat{\Psi} = \left(\begin{array}{c|ccc} I & x_1 & x_2 & x_3 \\ \hline i_1 & (0.4, 0.6, 0.4, 0.4) & (0.2, 0.7, 0.3, 0.5) & (0.3, 0.4, 0.7, 1) \\ i_2 & (1, 0.3, 0.5, 0.8) & (0.4, 0.6, 0.4, 0.6) & (0.9, 0.2, 0.4, 0.9) \\ i_3 & (0.7, 0.2, 0.4, 0.7) & (0.4, 0.5, 0.6, 0.6) & (0.8, 0.1, 0.5, 0.5) \end{array} \right).$$

Thus, $\hat{\Phi}_{(i_q)} \Subset \hat{\Psi}_{(i_q)}$ $(\forall i_q \in I)$.

Definition 8. *Let $\hat{\Phi}_{(i_q)}, \hat{\Psi}_{(i_q)} \in (\mathbb{SVNFS})^{XI}$ over $\mathbb{S}^{XI}$ and $\mu, \mu' \in [0,1]^X$, where*

$$\hat{\Phi}_{(i_q)} = \left\{ \frac{(T_{\hat{\Phi}_{(i_q)}}(x_p), I_{\hat{\Phi}_{(i_q)}}(x_p), F_{\hat{\Phi}_{(i_q)}}(x_p), \mu(x_p))}{x_p} \,\middle|\, i_q \in I,\ x_p \in X,\ 0 \le T_{\hat{\Phi}_{(i_q)}}(x_p) + I_{\hat{\Phi}_{(i_q)}}(x_p) + F_{\hat{\Phi}_{(i_q)}}(x_p) \le 3 \right\}$$

and

$$\hat{\Psi}_{(i_q)} = \left\{ \frac{(T'_{\hat{\Psi}_{(i_q)}}(x_p), I'_{\hat{\Psi}_{(i_q)}}(x_p), F'_{\hat{\Psi}_{(i_q)}}(x_p), \mu'(x_p))}{x_p} \,\middle|\, i_q \in I,\ x_p \in X,\ 0 \le T'_{\hat{\Psi}_{(i_q)}}(x_p) + I'_{\hat{\Psi}_{(i_q)}}(x_p) + F'_{\hat{\Psi}_{(i_q)}}(x_p) \le 3 \right\}.$$

Then, $\hat{\Phi}_{(i_q)} = \hat{\Psi}_{(i_q)}$ (i.e., $\hat{\Phi}_{(i_q)}$ is a single-valued neutrosophic fuzzy soft equal to $\hat{\Psi}_{(i_q)}$) if $\hat{\Phi}_{(i_q)} \Subset \hat{\Psi}_{(i_q)}$ and $\hat{\Phi}_{(i_q)} \Supset \hat{\Psi}_{(i_q)}$.

Definition 9. *Let $\hat{\Phi}_{(i_q)} \in (\mathbb{SVNFS})^{XI}$ over $\mathbb{S}^{XI}$ and $\mu \in [0,1]^X$, where*

$$\hat{\Phi}_{(i_q)} = \left\{ \frac{(T_{\hat{\Phi}_{(i_q)}}(x_p), I_{\hat{\Phi}_{(i_q)}}(x_p), F_{\hat{\Phi}_{(i_q)}}(x_p), \mu(x_p))}{x_p} \,\middle|\, i_q \in I,\ x_p \in X,\ 0 \le T_{\hat{\Phi}_{(i_q)}}(x_p) + I_{\hat{\Phi}_{(i_q)}}(x_p) + F_{\hat{\Phi}_{(i_q)}}(x_p) \le 3 \right\}$$

over $\mathbb{S}^{XI}$. Then,

(1) $\hat{\Phi}_{(i_q)}$ is called a single-valued neutrosophic fuzzy soft null set (denoted by $\hat{\emptyset}_{(i_q)}$), defined as

$$\hat{\emptyset}_{(i_q)} = \left\{ \frac{(0, 1, 1, 0)}{x_p} | i_q \in I, x_p \in X \right\}.$$

(2) $\hat{\Phi}_{(i_q)}$ is called a single-valued neutrosophic fuzzy soft universal set (denoted by $\hat{X}_{(i_q)}$), defined as

$$\hat{X}_{(i_q)} = \left\{ \frac{(1, 0, 0, 1)}{x_p} | i_q \in I, x_p \in X \right\}.$$

Example 4. *(Continued from Example 2). Then, $\hat{\emptyset}_{(i_q)}, \hat{X}_{(i_q)} \in (\mathbb{SVNFS})^{XI}$ are defined as follows:*

$$\hat{\emptyset} = \left(\begin{array}{c|ccc} I & x_1 & x_2 & x_3 \\ \hline i_1 & (0, 1, 1, 0) & (0, 1, 1, 0) & (0, 1, 1, 0) \\ i_2 & (0, 1, 1, 0) & (0, 1, 1, 0) & (0, 1, 1, 0) \\ i_3 & (0, 1, 1, 0) & (0, 1, 1, 0) & (0, 1, 1, 0) \end{array} \right)$$

and

$$\hat{X} = \left(\begin{array}{c|ccc} I & x_1 & x_2 & x_3 \\ \hline i_1 & (1, 0, 0, 1) & (1, 0, 0, 1) & (1, 0, 0, 1) \\ i_2 & (1, 0, 0, 1) & (1, 0, 0, 1) & (1, 0, 0, 1) \\ i_3 & (1, 0, 0, 1) & (1, 0, 0, 1) & (1, 0, 0, 1) \end{array} \right).$$

Definition 10. *Let $\hat{\Phi}_{(i_q)}, \hat{\Psi}_{(i_q)} \in (\mathbb{SVNFS})^{XI}$ over $\mathbb{S}^{XI}$ and $\mu, \mu' \in [0,1]^X$, where*

$$\hat{\Phi}_{(i_q)} = \left\{ \frac{(T_{\hat{\Phi}_{(i_q)}}(x_p), I_{\hat{\Phi}_{(i_q)}}(x_p), F_{\hat{\Phi}_{(i_q)}}(x_p), \mu(x_p))}{x_p} \,\middle|\, i_q \in I,\ x_p \in X,\ 0 \leq T_{\hat{\Phi}_{(i_q)}}(x_p) + I_{\hat{\Phi}_{(i_q)}}(x_p) + F_{\hat{\Phi}_{(i_q)}}(x_p) \leq 3 \right\}$$

and

$$\hat{\Psi}_{(i_q)} = \left\{ \frac{(T'_{\hat{\Psi}_{(i_q)}}(x_p), I'_{\hat{\Psi}_{(i_q)}}(x_p), F'_{\hat{\Psi}_{(i_q)}}(x_p), \mu'(x_p))}{x_p} \,\middle|\, i_q \in I,\ x_p \in X,\ 0 \leq T'_{\hat{\Psi}_{(i_q)}}(x_p) + I'_{\hat{\Psi}_{(i_q)}}(x_p) + F'_{\hat{\Psi}_{(i_q)}}(x_p) \leq 3 \right\}.$$

Then,

(1) The union $\hat{\Phi}_{(i_q)} \Cup \hat{\Psi}_{(i_q)}$ is defined as

$$\hat{\Phi}_{(i_q)} \Cup \hat{\Psi}_{(i_q)} = \left\{ \frac{(T_{\hat{\Phi}_{(i_q)}}(x_p) \circ T'_{\hat{\Psi}_{(i_q)}}(x_p), I_{\hat{\Phi}_{(i_q)}}(x_p) * I'_{\hat{\Psi}_{(i_q)}}(x_p), F_{\hat{\Phi}_{(i_q)}}(x_p) * F'_{\hat{\Psi}_{(i_q)}}(x_p), \mu(x_p) \circ \mu'(x_p))}{x_p} \,\middle|\, i_q \in I, x_p \in X \right\}.$$

(2) The intersection $\hat{\Phi}_{(i_q)} \Cap \hat{\Psi}_{(i_q)}$ is defined as

$$\hat{\Phi}_{(i_q)} \Cap \hat{\Psi}_{(i_q)} = \left\{ \frac{(T_{\hat{\Phi}_{(i_q)}}(x_p) * T'_{\hat{\Psi}_{(i_q)}}(x_p), I_{\hat{\Phi}_{(i_q)}}(x_p) \circ I'_{\hat{\Psi}_{(i_q)}}(x_p), F_{\hat{\Phi}_{(i_q)}}(x_p) \circ F'_{\hat{\Psi}_{(i_q)}}(x_p), \mu(x_p) * \mu'(x_p))}{x_p} \,\middle|\, i_q \in I, x_p \in X \right\}.$$

Example 5. *(Continued from Examples 2 and 3). For $\alpha, \beta \in [0,1]$, let the t-norm (i.e., given as $\alpha * \beta = \alpha \wedge \beta$) and the t-conorm (i.e., given as $\alpha \circ \beta = \alpha \vee \beta$). Then,*

$\hat{\Phi} \Cup \hat{\Psi} =$

I	x_1	x_2	x_3
i_1	(0.4, 0.6, 0.4, 0.4)	(0.2, 0.7, 0.3, 0.5)	(0.3, 0.4, 0.7, 1)
i_2	(1, 0.3, 0.5, 0.8)	(0.4, 0.6, 0.4, 0.6)	(0.9, 0.2, 0.4, 0.9)
i_3	(0.7, 0.2, 0.4, 0.7)	(0.4, 0.5, 0.6, 0.6)	(0.8, 0.1, 0.5, 0.5)

and

$\hat{\Phi} \Cap \hat{\Psi} =$

I	x_1	x_2	x_3
i_1	(0.3, 0.7, 0.5, 0.2)	(0.1, 0.8, 0.5, 0.5)	(0.2, 0.6, 0.8, 0.7)
i_2	(0.9, 0.4, 0.5, 0.7)	(0.3, 0.7, 0.5, 0.4)	(0.8, 0.2, 0.6, 0.8)
i_3	(0.6, 0.3, 0.5, 0.6)	(0.3, 0.5, 0.6, 0.4)	(0.7, 0.1, 0.6, 0.3)

Proposition 1. *Let $\hat{\emptyset}_{(i_q)}, \hat{X}_{(i_q)}, \hat{\Phi}_{(i_q)} \in (\mathbb{SVNFS})^{XI}$ over $\mathbb{S}^{XI}$ and $\mu \in [0,1]^X$. Then the following hold:*

(1) $\hat{\Phi}_{(i_q)} \Cup \hat{\Phi}_{(i_q)} = \hat{\Phi}_{(i_q)}$;
(2) $\hat{\Phi}_{(i_q)} \Cap \hat{\Phi}_{(i_q)} = \hat{\Phi}_{(i_q)}$;
(3) $\hat{\Phi}_{(i_q)} \Cup \hat{\emptyset}_{(i_q)} = \hat{\Phi}_{(i_q)}$;
(4) $\hat{\Phi}_{(i_q)} \Cap \hat{\emptyset}_{(i_q)} = \hat{\emptyset}_{(i_q)}$;
(5) $\hat{\Phi}_{(i_q)} \Cup \hat{X}_{(i_q)} = \hat{X}_{(i_q)}$;
(6) $\hat{\Phi}_{(i_q)} \Cap \hat{X}_{(i_q)} = \hat{\Phi}_{(i_q)}$.

Proof. Follows from Definitions 9 and 10. □

Proposition 2. *Let $\hat{\Phi}_{(i_q)}, \hat{\Psi}_{(i_q)}, \hat{\Gamma}_{(i_q)} \in (\mathbb{SVNFS})^{XI}$ over $\mathbb{S}^{XI}$ and $\mu, \mu', \mu'' \in [0,1]^X$. Then the following hold:*

(1) $\hat{\Phi}_{(i_q)} \Cup \hat{\Psi}_{(i_q)} = \hat{\Psi}_{(i_q)} \Cup \hat{\Phi}_{(i_q)}$;
(2) $\hat{\Phi}_{(i_q)} \Cap \hat{\Psi}_{(i_q)} = \hat{\Psi}_{(i_q)} \Cap \hat{\Phi}_{(i_q)}$;
(3) $\hat{\Phi}_{(i_q)} \Cup (\hat{\Psi}_{(i_q)} \Cup \hat{\Gamma}_{(i_q)}) = (\hat{\Phi}_{(i_q)} \Cup \hat{\Psi}_{(i_q)}) \Cup \hat{\Gamma}_{(i_q)}$;

(4) $\hat{\Phi}_{(i_q)} \Cap (\hat{\Psi}_{(i_q)} \Cap \hat{\Gamma}_{(i_q)}) = (\hat{\Phi}_{(i_q)} \Cap \hat{\Psi}_{(i_q)}) \Cap \hat{\Gamma}_{(i_q)}$;

(5) $\hat{\Phi}_{(i_q)} \Cap (\hat{\Psi}_{(i_q)} \Cup \hat{\Gamma}_{(i_q)}) = (\hat{\Phi}_{(i_q)} \Cap \hat{\Psi}_{(i_q)}) \Cup (\hat{\Phi}_{(i_q)} \Cap \hat{\Gamma}_{(i_q)})$;

(6) $\hat{\Phi}_{(i_q)} \Cup (\hat{\Psi}_{(i_q)} \Cap \hat{\Gamma}_{(i_q)}) = (\hat{\Phi}_{(i_q)} \Cup \hat{\Psi}_{(i_q)}) \Cap (\hat{\Phi}_{(i_q)} \Cup \hat{\Gamma}_{(i_q)})$.

Proof. Follows from Definition 10. □

Proposition 3. *Let $\hat{\Phi}_{(i_q)}, \hat{\Psi}_{(i_q)} \in (\mathbb{SVNFS})^{XI}$ over $\mathbb{S}^{XI}$, $\mu, \mu' \in [0,1]^X$, and $\hat{\Psi}_{(i_q)} \Subset \hat{\Phi}_{(i_q)}$. Then the following hold:*

(1) $\hat{\Phi}_{(i_q)} \Cup \hat{\Psi}_{(i_q)} = \hat{\Phi}_{(i_q)}$;

(2) $\hat{\Phi}_{(i_q)} \Cap \hat{\Psi}_{(i_q)} = \hat{\Psi}_{(i_q)}$.

Proof. Follows from Definitions 7 and 10. □

Next, we propose a definition, example, remark, and two propositions on the complement of $(\mathbb{SVNFS})^{XI}$ over $\mathbb{S}^{XI}$.

Definition 11. *Let $\hat{\Phi}_{(i_q)} \in (\mathbb{SVNFS})^{XI}$ over $\mathbb{S}^{XI}$ and $\mu \in [0,1]^X$, where*

$$\hat{\Phi}_{(i_q)} = \left\{ \frac{(T_{\Phi_{(i_q)}}(x_p), I_{\Phi_{(i_q)}}(x_p), F_{\Phi_{(i_q)}}(x_p), \mu(x_p))}{x_p} \,\middle|\, i_q \in I,\ x_p \in X,\ 0 \le T_{\Phi_{(i_q)}}(x_p) + I_{\Phi_{(i_q)}}(x_p) + F_{\Phi_{(i_q)}}(x_p) \le 3 \right\}.$$

Then, the complement $\hat{\Phi}^c_{(i_q)}$ of $\hat{\Phi}_{(i_q)}$ is defined as

$$\hat{\Phi}^c_{(i_q)} = \left\{ \frac{(F_{\Phi_{(i_q)}}(x_p), 1 - I_{\Phi_{(i_q)}}(x_p), T_{\Phi_{(i_q)}}(x_p), 1 - \mu(x_p))}{x_p} \,\middle|\, i_q \in I,\ x_p \in X \right\}.$$

Example 6. *(Continued from Example 2). The complement $\hat{\Phi}^c_{(i_q)}$ of $\hat{\Phi}_{(i_q)}$ is calculated by*

$\hat{\Phi}^c =$

I	x_1	x_2	x_3
i_1	(0.5, 0.3, 0.3, 0.8)	(0.5, 0.2, 0.1, 0.5)	(0.8, 0.4, 0.2, 0.3)
i_2	(0.5, 0.6, 0.9, 0.3)	(0.5, 0.3, 0.3, 0.6)	(0.6, 0.8, 0.8, 0.2)
i_3	(0.5, 0.7, 0.6, 0.4)	(0.6, 0.5, 0.3, 0.6)	(0.6, 0.9, 0.7, 0.7)

Proposition 4. *Let $\hat{\emptyset}_{(i_q)}, \hat{X}_{(i_q)}, \hat{\Phi}_{(i_q)} \in (\mathbb{SVNFS})^{XI}$ over $\mathbb{S}^{XI}$, and $\mu \in [0,1]^X$. Then, the following hold:*

(1) $\hat{\emptyset}^c_{(i_q)} = \hat{X}_{(i_q)}$;

(2) $\hat{X}^c_{(i_q)} = \hat{\emptyset}_{(i_q)}$;

(3) $(\hat{\Phi}^c_{(i_q)})^c = \hat{\Phi}^c_{(i_q)}$.

Proof. Follows from Definitions 9 and 11. □

Remark 1. *The equality of $\hat{\Phi}_{(i_q)} \Cup \hat{\Phi}^c_{(i_q)} = \hat{X}_{(i_q)}$ and $\hat{\Phi}_{(i_q)} \Cap \hat{\Phi}^c_{(i_q)} = \hat{\emptyset}_{(i_q)}$ does not hold by the following example.*

Example 7. *(Continued from Examples 2 and 6). Then, $\hat{\Phi}^c_{(i_q)}$ of $\hat{\Phi}_{(i_q)}$ is calculated by*

$\hat{\Phi} \Cup \hat{\Phi}^c =$

I	x_1	x_2	x_3
i_1	(0.5, 0.3, 0.3, 0.8)	(0.5, 0.2, 0.1, 0.5)	(0.8, 0.4, 0.2, 0.3)
i_1	(0.5, 0.6, 0.9, 0.3)	(0.5, 0.3, 0.3, 0.6)	(0.6, 0.8, 0.8, 0.2)
i_1	(0.5, 0.7, 0.6, 0.4)	(0.6, 0.5, 0.3, 0.6)	(0.6, 0.9, 0.7, 0.7)

and

$$\hat{\Phi} \Cap \hat{\Phi}^c = \left(\begin{array}{c|ccc} I & x_1 & x_2 & x_3 \\ \hline i_1 & (0.3, 0.7, 0.5, 0.2) & (0.1, 0.8, 0.5, 0.5) & (0.2, 0.6, 0.8, 0.7) \\ i_2 & (0.9, 0.4, 0.5, 0.7) & (0.3, 0.7, 0.5, 0.4) & (0.8, 0.2, 0.6, 0.8) \\ i_3 & (0.6, 0.3, 0.5, 0.6) & (0.3, 0.5, 0.6, 0.4) & (0.7, 0.1, 0.6, 0.3) \end{array}\right).$$

This shows that $\hat{\Phi}_{(i_q)} \Cup \hat{\Phi}^c_{(i_q)} \neq \hat{X}_{(i_q)}$ *and* $\hat{\Phi}_{(i_q)} \Cap \hat{\Phi}^c_{(i_q)} \neq \hat{\emptyset}_{(i_q)}$.

Proposition 5. *Let* $\hat{\Phi}_{(i_q)}, \hat{\Psi}_{(i_q)} \in (\mathbb{SVNFS})^{XI}$ *over* $\mathbb{S}^{XI}$ *and* $\mu, \mu' \in [0,1]^X$. *Then, the following hold:*

(1) $(\hat{\Phi}_{(i_q)} \Cup \hat{\Psi}_{(i_q)})^c = \hat{\Phi}^c_{(i_q)} \Cap \hat{\Psi}^c_{(i_q)}$;

(2) $(\hat{\Phi}_{(i_q)} \Cap \hat{\Psi}_{(i_q)})^c = \hat{\Phi}^c_{(i_q)} \Cup \hat{\Psi}^c_{(i_q)}$.

Proof. Consider $a * b = a \wedge b$ (t-norm) and $\alpha \circ \beta = \alpha \vee \beta$ (t-conorm) $(\forall \alpha, \beta \in [0,1])$. We have

(1) $(\hat{\Phi}_{(i_q)} \Cup \hat{\Psi}_{(i_q)})^c(x_p)$

$$= \left(\left\{\frac{(T_{\Phi_{(i_q)}}(x_p) \circ T'_{\Psi_{(i_q)}}(x_p), I_{\Phi_{(i_q)}}(x_p) * I'_{\Psi_{(i_q)}}(x_p), F_{\Phi_{(i_q)}}(x_p) * F'_{\Psi_{(i_q)}}(x_p), \mu(x_p) \circ \mu'(x_p))}{x_p}\bigg| i_q \in I, x_p \in X\right\}\right)^c$$

$$= \left\{\frac{(F_{\Phi_{(i_q)}}(x_p) * F'_{\Psi_{(i_q)}}(x_p), 1 - (I_{\Phi_{(i_q)}}(x_p) * I'_{\Psi_{(i_q)}}(x_p)), T_{\Phi_{(i_q)}}(x_p) \circ T'_{\Psi_{(i_q)}}(x_p), 1 - (\mu(x_p) \circ \mu'(x_p)))}{x_p}\bigg| i_q \in I, x_p \in X\right\}$$

$$= \left\{\frac{(F_{\Phi_{(i_q)}}(x_p) \wedge F'_{\Psi_{(i_q)}}(x_p), 1 - (I_{\Phi_{(i_q)}}(x_p) \wedge I'_{\Psi_{(i_q)}}(x_p)), T_{\Phi_{(i_q)}}(x_p) \vee T'_{\Psi_{(i_q)}}(x_p), 1 - (\mu(x_p) \vee \mu'(x_p)))}{x_p}\bigg| i_q \in I, x_p \in X\right\}$$

$$= \left\{\frac{(F_{\Phi_{(i_q)}}(x_p) \wedge F'_{\Psi_{(i_q)}}(x_p), 1 - I_{\Phi_{(i_q)}}(x_p) \vee 1 - I'_{\Psi_{(i_q)}}(x_p), T_{\Phi_{(i_q)}}(x_p) \vee T'_{\Psi_{(i_q)}}(x_p), 1 - \mu(x_p) \wedge 1 - \mu'(x_p))}{x_p}\bigg| i_q \in I, x_p \in X\right\}$$

$$= \left\{\frac{(F_{\Phi_{(i_q)}}(x_p) * F'_{\Psi_{(i_q)}}(x_p), 1 - I_{\Phi_{(i_q)}}(x_p) \circ 1 - I'_{\Psi_{(i_q)}}(x_p), T_{\Phi_{(i_q)}}(x_p) \circ T'_{\Psi_{(i_q)}}(x_p), 1 - \mu(x_p) * 1 - \mu'(x_p))}{x_p}\bigg| i_q \in I, x_p \in X\right\}$$

$$= \left\{\frac{(F_{\Phi_{(i_q)}}(x_p), 1 - I_{\Phi_{(i_q)}}(x_p), T_{\Phi_{(i_q)}}(x_p), 1 - \mu(x_p))}{x_p}\bigg| i_q \in I, x_p \in X\right\} \Cap \left\{\frac{(F'_{\Psi_{(i_q)}}(x_p), 1 - I'_{\Psi_{(i_q)}}(x_p), T'_{\Psi_{(i_q)}}(x_p), 1 - \mu'(x_p))}{x_p}\bigg| i_q \in I, x_p \in X\right\}$$

$$= \hat{\Phi}^c_{(i_q)}(x_p) \Cap \hat{\Psi}^c_{(i_q)}(x_p).$$

(2) $(\hat{\Phi}_{(i_q)} \Cap \hat{\Psi}_{(i_q)})^c(x_p)$

$$= \left(\left\{\frac{(T_{\Phi_{(i_q)}}(x_p) * T'_{\Psi_{(i_q)}}(x_p), I_{\Phi_{(i_q)}}(x_p) \circ I'_{\Psi_{(i_q)}}(x_p), F_{\Phi_{(i_q)}}(x_p) \circ F'_{\Psi_{(i_q)}}(x_p), \mu(x_p) * \mu'(x_p))}{x_p}\bigg| i_q \in I, x_p \in X\right\}\right)^c$$

$$= \left\{\frac{(F_{\Phi_{(i_q)}}(x_p) \circ F'_{\Psi_{(i_q)}}(x_p), 1 - (I_{\Phi_{(i_q)}}(x_p) \circ I'_{\Psi_{(i_q)}}(x_p)), T_{\Phi_{(i_q)}}(x_p) * T'_{\Psi_{(i_q)}}(x_p), 1 - (\mu(x_p) * \mu'(x_p)))}{x_p}\bigg| i_q \in I, x_p \in X\right\}$$

$$= \left\{\frac{(F_{\Phi_{(i_q)}}(x_p) \vee F'_{\Psi_{(i_q)}}(x_p), 1 - (I_{\Phi_{(i_q)}}(x_p) \vee I'_{\Psi_{(i_q)}}(x_p)), T_{\Phi_{(i_q)}}(x_p) \wedge T'_{\Psi_{(i_q)}}(x_p), 1 - (\mu(x_p) \wedge \mu'(x_p)))}{x_p}\bigg| i_q \in I, x_p \in X\right\}$$

$$= \left\{\frac{(F_{\Phi_{(i_q)}}(x_p) \vee F'_{\Psi_{(i_q)}}(x_p), 1 - I_{\Phi_{(i_q)}}(x_p) \wedge 1 - I'_{\Psi_{(i_q)}}(x_p), T_{\Phi_{(i_q)}}(x_p) \wedge T'_{\Psi_{(i_q)}}(x_p), 1 - \mu(x_p) \vee 1 - \mu'(x_p))}{x_p}\bigg| i_q \in I, x_p \in X\right\}$$

$$= \left\{\frac{(F_{\Phi_{(i_q)}}(x_p) \circ F'_{\Psi_{(i_q)}}(x_p), 1 - I_{\Phi_{(i_q)}}(x_p) * 1 - I'_{\Psi_{(i_q)}}(x_p), T_{\Phi_{(i_q)}}(x_p) * T'_{\Psi_{(i_q)}}(x_p), 1 - \mu(x_p) \circ 1 - \mu'(x_p))}{x_p}\bigg| i_q \in I, x_p \in X\right\}$$

$$= \left\{\frac{(F_{\Phi_{(i_q)}}(x_p), 1 - I_{\Phi_{(i_q)}}(x_p), T_{\Phi_{(i_q)}}(x_p), 1 - \mu(x_p))}{x_p}\bigg| i_q \in I, x_p \in X\right\} \Cup \left\{\frac{(F'_{\Psi_{(i_q)}}(x_p), 1 - I'_{\Psi_{(i_q)}}(x_p), T'_{\Psi_{(i_q)}}(x_p), 1 - \mu'(x_p))}{x_p}\bigg| i_q \in I, x_p \in X\right\}$$

$$= \hat{\Phi}^c_{(i_q)}(x_p) \Cup \hat{\Psi}^c_{(i_q)}(x_p).$$

□

4. Two Algorithms of Single-Valued Neutrosophic Fuzzy Soft Sets for Decision-Making

Depending on single-valued neutrosophic fuzzy soft sets, in the following, we introduce two new approaches for fuzzy decision-making problems.

Next, we construct Algorithm 1 as the first type for decision-making (i.e., the first application of a single-valued neutrosophic fuzzy soft set).

Algorithm 1: Determine the optimal decision based on a single-valued neutrosophic fuzzy soft set matrix.

First step: Input the single-valued neutrosophic fuzzy soft set $\hat{\Phi}_{(i_q)} \in (\mathbb{SVNFS})^{XI}$ as follows:

$$\hat{\Phi}_{(i_q)} = \left\{ \frac{(T_{\hat{\Phi}_{(i_q)}}(x_p), I_{\hat{\Phi}_{(i_q)}}(x_p), F_{\hat{\Phi}_{(i_q)}}(x_p), \mu(x_p))}{x_p} \middle| i_q \in I,\ x_p \in X,\ 0 \leq T_{\hat{\Phi}_{(i_q)}}(x_p) + I_{\hat{\Phi}_{(i_q)}}(x_p) + F_{\hat{\Phi}_{(i_q)}}(x_p) \leq 3 \right\},$$

to be evaluated by a group of experts n to element x on parameter i, where $T_{\hat{\Phi}_{(i_q)}}(x_p) \in [0,1]$ (i.e., the degree of truth membership), $I_{\hat{\Phi}_{(i_q)}}(x_p)$ (i.e., the degree of indeterminacy membership), $F_{\hat{\Phi}_{(i_q)}}(x_p)$ (i.e., the degree of falsity membership), and $\mu(x_p) \in [0,1]$.

Second step: Input the single-valued neutrosophic fuzzy soft set in matrix form (written as $\mathcal{M}_{q \times p}, p, q \in N$):

$$\mathcal{M}_{q\times p} = \begin{pmatrix} (T_{\hat{\Phi}_{(i_1)}}(x_1), I_{\hat{\Phi}_{(i_1)}}(x_1), F_{\hat{\Phi}_{(i_1)}}(x_1), \mu(x_1)) & (T_{\hat{\Phi}_{(i_1)}}(x_2), I_{\hat{\Phi}_{(i_1)}}(x_2), F_{\hat{\Phi}_{(i_1)}}(x_2), \mu(x_2)) & \cdots & (T_{\hat{\Phi}_{(i_2)}}(x_p), I_{\hat{\Phi}_{(i_2)}}(x_p), F_{\hat{\Phi}_{(i_2)}}(x_p), \mu(x_p)) \\ (T_{\hat{\Phi}_{(i_2)}}(x_1), I_{\hat{\Phi}_{(i_2)}}(x_1), F_{\hat{\Phi}_{(i_2)}}(x_1), \mu(x_1)) & (T_{\hat{\Phi}_{(i_2)}}(x_2), I_{\hat{\Phi}_{(i_2)}}(x_2), F_{\hat{\Phi}_{(i_2)}}(x_2), \mu(x_2)) & \cdots & (T_{\hat{\Phi}_{(i_2)}}(x_p), I_{\hat{\Phi}_{(i_2)}}(x_p), F_{\hat{\Phi}_{(i_2)}}(x_p), \mu(x_p)) \\ (T_{\hat{\Phi}_{(i_3)}}(x_1), I_{\hat{\Phi}_{(i_3)}}(x_1), F_{\hat{\Phi}_{(i_3)}}(x_1), \mu(x_1)) & (T_{\hat{\Phi}_{(i_3)}}(x_2), I_{\hat{\Phi}_{(i_3)}}(x_2), F_{\hat{\Phi}_{(i_3)}}(x_2), \mu(x_2)) & \cdots & (T_{\hat{\Phi}_{(i_3)}}(x_p), I_{\hat{\Phi}_{(i_3)}}(x_p), F_{\hat{\Phi}_{(i_3)}}(x_p), \mu(x_p)) \\ \vdots & \vdots & \ddots & \vdots \\ (T_{\hat{\Phi}_{(i_q)}}(x_1), I_{\hat{\Phi}_{(i_q)}}(x_1), F_{\hat{\Phi}_{(i_q)}}(x_1), \mu(x_1)) & (T_{\hat{\Phi}_{(i_q)}}(x_2), I_{\hat{\Phi}_{(i_q)}}(x_2), F_{\hat{\Phi}_{(i_q)}}(x_2), \mu(x_2)) & \cdots & (T_{\hat{\Phi}_{(i_q)}}(x_p), I_{\hat{\Phi}_{(i_q)}}(x_p), F_{\hat{\Phi}_{(i_q)}}(x_p), \mu(x_p)) \end{pmatrix}.$$

Third step: Calculate the center matrix (i.e., $\delta_{\hat{\Phi}_{(i_q)}}(x_p) = (T_{\hat{\Phi}_{(i_q)}}(x_p) + I_{\hat{\Phi}_{(i_q)}}(x_p) + F_{\hat{\Phi}_{(i_q)}}(x_p)) - \mu(x_p)$):

$$C_{q\times p} = \begin{pmatrix} \delta_{\hat{\Phi}_{(i_1)}}(x_1) & \delta_{\hat{\Phi}_{(i_1)}}(x_2) & \cdots, & \delta_{\hat{\Phi}_{(i_1)}}(x_p) \\ \delta_{\hat{\Phi}_{(i_2)}}(x_1) & \delta_{\hat{\Phi}_{(i_2)}}(x_2) & \cdots, & \delta_{\hat{\Phi}_{(i_2)}}(x_p) \\ \vdots & \vdots & \ddots & \vdots \\ \delta_{\hat{\Phi}_{(i_q)}}(x_1) & \delta_{\hat{\Phi}_{(i_q)}}(x_2) & \cdots, & \delta_{\hat{\Phi}_{(i_q)}}(x_p) \end{pmatrix}.$$

Fourth step: Calculate the $d^{max}(x_j)$ (maximum decision), $d^{min}(x_j)$ (minimum decision), and $S(x_j)$ (score) of elements x_j $(j = 1, 2, \cdots, p)$:

$$d^{max}(x_j) = \sum_{i=1}^{q} (1 - \delta_{\hat{\Phi}_{(i_q)}}(x_j))^2, \quad d^{min}(x_j) = \sum_{i=1}^{q} (\delta_{\hat{\Phi}_{(i_q)}}(x_j))^2$$
$$S(x_j) = d^{max}(x_j) + d^{min}(x_j).$$

(to understand the motivation behind this method, let ρ be the Euclidean metric on R^q, $\mathbf{0} = (0, \cdots, 0)^T \in R^q, \mathbf{1} = (1, \cdots, 1)^T \in R^q$, and $\boldsymbol{\theta}_j = (\theta_{1,x_j}, \theta_{2,x_j}, \cdots, \theta_{q,x_j})^T \in R^q$. Thus $S(x_j) = [\rho(\boldsymbol{\theta}_j, \mathbf{1})]^2 + [\rho(\boldsymbol{\theta}_j, \mathbf{0})]^2$ $(j = 1, 2, \cdots, p)$).

Fifth step: Obtain the decision p satisfying

$$x_p = \max\{S(x_1), S(x_2), \cdots, S(x_j)\}.$$

Now, we show the principle and steps of the above Algorithm 1 by using the following example.

Example 8. *An investment company wants to choose some investment projects to make full use of idle funds. There are five alternatives* $X = \{z_1, z_2, z_3, z_4, z_5\}$ *that can be selected: two internet education projects (denoted as* z_1 *and* z_2*) and three film studio investments (represented as* z_3, z_4, z_5*). According to the project investment books, the decision-makers evaluate the five alternatives from the following three parameters* $I = \{i_1, i_2, i_3\}$*, where* i_1 *is "human resources",* i_2 *is "social benefits", and* i_3 *is "expected benefits". The data of the single-valued neutrosophic fuzzy soft set* $\hat{\Phi}_{(i_q)} \in (\mathbb{SVNFS})^{XI}$ *is given by*

$$\hat{\Phi} = \left(\begin{array}{c|ccccc} I & z_1 & z_2 & z_3 & z_4 & z_5 \\ \hline i_1 & (0.3,0.7,0.5,0.2) & (0.1,0.8,0.5,0.5) & (0.2,0.6,0.8,0.7) & (0.5,0.6,0.5,0.2) & (0.4,0.7,0.9,0.1) \\ i_2 & (0.9,0.4,0.5,0.7) & (0.3,0.7,0.5,0.4) & (0.8,0.2,0.6,0.8) & (0.3,0.7,0.2,0.5) & (0.7,0.8,0.8,0.3) \\ i_3 & (0.6,0.3,0.5,0.6) & (0.3,0.5,0.6,0.4) & (0.7,0.1,0.6,0.3) & (0.8,0.9,0.6,0.4) & (0.7,0.8,0.9,0.6) \end{array} \right).$$

Now, we will explain the practical meaning of alternatives X by taking the alternative z_1 *as an example: the single-valued neutrosophic fuzzy soft set* $\hat{\Phi}_{(i_1)}(z_1) = (0.3, 0.7, 0.5, 0.2)$ *is the evaluation by four expert groups; the single-valued neutrosophic fuzzy soft value 0.3 (meaning that 30% say yes in the first expert group) in* $\hat{\Phi}_{(i_1)}(z_1)$*, the single-valued neutrosophic fuzzy soft value 0.7 (meaning 70% say no in the second expert group) in* $\hat{\Phi}_{(i_1)}(z_1)$*, the single-valued neutrosophic fuzzy soft value 0.5 (meaning 50% say yes in the third expert group) in* $\hat{\Phi}_{(i_1)}(z_1)$*, and fuzzy value 0.2 (meaning 20% say no in the fourth expert group) in* $\hat{\Phi}_{(i_1)}(z_1)$*. Then, the single-valued neutrosophic fuzzy soft set in matrix form* $\mathcal{M}_{3\times5}$ *in the second step of Algorithm 1 is given by*

$$\mathcal{M}_{3\times5} = \begin{pmatrix} (0.3,0.7,0.5,0.2) & (0.9,0.4,0.5,0.7) & (0.6,0.3,0.5,0.6) \\ (0.1,0.8,0.5,0.5) & (0.3,0.7,0.5,0.4) & (0.3,0.5,0.6,0.4) \\ (0.2,0.6,0.8,0.7) & (0.8,0.2,0.6,0.8) & (0.7,0.1,0.6,0.3) \\ (0.5,0.6,0.5,0.2) & (0.3,0.7,0.2,0.5) & (0.8,0.9,0.6,0.4) \\ (0.4,0.7,0.9,0.1) & (0.7,0.8,0.8,0.3) & (0.7,0.8,0.9,0.6) \end{pmatrix}.$$

Thus, we obtain the following center matrix $C_{3\times5}$ *of* $\mathcal{M}_{3\times5}$ *in the third step of Algorithm 1:*

$$C_{3\times5} = \begin{pmatrix} 1.3 & 1.1 & 0.8 \\ 0.9 & 1.1 & 1 \\ 0.9 & 0.8 & 1.1 \\ 1.4 & 0.7 & 1.9 \\ 1.9 & 2 & 1.8 \end{pmatrix}.$$

By calculating, we get $d^{max}(z_j)$*,* $d^{min}(z_j)$*, and* $S(z_j)$ *of elements* z_j $(j = 1, 2, 3, 4, 5)$*:*

$$d^{max}(z_1) = 0.14, d^{max}(z_2) = 0.02, d^{max}(z_3) = 0.06, d^1(z_4) = 1.06, d^{max}(z_5) = 2.45;$$

$$d^{min}(z_1) = 3.54, d^{min}(z_2) = 3.02, d^{min}(z_3) = 2.66, d^{min}(z_4) = 6.06, d^{min}(z_5) = 10.85;$$

$$S(z_1) = 3.68, S(z_2) = 3.04, S(z_3) = 2.72, S(z_4) = 7.12, S(z_5) = 13.3.$$

Finally, we can see from the fifth step that z_5 *is the best decision.*

Now, we present Algorithm 2 as a second type for a decision-making problem (i.e., a second application of the single-valued neutrosophic fuzzy soft set) as follows:

Algorithm 2: Determine the optimal decision based on AND operation of two single-valued neutrosophic fuzzy soft sets.

First step: Input the single-valued neutrosophic fuzzy soft sets $\hat{\Phi}_{(i_q)} \in (\mathbb{SVNFS})^{XI}$ and $\hat{\Psi}_{(j_q)} \in (\mathbb{SVNFS})^{XJ}$, defined, respectively, as follows:

$$\hat{\Phi}_{(i_q)} = \left\{ \frac{(T_{\hat{\Phi}_{(i_q)}}(x_p), I_{\hat{\Phi}_{(i_q)}}(x_p), F_{\hat{\Phi}_{(i_q)}}(x_p), \mu(x_p))}{x_p} \,\middle|\, i_q \in I,\ x_p \in X,\ 0 \le T_{\hat{\Phi}_{(i_q)}}(x_p) + I_{\hat{\Phi}_{(i_q)}}(x_p) + F_{\hat{\Phi}_{(i_q)}}(x_p) \le 3 \right\},$$

to be evaluated by a group of experts n to element x on parameter i, where $T_{\hat{\Phi}_{(i_q)}}(x_p) \in [0,1]$ (i.e., the degree of truth membership), $I_{\hat{\Phi}_{(i_q)}}(x_p)$ (i.e., the degree of indeterminacy membership), $F_{\hat{\Phi}_{(i_q)}}(x_p)$ (i.e., the degree of falsity membership), and $\mu(x_p) \in [0,1]$,

$$\hat{\Psi}_{(j_q)} = \left\{ \frac{(T'_{\hat{\Psi}_{(j_q)}}(x_p), I'_{\hat{\Psi}_{(j_q)}}(x_p), F'_{\hat{\Psi}_{(j_q)}}(x_p), \mu'(x_p))}{x_p} \,\middle|\, j_q \in J,\ x_p \in X,\ 0 \le T'_{\hat{\Psi}_{(j_q)}}(x_p) + I'_{\hat{\Psi}_{(j_q)}}(x_p) + F'_{\hat{\Psi}_{(j_q)}}(x_p) \le 3 \right\}$$

to be evaluated by a group of experts n to element x on parameter j, where $T_{\hat{\Psi}'_{(j_q)}}(x_p) \in [0,1]$ (i.e., the degree of truth membership), $I_{\hat{\Psi}'_{(j_q)}}(x_p)$ (i.e., the degree of indeterminacy membership), $F_{\hat{\Psi}'_{(j_q)}}(x_p)$ (i.e., the degree of falsity membership), and $\mu(x_p) \in [0,1]$.

Second step: Define and calculate the AND operation of two single-valued neutrosophic fuzzy soft sets $\hat{\Phi}_{(i_q)} \in (\mathbb{SVNFS})^{XI}$ and $\hat{\Psi}_{(j_q)} \in (\mathbb{SVNFS})^{XJ}$, denoted by $(\hat{\Phi}\overline{\wedge}\hat{\Psi})_{(i_q,j_q)}$ $(\forall i \in I, j \in J)$, defined as

$$(\hat{\Phi}\overline{\wedge}\hat{\Psi})_{(i_q,j_q)} = \left\{ \frac{(T_{\hat{\Phi}_{(i_q)}}(x_p) \wedge T'_{\hat{\Psi}_{(j_q)}}(x_p), I_{\hat{\Phi}_{(i_q)}}(x_p) \vee I'_{\hat{\Psi}_{(j_q)}}(x_p), F_{\hat{\Phi}_{(i_q)}}(x_p) \vee F'_{\hat{\Psi}_{(j_q)}}(x_p), \mu(x_p) \wedge \mu'(x_p))}{x_p} \,\middle|\, i_q \in I, j_q \in J, x_p \in X \right\}.$$

Third step: Define and write the truth membership $(\hat{\Phi}\overline{\wedge}\hat{\Psi})^T_{(i_q,j_q)}$, the indeterminacy membership $(\hat{\Phi}\overline{\wedge}\hat{\Psi})^I_{(i_q,j_q)}$, and the falsity membership $(\hat{\Phi}\overline{\wedge}\hat{\Psi})^F_{(i_q,j_q)}$, respectively, as follows:

$$(\hat{\Phi}\overline{\wedge}\hat{\Psi})^T_{(i_q,j_q)} = \left\{ \frac{(T_{\hat{\Phi}_{(i_q)}}(x_p) \wedge T'_{\hat{\Psi}_{(j_q)}}(x_p), \mu(x_p) \wedge \mu'(x_p))}{x_p} \,\middle|\, i_q \in I, j_q \in J, x_p \in X \right\},$$

$$(\hat{\Phi}\overline{\wedge}\hat{\Psi})^I_{(i_q,j_q)} = \left\{ \frac{(I_{\hat{\Phi}_{(i_q)}}(x_p) \vee I'_{\hat{\Psi}_{(j_q)}}(x_p), \mu(x_p) \wedge \mu'(x_p))}{x_p} \,\middle|\, i_q \in I, j_q \in J, x_p \in X \right\},$$

and

$$(\hat{\Phi}\overline{\wedge}\hat{\Psi})^F_{(i_q,j_q)} = \left\{ \frac{(F_{\hat{\Phi}_{(i_q)}}(x_p) \vee F'_{\hat{\Psi}_{(j_q)}}(x_p), \mu(x_p) \wedge \mu'(x_p))}{x_p} \,\middle|\, i_q \in I, j_q \in J, x_p \in X \right\}.$$

Fourth step: Define and compute the max-matrices of $(\hat{\Phi}\overline{\wedge}\hat{\Psi})^T_{(i_q,j_q)}$, $(\hat{\Phi}\overline{\wedge}\hat{\Psi})^I_{(i_q,j_q)}$, and $(\hat{\Phi}\overline{\wedge}\hat{\Psi})^F_{(i_q,j_q)}$, respectively, for every $x_p \in X$ as follows $(p = 1, 2, \cdots, N)$:

$$(\hat{\Phi}\overline{\wedge}\hat{\Psi})^T_{(i_q,j_q)}(x_p) = \frac{1}{2}\Big((T_{\hat{\Phi}_{(i_q)}}(x_p) \wedge T'_{\hat{\Psi}_{(j_q)}}(x_p)) + (\mu(x_p) \wedge \mu'(x_p)) \Big),$$

$$(\hat{\Phi}\overline{\wedge}\hat{\Psi})^I_{(i_q,j_q)}(x_p) = \Big((I_{\hat{\Phi}_{(i_q)}}(x_p) \vee I'_{\hat{\Psi}_{(j_q)}}(x_p)) \times (\mu(x_p) \wedge \mu'(x_p)) \Big),$$

and

$$(\hat{\Phi}\overline{\wedge}\hat{\Psi})^F_{(i_q,j_q)}(x_p) = \Big((F_{\hat{\Phi}_{(i_q)}}(x_p) \vee F'_{\hat{\Psi}_{(j_q)}}(x_p)) - (\mu(x_p) \wedge \mu'(x_p) \Big)^2.$$

Algorithm 2: *Cont.*

Fifth step: Calculate and write the max-decision τ_T (i.e., $\tau_T : X \to R$), τ_I (i.e., $\tau_I : X \to R$), and τ_F (i.e., $\tau_F : X \to R$) of $(\hat{\Phi}\overline{\wedge}\hat{\Psi})^T_{(i_q,j_q)}$, $(\hat{\Phi}\overline{\wedge}\hat{\Psi})^I_{(i_q,j_q)}$, and $(\hat{\Phi}\overline{\wedge}\hat{\Psi})^F_{(i_q,j_q)}$, respectively, for every $x_p \in X$ as follows ($p = 1, 2, \cdots, N$):

$$\tau_T(x_p) = \sum_{(i,j)\in I\times J} \delta_T(x_p)(i,j), \quad \tau_I(x_p) = \sum_{(i,j)\in I\times J} \delta_F(x_p)(i,j), \text{ and } \sum_{(i,j)\in I\times J} \delta_F(x_p)(i,j),$$

where

$$\delta_T(x_p)(i,j) = \begin{cases} (\hat{\Phi}\overline{\wedge}\hat{\Psi})^T_{(i_q,j_q)}(x_p), & (\hat{\Phi}\overline{\wedge}\hat{\Psi})^T_{(i_q,j_q)}(x_p) = \max\{(\hat{\Phi}\overline{\wedge}\hat{\Psi})^T_{(u_q,v_q)}(x_p) : (u,v) \in I \times J\} \\ 0, & \text{otherwise} \end{cases},$$

$$\delta_I(x_p)(i.j) = \begin{cases} (\hat{\Phi}\overline{\wedge}\hat{\Psi})^I_{(i_q,j_q)}(x_p), & (\hat{\Phi}\overline{\wedge}\hat{\Psi})^I_{(i_q,j_q)}(x_p) = \max\{(\hat{\Phi}\overline{\wedge}\hat{\Psi})^I_{(u_q,v_q)}(x_p) : (u,v) \in I \times J\} \\ 0, & \text{otherwise} \end{cases},$$

$$\delta_F(x_p)(i.j) = \begin{cases} (\hat{\Phi}\overline{\wedge}\hat{\Psi})^F_{(i_q,j_q)}(x_p), & (\hat{\Phi}\overline{\wedge}\hat{\Psi})^F_{(i_q,j_q)}(x_p) = \max\{(\hat{\Phi}\overline{\wedge}\hat{\Psi})^F_{(u_q,v_q)}(x_p) : (u,v) \in I \times J\} \\ 0, & \text{otherwise} \end{cases}.$$

Sixth step: Calculate the score $S(x_p)$ of element x_p as follows ($p = 1, 2, \cdots, N$):

$$S(x_p) = \tau_T(x_p) + \tau_I(x_p) + \tau_F(x_p).$$

Seventh step: Obtain the decision p satisfying

$$x_p = \max\{S(x_1), S(x_2), \cdots, S(x_j)\}.$$

Now, we show the principle and steps of the above Algorithm 2 using the following example.

Example 9. *(Continued from Example 11). Suppose that an investment company also adds three different parameters $J = \{j_1, j_2, j_3\}$, where j_1 is "marketing management", j_2 is "productivity of capital", and j_3 is "interest rates". The data of the single-valued neutrosophic fuzzy soft set $\hat{\Psi}_{(j_q)} \in (\mathbb{SVNFS})^{XJ}$ is given by*

$\hat{\Psi} =$

J	z_1	z_2	z_3	z_4	z_5
j_1	(0.5, 0.6, 0.7, 0.4)	(0.3, 0.2, 0.7, 0.8)	(0.6, 0.9, 0.4, 0.3)	(0.8, 0.8, 0.2, 0.1)	(0.9, 0.5, 0.4, 0.2)
j_2	(0.8, 0.4, 0.5, 0.2)	(0.7, 0.9, 0.2, 0.1)	(0.3, 0.3, 0.9, 0.4)	(0.9, 0.4, 0.5, 0.5)	(0.7, 0.8, 0.7, 0.2)
j_3	(0.9, 0.9, 0.5, 0.3)	(0.5, 0.9, 0.2, 0.1)	(0.6, 0.6, 0.1, 0.5)	(0.5, 0.7, 0.8, 0.8)	(0.6, 0.2, 0.4, 0.7)

Now, we explain the practical meaning of alternatives X by taking the alternative z_1 as an example: the single-valued neutrosophic fuzzy soft set $\hat{\Psi}_{(j_1)}(z_1) = (0.5, 0.6, 0.7, 0.4)$ is the evaluation by four expert groups; the single-valued neutrosophic fuzzy soft value 0.5 (meaning 50% say yes in the first expert group) in $\hat{\Psi}_{(j_1)}(z_1)$, the single-valued neutrosophic fuzzy soft value 0.6 (meaning 60% say no in the second expert group) in $\hat{\Psi}_{(j_1)}(z_1)$, the single-valued neutrosophic fuzzy soft value 0.7 (meaning 70% say yes in the third expert group) in $\hat{\Psi}_{(j_1)}(z_1)$, and fuzzy value 0.4 (meaning 40% say no in the fourth expert group) in $\hat{\Psi}_{(j_1)}(z_1)$. Then, by computing $(\hat{\Phi}\overline{\wedge}\hat{\Psi})_{(i_q,j_q)}$ ($q = 1, 2, 3$) in the second step of Algorithm 2, we obtain the following:

$\hat{\Phi}\overline{\wedge}\hat{\Psi}$	z_1	z_2	z_3	z_4	z_5
(i_1, j_1)	(0.3, 0.7, 0.7, 0.2)	(0.1, 0.8, 0.7, 0.5)	(0.2, 0.9, 0.8, 0.3)	(0.5, 0.8, 0.5, 0.1)	(0.4, 0.7, 0.9, 0.1)
(i_1, j_2)	(0.3, 0.7, 0.5, 0.2)	(0.1, 0.9, 0.5, 0.1)	(0.2, 0.6, 0.9, 0.4)	(0.5, 0.6, 0.5, 0.2)	(0.4, 0.8, 0.9, 0.1)
(i_1, j_3)	(0.3, 0.9, 0.5, 0.2)	(0.1, 0.9, 0.5, 0.1)	(0.2, 0.6, 0.8, 0.5)	(0.5, 0.7, 0.8, 0.2)	(0.4, 0.7, 0.9, 0.1)
(i_2, j_1)	(0.5, 0.6, 0.7, 0.4)	(0.3, 0.7, 0.7, 0.4)	(0.6, 0.9, 0.6, 0.3)	(0.3, 0.8, 0.2, 0.1)	(0.7, 0.8, 0.8, 0.2)
(i_2, j_2)	(0.8, 0.4, 0.5, 0.2)	(0.3, 0.9, 0.5, 0.1)	(0.3, 0.3, 0.9, 0.4)	(0.3, 0.7, 0.5, 0.5)	(0.7, 0.8, 0.8, 0.2)
(i_2, j_3)	(0.9, 0.9, 0.5, 0.3)	(0.3, 0.9, 0.5, 0.1)	(0.6, 0.6, 0.6, 0.5)	(0.3, 0.7, 0.8, 0.5)	(0.6, 0.8, 0.8, 0.3)
(i_3, j_1)	(0.5, 0.6, 0.7, 0.4)	(0.3, 0.5, 0.6, 0.4)	(0.7, 0.8, 0.6, 0.1)	(0.8, 0.9, 0.6, 0.1)	(0.7, 0.8, 0.9, 0.2)
(i_3, j_2)	(0.6, 0.4, 0.5, 0.2)	(0.3, 0.9, 0.6, 0.1)	(0.3, 0.3, 0.9, 0.3)	(0.8, 0.9, 0.6, 0.4)	(0.7, 0.8, 0.9, 0.2)
(i_3, j_3)	(0.6, 0.9, 0.5, 0.3)	(0.3, 0.9, 0.6, 0.1)	(0.6, 0.6, 0.6, 0.3)	(0.5, 0.9, 0.8, 0.4)	(0.6, 0.8, 0.9, 0.6)

By calculating in the third step of Algorithm 2, we get the truth membership $(\hat{\Phi}\overline{\wedge}\hat{\Psi})^T_{(i_q,j_q)}$, *the indeterminacy membership* $(\hat{\Phi}\overline{\wedge}\hat{\Psi})^I_{(i_q,j_q)}$, *and the falsity membership* $(\hat{\Phi}\overline{\wedge}\hat{\Psi})^F_{(i_q,j_q)}$, *respectively, as follows:* $(q = 1, 2, 3)$:

$(\hat{\Phi}\overline{\wedge}\hat{\Psi})^T$	z_1	z_2	z_3	z_4	z_5
(i_1, j_1)	(0.3, 0.2)	(0.1, 0.5)	(0.2, 0.3)	(0.5, 0.1)	(0.4, 0.1)
(i_1, j_2)	(0.3, 0.2)	(0.1, 0.1)	(0.2, 0.4)	(0.5, 0.2)	(0.4, 0.1)
(i_1, j_3)	(0.3, 0.2)	(0.1, 0.1)	(0.2, 0.5)	(0.5, 0.2)	(0.4, 0.1)
(i_2, j_1)	(0.5, 0.4)	(0.3, 0.4)	(0.6, 0.3)	(0.3, 0.1)	(0.7, 0.2)
(i_2, j_2)	(0.8, 0.2)	(0.3, 0.1)	(0.3, 0.4)	(0.3, 0.5)	(0.7, 0.2)
(i_2, j_3)	(0.9, 0.3)	(0.3, 0.1)	(0.6, 0.5)	(0.3, 0.5)	(0.6, 0.3)
(i_3, j_1)	(0.5, 0.4)	(0.3, 0.4)	(0.7, 0.1)	(0.8, 0.1)	(0.7, 0.2)
(i_3, j_2)	(0.6, 0.2)	(0.3, 0.1)	(0.3, 0.3)	(0.8, 0.4)	(0.7, 0.2)
(i_3, j_3)	(0.6, 0.3)	(0.3, 0.1)	(0.6, 0.3)	(0.5, 0.4)	(0.6, 0.6)

$(\hat{\Phi}\overline{\wedge}\hat{\Psi})^I$	z_1	z_2	z_3	z_4	z_5
(i_1, j_1)	(0.7, 0.2)	(0.8, 0.5)	(0.9, 0.3)	(0.8, 0.1)	(0.7, 0.1)
(i_1, j_2)	(0.7, 0.2)	(0.9, 0.1)	(0.6, 0.4)	(0.6, 0.2)	(0.8, 0.1)
(i_1, j_3)	(0.9, 0.2)	(0.9, 0.1)	(0.6, 0.5)	(0.7, 0.2)	(0.7, 0.1)
(i_2, j_1)	(0.6, 0.4)	(0.7, 0.4)	(0.9, 0.3)	(0.8, 0.1)	(0.8, 0.2)
(i_2, j_2)	(0.4, 0.2)	(0.9, 0.1)	(0.3, 0.4)	(0.7, 0.5)	(0.8, 0.2)
(i_2, j_3)	(0.9, 0.3)	(0.9, 0.1)	(0.6, 0.5)	(0.7, 0.5)	(0.8, 0.3)
(i_3, j_1)	(0.6, 0.4)	(0.5, 0.4)	(0.8, 0.1)	(0.9, 0.1)	(0.8, 0.2)
(i_3, j_2)	(0.4, 0.2)	(0.9, 0.1)	(0.3, 0.3)	(0.9, 0.4)	(0.8, 0.2)
(i_3, j_3)	(0.9, 0.3)	(0.9, 0.1)	(0.6, 0.3)	(0.9, 0.4)	(0.8, 0.6)

$(\hat{\Phi}\overline{\wedge}\hat{\Psi})^F$	z_1	z_2	z_3	z_4	z_5
(i_1, j_1)	(0.7, 0.2)	(0.7, 0.5)	(0.8, 0.3)	(0.5, 0.1)	(0.9, 0.1)
(i_1, j_2)	(0.5, 0.2)	(0.5, 0.1)	(0.9, 0.4)	(0.5, 0.2)	(0.9, 0.1)
(i_1, j_3)	(0.5, 0.2)	(0.5, 0.1)	(0.8, 0.5)	(0.8, 0.2)	(0.9, 0.1)
(i_2, j_1)	(0.7, 0.4)	(0.7, 0.4)	(0.6, 0.3)	(0.2, 0.1)	(0.8, 0.2)
(i_2, j_2)	(0.5, 0.2)	(0.5, 0.1)	(0.9, 0.4)	(0.5, 0.5)	(0.8, 0.2)
(i_2, j_3)	(0.5, 0.3)	(0.5, 0.1)	(0.6, 0.5)	(0.8, 0.5)	(0.8, 0.3)
(i_3, j_1)	(0.7, 0.4)	(0.6, 0.4)	(0.6, 0.1)	(0.6, 0.1)	(0.9, 0.2)
(i_3, j_2)	(0.5, 0.2)	(0.6, 0.1)	(0.9, 0.3)	(0.6, 0.4)	(0.9, 0.2)
(i_3, j_3)	(0.5, 0.3)	(0.6, 0.1)	(0.6, 0.3)	(0.8, 0.4)	(0.9, 0.6)

By calculating in the fourth step of Algorithm 2, we obtain the max-matrices of $(\hat{\Phi}\overline{\wedge}\hat{\Psi})^T_{(i_q,j_q)}$, $(\hat{\Phi}\overline{\wedge}\hat{\Psi})^I_{(i_q,j_q)}$, *and* $(\hat{\Phi}\overline{\wedge}\hat{\Psi})^F_{(i_q,j_q)}$ $(p = 1,2,3,4,5;\ q = 1,2,3)$, *respectively, for every* $z_p \in X$ *as follows:*

$$\left(\begin{array}{c|ccccc}
(\hat{\Phi}\overline{\wedge}\hat{\Psi})^T & z_1 & z_2 & z_3 & z_4 & z_5 \\ \hline
(i_1,j_1) & 0.25 & \underline{0.3} & 0.25 & \underline{0.3} & 0.25 \\
(i_1,j_2) & 0.25 & 0.1 & 0.3 & \underline{0.35} & 0.25 \\
(i_1,j_3) & 0.25 & 0.1 & \underline{0.35} & \underline{0.35} & 0.25 \\
(i_2,j_1) & \underline{0.45} & 0.35 & \underline{0.45} & 0.2 & \underline{0.45} \\
(i_2,j_2) & \underline{0.5} & 0.2 & 0.35 & 0.4 & 0.45 \\
(i_2,j_3) & \underline{0.6} & 0.2 & 0.55 & 0.4 & 0.45 \\
(i_3,j_1) & \underline{0.45} & 0.35 & 0.4 & \underline{0.45} & \underline{0.45} \\
(i_3,j_2) & 0.4 & 0.2 & 0.3 & \underline{0.6} & 0.45 \\
(i_3,j_3) & 0.45 & 0.2 & 0.45 & 0.45 & \underline{0.6}
\end{array}\right),$$

$$\left(\begin{array}{c|ccccc}
(\hat{\Phi}\overline{\wedge}\hat{\Psi})^I & z_1 & z_2 & z_3 & z_4 & z_5 \\ \hline
(i_1,j_1) & 0.14 & \underline{0.4} & 0.27 & 0.08 & 0.07 \\
(i_1,j_2) & 0.14 & 0.09 & \underline{0.24} & 0.12 & 0.08 \\
(i_1,j_3) & 0.18 & 0.09 & \underline{0.3} & 0.14 & 0.07 \\
(i_2,j_1) & 0.24 & \underline{0.28} & 0.27 & 0.08 & 0.16 \\
(i_2,j_2) & 0.08 & 0.08 & 0.12 & \underline{0.35} & 0.16 \\
(i_2,j_3) & 0.27 & 0.09 & 0.3 & \underline{0.35} & 0.24 \\
(i_3,j_1) & \underline{0.24} & 0.2 & 0.08 & 0.09 & 0.16 \\
(i_3,j_2) & 0.08 & 0.09 & 0.09 & \underline{0.36} & 0.16 \\
(i_3,j_3) & 0.27 & 0.09 & 0.18 & 0.36 & \underline{0.48}
\end{array}\right),$$

$$\left(\begin{array}{c|ccccc}
(\hat{\Phi}\overline{\wedge}\hat{\Psi})^F & x_1 & z_2 & z_3 & z_4 & z_5 \\ \hline
(i_1,j_1) & 0.25 & 0.04 & 0.25 & 0.16 & \underline{0.64} \\
(i_1,j_2) & 0.09 & 0.16 & 0.25 & 0.09 & \underline{0.64} \\
(i_1,j_3) & 0.09 & 0.16 & 0.09 & 0.36 & \underline{0.64} \\
(i_2,j_1) & 0.09 & 0.09 & 0.09 & 0.01 & \underline{0.36} \\
(i_2,j_2) & 0.09 & 0.16 & 0.25 & 0 & \underline{0.36} \\
(i_2,j_3) & 0.04 & 0.16 & 0.01 & 0.09 & \underline{0.25} \\
(i_3,j_1) & 0.09 & 0.04 & 0.25 & 0.25 & \underline{0.49} \\
(i_3,j_2) & 0.09 & 0.25 & 0.36 & 0.04 & \underline{0.49} \\
(i_3,j_3) & 0.04 & \underline{0.25} & 0.09 & 0.16 & 0.09
\end{array}\right).$$

By calculating in the fifth step of Algorithm 2, we obtain the max-decision τ_T, τ_I, *and* τ_F *of elements* z_p, *respectively, as follows* $(p = 1,2,3,4,5)$:

$$\tau_T(z_1) = 2,\ \tau_T(z_2) = 0.3,\ \tau_T(z_3) = 0.8,\ \tau_T(z_4) = 2.05,\ \tau_T(z_5) = 1.5;$$

$$\tau_I(z_1) = 0.24,\ \tau_I(z_2) = 0.68,\ \tau_I(z_3) = 0.54,\ \tau_I(z_4) = 1.06,\ \tau_I(z_5) = 0.48;$$

$$\tau_F(z_1) = 0,\ \tau_F(z_2) = 0.25,\ \tau_F(z_3) = 0,\ \tau_F(z_4) = 0,\ \tau_F(z_5) = 3.87.$$

By calculating in the sixth step of Algorithm 2, the scores $S(z_p)$ *of elements* $z_p (p = 1,2,3,4,5)$, *respectively, are as follows:*

$$S(z_1) = 2.24,\ S(z_2) = 1.23,\ S(z_3) = 1.34,\ S(z_4) = 3.11,\ S(z_5) = 5.85.$$

Finally, we know from the seventh step that z_5 has a high value. Therefore, the experts should select z_5 as the best choice.

Remark 2.

(1) By means of Algorithms 1 and 2, we can see that the final results are in agreement. Thus, x_5 is the most accurate and refinable.

(2) By comparing the steps in Algorithms 1 and 2, we can see that step 4 and step 5 in Algorithm 2 are complicated in their process compared to step 2 and step 3 in Algorithm 1, respectively. So, if we take the complexity of these steps into consideration, Algorithm 2 gives its decision concisely.

(3) Algorithms 1 and 2 that we have elaborated here arrive at their decisions by combining the concept of single-valued neutrosophic fuzzy set theory and soft set theory. As result, we can apply Algorithm 1 to picture fuzzy soft sets [29], generalized picture fuzzy soft sets [13], and interval-valued neutrosophic soft sets [12]. Further, Algorithm 2 can be applied to possibility m-polar fuzzy soft sets [15] and possibility multi-fuzzy soft sets [17].

5. Conclusions

We introduced the notion of the single-valued neutrosophic fuzzy soft set as a novel neutrosophic soft set model. We discussed the five operations of the single-valued neutrosophic fuzzy soft set, such as subset, equal, union, intersection, and complement. The structure properties of the single-valued neutrosophic fuzzy soft set are explained. Then, a novel approach (i.e., Algorithm 1) is presented as a single-valued neutrosophic fuzzy soft set decision method. Lastly, an application (i.e., Algorithm 2) of a single-valued neutrosophic fuzzy soft set for fuzzy decision-making is constructed, and the two approaches (i.e., Algorithms 1 and 2) introduce an important contribution to further research and relevant applications. Therefore, in the future, we will provide a real application with a real dataset or we will apply the two approaches (i.e., Algorithms 1 and 2) to lung cancer disease [30] and coronary artery disease [31]. In addition, we will describe in more detail in order to clarify if the methods (i.e., Algorithms 1 and 2) converge or diverge from standard approaches such as fuzzy sets [1], intuitionistic fuzzy sets [2], picture fuzzy sets [3].

Author Contributions: Conceptualization and data curation, A.M.K.; Methodology, A.A.A. and W.A.; Writing—original draft preparation, F.S.; Funding acquisition, D.C. All authors have read and agreed to the published version of the manuscript.

Funding: This research received no external funding.

Acknowledgments: The authors thank the editors and the anonymous reviewers for their insightful comments which improved the quality of the paper. The authors wish to thank the Deanship of Scientific Research at Prince Sattam Bin Abdulaziz University, Alkharj 11942, Saudi Arabia, for their support of their research.

Conflicts of Interest: The authors declare no conflicts of interest.

References

1. Zadeh, L.A. Fuzzy sets. *Inf. Contr.* **1965**, *8*, 338–353. [CrossRef]
2. Atanassov, K.T. Intuitionistic fuzzy sets. *Fuzzy Sets Syst.* **1986**, *20*, 87–96. [CrossRef]
3. Cuong, B.C. Picture fuzzy sets. *J. Comput. Sci. Cybern.* **2014**, *30*, 409–420.
4. Molodtsov, Soft set theory-first results. *Comput. Math. Appl.* **1999**, *37*, 19–31. [CrossRef]
5. Maji, P.K.; Biswas, R.; Roy, A.R. Soft set theory. *Comput. Math. Appl.* **2003**, *44*, 555–562. [CrossRef]
6. Maji, P.K.; Roy, A.R.; Biswas, R. An application of soft sets in a decision making problem. *Comput. Math. Appl.* **2002**, 44, 1077–1083. [CrossRef]
7. Maji, P.K.; Biswas, R.; Roy, A.R. Fuzzy soft sets. *J. Fuzzy Math.* **2001**, *9*, 589–602.
8. Khalil, A.M.; Hassan N. Inverse fuzzy soft set and its application in decision making. *Int. J. Inf. Decis. Sci.* **2019**, *11*, 73–92.
9. Vijayabalaji, S.; Ramesh, A. Belief interval-valued soft set. *Expert Syst. Appl.* **2019**, *119*, 262–271. [CrossRef]

10. Jiang, Y.; Tang, Y.; Chen, Q.; Liu, H.; Tang, J. Interval-valued intuitionistic fuzzy soft sets and their properties. *Comput. Math. Appl.* **2010**, *60*, 906–918. [CrossRef]
11. Khalil, A.M.; Li, S.; Garg, H.; Li, H.; Ma, S. New operations on interval-valued picture fuzzy set, interval-valued picture fuzzy soft set and their applications. *IEEE Access* **2019**, *7*, 51236–51253. [CrossRef]
12. Deli, I. Interval-valued neutrosophic soft sets and its decision making. *Int. J. Mach. Learn. Cybern.* **2017**, *8*, 665–676. [CrossRef]
13. Khan, M.J.; Kumam, P.; Ashraf, S.; Kumam, W. Generalized picture fuzzy soft sets and their application in decision support systems. *Symmetry* **2019**, *11*, 415. [CrossRef]
14. Hua, D.J.; Zhang, H.D.; He, Y. Possibility Pythagorean fuzzy soft set and its application. *J. Intell. Fuzzy Syst.* **2019**, *36*, 413–421.
15. Khalil, A.M.; Li, S.; Li, H.; Ma, S. Possibility m-polar fuzzy soft sets and its application in decision-making problems. *J. Intell. Fuzzy Syst.* **2019**, *37*, 929–940. [CrossRef]
16. Karaaslan, F. Possibility neutrosophic soft sets and PNS-decision making method. *Appl. Soft Comput.* **2017**, *54*, 403–414. [CrossRef]
17. Zhang, H.D.; Shu, L. Possibility multi-fuzzy soft set and its application in decision making. *J. Intell. Fuzzy Syst.* **2014**, *27*, 2115–2125. [CrossRef]
18. Karaaslan, F.; Hunu, F. Type-2 single-valued neutrosophic sets and their applications in multi-criteria group decision making based on TOPSIS method. *J. Ambient Intell. Human. Comput.* **2020**. [CrossRef]
19. Al-Quran, A.; Hassan, N.; Alkhazaleh, S. Fuzzy parameterized complex neutrosophic soft expert set for decision under uncertainty. *Symmetry* **2019**, *11*, 382. [CrossRef]
20. Abu Qamar, M.; Hassan, N. An approach toward a Q-neutrosophic soft set and its application in decision making. *Symmetry* **2019**, *11*, 119. [CrossRef]
21. Abu Qamar, M.; Hassan, N. Generalized Q-neutrosophic soft expert set for decision under uncertainty. *Symmetry* **2018**, *10*, 621. [CrossRef]
22. Uluçay, V.; Şahin, M.; Hassan, N. Generalized neutrosophic soft expert set for multiple-criteria decision-making. *Symmetry* **2018**, *10*, 437. [CrossRef]
23. Zhang, X.; Bo, C.; Smarandache, F.; Park, C. New operations of totally dependent-neutrosophic sets and totally dependent-neutrosophic soft sets. *Symmetry* **2018**, *10*, 187. [CrossRef]
24. Smarandache, F. Extension of soft set to hypersoft set, and then to Plithogenic hypersoft set. *Neutrosophic Sets Syst.* **2018**, *22*, 168–170.
25. Smarandache, F. *Neutrosophy, Neutrosophic Probability, Set, and Logic*; American Research Press: Rehoboth, DE, USA, 1998.
26. Wang, H.; Smarandache, F.; Zhang, Y.; Sunderraman, R. Single valued neutrosophic sets. *Multissp. Multistruct.* **2010**, *4*, 410–413.
27. Das, S.; Roy, B.K.; Kar, M.B.; Pamučar, D. Neutrosophic fuzzy set and its application in decision making. *J. Ambient Intell. Human. Comput.* **2020**. [CrossRef]
28. Maji, P.K. Neutrosophic soft set. *Ann. Fuzzy Math. Inform.* **2013**, *5*, 157–168.
29. Yang, Y.; Liang, C.; Ji, S.; Liu, T. Adjustable soft discernibility matrix based on picture fuzzy soft sets and its application in decision making. *J. Intell. Fuzzy Syst.* **2015**, *29*, 1711–1722. [CrossRef]
30. Khalil, A.M.; Li, S.G.; Lin, Y.; Li, H.X; Ma, S.G. A new expert system in prediction of lung cancer disease based on fuzzy soft sets. *Soft Comput.* **2020**. [CrossRef]
31. Hassan, N.; Sayed, O.R.; Khalil, A.M.; Ghany M.A. Fuzzy soft expert system in prediction of coronary artery disease. *Int. J. Fuzzy Syst.* **2017**, *19*, 1546–1559. [CrossRef]

symmetry

Article

Single-Valued Neutrosophic Set Correlation Coefficient and Its Application in Fault Diagnosis

Shchur Iryna [1], Yu Zhong [1], Wen Jiang [1,2,*], Xinyang Deng [1] and Jie Geng [1]

[1] School of Electronics and Information, Northwestern Polytechnical University, Xi'an 710072, China; irynashchur93@mail.nwpu.edu.cn (S.I.); yuzhong@mail.nwpu.edu.cn (Y.Z.); xinyang.deng@nwpu.edu.cn (X.D.); gengjie@nwpu.edu.cn (J.G.)

[2] Peng Cheng Laboratory, Shenzhen 518055, China

* Correspondence: jiangwen@nwpu.edu.cn; Tel.: +86-29-8843-1267

Received: 3 July 2020; Accepted: 11 August 2020; Published: 17 August 2020

Abstract: With the increasing automation of mechanical equipment, fault diagnosis becomes more and more important. However, the factors that cause mechanical failures are becoming more and more complex, and the uncertainty and coupling between the factors are getting higher and higher. In order to solve the given problem, this paper proposes a single-valued neutrosophic set ISVNS algorithm for processing of uncertain and inaccurate information in fault diagnosis, which generates neutrosophic set by triangular fuzzy number and introduces the formula of the improved weighted correlation coefficient. Since both the single-valued neutrosophic set data and the ideal neutrosophic set data are considered, the proposed method solves the fault diagnosis problem more effectively. Finally, experiments show that the algorithm can significantly improve the accuracy degree of fault diagnosis, and can better satisfy the diagnostic requirements in practice.

Keywords: neutrosophic set; fault diagnosis; triangle fuzzy number; weighted correlation coefficient

---

1. Introduction

With the development of automation technology, these mechanical machines gradually came into the stage of fully automated control operation [1–5]. In this way, people's hands are comparatively free, and machines are more intelligent and comprehensive; however, this kind of full automation greatly increases the probability of mechanical equipment failure as well [6–11]. If the mechanical equipment has faults, the quality of the manufactured products will not pass the standard, which will affect the economic benefits of the enterprise [12–15]; additionally, it will bring potential danger to personal safety [16–19]. In order to solve this problem, it is necessary to carry out fault diagnosis on mechanical equipment on a regular basis to detect and repair mechanical equipment and ensure its normal operation.

1.1. Research Status

Therefore, the fault diagnosis of mechanical equipment has been widely concerned by many scholars, and has been applied in the military [20,22–24], medical [25–28], economic [29–32], and other fields. In ref. [33], looking at the problem of low efficiency of fault diagnosis of automobile exhaust system, based on a cold test, a fault diagnosis method is proposed for port vehicle exhaust system based on the principal component analysis. The variance contribution rate of principal component model is analyzed by the change of each variable of measurement data, and the fault diagnosis is achieved; In ref. [34], aiming at the problem of fault diagnosis of the data-driven system, a new diagnosis method based on Bayesian network (BN) combined with fault frequency is proposed to realize fault diagnosis; In ref. [35], based on the particle filter (PF) program, a dual estimation method is applied to fault diagnosis; In ref. [36], for the problem of bearing diagnosis under the condition of variable speed, the method of support vector machine and neural network is used for bearing fault diagnosis; On the basis of machine learning technology, a new depth neural network model with domain self-adaptability is proposed in ref. [37], which realizes fault diagnosis, but the selection of the best parameters of the model is random, and the application of the model is limited.

In the actual operating environment of mechanical equipment, fault information is usually inaccurate, incomplete and uncertain. It is difficult to use the above fault diagnosis method for deterministic analysis and processing of fault information. In order to deal with the uncertain and inaccurate information in the fault information, so as to better handle the fault information and get more accurate fault diagnosis results, Smarandache [38] proposed the theory of neutrosophic set from a philosophical point of view. It describes the uncertainty, imprecision, and inconsistent information in the objective world much better. The literature [39] introduced the theory of interval neutrosophic set and single-valued neutrosophic set; The literature [40] proposed the theory of simplified neutrosophic set; Literature [41] proposed a single-valued neutrosophic set SVNS method with a weighted correlation coefficient to realize fault diagnosis. However, the correlation coefficient does not comprehensively consider the single-valued neutrosophic set and ideal neutrosophic set data under various faults [42–46], only the maximum value between them is considered [47–49]. This does not completely deal with the uncertain and inaccurate information in the fault information, and may lead to an incorrect diagnosis.

1.2. Contribution of This Work

Based on the problems above, properly handling the uncertain information in the fault diagnosis process is an important goal to be achieved, however, complicated and changeable environmental information, the mutual influence between the factors causing the failure are difficult to handle. Due to neutrosophic set's outstanding performance in handling uncertain information issues, this paper proposes a single-valued neutrosophic set ISVNS algorithm, which generates neutrosophic set by triangular fuzzy number and introduces the formula of the improved weighted correlation coefficient. In addition, the ISVNS algorithm comprehensively considers the single-valued neutrosophic set and ideal neutrosophic set data of various faults, so make it possible to analyze the data more comprehensively and make more accurate judgments. Finally, an example was used to diagnose the fault; the degree of accuracy of the fault diagnosis was calculated; the excellent productivity of the improved method, proposed in this paper was obtained by comparison. For the current difficulties in dealing with some uncertain issues, this method may have some enlightenment.

Due to the Overall Equipment Effectiveness (OEE) and the Overall labor effectiveness (OLE) are simple and practical production management tool, which has been widely used in European and American manufacturing and Chinese multinational companies. The global equipment efficiency index has become an important standard for measuring the production efficiency of enterprises, so it is also important to consider the proposed method's impact on OEE and OLE. During the simulation process in the laboratory, the ratio of the operating hours and the planned working hours is relatively high; therefore, the fault can be repaired in a more timely manner based on the diagnosis result, and the OEE and the OLE can be improved.

The remainder of this paper is organized as follows. Section 2 briefly introduced triangular fuzzy numbers and single-valued neutrosophic sets. Section 3 proposed improved correlation coefficient between single-valued neutrosophic sets. In Section 4, a numerical example is given to fault diagnosis and fault diagnosis accuracy based on the proposed approach. Some conclusions are shown in Section 5.

2. Preliminaries

2.1. Triangular Fuzzy Numbers

About data sets $D = \{d_1, d_2, \cdots, d_n\}$, take its minimum value $l = \min\{d_1, d_2, \cdots, d_n\}$, average value $m = mean\{d_1, d_2, \cdots, d_n\}$, maximum value $u = \max\{d_1, d_2, \cdots, d_n\}$, so $s = [l; m; u]$ named as triangular fuzzy number [9] of this data set D, as shown in Figure 1:

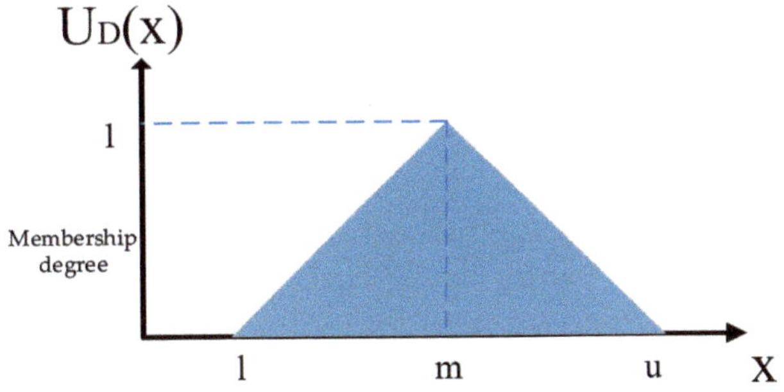

Figure 1. Geometric Interpretation of Triangular Fuzzy Number.

It can be seen from the geometric interpretation of the triangular fuzzy number in the figure above:
If $x = m$, so x completely belongs to D;
If $l \leq x \leq u$, so x takes a certain degree of data set D;
If $x < l$ or $x > u$, so x completely does not belong to the data set D.
And, for two triangle fuzzles $M_1 = (l_1, m_1, u_1)$ and $M_2 = (l_2, m_2, u_2)$, have the following counting method:

$$\begin{aligned} M_1 \oplus M_2 &= (l_1 + l_2, m_1 + m_2, u_1 + u_2) \\ M_1 \otimes M_2 &\approx (l_1 l_2, m_1 m_2, u_1 u_2) \\ \lambda \otimes M_1 &\approx (\lambda l_1, \lambda m_1, \lambda u_1) \\ \tfrac{1}{M_1} &\approx (\tfrac{1}{u_1}, \tfrac{1}{m_1}, \tfrac{1}{l_1}) \end{aligned}$$

2.2. Single-Valued Neutrosophic Sets

Neutrosophic set introduced by Smarandache [38] is an effective tool for solving the problems under complex environment. The definition of neutrosophic set is as follows.

Definition 1. *X denote a space of points or objects, and each element of it is denoted as x. A neutrosophic set A in X is characterized by a truth-membership function T_A, an indeterminacy-membership function I_A and a falsity-membership function F_A. $T_A(x)$, $I_A(x)$ and $F_A(x)$ are real standard or non-standard subsets of $]0^-, 1^+[$. That is:*

$$\begin{aligned} T_A &: X \mapsto]0^-, 1^+[\\ I_A &: X \mapsto]0^-, 1^+[\\ F_A &: X \mapsto]0^-, 1^+[\end{aligned} \tag{1}$$

For Formula (1), $0^- \leq supT_A(x) + supI_A(x) + supF_A(x) \leq 3^+$.

To facilitate the application of neutrosophic set in real scientific and engineering problems, the notion of SVNS was defined as follows.

Definition 2. *X denote a space of points or objects, and each element of it is denoted as x. A neutrosophic set A in X is characterized by a truth-membership function $T_A(x)$, a indeterminacy-membership function $I_A(x)$*

and a falsity-membership function $F_A(x)$*. If* $T_A(x) : X \to [0,1]$, $I_A(x) : X \to [0,1]$ *and* $F_A(x) : X \to [0,1]$ *satisfied:*

$$\begin{array}{c} x \in X \mapsto T_A(x) \in [0,1] \\ x \in X \mapsto I_A(x) \in [0,1] \\ x \in X \mapsto F_A(x) \in [0,1] \\ 0 \leq T_A(x) + I_A(x) + F_A(x) \leq 3 \end{array} \tag{2}$$

then an SVNS A in X can be denoted as:

$$A = \{\langle x,\ T_A(x),\ I_A(x),\ F_A(x)\rangle | x \in X\} \tag{3}$$

which is called an SVNS. Especially, if X includes only one element, $N = \langle T_A(x),\ I_A(x),\ F_A(x)\rangle$ *is called a single-valued neutrosophic number (SVN).*

For any two SVNSs ($A = \langle T_A(x),\ I_A(x),\ F_A(x)\rangle$, $B = \langle T_B(x),\ I_B(x),\ F_B(x)\rangle$ operational relations are defined as follows:

$$\begin{array}{l} (1)\ A + B = \langle T_A(x) + T_B(x) - T_A(x)T_B(x),\ I_A(x) + I_B(x) - I_A(x)I_B(x), \\ \qquad F_A(x) + F_B(x) - F_A(x)F_B(x)\rangle \\ (2)\ A \times B = \langle T_A(x)T_B(x), I_A(x)I_B(x), F_A(x)F_B(x)\rangle \\ (3)\ \lambda A = \left\langle 1 - (1 - T_A(x))^{\lambda}, 1 - (1 - I_A(x))^{\lambda}, 1 - (1 - F_A(x))^{\lambda}\right\rangle, \quad \lambda > 0 \\ (4)\ A^{\lambda} = \left\langle T_A(x)^{\lambda}, I_A(x), F_A(x)\right\rangle, \quad \lambda > 0 \end{array} \tag{4}$$

These are a series of common laws in operation for SVNSs.

Moreover, the assumption and operation requirements are as follows: Because ISVNS algorithm generates neutral set by triangular fuzzy number, the fault template data of each fault must be greater than or equal to 3.

3. The Proposed Method

3.1. Correlation Coefficient between Single-Valued Neutrosophic Sets

For any two neutrosophic sets $A = \{T_A, F_A, I_A\}$ and $B = \{T_B, F_B, I_B\}$, the improved correlation coefficient is defined as follows:

$$\begin{aligned} W(A,B) &= \frac{2 \cdot C(A,B)}{C(A,A) + C(B,B)} \\ &= \frac{2 \cdot \sum\limits_{i=1}^{n} [T_A(x_i) \cdot T_B(x_i) + F_A(x_i) \cdot F_B(x_i) + I_A(x_i) \cdot I_B(x_i)]}{\sum\limits_{i=1}^{n} [T_A{}^2(x_i) + F_A{}^2(x_i) + I_A{}^2(x_i)] + \sum\limits_{i=1}^{n} [T_B{}^2(x_i) + F_B{}^2(x_i) + I_B{}^2(x_i)]} \end{aligned} \tag{5}$$

In addition, the correlation coefficient for any $A = \{T_A, F_A, I_A\}$ and $B = \{T_B, F_B, I_B\}$ must satisfy the following three mathematical rules10:

$$\begin{cases} W(A,B) = W(B,A) \\ 0 \leq W(A,B) \leq 1 \\ if\ A = B,\ W(A,B) = 1 \end{cases} \tag{6}$$

For Formula (5), prove separately as follows:

(1) According to the structural symmetry of the Formula (5), the condition $W(A,B) = W(B,A)$ is satisfied.

(2) For each element in the Formula (5), they are satisfied ≥ 0, so obviously $W(A,B) \geq 0$; The proof of inequality $W(A,B) \leq 1$ as follows:

$$\begin{aligned} C(A,B) &= \sum_{i=1}^{n} [T_A(x_i) \cdot T_B(x_i) + F_A(x_i) \cdot F_B(x_i) + I_A(x_i) \cdot I_B(x_i)] \\ &= T_A(x_1) \cdot T_B(x_1) + T_A(x_2) \cdot T_B(x_2) + \cdots + T_A(x_n) \cdot T_B(x_n) \\ &+ F_A(x_1) \cdot F_B(x_1) + F_A(x_2) \cdot F_B(x_2) + \cdots + F_A(x_n) \cdot F_B(x_n) \\ &+ I_A(x_1) \cdot I_B(x_1) + I_A(x_2) \cdot I_B(x_2) + \cdots + I_A(x_n) \cdot I_B(x_n) \end{aligned}$$

And because of the inequality:

$$ab \leq \frac{a^2 + b^2}{2}$$

Therefore, we can get:

$$\begin{aligned} C(A,B) &= \sum_{i=1}^{n} [T_A(x_i) \cdot T_B(x_i) + F_A(x_i) \cdot F_B(x_i) + I_A(x_i) \cdot I_B(x_i)] \\ &= T_A(x_1) \cdot T_B(x_1) + T_A(x_2) \cdot T_B(x_2) + \cdots + T_A(x_n) \cdot T_B(x_n) \\ &+ F_A(x_1) \cdot F_B(x_1) + F_A(x_2) \cdot F_B(x_2) + \cdots + F_A(x_n) \cdot F_B(x_n) \\ &+ I_A(x_1) \cdot I_B(x_1) + I_A(x_2) \cdot I_B(x_2) + \cdots + I_A(x_n) \cdot I_B(x_n) \\ &\leq \frac{T_A{}^2(x_1) + T_B{}^2(x_1)}{2} + \frac{T_A{}^2(x_2) + T_B{}^2(x_2)}{2} + \cdots + \frac{T_A{}^2(x_n) + T_B{}^2(x_n)}{2} \\ &+ \frac{F_A{}^2(x_1) + F_B{}^2(x_1)}{2} + \frac{F_A{}^2(x_2) + F_B{}^2(x_2)}{2} + \cdots + \frac{F_A{}^2(x_n) + F_B{}^2(x_n)}{2} \\ &+ \frac{I_A{}^2(x_1) + I_B{}^2(x_1)}{2} + \frac{I_A{}^2(x_2) + I_B{}^2(x_2)}{2} + \cdots + \frac{I_A{}^2(x_n) + I_B{}^2(x_n)}{2} \\ &= \frac{1}{2}\left\{ \sum_{i=1}^{n} [T_A{}^2(x_i) + F_A{}^2(x_i) + I_A{}^2(x_i)] + \sum_{i=1}^{n} [T_B{}^2(x_i) + F_B{}^2(x_i) + I_B{}^2(x_i)] \right\} \\ &= \frac{1}{2}[C(A,A) + C(B,B)] \end{aligned}$$

Therefore:

$$C(A,B) \leq \frac{1}{2}[C(A,A) + C(B,B)]$$

There is:

$$2 \cdot C(A,B) \leq C(A,A) + C(B,B)$$

Finally, contacting the previous types, there are:

$$W(A,B) = \frac{2 \cdot C(A,B)}{C(A,A) + C(B,B)} \leq 1$$

In summary, the condition $0 \leq W(A,B) \leq 1$ is satisfied;

(3) If $A = B$, so for any $x_i \in X(i = 1, 2, \cdots, n)$, all $T_A(x_i) = T_B(x_i)$, $F_A(x_i) = F_B(x_i)$, $I_A(x_i) = I_B(x_i)$, we can see from the structure of Formula (5), $W(A,B) = 1$.

In practical application, it is usually necessary to consider the weight of neutrosophic sets, so the weighted correlation coefficient of neutrosophic sets is given:

$$W(A,B) = \frac{2 \cdot \sum_{i=1}^{n} w_i [T_A(x_i) \cdot T_B(x_i) + F_A(x_i) \cdot F_B(x_i) + I_A(x_i) \cdot I_B(x_i)]}{\sum_{i=1}^{n} w_i [T_A{}^2(x_i) + F_A{}^2(x_i) + I_A{}^2(x_i)] + \sum_{i=1}^{n} w_i [T_B{}^2(x_i) + F_B{}^2(x_i) + I_B{}^2(x_i)]} \tag{7}$$

Among them, $w_j(j = 1, 2, \cdots, n)$ represents the weight of the i fault template, and $\sum_{i=1}^{n} w_i = 1$. In addition, the formula satisfies the three conditions of the formula and proves to be the same as the Formula (5), which is not repeated here.

3.2. Fault Diagnosis Method

Based on the above analysis, properly handling the uncertain information in the fault diagnosis process is an important goal to be achieved, however, complicated and changeable environmental information, the mutual influence between the factors causing the failure are difficult to handle. Due to neutrosophic set's outstanding performance in handling uncertain information issues, this paper proposes a single-valued neutrosophic set ISVNS algorithm, which generates neutrosophic set by triangular fuzzy number and introduces the formula of the improved weighted correlation coefficient.

The objectives of the proposed algorithm are to rationally process the uncertain information in the diagnosis process, and obtain correct and reasonable fault diagnosis results from the fault data. Therefore, the laboratory simulation of the algorithm is carried out under the assumption that the actual fault is one of several known fault templates, the collected fault data is reasonable, and there is no major abnormality. The operating requirement is to collect fault data in a stable and equal time interval way. The detailed flow chart of the fault diagnosis method is shown in Figure 2:

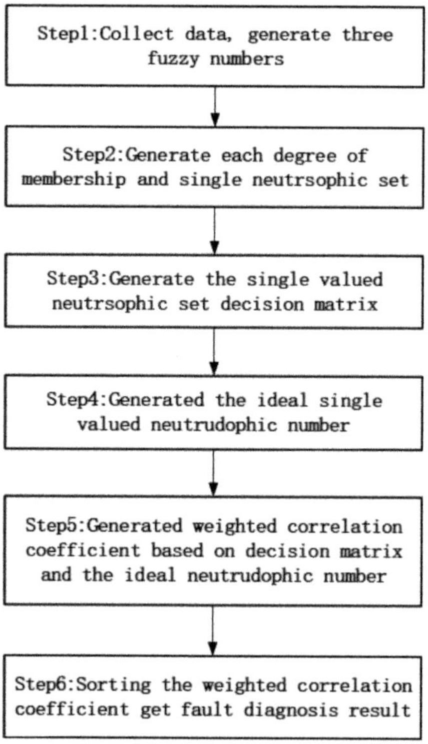

Figure 2. The detailed flow chart of the fault diagnosis method.

Step 1: For fault template set $A = \{A_1, A_2, \cdots, A_m\}$, and test sample set $C = \{C_1, C_2, \cdots, C_n\}$.

Firstly, three fuzzy numbers of fault template data and test sample data are generated, and the calculation method is as follows:

In a group number, the largest value is the right end value of the triangle; the minimum value is the left end-point of the triangle; the average value is the upper-end value of the triangle; the height of the triangle is 1, as shown in Figure 1.

Step 2: By comparing the three fuzzy number of each attribute of the test sample and the three fuzzy number of the same attribute of the fault template, the degree of determinacy-membership $T_{A_i}(C_j)$, the degree of non-membership $F_{A_i}(C_j)$, and the degree of indeterminacy-membership $I_{A_i}(C_j)$ are obtained, as shown in Figure 3, and the calculation method is as follows:

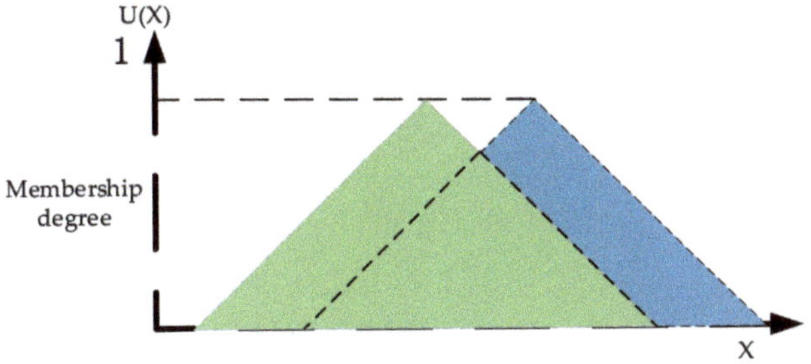

Figure 3. Each degree of membership and neutrsophic set generate a schematic diagram.

$T_{A_i}(C_j)$: The ratio of the overlapping area of the analyzed sample triangular fuzzy number and the triangular fuzzy number of the fault template under a certain attribute, and the analyzed sample total area of the triangular fuzzy number;

$F_{A_i}(C_j)$: The ratio of the area of the part not overlapped with the fault template in the analyzed sample triangular fuzzy number to the total area of the analyzed triangular fuzzy number. Moreover, for each attribute, the sum of the non-membership degree and the determined membership degree is 1;

$I_{A_i}(C_j)$: The calculation of the indeterminacy-membership degree is as follows:

$$(R_1 + R_2)/2 \tag{8}$$

Among them, $R_1 = 1 - S_1/S_k$, $R_2 = S_k/\max\{S_A, S_B, S_C\}$, S_1 is the overlapping area of the analyzed sample triangular fuzzy number and the triangular fuzzy number of the fault template under the current attributes, $\{S_k | k = A, B, C\}$ is the area of the k fault triangular fuzzy number of the under the current attribute.

Finally, each degree of membership, the relationship of $A_i(i = 1, 2, \cdots m)$ and $C_j(j = 1, 2, \cdots n)$ is shown below:

$$A_i = \left\{ < T_{A_i}(C_j), F_{A_i}(C_j), I_{A_i}(C_j) > \middle| C_j \in C, j = 1, 2, \cdots, n \right\} \tag{9}$$

Among them $T_{A_i}(C_j), F_{A_i}(C_j), I_{A_i}(C_j) \in [0, 1]$ respectively denote the degree of determinacy-membership, the degree of non-membership, and the degree of indeterminacy-membership between $A_i(i = 1, 2, \cdots m)$ and $C_j(j = 1, 2, \cdots n)$, also $0 \leq T_{A_i}(C_j), F_{A_i}(C_j), I_{A_i}(C_j) \leq 3$.

Step 3: $T_{A_i}(C_j), F_{A_i}(C_j), I_{A_i}(C_j)$ can be expressed as single-valued neutrosophic set $a_{ij} =< t_{ij}, f_{ij}, i_{ij} >$, at this point, a single-valued neutrsophic set decision matrix can be generated as follows:

$$\begin{aligned} D &= (a_{ij})_{m \times n} \\ &= \begin{bmatrix} < t_{11}, f_{11}, i_{11} > & < t_{12}, f_{12}, i_{12} > & \cdots & < t_{in}, f_{ij}, i_{ij} > \\ < t_{21}, f_{21}, i_{21} > & < t_{22}, f_{22}, i_{22} > & \cdots & < t_{2n}, f_{2n}, i_{2n} > \\ \vdots & \vdots & & \vdots \\ < t_{m1}, f_{m1}, i_{m1} > & < t_{m2}, f_{m2}, i_{m2} > & \cdots & < t_{mn}, f_{mn}, i_{mn} > \end{bmatrix} \end{aligned} \tag{10}$$

Step 4: After obtaining the single-valued neutrosophic set decision matrix D, the ideal single-valued neutrudophic number for attribute $j(j = 1, 2, \cdots, n)$ can be generated by column as follows:

$$a^*_j = < t^*_j, f^*_j, i^*_j > = < \max_i(t_{ij}), \min_i(f_{ij}), \min_i(i_{ij}) > \tag{11}$$

Among them, $\max_i(t_{ij}), \min_i(f_{ij}), \min_i(i_{ij})$ respectively denote the maximum value of the jth column in t_{ij}, the minimum value of the jth column in f_{ij}, the minimum value of the jth column in i_{ij}.

Step 5: According to Formula (7), generated weighted correlation coefficient based on single-valued neutrsophic set decision matrix D and the ideal single-valued neutrudophic number a^*, the calculation formula is as follows:

$$W(A_i, B) = \frac{2 \cdot \sum_{j=1}^{n} w_j [t_{ij} \cdot t^*_j + f_{ij} \cdot f^*_j + i_{ij} \cdot i^*_j]}{\sum_{j=1}^{n} w_j [t_{ij}^2 + f_{ij}^2 + i_{ij}^2] + \sum_{j=1}^{n} w_j [t^{*2}_j + f^{*2}_j + i^{*2}_j]} \tag{12}$$

Among them, $w_j(j = 1, 2, \cdots, n)$ represents the weight of the j attribute, and $\sum_{j=1}^{n} w_j = 1$; $a_{ij} = < t_{ij}, f_{ij}, i_{ij} >$ denote single-valued neutrosophic set for attribute j from decision matrix D, $a^*_j = < t^*_j, f^*_j, i^*_j >$ denote the ideal single-valued neutrudophic number for attribute j.

Step 6: Finally, sorting the $W(A_i, B)$ of each analyzed sample, the largest value indicates that the template data belongs to this kind of fault.

4. Illustrative Example and Discussion

In this section, in order to demonstrate the validity and accuracy rate of the proposed method, an example of a motor rotor is used. The data in this paper is originated from ref. [11], and the data analysis software is the LABVIEW environment12.

4.1. Fault Diagnosis

The specific steps for fault diagnosis using the ISVNS method proposed in this paper are as follows:

(i). According to the fault template data, the triangular fuzzy numbers under various attributes are obtained, in turn, as shown in Table 1: According to the analyzed sample data, the triangular fuzzy numbers under various attributes are obtained, in turn, as shown in Table 2: For the analyzed sample Xk ($k = 1, 2, 3, 4$ represents the k attribute), Xk and G_{k1-k5} (where $G = X, Y, Z$ represent A, B, C three kinds of faults) are used for matching, respectively. The neutrosophic numbers (T, F, I) statistics generated by the determined-membership degree T, non-membership degree F, and indeterminacy-membership degree I, are calculated, as shown in Table 4:

(ii). Next, for the same fault template, neutrosophic sets with different attributes under fuzzy sample X, we can get the single-valued neutrosophic decision matrix, as shown in Table 3:

(iii). According to the single-valued neutrosophic set decision matrix and Formula (11) under sample X in Table 3, the ideal neutrosophic set B_X can be obtained as follows:

$$\begin{aligned} B_X = \ & [< 0.9612, 0.0388, 0.6747 >, < 0.7540, 0.2460, 0.5972 >, \\ & < 0.9836, 0.0164, 0.6451 >, < 0.9966, 0.0034, 0.5757 >] \end{aligned} \tag{13}$$

(iv). The weights of attributes $j(j = 1, 2, 3, 4)$ are all the same, that is the weight matrix w is as follows:

$$w = [0.25, 0.25, 0.25, 0.25] \tag{14}$$

Next, according to Table 3, Formula (7), (13), (14), for the fault template type A_i (A_1= X11 − X45,A_2= Y11 − Y45,A_3= Z11 − Z45) and the ideal single-valued neutrosophic set B_X, calculate the improved weight correlation coefficient as follows:

$$\begin{cases} W[A_1, B_X] = 0.8126 \\ W[A_2, B_X] = 0.4133 \\ W[A_3, B_X] = 0.5398 \end{cases} \quad (15)$$

(v). Finally, according to Formula (15), $A_1 > A_3 > A_2$, it can be seen that the analyzed samples X1–X4 belong to the first type of fault, namely, the X fault.

Table 1. Triangle fuzzy number of fault template.

		Min Value	Average Value	Max Value	Area
X	X11-X15	0.0661	0.1614605	0.2006	0.06725
	X21-X25	0.121	0.149226	0.3468	0.1129
	X31-X35	0.0899	0.1123885	0.1296	0.01985
	X41-X45	0.357	4.3256515	4.666	2.1545
Y	Y11-Y15	0.1567	0.181797	0.2038	0.02355
	Y21-Y25	0.3071	0.329311	0.351	0.02195
	Y31-Y35	0.1865	0.242014	0.3218	0.06765
	Y41-Y45	4.094	4.715255	8.896	2.401
Z	Z11-Z15	0.3006	0.3294004	0.3476	0.0235
	Z21-Z25	0.2801	0.343854	0.3647	0.0423
	Z31-Z35	0.1151	0.136169	0.1864	0.03565
	Z41-Z45	9.385	9.810633	10.112	0.3635

Table 2. Triangular fuzzy numbers data of the analyzed sample.

		Min Value	Average Value	Max Value	Area
X	X1	0.1416	0.14265	0.144	0.0012
	X2	0.1028	0.11092	0.3058	0.1015
	X3	0.1279	0.133655	0.1378	0.00495
	X4	4.06	4.0938	4.18	0.06

Table 3. Single-valued neutrosophic set decision matrix under sample X.

Diagnosis Fault	X1	X2	X3	X4
X11-X45	(0.9612,0.0388,0.9914)	(0.7540,0.2460,0.6610)	(0.0127,0.9873,0.6451)	(0.9966,0.0034,0.9348)
Y11-Y45	(0,1,0.6751)	(0,1,0.5972)	(0,1,1)	(0.0871,0.9129,0.9989)
Z11-Z45	(0,1,0.6747)	(0.0126,0.9874,0.6722)	(0.9836,0.0164,0.6952)	(0,1,0.5757)

Table 4. The calculation result of the membership degree of the analyzed sample X.

Analyzed Sample	Fault Template	Neutrosophic Number
X1	X11-X15	(0.9612,0.0388,0.9914)
	Y11-Y15	(0,1,0.6751)
	Z11-Z15	(0,1,0.6747)
X2	X21-X25	(0.7540,0.2460,0.6610)
	Y21-Y25	(0,1,0.5972)
	Z21-Z25	(0.0126,0.9874,0.6722)
X3	X31-X35	(0.0127,0.9873,0.6451)
	Y31-Y35	(0,1,1)
	Z31-Z35	(0.9836,0.0164,0.6952)
X4	X41-X45	(0.9966,0.0034,0.9348)
	Y41-Y45	(0.0871,0.9129,0.9989)
	Z41-Z45	(0,1,0.5757)

4.2. Fault Diagnosis Accuracy

To verify the accuracy of fault diagnosis, separately extract arbitrary 40 faults data from the three faults template [18–42]; These 120 faults data are used as a diagnosis template. Diagnose the fault type to which it belongs. Finally, compare each test sample with its original fault template [43,44], Calculate the overall accuracy of fault diagnosis.

The SVNPWA algorithm in ref. [44] is used for verifying these 120 unknown fault samples, and the diagnosis accuracy is 98.33%. Moveover, the ISVNS algorithm proposed in this paper also applied for diagnosing the same 120 unknown fault samples, and the diagnosis accuracy is 99.16%. The diagnosis results are shown in Table 5.

Table 5. Diagnosis results of applying the SVNPWA and proposed algorithm.

Unknow Fault	SVNPWA		The Proposed Algorithm	
	Times of Right	Times of Error	Times of Right	Times of Error
X	38	2	40	0
Y	40	0	39	1
Z	40	0	40	0

It can be seen from Table 5: Compared with the SVNPWA algorithm, the fault diagnostic accuracy rate of the ISVNS algorithm, proposed in this paper, is improved by 0.83%. That is, the ISVNS algorithm can better satisfy the diagnostic needs than the basic SVNPWA algorithm.

5. Conclusions

This paper proposes an ISVNS algorithm, which introduces the improved weighted correlation coefficient formula, and more comprehensively considers both single-valued neutrosophic set and ideal neutrosophic set under various faults, effectively solved the problem of fault diagnosis. An example of a motor rotor is illustrated that the ISVNS algorithm could improve the diagnostic accuracy rate compared with the SVNPWA algorithm. In conclusion, the ISVNS algorithm can obtain better fault diagnosis accuracy, and satisfy the fault diagnosis needs in practice.

Since the collection and aggregation of data on technical faults is a laborious process. Moreover, the parameters in the diagnosis process are automatically generated in the laboratory simulation, without human intervention—therefore, the proposed method could be automating in practical application.

Therefore, the next step of the work will focus on how to automating the proposed method in order to scale the use of the proposed algorithm.

Author Contributions: Conceptualization, W.J. and S.I.; methodology, W.J., S.I. and Y.Z.; validation, S.I.; formal analysis, W.J., S.I., Y.Z., X.D. and J.G.; writing—original draft preparation, S.I.; writing—review and editing, W.J., S.I., Y.Z., X.D. and J.G.; funding acquisition, W.J. All authors have read and agreed to the published version of the manuscript.

Funding: The work is partially supported by National Natural Science Foundation of China (Program No. 61703338), Equipment Pre-Research Fund (Program No. 61400010109), National Science and Technology Major Project (2017-I-0001-0001).

Conflicts of Interest: The authors declare no conflict of interest.

References

1. Wang, H.; Deng, X.; Zhang, Z.; Jiang, W. A new failure mode and effects analysis method based on dempster–shafer theory by integrating evidential network. *IEEE Access* **2019**, *7*, 79579–79591.
2. Ma, Y.; Wang, J.; Wang, J.; Wu, X. An interval neutrosophic linguistic multi-criteria group decision-making method and its application in selecting medical treatment options. *Neural Comput. Appl.* **2017**, *28*, 2745–2765.
3. Wei, G.; Zhang, Z. Some single-valued neutrosophic bonferroni power aggregation operators in multiple attribute decision making. *J. Ambient. Intell. Humaniz. Comput.* **2018**, *10*, 1–20.
4. Zadeh, L.A. Fuzzy sets. *Inf. Control.* **1965**, *8*, 338–353.
5. Zhang, Y.; Jiang, W.; Deng, X. Fault diagnosis method based on time domain weighted data aggregation and information fusion. *Int. J. Distrib. Sens. Netw.* **2019**, *15*. [CrossRef]
6. Deng, X.; Jiang, W. A total uncertainty measure for d numbers based on belief intervals. *Int. J. Intell. Syst.* **2019**. [CrossRef]
7. Han, C.; Shih, R.; Lee, L. Quantifying signed directed graphs with the fuzzy set for fault diagnosis resolution improvement. *Ind. Eng. Chem. Res.* **2010**, *33*, 1943–1954.
8. Li, Y.; Shu, N. Transformer fault diagnosis based on fuzzy clustering and complete binary tree support vector machine. *Trans. China Electrotech. Soc.* **2016**, *31*, 64–70.
9. Jiang, W.; Zhong, Y.; Deng, X. A Neutrosophic Set Based Fault Diagnosis Method Based on Multi-Stage Fault Template Data. *Symmetry* **2018**, *10*, 346.
10. Fu, C.; Chang, W.; Liu, W.; Yang, S. Data-driven group decision making for diagnosis of thyroid nodule. *Sci. China Inf. Sci.* **2019**, *62*, 212205.
11. Xu, X.; Wen, C. *Theory and Application of Multi-Source and Uncertain Information Fusion*; Science Press: Beijing, China, 2012; pp. 98–108.
12. Geng, J.; Ma, X.; Zhou, X.; Wang, S.; Yang, S.; Jiao, L. Saliency-Guided Deep Neural Networks for SAR Image Change Detection. IEEE Transactions on Geoence and Remote Sensing. *Remote. Sens. Lett.* **2019**, *99*, 1–13. [CrossRef]
13. Sun, C.; Li, S.; Den, Y. Determining Weights in Multi-Criteria Decision Making Based on Negation of Probability Distribution under Uncertain Environment. *Mathematics* **2020**, *8*, 191. [CrossRef]
14. Kandasamy, W.B.V.; Smarandache, I.; Neutrosophic, F. Components Semigroups and Multiset Neutrosophic Components Semigroups. *Symmetry* **2020**, *12*, 818.
15. Yang, W.; Cai, L.; Edalatpanah, S.A.; Smarandache, F. Triangular Single Valued Neutrosophic Data Envelopment Analysis: Application to Hospital Performance Measurement. *Symmetry* **2020**, *12*, 588.
16. Zhou, Q.; Mo, H.; Deng, Y. A New Divergence Measure of Pythagorean Fuzzy Sets Based on Belief Function and Its Application in Medical Diagnosis. *Mathematics* **2020**, *8*, 142. [CrossRef]
17. Zhou, X.; Li, P.; Smarandache, F.; Khalil, A.M. New Results on Neutrosophic Extended Triplet Groups Equipped with a Partial Order. *Symmetry* **2019**, *11*, 1514. [CrossRef]
18. Saber, Y.; Alsharari, F.; Smarandache, F. On Single-Valued Neutrosophic Ideals in Šostak Sense. *Symmetry* **2020**, *12*, 193. [CrossRef]
19. Liu, B.; Deng, Y. Risk Evaluation in Failure Mode and Effects Analysis Based on D Numbers Theory. *Int. J. Comput. Commun. Control.* **2019**, *14*, 672–691.
20. Caliskan, F.; Zhang, Y.; Wu, N.E.; Shin, J.-Y. Actuator Fault Diagnosis in a Boeing 747 Model via Adaptive Modified Two-Stage Kalman Filter. *Int. J. Aerosp. Eng.* **2014**, *2014*, 10. [CrossRef]

21. Deng, X.; Jiang, W. Evaluating Green Supply Chain Management Practices under Fuzzy Environment: A Novel Method Based on D Number Theory. *Int. J. Fuzzy Syst.* **2019**, *21*, 1389–1402. [CrossRef]
22. Deng, X.; Jiang, W. D number theory based game-theoretic framework in adversarial decision making under a fuzzy environment. *Int. J. Approx. Reason.* **2019**, *106*, 194–213. [CrossRef]
23. Deng, X.; Jiang, W.; Wang, Z. Zero-sum polymatrix games with link uncertainty: A Dempster-Shafer theory solution. *Appl. Math. Comput.* **2019**, *340*, 101–112.
24. Zieja, M.; Golda, P.; Zokowski, M.; Majewski, P. Vibroacoustic technique for the fault diagnosis in a gear transmission of a military helicopter. *J. Vibroengineering* **2017**, *19*, 1039–1049. [CrossRef]
25. Zhang, C.; Li, D.; Broumi, S.; Kumar, A. Medical Diagnosis Based on Single-Valued Neutrosophic Probabilistic Rough Multisets over Two Universes. *Symmetry* **2018**, *10*, 213. [CrossRef]
26. Jiang, W.; Cao, Y.; Deng, X. A Novel Z-network Model Based on Bayesian Network and Z-number. *IEEE Trans. Fuzzy Syst.* **2019**. [CrossRef]
27. Jiang, W. A correlation coefficient for belief functions. *Int. J. Approx. Reason.* **2018**, *103*, 94–106. [CrossRef]
28. Oliveira, C.C.; da Silva, J.M. Fault Diagnosis in Highly Dependable Medical Wearable Systems. *J. Electron. Test.* **2016**, *32*, 467–479. [CrossRef]
29. Strydom, J.J.; Miskin, J.J.; Mccoy, J.T.; Auret, L.; Dorfling, C. Fault diagnosis and economic performance evaluation for a simulated base metal leaching operation. *Miner. Eng.* **2018**. [CrossRef]
30. Huang, Z.; Yang, L.; Jiang, W. Uncertainty measurement with belief entropy on the interference effect in the quantum-like Bayesian Networks. *Appl. Math. Comput.* **2019**, *347*, 417–428.
31. He, Z.; Jiang, W. An evidential Markov decision making model. *Inf. Sci.* **2018**, *467*, 357–372. [CrossRef]
32. Gong, X.; Qiao, W. Bearing Fault Diagnosis for Direct-Drive Wind Turbines via Current-Demodulated Signals. *Ind. Electron. IEEE Trans.* **2013**, *60*, 3419–3428.
33. Ji, W.; Li, B. Fault Diagnosis Method of Exhaust System of Port Vehicle. *J. Coast. Res.* **2018**, *83*, 469–473.
34. Askarian, M.; Zarghami, R.; Jalali-Farahani, F.; Navid, M. Fault Diagnosis of Chemical Processes Considering Fault Frequency via Bayesian Network. *Can. J. Chem. Eng.* **2016**, *94*, 2315–2325. [CrossRef]
35. Daroogheh, N.; Meskin, N.; Khorasani, K. Particle Filter-Based Fault Diagnosis of Nonlinear Systems Using a Dual Particle Filter Scheme. *IEEE Trans. Control Syst. Technol.* **2016**. [CrossRef]
36. Tra, V.; Kim, J.; Khan, S.A.; KIM, J.-M. Incipient fault diagnosis in bearings under variable speed conditions using multiresolution analysis and a weighted committee machine. *J. Acoust. Soc. Am.* **2017**, *142*, EL35. [CrossRef]
37. Lu, W.; Liang, B.; Cheng, Y.; Meng, D.; Yang, J.; Zhang, T. Deep Model Based Domain Adaptation for Fault Diagnosis. *IEEE Trans. Ind. Electron.* **2016**, *64*, 2296–2305. [CrossRef]
38. Smarandache, F. *Neutrosophy: Neutrosophic Probability, Set, and Logic: Analytic Synthesis and Synthetic Analysis*; American Research Press: Rehoboth, MI, USA, 1998.
39. Wang, H.; Smarandache, F.; Zhang, Y.; Sunderraman, R. Single valued neutrosophic sets. In Proceedings of the 8th Joint Conference on Information Sciences. Joint Conference Information Science, Salt Lake City, UT, USA, 21–26 July 2005; pp. 94–97.
40. Ye, J. A multicriteria decision-making method using aggregation operators for simplified neutrosophic sets. *J. Intell. Fuzzy Syst.* **2014**, *26*, 2459–2466. [CrossRef]
41. Ye, J. Another Form of Correlation Coefficient between Single Valued Neutrosophic Sets and Its Multiple Attribute Decision-Making Method. *Neutrosophic Sets Syst.* **2013**. [CrossRef]
42. Zhang, H.; Deng, Y. Weighted belief function of sensor data fusion in engine fault diagnosis. *Soft Comput.* **2020**, *24*, 2329–2339. [CrossRef]
43. Xiao, F.; Ding, W. Divergence measure of Pythagorean fuzzy sets and its application in medical diagnosis. *Appl. Soft Comput.* **2019**, *79*, 254–267.
44. Xiao, F. Multi-sensor data fusion based on the belief divergence measure of evidences and the belief entropy. *Inf. Fusion* **2019**, *46*, 23–32. [CrossRef]
45. Gou, L.; Zhong, Y. A New Fault Diagnosis Method Based on Attributes Weighted Neutrosophic Set. *IEEE Access* **2019**, *7*, 117740–117748. [CrossRef]
46. Jiang, W.; Huang, C.; Deng, X. A new probability transformation method based on a correlation coefficient of belief functions. *Int. J. Intell. Syst.* **2019**, *34*, 1337–1347. [CrossRef]

47. Broumi, S.; Nagarajan, D.; Bakali, A.; Talea, M.; Smarandache, F.; Lathamaheswari, M. The shortest path problem in interval valued trapezoidal and triangular neutrosophic environment. *Complex Intell. Syst.* **2019**, *5*, 391–402. [CrossRef]
48. Broumi, S.; Nagarajan, D.; Bakali, A.; Talea, M.; Smarandache, F.; Lathamaheswari, M.; Kavikumar, J. Implementation of neutrosophic function memberships using matlab program. *Neutrosophic Sets Syst.* **2019**, *27*, 44–52.
49. Chakraborty, A. A new score function of pentagonal neutrosophic number and its application in networking problem. *Int. J. Neutrosophic Sci.* **2020**, *1*, 40–51.

symmetry

MDPI

Article

A New Multi-Sensor Fusion Target Recognition Method Based on Complementarity Analysis and Neutrosophic Set

Yuming Gong [1], Zeyu Ma [1], Meijuan Wang [1], Xinyang Deng [1,*] and Wen Jiang [1,2,*]

[1] School of Electronics and Information, Northwestern Polytechnical University, Xi'an 710072, China; gongyuming@mail.nwpu.edu.cn (Y.G.); mazeyu@mail.nwpu.edu.cn (Z.M.); wangmeijuan@mail.nwpu.edu.cn (M.W.)

[2] Peng Cheng Laboratory, Shenzhen 518055, China

* Correspondence: xinyang.deng@nwpu.edu.cn (X.D.); jiangwen@nwpu.edu.cn (W.J.)

Received: 1 August 2020; Accepted: 27 August 2020; Published: 31 August 2020

check for updates

Abstract: To improve the efficiency, accuracy, and intelligence of target detection and recognition, multi-sensor information fusion technology has broad application prospects in many aspects. Compared with single sensor, multi-sensor data contains more target information and effective fusion of multi-source information can improve the accuracy of target recognition. However, the recognition capabilities of different sensors are different during target recognition, and the complementarity between sensors needs to be analyzed during information fusion. This paper proposes a multi-sensor fusion recognition method based on complementarity analysis and neutrosophic set. The proposed method mainly has two parts: complementarity analysis and data fusion. Complementarity analysis applies the trained multi-sensor to extract the features of the verification set into the sensor, and obtain the recognition result of the verification set. Based on recognition result, the multi-sensor complementarity vector is obtained. Then the sensor output the recognition probability and the complementarity vector are used to generate multiple neutrosophic sets. Next, the generated neutrosophic sets are merged within the group through the simplified neutrosophic weighted average (SNWA) operator. Finally, the neutrosophic set is converted into crisp number, and the maximum value is the recognition result. The practicality and effectiveness of the proposed method in this paper are demonstrated through examples.

Keywords: neutrosophic set; target recognition; complementarity analysis; data fusion

1. Introduction

In daily life, target recognition involves all aspects of our lives, such as intelligent video surveillance and face recognition. These applications also make target recognition technology more popular. The development of related technologies has greatly enriched the application scenarios of target recognition and tracking theories. Research on related theoretical methods has also received extensively attention. Target recognition involves image processing, calculation computer vision, pattern recognition and other subjects.

Generalized target recognition includes two stages, feature extraction, and classifier classification. Through feature extraction, image, video, and other target observation data are preprocessed to extract feature information, and then the classifier algorithm implements target classification based on the feature information [1]. Common image features can be divided into color gray statistical feature, texture edge feature, algebraic feature, and variation coefficient feature. The feature extraction methods corresponding to the above features are color histogram, gray-level co-occurrence matrix method, principal component analysis method, wavelet transform [2]. The classic target classification algorithms include decision tree, support vector machine [3], neural network [4], logistic regression, and naive Bayes classification [5–8]. On the basis of a single classifier, the ensemble classifier integrates the classification results of a series of weak classifiers through an ensemble learning method, so as to obtain a better classification effect than a single classifier, mainly including Bagging and Boosting [9]. With the development of high-performance computing equipment and the enlargement in the amount of available data, deep learning related theories and methods have developed rapidly, and the application of deep learning in the direction of target recognition has also made it break through the limitations of traditional methods based on deep neural networks [10–12]. Classification network models include LeNet, AlexNet, VggNet.

In practical applications, multi-source sensor data is often processed in decision analysis [13–19] as the number of sensors increases. For the problem of uncertain information processing, there are many theoretical methods such as Dempster-Shafer evidence theory [20–26], fuzzy set theory [27,28], D number [29–32] and rough set theory [33].

Smarandache [34] firstly generalized the concepts of fuzzy sets [35], intuitionistic fuzzy sets (IFS) [36] and interval-valued intuitionistic fuzzy sets (IVIFS) [37], and proposed the neutrosophic set. It is very suitable to use the neutrosophic set to deal with uncertain and inconsistent information in the real world. However, its authenticity, uncertainty, and false membership function are defined in the real number standard or non-standard subset. Therefore, non-standard intervals are not suitable for scientific and engineering applications. Therefore, Ye [38] introduced a simplified neutrosophic set (SNS), which limits the true value, uncertainty, and false membership function to the actual standard interval $[0,1]$. In addition, SNS also includes single value neutrosophic set (SVNS) [39–41] and interval neutrosophic set (INS) [42].

As a new kind of fuzzy set, neutrosophic set [43,44] have been used in many fields, such as decision-making [45–48], data analysis [49,50], fault diagnosis [51], the shortest path problem [52]. There is also a lot of progress in the related theoretical research of neutrosophic set. For example, score function of pentagonal neutrosophic set [53,54].

Existing multi-sensor fusion methods, such as evidence theory, have complex calculation and long calculation time [55], and there are a few methods that use neutrosophic set in multi-sensor fusion. Therefore, this paper proposes a multi-sensor fusion based on neutrosophic set. First, the complementarity vectors between multiple sensors are calculated. Then these complementarity vectors and the probability of sensor output are used to form a group of neutrosophic sets, and generated neutrosophic sets are fused through the SNWA operator. Finally, the neutrosophic set is converted to the crisp number, and the maximum value is the recognition result. The proposed method has simple calculation and fast operation, and can effectively improve the accuracy of target recognition.

The rest of this article is organized as follows: Section 2 introduces some necessary concepts, such as neutrosophic set and multi-category evaluation standard. The proposed multi-sensor fusion recognition method is listed step by step in Section 3. In Section 4, an example is used to illustrate and explain the effective of proposed method. Some results discussion are shown in Section 5.

2. Preliminaries

2.1. Neutrosophic Set

Definition 1. *The the simplified neutrosophic set (SNS) is defined as follows [38]:*

X is a finite set, with a element of X denoted by x. A neutrosophic set (A) in X contains three parts: a truth-membership function (T_p), an indeterminacy-membership function (I_p), and a falsity-membership function (F_p).

$$0 \leq T_P(x) \leq 1 \tag{1}$$
$$0 \leq I_P(x) \leq 1 \tag{2}$$
$$0 \leq F_P(x) \leq 1 \tag{3}$$
$$0 \leq T_P(x) + I_P(x) + F_P(x) \leq 3 \tag{4}$$

A single-valued neutrosophic set P on X is defined as:

$$P = \{\langle x, T_P(x), I_P(x), F_P(x)\rangle | x \in X\} \tag{5}$$

This is called a SNS. In particular, if X includes only one element, $N = \langle x, T_P(x), I_P(x), F_P(x)\rangle$ *is called a SNN (the simplified neutrosophic number) and is denoted by* $\alpha = \langle \mu, \pi, \nu\rangle$*. The numbers* μ, π, ν *denote the degree of membership, the degree of indeterminacy-membership, and the degree of non-membership.*

Definition 2. *The crisp number of each SNN is deneutrosophicated and calculated as follows [56]:*

$$S_i = \mu_i + (\pi_i) \times (\frac{\mu_i}{\mu_i + \nu_i}) \tag{6}$$

S_i *can be regarded as the score of SNN, so SNN can be sorted according to crisp number* S_i*.*

2.2. Commonly Used Evaluation Indicators for Multi-Classification Problems

The index for evaluating the performance of a classifier is generally the accuracy of the classifier, which is defined as the ratio of the number of samples correctly classified by the classifier to the total number of samples for a given test data set [57].

The commonly used evaluation indicators for classification problems are precision and recall. Usually, the category of interest is regarded as the positive category, and the other categories are regarded as the negative category. The prediction of the classifier on the test data set is correct or incorrect. There are four situations as follows:

- True Positive(TP): The true category is a positive example, and the predicted category is a positive example.
- False Positive (FP): The true category is negative, and the predicted category is positive.
- False Negative (FN): The true category is positive, and the predicted category is negative.
- True Negative (TN): The true category is negative, and the predicted category is negative.

Based on the above basic concepts, the commonly used evaluation indicators for multi-classification problems are as follows.

1. Precision or precision rate, also known as precision (P):

$$P = \frac{TP}{TP + FP} \tag{7}$$

2. Recall rate, also known as recall rate (R):

$$R = \frac{TP}{TP + FN} \tag{8}$$

3. F1 score is an index used to measure the accuracy of the classification model. It also takes into account the accuracy and recall of the classification model. The score can be regarded as a harmonic average of model accuracy and recall. Its maximum value is 1 and its minimum value is 0:

$$F_1 = \frac{2 \times P \times R}{P + R} \tag{9}$$

2.3. AdaBoost Algorithm

AdaBoost is essentially an iterative algorithm [58], and its core idea is to train some weak classifier h_i based on the initial sample using the decision tree algorithm. Use the classifier to detect the sample set. For each training sample point, adjust its weight according to whether the result of its classification is accurate: if h_i makes it classified correctly, reduce the weight of the sample point; otherwise, increase the sample the weight of the point. The adjusted weight is calculated according to the accuracy of the detection result. The sample set after adjusting the weight constitutes the sample set to be trained at the next level, which is used to train the next level classifier. In this way, iterate step by step to obtain a new classifier until the classifier h_m is obtained, and the sample detection error rate is 0.

Combine $h_1, h_2, \ldots, h_m$ according to the error rate of the sample detection: make the weak classifier with the larger error account for the smaller weight in the combined classifier, and the weak classifier with the smaller error account for the larger weight to obtain a combined classifier.

The algorithm is essentially a comprehensive improvement of the weak classifier trained by the basic decision tree algorithm. Through continuous training of samples and weight adjustment, multiple classifiers are obtained, and the classifiers are combined by weight to obtain a comprehensive classifier that improves the ability of data classification. The whole process is as follows:

- Train weak classifiers with sample sets.
- Calculate the error rate of the weak classifier, and obtain the correct and incorrect sample sets.
- Adjust the sample set weight according to the classification result to obtain a redistributed sample set.

After M cycles, M weak classifiers are obtained, and the joint weight of the classifier is calculated according to the detection accuracy of each weak classifier, and finally a strong classifier is obtained.

2.4. HOG Feature

Histogram of oriented gradient (HOG), which is a feature descriptor for target detection. This technology counts the number of directional gradients that appear locally in the image. This method is similar to the histogram of edge orientation and scale-invariant feature transform, but the difference is hog calculate the density matrix based on the uniform space to improve accuracy. Navneet Dalal and Bill Triggs first proposed HOG in 2005 for pedestrian detection in static images or videos [59].

The core idea of HOG is that the shape of the detected local object can be described by the light intensity gradient or the distribution of the edge direction. By dividing the entire image into small connected areas (called cells). Each cell generates a directional gradient histogram or the edge direction of the pixel in the cell, and the descriptor is represented by combining the histogram. To improve the accuracy, the local histogram can be compared and standardized by calculating the light intensity of a larger area (called block) in the image as a measure, and then using this value (measure) to normalize all cells in the block. This normalization process completes better illumination/shadow invariance. Compared with other descriptors, the descriptors obtained by HOG maintain the invariance of geometric and optical transformations (unless the object orientation changes).

2.5. Gabor Feature

Gabor feature [60] is a feature that can be used to describe the texture information of an image. The frequency and direction of the Gabor filter are similar to the human visual system, and it is particularly suitable for texture representation and discrimination. The Gabor feature mainly relies on the Gabor kernel to window the signal in the frequency domain, so as to describe the local frequency information of the signal.

In terms of feature extraction, Gabor wavelet transform is compared with other methods: on the one hand, it processes less data and can meet the real-time requirements of the system; on the other hand, wavelet transform is insensitive to changes in illumination and can tolerate a certain degree of when image rotation and deformation are used for recognition based on Euclidean distance, the feature pattern and the feature to be measured do not need to correspond strictly, so the robustness of the system can be improved.

2.6. D-AHP Theory

Analytic Hierarchy Process (AHP) [61] is a systematic and hierarchical analysis method that combines qualitative and quantitative analysis. The characteristic of this method is that on the basis of in-depth research on the nature, influencing factors and internal relations of complex decision-making problems, it uses less quantitative information to mathematicize the thinking process of decision-making, thereby providing multi-objective, multi-criteria or complex decision-making problems with no structural characteristics provide simple decision-making methods.

The D-AHP method extends the traditional AHP method in theory. In the D-AHP method [62], the derived results about the ranking and priority weights of alternatives are impacted by the credibility of providing information. A parameter λ is used to express the credibility of information, and its value is associated with the cognitive ability of experts. If the comparison information used in the decision-making process is provided by an authoritative expert, λ will take a smaller value. If the comparison information comes from an expert whose judgment is with low belief, λ takes a higher value.

3. The Proposed Method

In general sensor recognition, the training set is inputting to train the sensor by extracting feature, and then the test set is inputting to test and get the recognition result. To improve the accuracy of multi-sensor fusion recognition, this paper proposes a fusion recognition method based on neutrosophic set. The proposed method in this paper is mainly divided into two parts: complementarity analysis and data fusion. The main steps of the method proposed in this paper are shown in Figure 1.

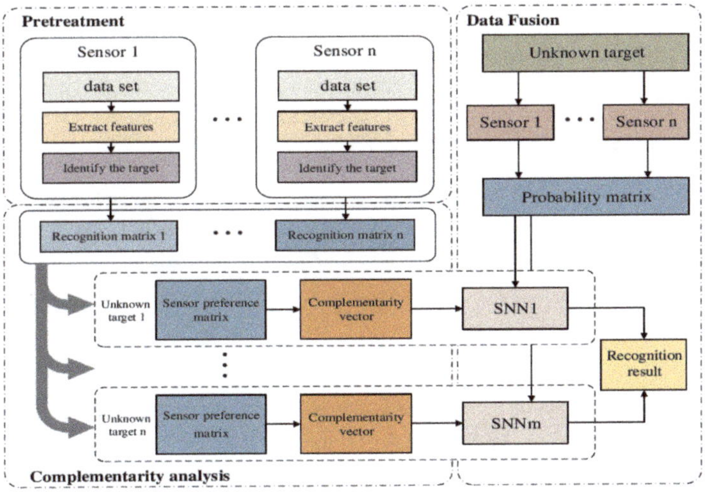

Figure 1. Target fusion recognition based on sensor complementarity and neutrosophic set.

The essence of complementarity is to calculate the weight of the recognition ability of the base sensor in different categories. Based on this, the data set is divided into training set, validation set, and test set. First, the base sensor preference matrix is obtained from the recognition matrix of the trained base sensor on the verification set, and then the sensor complementarity vector is calculated in each category.

Data fusion aims at different target types, based on the sensor complementarity vector, the recognition results of different sensors are generated by the neutrosophic set. Then the fusion of the neutrosophic set can be obtained in different categories, and finally the neutrosophic set is converted into crisp number, and the maximum value is taken as the recognition result.

3.1. *Complementarity*

The main steps of the multi-sensor complementarity analysis are proposed as follows:

- According to the data test result, the sensor recognition matrix can be obtained.
- The sensor preference matrix for different target types is obtained from multiple sensor recognition matrices.
- The sensor complementarity vector is gotten from the sensor preference relationship matrix.

3.1.1. Sensor Recognition Matrix

Suppose there are n types of sensors $X_1, X_2, X_3, \cdots, X_n$, and m types of target types $Y_1, Y_2, Y_3, \cdots, Y_m$. For a certain data set i, the recognition results of all its samples can be represented by the following recognition matrix R_i.

$$R_i = \begin{bmatrix} r^i{}_{11} & r^i{}_{12} & \cdots & r^i{}_{1m} \\ r^i{}_{21} & r^i{}_{22} & \cdots & r^i{}_{2m} \\ \vdots & \vdots & \ddots & \vdots \\ r^i{}_{n1} & r^i{}_{n2} & \cdots & r^i{}_{nm} \end{bmatrix} \tag{10}$$

3.1.2. Sensor Preference Matrix

To obtain the preference relationship matrix between sensors, the preference between sensors needs to be defined. For two sensors, if the recognition performance of the sensor X_1 on the target Y_j is better than the recognition performance of the sensor X_2, then for the target Y_j, the sensor X_1 is better than the sensor X_2. The recognition results of a certain sensor on the samples in the data set i can be organized in the form of a recognition matrix, but the rows and columns become the recognized category of the sample and the true category of the sample. Furthermore, for the target of category Y_j, according to the recognition matrix, we can get Table 1:

Table 1. R^i_j: The recognition situation of a certain sensor to the target of category Y_j.

Real-Recognition	Y_j	**Non-**Y_j
Y_j	$r^i{}_{jj}$	$\sum\limits_{k\neq j} r^i{}_{jk}$
non-Y_j	$\sum\limits_{k\neq j} r^i{}_{kj}$	$\sum\limits_{l\neq j}\sum\limits_{k\neq j} r^i{}_{kl}$

Record the above matrix as R^i_j in Table 1, where $r^i{}_{jj}$ is the number of correct recognition of the category Y_j by the sensor, $\sum\limits_{k\neq j} r^i{}_{kj}$ is the number of samples that the sensor misrecognizes non-targets. $\sum\limits_{k\neq j} r^i{}_{jk}$ is the number of misrecognized category Y_j samples into other categories. $\sum\limits_{l\neq j}\sum\limits_{k\neq j} r^i{}_{kl}$ is the number of samples other than the above three cases number. If the optimal performance of sensor recognition is expressed as a matrix:

$$I_j = \begin{bmatrix} \sum_{l=1}^{h} r^i_{jl} & 0 \\ 0 & \sum_{k \neq j} \sum_{l=1}^{h} r^i_{kl} \end{bmatrix} \tag{11}$$

Which means the category is fully recognized correctly. Then the recognition performance of this sensor to the category Y_j is defined as $\frac{1}{\|R^i_j - I_j\|}$. Therefore, if there are two sensors X_k and X_l, the preference value of the recognition accuracy rate of X_k versus X_l in the category (representing the priority of the recognition ability of X_k over X_l in the category of Y_j) is defined as:

$$p^j{}_{kl} = \frac{\frac{1}{\|R^k{}_j - I_j\|}}{\frac{1}{\|R^k{}_j - I_j\|} + \frac{1}{\|R^l{}_j - I_j\|}} = \frac{\|R^l{}_j - I_j\|}{\|R^k{}_j - I_j\| + \|R^l{}_j - I_j\|} \tag{12}$$

In the same way, the recognition accuracy preference value of sensor X_l vs X_k on category Y_j is defined as follows:

$$p^j{}_{lk} = \frac{\|R^k{}_j - I_j\|}{\|R^k{}_j - I_j\| + \|R^l{}_j - I_j\|} \tag{13}$$

It is easy to get from the above two formulas that $p^j{}_{kl} + p^j{}_{lk} = 1$, which is the sum of the two preference values is 1. If $l = k$, then $p^j{}_{kl} = 0.5$. For the case where multiple sensors recognize at the same time, the preference relationship matrix P^j on the category Y_j can be obtained. For each target category in the data set i, a corresponding preference relationship matrix can be obtained.

3.1.3. Sensor Complementarity Vector

Next, using the method in D-AHP theory [62], the complementarity vector can be calculated by the preference relationship matrix.

According to the classifier preference relation matrix P_j of category Y_j.

$$P^j = \begin{bmatrix} P^j{}_{11} & P^j{}_{12} & \cdots & P^j{}_{1n} \\ P^j{}_{21} & P^j{}_{22} & \cdots & P^j{}_{2n} \\ \vdots & \vdots & \ddots & \vdots \\ P^j{}_{n1} & P^j{}_{n2} & \cdots & P^j{}_{nn} \end{bmatrix} \tag{14}$$

It is calculated that for each sensor complementarity vector C^j of category Y_j, C^j is a $1 \times n$ dimensional vector. The flowchart of the C^j calculation is presented as Figure 2, the calculation steps are as follows:

1. Express the importance of the index relative to the evaluation target through the preference relationship, and construct the D number preference matrix R_D.
2. According to the integrated representation of the D number, transform the D number preference matrix into a certain number matrix R_I.
3. Construct a probability matrix R_I based on the deterministic number matrix R_P, and calculate the preference probability between the indicators compared in pairs.
4. Convert the probability matrix R_P into a triangularized probability matrix $R^T{}_P$, and sort the indicators according to their importance.
5. According to the index sorting result, the deterministic number matrix R_I is expressed as a matrix $R^T{}_I$, finally C^j is obtained.

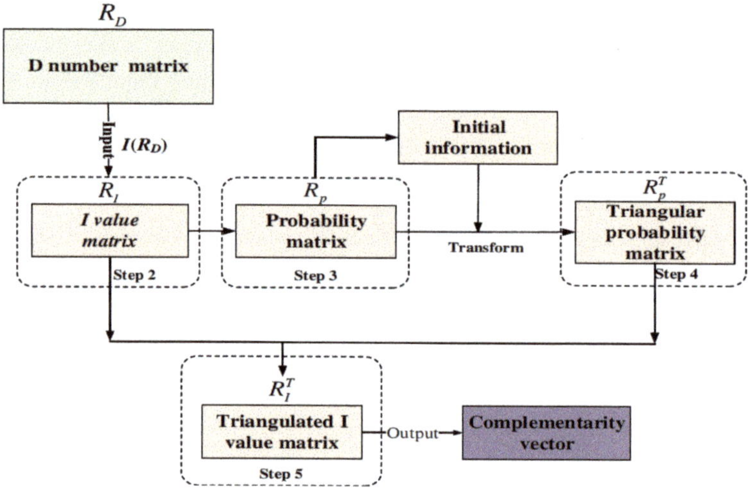

Figure 2. Complementary vector generation process [62].

3.2. Data Fusion

For a picture with unknown target type, the probability matrix is formed via the recognition result vectors of each sensor.

$$Q = \begin{bmatrix} q_{11} & \cdots & q_{1m} \\ \vdots & \ddots & \vdots \\ q_{n1} & \cdots & q_{nm} \end{bmatrix} \tag{15}$$

where q_{ij} is the probability that the sensor X_i considers the unknown target as the target type Y_j.

At this time, the complementary vector is normalized to obtain the weight coefficient:

$$H^j = C^j / (C^j{}_{min} + C^j{}_{max}) \tag{16}$$

If the target type is Y_j, the complementarity vector H^j and Q can be combined to obtain a neutrosophic set $\alpha = \langle \mu, \pi, \nu \rangle$. Since there are n sensors, n groups neutrosophic sets can be obtained according to Q and Y_j as follows:

$$\alpha_i = [\mu = q_{ij} \times H_j, \pi = (1 - q_{ij}) \times H_j, \nu = 1 - H_j] \tag{17}$$

Combining n groups of neutrosophic sets, a group of fused neutrosophic set can be obtianed, and the target recognition neutrosophic set is calculated under the imaginary target type Y_j. Since there are m types of target, we can finally use the SNWA operator [63] to get m neutrosophic sets:

$$\alpha_{Ti} = W_1 \times \alpha_1 + W_2 \times \alpha_2 + \cdots + W_n \times \alpha_n = [1 - \prod_{i=1}^{n} (1-\mu_i)^{W_i}, \prod_{i=1}^{n} (\pi_i)^{W_i}, \prod_{i=1}^{n} (\nu_i)^{W_i}] \tag{18}$$

Finally, convert these m SNNs into crisp numbers, and take the maximum value as the recognition result.

$$RS = Max(S) = Max[S_1 = \mu_{T_1} + (\pi_{T_1}) \times (\frac{\mu_{T_1}}{\mu_{T_1} + \nu_{T_1}}), \ldots, S_m = \mu_{T_m} + (\pi_{T_m}) \times (\frac{\mu_{T_m}}{\mu_{T_m} + \nu_{T_m}})] \tag{19}$$

4. Simulation

4.1. Data Set

The data source type of the experiment consists of two types: visible light image and infrared light image. There are four target types: sailboat (1), cargo ship (2), speed boat (3), and fishing boat (4). The structure of the experimental data is shown in Tables 2 and 3. The data consists of a train set, a verification set, and a test set. The verification set and the test set are the same pictures. Since the information of visible light and infrared sensors needs to be fused in the verification set and test set, two visible and infrared images taken at the same location are required for the same target, as shown in Figures 3–6. Due to the lack of data, K = 8 cross-validation is used for verificating and testing. This means that the validation set (test set) is divided into eight groups, and when a certain one group is tested, the remaining seven groups are used as the validation set. The image features use Gabor and HOG, and the classifier uses AdaBoost.

Table 2. Target recognition image of infrared data.

Infrared Light Data	Train Set	Validation Set	Test Set
Sailboat	65	28	28
Cargo ship	68	35	35
Speed boat	63	25	25
Fishing boat	79	35	35

Table 3. Target recognition image of visible light data.

Visible Light Data	Train Set	Validation Set	Test Set
Sailboat	65	28	28
Cargo ship	79	35	35
Speed boat	70	25	25
Fishing boat	78	35	35

Figure 3. Visible light image and infrared light image for the same sailboat.

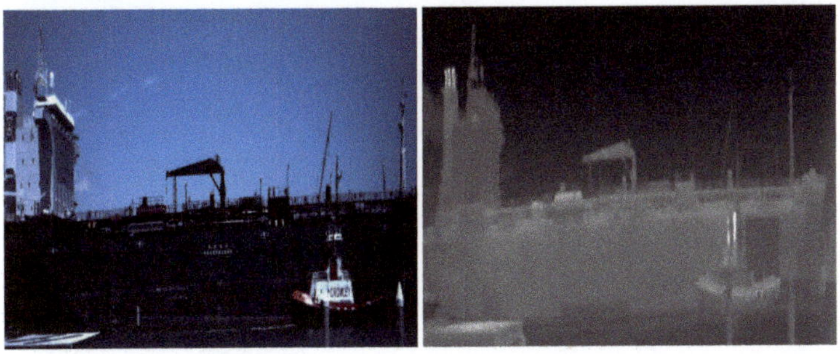

Figure 4. Visible light image and infrared light image for the same cargo ship.

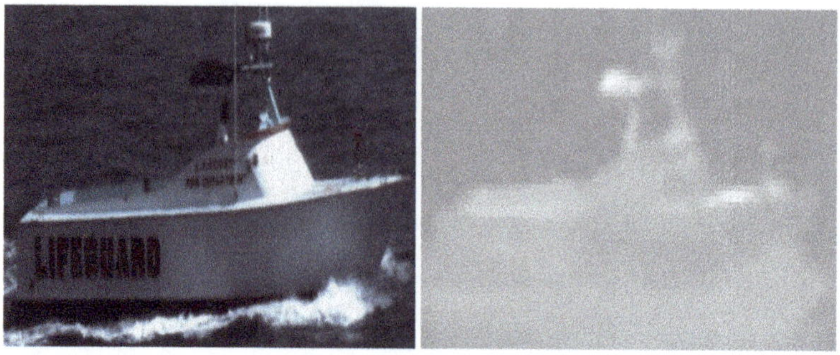

Figure 5. Visible light image and infrared light image for the same speedboat.

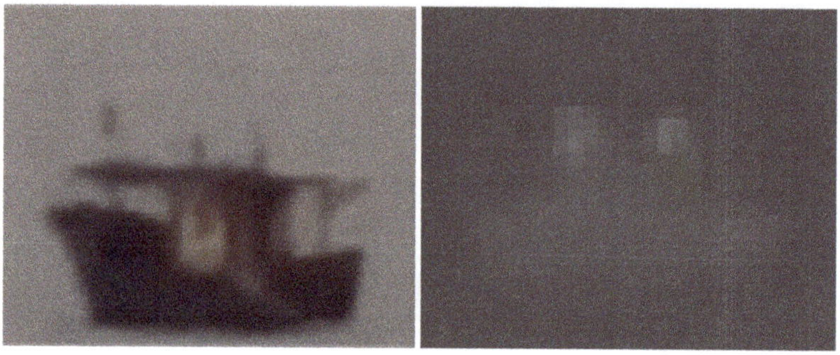

Figure 6. Visible light image and infrared light image for the same fishing boat.

4.2. Sensor

Two data sources (visible light, infrared), two image features (HOG, Gabor), and the classification algorithm AdaBoost can be combined separately to obtain 4 classifiers, as shown in Table 4. Since the background of this research is target recognition, these classifiers are regarded as different sensors, so as to recognize the target and generate their respective recognition results. The specific process of recognition is shown in Figure 7.

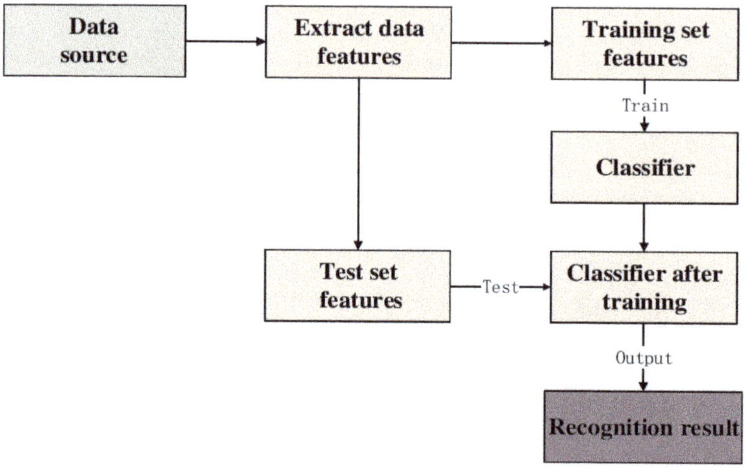

Figure 7. The work process of sensor.

Table 4. Base sensor recognition confusion matrix.

Sensor 1	**Sensor 2**
Visible light + HOG + AdaBoost	Infrared light + HOG + AdaBoost
Sensor 3	**Sensor 4**
Visible light + GABOR + AdaBoost	Infrared light + GABOR + AdaBoost

4.3. Base Sensor Recognition Confusion Matrix

The verification set is input into the trained base sensor, and the recognition confusion matrix of the base sensor can be obtained according to the recognition result of the sensor. The Tables 5–8 are the identification confusion matrixs of the base sensors on the verification set.

Table 5. Recognition confusion matrix of sensor 1.

Real Category/Identify Category	Sailboat	Cargo Ship	Speedboat	Fishing Boat
Sailboat	23	0	2	0
Cargo ship	0	29	3	0
Speed boat	1	1	20	0
Fishing boat	0	0	0	31

Table 6. Recognition confusion matrix of sensor 2.

Real Category/Identify Category	Sailboat	Cargo Ship	Speedboat	Fishing Boat
Sailboat	16	3	6	0
Cargo ship	1	13	17	1
Speed boat	0	0	18	4
Fishing boat	0	1	9	21

Table 7. Recognition confusion matrix of sensor 3.

Real Category/Identify Category	Sailboat	Cargo Ship	Speedboat	Fishing Boat
Sailboat	16	1	8	0
Cargo ship	5	17	10	0
Speed boat	0	0	22	0
Fishing boat	0	0	0	31

Table 8. Recognition confusion matrix of sensor 4.

Real Category/Identify Category	Sailboat	Cargo Ship	Speedboat	Fishing Boat
Sailboat	25	0	0	0
Cargo ship	5	25	0	2
Speed boat	0	10	1	11
Fishing boat	0	0	6	25

4.4. Preference Matrix

Tables 9–12 show the preference comparison matrix of the four sensors for the four types of target.

Table 9. Preference matrix $P(1,1)$ of Sailboat.

$P(1,1)$	Sensor 1	Sensor 2	Sensor 3	Sensor 4
Sensor 1	0.500	0.801	0.821	0.690
Sensor 2	0.198	0.500	0.532	0.355
Sensor 3	0.178	0.467	0.500	0.326
Sensor 4	0.309	0.644	0.673	0.500

Table 10. Preference matrix $P(1,2)$ of Cargo ship.

$P(1,2)$	Sensor 1	Sensor 2	Sensor 3	Sensor 4
Sensor 1	0.500	0.859	0.826	0.794
Sensor 2	0.140	0.500	0.436	0.386
Sensor 3	0.173	0.563	0.500	0.448
Sensor 4	0.205	0.614	0.551	0.500

Table 11. Preference matrix $P(1,3)$ of Speedboat.

$P(1,3)$	Sensor 1	Sensor 2	Sensor 3	Sensor 4
Sensor 1	0.500	0.856	0.769	0.802
Sensor 2	0.143	0.500	0.358	0.403
Sensor 3	0.230	0.641	0.500	0.548
Sensor 4	0.197	0.596	0.451	0.500

Table 12. Preference matrix $P(1,4)$ of Fishing boat.

$P(1,4)$	Sensor 1	Sensor 2	Sensor 3	Sensor 4
Sensor 1	0.500	1.000	0.500	1.000
Sensor 2	0	0.500	0	0.561
Sensor 3	0.500	1.000	0.500	1.000
Sensor 4	0	0.438	0	0.500

4.5. Complementarity Vector

According to the previous research, the complementarity vector can be obtained from the preference matrix, as shown in Table 13, which reflects the complementarity between the 4 sensors if the target to be identified is the first type of target sailboat. These information also reflect the importance of each sensor, so the complementary vector is used as the weight to generate the neutrosophic set when fusing the neutrosophic set.

Table 13. Complementarity vector: C.

	Sensor 1	Sensor 2	Sensor 3	Sensor 4
$C(1,1)$	0.473	0.138	0.106	0.283
$C(1,2)$	0.513	0.102	0.166	0.219
$C(1,3)$	0.500	0.086	0.231	0.183
$C(1,4)$	0.382	0.132	0.382	0.104

4.6. Data Fusion

When identifying an unknown target, the four sensors can obtain the probability of the category through the trained classifier, as shown in Table 14.

Table 14. Probability of the 4 sensors to recognize the unknown target.

Type	Sailboat	Cargo Ship	Speedboat	Fishing Boat
Sensor 1	0.256	0.122	0.180	0.442
Sensor 2	0.136	0.237	0.315	0.312
Sensor 3	0.078	0.107	0.352	0.463
Sensor 4	0.099	0.162	0.286	0.453

According to these probabilities, by Equation (16), the complementarity vectors are converted into weight vectors, as shown in Table 15.

Table 15. Weight vector: *H*.

	Sensor 1	Sensor 2	Sensor 3	Sensor 4
$W(1,1)$	0.817	0.238	0.183	0.487
$W(1,2)$	0.833	0.167	0.262	0.355
$W(1,3)$	0.853	0.147	0.393	0.311
$W(1,4)$	0.789	0.274	0.789	0.210

Use Equation (17) to combine P and W and get four neutrosophic sets in one category. The current recognition framework has four categories, so four groups of neutrosophic set are obtained, as follows:

$$\alpha_1 = \begin{bmatrix} \alpha_{11} = [0.209, 0.607, 0.184] \\ \alpha_{12} = [0.032, 0.205, 0.761] \\ \alpha_{13} = [0.014, 0.168, 0.817] \\ \alpha_{14} = [0.048, 0.439, 0.513] \end{bmatrix}$$

$$\alpha_2 = \begin{bmatrix} \alpha_{21} = [0.102, 0.731, 0.167] \\ \alpha_{22} = [0.039, 0.127, 0.833] \\ \alpha_{23} = [0.029, 0.242, 0.729] \\ \alpha_{24} = [0.057, 0.297, 0.645] \end{bmatrix}$$

$$\alpha_3 = \begin{bmatrix} \alpha_{31} = [0.154, 0.699, 0.147] \\ \alpha_{32} = [0.046, 0.101, 0.853] \\ \alpha_{33} = [0.138, 0.255, 0.607] \\ \alpha_{34} = [0.089, 0.222, 0.689] \end{bmatrix}$$

$$\alpha_4 = \begin{bmatrix} \alpha_{41} = [0.348, 0.442, 0.210] \\ \alpha_{42} = [0.085, 0.189, 0.726] \\ \alpha_{43} = [0.365, 0.425, 0.210] \\ \alpha_{44} = [0.095, 0.116, 0.789] \end{bmatrix}$$

Combine these four groups of neutrosophic sets according to Equation (18) to obtain 4 neutrosophic sets:

$$\begin{aligned} \alpha_1 &= 0.25 \times \alpha_{11} + 0.25 \times \alpha_{12} + 0.25 \times \alpha_{13} + 0.25 \times \alpha_{14} = [0.079, 0.492, 0.310] \\ \alpha_2 &= 0.25 \times \alpha_{21} + 0.25 \times \alpha_{22} + 0.25 \times \alpha_{23} + 0.25 \times \alpha_{24} = [0.058, 0.506, 0.286] \\ \alpha_3 &= 0.25 \times \alpha_{31} + 0.25 \times \alpha_{32} + 0.25 \times \alpha_{33} + 0.25 \times \alpha_{34} = [0.108, 0.479, 0.251] \\ \alpha_4 &= 0.25 \times \alpha_{41} + 0.25 \times \alpha_{42} + 0.25 \times \alpha_{43} + 0.25 \times \alpha_{44} = [0.234, 0.399, 0.253] \end{aligned}$$

Furthermore, the four neutrosophic sets are transformed into crisp numbers by Equation (19).

$$RS = 4 = Max(S) = Max[S_1 = 0.181, S_2 = 0.142, S_3 = 0.252, S_4 = 0.427]$$

Therefore, the recognition result of the unknown target is fishing boat.

4.7. Recognition Result

After all the test sets are finally tested, the results of the method proposed in this paper are shown in Table 16. And the results of two other fusion methods such as simple fusion (Table 17) and D-S fusion (Table 18) are given to compare with the proposed method.

Table 16. The proposed method recognition result.

Real Category/Identify Category	Sailboat	Cargo Ship	Speedboat	Fishing Boat
Sailboat	27	0	1	0
Cargo ship	1	31	3	0
Speed boat	0	0	25	0
Fishing boat	0	0	0	35

Table 17. Simple fusion result.

Real Category/Identify Category	Sailboat	Cargo Ship	Speedboat	Fishing Boat
Sailboat	27	0	1	0
Cargo ship	1	33	1	0
Speed boat	1	1	21	2
Fishing boat	0	0	0	35

Table 18. D-S fusion result.

Real Category/Identify Category	Sailboat	Cargo Ship	Speedboat	Fishing Boat
Sailboat	26	1	1	0
Cargo ship	2	28	5	0
Speed boat	0	0	25	0
Fishing boat	0	0	2	33

The recognition results of the 4 base sensors on the test set are shown in Tables 19–22:

Table 19. Sensor 1 recognition result.

Real Category/Identify Category	Sailboat	Cargo Ship	Speedboat	Fishing Boat
Sailboat	25	0	3	0
Cargo ship	0	31	4	0
Speed boat	1	1	23	0
Fishing boat	0	0	0	35

Table 20. Sensor 2 recognition result.

Real Category/Identify Category	Sailboat	Cargo Ship	Speedboat	Fishing Boat
Sailboat	17	3	8	0
Cargo ship	1	14	18	2
Speed boat	0	1	19	5
Fishing boat	0	1	10	24

Table 21. Sensor 3 recognition result.

Real Category/Identify Category	Sailboat	Cargo Ship	Speedboat	Fishing Boat
Sailboat	17	1	10	0
Cargo ship	6	18	11	0
Speed boat	0	0	25	0
Fishing boat	0	0	0	35

Table 22. Sensor 4 recognition result.

Real Category/Identify Category	Sailboat	Cargo Ship	Speedboat	Fishing Boat
Sailboat	28	0	0	0
Cargo ship	6	27	0	2
Speed boat	0	12	1	12
Fishing boat	0	0	6	29

The multi-category evaluation criteria is used in the previous article to evaluate the classification results, shown in Table 23. It can be seen that the method proposed in this paper has improved in the recall rate, accuracy rate, F1 score and other indicators compared with these four sensors. After multi-sensor fusion recognition, the accuracy rate of a single sensor is increased by 3.25% at the lowest and 35.77% at the highest, and the average performance of a single sensor is improved by 21.13%. At the same time, compared with the other two fusion methods, the method proposed in this parper also performs better.It can be seen that the accuracy of fusion recognition can be significantly improved.

Table 23. Recognition result analysis.

	Category	Accuracy Rate	Recall Rate	F1 Score	Count Time	Correct Rate
Sensor 1	Sailboat	0.962	0.893	0.926	54*S*	92.68%
	Cargo ship	0.969	0.886	0.925		
	Speedboat	0.767	0.920	0.836		
	Fishing boat	1.000	1.000	1.000		
Sensor 2	Sailboat	0.944	0.607	0.739	52*S*	60.16%
	Cargo ship	0.737	0.400	0.519		
	Speedboat	0.345	0.760	0.475		
	Fishing boat	0.774	0.686	0.727		
Sensor 3	Sailboat	0.739	0.607	0.667	162*S*	77.24%
	Cargo ship	0.947	0.514	0.667		
	Speedboat	0.543	1.000	0.704		
	Fishing boat	1.000	1.000	1.000		
Sensor 4	Sailboat	0.824	1.000	0.903	158*S*	69.11%
	Cargo ship	0.692	0.771	0.730		
	Speedboat	0.143	0.040	0.063		
	Fishing boat	0.674	0.829	0.744		
D-S fusion	Sailboat	0.929	0.929	0.929	434*S*	91.56%
	Cargo ship	0.966	0.800	0.875		
	Speedboat	0.758	1.000	0.862		
	Fishing boat	1.000	0.943	0.971		
Simple fusion	Sailboat	0.931	0.964	0.947	413*S*	94.30%
	Cargo ship	0.971	0.943	0.957		
	Speedboat	0.913	0.840	0.875		
	Fishing boat	1.000	0.946	0.972		
Proposed method	Sailboat	0.964	0.964	0.964	398*S*	95.93%
	Cargo ship	1.000	0.886	0.940		
	Speedboat	0.862	1.000	0.926		
	Fishing boat	1.000	1.000	1.000		

5. Results

Aiming at the problem of multi-sensor target recognition, this paper proposes a new method based on the complementary characteristics of sensors in the fusion of neutrosophic set, which improves the accuracy of target type recognition. Using the identification of the sea surface vessel type as the verification scenario, the category-oriented sensor complementarity vector is constructed through feature extraction, sensor training of the target's infrared and visible image training data.

The multi-sensor neutrosophic set model is performed on the target to be recognized to realize the multi-sensor. Compared with other methods, the method proposed in this paper performs better in recognition accuracy, Compared with other fuzzy mathematics theories, the neutrosophic set theory is more helpful for us to deal with the complementary information between sensors. At the same time, the three sets of functions included in the neutrosophic set allow us to flexibly adjust the weight and other parameters, and the calculation of the neutrosophic set is simple, it takes less time to run the program. Further research will mostly concentrate on the the proposed method can be used to more complicated study to further demonstrate its efficiency.

Author Contributions: Y.G. wrote this paper, W.J., X.D. and M.W. reviewed and improved this article, Y.G. and Z.M. discussed and analyzed the numerical results. All authors have read and agreed to the published version of the manuscript.

Funding: The work is partially supported by Equipment Pre-Research Fund (Program No. 61400010109), the Seed Foundation of Innovation and Creation for Graduate Students in Northwestern Polytechnical University (Program No. CX2020151).

Conflicts of Interest: The authors declare that they have no competing interest.

References

1. Akula, A.; Singh, A.; Ghosh, R.; Kumar, S.; Sardana, H.K. *Target Recognition in Infrared Imagery Using Convolutional Neural Network*; Springer: Singapore, 2017; pp. 25–34.
2. Liu, J.; Tang, Y.Y. An evolutionary autonomous agents approach to image feature extraction. *IEEE Trans. Evol. Comput.* **1997**, *1*, 141–158.
3. Hearst, M.A.; Dumais, S.T.; Osman, E.; Platt, J.; Scholkopf, B. Support vector machines. *IEEE Intell. Syst. Their Appl.* **1998**, *13*, 18–28. [CrossRef]
4. Sutskever, I.; Vinyals, O.; Le, Q. Sequence to Sequence Learning with Neural Networks. In Proceedings of the Advances in Neural Information Processing Systems 27 (NIPS 2014), Montreal, QC, Canada, 8–13 December 2014; pp. 3104–3112.
5. Calders, T.; Verwer, S. Three naive Bayes approaches for discrimination-free classification. *Data Min. Knowl. Discov.* **2010**, *21*, 277–292. [CrossRef]
6. Jiang, W.; Cao, Y.; Deng, X. A Novel Z-network Model Based on Bayesian Network and Z-number. *IEEE Trans. Fuzzy Syst.* **2020**, *28*, 1585–1599. [CrossRef]
7. Zhang, L.; Wu, X.; Qin, Y.; Skibniewski, M.J.; Liu, W. Towards a Fuzzy Bayesian Network Based Approach for Safety Risk Analysis of Tunnel-Induced Pipeline Damage. *Risk Anal.* **2016**, *36*, 278–301. [CrossRef] [PubMed]
8. Huang, Z.; Yang, L.; Jiang, W. Uncertainty measurement with belief entropy on the interference effect in the quantum-like Bayesian Networks. *Appl. Math. Comput.* **2019**, *347*, 417–428. [CrossRef]
9. Wood, D.; Underwood, J.; Avis, P. Integrated learning systems in the classroom. *Comput. Educ.* **1999**, *33*, 91–108. [CrossRef]
10. Geng, J.; Deng, X.; Ma, X.; Jiang, W. Transfer Learning for SAR Image Classification Via Deep Joint Distribution Adaptation Networks. *IEEE Trans. Geosci. Remote Sens.* **2020**, *58*, 5377–5392. [CrossRef]
11. Geng, J.; Jiang, W.; Deng, X. Multi-scale deep feature learning network with bilateral filtering for SAR image classification. *ISPRS J. Photogramm. Remote Sens.* **2020**, *167*, 201–213. [CrossRef]
12. Jiang, W.; Huang, K.; Geng, J.; Deng, X. Multi-Scale Metric Learning for Few-Shot Learning. *IEEE Trans. Circuits Syst. Video Technol.* **2020**. [CrossRef]
13. He, Z.; Jiang, W. An evidential dynamical model to predict the interference effect of categorization on decision making. *Knowl.-Based Syst.* **2018**, *150*, 139–149. [CrossRef]
14. He, Z.; Jiang, W. An evidential Markov decision making model. *Inf. Sci.* **2018**, *467*, 357–372. [CrossRef]
15. Fu, C.; Chang, W.; Xue, M.; Yang, S. Multiple criteria group decision making with belief distributions and distributed preference relations. *Eur. J. Oper. Res.* **2019**, *273*, 623–633. [CrossRef]
16. Sun, C.; Li, S.; Deng, Y. Determining Weights in Multi-Criteria Decision Making Based on Negation of Probability Distribution under Uncertain Environment. *Mathematics* **2020**, *8*, 191. [CrossRef]

17. Fu, C.; Chang, W.; Liu, W.; Yang, S. Data-driven group decision making for diagnosis of thyroid nodule. *Sci.-China-Inf. Sci.* **2019**, *62*, 212205. [CrossRef]
18. Xiao, F. GIQ: A generalized intelligent quality-based approach for fusing multi-source information. *IEEE Trans. Fuzzy Syst.* **2020**. [CrossRef]
19. Xiao, F.; Cao, Z.; Jolfaei, A. A novel conflict measurement in decision making and its application in fault diagnosis. *IEEE Trans. Fuzzy Syst.* **2020**. [CrossRef]
20. Li, S.; Liu, G.; Tang, X.; Lu, J.; Hu, J. An Ensemble Deep Convolutional Neural Network Model with Improved D-S Evidence Fusion for Bearing Fault Diagnosis. *Sensors* **2017**, *17*, 1729. [CrossRef]
21. Jiang, W. A correlation coefficient for belief functions. *Int. J. Approx. Reason.* **2018**, *103*, 94–106. [CrossRef]
22. Deng, X.; Jiang, W.; Wang, Z. Zero-sum polymatrix games with link uncertainty: A Dempster-Shafer theory solution. *Appl. Math. Comput.* **2019**, *340*, 101–112. [CrossRef]
23. Deng, X.; Jiang, W.; Wang, Z. Weighted belief function of sensor data fusion in engine fault diagnosis. *Soft Comput.* **2020**, *24*, 2329–2339.
24. Fei, L.; Feng, Y.; Liu, L. Evidence combination using OWA-based soft likelihood functions. *Int. J. Intell. Syst.* **2019**, *34*, 2269–2290. [CrossRef]
25. Fei, L.; Lu, J.; Feng, Y. An extended best-worst multi-criteria decision-making method by belief functions and its applications in hospital service evaluation. *Comput. Ind. Eng.* **2020**, *142*, 106355. [CrossRef]
26. Mao, S.; Han, Y.; Deng, Y.; Pelusi, D. A hybrid DEMATEL-FRACTAL method of handling dependent evidences. *Eng. Appl. Artif. Intell.* **2020**, *91*, 103543. [CrossRef]
27. Zimmermann, H.J. Fuzzy set theory. *Wiley Interdiscip. Rev. Comput. Stat.* **2010**, *2*, 317–332. [CrossRef]
28. Xiao, F. A distance measure for intuitionistic fuzzy sets and its application to pattern classification problems. *IEEE Trans. Syst. Man Cybern. Syst.* **2019**. [CrossRef]
29. Deng, X.; Jiang, W. D number theory based game-theoretic framework in adversarial decision making under a fuzzy environment. *Int. J. Approx. Reason.* **2019**, *106*, 194–213. [CrossRef]
30. Deng, X.; Jiang, W. Evaluating green supply chain management practices under fuzzy environment: A novel method based on D number theory. *Int. J. Fuzzy Syst.* **2019**, *21*, 1389–1402. [CrossRef]
31. Deng, X.; Jiang, W. A total uncertainty measure for D numbers based on belief intervals. *Int. J. Intell. Syst.* **2019**, *34*, 3302–3316. [CrossRef]
32. Liu, B.; Deng, Y. Risk Evaluation in Failure Mode and Effects Analysis Based on D Numbers Theory. *Int. J. Comput. Commun. Control* **2019**, *14*, 672–691.
33. Qian, Y.; Liang, J.; Pedrycz, W.; Dang, C. Positive approximation: An accelerator for attribute reduction in rough set theory. *Artif. Intell.* **2010**, *174*, 597–618. [CrossRef]
34. Smarandache, F. *Neutrosophic Probability, Set, and Logic (First Version)*; American Research Press: Rehoboth, DE, USA, 2016; pp. 41–48.
35. Zadeh, L.A. Fuzzy sets. *Inf. Control* **1965**, *8*, 338–353. [CrossRef]
36. Atanassov, K.T. Intuitionistic fuzzy sets. *Fuzzy Sets Syst.* **1986**, *20*, 87–96. [CrossRef]
37. Atanassov, K.; Gargov, G. Interval valued intuitionistic fuzzy sets. *Fuzzy Sets Syst.* **1989**, *31*, 343–349. [CrossRef]
38. Ye, J. A multicriteria decision-making method using aggregation operators for simplified neutrosophic sets. *J. Intell. Fuzzy Syst.* **2014**, *26*, 2459–2466. [CrossRef]
39. Haibin, W.; Smarandache, F.; Zhang, Y.; Sunderraman, R. Single Valued Neutrosophic Sets. 2010. Available online: http://fs.unm.edu/SingleValuedNeutrosophicSets.pdf (accessed on 31 July 2020).
40. Zhang, C.; Li, D.; Broumi, S.; Sangaiah, A.K. Medical Diagnosis Based on Single-Valued Neutrosophic Probabilistic Rough Multisets over Two Universes. *Symmetry* **2018**, *10*, 213. [CrossRef]
41. Saber, Y.; Alsharari, F.; Smarandache, F. On Single-Valued Neutrosophic Ideals in Šostak Sense. *Symmetry* **2020**, *12*, 193. [CrossRef]
42. Haibin, W.; Smarandache, F.; Zhang, Y.Q.; Sunderraman, R. Interval Neutrosophic Sets and Logic: Theory and Applications in Computing. 2012. Available online: https://arxiv.org/abs/cs/0505014 (accessed on 31 July 2020).
43. Smarandache, F. *Neutrosophy: Neutrosophic Probability, Set, and Logic: Analytic Synthesis and Synthetic Analysis*; American Research Press: Rehoboth, DE, USA, 1998.
44. Zhou, X.; Li, P.; Smarandache, F.; Khalil, A.M. New Results on Neutrosophic Extended Triplet Groups Equipped with a Partial Order. *Symmetry* **2019**, *14*, 1514. [CrossRef]

45. Ma, Y.X.; Wang, J.-Q.; Wang, J.; Wu, X.-H. An interval neutrosophic linguistic multi-criteria group decision-making method and its application in selecting medical treatment options. *Neural Comput. Appl.* **2017**, *28*, 2745–2765. [CrossRef]
46. Wei, G.; Zhang, Z. Some single-valued neutrosophic Bonferroni power aggregation operators in multiple attribute decision making. *J. Ambient. Intell. Humaniz. Comput.* **2018**, *10*, 863–882. [CrossRef]
47. Ye, J. Another Form of Correlation Coefficient between Single Valued Neutrosophic Sets and Its Multiple Attribute Decision- Making Method. *Neutrosophic Sets Syst.* **2013** *1*, 8–12. [CrossRef]
48. Ye, J. Trapezoidal neutrosophic set and its application to multiple attribute decision-making. *Neural Comput. Appl.* **2015**, *26*, 1157–1166. [CrossRef]
49. Kandasamy, V.W.B.; Kandasamy, I.; Smarandache, F. Neutrosophic Components Semigroups and Multiset Neutrosophic Components Semigroups. *Symmetry* **2020**, *14*, 818.
50. Yang, W.; Cai, L.; Edalatpanah, S.A.; Smarandache, F. Triangular Single Valued Neutrosophic Data Envelopment Analysis: Application to Hospital Performance Measurement. *Symmetry* **2020**, *14*, 588. [CrossRef]
51. Gou, L.; Zhong, Y. A New Fault Diagnosis Method Based on Attributes Weighted Neutrosophic Set. *IEEE Access* **2019**, *7*, 117740–117748. [CrossRef]
52. Broumi, S.; Nagarajan, D.; Bakali, A.; Talea, M.; Smarandache, F.; Lathamaheswari, M. The shortest path problem in interval valued trapezoidal and triangular neutrosophic environment. *Complex Intell. Syst.* **2019**, *5*, 391–402. [CrossRef]
53. Chakraborty, A. A new score function of pentagonal neutrosophic number and its application in networking problem. *Int. J. Neutrosophic Sci.* **2020**, *1*, 40–51.
54. Broumi, S.; Nagarajan, D.; Bakali, A.; Talea, M.; Smarandache, F.; Lathamaheswari, M.; Kavikumar, J. Implementation of neutrosophic function memberships using matlab program. *Neutrosophic Sets Syst.* **2019**, *14*, 44–52.
55. Zhang, L.; Wu, X.; Zhu, H.; AbouRizk, S.M. Perceiving safety risk of buildings adjacent to tunneling excavation: An information fusion approach. *Autom. Constr.* **2017**, *73*, 88–101. [CrossRef]
56. Kavita; Yadav, S.P.; Kumar, S. *A Multi-Criteria Interval-Valued Intuitionistic Fuzzy Group Decision Making for Supplier Selection with TOPSIS Method*; Springer: Berlin/Heidelberg, Germany, 2009; Volume 5908.
57. Li, H. *Statistical Learning Methods*; Tsinghua University Press: Beijing, China, 2012.
58. Rtsch, G.; Onoda, T.; Muller, K.R. Soft Margins for AdaBoost. *Mach. Learn.* **2001**, *42*, 287–320. [CrossRef]
59. Dalal, N.; Triggs, B. Histograms of Oriented Gradients for Human Detection. In Proceedings of the IEEE Conference on Computer Vision and Pattern Recognition (CVPR 2005), San Diego, CA, USA, 20–25 June 2005; Volume 2.
60. Liu, C.; Koga, M.; Fujisawa, H. Gabor feature extraction for character recognition: Comparison with gradient feature. In Proceedings of the Eighth International Conference on Document Analysis and Recognition, Seoul, Korea, 31 August–1 September 2005; Volumes 1 and 2, pp. 121–125.
61. Saaty, T.L. Analytic Hierarchy Process. 2013. Available online: https://onlinelibrary.wiley.com/doi/abs/10.1002/0470011815.b2a4a002 (accessed on 31 July 2020).
62. Deng, X.; Deng, Y. D-AHP method with different credibility of information. *Soft Comput.* **2019**, *23*, 683–691. [CrossRef]
63. Peng, J.-J.; Wang, J.-Q.; Wang, J.; Zhang, H.-Y.; Chen, X.-H. Simplified neutrosophic sets and their applications in multi-criteria group decision-making problems. *Int. J. Syst. Sci.* **2016**, *47*. [CrossRef]

symmetry

MDPI

Article

On Single-Valued Neutrosophic Ideals in Šostak Sense

Yaser Saber [1,2], Fahad Alsharari [1] and Florentin Smarandache [3,*]

1 Department of Mathematics, College of Science and Human Studies, Hotat Sudair, Majmaah University, Majmaah 11952, Saudi Arabia; y.sber@mu.edu.sa or m.ah75@yahoo.com (Y.S.); f.alsharari@mu.edu.sa (F.A.)
2 Department of Mathematics, Faculty of Science Al-Azhar University, Assiut 71524, Egypt
3 Department of Mathematics, University of New Mexico, Gallup, NM 87301, USA
* Correspondence: smarand@unm.edu

Received: 15 December 2019; Accepted: 6 January 2020; Published: 25 January 2020

check for updates

Abstract: Neutrosophy is a recent section of philosophy. It was initiated in 1980 by Smarandache. It was presented as the study of origin, nature, and scope of neutralities, as well as their interactions with different ideational spectra. In this paper, we introduce the notion of single-valued neutrosophic ideals sets in Šostak's sense, which is considered as a generalization of fuzzy ideals in Šostak's sense and intuitionistic fuzzy ideals. The concept of single-valued neutrosophic ideal open local function is also introduced for a single-valued neutrosophic topological space. The basic structure, especially a basis for such generated single-valued neutrosophic topologies and several relations between different single-valued neutrosophic ideals and single-valued neutrosophic topologies, are also studied here. Finally, for the purpose of symmetry, we also define the so-called single-valued neutrosophic relations.

Keywords: single-valued neutrosophic closure; single-valued neutrosophic ideal; single-valued neutrosophic ideal open local function; single-valued neutrosophic ideal closure; single-valued neutrosophic ideal interior; single-valued neutrosophic ideal open compatible

1. Introduction

The notion of fuzzy sets, employed as an ordinary set generalization, was introduced in 1965 by Zadeh [1]. Later on, using fuzzy sets through the fuzzy topology concept was initially introduced in 1968 by Chang [2]. Afterwards, many properties in fuzzy topological spaces have been explored by various researchers [3–13]

Paradoxically, it is to be emphasized that being fuzzy or what is termed as fuzzy topology in fuzzy openness concept is not highlighted and well-studied. Meanwhile, Samanta et al. [14,15] introduced what is called the graduation of openness of fuzzy sets. Later on, Ramadan [16] introduced smooth continuity, a number of their properties, and smooth topology. Demirci [17] investigated properties and systems of smooth Q-neighborhood and smooth neighborhood alike. It is worth mentioning that Chattopadhyay and Samanta [18] have initiated smooth connectedness and smooth compactness. On the other hand, Peters [19] tackled the notion of primary fuzzy smooth characteristics and structures together with smooth topology in Lowen sense. He [20] further evidenced that smooth topologies collection constitutes a complete lattice. Furthermore, Onassanya and Hošková-Mayerová [21] inspected certain features of subsets of α-level as an integral part of a fuzzy subset topology. Likewise, more specialists in the field like Çoker and Demirci [22], in addition to Samanta and Mondal [23,24], have provided definitions to the concept of graduation intuitionistic openness of fuzzy sets based on Šostak's sense [25] according to Atanassov's [26] intuitionistic fuzzy sets. Essentially, they focused on intuitionistic gradation of openness in light of Chang. On the other hand, Lim et al. [27] examined

Lowen's framework smooth intuitionistic topological spaces. In recent times, Kim et al. [28] considered systems of neighborhood and continuities within smooth intuitionistic topological spaces. Moreover, Choi et al. [29] scrutinized smooth interval-valued topology through graduation of the concept of interval-valued openness of fuzzy sets, as suggested by Gorzalczany [30] and Zadeh [31], respectively. Ying [32] put forward a topology notion termed as fuzzifying topology, taking into consideration the extent of ordinary subset of a set openness. General properties in ordinary smooth topological spaces were elaborated in 2012 by Lim et al. [33]. In addition, they [34–36] inspected compactness, interiors, and closures within normal smooth topological spaces. In 2014, Saber et al. [37] shaped the notion of fuzzy ideal and r-fuzzy open local function in fuzzy topological spaces in view of the definition of Šostak. In addition, they [38,39] inspected intuitionistic fuzzy ideals, fuzzy ideals and fuzzy open local function in fuzzy topological spaces in view of the definition of Chang.

Smarandache [40] determined the notion of a neutrosophic set as intuitionistic fuzzy set generalization. Meanwhile, Salama et al. [41,42] familiarized the concepts of neutrosophic crisp set and neutrosophic crisp relation neutrosophic set theory. Correspondingly, Hur et al. [43,44] initiated classifications NSet(H) and NCSet including neutrosophic crisp and neutrosophic sets, where they examined them in a universe topological position. Furthermore, Salama and Alblowi [45] presented neutrosophic topology as they claimed a number of its characteristics. Salama et al. [46] defined a neutrosophic crisp topology and studied some of its properties. Others, such as Wang et al. [47], defined the single-valued neutrosophic set concept. Currently, Kim et al. [48] has come to grips with a neutrosophic partition single-value, neutrosophic equivalence relation single-value, and neutrosophic relation single-value.

Preliminaries of single-value neutrosophic sets and single-valued neutrosophic topology are reviewed in Section 2. Section 3 is devoted to the concepts of single-valued neutrosophic closure space and single-valued neutrosophic ideal. Some of their characteristic properties are considered. Finally, the concepts of single-valued neutrosophic ideal open local function has been introduced and studied. Several preservation properties and some characterizations concerning single-valued neutrosophic ideal open compatible have been obtained.

2. Preliminaries

In this section, we attempt to cover enough of the fundamental concepts and definitions.

Definition 1 ([49]). *A neutrosophic set $\mathcal{H}$ (**NS**, for short) on a nonempty set $\mathcal{S}$ is defined as*

$$\mathcal{H} = \langle \kappa, T_{\mathcal{H}}, I_{\mathcal{H}}, F_{\mathcal{H}} : \kappa \in \mathcal{S} \rangle,$$

where

$$T_{\mathcal{H}} : \mathcal{S} \to]^{-}0, 1^{+}[, \quad I_{\mathcal{H}} : \mathcal{S} \to]^{-}0, 1^{+}[, \quad F_{\mathcal{H}} : \mathcal{S} \to]^{-}0, 1^{+}[$$

and

$$^{-}0 \leq T_{\mathcal{H}}(\kappa) + I_{\mathcal{H}}(\kappa) + F_{\mathcal{H}}(\kappa) \leq 3^{+},$$

representing the degree of membership (namely, $T_{\mathcal{H}}(\kappa)$), the degree of indeterminacy (namely, $I_{\mathcal{H}}(\kappa)$), and the degree of nonmembership (namely, $F_{\mathcal{H}}(\kappa)$); for all $\kappa \in \mathcal{S}$ to the set $\mathcal{H}$.

Definition 2 ([49]). *Let $\mathcal{H}$ and $\mathcal{R}$ be fuzzy neutrosophic sets in $\mathcal{S}$. Then, $\mathcal{H}$ is a subset of $\mathcal{R}$ if, for each $\kappa \in \mathcal{S}$,*

$$\inf T_{\mathcal{H}}(x) \leq \inf T_{\mathcal{R}}(\kappa), \quad \inf I_{\mathcal{H}}(x) \geq \inf I_{\mathcal{R}}(\kappa), \quad \inf F_{\mathcal{H}}(x) \geq \inf F_{\mathcal{R}}(\kappa)$$

and

$$\sup T_{\mathcal{H}}(\kappa) \leq \sup T_{\mathcal{R}}(\kappa), \quad \sup I_{\mathcal{H}}(\kappa) \geq \sup I_{\mathcal{R}}(\kappa), \quad \sup F_{\mathcal{H}}(\kappa) \geq \sup F_{\mathcal{R}}(\kappa).$$

Definition 3 ([47]). *Let $\mathcal{H}$ be a space of points (objects) with a generic element in $\mathcal{S}$ denoted by κ. Then, $\mathcal{H}$ is called a single-valued neutrosophic set (in short,* **SVNS**) *in $\mathcal{S}$ if $\mathcal{H}$ has the form $\mathcal{H} = \langle T_{\mathcal{H}}, I_{\mathcal{H}}, F_{\mathcal{H}} \rangle$, where $T_{\mathcal{H}}, I_{\mathcal{H}}, F_{\mathcal{H}} : \mathcal{S} \to [0,1]$.*

In this case, $T_{\mathcal{H}}, I_{\mathcal{H}}, F_{\mathcal{H}}$ are called truth-membership function, indeterminacy-membership function, and falsity-membership function, respectively, and we will denote the set of all **SVNS's** *in $\mathcal{S}$ as* $\mathbf{SVNS}(\mathcal{S})$.

Moreover, we will refer to the Null (empty) **SVNS** *(or the absolute (universe)* **SVNS**) *in $\mathcal{S}$ as 0_N (or 1_N) and define by $0_N = (0,1,1)$ (or $1_N = (1,0,0)$) for each $\kappa \in \mathcal{S}$.*

Definition 4 ([47]). *Let $\mathcal{H} = \langle T_{\mathcal{H}}, I_{\mathcal{H}}, F_{\mathcal{H}} \rangle$ be an* **SVNS** *on $\mathcal{S}$. The complement of the set $\mathcal{H}$ ($\mathcal{H}^c$, for short) and is defined as follows: for every $\kappa \in \mathcal{S}$,*

$$T_{\mathcal{H}^c}(\kappa) = F_{\mathcal{H}}(\kappa), \quad I_{\mathcal{H}^c}(\kappa) = 1 - I_{\mathcal{H}}(\kappa), \quad F_{\mathcal{H}^c}(\kappa) = T_{\mathcal{H}}(\kappa).$$

Definition 5 ([50]). *Suppose that $\mathcal{H} \in$* $\mathbf{SVNS}(\mathcal{S})$. *Then,*

(i) $\mathcal{H}$ is said to be contained in $\mathcal{R}$, denoted by $\mathcal{H} \subseteq \mathcal{R}$, if, for every $\kappa \in \mathcal{S}$,

$$T_{\mathcal{H}}(\kappa) \leq T_{\mathcal{R}}(\kappa), \quad I_{\mathcal{H}}(\kappa) \geq I_{\mathcal{R}}(\kappa), \quad F_{\mathcal{H}}(\kappa) \geq F_{\mathcal{R}}(\kappa);$$

(ii) $\mathcal{H}$ is said to be equal to $\mathcal{R}$, denoted by $\mathcal{H} = \mathcal{H}$, if $\mathcal{R} \subseteq \mathcal{R}$ and $\mathcal{H} \supseteq \mathcal{R}$.

Definition 6 ([51]). *Suppose that $\mathcal{H}, \mathcal{R} \in$* $\mathbf{SVNS}(\mathcal{S})$. *Then,*

(i) the union of $\mathcal{H}$ and $\mathcal{R}$ ($\mathcal{H} \cup \mathcal{R}$, for short) is an **SVNS** *in $\mathcal{S}$ defined as*

$$\mathcal{H} \cup \mathcal{R} = (T_{\mathcal{H}} \cup T_{\mathcal{R}}, I_{\mathcal{H}} \cap I_{\mathcal{R}}, F_{\mathcal{H}} \cap F_{\mathcal{R}}),$$

where $(T_{\mathcal{H}} \cup T_{\mathcal{R}})(\kappa) = T_{\mathcal{H}}(\kappa) \cup T_{\mathcal{R}}(\kappa)$ and $(F_{\mathcal{H}} \cap F_{\mathcal{R}})(\kappa) = F_{\mathcal{H}}(\kappa) \cap F_{\mathcal{R}}(\kappa)$, for each $\kappa \in \mathcal{S}$;

(ii) the intersection of $\mathcal{H}$ and $\mathcal{R}$, ($\mathcal{H} \cap \mathcal{R}$, for short), is an **SVNS** *in $\mathcal{S}$ defined as*

$$\mathcal{H} \cap \mathcal{R} = (T_{\mathcal{H}} \cap T_{\mathcal{R}}, I_{\mathcal{H}} \cup I_{\mathcal{R}}, F_{\mathcal{H}} \cup F_{\mathcal{R}}).$$

Definition 7 ([45]). *Let $\mathcal{H} \in$* $\mathbf{SVNS}(\mathcal{S})$. *Then,*

(i) the union of $\{\mathcal{H}_i\}_{i \in J}$ ($\bigcup_{i \in J} \mathcal{H}_i$, for short) is an **SVNS** *in $\mathcal{S}$ defined as follows: for every $\kappa \in \mathcal{S}$,*

$$(\bigcup_{i \in J} \mathcal{H}_i)(\kappa) = (\bigcup_{i \in J} T_{\mathcal{H}_i}(\kappa), \bigcap_{i \in J} I_{\mathcal{H}_i}(\kappa), \bigcap_{i \in J} F_{\mathcal{H}_i}(\kappa);$$

(ii) the intersection of $\{\mathcal{H}_i\}_{i \in J}$ ($\bigcap_{i \in J} \mathcal{H}_i$, for short) is an **SVNS** *in $\mathcal{S}$ defined as follows: for every $\kappa \in \mathcal{S}$,*

$$(\bigcap_{i \in J} \mathcal{H}_i)(\kappa) = (\bigcap_{i \in J} T_{\mathcal{H}_i}(\kappa), \bigcup_{i \in J} I_{\mathcal{H}_i}(\kappa), \bigcup_{i \in J} F_{\mathcal{H}_i}(\kappa).$$

Definition 8 ([52]). *A single-valued neutrosophic topology on $\mathcal{S}$ is a map $(\tau^T, \tau^I, \tau^F) : I^{\mathcal{S}} \to I$ satisfying the following three conditions:*

(SVNT1) $\tau^T(\underline{0}) = \tau^T(\underline{1}) = 1$ *and* $\tau^I(\underline{0}) = \tau^I(\underline{1}) = \tau^F(\underline{0}) = \tau^F(\underline{1}) = 0$,

(SVNT2) $\tau^T(\mathcal{H} \cap \mathcal{R}) \geq \tau^T(\mathcal{H}) \cap \tau^T(\mathcal{R}), \quad \tau^I(\mathcal{H} \cap \mathcal{R}) \leq \tau^I(\mathcal{H}) \cup \tau^I(\mathcal{R})$, $\tau^F(\mathcal{H} \cap \mathcal{R}) \leq \tau^F(\mathcal{H}) \cup \tau^F(\mathcal{R})$, *for any* $\mathcal{H}, \mathcal{R} \in I^{\mathcal{S}}$,

(SVNT3) $\tau^T(\cup_{i\in j}\mathcal{H}_i) \geq \cap_{i\in j}\tau^T(\mathcal{H}_i)$, $\tau^I(\cup_{i\in j}\mathcal{H}_i) \leq \cup_{i\in j}\tau^I(\mathcal{H}_i)$, $\tau^F(\cup_{i\in j}\mathcal{H}_i) \leq \cup_{i\in j}\tau^F(\mathcal{H}_i)$, *for any* $\{\mathcal{H}_i\}_{i\in J} \in I^{\mathcal{S}}$.

The pair $(X, \tau^T, \tau^I, \tau^F)$ is called single-valued neutrosophic topological spaces (**SVNTS**, *for short*). We will occasionally write τ^{TIF} for (τ^T, τ^I, τ^F) and it will cause no ambiguity.

3. Single-Valued Neutrosophic Closure Space and Single-Valued Neutrosophic Ideal in Šostak Sense

This section deals with the definition of single-valued neutrosophic closure space. The researchers examine the connection between single-valued neutrosophic closure space and **SVNTS** based in Šostak sense. Moreover, the researchers focused on the single-valued neutrosophic ideal notion where they obtained fundamental properties. Based on Šostak's sense, where a single-valued neutrosophic ideal takes the form $(\mathcal{S}, \mathcal{L}^T, \mathcal{L}^I, \mathcal{L}^F)$ and the mappings $\mathcal{L}^T, \mathcal{L}^I, \mathcal{L}^F : I^{\mathcal{S}} \to I$, where $(\mathcal{L}^T, \mathcal{L}^I, \mathcal{L}^F)$ are the degree of openness, the degree of indeterminacy, and the degree of non-openness, respectively.

In this paper, $\mathcal{S}$ is used to refer to nonempty sets, whereas I is used to refer to closed interval $[0,1]$ and I_o is used to refer to the interval $(0,1]$. Concepts and notations that are not described in this paper are standard, instead, $\mathcal{S}$ is usually used.

Definition 9. *A mapping* $\mathbb{C} : I^{\mathcal{S}} \times I_0 \to I^{\mathcal{S}}$ *is called a single-valued neutrosophic closure operator on* $\mathcal{S}$ *if, for every* $\mathcal{H}, \mathcal{R} \in I^{\mathcal{S}}$ *and* $r, s \in I_0$*, the following axioms are satisfied:*

($\mathbb{C}_1$) $\mathbb{C}((0.1.1), s) = (0.1.1)$,
($\mathbb{C}_2$) $\mathcal{H} \leq \mathbb{C}(\mathcal{H}, s)$,
($\mathbb{C}_3$) $\mathbb{C}(\mathcal{H}, s) \vee \mathbb{C}(\mathcal{R}, s) = \mathbb{C}(\mathcal{H} \vee \mathcal{R}, s)$,
($\mathbb{C}_4$) $\mathbb{C}(\mathcal{H}, s) \leq \mathbb{C}(\mathcal{H}, r)$ *if* $s \leq r$,
($\mathbb{C}_5$) $\mathbb{C}(\mathbb{C}(\mathcal{H}, s), s) = \mathbb{C}(\mathcal{H}, s)$.

The pair $(X, \mathbb{C})$ is a single-valued neutrosophic closure space ($\mathcal{SVNCS}$, *for short*).

Suppose that $\mathbb{C}_1$ and $\mathbb{C}_2$ are single-valued neutrosophic closure operators on $\mathcal{S}$. Then, $\mathbb{C}_1$ is finer than $\mathbb{C}_2$, denoted by $\mathbb{C}_2 \leq \mathbb{C}_1$ iff $\mathbb{C}_1(\mathcal{H}, s) \leq \mathbb{C}_2(\mathcal{H}, s)$, for every $\mathcal{H} \in I^{\mathcal{S}}$ and $s \in I_0$.

Theorem 1. *Let* $(\mathcal{S}, \tau^{TIF})$ *be an* **SVNTS**. *Then, for any* $\mathcal{H} \in I^{\mathcal{S}}$ *and* $s \in I_0$*, we define an operator* $\mathbb{C}_{\tau^{TIF}} : I^{\mathcal{S}} \times I_0 \to I^{\mathcal{S}}$ *as follows:*

$$\mathbb{C}_{\tau^{TIF}}(\mathcal{H}, s) = \bigwedge\{\mathcal{R} \in I^X : \mathcal{H} \leq \mathcal{R}, \quad \tau^T(\underline{1} - \mathcal{R}) \geq s, \quad \tau^I(\underline{1} - \mathcal{R}) \leq 1 - s, \quad \tau^F(\underline{1} - \mathcal{R}) \leq 1 - s\}.$$

Then, $(\mathcal{S}, \mathbb{C}_{\tau^{TIF}})$ *is an* $\mathcal{SVNCS}$.

Proof. Suppose that $(\mathcal{S}, \tau^{TIF})$ is an **SVNTS**. Then, $\mathbb{C}_1$, ($\mathbb{C}_2$) and ($\mathbb{C}_4$) follows directly from the definition of $\mathbb{C}_{\tau^{TIF}}$.

($\mathbb{C}_3$) Since $\mathcal{R}, \mathcal{H} \leq \mathcal{H} \cup \mathcal{R}$, $\mathbb{C}_{\tau^{TIF}}(\mathcal{R}, s) \leq \mathbb{C}_{\tau^{TIF}}(\mathcal{H} \cup \mathcal{R}, s)$ and $\mathbb{C}_{\tau^{TIF}}(\mathcal{H}, s) \leq \mathbb{C}_{\tau^{TIF}}(\mathcal{H} \cup \mathcal{R}, s)$, therefore,

$$\mathbb{C}_{\tau^{TIF}}(\mathcal{H}, s) \cup \mathbb{C}_{\tau^{TIF}}(\mathcal{R}, s) \leq \mathbb{C}_{\tau^{TIF}}(\mathcal{H} \cup \mathcal{R}, s).$$

Let (X, τ^{TIF}) be an **SVNTS**. From ($\mathbb{C}_2$), we have

$$\mathcal{H} \leq \mathbb{C}_{\tau^{TIF}}(\mathcal{H}, s), \quad \tau^T(\underline{1} - \mathbb{C}_{\tau^{TIF}}(\mathcal{H}, s)) \geq s, \ \tau^I(\underline{1} - \mathbb{C}_{\tau^{TIF}}(\mathcal{H}, s)) \leq 1 - s$$
$$\text{and } \tau^F(\underline{1} - \mathbb{C}_{\tau^{TIF}}(\mathcal{H}, s)) \leq 1 - s,$$

$$\mathcal{R} \leq \mathbb{C}_{\tau^{TIF}}(\mathcal{R},s), \quad \tau^T(\underline{1} - \mathbb{C}_{\tau^{TIF}}(\mathcal{R},s)) \geq s, \ \tau^I(\underline{1} - \mathbb{C}_{\tau^{TIF}}(\mathcal{R},s)) \leq 1-s$$
$$and \ \tau^F(\underline{1} - \mathbb{C}_{\tau^{TIF}}(\mathcal{R},s)) \leq 1-s.$$

It implies that $\mathcal{H} \cup \mathcal{R} \leq \mathbb{C}_{\tau^{TIF}}(\mathcal{H},s) \cup \mathbb{C}_{\tau^{TIF}}(\mathcal{R},s)$,

$$\begin{aligned}\tau^T(\underline{1} - (\mathbb{C}_{\tau^{TIF}}(\mathcal{H},s) \cup \mathbb{C}_{\tau^{TIF}}(\mathcal{R},s))) &= \tau^T((\underline{1} - \mathbb{C}_{\tau^{TIF}}(\mathcal{H},s)) \cap (\underline{1} - \mathbb{C}_{\tau^{TIF}}(\mathcal{R},s))) \\ &\geq \tau^T(\underline{1} - \mathbb{C}_{\tau^{TIF}}(\mathcal{H},s)) \cap \tau^T(\underline{1} - \mathbb{C}_{\tau^{TIF}}(\mathcal{R},s)) \geq s,\end{aligned}$$

$$\begin{aligned}\tau^I(\underline{1} - (\mathbb{C}_{\tau^{TIF}}(\mathcal{H},s) \cup \mathbb{C}_{\tau^{TIF}}(\mathcal{R},s))) &= \tau^I((\underline{1} - \mathbb{C}_{\tau^{TIF}}(\mathcal{H},s)) \cap (\underline{1} - \mathbb{C}_{\tau^{TIF}}(\mathcal{R},s))) \\ &\leq \tau^I((\underline{1} - \mathbb{C}_{\tau^{TIF}}(\mathcal{H},s)) \cup \tau^I(\underline{1} - \mathbb{C}_{\tau^{TIF}}(\mathcal{R},s)) \leq 1-s,\end{aligned}$$

$$\begin{aligned}\tau^F(\underline{1} - (\mathbb{C}_{\tau^{TIF}}(\mathcal{H},s) \cup \mathbb{C}_{\tau^{TIF}}(\mathcal{R},s))) &= \tau^F((\underline{1} - \mathbb{C}_{\tau^{TIF}}(\mathcal{H},s)) \cap (\underline{1} - \mathbb{C}_{\tau^{TIF}}(\mathcal{R},s))) \\ &\leq \tau^F(\underline{1} - \mathbb{C}_{\tau^{TIF}}(\mathcal{H},s)) \cup \tau^F(\underline{1} - \mathbb{C}_{\tau^{TIF}}(\mathcal{R},s)) \leq 1-s.\end{aligned}$$

Hence, $\mathbb{C}_{\tau^{TIF}}(\mathcal{H},s) \cup \mathbb{C}_{\tau^{TIF}}(\mathcal{H} \cup \mathcal{R},s) \geq \mathbb{C}_{\tau^{TIF}}(\mathcal{H} \cup \mathcal{R},s)$. Therefore,

$$\mathbb{C}_{\tau^{TIF}}(\mathcal{H},s) \cup \mathbb{C}_{\tau^{TIF}}(\mathcal{H} \cup \mathcal{R},s) = \mathbb{C}_{\tau^{TIF}}(\mathcal{H} \cup \mathcal{R},s).$$

($\mathbb{C}_5$) Suppose that there exists $s \in I_0$, $\mathcal{H} \in I^{\mathcal{S}}$, and $\kappa \in \mathcal{S}$ such that

$$\mathbb{C}_{\tau^{TIF}}(\mathbb{C}_{\tau^{TIF}}(\mathcal{H},s),s)(\kappa) > \mathbb{C}_{\tau^{TIF}}(\mathcal{H},s)(\kappa).$$

By the definition of $\mathbb{C}_{\tau^{TIF}}$, there exists $\mathcal{D} \in I^{\mathcal{S}}$ with $\mathcal{D} \geq \mathcal{H}$, and $\tau^T(\underline{1} - \mathcal{D}) \geq s$, $\tau^I(\underline{1} - \mathcal{D}) \leq 1-s$ and $\tau^F(\underline{1} - \mathcal{D}) \leq 1-s$ such that

$$\mathbb{C}_{\tau^{TIF}}(\mathbb{C}_{\tau^{TIF}}(\mathcal{H},s),s)(\kappa) > \mathcal{D}(\kappa) \geq \mathbb{C}_{\tau^{TIF}}(\mathcal{H},s)(\kappa).$$

Since $\mathbb{C}_{\tau^{TIF}}(\mathcal{H},s) \leq \mathcal{D}$ and $\tau^T(\underline{1} - \mathcal{D}) \geq s$, $\tau^I(\underline{1} - \mathcal{D}) \leq 1-s$, and $\tau^F(\underline{1} - \mathcal{D}) \leq 1-s$, by the definition of $\mathbb{C}_{\tau^{TIF}}(\mathbb{C}_{\tau^{TIF}})$, we have

$$\mathbb{C}_{\tau^{TIF}}(\mathbb{C}_{\tau^{TIF}}(\mathcal{H},s),s) \leq \mathcal{D}.$$

It is a contradiction. Thus, $\mathbb{C}_{\tau^{TIF}}(\mathbb{C}_{\tau^{TIF}}(\mathcal{H},s),s) = \mathbb{C}_{\tau^{TIF}}(\mathcal{H},s)$. Hence, $\mathbb{C}_{\tau^{TIF}}$ is a single-valued neutrosophic closure operator on $\mathcal{S}$. □

Theorem 2. *Let $(\mathcal{S},\mathbb{C})$ be an $\mathcal{SVNCS}$ and $\mathcal{H} \in \mathcal{S}$. Define the mapping $\tau_{\mathbb{C}}^{TIF} : I^{\mathcal{S}} \to I$ on $\mathcal{S}$ by*

$$\tau_{\mathbb{C}}^T(\mathcal{H}) = \bigcup\{s \in I_0 \mid \mathbb{C}(\overline{1} - \mathcal{H},s) = \overline{1} - \mathcal{H}\},$$

$$\tau_{\mathbb{C}}^I(\mathcal{H}) = \bigcap\{1-s \in I_0 \mid \mathbb{C}(\overline{1} - \mathcal{H},s) = \overline{1} - \mathcal{H}\},$$

$$\tau_{\mathbb{C}}^F(\mathcal{H}) = \bigcap\{1-s \in I_0 \mid \mathbb{C}(\overline{1} - \mathcal{H},s) = \overline{1} - \mathcal{H}\},$$

Then,

(1) *$\tau_{\mathbb{C}}^{TIF}$ is an* ***SVNTS*** *on $\mathcal{S}$;*
(2) *$\mathbb{C}_{\tau_{\mathbb{C}}^{TIF}}$ is finer than $\mathbb{C}$.*

Proof. (SVNT1) Let $(\mathcal{S},\mathbb{C})$ be an $\mathcal{SVNCS}$. Since $\mathbb{C}((0.1.1),r)=(0.1.1)$ and $\mathbb{C}(1,0,0),r)=(1,0,0)$ for every $s\in I_0$, (SVNT1).

(SVNT2) Let $(\mathcal{S},\mathbb{C})$ be an $\mathcal{SVNCS}$. Suppose that there exists $\mathcal{H}_1,\mathcal{H}_2\in I^{\mathcal{S}}$ such that

$$\tau^T_{\mathbb{C}}(\mathcal{H}_1\cap\mathcal{H}_2)<\tau^T_{\mathbb{C}}(\mathcal{H}_1)\cap\tau^T_{\mathbb{C}}(\mathcal{H}_2),\quad \tau^I_{\mathbb{C}}(\mathcal{H}_1\cap\mathcal{H}_2)>\tau^I_{\mathbb{C}}(\mathcal{H}_1)\cup\tau^I_{\mathbb{C}}(\mathcal{H}_2),$$

$$\tau^F_{\mathbb{C}}(\mathcal{H}_1\cap\mathcal{H}_2)>\tau^F_{\mathbb{C}}(\mathcal{H}_1)\cup\tau^F_{\mathbb{C}}(\mathcal{H}_2).$$

There exists $s\in I_0$ such that

$$\tau^T_{\mathbb{C}}(\mathcal{H}_1\cap\mathcal{H}_2)<s<\tau^T_{\mathbb{C}}(\mathcal{H}_1)\cap\tau^T_{\mathbb{C}}(\mathcal{H}_2),\qquad \tau^I_{\mathbb{C}}(\mathcal{H}_1\cap\mathcal{H}_2)>1-s>\tau^I_{\mathbb{C}}(\mathcal{H}_1)\cup\tau^I_{\mathbb{C}}(\mathcal{H}_2),$$

$$\tau^F_{\mathbb{C}}(\mathcal{H}_1\cap\mathcal{H}_2)>1-s>\tau^F_{\mathbb{C}}(\mathcal{H}_1)\cup\tau^F_{\mathbb{C}}(\mathcal{H}_2).$$

For each $i\in\{1,2\}$, there exists $s\in I_0$ with $\mathbb{C}(\mathcal{H}_i,s_i)=\overline{1}-\mathcal{H}_i$ such that

$$s<s_i\le\tau^T_{\mathbb{C}}(\mathcal{H}_i),\quad \tau^I_{\mathbb{C}}(\mathcal{H}_i)\le 1-s_i<1-s,\quad \tau^F_{\mathbb{C}}(\mathcal{H}_i)\le 1-s_i<1-s.$$

In addition, since $(\overline{1}-\mathcal{H}_i,r)=\overline{1}-\mathcal{H}_i$ by $\mathbb{C}_2$ and $\mathbb{C}_4$ of Definition 9, for any $i\in\{1,2\}$,

$$\mathbb{C}((\overline{1}-\mathcal{H}_1)\cup(\overline{1}-\mathcal{H}_2),s)=(\overline{1}-\mathcal{H}_1)\cup(\overline{1}-\mathcal{H}_2).$$

It follows that $\tau^T_{\mathbb{C}}(\mathcal{H}_1\cap\mathcal{H}_2)\ge s$, $\tau^I_{\mathbb{C}}(\mathcal{H}_1\cap\mathcal{H}_2)\le 1-s$, and $\tau^F_{\mathbb{C}}(\mathcal{H}_1\cap\mathcal{H}_2)\le 1-s$. It is a contradiction. Thus, for every $\mathcal{H},\mathcal{R}\in I^{\mathcal{S}}$, $\tau^T_{\mathbb{C}}(\mathcal{H}\cap\mathcal{R})\ge\tau^T_{\mathbb{C}}(\mathcal{H})\cap\tau^T_{\mathbb{C}}(\mathcal{B})$, $\tau^I_{\mathbb{C}}(\mathcal{H}\cap\mathcal{R})\le\tau^I_{\mathbb{C}}(\mathcal{H})\cup\tau^I_{\mathbb{C}}(\mathcal{R})$, and $\tau^F_{\mathbb{C}}(\mathcal{H}\cap\mathcal{R})\le\tau^F_{\mathbb{C}}(\mathcal{H})\cup\tau^F_{\mathbb{C}}(\mathcal{R})$.

(SVNT3) Suppose that there exists $\mathcal{H}=\bigcup_{i\in I}\mathcal{H}_i\in I^{\mathcal{S}}$ such that

$$\tau^T_{\mathbb{C}}(\mathcal{H})<\bigcup_{i\in I}\tau^T_{\mathbb{C}}(\mathcal{H}_i),\quad \tau^I_{\mathbb{C}}(\mathcal{H})>\bigcup_{i\in I}\tau^I_{\mathbb{C}}(\mathcal{H}_i),\quad \tau^F_{\mathbb{C}}(\mathcal{H})>\bigcup_{i\in I}\tau^F_{\mathbb{C}}(\mathcal{H}_i).$$

There exists $s_0\in I_0$ such that

$$\tau^T_{\mathbb{C}}(\mathcal{H})<s_0<\bigcup_{i\in I}\tau^T_{\mathbb{C}}(\mathcal{H}_i),\quad \tau^I_{\mathbb{C}}(\mathcal{H})>1-s_0>\bigcup_{i\in I}\tau^I_{\mathbb{C}}(\mathcal{H}_i),\quad \tau^F_{\mathbb{C}}(\mathcal{H})>1-s_0>\bigcup_{i\in I}\tau^F_{\mathbb{C}}(\mathcal{H}_i).$$

For every $i\in I$, there exists $\mathbb{C}(\mathcal{H}_i,s_i)=\overline{1}-\mathcal{H}_i$ and $s_i\in I_0$ such that

$$s_0<s_i\le\tau^T_{\mathbb{C}}(\mathcal{H}_i),\quad 1-s_0>1-s_i\ge\tau^I_{\mathbb{C}}(\mathcal{H}_i),\quad 1-s_i>1-s_0\ge\tau^F_{\mathbb{C}}(\mathcal{H}_i).$$

In addition, since $\mathbb{C}(\overline{1}-\mathcal{H}_i,r_0)\le\mathbb{C}(\overline{1}-\mathcal{H}_i,s_i)=\overline{1}-\mathcal{H}_i$, by $\mathbb{C}_2$ of Definition 9,

$$\mathbb{C}(\overline{1}-\mathcal{H}_i,s_0)=\overline{1}-\mathcal{H}_i.$$

It implies, for all $i\in I$,

$$\mathbb{C}(\overline{1}-\mathcal{H},s_0)\le\mathbb{C}(\overline{1}-\mathcal{H}_i,s_0)=\overline{1}-\mathcal{H}_i.$$

It follows that

$$\mathbb{C}(\overline{1}-\mathcal{H},r_0)\le\bigcap_{i\in J}(\overline{1}-\mathcal{H}_i)=\overline{1}-\mathcal{H}.$$

Thus, $\mathbb{CI}(\overline{1}-\mathcal{H},s_0)=\overline{1}-\mathcal{H}$, that is, $\tau^T_{\mathbb{C}}(\mathcal{H})\geq s_0$, $\tau^I_{\mathbb{C}}(\mathcal{H})\leq 1-s_0$, and $\tau^F_{\mathbb{C}}(\mathcal{H})\leq 1-s_0$. It is a contradiction. Hence, $\tau^{TIF}_{\mathbb{C}}$ is an **SVNTS** on $\mathcal{S}$.

(2) Since $\mathcal{H}\leq\mathbb{C}(\mathcal{H},r)$,

$$\tau^T_{\mathbb{C}}(\overline{1}-\mathbb{C}(\mathcal{H},s))\geq s,\ \tau^I_{\mathbb{C}}(\overline{1}-\mathbb{C}(\mathcal{H},s))\leq 1-s,\ \tau^F_{\mathbb{C}}(\overline{1}-\mathbb{C}(\mathcal{H},s))\leq 1-s.$$

From $\mathbb{C}_5$ of Definition 9, we have $\mathbb{C}_{\tau^{TIF}_{\mathbb{C}}}(\mathcal{H},s)\leq\mathbb{C}(\mathcal{H},s)$. Thus, $\mathbb{C}_{\tau^{TIF}_{\mathbb{C}}}$ is finer than $\mathbb{C}$. □

Example 1. *Let $\mathcal{S}=\{a,b\}$. Define $\mathcal{B},\mathcal{H},\mathcal{A}\in I^{\mathcal{S}}$ as follows:*

$$\mathcal{B}=\langle(0.2,0.2),(0.3,0.3),(0.3,0.3)\rangle;\mathcal{H}=\langle(0.5,0.5),(0.1,0.1),(0.1,0.1)\rangle.$$

We define the mapping $\mathbb{C}:I^{\mathcal{S}}\times I_0\to I^{\mathcal{S}}$ as follows:

$$\mathbb{C}(\mathcal{A},s)=\begin{cases}(0.1.1), & \text{if } \mathcal{A}=(0.1.1),\quad s\in I_0,\\ \mathcal{B}\cap\mathcal{H}, & \text{if } 0\neq\mathcal{A}\leq\mathcal{B}\cap\mathcal{H},\ 0<r<\frac{1}{2},\\ \mathcal{B}, & \text{if } \mathcal{A}\leq\mathcal{B},\mathcal{A}\nleq\mathcal{H},\ 0<r<\frac{1}{2},\\ & \text{or } 0\neq\mathcal{A}\leq\mathcal{B}\quad \frac{1}{2}<r<\frac{2}{3},\\ \mathcal{H}, & \text{if } \mathcal{A}\leq\mathcal{H},\mathcal{A}\nleq\mathcal{B},\ 0<r<\frac{1}{2},\\ \mathcal{B}\cup\mathcal{H}, & \text{if } 0\neq\mathcal{A}\leq\mathcal{B}\cup\mathcal{H},\ 0<r<\frac{1}{2},\\ \overline{1}, & \text{otherwise.}\end{cases}$$

Then, $\mathbb{C}$ is a single-valued neutrosophic closure operator.
From Theorem 2, we have a single-valued neutrosophic topology $(\tau^T_{\mathbb{C}},\tau^I_{\mathbb{C}},\tau^F_{\mathbb{C}})$ on $\mathcal{S}$ as follows:

$$\tau^T_{\mathbb{C}}(\mathcal{A})=\begin{cases}1, & \text{if } \mathcal{A}=(1,0,0) \text{ or } (0,1,1),\\ \frac{2}{3}, & \text{if } \mathcal{A}=\mathcal{B}^c,\\ \frac{1}{2}, & \text{if } \mathcal{A}=\mathcal{H}^c,\\ \frac{1}{2}, & \text{if } \mathcal{A}=\mathcal{B}^c\cup\mathcal{H}^c,\\ \frac{1}{2}, & \text{if } \mathcal{A}=\mathcal{B}^c\cap\mathcal{H}^c,\\ 0, & \text{otherwise.}\end{cases}$$

$$\tau^I_{\mathbb{C}}(\mathcal{A})=\begin{cases}0, & \text{if } \mathcal{A}=(1,0,0) \text{ or } (0,1,1),\\ \frac{1}{3}, & \text{if } \mathcal{A}=\mathcal{B}^c,\\ \frac{1}{2}, & \text{if } \mathcal{A}=\mathcal{H}^c,\\ \frac{1}{2}, & \text{if } \mathcal{A}=\mathcal{B}^c\cup\mathcal{H}^c,\\ \frac{1}{2}, & \text{if } \mathcal{A}=\mathcal{B}^c\cap\mathcal{H}^c,\\ 1, & \text{otherwise.}\end{cases}$$

$$\tau^F_{\mathbb{C}}(\mathcal{A})=\begin{cases}0, & \text{if } \mathcal{A}=(1,0,0) \text{ or } (0,1,1),\\ \frac{1}{3}, & \text{if } \mathcal{A}=\mathcal{B}^c,\\ \frac{1}{2}, & \text{if } \mathcal{A}=\mathcal{H}^c,\\ \frac{1}{2}, & \text{if } \mathcal{A}=\mathcal{B}^c\cup\mathcal{H}^c,\\ \frac{1}{2}, & \text{if } \mathcal{A}=\mathcal{B}^c\cap\mathcal{H}^c,\\ 1, & \text{otherwise.}\end{cases}$$

Thus, the $\tau^{TIF}_{\mathbb{C}}$ is a single-valued neutrosophic topology on $\mathcal{S}$.

Definition 10. *A single-valued neutrosophic ideal (**SVNI**) on $\mathcal{S}$ in Šostak's sense on a nonempty set $\mathcal{S}$ is a family $\mathcal{L}^T,\mathcal{L}^I,\mathcal{L}^F$ of single-valued neutrosophic sets in $\mathcal{S}$ satisfying the following axioms:*

(L_1) $\mathcal{L}^T(\underline{0}) = 1$ *and* $\mathcal{L}^I(\underline{0}) = \mathcal{L}^F(\underline{0}) = 0$.

(L_2) *If* $\mathcal{H} \leq \mathcal{B}$, *then* $\mathcal{L}^T(\mathcal{R}) \leq \mathcal{L}^T(\mathcal{H})$, $\mathcal{L}^I(\mathcal{R}) \geq \mathcal{L}^I(\mathcal{H})$, *and* $\mathcal{L}^F(\mathcal{R}) \geq \mathcal{L}^F(\mathcal{H})$, *for each single-valued neutrosophic set* $\mathcal{R}, \mathcal{H}$ *in* $I^{\mathcal{S}}$.

(L_3) $\mathcal{L}^T(\mathcal{R} \cup \mathcal{H}) \geq \mathcal{L}^T(\mathcal{R}) \cap \mathcal{L}^T(\mathcal{H})$, $\mathcal{L}^I(\mathcal{R} \cup \mathcal{H}) \leq \mathcal{L}^I(\mathcal{R}) \cup \mathcal{L}^I(\mathcal{H})$, *and* $\mathcal{L}^F(\mathcal{R} \cup \mathcal{H}) \leq \mathcal{L}^F(\mathcal{R}) \cup \mathcal{L}^F(\mathcal{H})$, *for each single-valued neutrosophic set* $\mathcal{R}, \mathcal{H}$ *in* $I^{\mathcal{S}}$.

If $\mathcal{L}_1$ *and* $\mathcal{L}_2$ *are* **SVNI** *on* $\mathcal{S}$, *we say that* $\mathcal{L}_1$ *is finer than* $\mathcal{L}_2$, *denoted by* $\mathcal{L}_1 \leq \mathcal{L}_2$, *iff* $\mathcal{L}_1^T(\mathcal{H}) \leq \mathcal{L}_2^T(\mathcal{H})$, $\mathcal{L}_1^I(\mathcal{H}) \geq \mathcal{L}_2^I(\mathcal{A})$, *and* $\mathcal{L}_1^F(\mathcal{H}) \geq \mathcal{L}_2^F(\mathcal{H})$, *for* $\mathcal{H} \in I^{\mathcal{S}}$.

The triable $(X, (\tau^T, \tau^I, \tau^F), (\mathcal{L}^T, \mathcal{L}^I, \mathcal{L}^F)$ *is called a single-valued neutrosophic ideal topological space in Šostak sense (***SVNITS***, for short).*

We will occasionally write $\mathcal{L}^{TIF}$, $\mathcal{L}_i^{TIF}$, *and* $\mathcal{L}^{TIF} : I^X \to I$ *for* $(\mathcal{L}^T, \mathcal{L}^I, \mathcal{L}^F)$, $(\mathcal{L}_i^T, \mathcal{L}_i^I, \mathcal{L}_i^F)$, *and* $\mathcal{L}^T, \mathcal{L}^I, \mathcal{L}^F : I^{\mathcal{S}} \to I$, *respectively.*

Remark 1. *The conditions* (L_2) *and* (L_3), *which are given in Definition 10, are equivalent to the following axioms:* $\mathcal{L}^T(\mathcal{H} \cup \mathcal{R}) = \mathcal{L}^T(\mathcal{H}) \cap \mathcal{L}^T(\mathcal{R})$, $\mathcal{L}^I(\mathcal{H} \cup \mathcal{R}) \neq \mathcal{L}^I(\mathcal{H}) \cup \mathcal{L}^I(\mathcal{R})$, *and* $\mathcal{L}^F(\mathcal{H} \cup \mathcal{R}) \neq \mathcal{L}^F(\mathcal{H}) \cup \mathcal{L}^F(\mathcal{R})$, *for every* $\mathcal{R}, \mathcal{H} \in I^{\mathcal{S}}$.

Example 2. *Let* $\mathcal{S} = \{a, b\}$. *Define the single-valued neutrosophic sets* $\mathcal{R}, \mathcal{C}, \mathcal{H}, \mathcal{A}$ *and* $(\mathcal{L}^T, \mathcal{L}^T, \mathcal{L}^T) : I^{\mathcal{S}} \to I$ *as follows:*

$$\mathcal{R} = \langle (0.3, 0.5), (0.4, 0.5), (0.5, 0.5) \rangle; \quad \mathcal{C} = \langle (0.3, 0.4), (0.5, 0.5), (0.3, 0.4) \rangle,$$

$$\mathcal{H} = \langle (0.1, 0.2), (0.5, 0.5), (0.5, 0.5) \rangle.$$

$$\mathcal{L}^T(\mathcal{A}) = \begin{cases} 1, & \text{if } \mathcal{B} = (0.1.1), \\ \frac{1}{2}, & \text{if } \mathcal{A} = \mathcal{R}, \\ \frac{2}{3}, & \text{if } (0.1.1) < \mathcal{A} < \mathcal{R}, \\ 0, & \text{otherwise.} \end{cases}$$

$$\mathcal{L}^I(\mathcal{A}) = \begin{cases} 0, & \text{if } \mathcal{A} = (0.1.1), \\ \frac{1}{2}, & \text{if } \mathcal{A} = \mathcal{C}, \\ \frac{1}{4}, & \text{if } (0.1.1) < \mathcal{A} < \mathcal{C}, \\ 1, & \text{otherwise.} \end{cases}$$

$$\mathcal{L}^T(\mathcal{B}) = \begin{cases} 0, & \text{if } \mathcal{A} = (0, 1, 1), \\ \frac{1}{2}, & \text{if } \mathcal{A} = \mathcal{H}, \\ \frac{1}{4}, & \text{if } (0.1.1) < \mathcal{A} < \mathcal{H}, \\ 1, & \text{otherwise.} \end{cases}$$

Then, $\mathcal{L}^{TIF}$ *is an* **SVNI** *on* $\mathcal{S}$.

Remark 2. *(i) If* $\mathcal{L}^T(\underline{1}) = 1$, $\mathcal{L}^I(\underline{1}) = 0$, *and* $\mathcal{L}^F(\underline{1}) = 0$, *then* $\mathcal{L}^{TIF}$ *is called a single-valued neutrosophic proper ideal.*

(ii) If $\mathcal{L}^T(\underline{1}) = 0$, $\mathcal{L}^I(\underline{1}) = 1$, *and* $\mathcal{L}^F(\underline{1}) = 1$, *then* $\mathcal{L}^{TIF}$ *is called a single-valued neutrosophic improper ideal.*

Proposition 1. *Let* $\{\mathcal{L}_i^{TIF}\}_{i \in J}$ *be a family of* **SVNI** *on* $\mathcal{S}$. *Then, their intersection* $\bigcap_{i \in J} \mathcal{L}_i^{TIF}$ *is also* **SVNI**.

Proof. Directly from Definition 7. □

Proposition 2. *Let* $\{\mathcal{L}_i^{TIF}\}_{i\in J}$ *be a family of* **SVNI** *on* $\mathcal{S}$. *Then, their union* $\bigcup_{i\in J}\mathcal{L}_i^{TIF}$ *is also an* **SVNI**.

Proof. Directly from Definition 7. □

4. Single-Valued Neutrosophic Ideal Open Local Function in Šostak Sense

In this section, we study the single-valued neutrosophic ideal open local function in Šostak's sense and present some of their properties. Additionally, properties preserved by single-valued neutrosophic ideal open compatible are examined.

Definition 11. *Let* $s,t,p \in I_0$ *and* $s+t+p \leq 3$. *A single-valued neutrosophic point* $x_{s,t,r}$ *of* $\mathcal{S}$ *is the single-valued neutrosophic set in* $I^{\mathcal{S}}$ *for each* $\kappa \in \mathcal{H}$, *defined by*

$$x_{s,t,p}(\kappa) = \begin{cases} (s,t,p), & \text{if } x = \kappa, \\ (0,1,1), & \text{if } x \neq \kappa. \end{cases}$$

A single-valued neutrosophic point $x_{s,t,p}$ *is said to belong to a single-valued neutrosophic set* $\mathcal{H} = \langle T_{\mathcal{H}}, I_{\mathcal{H}}, F_{\mathcal{H}}\rangle \in I^{\mathcal{S}}$, *denoted by* $x_{s,t,p} \in \mathcal{H}$ *iff* $s < T_{\mathcal{H}}$, $t \geq I_{\mathcal{H}}$ *and* $p \geq F_{\mathcal{H}}$. *1. We indicate the set of all single-valued neutrosophic points in* $\mathcal{S}$ *as* **SVNP**$(\mathcal{S})$.

For every $x_{s,t,p} \in$ **SVNP**$(\mathcal{S})$ and $\mathcal{H} \in I^{\mathcal{S}}$ we shall write $x_{s,t,p}$ quasi-coincident with $\mathcal{H}$, denoted by $x_{s,t,p}q\mathcal{H}$, if

$$s + T_{\mathcal{H}}(\kappa) > 1, \qquad t + I_{\mathcal{H}}(\kappa) \leq 1, \qquad p + F_{\mathcal{H}}(\kappa) \leq 1.$$

For every $\mathcal{R}, \mathcal{H} \in \mathcal{S}$ we shall write $\mathcal{H}\bar{q}\mathcal{R}$ to mean that $\mathcal{H}$ is quasi-coincident with $\mathcal{R}$ if there exists $\kappa \in \mathcal{S}$ such that

$$T_{\mathcal{H}}(\kappa) + T_{\mathcal{R}}(\kappa) > 1, \quad I_{\mathcal{H}}(\kappa) + I_{\mathcal{R}}(\kappa) \leq 1, \quad F_{\mathcal{H}}(\kappa) + F_{\mathcal{R}}(\kappa) \leq 1.$$

Definition 12. *Let* $(\mathcal{S}, \tau^{TIF})$ *be an* **SVNTS**. *For each* $r \in I_0$, $\mathcal{H} \in I^{\mathcal{S}}$, $x_{s,t,p} \in$ **SVNP**$(\mathcal{S})$, *a single-valued neutrosophic open* $Q_{\tau^{TIF}}$*-neighborhood of* $x_{s,t,p}$ *is defined as follows:*

$$Q_{\tau^{TIF}}(x_{s,t,p}, r) = \{\mathcal{H} | (x_{s,t,p})q\mathcal{H}, \qquad \tau^{T}(\mathcal{H}) \geq r, \qquad \tau^{I}(\mathcal{H}) \leq 1-r, \qquad \tau^{F}(\mathcal{H}) \leq 1-r\}.$$

Lemma 1. *A single-valued neutrosophic point* $x_{s,t,p} \in \mathbb{C}_{\tau^{TIF}}(\mathcal{R}, r)$ *iff every single-valued neutrosophic open* $Q_{\tau^{TIF}}$*-neighborhood of* $x_{s,t,p}$ *is quasi-coincident with* $\mathcal{H}$.

Definition 13. *Let* $(\mathcal{S}, \tau^{TIF})$ *be an* **SVNTS** *for each* $\mathcal{H} \in I^{\mathcal{S}}$. *Then, the single-valued neutrosophic ideal open local function* $\mathcal{H}_r^{\star}(\tau^{TIF}, \mathcal{L}^{TIF})$ *of* $\mathcal{H}$ *is the union of all single-valued neutrosophic points* $x_{s,t,p}$ *such that if* $\mathcal{R} \in Q_{\tau^{TIF}}(x_{s,t,p}, r)$ *and* $\mathcal{L}^{T}(\mathcal{C}) \geq r$, $\mathcal{L}^{I}(\mathcal{C}) \leq 1-r$, $\mathcal{L}^{F}(\mathcal{C}) \leq 1-r$, *then there is at least one* $\kappa \in \mathcal{S}$ *for which* $T_{\mathcal{R}}(\kappa) + T_{\mathcal{H}}(\kappa) - 1 > T_{\mathcal{C}}(\kappa)$, $I_{\mathcal{R}}(\kappa) + I_{\mathcal{H}}(\kappa) - 1 \leq I_{\mathcal{C}}(\kappa)$, *and* $F_{\mathcal{R}}(\kappa) + F_{\mathcal{H}}(\kappa) - 1 \leq F_{\mathcal{C}}(\kappa)$.

Occasionally, we will write $\mathcal{H}_r^{\star}$ for $\mathcal{H}_r^{\star}(\tau^{TIF}, \mathcal{L}^{TIF})$ and it will have no ambiguity.

Example 3. *Let* $(\mathcal{S}, \tau^{TIF}, \mathcal{L}^{TIF})$ *be an* **SVNITS**. *The simplest single-valued neutrosophic ideal on* $\mathcal{S}$ *is* $\mathcal{L}_0^{TIF} : I^{\mathcal{S}} \to I$, *where*

$$\mathcal{L}_0^{TIF}(\mathcal{R}) = \begin{cases} 1, & \text{if } \mathcal{R} = (1,0,0), \\ 0, & \text{otherwise.} \end{cases}$$

If we take $\mathcal{L}^{TIF} = \mathcal{L}_0^{TIF}$, *for each* $\mathcal{H} \in I^{\mathcal{S}}$ *we have* $\mathcal{H}_r^{\star} = \mathbb{C}_{\tau^{TIF}}(\mathcal{H}, r)$.

Theorem 3. *Let* $(\mathcal{S}, \tau^{TIF})$ *be an* **SVNTS** *and* $\mathcal{L}_1^{TIF}, \mathcal{L}_2^{TIF} \in$ **SVNI**$(\mathcal{S})$*. Then, for any* $\mathcal{H}, \mathcal{R} \in I^{\mathcal{S}}$ *and* $r \in I_0$*, we have*

(1) *If* $\mathcal{H} \leq \mathcal{R}$*, then* $\mathcal{H}_r^\star \leq \mathcal{R}_r^\star$*;*
(2) *If* $\mathcal{L}_1^T \leq \mathcal{L}_2^T$*,* $\mathcal{L}_1^I \geq \mathcal{L}_2^I$ *and* $\mathcal{L}_1^F \geq \mathcal{L}_2^F$*, then* $\mathcal{H}_r^\star(\mathcal{L}_1^{TIF}, \tau^{TIF}) \geq \mathcal{H}_r^\star((\mathcal{L}_2^{TIF}, \tau^{TIF})$*;*
(3) $\mathcal{H}_r^\star = C_{\tau^{TIF}}(\mathcal{A}_r^\star, r) \leq \mathbb{C}_{\tau^{TIF}}(\mathcal{H}, r)$*;*
(4) $(\mathcal{H}_r^\star)_r^\star \leq \mathcal{H}_r^\star$*;*
(5) $(\mathcal{H}_r^\star \vee \mathcal{R}_r^\star) = (\mathcal{H} \vee \mathcal{R})_r^\star$*;*
(6) *If* $\mathcal{L}^T(\mathcal{H}) \geq r$*,* $\mathcal{L}^I(\mathcal{R}) \leq 1 - r$*, and* $\mathcal{L}^F(\mathcal{R}) \leq 1 - r$ *then* $(\mathcal{H} \vee \mathcal{R})_r^\star = \mathcal{A}_r^\star \vee \mathcal{R}_r^\star = \mathcal{H}_r^\star$*;*
(7) *If* $\tau^T(\mathcal{R}) \geq r$*,* $\tau^I(\mathcal{R}) \leq 1 - r$*, and* $\tau^F(\mathcal{R}) \leq 1 - r$*, then* $(\mathcal{R} \wedge \mathcal{H}_r^\star) \leq (\mathcal{R} \wedge \mathcal{H})_r^\star$*;*
(8) $(\mathcal{H}_r^\star \wedge \mathcal{R}_r^\star) \geq (\mathcal{H} \wedge \mathcal{R})_r^\star$*.*

Proof. (1) Suppose that $\mathcal{H} \in I^{\mathcal{S}}$ and $\mathcal{H}_r^\star \not\leq \mathcal{R}_r^\star$. Then, there exists $\kappa \in \mathcal{S}$ and $s, t, p \in I_0$ such that

$$T_{\mathcal{H}_r^\star}(\kappa) \geq s > T_{\mathcal{R}_r^\star}(\kappa), \quad I_{\mathcal{H}_r^\star}(\kappa) < t \leq I_{\mathcal{R}_r^\star}(\kappa), \quad F_{\mathcal{H}_r^\star}(\kappa) < p \leq F_{\mathcal{R}_r^\star}(\kappa). \tag{1}$$

Since $T_{\mathcal{R}_r^\star}(\kappa) < s$, $I_{\mathcal{R}_r^\star}(\kappa) \geq t$, and $F_{\mathcal{R}_r^\star}(\kappa) \geq p$. Then, there exists $\mathcal{D} \in Q_{(\tau^{TIF})}(x_{s,t,p}, r)$, $\mathcal{L}^T(\mathcal{C}) \geq r$, $\mathcal{L}^I(\mathcal{C}) \leq 1 - r$, and $\mathcal{L}^F(\mathcal{C}) \leq 1 - r$ such that for any $\kappa_1 \in \mathcal{S}$,

$$T_{\mathcal{D}}(\kappa_1) + T_{\mathcal{R}}(\kappa_1) - 1 \leq T_{\mathcal{C}}(\kappa_1), \quad I_{\mathcal{D}}(\kappa_1) + I_{\mathcal{R}}(\kappa_1) - 1 > I_{\mathcal{C}}(\kappa_1), \quad F_{\mathcal{D}}(\kappa_1) + F_{\mathcal{R}}(\kappa_1) - 1 > F_{\mathcal{C}}(\kappa_1).$$

Since $\mathcal{H} \leq \mathcal{R}$,

$$T_{\mathcal{D}}(\kappa_1) + T_{\mathcal{H}}(\kappa_1) - 1 \leq T_{\mathcal{C}}(\kappa_1), \quad I_{\mathcal{D}}(\kappa_1) + I_{\mathcal{H}}(\kappa_1) - 1 > I_{\mathcal{C}}(\kappa_1), \quad F_{\mathcal{D}}(\kappa_1) + F_{\mathcal{H}}(\kappa_1) - 1 > F_{\mathcal{C}}(\kappa_1).$$

So, $T_{\mathcal{H}_r^\star}(\kappa) < s$, $I_{\mathcal{H}_r^\star}(\kappa) \geq t$, and $F_{\mathcal{H}_r^\star}(\kappa) \geq p$ and we arrive at a contradiction for Equation (1). Hence, $\mathcal{H}_r^\star \leq \mathcal{R}_r^\star$.

(2) Suppose $\mathcal{H}_r^\star(\mathcal{L}_1^{TIF}, \tau^{TIF}) \not\geq \mathcal{H}_r^\star(\mathcal{L}_2^{TIF}, \tau^{TIF})$. Then, there exists $s, t, p \in I_0$ and $\kappa \in \mathcal{S}$ such that

$$T_{\mathcal{H}_r^\star(\mathcal{L}_1^{TIF}, \tau^{TIF})}(\kappa) < s \leq T_{\mathcal{H}_r^\star(\mathcal{L}_2^{TIF}, \tau^{TIF})}(\kappa),$$

$$I_{\mathcal{H}_r^\star(\mathcal{L}_1^{TIF}, \tau^{TIF})}(\kappa) \geq t > I_{\mathcal{H}_r^\star(\mathcal{L}_2^{TIF}, \tau^{TIF})}(\kappa), \tag{2}$$

$$F_{\mathcal{H}_r^\star(\mathcal{L}_1^{TIF}, \tau^{TIF})}(\kappa) \geq p > F_{\mathcal{H}_r^\star(\mathcal{L}_2^{TIF}, \tau^{TIF})}(\kappa).$$

Since $T_{\mathcal{H}_r^\star(\mathcal{L}_1^{TIF}, \tau^{TIF})}(\kappa) < s$, $I_{\mathcal{H}_r^\star(\mathcal{L}_1^{TIF}, \tau^{TIF})}(\kappa) \geq t$, and $F_{\mathcal{H}_r^\star(\mathcal{L}_1^{TIF}, \tau^{TIF})}(\kappa) \geq p$, $\mathcal{D} \in Q_{\tau^{TIF}}(x_{s,t,p}, r)$ with $\mathcal{L}_1^T(\mathcal{C}) \geq r$, $\mathcal{L}_1^I(\mathcal{C}) \leq 1 - r$ and $\mathcal{L}_1^F(\mathcal{C}) \leq 1 - r$. Thus, for every $\kappa_1 \in \mathcal{S}$,

$$T_{\mathcal{D}}(\kappa_1) + T_{\mathcal{H}}(\kappa_1) - 1 \leq T_{\mathcal{C}}(\kappa_1), \quad I_{\mathcal{D}}(\kappa_1) + I_{\mathcal{H}}(\kappa_1) - 1 > I_{\mathcal{C}}(\kappa_1), \quad F_{\mathcal{D}}(\kappa_1) + F_{\mathcal{H}}(\kappa_1) - 1 > F_{\mathcal{C}}(\kappa_1).$$

Since $\mathcal{L}_2^T(\mathcal{C}) \geq \mathcal{L}_1^T(\mathcal{C})) \geq r$, $\mathcal{L}_2^I(\mathcal{C}) \leq \mathcal{L}_1^I(\mathcal{C})) \leq 1 - r$, and $\mathcal{L}_2^F(\mathcal{C}) \leq \mathcal{L}_1^F(\mathcal{C})) \leq 1 - r$,

$$T_{\mathcal{D}}(\kappa_1) + T_{\mathcal{H}}(\kappa_1) - 1 \leq T_{\mathcal{C}}(\kappa_1), \quad I_{\mathcal{D}}(\kappa_1) + I_{\mathcal{H}}(\kappa_1) - 1 > I_{\mathcal{C}}(\kappa_1), \quad F_{\mathcal{D}}(\kappa_1) + F_{\mathcal{H}}(\kappa_1) - 1 > F_{\mathcal{C}}(\kappa_1).$$

Thus, $T_{\mathcal{H}_r^\star(\mathcal{L}_2^{TIF}, \tau^{TIF})}(\kappa) < s$, $I_{\mathcal{H}_r^\star(\mathcal{L}_2^{TIF}, \tau^{TIF})}(\kappa) \geq t$, and $F_{\mathcal{H}_r^\star(\mathcal{L}_2^{TIF}, \tau^{TIF})}(\kappa) \geq p$. This is a contradiction for Equation (2). Hence, $\mathcal{H}_r^\star((\mathcal{L}_1^{TIF}, \tau^{TIF})) \geq \mathcal{H}_r^\star((\mathcal{L}_2^{TIF}, \tau^{TIF}))$.

(3)($\Rightarrow$) Suppose $\mathcal{H}_r^\star \not\leq \mathbb{C}_{\tau^{TIF}}(\mathcal{H}, r)$. Then, there exists $s, t, p \in I_0$ and $\kappa \in \mathcal{S}$ such that

$$T_{\mathcal{H}_r^\star}(\kappa) \geq s > T_{\mathbb{C}_{\tau^{TIF}}(\mathcal{H},r)}(\kappa), \quad I_{\mathcal{H}_r^\star}(\kappa) < t \leq I_{\mathbb{C}_{\tau^{TIF}}(\mathcal{H},r)}(\kappa), \quad F_{\mathcal{H}_r^\star}(\kappa) < p \leq F_{\mathbb{C}_{\tau^{TIF}}(\mathcal{H},r)}(\kappa). \tag{3}$$

Since $T_{\mathcal{H}_r^\star}(\kappa) \geq s$, $I_{\mathcal{H}_r^\star}(\kappa) < t$ and $F_{\mathcal{H}_r^\star}(\kappa) < p$, $x_{s,t,p} \in \mathcal{H}_r^\star$. So there is at least one $\kappa_1 \in \mathcal{S}$ for every $\mathcal{D} \in Q_{\tau TIF}(x_{s,t,p}, r)$ with $\mathcal{L}_1^T(\mathcal{C}) \geq r$, $\mathcal{L}_1^I(\mathcal{C}) \leq 1-r$, $\mathcal{L}_1^F(\mathcal{C}) \leq 1-r$ such that

$$T_{\mathcal{D}}(\kappa_1) + T_{\mathcal{H}}(\kappa_1) > T_{\mathcal{C}}(\kappa_1) + 1, \quad I_{\mathcal{D}}(\kappa_1) + I_{\mathcal{H}}(\kappa_1) \leq I_{\mathcal{C}}(\kappa_1) + 1, \quad F_{\mathcal{D}}(\kappa_1) + F_{\mathcal{H}}(\kappa_1) \leq F_{\mathcal{C}}(\kappa_1) + 1.$$

Therefore, by Lemma 1, $x_{s,t,p} \in \mathbb{C}_{\tau TIF}(\mathcal{H}, r)$ which is a contradiction for Equation (3). Hence, $\mathcal{H}_r^\star \leq \mathbb{C}_{\tau TIF}(\mathcal{H}, r)$.

($\Leftarrow$) Suppose $\mathcal{H}_r^\star \not\geq \mathbb{C}_{\tau TIF}(\mathcal{H}_r^\star, r)$. Then, there exists $s, t, p \in I_0$ and $\kappa \in \mathcal{S}$ such that

$$T_{\mathcal{H}_r^\star}(\kappa) < s \leq T_{\mathbb{C}_{\tau TIF}(\mathcal{H}_r^\star, r)}(\kappa), \quad I_{\mathcal{H}_r^\star}(\kappa) \geq t > I_{\mathbb{C}_{\tau TIF}(\mathcal{H}_r^\star, r)}(\kappa), \quad F_{\mathcal{H}_r^\star}(\kappa) \geq p > F_{\mathbb{C}_{\tau TIF}(\mathcal{H}_r^\star, r)}(\kappa). \tag{4}$$

Since $T_{\mathbb{C}_{\tau TIF}(\mathcal{H}_r^\star, r)}(\kappa) \geq t$, $I_{\mathbb{C}_{\tau TIF}(\mathcal{H}_r^\star, r)}(\kappa) < s$, $\mathbb{C}_{\tau TIF}(\mathcal{H}_r^\star, r)(\kappa) < p$ we have $x_{s,t,p} \in \mathbb{C}_{\tau TIF}(\mathcal{H}_r^\star, r)$. So, there is at least one $\kappa_1 \in \mathcal{S}$ with $\mathcal{R} \in Q_{\tau TIF}(x_{s,t,p}, r)$ such that

$$T_{\mathcal{R}}(\kappa_1) + T_{\mathcal{H}_r^\star}(\kappa_1) > 1, \quad I_{\mathcal{R}}(\kappa_1) + I_{\mathcal{H}_r^\star}(\kappa_1) \leq 1, \quad F_{\mathcal{R}}(\kappa_1) + F_{\mathcal{H}_r^\star}(\kappa_1) \leq 1.$$

Therefore, $\mathcal{H}_r^\star(\kappa_1) \neq 0$. Let $s_1 = T_{\mathcal{H}_r^\star}(\kappa_1)$, $t_1 = I_{\mathcal{H}_r^\star}(\kappa_1)$, and $p_1 = F_{\mathcal{H}_r^\star}(\kappa_1)$. Then, $(\kappa_1)_{s_1,t_1,p_1} \in \mathcal{H}_r^\star$ and $s_1 + T_{\mathcal{R}}(\kappa_1) > 1$, $t_1 + I_{\mathcal{R}}(\kappa_1) \leq 1$, and $p_1 + F_{\mathcal{R}}(\kappa_1) \leq 1$ so that $\mathcal{R} \in Q_{\tau TIF}((\kappa_1)_{s_1,t_1,p_1}, r)$. Now, $(\kappa_1)_{s_1,t_1,p_1} \in \mathcal{H}_r^\star$ implies there is at least one $\kappa' \in \mathcal{S}$ such that $T_{\mathcal{D}}(\kappa') + T_{\mathcal{H}}(\kappa') - 1 > T_{\mathcal{C}}(\kappa')$, $I_{\mathcal{D}}(\kappa') + I_{\mathcal{H}}(\kappa') - 1 \leq I_{\mathcal{C}}(\kappa')$, and $F_{\mathcal{D}}(\kappa') + F_{\mathcal{H}}(\kappa') - 1 \leq F_{\mathcal{C}}(\kappa')$, for all $\mathcal{L}^T(\mathcal{C}) \geq r$, $\mathcal{L}^I(\mathcal{C}) \leq 1-r$, $\mathcal{L}^F(\mathcal{C}) \leq 1-r$, and $\mathcal{D} \in Q_{\tau TIF}((\kappa_1)_{s_1,t_1,p_1}, r)$. That is also true for $\mathcal{R}$. So there is at least one $\kappa'' \in \mathcal{S}$ such that $T_{\mathcal{R}}(\kappa'') + T_{\mathcal{H}}(\kappa'') - 1 > T_{\mathcal{C}}(\kappa'')$, $I_{\mathcal{R}}(\kappa'') + I_{\mathcal{H}}(\kappa'') - 1 \leq I_{\mathcal{C}}(\kappa'')$, and $F_{\mathcal{R}}(x'') + F_{\mathcal{H}}(\kappa'') - 1 \leq F_{\mathcal{C}}(\kappa'')$. Since $\mathcal{R} \in Q_{\tau TIF}(\kappa_{s,t,p}, r)$ and $\mathcal{R}$ is arbitrary; then $T_{\mathcal{H}_r^\star}(\kappa) > s$, $I_{\mathcal{H}_r^\star}(\kappa) \leq t$ and $T_{\mathcal{H}_r^\star}(\kappa) \leq p$. It is a contradiction for (4). Thus, $\mathcal{H}_r^\star \geq \mathbb{C}_{\tau TIF}(\mathcal{H}_r^\star, r)$.

(4) ($\Rightarrow$) Can be easily established using standard technique.

(5) ($\Rightarrow$) Since $\mathcal{H}, \mathcal{R} \leq \mathcal{H} \cup \mathcal{R}$. By (1), $\mathcal{H}_r^\star \leq (\mathcal{H} \cup \mathcal{R})_r^\star$ and $\mathcal{R}_r^\star \leq (\mathcal{H} \cup \mathcal{R})_r^\star$. Hence, $\mathcal{H}_r^\star \cup \mathcal{B}_r^\star \leq (\mathcal{H} \cup \mathcal{R})_r^\star$.

($\Leftarrow$) Suppose $(\mathcal{H}_r^\star \cup \mathcal{R}_r^\star) \not\geq (\mathcal{H} \cup \mathcal{R})_r^\star$. Then, there exists $s, t, p \in I_0$ and $\kappa \in \mathcal{S}$ such that

$$T_{(\mathcal{H}_r^\star \cup \mathcal{R}_r^\star)}(\kappa) < s \leq T_{(\mathcal{H} \cup \mathcal{R})_r^\star}(\kappa), \; I_{(\mathcal{H}_r^\star \cup \mathcal{R}_r^\star)}(\kappa) \geq t > I_{(\mathcal{H} \cup \mathcal{R})_r^\star}(\kappa), \; F_{(\mathcal{H}_r^\star \cup \mathcal{R}_r^\star)}(\kappa) \geq p > F_{(\mathcal{H} \cup \mathcal{R})_r^\star}(\kappa). \tag{5}$$

Since $T_{(\mathcal{H}_r^\star \cup \mathcal{R}_r^\star)}(\kappa) < s$, $I_{(\mathcal{H}_r^\star \cup \mathcal{R}_r^\star)}(\kappa) \geq t$, and $F_{(\mathcal{H}_r^\star \cup \mathcal{R}_r^\star)}(\kappa) \geq p$, we have $T_{\mathcal{H}_r^\star}(\kappa) < s$, $I_{\mathcal{H}_r^\star}(\kappa) \geq t$, $F_{\mathcal{H}_r^\star}(\kappa) \geq p$ or $T_{\mathcal{R}_r^\star}(\kappa) < t$, $I_{\mathcal{R}_r^\star}(\kappa) \geq t$, $F_{\mathcal{R}_r^\star}(\kappa) \geq t$. So, there exists $\mathcal{D}_1 \in Q_{\tau TIF}(x_{s,t,p}, r)$ such that for every $\kappa_1 \in \mathcal{S}$ and for some $\mathcal{L}^T(\mathcal{C}_1) \geq r$, $\mathcal{L}^I(\mathcal{C}_1) \leq 1-r$, $\mathcal{L}^F(\mathcal{C}_1) \leq 1-r$, we have

$$T_{\mathcal{D}_1}(\kappa_1) + T_{\mathcal{H}}(\kappa_1) - 1 \leq T_{\mathcal{C}_1}(\kappa_1), \quad I_{\mathcal{D}_1}(\kappa_1) + I_{\mathcal{H}}(\kappa_1) - 1 > I_{\mathcal{C}_1}(\kappa_1), \quad F_{\mathcal{D}_1}(\kappa_1) + F_{\mathcal{H}}(\kappa_1) - 1 > F_{\mathcal{C}_1}(\kappa_1).$$

Similarly, there exists $\mathcal{D}_2 \in Q_{\tau TIF}(x_{s,t,p}, r)$ such that for every $\kappa_1 \in \mathcal{S}$ and for some $\mathcal{L}^T(\mathcal{C}_2) \geq r$, $\mathcal{L}^I(\mathcal{C}_2) \leq 1-r$, $\mathcal{L}^F(\mathcal{C}_2) \leq 1-r$, we have

$$T_{\mathcal{D}_2}(\kappa_1) + T_{\mathcal{H}}(\kappa_1) - 1 \leq T_{\mathcal{C}_2}(\kappa_1), \quad I_{\mathcal{D}_2}(\kappa_1) + I_{\mathcal{H}}(\kappa_1) - 1 > I_{\mathcal{C}_2}(\kappa_1), \quad F_{\mathcal{D}_2}(\kappa_1) + F_{\mathcal{H}}(\kappa_1) - 1 > F_{\mathcal{C}_2}(\kappa_1).$$

Since $\mathcal{D} = \mathcal{D}_1 \wedge \mathcal{D}_2 \in Q_{\tau TIF}(x_{s,t,p}, r)$ and by (L_3), $\mathcal{L}^T(\mathcal{C}_1 \cup \mathcal{C}_2) \geq \mathcal{L}^T(\mathcal{C}_1) \cap \mathcal{L}^T(\mathcal{C}_2) \geq r$, $\mathcal{L}^I(\mathcal{C}_1 \cup \mathcal{C}_2) \leq \mathcal{L}^I(\mathcal{C}_1) \cup \mathcal{L}^I(\mathcal{C}_2) \leq 1-r$, and $\mathcal{L}^F(\mathcal{C}_1 \cup \mathcal{C}_2) \leq \mathcal{L}^T(\mathcal{C}_1) \cup \mathcal{L}^T(\mathcal{C}_2) \leq 1-r$. Thus, for every $\kappa_1 \in \mathcal{S}$,

$$\begin{aligned} T_{\mathcal{D}}(\kappa_1) + T_{\mathcal{R}\cup\mathcal{H}}(\kappa_1) - 1 &\leq T_{\mathcal{C}_1\cup\mathcal{C}_2}(\kappa_1), \\ I_{\mathcal{D}}(\kappa_1) + I_{\mathcal{R}\cup\mathcal{H}}(\kappa_1) - 1 &\geq I_{\mathcal{C}_1\cup\mathcal{C}_2}(\kappa_1), \\ F_{\mathcal{D}}(\kappa_1) + F_{\mathcal{R}\cup\mathcal{H}}(\kappa_1) &\geq F_{\mathcal{C}_1\cup\mathcal{C}_2}(\kappa_1). \end{aligned}$$

Therefore, $T_{(\mathcal{H}\cup\mathcal{R})_r^\star}(\kappa) < s$, $I_{(\mathcal{H}\cup\mathcal{R})_r^\star}(\kappa) \geq t$, and $F_{(\mathcal{H}\cup\mathcal{R})_r^\star}(\kappa) \geq p$. So, we arrive at a contradiction for (5). Hence, $(\mathcal{H}_r^\star \cup \mathcal{R}_r^\star) \geq (\mathcal{H} \cup \mathcal{R})_r^\star$.

(6), (7), and (8) can be easily established using the standard technique. □

Example 4. *Let $\mathcal{S} = \{a, b\}$. Define $\mathcal{R}, \mathcal{C}, \mathcal{H} \in \mathcal{S}$ as follows:*

$\mathcal{R}_1 = \langle(0.5, 0.5, 0.5), (0.5, 0.5, 0.5), (0.5, 0.5, 0.5)\rangle$; $\mathcal{R}_2 = \langle(0.4, 0.4, 0.4), (0.1, 0.1, 0.1), (0.1, 0.1, 0.1)\rangle$;

$\mathcal{R}_3 = \langle(0.3, 0.3, 0.3), (0.1, 0.1, 0.1), (0.1, 0.1, 0.1)$; $\mathcal{C}_1 = \langle(0.3, 0.3, 0.3), (0.3, 0.3, 0.3), (0.1, 0.1, 0.1)\rangle$;

$\mathcal{C}_2 = \langle(0.2, 0.2, 0.2), (0.2, 0.2, 0.2), (0.1, 0.1, 0.1)$; $\mathcal{C}_3 = \langle(0.1, 0.1, 0.1), (0.1, 0.1, 0.1), (0.1, 0.1, 0.1)\rangle$.

Define $\tau^{TIF}, \mathcal{L}^{TIF} : I^X \to I$ as follows:

$$\tau^T(\mathcal{H}) = \begin{cases} 1, & \text{if } \mathcal{H} = (0,1,1), \\ 1, & \text{if } \mathcal{H} = (1,0,0), \\ \frac{1}{2}, & \text{if } \mathcal{H} = \mathcal{R}_1; \end{cases} \qquad \mathcal{L}^T(\mathcal{H}) = \begin{cases} 1, & \text{if } \mathcal{H} = (0,1,1), \\ \frac{1}{2}, & \text{if } \mathcal{H} = \mathcal{C}_1, \\ \frac{2}{3}, & \text{if } \underline{0} < \mathcal{H} < \mathcal{C}_1; \end{cases}$$

$$\tau^I(\mathcal{H}) = \begin{cases} 0, & \text{if } \mathcal{H} = (0,1,1), \\ 0, & \text{if } \mathcal{H} = (1,0,0), \\ \frac{1}{2}, & \text{if } \mathcal{H} = \mathcal{R}_2; \end{cases} \qquad \mathcal{L}^I(\mathcal{R}) = \begin{cases} 0, & \text{if } \mathcal{H} = (0,1,1), \\ \frac{1}{2}, & \text{if } \mathcal{H} = \mathcal{C}_2, \\ \frac{1}{4}, & \text{if } \underline{0} < \mathcal{H} < \mathcal{C}_2; \end{cases}$$

$$\tau^F(\mathcal{H}) = \begin{cases} 0, & \text{if } \mathcal{H} = (0,1,1), \\ 0, & \text{if } \mathcal{H} = (1,0,0), \\ \frac{1}{2}, & \text{if } \mathcal{H} = \mathcal{R}_3; \end{cases} \qquad \mathcal{L}^F(\mathcal{H}) = \begin{cases} 0, & \text{if } \mathcal{H} = (0,1,1), \\ \frac{1}{2}, & \text{if } \mathcal{H} = \mathcal{C}_3, \\ \frac{1}{4}, & \text{if } \underline{0} < \mathcal{H} < \mathcal{C}_3. \end{cases}$$

Let $\mathcal{G} = \langle(0.4, 0.4, 0.4), (0.4, 0.4, 0.4), (0.4, 0.4, 0.4)\rangle$. Then, $\mathcal{G}^{\star}_{\frac{1}{2}} = \mathcal{R}_1$.

Theorem 4. *Let $\{\mathcal{H}_i\}_{i \in J} \subset I^{\mathcal{S}}$ be a family of single-valued neutrosophic sets on $\mathcal{S}$ and $(\mathcal{S}, \tau^{TIF}, \mathcal{L}^{TIF})$ be an* **SVNITS**. *Then,*

(1) $(\bigcup(\mathcal{H}_i)^{\star}_r : i \in J) \leq (\bigcup \mathcal{H}_i : i \in J)^{\star}_r$;
(2) $(\bigcap(\mathcal{H}_i)^{\star}_r : i \in J) \geq (\bigcap \mathcal{H}_i : i \in J)^{\star}_r$.

Proof. (1) Since $\mathcal{H}_i \leq \bigcup \mathcal{H}_i$ for all $i \in J$, and by Theorem 3 (1), we obtain $(\bigcup(\mathcal{H}_i)^{\star}_r, i \in J) \leq (\bigcup \mathcal{H}_i, i \in J)^{\star}_r$. Then, (1) holds.

(2) Easy, so omitted. □

Remark 3. *Let $(\mathcal{S}, \tau^{TIF}, \mathcal{L}^{TIF})$ be an* **SVNITS** *and $\mathcal{H} \in I^{\mathcal{S}}$, we can define*

$$\mathbb{C}^{\star}_{\tau^{TIF}}(\mathcal{H}, r) = \mathcal{H} \cup \mathcal{H}^{\star}_r, \qquad int^{\star}_{\tau^{TIF}}(\mathcal{H}, r) = \mathcal{H} \wedge [\underline{1} - (\underline{1} - \mathcal{H})^{\star}_r].$$

It is clear, $\mathbb{C}^{\star}_{\tau^{TIF}}$ is a single-valued neutrosophic closure operator and $(\tau^{T\star}(\mathcal{L}^T), \tau^{I\star}(\mathcal{L}^I), \tau^{F\star}(\mathcal{L}^F)$ is the single-valued neutrosophic topology generated by $\mathbb{C}^{\star}_{\tau^{TIF}}$, i.e.,

$$\tau^{\star}(\mathcal{I})(\mathcal{H}) = \bigcup\{r \mid \mathbb{C}^{\star}_{\tau^{TIF}}(\underline{1} - \mathcal{H}, r) = \underline{1} - \mathcal{H}\}.$$

Now, if $\mathcal{L}^{TIF} = \mathcal{L}^{TIF}_0$, then, $\mathbb{C}^{\star}_{\tau^{TIF}}(\mathcal{H}, r) = \mathcal{H}^{\star}_r \cup \mathcal{H} = \mathbb{C}^{\star}_{\tau^{TIF}}(\mathcal{H}, r) \cup \mathcal{H} = \mathbb{C}_{\tau^{TIF}}(\mathcal{H}, r)$, for $\mathcal{H} \in I^{\mathcal{S}}$. So, $\tau^{TIF\star}(\mathcal{L}^{TIF}_0) = \tau^{TIF}$.

Proposition 3. *Let $(\mathcal{S}, \tau^{TIF}, \mathcal{L}^{TIF})$ be an* **SVNITS**, *$r \in I_0$, and $\mathcal{H} \in I^{\mathcal{S}}$. Then,*

(1) $\mathbb{C}^{\star}_{\tau^{TIF}}(\underline{1}, r) = \underline{1}$;

(2) $\mathbb{C}^{\star}_{\tau^{TIF}}(\underline{0}, r) = \underline{0}$;
(3) $int^{\star}_{\tau^{TIF}}(\mathcal{H} \cup \mathcal{R}, r) \leq int^{\star}_{\tau^{TIF}}(\mathcal{H}, r) \cup int^{\star}_{\tau^{TIF}}(\mathcal{R}, r)$;
(4) $int^{\star}_{\tau^{TIF}}(\mathcal{H}, r) \leq \mathcal{H} \leq \mathbb{C}^{\star}_{\tau^{TIF}}(\mathcal{H}, r) \leq \mathbb{C}_{\tau^{TIF}}(\mathcal{H}, r)$;
(5) $\mathbb{C}^{\star}_{\tau^{TIF}}(\underline{1} - \mathcal{H}, r) = \underline{1} - int^{\star}_{\tau^{TIF}}(\mathcal{H}, r)$ and $\underline{1} - \mathbb{C}^{\star}_{\tau^{TIF}}\star(\mathcal{H}, r) = int^{\star}_{\tau^{TIF}}(\underline{1} - \mathcal{H}, r)$;
(6) $int^{\star}_{\tau^{TIF}}(\mathcal{H} \cap \mathcal{R}, r) = int^{\star}_{\tau^{TIF}}(\mathcal{H}, r) \cap int^{\star}_{\tau^{TIF}}(\mathcal{R}, r)$.

Proof. Follows directly from definitions of $\mathbb{C}^{\star}_{\tau^{TIF}}$, $int^{\star}_{\tau^{TIF}}$, $\mathbb{C}_{\tau^{TIF}}$, and Theorem 3 (5). □

Theorem 5. *Let $(\mathcal{S}, \tau_1^{TIF}, \mathcal{L}^{TIF})$ and $(\mathcal{S}, \tau_2^{TIF}, \mathcal{L}^{TIF})$ be* **SVNTS's** *and $\tau_1^{TIF} \leq \tau_2^{TIF}$. Then, $\mathcal{H}^{\star}_r(\tau_2^{TIF}, \mathcal{L}^{TIF}) \leq \mathcal{H}^{\star}_r(\tau_1^{TIF}, \mathcal{L}^{TIF})$.*

Proof. Suppose $\mathcal{H}^{\star}_r(\tau_2^{TIF}, \mathcal{L}^{TIF}) \not\leq \mathcal{H}^{\star}_r(\tau_1^{TIF}, \mathcal{L}^{TIF})$. Then, there exists $s, t, p \in I_0$, $\kappa \in \mathcal{S}$ such that

$$T_{\mathcal{H}^{\star}_r(\tau_2^{TIF}, \mathcal{L}^{TIF})}(\kappa) \geq s > T_{\mathcal{H}^{\star}_r(\tau_1^{TIF}, \mathcal{L}^{TIF})}(\kappa),$$

$$I_{\mathcal{H}^{\star}_r(\tau_2^{TIF}, \mathcal{L}^{TIF})}(\kappa) < t \leq I_{\mathcal{H}^{\star}_r(\tau_1^{TIF}, \mathcal{L}^{TIF})}(\kappa), \tag{6}$$

$$F_{\mathcal{H}^{\star}_r(\tau_2^{TIF}, \mathcal{L}^{TIF})}(\kappa) < t \leq F_{\mathcal{H}^{\star}_r(\tau_1^{TIF}, \mathcal{L}^{TIF})}(\kappa).$$

Since $T_{\mathcal{H}^{\star}_r(\tau_1^{TIF}, \mathcal{L}^{TIF})}(\kappa) < s$, $I_{\mathcal{H}^{\star}_r(\tau_1^{TIF}, \mathcal{L}^{TIF})}(\kappa) \geq t$, $F_{\mathcal{H}^{\star}_r(\tau_1^{TIF}, \mathcal{L}^{TIF})}(\kappa) \geq p$, there exists $\mathcal{D} \in Q_{\tau_1^{TIF}}(x_{s,t,p}, r)$ with $\mathcal{L}^T(\mathcal{C}_1) \geq r$, $\mathcal{L}^I(\mathcal{C}_1) \leq 1 - r$ and $\mathcal{L}^F(\mathcal{C}_1) \leq 1 - r$, such that for any $\kappa_1 \in \mathcal{S}$,

$$T_{\mathcal{D}}(\kappa_1) + T_{\mathcal{H}}(\kappa_1) - 1 \leq T_{\mathcal{C}}\kappa_1), \quad I_{\mathcal{D}}(\kappa_1) + I_{\mathcal{H}}(\kappa_1) - 1 > I_{\mathcal{C}}(\kappa_1), \quad F_{\mathcal{D}}(\kappa_1) + F_{\mathcal{H}}(\kappa_1) - 1 > F_{\mathcal{C}}(\kappa_1).$$

Since $\tau_1^{TIF} \leq \tau_2^{TIF}$, $\mathcal{D} \in Q_{\tau_2^{TIF}}(x_{s,t,p}, r)$. Thus, $T_{\mathcal{H}^{\star}_r(\tau_2^{TIF}, \mathcal{L}^{TIF})}(\kappa) < s$, $I_{\mathcal{H}^{\star}_r(\tau_2^{TIF}, \mathcal{L}^{TIF}))}(\kappa) \geq t$, $F_{\mathcal{H}^{\star}_r(\tau_2^{TIF}, \mathcal{L}^{TIF})}(\kappa) \geq p$. It is a contradiction for Equation (6). □

Theorem 6. *Let $(\mathcal{S}, \tau^{TIF}, \mathcal{L}_1^{TIF})$ and $(\mathcal{S}, \tau^{TIF}, \mathcal{L}_2^{TIF})$ be* **SVNTS's** *and $\mathcal{L}_1^{TIF} \leq \mathcal{L}_2^{TIF}$. Then, $\mathcal{H}^{\star}_r(\mathcal{L}_1^{TIF}, \tau^{TIF}) \geq \mathcal{H}^{\star}_r(\mathcal{L}_2^{TIF}, \tau^{TIF})$.*

Proof. Clear. □

Definition 14. *Let Θ be a subset of $I^{\mathcal{S}}$, and $\underline{0} \notin \Theta$. A mapping $\beta^T, \beta^I, \beta^F : \Theta \to I$ is called a single-valued neutrosophic base on $\mathcal{S}$ if it satisfies the following conditions:*

(1) $\beta^T(\underline{1}) = 1$ and $\beta^I(\underline{1}) = \beta^F(\underline{1}) = 0$;
(2) For all $\mathcal{H}, \mathcal{R} \in \Theta$,

$$\beta^T(\mathcal{H} \cap \mathcal{R}) \geq \beta^T(\mathcal{H}) \cap \beta^T(\mathcal{R}), \quad \beta^I(\mathcal{H} \cap \mathcal{R}) \leq \beta^I(\mathcal{H}) \cup \beta^I(\mathcal{R}), \quad \beta^F(\mathcal{H} \cap \mathcal{R}) \leq \beta^F(\mathcal{H}) \cup \beta^F(\mathcal{R}).$$

Theorem 7. *Define a mapping $\beta : \Theta \to I$ on $\mathcal{S}$ by*

$$\beta^I(\mathcal{H}) = \bigcup\{\tau^T(\mathcal{R}) \cap \mathcal{I}^T(\mathcal{C}) |\ \mathcal{H} = \mathcal{R} \cap (\underline{1} - \mathcal{C})\},$$

$$\beta^I(\mathcal{H}) = \bigcap\{\tau^I(\mathcal{R}) \cup \mathcal{I}^I(\mathcal{C}) |\ \mathcal{H} = \mathcal{R} \cap (\underline{1} - \mathcal{C})\},$$

$$\beta^F(\mathcal{H}) = \bigcap\{\tau^F(\mathcal{R}) \cup \mathcal{I}^F(\mathcal{C}) |\ \mathcal{H} = \mathcal{R} \cap (\underline{1} - \mathcal{C})\}.$$

Then, β^{TIF} is a base for the single-valued neutrosophic topology $\tau^{TIF\star}$.

Proof. (1) Since $\mathcal{L}^T(\underline{0}) = 1$ and $\mathcal{L}^I(\underline{0}) = \mathcal{L}^F(\underline{0}) = 0$, we have $\beta^T(\underline{1}) = 1$ and $\beta^I(\underline{1}) = \beta^F(\underline{1}) = 0$;
(2) Suppose that there exists $\mathcal{H}_1, \mathcal{H}_2 \in \Theta$ such that

$$\begin{aligned}\beta^T(\mathcal{H}_1 \cap \mathcal{H}_2) &\ngeq \beta^T(\mathcal{H}_1) \cap \beta^T(\mathcal{H}_2),\\ \beta^I(\mathcal{H}_1 \cap \mathcal{H}_2) &\nleq \beta^I(\mathcal{H}_1) \cup \beta^I(\mathcal{H}_2),\\ \beta^F(\mathcal{H}_1 \cap \mathcal{H}_2) &\nleq \beta^F(\mathcal{H}_1) \cup \beta^F(\mathcal{H}_2).\end{aligned}$$

There exists $s, t, p \in I_0$ and $\kappa \in \mathcal{S}$ such that

$$\beta^T(\mathcal{H}_1 \cap \mathcal{H}_2)(\kappa) < s \leq \beta^T(\mathcal{H}_1)(x) \cap \beta^T(\mathcal{H}_2)(\kappa),$$

$$\beta^I(\mathcal{H}_1 \cap \mathcal{H}_2)(\kappa) \geq t > \beta^I(\mathcal{H}_1)(\kappa) \cap \beta^I(\mathcal{H}_2)(\kappa), \tag{7}$$

$$\beta^F(\mathcal{H}_1 \cap \mathcal{H}_2)(\kappa) \geq p > \beta^F(\mathcal{H}_1)(\kappa) \cup \beta^F(\mathcal{H}_2)(\kappa).$$

Since $\beta^T(\mathcal{H}_1)(\kappa) \geq s$, $\beta^I(\mathcal{H}_1)(\kappa) < t$, $\beta^F(\mathcal{H}_1)(\kappa) < p$, and $\beta^T(\mathcal{H}_2)(\kappa) \geq s$, $\beta^I(\mathcal{H}_2)(\kappa) < t$, $\beta^F(\mathcal{H}_2)(\kappa) < p$, then there exists $\mathcal{R}_1, \mathcal{R}_1, \mathcal{C}_1, \mathcal{C}_2 \in \Theta$ with $\mathcal{H}_1 = \mathcal{R}_1 \cap (\underline{1} - \mathcal{C}_1)$ and $\mathcal{H}_2 = \mathcal{R}_2 \cap (\underline{1} - \mathcal{C}_2)$, such that $\beta^T(\mathcal{H}_1) \geq \tau^T(\mathcal{R}_1) \cap \mathcal{L}^T(\mathcal{C}_1) \geq s$, $\beta^I(\mathcal{H}_1) \leq \tau^I(\mathcal{R}_1) \cup \mathcal{L}^I(\mathcal{C}_1) < t$, $\beta^F(\mathcal{H}_1) \leq \tau^F(\mathcal{R}_1) \cup \mathcal{L}^F(\mathcal{C}_1) < p$, and $\beta^T(\mathcal{H}_2) \geq \tau^T(\mathcal{R}_2) \cap \mathcal{L}^T(\mathcal{C}_2) \geq s$, $\beta^I(\mathcal{H}_2) \leq \tau^I(\mathcal{R}_2) \cup \mathcal{L}^I(\mathcal{C}_2) < t$, $\beta^F(\mathcal{H}_2) \leq \tau^F(\mathcal{R}_2) \cup \mathcal{L}^F(\mathcal{C}_2) < p$. Therefore,

$$\begin{aligned}\mathcal{H}_1 \cap \mathcal{H}_2 &= (\mathcal{R}_1 \cap (\underline{1} - \mathcal{C}_1)) \cap (\mathcal{R}_2 \cap (\underline{1} - \mathcal{C}_2))\\ &= (\mathcal{R}_1 \cap \mathcal{R}_2) \cap ((\underline{1} - \mathcal{C}_1) \cap (\underline{1} - \mathcal{C}_2))\\ &= (\mathcal{R}_1 \cap \mathcal{R}_2) \cap (\underline{1} - (\mathcal{C}_1 \cup \mathcal{C}_2)).\end{aligned}$$

Hence, from Definition 14, we have

$$\begin{aligned}\beta^T(\mathcal{H}_1 \cap \mathcal{H}_2) &\geq \tau^T(\mathcal{R}_1 \cap \mathcal{R}_2) \cap \mathcal{L}^T(\mathcal{C}_1 \cup \mathcal{C}_2)\\ &\geq \tau^T(\mathcal{R}_1) \cap \tau^T(\mathcal{R}_2) \cap \mathcal{L}^T(\mathcal{C}_1) \cap \mathcal{L}^T(\mathcal{C}_2)\\ &= (\tau^T(\mathcal{R}_1) \cap \mathcal{L}^T(\mathcal{C}_1)) \cap (\tau^T(\mathcal{R}_2) \cap \mathcal{L}^T(\mathcal{C}_2)) \geq s,\end{aligned}$$

$$\begin{aligned}\beta^I(\mathcal{H}_1 \cap \mathcal{H}_2) &\leq \tau^I(\mathcal{R}_1 \cap \mathcal{R}_2) \cup \mathcal{L}^I(\mathcal{C}_1 \cup \mathcal{C}_2)\\ &\leq \tau^I(\mathcal{R}_1) \cup \tau^I(\mathcal{R}_2) \cup \mathcal{L}^I(\mathcal{C}_1) \cup \mathcal{L}^I(\mathcal{C}_2)\\ &= (\tau^I(\mathcal{R}_1) \cup \mathcal{L}^F(\mathcal{C}_1)) \cup (\tau^I(\mathcal{R}_2) \cup \mathcal{L}^I(\mathcal{C}_2)) < t,\end{aligned}$$

$$\begin{aligned}\beta^F(\mathcal{H}_1 \cap \mathcal{H}_2) &\leq \tau^F(\mathcal{R}_1 \cap \mathcal{R}_2) \cup \mathcal{L}^F(\mathcal{C}_1 \cup \mathcal{C}_2)\\ &\leq \tau^F(\mathcal{R}_1) \cup \tau^F(\mathcal{R}_2) \cup \mathcal{L}^F(\mathcal{C}_1) \cup \mathcal{L}^F(\mathcal{C}_2)\\ &= (\tau^F(\mathcal{R}_1) \cup \mathcal{L}^F(\mathcal{C}_1)) \cup (\tau^F(\mathcal{R}_2) \cup \mathcal{L}^F(\mathcal{C}_2)) < p.\end{aligned}$$

It is a contradiction for Equation (7). Thus,

$$\beta^T(\mathcal{H}_1 \cap \mathcal{H}_2) \geq \beta^T(\mathcal{H}_1) \cap \beta^T(\mathcal{H}_2), \beta^I(\mathcal{H}_1 \cap \mathcal{H}_2) \leq \beta^I(\mathcal{H}_1) \cup \beta^I(\mathcal{H}_2), \beta^F(\mathcal{H}_1 \cap \mathcal{H}_2) \leq \beta^F(\mathcal{H}_1) \cup \beta^F(\mathcal{H}_2).$$

□

Theorem 8. *Let $(\mathcal{S}, \tau^{TIF})$ be an* **SVNTS**, *and $\mathcal{L}_1^{TIF}$ and $\mathcal{L}_1^{TIF}$ be two single-valued neutrosophic ideals on $\mathcal{S}$. Then, for every $r \in I_0$ and $\mathcal{H} \in I^{\mathcal{S}}$,*

(1) $\mathcal{H}_r^\star(\mathcal{L}_1^{TIF} \cap \mathcal{L}_2^{TIF}, \tau^{TIF}) = \mathcal{H}_r^\star(\mathcal{L}_1^{TIF}, \tau^{TIF}) \cup \mathcal{H}_r^\star(\mathcal{L}_2^{TIF}, \tau^{TIF})$,
(2) $\mathcal{H}_r^\star(\mathcal{L}_1^{TIF} \cup \mathcal{L}_2^{TIF}, \tau) = \mathcal{H}_r^\star(\mathcal{L}_1^{TIF}, \tau^{T\star}(\mathcal{L}_2^{TIF})) \cap \mathcal{H}^\star(\mathcal{L}_2^{TIF}, \tau^{T\star}(\mathcal{L}_1^{TIF}))$.

Proof. (1) Suppose that $\mathcal{H}_r^\star(\mathcal{L}_1^{TIF} \cap \mathcal{L}_2^{TIF}, \tau^{TIF}) \not\leq \mathcal{H}_r^\star(\mathcal{L}_1^{TIF}, \tau^{TIF}) \cup \mathcal{H}_r^\star(\mathcal{L}_2^{TIF}, \tau^{TIF})$, there exists $\kappa \in \mathcal{S}$ and $s, t, p \in I_0$ such that

$$T_{\mathcal{H}_r^\star(\mathcal{L}_1^T \cap \mathcal{L}_2^T, \tau^T)}(\kappa) \geq s > T_{\mathcal{H}_r^\star(\mathcal{L}_1^T, \tau^T)}(\kappa) \cup T_{\mathcal{H}_r^\star(\mathcal{L}_2^T, \tau^T)}(\kappa), \tag{8}$$
$$I_{\mathcal{H}_r^\star(\mathcal{L}_1^I \cap \mathcal{L}_2^I, \tau^I)}(\kappa) < t \leq I_{\mathcal{H}_r^\star(\mathcal{L}_1^I, \tau^I)}(\kappa) \cup I_{\mathcal{H}_r^\star(\mathcal{L}_2^I, \tau^I)}(\kappa),$$

$$F_{\mathcal{H}_r^\star(\mathcal{L}_1^F \cap \mathcal{L}_2^F, \tau^F)}(\kappa) < p \leq F_{\mathcal{H}_r^\star(\mathcal{L}_1^F, \tau^F)}(\kappa) \cap F_{\mathcal{H}_r^\star(\mathcal{L}_2^F, \tau^F)}(\kappa).$$

Since $T_{\mathcal{H}_r^\star(\mathcal{L}_1^T, \tau^T)}(\kappa) \cup T_{\mathcal{H}_r^\star(\mathcal{L}_2^T, \tau^T)}(\kappa) < s$, $I_{\mathcal{H}_r^\star(\mathcal{L}_1^I, \tau^I)}(\kappa) \cap I_{\mathcal{H}_r^\star(\mathcal{L}_2^I, \tau^I)}(\kappa) \geq t$, $F_{\mathcal{H}_r^\star(\mathcal{L}_1^F, \tau^F)}(\kappa) \cap F_{\mathcal{H}_r^\star(\mathcal{L}_2^F, \tau^F)}(\kappa) \geq p$, we have, $T_{\mathcal{H}_r^\star(\mathcal{L}_1^T, \tau^T)}(\kappa) < s$, $I_{\mathcal{H}_r^\star(\mathcal{L}_1^I, \tau^I)}(\kappa) \geq t$, $F_{\mathcal{H}_r^\star(\mathcal{L}_1^F, \tau^F)}(\kappa) \geq p$, and $I_{\mathcal{H}_r^\star(\mathcal{L}_2^I, \tau^I)}(\kappa) < s$, $I_{\mathcal{H}_r^\star(\mathcal{L}_2^I, \tau^I)}(\kappa) \geq t$, $F_{\mathcal{H}_r^\star(\mathcal{L}_2^F, \tau^F)}(\kappa) \geq p$.

Now, $T_{\mathcal{H}_r^\star(\mathcal{L}_1^T, \tau^T)}(\kappa) < s$, $I_{\mathcal{H}_r^\star(\mathcal{L}_1^I, \tau^I)}(\kappa) \geq t$, $F_{\mathcal{H}_r^\star(\mathcal{L}_1^F, \tau^F)}(\kappa) \geq p$ implies that there exists $\mathcal{D}_1 \in Q_{\tau^{TIF}}(x_{s,t,p}, r)$ and for some $\mathcal{L}_1^T(\mathcal{C}_1) \geq r$, $\mathcal{L}_1^I(\mathcal{C}_1) \leq 1 - r$ and $\mathcal{L}_1^F(\mathcal{C}_1) \leq 1 - r$ such that for every $\kappa_1 \in \mathcal{S}$,

$$T_{\mathcal{D}_1}(\kappa_1) + T_{\mathcal{H}}(\kappa_1) - 1 \leq T_{\mathcal{C}_1}(\kappa_1), \quad I_{\mathcal{D}_1}(\kappa_1) + I_{\mathcal{H}}(\kappa_1) - 1 \geq I_{\mathcal{C}_1}(\kappa_1), \quad F_{\mathcal{D}_1}(\kappa_1) + F_{\mathcal{H}}(\kappa_1) - 1 \geq F_{\mathcal{C}_1}(\kappa_1).$$

Once again, $T_{\mathcal{H}_r^\star(\mathcal{L}_2^T, \tau^T)}(\kappa) < s$, $I_{\mathcal{H}_r^\star(\mathcal{L}_2^I, \tau^I)}(\kappa) \geq t$, $F_{\mathcal{H}_r^\star(\mathcal{L}_2^F, \tau^F)}(\kappa) \geq p$, implies there exists $\mathcal{D}_2 \in Q_{\tau^{TIF}}(x_{s,t,p}, r)$ and for some $\mathcal{L}_2^T(\mathcal{C}_2) \geq r$, $\mathcal{L}_2^I(\mathcal{C}_2) \leq 1 - r$ and $\mathcal{L}_2^F(\mathcal{C}_2) \leq 1 - r$, such that for $\kappa_1 \in \mathcal{S}$,

$$T_{\mathcal{D}_2}(\kappa_1) + T_{\mathcal{H}}(\kappa_1) - 1 \leq T_{\mathcal{C}_2}(\kappa_1), \quad I_{\mathcal{D}_2}(\kappa_1) + I_{\mathcal{H}}(\kappa_1) - 1 \geq I_{\mathcal{C}_2}(\kappa), \quad F_{\mathcal{D}_2}(\kappa_1) + F_{\mathcal{H}}(\kappa_1) - 1 \geq F_{\mathcal{C}_2}(\kappa_1),$$

Therefore, for every $\kappa_1 \in \mathcal{S}$, we have

$$T_{\mathcal{D}_1 \cap \mathcal{D}_2}(\kappa_1) + T_{\mathcal{H}}(\kappa_1) - 1 \leq T_{\mathcal{C}_1 \cap \mathcal{C}_2}(\kappa_1), \quad I_{\mathcal{D}_1 \cup \mathcal{D}_2}(\kappa_1) + I_{\mathcal{H}}(\kappa_1) - 1 \geq I_{\mathcal{C}_1 \cup \mathcal{C}_2}(\kappa_1),$$

$$F_{\mathcal{D}_1 \cup \mathcal{D}_2}(\kappa_1) + F_{\mathcal{H}}(\kappa_1) - 1 \geq F_{\mathcal{C}_1 \cup \mathcal{C}_2}(\kappa_1).$$

Since $(\mathcal{D}_1 \wedge \mathcal{D}_2) \in Q_{\tau^{TIF}}(x_{s,t,p}, r)$ and $(\mathcal{L}_1^T \cap \mathcal{L}_2^T)(\mathcal{C}_1 \cap \mathcal{C}_2) \geq r$, $(\mathcal{L}_1^I \cap \mathcal{L}_2^I)(\mathcal{C}_1 \cup \mathcal{C}_2) \leq 1 - r$, and $(\mathcal{L}_1^F \cap \mathcal{L}_2^F)(\mathcal{C}_1 \cup \mathcal{C}_2) \geq 1 - r$ we have $T_{\mathcal{H}_r^\star(\mathcal{L}_1^T \cap \mathcal{L}_2^T, \tau^T)}(\kappa) \leq s$, $I_{\mathcal{H}_r^\star(\mathcal{L}_1^I \cap \mathcal{L}_2^I, \tau^I)}(\kappa) > t$, and $F_{\mathcal{H}_r^\star(\mathcal{L}_1^F \cap \mathcal{L}_2^F, \tau^F)}(\kappa) > t$ and this is a contradiction for Equation (8). So that

$$\mathcal{H}_r^\star(\mathcal{L}_1^{TIF} \cap \mathcal{L}_2^{TIF}, \tau^{TIF}) \leq \mathcal{H}_r^\star(\mathcal{L}_1^{TIF}, \tau^{TIF}) \cup \mathcal{H}_r^\star(\mathcal{L}_2^{TIF}, \tau^{TIF}).$$

On the opposite direction, $\mathcal{L}_1^{TIF} \geq \mathcal{L}_1^{TIF} \cap \mathcal{L}_2^{TIF}$ and $\mathcal{L}_2^{TIF} \geq \mathcal{L}_1^{TIF} \cap \mathcal{L}_2^{TIF}$, so by Theorem 3 (2),

$$\mathcal{H}_r^\star(\mathcal{L}_1^{TIF} \cap \mathcal{L}_2^{TIF}, \tau^T) \geq \mathcal{H}_r^\star(\mathcal{L}_1^{TIF}, \tau^{TIF}) \cup \mathcal{H}_r^\star(\mathcal{L}_2^{TIF}, \tau^{TIF}).$$

Then,

$$\mathcal{H}_r^\star(\mathcal{L}_1^{TIF} \cap \mathcal{L}_2^{TIF}, \tau^{TIF}) = \mathcal{H}_r^\star(\mathcal{L}_1^{TIF}, \tau^{TIF}) \cup \mathcal{H}_r^\star(\mathcal{L}_2^{TIF}, \tau^{TIF}).$$

(2) Straightforward. □

The above theorem results in an important consequence. $\tau^{TIF\star}(\mathcal{L}^{TIF})$ and $[\tau^{TIF\star}(\mathcal{L}^{TIF})]^\star(\mathcal{L}^{TIF})$ (in short $\tau^{\star\star}$) are equal for any single-valued neutrosophic ideal on $\mathcal{S}$.

Corollary 1. *Let $(\mathcal{S}, \tau^{TIF}, \mathcal{L}^{TIF})$ be an* **SVNITS**. *For every $r \in I_0$ and $\mathcal{H} \in I^X$, $\mathcal{H}_r^\star(\mathcal{L}^{TIF}) = \mathcal{H}_r^\star(\mathcal{L}^{TIF}, \tau^{TIF\star})$ and $\tau^{TIF\star}(\mathcal{L}^{TIF}) = \tau^{TIF\star\star}$.*

Proof. Putting $\mathcal{L}_1^{TIF} = \mathcal{L}_2^{TIF}$ in Theorem 8 (2), we have the required result. □

Corollary 2. *Let $(\mathcal{S}, \tau^{TIF})$ be an* **SVNTS**, *and $\mathcal{L}_1^{TIF}$ and $\mathcal{L}_1^{TIF}$ be two single-valued neutrosophic ideals on $\mathcal{S}$. Then, for any $\mathcal{H} \in I^{\mathcal{S}}$ and $r \in I_0$,*

(1) $\tau^{T\star}(\mathcal{L}_1^{TIF} \cup \mathcal{I}_2^{TIF}) = (\tau^{TIF\star}(\mathcal{L}_2^{TIF}))^\star(\mathcal{L}_1^T) = (\tau^{TIF\star}(\mathcal{L}_1^{TIF}))^\star(\mathcal{L}_2^T)$,

(2) $\tau^{T\star}(\mathcal{L}_1^{TIF} \cap \mathcal{L}_2^{TIF}) = \tau^{TIF\star}(\mathcal{L}_1^{TIF}) \cap \tau^{T\star}(\mathcal{L}_2^{TIF})$.

Proof. Straightforward. □

Definition 15. *For an* **SVNTS** *$(\mathcal{S}, \tau^{TIF})$ with a single-valued neutrosophic ideal $\mathcal{I}^{TIF}$, τ^{TIF} is said to be single-valued neutrosophic ideal open compatible with $\mathcal{I}^{TIF}$, denoted by $\tau^{TIF} \sim \mathcal{L}^{TIF}$, if for each $\mathcal{H}, \mathcal{C} \in I^{\mathcal{S}}$ and $x_{s,t,p} \in \mathcal{H}$ with $\mathcal{L}^T(\mathcal{C}) \geq r$, $\mathcal{L}^I(\mathcal{C}) \leq 1 - r$, and $\mathcal{L}^F(\mathcal{C}) \leq 1 - r$, there exists $\mathcal{D} \in Q_{\tau^{TIF}}(x_t, r)$ such that $T_{\mathcal{D}}(\kappa) + T_{\mathcal{H}}(\kappa) - 1 \leq T_{\mathcal{C}}(\kappa)$, $I_{\mathcal{D}}(\kappa) + I_{\mathcal{H}}(\kappa) - 1 > I_{\mathcal{C}}(\kappa)$, and $F_{\mathcal{D}}(\kappa) + F_{\mathcal{H}}(\kappa) - 1 > F_{\mathcal{C}}(\kappa)$ holds for any $\kappa \in \mathcal{S}$, then $\mathcal{L}^T(\mathcal{H}) \geq r$, $\mathcal{L}^I(\mathcal{H}) \leq 1 - r$ and $\mathcal{L}^F(\mathcal{H}) \leq 1 - r$.*

Definition 16. *Let $\{\mathcal{R}_j\}_{j \in J}$ be an indexed family of a single-valued neutrosophic set of $\mathcal{S}$ such that $\mathcal{R}_j q \mathcal{H}$ for each $j \in J$, where $\mathcal{H} \in I^{\mathcal{S}}$. Then, $\{\mathcal{R}_j\}_{j \in J}$ is said to be a single-valued neutrosophic quasi-cover of $\mathcal{H}$ iff $T_{\mathcal{H}}(\kappa) + T_{\bigvee_{j \in J}(\mathcal{R}_j)}(\kappa) \geq 1$, $I_{\mathcal{H}}(\kappa) + I_{\bigvee_{j \in J}(\mathcal{R}_j)}(\kappa) < 1$, and $F_{\mathcal{H}}(\kappa) + F_{\bigvee_{j \in J}(\mathcal{R}_j)}(\kappa) < 1$, for every $\kappa \in \mathcal{S}$.*

Further, let $(\mathcal{S}, \tau^{TIF})$ be an **SVNTS**, for each $\tau^T(\mathcal{R}_j) \geq r$, $\tau^I(\mathcal{R}_j) \leq 1 - r$, and $\tau^F(\mathcal{R}_j) \leq 1 - r$. Then, any single-valued neutrosophic quasi-cover will be called single-valued neutrosophic quasi open-cover of $\mathcal{H}$.

Theorem 9. *Let $(\mathcal{S}, \tau^{TIF})$ be an* **SVNTS** *with single-valued neutrosophic ideal $\mathcal{L}^{TIF}$ on $\mathcal{S}$. Then, the following conditions are equivalent:*

(1) $\tau \sim \mathcal{L}$.

(2) *If for every $\mathcal{H} \in I^{\mathcal{S}}$ has a single-valued neutrosophic quasi open-cover of $\{\mathcal{R}_j\}_{j \in J}$ such that for each j, $T_{\mathcal{H}}(\kappa) + T_{R_j}(\kappa) - 1 \leq T_{\mathcal{C}}(\kappa)$, $I_{\mathcal{H}}(\kappa) + I_{R_j}(\kappa) - 1 > I_{\mathcal{C}}(\kappa)$, and $F_{\mathcal{H}}(\kappa) + F_{R_j}(\kappa) - 1 > F_{\mathcal{C}}(\kappa)$ for every $\kappa \in \mathcal{S}$ and for some $\mathcal{L}^T(\mathcal{C}) \geq r$, $\mathcal{L}^I(\mathcal{C}) \leq 1 - r$, and $\mathcal{L}^F(\mathcal{C}) \leq 1 - r$, then $\mathcal{L}^T(\mathcal{H}) \geq r$, $\mathcal{L}^I(\mathcal{H}) \leq 1 - r$, and $\mathcal{L}^F(\mathcal{H}) \leq 1 - r$,*

(3) *For every $\mathcal{H} \in I^{\mathcal{S}}$, $\mathcal{H} \wedge \mathcal{H}_r^\star = (0,1,1)$ implies $\mathcal{L}^T(\mathcal{H}) \geq r$, $\mathcal{L}^I(\mathcal{H}) \leq 1 - r$, and $\mathcal{L}^F(\mathcal{H}) \leq 1 - r$,*

(4) *For every $\mathcal{H} \in I^{\mathcal{S}}$, $\mathcal{L}^T(\widetilde{\mathcal{H}}) \geq r$, $\mathcal{L}^I(\widetilde{\mathcal{H}}) \leq 1 - r$, and $\mathcal{L}^F(\widetilde{\mathcal{H}}) \leq 1 - r$, where $\widetilde{\mathcal{H}} = \bigvee x_{s,t,p}$ such that $x_{s,t,p} \in \mathcal{H}$ but $x_{s,t,p} \notin \mathcal{H}_r^\star$,*

(5) *For every $\tau^{T\star}(\underline{1} - \mathcal{H}) \geq r$, $\tau^{I\star}(\underline{1} - \mathcal{H}) \leq 1 - r$, and $\tau^{F\star}(\underline{1} - \mathcal{H}) \leq 1 - r$ we have $\mathcal{L}^T(\widetilde{\mathcal{H}}) \geq r$, $\mathcal{L}^I(\widetilde{\mathcal{H}}) \leq 1 - r$, and $\mathcal{L}^F(\widetilde{\mathcal{H}}) \leq 1 - r$,*

(6) *For every $\mathcal{H} \in I^{\mathcal{S}}$, if $\mathcal{A}$ contains no $\mathcal{R} \neq (0,1,1)$ with $\mathcal{R} \leq \mathcal{R}_r^\star$, then $\mathcal{L}^T(\mathcal{H}) \geq r$, $\mathcal{L}^I(\mathcal{H}) \leq 1 - r$, and $\mathcal{L}^F(\mathcal{H}) \leq 1 - r$.*

Proof. It is proved that most of the equivalent conditions ultimately prove the all the equivalence.

(1)⇒(2): Let $\{\mathcal{R}_j\}_{j \in J}$ be a single-valued neutrosophic quasi open-cover of $\mathcal{H} \in I^{\mathcal{S}}$ such that for $j \in J$, $T_{\mathcal{H}}(\kappa) + T_{R_j}(\kappa) - 1 \leq T_{\mathcal{C}}(\kappa)$, $I_{\mathcal{H}}(\kappa) + I_{R_j}(\kappa) - 1 > I_{\mathcal{C}}(\kappa)$, and $F_{\mathcal{H}}(\kappa) + F_{R_j}(\kappa) - 1 > F_{\mathcal{C}}(\kappa)$ for every $\kappa \in \mathcal{R}$ and for some $\mathcal{L}^T(\mathcal{C}) \geq r$, $\mathcal{L}^I(\mathcal{C}) \leq 1 - r$, and $\mathcal{L}^F(\mathcal{C}) \leq 1 - r$. Therefore, as $\{\mathcal{R}_j\}_{j \in J}$ is a single-valued neutrosophic quasi open-cover of $\mathcal{R}$, for each $x_{s,t,p} \in \mathcal{H}$, there exists at least one $\mathcal{R}_{j_o}$ such that $x_{s,t,p} q \mathcal{R}_{j_o}$ and for every $\kappa \in \mathcal{S}$, $T_{\mathcal{H}}(\kappa) + T_{R_{j_o}}(\kappa) - 1 \leq T_{\mathcal{C}}(\kappa)$, $I_{\mathcal{H}}(\kappa) + I_{R_{j_o}}(\kappa) - 1 > I_{\mathcal{C}}(\kappa)$, and $F_{\mathcal{H}}(\kappa) + F_{R_{jo}}(\kappa) - 1 > F_{\mathcal{C}}(\kappa)$ for every $\kappa \in \mathcal{S}$ and for some $\mathcal{L}^T(\mathcal{C}) \geq r$, $\mathcal{L}^I(\mathcal{C}) \leq 1-r$ and $\mathcal{L}^F(\mathcal{C}) \leq 1-r$. Obviously, $\mathcal{R}_{jo} \in Q_{\tau^{TIF}}(x_{s,t,p}, r)$. By (1), we have $\mathcal{L}^T(\mathcal{H}) \geq r$, $\mathcal{L}^I(\mathcal{H}) \leq 1-r$, and $\mathcal{L}^F(\mathcal{H}) \leq 1-r$.

(2)⇒(1): Clear from the fact that a collection of $\{\mathcal{R}_j\}_{j\in J}$, which contains at least one $\mathcal{R}_{jo} \in Q_{\tau^{TIF}}(x_{s,t,p}, r)$ of each single-valued neutrosophic point of $\mathcal{H}$, constitutes a single-valued neutrosophic quasi-open cover of $\mathcal{H}$.

(1)⇒(3): Let $\mathcal{H} \cap \mathcal{H}_r^\star = (0,1,1)$, for every $\kappa \in \mathcal{S}$, $x_t \in \mathcal{H}$ implies $x_{s,t,p} \notin \mathcal{H}_r^\star$. Then, there exists $\mathcal{D} \in Q_{\tau^{TIF}}(x_{s,t,p}, r)$ and $\mathcal{L}^T(\mathcal{C}) \geq r$, $\mathcal{L}^I(\mathcal{C}) \leq 1-r$, $\mathcal{L}^F(\mathcal{C}) \leq 1-r$ such that for every $\kappa \in \mathcal{S}$, $T_{\mathcal{D}}(\kappa) + T_{\mathcal{H}}(\kappa) - 1 \leq T_{\mathcal{C}}(\kappa)$, $I_{\mathcal{D}}(\kappa) + I_{\mathcal{H}}(\kappa) - 1 > I_{\mathcal{C}}(\kappa)$, and $F_{\mathcal{D}}(\kappa) + F_{\mathcal{H}}(\kappa) - 1 > F_{\mathcal{C}}(\kappa)$. Since $\mathcal{D} \in Q_{\tau^{TIF}}(x_{s,t,p}, r)$, By (1), we have $\mathcal{L}^T(\mathcal{H}) \geq r$, $\mathcal{L}^I(\mathcal{H}) \leq 1-r$, and $\mathcal{L}^F(\mathcal{H}) \leq 1-r$.

(3)⇒(1): For every $x_{s,t,p} \in \mathcal{H}$, there exists $\mathcal{D} \in Q_{\tau^{TIF}}(x_{s,t,p}, r)$ such that for every $\kappa \in \mathcal{S}$, $T_{\mathcal{D}}(\kappa) + T_{\mathcal{H}}(\kappa) - 1 \leq T_{\mathcal{C}}(\kappa)$, $I_{\mathcal{D}}(\kappa) + I_{\mathcal{H}}(\kappa) - 1 > I_{\mathcal{C}}(\kappa)$, and $F_{\mathcal{D}}(\kappa) + F_{\mathcal{H}}(\kappa) - 1 > F_{\mathcal{C}}(\kappa)$, for some $\mathcal{L}^T(\mathcal{C}) \geq r$, $\mathcal{L}^I(\mathcal{C}) \leq 1-r$, $\mathcal{L}^F(\mathcal{C}) \leq 1-r$. This implies $x_{s,t,p} \notin \mathcal{H}_r^\star$. Now, there are two cases: either $\mathcal{H}_r^\star = (0,1,1)$ or $\mathcal{H}_r^\star \neq (0,1,1)$ but $s > T_{\mathcal{H}_r^\star}(\kappa) \neq 0$, $t \leq I_{\mathcal{H}_r^\star}(\kappa) \neq 1$, and $p \leq F_{\mathcal{H}_r^\star}(\kappa) \neq 1$. Let, if possible, $x_{s,t,p} \in \mathcal{H}$ such that $t > T_{\mathcal{H}_r^\star}(\kappa) \neq 0$, $t \leq I_{\mathcal{H}_r^\star}(\kappa) \neq 1$, and $t \leq F_{\mathcal{H}_r^\star}(\kappa) \neq 1$. Let $s' = T_{\mathcal{H}_r^\star}(\kappa) \neq 0$, $t' = I_{\mathcal{H}_r^\star}(\kappa) \neq 1$, and $p' = F_{\mathcal{H}_r^\star}(\kappa) \neq 1$. Then, $x_{s',t',p'} \in \mathcal{H}_r^*(\kappa)$. In addition, $x_{s',t',p'} \in \mathcal{H}$. Thus, for every $\mathcal{V} \in Q_{\tau^{TIF}}(x_{s,t,p}, r)$, for every $\mathcal{L}^T(\mathcal{C}) \geq r$, $\mathcal{L}^I(\mathcal{C}) \leq 1-r$, and $\mathcal{L}^F(\mathcal{C}) \leq 1-r$, there is at least one $\kappa \in \mathcal{S}$ such that $T_{\mathcal{V}}(\kappa) + T_{\mathcal{H}}(\kappa) - 1 > T_{\mathcal{C}}(\kappa)$, $I_{\mathcal{V}}(\kappa) + I_{\mathcal{H}}(\kappa) - 1 \leq I_{\mathcal{C}}(\kappa)$, and $F_{\mathcal{V}}(\kappa) + F_{\mathcal{H}}(\kappa) - 1 \leq F_{\mathcal{C}}(\kappa)$. Since $x_{s,t,p} \in \mathcal{H}$, this contradicts the assumption for every single-valued neutrosophic point of $\mathcal{H}$. So, $\mathcal{H}_r^\star = (0,1,1)$. That means $x_{s,t,p} \in \mathcal{H}$ implies $x_{s,t,p} \notin \mathcal{H}_r^\star$. Now this is true for every $\mathcal{H} \in I^{\mathcal{S}}$. So, for any $\mathcal{H} \in I^{\mathcal{S}}$, $\mathcal{H} \cap \mathcal{H}_r^\star = (0,1,1)$. Hence, by (3), we have $\mathcal{L}^T(\mathcal{H}) \geq r$, $\mathcal{L}^I(\mathcal{H}) \leq 1-r$, $\mathcal{L}^F(\mathcal{H}) \leq 1-r$, which implies $\tau^{TIF} \sim \mathcal{L}^{TIF}$.

(3)⇒(4): Let $x_{s,t,p} in \widetilde{\mathcal{H}}$. Then, $x_{s,t,p} \in \mathcal{H}$ but $x_{s,t,p} \notin \mathcal{H}_r^\star$. So, there exists a $\mathcal{D} \in Q_{\tau^{TIF}}(x_{s,t,p}, r)$ such that for every $\kappa \in \mathcal{S}$, $T_{\mathcal{D}}(\kappa) + T_{\mathcal{H}}(\kappa) - 1 \leq T_{\mathcal{C}}(\kappa)$, $I_{\mathcal{D}}(\kappa) + I_{\mathcal{H}}(\kappa) - 1 > I_{\mathcal{C}}(\kappa)$, and $F_{\mathcal{D}}(\kappa) + F_{\mathcal{H}}(\kappa) - 1 > F_{\mathcal{C}}(\kappa)$, for some $\mathcal{L}^T(\mathcal{C}) \geq r$, $\mathcal{L}^I(\mathcal{C}) \leq 1-r$, $\mathcal{L}^F(\mathcal{C}) \leq 1-r$. Since $\widetilde{\mathcal{H}} \leq \mathcal{H}$, for every $\kappa \in \mathcal{S}$, $T_{\mathcal{D}}(\kappa) + T_{\widetilde{\mathcal{H}}}(\kappa) - 1 \leq T_{\mathcal{C}}(\kappa)$, $I_{\mathcal{D}}(\kappa) + I_{\widetilde{\mathcal{H}}}(\kappa) - 1 > I_{\mathcal{C}}(\kappa)$, and $F_{\mathcal{D}}(\kappa) + F_{\widetilde{\mathcal{H}}}(\kappa) - 1 > F_{\mathcal{C}}(\kappa)$, for some $\mathcal{L}^T(\mathcal{C}) \geq r$, $\mathcal{L}^I(\mathcal{C}) \leq 1-r$ and $\mathcal{L}^F(\mathcal{C}) \leq 1-r$. Therefore, $x_{s,t,p} \notin \widetilde{\mathcal{H}}_r^\star$ implies that $\widetilde{\mathcal{H}}_r^\star = (0,1,1)$ or $\widetilde{\mathcal{H}}_r^\star \neq (0,1,1)$ but $s > T_{\widetilde{\mathcal{H}}_r^\star}$, $t \leq I_{\widetilde{\mathcal{H}}_r^\star}$, and $p \leq F_{\widetilde{\mathcal{H}}_r^\star}$. Let $x_{s',t',p'}$ in $SVNP(\mathcal{S})$ such that $s' \leq T_{\widetilde{\mathcal{H}}_r^\star}(\kappa) < s$, $t \leq I_{\widetilde{\mathcal{A}}_r^\star}(\kappa) < t'$, and $p \leq F_{\widetilde{\mathcal{H}}_r^\star}(\kappa) < p'$, i.e., $x_{s',t',p'} \in \widetilde{\mathcal{H}}_r^\star$. Then, for each $\mathcal{V} \in Q_{\tau^{TIF}}(x_{s',t',p'}, r)$ and for each $\mathcal{L}^T(\mathcal{C}) \geq r$, $\mathcal{L}^I(\mathcal{C}) \leq 1-r$, $\mathcal{L}^F(\mathcal{C}) \leq 1-r$, there is at least one $\kappa \in \mathcal{S}$ such that $T_{\mathcal{V}}(\kappa) + T_{\widetilde{\mathcal{H}}}(\kappa) - 1 > T_{\mathcal{C}}(\kappa)$, $I_{\mathcal{V}}(\kappa) + I_{\widetilde{\mathcal{H}}}(\kappa) - 1 \leq I_{\mathcal{C}}(\kappa)$, and $F_{\mathcal{V}}(\kappa) + F_{\widetilde{\mathcal{H}}}(\kappa) - 1 \leq F_{\mathcal{C}}(\kappa)$. Since $\widetilde{\mathcal{H}} \leq \mathcal{H}$, then for each $\mathcal{V} \in Q_{\tau^{TIF}}(x_{s',t',p'}, r)$ and for each $\mathcal{L}^T(\mathcal{C}) \geq r$, $\mathcal{L}^I(\mathcal{C}) \leq 1-r$, $\mathcal{L}^F(\mathcal{C}) \leq 1-r$, there is at least one $\kappa \in \mathcal{S}$ such that $T_{\mathcal{V}}(\kappa) + T_{\mathcal{H}}(\kappa) - 1 > T_{\mathcal{C}}(\kappa)$, $I_{\mathcal{V}}(\kappa) + I_{\mathcal{H}}(\kappa) - 1 \leq I_{\mathcal{C}}(\kappa)$, and $F_{\mathcal{V}}(\kappa) + F_{\mathcal{H}}(\kappa) - 1 \leq F_{\mathcal{C}}(\kappa)$. This implies $x_{s',t',p'} \in \mathcal{H}_r^\star$. But as $s' < s$, $t' < t$, and $p' < p$, $x_{s,t,p} \in \widetilde{\mathcal{H}}$ implies $x_{s',t',p'} \in \widetilde{\mathcal{H}}$, and therefore, $x_{s',t',p'} \notin \mathcal{H}_r^\star$. This is a contradiction. Hence, $\mathcal{H}_r^\star = (0,1,1)$, so that $x_{s,t,p} \in \widetilde{\mathcal{H}}$ implies $x_{s,t,p} \notin \widetilde{\mathcal{H}}_r^\star$ with $\widetilde{\mathcal{H}}_r^\star = (0,1,1)$. Thus, $\widetilde{\mathcal{H}} \cap \widetilde{\mathcal{H}}_r^\star = \underline{0}$, for every $\mathcal{H} \in I^X$. Hence, by (3), $\mathcal{L}^T(\widetilde{\mathcal{H}}) \geq r$, $\mathcal{L}^I(\widetilde{\mathcal{H}}) \leq 1-r$, and $\mathcal{L}^F(\widetilde{\mathcal{H}}) \leq 1-r$.

(4)⇒(5): Straightforward.

(4)⇒(6): Let $\mathcal{H} \in I^{\mathcal{S}}$ and $\mathcal{H} \leq R \neq (0,1,1)$ with $\mathcal{R} \leq \mathcal{R}_r^\star$. Then, for any $\mathcal{H} \in I^{\mathcal{S}}$, $\mathcal{H} = \widetilde{\mathcal{H}} \cup (\mathcal{H} \cap \mathcal{H}_r^\star)$. Therefore, $\mathcal{H}_r^\star = (\widetilde{\mathcal{A}} \cup (\mathcal{H} \cap \mathcal{H}_r^\star))_r^\star = \widetilde{\mathcal{H}}_r^\star \cup (\mathcal{H} \cap \mathcal{H}_r^\star)_r^\star$. by Theorem 3 (5).

Now, by (4), we have $\mathcal{L}^T(\widetilde{\mathcal{H}}) \geq r$, $\mathcal{L}^I(\widetilde{\mathcal{H}}) \leq 1-r$, and $\mathcal{L}^F(\widetilde{\mathcal{H}}) \leq 1-r$, then $\widetilde{\mathcal{H}}_r^\star = (0,1,1)$. Hence, $(\mathcal{H} \cap \mathcal{H}_r^\star)_r^\star = \mathcal{H}_r^\star$ but $\mathcal{H} \cap \mathcal{H}_r^\star \leq \mathcal{H}_r^\star$, then $\mathcal{H} \cap \mathcal{A}_r^\star \leq (\mathcal{H} \cap \mathcal{H}_r^\star)_r^\star$. This contradicts the hypothesis about every single-valued neutrosophic set $\mathcal{H} \in I^{\mathcal{S}}$, if $(0,1,1) \neq \mathcal{R} \leq \mathcal{H}$ with $\mathcal{R} \leq \mathcal{R}_r^\star$. Therefore, $\mathcal{H} \cap \mathcal{H}_r^\star = (0,1,1)$, so that $\mathcal{H} = \widetilde{\mathcal{H}}$ by (4), we have $\mathcal{L}^T(\mathcal{H}) \geq r$, $\mathcal{L}^I(\mathcal{H}) \leq 1-r$, and $\mathcal{L}^F(\mathcal{H}) \leq 1-r$.

(6)⇒(4): Since, for every $\mathcal{H} \in I^{\mathcal{S}}$, $\mathcal{H} \cap \mathcal{H}_r^\star = (0,1,1)$. Therefore, by (6), as $\mathcal{H}$ contains no non-empty single-valued neutrosophic subset $\mathcal{R}$ with $\mathcal{R} \leq \mathcal{R}_r^\star$, $\mathcal{L}^T(\mathcal{H}) \geq r$, $\mathcal{L}^I(\mathcal{H}) \leq 1-r$, and $\mathcal{L}^F(\mathcal{H}) \leq 1-r$.

(5)⇒(1): For every $\mathcal{H} \in I^{\mathcal{S}}$, $x_{s,t,p} \in \mathcal{H}$, there exists an $\mathcal{D} \in Q_{\tau^{TIF}}(x_{s,t,p}, r)$ such that $T_{\mathcal{D}}(\kappa) + T_{\mathcal{H}}(\kappa) - 1 \leq T_{\mathcal{C}}(\kappa)$, $I_{\mathcal{D}}(\kappa) + I_{\mathcal{H}}(\kappa) - 1 > I_{\mathcal{C}}(\kappa)$, and $F_{\mathcal{D}}(\kappa) + F_{\mathcal{H}}(\kappa) - 1 > F_{\mathcal{C}}(\kappa)$ holds for every $\kappa \in \mathcal{S}$ and for some $\mathcal{L}^T(\mathcal{H}) \geq r$, $\mathcal{L}^I(\mathcal{H}) \leq 1-r$, and $\mathcal{L}^F(\mathcal{H}) \leq 1-r$. This implies $x_{s,t,p} \notin \mathcal{H}_r^\star$. Let $\mathcal{R} =$

$\mathcal{H} \cup \mathcal{H}_r^*$. Then, $\mathcal{R}_r^* = (\mathcal{H} \cup \mathcal{H}_r^*)_r^* = \mathcal{H}_r^* \cup (\mathcal{H}_r^*)_r^* = \mathcal{H}_r^*$ by Theorem 3(4). So, $C^*_{\tau^{TIF}}(\mathcal{R}, r) = \mathcal{R} \cup \mathcal{R}_r^* = \mathcal{R}$. That means $\tau^{T\star}(\underline{1} - \mathcal{R}) \geq r$, $\tau^{I\star}(\underline{1} - \mathcal{R}) \leq 1 - r$, and $\tau^{F\star}(\underline{1} - \mathcal{R}) \leq 1 - r$. Therefore, by (5), we have $\mathcal{L}^T(\mathcal{R}) \geq r$, $\mathcal{L}^I(\mathcal{R}) \leq 1 - r$, and $\mathcal{L}^F(\mathcal{R}) \leq 1 - r$.

Once again, for any $x_{s,t,p}$ in $SVNP(X)$, $x_{s,t,p} \notin \widetilde{\mathcal{R}}_r^\star$ implies $x_{s,t,p} \in \mathcal{R}$ but $x_{s,t,p} \notin \mathcal{R}_r^\star = \mathcal{H}_r^\star$ So, as $\mathcal{B} = \mathcal{H} \vee \mathcal{H}_r^\star$, $x_{s,t,p} \in \mathcal{H}$. Now, by hypothesis about $\mathcal{H}$. Then, for any $x_{s,t,p} \in \mathcal{H}_r^\star$. So, $\widetilde{\mathcal{R}} = \mathcal{H}$. Hence, $\mathcal{L}^T(\mathcal{H}) \geq r$, $\mathcal{L}^I(\mathcal{H}) \leq 1 - r$, and $\mathcal{L}^F(\mathcal{H}) \leq 1 - r$, i.e., $\tau^{TIF} \sim \mathcal{L}^{TIF}$. □

Theorem 10. *Let $(\mathcal{S}, \tau^{TIF})$ be an* **SVNTS** *with single-valued neutrosophic ideal $\mathcal{L}^{TIF}$ on $\mathcal{S}$. Then, the following are equivalent and implied by $\tau \sim \mathcal{L}$.*

(1) *For every $\mathcal{H} \in I^{\mathcal{S}}$, $\mathcal{H} \wedge \mathcal{H}_r^* = (0,1,1)$ implies $\mathcal{H}_r^* = (0,1,1)$;*
(2) *For any $\mathcal{H} \in I^{\mathcal{S}}$, $\widetilde{\mathcal{H}}_r^\star = (0,1,1)$;*
(3) *For every $\mathcal{H} \in I^{\mathcal{S}}$, $\mathcal{H} \wedge \mathcal{H}_r^\star = \mathcal{H}_r^\star$.*

Proof. Clear from Theorem 9. □

The following corollary is an important consequence of Theorem 10.

Corollary 3. *Let $\tau^{TIF} \sim \mathcal{L}^{TIF}$. Then, $\beta(\tau^{TIF}, \mathcal{L}^{TIF})$ is a base for $\tau^{TIF\star}$ and also $\beta(\tau^{TIF}, \mathcal{L}^{TIF}) = \tau^{TIF\star}$.*

Definition 17. *Let $\mathcal{H}, \mathcal{R} \in$* **SVNS** *on $\mathcal{S}$. If $\mathcal{H}$ is a single-valued neutrosophic relation on a set $\mathcal{S}$, then $\mathcal{H}$ is called a single-valued neutrosophic relation on $\mathcal{B}$ if, for every $\kappa, \kappa_1 \in \mathcal{S}$,*
$T_{\mathcal{R}}(\kappa, \kappa_1) \leq \min(T_{\mathcal{H}}(\kappa), T_{\mathcal{H}}(\kappa_1))$,
$I_{\mathcal{R}}(\kappa, \kappa_1) \geq \max(I_{\mathcal{H}}(\kappa), I_{\mathcal{H}}(\kappa_1))$, *and*
$F_{\mathcal{R}}(\kappa, \kappa_1) \geq \max(F_{\mathcal{H}}(\kappa), F_{\mathcal{H}}(\kappa_1))$.

A single-valued neutrosophic relation $\mathcal{H}$ on $\mathcal{S}$ is called symmetric if, for every $\kappa, \kappa_1 \in \mathcal{S}$,

$$T_{\mathcal{H}}(\kappa, \kappa_1) = T_{\mathcal{H}}(\kappa_1, \kappa), \quad I_{\mathcal{H}}(\kappa, \kappa_1) = I_{\mathcal{H}}(\kappa_1, \kappa), \quad F_{\mathcal{H}}(\kappa, \kappa_1) = F_{\mathcal{H}}(\kappa_1, \kappa); \text{ and}$$

$$T_{\mathcal{R}}(\kappa, \kappa_1) = T_{\mathcal{R}}(\kappa_1, \kappa) \quad I_{\mathcal{R}}(\kappa, \kappa_1) = I_{\mathcal{R}}(\kappa_1, \kappa), \quad F_{\mathcal{R}}(\kappa, \kappa_1) = F_{\mathcal{R}}(\kappa_1, \kappa).$$

In the purpose of symmetry, we can replace Definition 3 with Definition 17.

5. Conclusions

In this paper, we defined a single-valued neutrosophic closure space and single-valued neutrosophic ideal to study some characteristics of neutrosophic sets and obtained some of their basic properties. Next, the single-valued neutrosophic ideal open local function, single-valued neutrosophic ideal closure, single-valued neutrosophic ideal interior, single-valued neutrosophic ideal open compatible, and ordinary single-valued neutrosophic base were introduced and studied.

Discussion for further works:
We can apply the following ideas to the notion of single-valued ideal topological spaces.

(a) The collection of bounded single-valued sets [53];
(b) The concept of fuzzy bornology [54];
(c) The notion of boundedness in topological spaces. [54].

Author Contributions: This paper was organized by the idea of Y.S., F.A. analyzed the related papers with this research, and F.S. checked the overall contents and mathematical accuracy. All authors have read and agreed to the published version of the manuscript.

Funding: This research was supported by Majmaah University.

Acknowledgments: The authors would like to thank Deanship of Scientific Research at Majmaah University for supporting this work under Project Number No: R-1441-62. The authors would also like to express their sincere thanks to the referees for their useful suggestions and comments.

Conflicts of Interest: The authors declare no conflict of interest.

References

1. Zadeh, L.A. Fuzzy sets. *Inf. Control* **1965**, *8*, 338–353.
2. Chang, C.L. Fuzzy topological spaces. *J. Math. Anal. Appl.* **1968**, *24*, 182–190.
3. El-Gayyar, M.K.; Kerre, E.E.; Ramadan, A.A. On smooth topological space II: Separation axioms. *Fuzzy Sets Syst.* **2001**, *119*, 495–504.
4. Ghanim, M.H.; Kerre, E.E.; Mashhour, A.S. Separation axioms, subspaces and sums in fuzzy topology. *J. Math. Anal. Appl.* **1984**, *102*, 189–202.
5. Kandil, A.; El Etriby, A.M. On separation axioms in fuzzy topological space. *Tamkang J. Math.* **1987**, *18*, 49–59.
6. Kandil, A.; Elshafee, M.E. Regularity axioms in fuzzy topological space and FRi-proximities. *Fuzzy Sets Syst.* **1988**, *27*, 217–231.
7. Kerre, E.E. Characterizations of normality in fuzzy topological space. *Simon Steven* **1979**, *53*, 239–248.
8. Lowen, R. Fuzzy topological spaces and fuzzy compactness. *J. Math. Anal. Appl.* **1976**, *56*, 621–633.
9. Lowen, R. A comparison of different compactness notions in fuzzy topological spaces. *J. Math. Anal.* **1978**, *64*, 446–454.
10. Lowen, R. Initial and final fuzzy topologies and the fuzzy Tychonoff Theorem. *J. Math. Anal.* **1977**, *58*, 11–21.
11. Pu, P.M.; Liu, Y.M. Fuzzy topology I. Neighborhood structure of a fuzzy point. *J. Math. Anal. Appl.* **1982**, *76*, 571–599.
12. Pu, P.M.; Liu, Y.M. Fuzzy topology II. Products and quotient spaces. *J. Math. Anal. Appl.* **1980**, *77*, 20–37.
13. Yalvac, T.H. Fuzzy sets and functions on fuzzy spaces. *J. Math. Anal.* **1987**, *126*, 409–423.
14. Chattopadhyay, K.C.; Hazra, R.N.; Samanta, S.K. Gradation of openness: Fuzzy topology. *Fuzzy Sets Syst.* **1992**, *49*, 237–242.
15. Hazra, R.N.; Samanta, S.K.; Chattopadhyay, K.C. Fuzzy topology redefined. *Fuzzy Sets Syst.* **1992**, *45*, 79–82.
16. Ramaden, A.A. Smooth topological spaces. *Fuzzy Sets Syst.* **1992**, *48*, 371–375.
17. Demirci, M. Neighborhood structures of smooth topological spaces. *Fuzzy Sets Syst.* **1997**, *92*, 123–128.
18. Chattopadhyay, K.C.; Samanta, S.K. Fuzzy topology: Fuzzy closure operator, fuzzy compactness and fuzzy connectedness. *Fuzzy Sets Syst.* **1993**, *54*, 207–212.
19. Peeters, W. Subspaces of smooth fuzzy topologies and initial smooth fuzzy structures. *Fuzzy Sets Syst.* **1999**, *104*, 423–433.
20. Peeters, W. The complete lattice $(S(X), \preceq)$ of smooth fuzzy topologies. *Fuzzy Sets Syst.* **2002**, *125*, 145–152.
21. Onasanya, B.O.; Hošková-Mayerová, Š. Some topological and algebraic properties of α-level subsets' topology of a fuzzy subset. *An. Univ. Ovidius Constanta* **2018**, *26*, 213–227.
22. Çoker, D.; Demirci, M. An introduction to intuitionistic fuzzy topological spaces in Šostak's sense. *Busefal* **1996**, *67*, 67–76.
23. Samanta, S.K.; Mondal, T.K. Intuitionistic gradation of openness: Intuitionistic fuzzy topology. *Busefal* **1997**, *73*, 8–17.
24. Samanta, S.K.; Mondal, T.K. On intuitionistic gradation of openness. *Fuzzy Sets Syst.* **2002**, *131*, 323–336.
25. Šostak, A. On a fuzzy topological structure. In *Circolo Matematico di Palermo, Palermo; Rendiconti del Circolo Matematico di Palermo, Proceedings of the 13th Winter School on Abstract Analysis, Section of Topology, Srni, Czech Republic, 5–12 January 1985* Circolo Matematico di Palermo: Palermo, Italy, 1985; pp. 89–103.
26. Atanassov, K. Intuitionistic fuzzy sets. *Fuzzy Sets Syst.* **1986**, *20*, 87–96.
27. Lim, P.K.; Kim, S.R.; Hur, K. Intuitionisic smooth topological spaces. *J. Korean Inst. Intell. Syst.* **2010**, *20*, 875–883.
28. Kim, S.R.; Lim, P.K.; Kim, J.; Hur, K. Continuities and neighborhood structures in intuitionistic fuzzy smooth topological spaces. *Ann. Fuzzy Math. Inform.* **2018**, *16*, 33–54.
29. Choi, J.Y.; Kim, S.R.; Hur, K. Interval-valued smooth topological spaces. *Honam Math. J.* **2010**, *32*, 711–738.

30. Gorzalczany, M.B. A method of inference in approximate reasoning based on interval-valued fuzzy sets. *Fuzzy Sets Syst.* **1987**, *21*, 1–17.
31. Zadeh, L.A. The concept of a linguistic variable and its application to approximate reasoning I. *Inform. Sci.* **1975**, *8*, 199–249.
32. Ying, M.S. A new approach for fuzzy topology(I). *Fuzzy Sets Syst.* **1991**, *39*, 303–321.
33. Lim, P.K.; Ryou, B.G.; Hur, K. Ordinary smooth topological spaces. *Int. J. Fuzzy Log. Intell. Syst.* **2012**, *12*, 66–76.
34. Lee, J.G.; Lim, P.K.; Hur, K. Some topological structures in ordinary smooth topological spaces. *J. Korean Inst. Intell. Syst.* **2012**, *22*, 799–805.
35. Lee, J.G.; Lim, P.K.; Hur, K. Closures and interiors redefined, and some types of compactness in ordinary smooth topological spaces. *J. Korean Inst. Intell. Syst.* **2013**, *23*, 80–86.
36. Lee, J.G.; Hur, K.; Lim, P.K. Closure, interior and compactness in ordinary smooth topological spaces. *Int. J. Fuzzy Log. Intell. Syst.* **2014**, *14*, 231–239.
37. Saber, Y.M.; Abdel-Sattar, M.A. Ideals on Fuzzy Topological Spaces. *Appl. Math. Sci.* **2014**, *8*, 1667–1691.
38. Salama, A.A.; Albalwi, S.A. Intuitionistic Fuzzy Ideals Topological Spaces. *Adv. Fuzzy Math.* **2012**, *7*, 51–60.
39. Sarkar, D. Fuzzy ideal theory fuzzy local function and generated fuzzy topology fuzzy topology. *Fuzzy Sets Syst.* **1997**, *87*, 117–123.
40. Smarandache, F. *Neutrosophy, Neutrisophic Property, Sets, and Logic*; American Research Press: Rehoboth, DE, USA, 1998.
41. Salama, A.A.; Broumi, S.; Smarandache, F. Some types of neutrosophic crisp sets and neutrosophic crisp relations. *I. J. Inf. Eng. Electron. Bus.* **2014**. Available online: http://fs.unm.edu/Neutro-SomeTypeNeutrosophicCrisp.pdf (accessed on 10 February 2019).
42. Salama, A.A.; Smarandache, F. *Neutrosophic Crisp Set Theory*; The Educational Publisher Columbus: Columbus, OH, USA, 2015.
43. Hur, K.; Lim, P.K.; Lee, J.G.; Kim, J. The category of neutrosophic crisp sets. *Ann. Fuzzy Math. Inform.* **2017**, *14*, 43–54.
44. Hur, K.; Lim, P.K.; Lee, J.G.; Kim, J. The category of neutrosophic sets. *Neutrosophic Sets Syst.* **2016**, *14*, 12–20.
45. Salama, A.A.; Alblowi, S.A. Neutrosophic set and neutrosophic topological spaces. *IOSR J. Math.* **2012**, *3*, 31–35.
46. Salama, A.A.; Smarandache, F.; Kroumov, V. Neutrosophic crisp sets and neutrosophic crisp topological spaces. *Neutrosophic Sets Syst.* **2014**, *2*, 25–30.
47. Wang, H.; Smarandache, F.; Zhang, Y.Q.; Sunderraman, R. Single valued neutrosophic sets. *Multispace Multistruct.* **2010**, *4*, 410–413.
48. Kim, J.; Lim, P.K.; Lee, J.G.; Hur, K. Single valued neutrosophic relations. *Ann. Fuzzy Math. Inform.* **2018**, *16*, 201–221.
49. Smarandache, F. *A Unifying Field in Logics: Neutrosophic Logic. Neutrosophy, Neutrosophic Set, Neutrosophic Probability and Statistics*, 6th ed.; InfoLearnQuest: Ann Arbor, MI, USA, 2007.
50. Ye, J. A multicriteria decision-making method using aggregation operators for simplified neutrosophic sets. *J. Intell. Fuzzy Syst.* **2014**, *26*, 2450–2466.
51. Yang, H.L.; Guo, Z.L.; Liao, X. On single valued neutrosophic relations. *J. Intell. Fuzzy Syst.* **2016**, *30*, 1045–1056.
52. El-Gayyar, M. Smooth Neutrosophic Topological Spaces. *Neutrosophic Sets Syst.* **2016**, *65*, 65–72.
53. Yan, C.H.; Wu, C.X. Fuzzy L-bornological spaces. *Information Sciences.* **2005**, *173*, 1–10.
54. Lambrinos, P. A. A topological notion of boundedness. *Manuscripta Math.* **1973**, *10*, 289–296.

symmetry

MDPI

Article

Connectedness and Stratification of Single-Valued Neutrosophic Topological Spaces

Yaser Saber [1,2], Fahad Alsharari [1,*], Florentin Smarandache [3] and Mohammed Abdel-Sattar [4,5]

1 Department of Mathematics, College of Science and Human Studies, Hotat Sudair, Majmaah University, Majmaah 11952, Saudi Arabia; y.sber@mu.edu.sa
2 Department of Mathematics, Faculty of Science Al-Azhar University, Assiut 71524, Egypt
3 Department of Mathematics, University of New Mexico, Gallup, NM 87301, USA; smarand@unm.edu
4 Department of Mathematics, College of Science and Arts King Khaled University, Mhayal Asier 61913, Saudi Arabia; mabdulhafiz@kku.edu.sa
5 Department of Mathematics and Computer Science, Faculty of Science, Beni-Suef University, Beni-Suef 62511, Egypt
* Correspondence: f.alsharari@mu.edu.sa

Received: 26 August 2020; Accepted: 4 September 2020; Published: 7 September 2020

check for updates

Abstract: This paper aims to introduce the notion of r-single-valued neutrosophic connected sets in single-valued neutrosophic topological spaces, which is considered as a generalization of r-connected sets in Šostak's sense and r-connected sets in intuitionistic fuzzy topological spaces. In addition, it introduces the concept of r-single-valued neutrosophic separated and obtains some of its basic properties. It also tries to show that every r-single-valued neutrosophic component in single-valued neutrosophic topological spaces is an r-single-valued neutrosophic component in the stratification of it. Finally, for the purpose of symmetry, it defines the so-called single-valued neutrosophic relations.

Keywords: stratification of single-valued neutrosophic topological spaces; r-single-valued neutrosophic separated; r-single-valued neutrosophic connected and r-single-valued neutrosophic component

---

1. Introduction

Under a neutrosophic environment, Smarandache had established a generalization of intuitionistic fuzzy sets. His neutrosophic framework has a very large impact of constant applications for different fields in applied and pure sciences. In 1965, Zadeh [1] defined the so-called fuzzy sets ($\mathcal{FS}$) and, later on, Atanassov [2] defined the intuitionistic fuzzy sets ($\mathcal{IFS}$) in 1983. Topology is, of course, a cornerstone notion of mathematics, especially for ordinary subjects. The main concept of fuzzy topology ($\mathcal{FT}$) was defined by Chang [3]. Moreover, Lowen [4] gave the introduction to the concept of stratified fuzzy topology in the sense of Chang's fuzzy topology. Lee et al. and Liu et al. in their papers [5,6] investigated fuzzy connectedness ($\mathcal{F}$-*connected*) in fuzzy topological spaces. Again, researchers in [7–10] have studied the concept of ($\mathcal{F}$-*connected*). Sostak [11], however, also introduced the concept of smooth topology as an extension of Lowen and Chang's work.

In his paper [12], Smarandache characterized the neutrosophic set into three segment neutrosophic sets (F-Falsehood, I-Indeterminacy, T-Truth), and neutrosophic topological spaces ($\mathcal{SVNT}$) presented by Salama et al. [13,14]. Single valued neutrosophic sets (in sort, $\mathcal{SVN}$) were proposed by Wang et al. [15]. Meanwhile, Kim et al. [16] inspected the single valued neutrosophic relations ($\mathcal{SVNR}s$) and symmetric closure of $\mathcal{SVNR}$, respectively. In recent times, Saber et al. [17] familiarized the concepts of single-valued neutrosophic ideal open local function and single-valued neutrosophic topological space.

In this paper, we introduce the concept of r-single-valued neutrosophic connected sets and r-single-valued neutrosophic component in single-valued neutrosophic topological spaces. We then define the stratification of the single-valued neutrosophic topological spaces and show that every r-single-valued neutrosophic component in a single-valued neutrosophic is an r-single-valued neutrosophic component in the stratification of it. We have performed distinguished definitions, theorems, and counterexamples in-depth analysis to investigate some of their significant properties and to find out the best results and consequences. It can be said that different crucial notions in single valued neutrosophic topology were developed and generalized in this article. Different attributes like connectedness and stratification which have a significant impact on the overall topology's notions were also studied.

Innovative aspects and benefits of this article compared to relevant recent research on groups related to it are very useful. This paper studies connectedness and stratification of single-valued neutrosophic topological spaces. What makes this paper interesting is the introduction of the concept of r-single-valued neutrosophic separated. The authors obtain some of its basic properties. They show that every r-single-valued neutrosophic component in single-valued neutrosophic topological spaces is an r-single-valued neutrosophic component in the stratification of it.

A neutrosophic set is a power general formal framework, which generalizes the concept of the classic set, fuzzy set, interval valued fuzzy set, intuitionistic fuzzy set, and interval intuitionistic fuzzy set from a philosophical point of view. The applications aspects of these kinds of sets can be further noted. It can be seen In Geographical Information Systems (GIS) where there is a need to model spatial regions with indeterminate boundary and under indeterminacy (see [18]). In addition, possible applications to superstrings and ζ^{∞} space–time are touched upon (see [19]). It can also be applicable to control engineering in average consensus in multi-agent systems with uncertain topologies, multiple time-varying delays, and emergence in random noisy environments (see [20]).

In this work, $\tilde{X}$ is assumed to be a nonempty set, $\zeta = [0,1]$ and $\zeta_0 = (0,1]$. For $\alpha \in \zeta$, $\tilde{\alpha}(x) = \alpha$ for all $x \in \tilde{X}$. The family of all single-valued neutrosophic sets on $\tilde{X}$ is denoted by $\zeta^{\tilde{X}}$.

2. Preliminaries

This section is devoted to bring a complete survey and previous studies and important related notions and ideas.

Definition 1 ([21]). *Let $\tilde{X}$ be a non-empty set. A neutrosophic set (briefly, $\mathcal{NS}$) in $\tilde{X}$ is an object having the form*

$$\mathcal{S} = \{\langle x, \tilde{\gamma}_{\mathcal{S}}, \tilde{\eta}_{\mathcal{S}}, \tilde{\mu}_{\mathcal{S}} \rangle : x \in \tilde{X}\},$$

where $\tilde{\gamma}_{\mathcal{S}}, \tilde{\eta}_{\mathcal{S}}, \tilde{\mu}_{\mathcal{S}}$ and the degree of membership (namely $\tilde{\gamma}_{\mathcal{S}}(x)$), the degree of indeterminacy (namely $\tilde{\eta}_{\mathcal{S}}(x)$), and the degree of non-membership (namely $\tilde{\mu}_{\mathcal{S}}(x)$); for all $x \in \tilde{X}$ to the set $\mathcal{S}$. A neutrosophic set $\mathcal{S} = \{\langle x, \tilde{\gamma}_{\mathcal{S}}, \tilde{\eta}_{\mathcal{S}}, \tilde{\mu}_{\mathcal{S}} \rangle : x \in \tilde{X}\}$ can be identified as $\tilde{\gamma}_{\mathcal{S}}, \tilde{\eta}_{\mathcal{S}}, \tilde{\mu}_{\mathcal{S}}$ in $\rfloor^{-}0, 1^{+}\lfloor$ in $\tilde{X}$.

Definition 2 ([22]). *Suppose that $\mathcal{S}$ and $\mathcal{E}$ are $\mathcal{NS}'$s of the form $\mathcal{S} = \{\langle x, \tilde{\gamma}_{\mathcal{S}}, \tilde{\eta}_{\mathcal{S}}, \tilde{\mu}_{\mathcal{S}} \rangle : x \in \tilde{X}\}$ and $\mathcal{E} = \{\langle x, \tilde{\gamma}_{\mathcal{S}}, \tilde{\eta}_{\mathcal{S}}, \tilde{\mu}_{\mathcal{S}} \rangle : x \in \tilde{X}\}$ Then, $\mathcal{S} \subseteq \mathcal{E}$, iff for every $x \in \tilde{X}$,*

$$\inf\tilde{\gamma}_{\mathcal{S}}(x) \leq \inf\tilde{\gamma}_{\mathcal{E}}(x), \quad \inf\tilde{\eta}_{\mathcal{S}}(x) \geq \inf\tilde{\eta}_{\mathcal{E}}(x), \quad \inf\tilde{\mu}_{\mathcal{S}}(x) \geq \inf\tilde{\mu}_{\mathcal{E}}(x),$$

$$\sup\tilde{\gamma}_{\mathcal{S}}(x) \leq \sup\tilde{\gamma}_{\mathcal{E}}(x), \quad \sup\tilde{\eta}_{\mathcal{S}}(x) \geq \sup\tilde{\eta}_{\mathcal{E}}(x), \quad \sup\tilde{\mu}_{\mathcal{S}}(x) \geq \sup\tilde{\mu}_{\mathcal{E}}(x).$$

Definition 3 ([15]). *Let $\tilde{X}$ be a space of points (objects), with a generic element in $\tilde{X}$ denoted by x. Then, $\mathcal{S}$ is called a single valued neutrosophic set (briefly, $\mathcal{SVNS}$) in $\tilde{X}$, if $\mathcal{S}$ has the form $\mathcal{S} = \{\langle x, \tilde{\gamma}_{\mathcal{S}}, \tilde{\eta}_{\mathcal{S}}, \tilde{\mu}_{\mathcal{S}} \rangle : x \in \tilde{X}\}$, where $\tilde{\gamma}_{\mathcal{S}}, \tilde{\eta}_{\mathcal{S}}, \tilde{\mu}_{\mathcal{S}} : \tilde{X} \to [0,1]$.*

In this case, $\tilde{\gamma}_{\mathcal{S}}, \tilde{\eta}_{\mathcal{S}}, \tilde{\mu}_{\mathcal{S}}$ *are called truth-membership, indeterminacy-membership, falsify-membership mappings, respectively, and we will denote the set of all* $\mathcal{SVNS}'s$ *in* $\tilde{X}$ *as* $I^{\tilde{X}}$*. Moreover, we will refer to the Null (empty)* $\mathcal{SVNS}$ *(resp. the absolute (universe)* $\mathcal{SVNS}$*) in* $\tilde{X}$ *as* $\tilde{0}$ *(resp.* $\tilde{1}$*) and defined by* $\tilde{0} = \langle 0,1,1 \rangle$ *(resp.* $\tilde{1} = \langle 1,0,0 \rangle$*) for each* $x \in \tilde{X}$*.*

Definition 4 ([15])**.** *Let* $\mathcal{S} = \{\langle x, \tilde{\gamma}_{\mathcal{S}}, \tilde{\eta}_{\mathcal{S}}, \tilde{\mu}_{\mathcal{S}} \rangle : x \in \tilde{X}\}$ *be an* $\mathcal{SVNS}$ *on* $\tilde{X}$*. The complement of the set* $\mathcal{S}$ *(briefly* $\mathcal{S}^c$*) is defined as follows:*

$$\tilde{\gamma}_{\mathcal{S}^c}(x) = \tilde{\mu}_{\mathcal{S}}(x), \quad \tilde{\eta}_{\mathcal{S}^c}(x) = 1 - \tilde{\eta}_{\mathcal{S}}(x), \quad \tilde{\mu}_{\mathcal{S}^c}(x) = \tilde{\gamma}_{\mathcal{S}}(x),$$

for every $x \in \tilde{X}$*.*

Definition 5 ([23])**.** *Let* $\mathcal{S} = \{\langle x, \tilde{\gamma}_{\mathcal{S}}, \tilde{\eta}_{\mathcal{S}}, \tilde{\mu}_{\mathcal{S}} \rangle : x \in \tilde{X}\}$ *and* $\mathcal{E} = \{\langle x, \tilde{\gamma}_{\mathcal{E}}, \tilde{\eta}_{\mathcal{E}}, \tilde{\mu}_{\mathcal{E}} \rangle : x \in \tilde{X}\}$ *be an* $\mathcal{SVNS}$*. Then,*

(i) *A* $\mathcal{SVNS}$ $\mathcal{S}$ *is contained in the other* $\mathcal{SVNS}$ $\mathcal{E}$ *(briefly,* $\mathcal{S} \subseteq \mathcal{E}$*), if and only if*

$$\tilde{\gamma}_{\mathcal{S}}(x) \leq \tilde{\gamma}_{\mathcal{E}}(x), \quad \tilde{\eta}_{\mathcal{S}}(x) \geq \tilde{\eta}_{\mathcal{E}}(x), \quad \tilde{\mu}_{\mathcal{S}}(x) \geq \tilde{\mu}_{\mathcal{E}}(x)$$

for every $\omega \in \tilde{X}$*,*

(ii) *we say that* $\mathcal{S}$ *is equal to* $\mathcal{E}$*, denoted by* $\mathcal{S} = \mathcal{E}$*, if* $\mathcal{S} \subseteq \mathcal{E}$ *and* $\mathcal{S} \supseteq \mathcal{E}$*.*

Definition 6 ([22])**.** *Let* $\mathcal{S} = \{\langle x, \tilde{\gamma}_{\mathcal{S}}, \tilde{\eta}_{\mathcal{S}}, \tilde{\mu}_{\mathcal{S}} \rangle : x \in \tilde{X}\}$ *and* $\mathcal{E} = \{\langle x, \tilde{\gamma}_{\mathcal{E}}, \tilde{\eta}_{\mathcal{E}}, \tilde{\mu}_{\mathcal{E}} \rangle : x \in \tilde{X}\}$ *be an* $\mathcal{SVNS}$*. Then,*

(i) *the intersection of* $\mathcal{S}$ *and* $\mathcal{E}$ *(briefly,* $\mathcal{S} \cap \mathcal{E}$*) is a* $\mathcal{SVNS}$ *in* $\tilde{X}$ *defined as:*

$$\mathcal{S} \cap \mathcal{E} = (\tilde{\gamma}_{\mathcal{S}} \cap \tilde{\gamma}_{\mathcal{E}}, \tilde{\eta}_{\mathcal{S}} \cup \tilde{\eta}_{\mathcal{E}}, \tilde{\mu}_{\mathcal{S}} \cup \tilde{\mu}_{\mathcal{E}})$$

where $(\tilde{\mu}_{\mathcal{S}} \cup \tilde{\mu}_{\mathcal{E}})(x) = \tilde{\mu}_{\mathcal{S}}(x) \cup \tilde{\mu}_{\mathcal{E}}(x)$ *and* $(\tilde{\gamma}_{\mathcal{S}} \cap \tilde{\gamma}_{\mathcal{E}})(x) = \tilde{\gamma}_{\mathcal{S}}(x) \cap \tilde{\gamma}_{\mathcal{E}}(x)$*, for all* $x \in \tilde{X}$*,*

(ii) *the union of* $\mathcal{S}$ *and* $\mathcal{E}$ *(briefly,* $\mathcal{S} \cup \mathcal{E}$*) is an* $\mathcal{SVNS}$ *on* $\tilde{X}$ *defined as:*

$$\mathcal{S} \cup \mathcal{E} = (\tilde{\gamma}_{\mathcal{S}} \cup \tilde{\gamma}_{\mathcal{E}}, \tilde{\eta}_{\mathcal{S}} \cap \tilde{\eta}_{\mathcal{E}}, \tilde{\mu}_{\mathcal{S}} \cap \tilde{\mu}_{\mathcal{E}}).$$

Definition 7 ([13])**.** *Let* $\{\mathcal{S}_j, j \in \Gamma\}$ *be an arbitrary family of* $\mathcal{SVNS}'s$ *on* $\tilde{X}$*. Then,*

(i) *the intersection of* $\{\mathcal{S}_j, j \in \Gamma\}$ *(briefly,* $\bigcap_{j \in \Gamma} \mathcal{S}_j$*) is* $\mathcal{SVNS}$ *over* $\tilde{X}$ *defined as:*

$$(\bigcap_{J \in \Gamma} \mathcal{S}_j)(x) = (\bigcap_{j \in \Gamma} \tilde{\gamma}_{\mathcal{S}_j}(x), \quad \bigcup_{j \in \Gamma} \tilde{\eta}_{\mathcal{S}_j}(x), \quad \bigcup_{j \in \Gamma} \tilde{\mu}_{\mathcal{S}_j}(x)),$$

for all $x \in \tilde{X}$*,*

(ii) *the union of* $\{\mathcal{S}_j, j \in \Gamma\}$ *(briefly,* $\bigcup_{j \in \Gamma} \mathcal{S}_j$*) is* $\mathcal{SVNS}$ *over* $\tilde{X}$ *defined as:*

$$(\bigcup_{j \in \Gamma} \mathcal{S}_j)(x) = (\bigcup_{j \in \Gamma} \tilde{\gamma}_{\mathcal{S}_j}(x), \bigcap_{j \in \Gamma} \tilde{\eta}_{\mathcal{S}_j}(x), \bigcap_{j \in \Gamma} \tilde{\mu}_{\mathcal{S}_j}(x)),$$

for all $x \in \tilde{X}$*.*

Definition 8 ([24])**.** *A single-valued neutrosophic topology (*$\mathcal{SVNT}$*) on* $\tilde{X}$ *is an ordered triple* $(\tilde{\tau}^{\tilde{\gamma}}, \tilde{\tau}^{\tilde{\eta}}, \tilde{\tau}^{\tilde{\mu}})$ *as mappings from* $\zeta^{\tilde{X}}$ *to* ζ *such that:*

(SVNT1) $\tilde{\tau}^{\tilde{\gamma}}(\tilde{0}) = \tilde{\tau}^{\tilde{\gamma}}(\tilde{1}) = 1$ *and* $\tilde{\tau}^{\tilde{\eta}}(\tilde{0}) = \tilde{\tau}^{\tilde{\eta}}(\tilde{1}) = \tilde{\tau}^{\tilde{\mu}}(\tilde{0}) = \tilde{\tau}^{\tilde{\mu}}(\tilde{1}) = 0,$

(SVNT2) $\tilde{\tau}^{\tilde{\gamma}}(\mathcal{S} \cap \mathcal{E}) \geq \tilde{\tau}^{\tilde{\gamma}}(\mathcal{S}) \cap \tilde{\tau}^{\tilde{\gamma}}(\mathcal{E}), \qquad \tilde{\tau}^{\tilde{\eta}}(\mathcal{S} \cap \mathcal{E}) \leq \tau^{\tilde{\eta}}(\mathcal{S}) \cup \tilde{\tau}^{\tilde{\eta}}(\mathcal{E}),$

$$\tilde{\tau}^{\tilde{\mu}}(\mathcal{S} \cap \mathcal{E}) \leq \tilde{\tau}^{\tilde{\mu}}(\mathcal{S}) \cup \tilde{\tau}^{\tilde{\mu}}(\mathcal{E}), \text{for all } \mathcal{S}, \mathcal{E} \in \zeta^{\tilde{X}},$$

(SVNT3) $\tilde{\tau}^{\tilde{\gamma}}(\cup_{j\in\Gamma}\mathcal{S}_j) \geq \cap_{j\in\Gamma}\tilde{\tau}^{\tilde{\gamma}}(\mathcal{S}_j), \quad \tilde{\tau}^{\tilde{\eta}}(\cup_{i\in\Gamma}\mathcal{S}_j) \leq \cup_{j\in\Gamma}\tilde{\tau}^{\tilde{\eta}}(\mathcal{S}_j),$

$$\tilde{\tau}^{\tilde{\mu}}(\cup_{j\in\Gamma}\mathcal{S}_j) \leq \cup_{j\in\Gamma}\tilde{\tau}^{\tilde{\mu}}(\mathcal{S}_j) \text{ for all } \{\mathcal{S}_j, j \in \Gamma\} \in \zeta^{\tilde{X}}.$$

The quadruple $(\tilde{X}, \tilde{\tau}^{\tilde{\gamma}}, \tilde{\tau}^{\tilde{\eta}}, \tilde{\tau}^{\tilde{\mu}})$ is called $\mathcal{SVNTS}$. $\tilde{\tau}^{\tilde{\gamma}}$, $\tilde{\tau}^{\tilde{\eta}}$ and $\tilde{\tau}^{\tilde{\mu}}$ may be interpreted as the degree of openness, the degree of indeterminacy, and the degree of non-openness, respectively, and any single valued neutrosophic (briefly, $\mathcal{SVNS}$) set in $\tilde{X}$ is known as a single valued neutrosophic open set (briefly, r-**SVNO**) set in $\tilde{X}$. The elements of $\tilde{\tau}^{\tilde{\gamma}}, \tilde{\tau}^{\tilde{\eta}}, \tilde{\tau}^{\tilde{\mu}}$ are called open single valued neutrosophic sets (such that, for any $\mathcal{SVNS}$ $\mathcal{S} \in \tilde{X}$ and $r \in I_0$, we obtain $\tilde{\tau}^{\tilde{\gamma}}(\mathcal{S}) \geq r$, $\tilde{\tau}^{\tilde{\eta}}(\mathcal{S}) \leq 1-r$, and $\tilde{\tau}^{\tilde{\eta}}(\mathcal{S}) \leq 1-r$]. Then, the complement of r-**SVNO** is a single valued neutrosophic closed set (briefly, r-**SVNC**), and this will cause no ambiguity. Occasionally, we will write $\tilde{\tau}^{\tilde{\gamma}\tilde{\eta}\tilde{\mu}}$ for $(\tilde{\tau}^{\tilde{\gamma}}, \tilde{\tau}^{\tilde{\eta}}, \tilde{\tau}^{\tilde{\mu}})$, and it will be no ambiguity.

Definition 9 ([17]). *Let $(\tilde{X}, \tilde{\tau}^{\tilde{\gamma}}, \tilde{\tau}^{\tilde{\eta}}, \tilde{\tau}^{\tilde{\mu}})$ be an $\mathcal{SVNTS}$. A mapping $C : \zeta^{\tilde{X}} \times \zeta_0 \to \zeta^{\tilde{X}}$ is called a single-valued neutrosophic closure operator if, for every $\mathcal{S}, \mathcal{E} \in \zeta^{\tilde{X}}$ and $r, s \in \zeta_0$, it satisfies the following conditions:*

(C_1) $C(\tilde{0}, r) = \tilde{0}$,
(C_2) $\mathcal{S} \leq C(\mathcal{S}, r)$,
(C_3) $C(\mathcal{S}, r) \cup C(\mathcal{E}, r) = C(\mathcal{S} \cup \mathcal{E}, r)$,
(C_4) $C(\mathcal{S}, r) \leq C(\mathcal{S}, s)$ *if* $r \leq s$.
(C_5) $C(C(\mathcal{S}, r), r) = C(\mathcal{S}, r)$.

The pair $(\tilde{X}, C)$ is a single-valued neutrosophic closure space (briefly, $\mathcal{SVNCS}$).

Theorem 1 ([17]). *Let $C_{\tilde{\tau}^{\tilde{\gamma}}, \tilde{\tau}^{\tilde{\eta}}, \tilde{\tau}^{\tilde{\mu}}}$ be an single-valued neutrosophic closure operator on $\tilde{X}$. Define the mappings $\tilde{\tau}^{\tilde{\gamma}}_{C_{\tilde{\tau}^{\tilde{\gamma}}}}, \tilde{\tau}^{\tilde{\eta}}_{C_{\tilde{\tau}^{\tilde{\eta}}}}, \tilde{\tau}^{\tilde{\mu}}_{C_{\tilde{\tau}^{\tilde{\mu}}}} : \zeta^{\tilde{X}} \to \zeta$ by*

$$\tilde{\tau}^{\tilde{\gamma}}_{C_{\tilde{\tau}^{\tilde{\gamma}}}}(\mathcal{S}) = \bigcup\{r \in \zeta_0 \mid C_{\tilde{\tau}^{\tilde{\gamma}}}(\mathcal{S}^c, r) = \mathcal{S}^c\}, \quad \tilde{\tau}^{\tilde{\eta}}_{C_{\tilde{\tau}^{\tilde{\eta}}}}(\mathcal{S}) = \bigcap\{1 - r \in \zeta_0 \mid C_{\tilde{\tau}^{\tilde{\eta}}}(\mathcal{S}^c, r) = \mathcal{S}^c\},$$

$$\tilde{\tau}^{\tilde{\mu}}_{C_{\tilde{\tau}^{\tilde{\mu}}}}(\mathcal{S}) = \bigcap\{1 - r \in \zeta_0 \mid C_{\tilde{\tau}^{\tilde{\mu}}}(\mathcal{S}^c, r) = \mathcal{S}^c\}.$$

Then, $(\tau^{\tilde{\gamma}}_{C_{\tau^{\tilde{\gamma}}}}, \tau^{\tilde{\eta}}_{C_{\tau^{\tilde{\eta}}}}, \tau^{\tilde{\mu}}_{C_{\tau^{\tilde{\mu}}}})$ is an $\mathcal{SVNT}$ on $\tilde{X}$.

Definition 10 ([25]). *Let $f : (\tilde{X}, \tilde{\tau}_1^{\tilde{\gamma}}, \tilde{\tau}_1^{\tilde{\eta}}, \tilde{\tau}_1^{\tilde{\mu}}) \to (\tilde{Y}, \tilde{\tau}_2^{\tilde{\gamma}}, \tilde{\tau}_2^{\tilde{\eta}}, \tilde{\tau}_2^{\tilde{\mu}})$ be a mapping and $r \in \zeta_0$. Then, f is said to be $\mathcal{SVN}$-continuous if $\tilde{\tau}_2^{\tilde{\gamma}}(\mathcal{S}) \leq \tilde{\tau}_1^{\tilde{\gamma}}(f^{-1}(\mathcal{S}))$, $\tilde{\tau}_2^{\tilde{\eta}}(\mathcal{S}) \geq \tilde{\tau}_1^{\tilde{\eta}}(f^{-1}(\mathcal{S}))$, and $\tilde{\tau}_2^{\tilde{\mu}}(\mathcal{S}) \geq \tilde{\tau}_1^{\tilde{\mu}}(f^{-1}(\mathcal{S}))$ for all $\mathcal{S} \in \zeta^{\tilde{Y}}$.*

3. Connectedness in Single-Valued Neutrosophic Topological Spaces

The aim of this section is to introduce the r-single-valued neutrosophic separated (briefly, r-$\mathcal{SVNSEP}$), r-single-valued neutrosophic connected (briefly, r-$\mathcal{SVNCON}$), and r-single-valued neutrosophic component (briefly, r-$\mathcal{SVNCOM}$).

Definition 11. *Let $(\tilde{X}, \tilde{\tau}^{\tilde{\gamma}}, \tilde{\tau}^{\tilde{\eta}}, \tilde{\tau}^{\tilde{\mu}})$ be an $\mathcal{SVNTS}$. For every $\mathcal{S}, \mathcal{E}, \mathcal{R} \in \zeta^{\tilde{X}}$, $\mathcal{S}$ and $\mathcal{E}$ are called r-single-valued neutrosophic separated (briefly, r-$\mathcal{SVNSEP}$) if for $r \in \zeta_0$,*

$$C_{\tilde{\tau}^{\tilde{\gamma}}, \tilde{\tau}^{\tilde{\eta}}, \tilde{\tau}^{\tilde{\mu}}}(\mathcal{S}, r) \cap \mathcal{E} = C_{\tilde{\tau}^{\tilde{\gamma}}, \tilde{\tau}^{\tilde{\eta}}, \tilde{\tau}^{\tilde{\mu}}}(\mathcal{E}, r) \cap \mathcal{S} = \tilde{0}$$

A $\mathcal{SVNS}$, $\mathcal{R}$ is called r-single-valued neutrosophic connected (briefly, r-$\mathcal{SVNCON}$) if r-$\mathcal{SVNSEP}$ $\mathcal{S},\mathcal{E} \in \zeta^{\tilde{X}} - \{\tilde{0}\}$ such that $\mathcal{R} = \mathcal{S} \cup \mathcal{E}$ does not exist. A $\mathcal{SVNS}$ $\mathcal{R}$ is said to be $\mathcal{SVNCON}$ if it is r-$\mathcal{SVNCON}$ for any $r \in \zeta_0$. A quadruple $(\tilde{X}, \tilde{\tau}^{\tilde{\gamma}}, \tilde{\tau}^{\tilde{\eta}}, \tilde{\tau}^{\tilde{\mu}})$ is said to be r-$\mathcal{SVNCON}$ if $\tilde{1}$ is r-$\mathcal{SVNCON}$.

Remark 1. *Let $\mathcal{S}$ and $\mathcal{E}$ be r-$\mathcal{SVNSEP}$. Then for every $\mathcal{R} \in \zeta^{\tilde{X}}$ and $r_1 \le r$. We have $C_{\tilde{\tau}^{\tilde{\gamma}},\tilde{\tau}^{\tilde{\eta}},\tilde{\tau}^{\tilde{\mu}}}(\mathcal{R}, r_1) \le C_{\tilde{\tau}^{\tilde{\gamma}},\tilde{\tau}^{\tilde{\eta}},\tilde{\tau}^{\tilde{\mu}}}(\mathcal{R}, r)$, and $\mathcal{S}$ and $\mathcal{E}$ are said to be r_1-$\mathcal{SVNSEP}$. Conversely, from this fact, if $\mathcal{R}$ is r_1-$\mathcal{SVNCON}$ and $r \ge r_1$, then $\mathcal{R}$ is called r-$\mathcal{SVNCON}$.*

Example 1. *Let $\tilde{X} = \{a, b, c\}$ be a set. Define $\mathcal{E}_1, \mathcal{E}_2 \in \zeta^{\tilde{X}}$ as follows:*

$$\mathcal{E}_1 = \langle (1,1,0),(1,1,0),(1,1,0) \rangle; \quad \mathcal{E}_2 = \langle (0,0,1),(0,0,1),(0,0,1) \rangle.$$

We define an $\mathcal{SVNT}$ $(\tilde{\tau}^{\tilde{\gamma}}, \tilde{\tau}^{\tilde{\eta}}, \tilde{\tau}^{\tilde{\mu}})$ on $\tilde{X}$ as follows: for each $\mathcal{S} \in \zeta^{\tilde{X}}$,

$$\tilde{\tau}^{\tilde{\gamma}}(\mathcal{S}) = \begin{cases} 1, \text{if } \mathcal{S} = \tilde{0}, \\ 1, \text{if } \mathcal{S} = \tilde{1}, \\ \frac{1}{3}, \text{if } \mathcal{S} = \mathcal{E}_1, \\ \frac{1}{2}, \text{if } \mathcal{S} = \mathcal{E}_2, \\ 0, \quad \text{otherwise}, \end{cases} \qquad \tilde{\tau}^{\tilde{\eta}}(\mathcal{S}) = \begin{cases} 0, \text{if } \mathcal{S} = \tilde{0}, \\ 0, \text{if } \mathcal{S} = \tilde{1}, \\ \frac{2}{3}, \text{if } \mathcal{S} = \mathcal{E}_1, \\ \frac{1}{2}, \text{if } \mathcal{S} = \mathcal{E}_2, \\ 1, \quad \text{otherwise}, \end{cases}$$

$$\tilde{\tau}^{\tilde{\mu}}(\mathcal{S}) = \begin{cases} 0, \text{if } \mathcal{S} = \tilde{0}, \\ 0, \text{if } \mathcal{S} = \tilde{0}, \\ \frac{2}{3}, \text{if } \mathcal{S} = \mathcal{E}_1, \\ \frac{1}{2}, \text{if } \mathcal{S} = \mathcal{E}_2, \\ 1, \quad \text{otherwise}. \end{cases}$$

We thus obtain

$$C_{\tilde{\tau}^{\tilde{\gamma}},\tilde{\tau}^{\tilde{\eta}},\tilde{\tau}^{\tilde{\mu}}}(\mathcal{S}, r) = \begin{cases} \tilde{0}, \text{ if } \mathcal{S} = \tilde{0}, \ r \in \zeta_0, \\ \mathcal{E}_2^c, \text{if } \mathcal{S} \le \mathcal{E}_1, r \le \frac{1}{2}, 1 - r \ge \frac{1}{2}, \\ \mathcal{E}_1^c, \text{if } \mathcal{S} \le \mathcal{E}_2, r \le \frac{1}{3}, 1 - r \ge \frac{2}{3}, \\ \tilde{0}, \quad \text{otherwise}. \end{cases}$$

If $r \le \frac{1}{3}$ and $1 - r \ge \frac{2}{3}$, then $\mathcal{E}_2^c = C_{\tilde{\tau}^{\tilde{\gamma}},\tilde{\tau}^{\tilde{\eta}},\tilde{\tau}^{\tilde{\mu}}}(\mathcal{E}_1, r) \cap \mathcal{E}_2 = \tilde{0}$ and $\mathcal{E}_1^c = C_{\tilde{\tau}^{\tilde{\gamma}},\tilde{\tau}^{\tilde{\eta}},\tilde{\tau}^{\tilde{\mu}}}(\mathcal{E}_2, r) \cap \mathcal{E}_1 = \tilde{0}$. Thus, $\mathcal{E}_1 \cup \mathcal{E}_2 = \tilde{1}$ is not r-$\mathcal{SVNCON}$ for $r \le \frac{1}{3}$ and $1 - r \ge \frac{2}{3}$. If $r > \frac{1}{3}$ and $1 - r < \frac{2}{3}$, $(\tilde{X}, \tilde{\tau}^{\tilde{\gamma}}, \tilde{\tau}^{\tilde{\eta}}, \tilde{\tau}^{\tilde{\mu}})$ is r-$\mathcal{SVNCON}$.

Before we proceed further, we need to recall the following theorem given in [17] and prove its second part.

Theorem 2 ([17]). *Suppose that $(\tilde{X}, \tilde{\tau}^{\tilde{\gamma}}, \tilde{\tau}^{\tilde{\eta}}, \tilde{\tau}^{\tilde{\mu}})$ is an $\mathcal{SVNTS}$. For every $r \in \zeta_0$ and $\mathcal{S} \in \zeta^{\tilde{X}}$. Define an operator $C_{\tilde{\tau}^{\tilde{\gamma}},\tilde{\tau}^{\tilde{\eta}},\tilde{\tau}^{\tilde{\mu}}} : \zeta^{\tilde{X}} x \zeta_0 \to \zeta$ as follows:*

$$C_{\tilde{\tau}^{\tilde{\gamma}},\tilde{\tau}^{\tilde{\eta}},\tilde{\tau}^{\tilde{\mu}}}(\mathcal{S}, r) = \bigcap \{\mathcal{E} \in \zeta^{\tilde{X}} : \mathcal{E} \le \mathcal{S}, \ \tilde{\tau}^{\tilde{\gamma}}(\mathcal{E}^c) \ge r, \ \tilde{\tau}^{\tilde{\eta}}(\mathcal{E}^c) \le 1 - r, \ \tilde{\tau}^{\tilde{\mu}}(\mathcal{E}^c) \le 1 - r\}.$$

Then,

(1) $(\tilde{X}, C_{\tilde{\tau}^{\tilde{\gamma}},\tilde{\tau}^{\tilde{\eta}},\tilde{\tau}^{\tilde{\mu}}})$ *is an $\mathcal{SVNCS}$,*
(2) $\tilde{\tau}^{\tilde{\gamma}}_{C_{\tilde{\tau}^{\tilde{\gamma}},\tilde{\tau}^{\tilde{\eta}},\tilde{\tau}^{\tilde{\mu}}}} = \tilde{\tau}^{\tilde{\gamma}}$, $\tilde{\tau}^{\tilde{\eta}}_{C_{\tilde{\tau}^{\tilde{\gamma}},\tilde{\tau}^{\tilde{\eta}},\tilde{\tau}^{\tilde{\mu}}}} = \tilde{\tau}^{\tilde{\eta}}$ *and* $\tilde{\tau}^{\tilde{\mu}}_{C_{\tilde{\tau}^{\tilde{\gamma}},\tilde{\tau}^{\tilde{\eta}},\tilde{\tau}^{\tilde{\mu}}}} = \tilde{\tau}^{\tilde{\mu}}$

Proof. (1) It has been proven in [17].

(2) Suppose that $\tilde{\tau}^{\tilde{\gamma}}(\mathcal{E}) = r$, $\tilde{\tau}^{\tilde{\eta}}(\mathcal{E}) = 1-r$ and $\tilde{\tau}^{\tilde{\mu}}(\mathcal{R}) = 1-r$. Then, $C_{\tilde{\tau}^{\tilde{\gamma}},\tilde{\tau}^{\tilde{\eta}},\tilde{\tau}^{\tilde{\mu}}}(\mathcal{E}^c,r) = \mathcal{E}^c$. Therefore, $\tilde{\tau}^{\tilde{\gamma}}_{C_{\tilde{\tau}^{\tilde{\gamma}},\tilde{\tau}^{\tilde{\eta}},\tilde{\tau}^{\tilde{\mu}}}} \geq \tilde{\tau}^{\tilde{\gamma}}$, $\tilde{\tau}^{\tilde{\eta}}_{C_{\tilde{\tau}^{\tilde{\gamma}},\tilde{\tau}^{\tilde{\eta}},\tilde{\tau}^{\tilde{\mu}}}} \leq \tilde{\tau}^{\tilde{\eta}}$ and $\tilde{\tau}^{\tilde{\mu}}_{C_{\tilde{\tau}^{\tilde{\gamma}},\tilde{\tau}^{\tilde{\eta}},\tilde{\tau}^{\tilde{\mu}}}} \leq \tilde{\tau}^{\tilde{\mu}}$. Suppose that

$$\tilde{\tau}^{\tilde{\gamma}}_{C_{\tilde{\tau}^{\tilde{\gamma}},\tilde{\tau}^{\tilde{\eta}},\tilde{\tau}^{\tilde{\mu}}}} \not\leq \tilde{\tau}^{\tilde{\gamma}}, \quad \tilde{\tau}^{\tilde{\eta}}_{C_{\tilde{\tau}^{\tilde{\gamma}},\tilde{\tau}^{\tilde{\eta}},\tilde{\tau}^{\tilde{\mu}}}} \not\geq \tilde{\tau}^{\tilde{\eta}}, \quad \tilde{\tau}^{\tilde{\mu}}_{C_{\tilde{\tau}^{\tilde{\gamma}},\tilde{\tau}^{\tilde{\eta}},\tilde{\tau}^{\tilde{\mu}}}} \not\geq \tilde{\tau}^{\tilde{\mu}}.$$

Then, there exists $\mathcal{E}$ with $C_{\tilde{\tau}^{\tilde{\gamma}},\tilde{\tau}^{\tilde{\eta}},\tilde{\tau}^{\tilde{\mu}}}(\mathcal{E}^c,r) = \mathcal{E}^c$ such that

$$\tilde{\tau}^{\tilde{\gamma}}_{C_{\tilde{\tau}^{\tilde{\gamma}},\tilde{\tau}^{\tilde{\eta}},\tilde{\tau}^{\tilde{\mu}}}}(\mathcal{E}) \geq r > \tilde{\tau}^{\tilde{\gamma}}(\mathcal{E}), \quad \tilde{\tau}^{\tilde{\eta}}_{C_{\tilde{\tau}^{\tilde{\gamma}},\tilde{\tau}^{\tilde{\eta}},\tilde{\tau}^{\tilde{\mu}}}}(\mathcal{E}) \leq 1-r < \tilde{\tau}^{\tilde{\eta}}(\mathcal{E}), \quad \tilde{\tau}^{\tilde{\mu}}_{C_{\tilde{\tau}^{\tilde{\gamma}},\tilde{\tau}^{\tilde{\eta}},\tilde{\tau}^{\tilde{\mu}}}}(\mathcal{E}) \leq 1-r < \tilde{\tau}^{\tilde{\mu}}(\mathcal{E}). \tag{1}$$

By the definition of $C_{\tilde{\tau}^{\tilde{\gamma}},\tilde{\tau}^{\tilde{\eta}},\tilde{\tau}^{\tilde{\mu}}}$, we have $\tilde{\tau}^{\tilde{\gamma}}(\mathcal{E}) \geq r$, $\tilde{\tau}^{\tilde{\eta}}(\mathcal{E}) \leq 1-r$ and $\tilde{\tau}^{\tilde{\mu}}(\mathcal{E}) \leq 1-r$. It is a contradiction for Equation (1). □

Example 2. *Let $\tilde{X} = \{a,b\}$ be a set. Define $\mathcal{E}_1, \mathcal{E}_2 \in \zeta^{\tilde{X}}$.*

$$\mathcal{E}_1 = \langle (0.2,0.2),(0.3,0.3),(0.3,0.3)\rangle; \mathcal{E}_2 = \langle (0.5,0.5),(0.1,0.1),(0.1,0.1)\rangle.$$

We define the mapping $C : \zeta^{\tilde{X}} \times \zeta_0 \to \zeta^{\tilde{X}}$ as follows:

$$C(\mathcal{S},r) = \begin{cases} \tilde{0}, & \text{if } \mathcal{S} = \tilde{0}, \quad r \in I_0, \\ \mathcal{E}_1 \cap \mathcal{E}_2, & \text{if } 0 \neq \mathcal{S} \leq \mathcal{E}_1 \cap \mathcal{E}_2, \ 0 < r < \frac{1}{2}, \\ \mathcal{E}_1, & \text{if } \mathcal{S} \leq \mathcal{E}_1, \mathcal{S} \not\leq \mathcal{E}_2, \ 0 < r < \frac{1}{2}, \\ & \text{or } 0 \neq \mathcal{S} \leq \mathcal{E}_1 \ \ \frac{1}{2} < r < \frac{2}{3}, \\ \mathcal{E}_2, & \text{if } \mathcal{S} \leq \mathcal{E}_2, \mathcal{S} \not\leq \mathcal{E}_1, \ 0 < r < \frac{1}{2}, \\ \mathcal{E}_1 \cup \mathcal{E}_2, & \text{if } 0 \neq \mathcal{S} \leq \mathcal{E}_1 \cup \mathcal{E}_2, \ 0 < r < \frac{1}{2}, \\ \tilde{1}, & \text{otherwise.} \end{cases}$$

Then, C is a single-valued neutrosophic closure operator.

From Theorem 1, we have a single-valued neutrosophic topology $(\tau_C^{\tilde{\gamma}}, \tau_C^{\tilde{\eta}}, \tau_C^{\tilde{\mu}})$ on $\tilde{X}$ as follows:

$$\tau_C^{\tilde{\gamma}}(\mathcal{S}) = \begin{cases} 1, & \text{if } \mathcal{S} = \tilde{1} \text{ or } \tilde{0}, \\ \frac{2}{3}, & \text{if } \mathcal{S} = \mathcal{E}_1^c, \\ \frac{1}{2}, & \text{if } \mathcal{S} = \mathcal{E}_2^c, \\ \frac{1}{2}, & \text{if } \mathcal{S} = \mathcal{E}_1^c \cup \mathcal{H}^c, \\ \frac{1}{2}, & \text{if } \mathcal{S} = \mathcal{E}_1^c \cap \mathcal{E}_2^c, \\ 0, & \text{otherwise.} \end{cases}$$

$$\tau_C^{\tilde{\eta}}(\mathcal{S}) = \begin{cases} 0, & \text{if } \mathcal{S} = \tilde{1} \text{ or } \tilde{0}, \\ \frac{1}{3}, & \text{if } \mathcal{S} = \mathcal{E}_1^c, \\ \frac{1}{2}, & \text{if } \mathcal{S} = \mathcal{E}_2^c, \\ \frac{1}{2}, & \text{if } \mathcal{S} = \mathcal{E}_1^c \cup \mathcal{E}_2^c, \\ \frac{1}{2}, & \text{if } \mathcal{S} = \mathcal{E}_1^c \cap \mathcal{E}_2^c, \\ 1, & \text{otherwise.} \end{cases}$$

$$\tau_C^{\tilde{\mu}}(\mathcal{S}) = \begin{cases} 0, & \text{if } \mathcal{S} = \tilde{1} \text{ or } \tilde{0}, \\ \frac{1}{3}, & \text{if } \mathcal{S} = \mathcal{E}_1^c, \\ \frac{1}{2}, & \text{if } \mathcal{S} = \mathcal{E}_2^c, \\ \frac{1}{2}, & \text{if } \mathcal{S} = \mathcal{E}_1^c \cup \mathcal{E}_2^c, \\ \frac{1}{2}, & \text{if } \mathcal{S} = \mathcal{E}_1^c \cap \mathcal{E}_2^c, \\ 1, & \text{otherwise.} \end{cases}$$

Thus, the $(\tau_C^{\tilde{\gamma}}, \tau_C^{\tilde{\eta}}, \tau_C^{\tilde{\mu}})$ *is a single-valued neutrosophic topology on* $\tilde{X}$.

Theorem 3. *Let* $(\tilde{X}, \tilde{\tau}^{\tilde{\gamma}}, \tilde{\tau}^{\tilde{\eta}}, \tilde{\tau}^{\tilde{\mu}})$ *be an* $\mathcal{SVNTS}$. *Then, the following are equivalent.*

(1) $(\tilde{X}, \tilde{\tau}^{\tilde{\gamma}}, \tilde{\tau}^{\tilde{\eta}}, \tilde{\tau}^{\tilde{\mu}})$ *is r-*$\mathcal{SVNCON}$.

(2) *if* $\mathcal{S} \cup \mathcal{E} = \tilde{1}$ *and* $\mathcal{S} \cap \mathcal{E} = \tilde{0}$ *for* $(\tilde{\tau}^{\tilde{\gamma}}(\mathcal{E}) \geq r, \tilde{\tau}^{\tilde{\eta}}(\mathcal{E}) \leq 1-r, \tilde{\tau}^{\tilde{\mu}}(\mathcal{E}) \leq 1-r)$ *and* $(\tilde{\tau}^{\tilde{\gamma}}(\mathcal{S}) \geq r, \tilde{\tau}^{\tilde{\eta}}(\mathcal{S}) \leq 1-r, \tilde{\tau}^{\tilde{\mu}}(\mathcal{S}) \leq 1-r)$, *then* $\mathcal{E} = \tilde{0}$ *or* $\mathcal{S} = \tilde{0}$,

(3) *if* $\mathcal{S} \cup \mathcal{E} = \tilde{1}$, $\mathcal{E}_1 \cap \mathcal{E}_2 = \tilde{0}$ *for* $(\tilde{\tau}^{\tilde{\gamma}}(\mathcal{E}^c) \geq r, \tilde{\tau}^{\tilde{\eta}}(\mathcal{E}^c) \leq 1-r, \tilde{\tau}^{\tilde{\mu}}(\mathcal{E}^c) \leq 1-r)$ *and* $(\tilde{\tau}^{\tilde{\gamma}}(\mathcal{S}^c) \geq r, \tilde{\tau}^{\tilde{\eta}}(\mathcal{S}^c) \leq 1-r, \tilde{\tau}^{\tilde{\mu}}(\mathcal{S}^c) \leq 1-r)$, *then* $\mathcal{E} = \tilde{0}$ *or* $\mathcal{S} = \tilde{0}$.

Proof. (1)⇒(2): Let there exist $\mathcal{S}, \mathcal{E} \in \zeta^{\tilde{X}} - \{\tilde{0}\}$ such that for every $(\tilde{\tau}^{\tilde{\gamma}}(\mathcal{E}) \geq r, \tilde{\tau}^{\tilde{\eta}}(\mathcal{E}) \leq 1-r, \tilde{\tau}^{\tilde{\mu}}(\mathcal{E}) \leq 1-r)$ and $(\tilde{\tau}^{\tilde{\gamma}}(\mathcal{S}) \geq r, \tilde{\tau}^{\tilde{\eta}}(\mathcal{S}) \leq 1-r, \tilde{\tau}^{\tilde{\mu}}(\mathcal{S}) \leq 1-r)$, $\mathcal{S} \cup \mathcal{E} = \tilde{1}, \mathcal{S} \cap \mathcal{E} = \tilde{0}$. It implies

$$\mathcal{S}^c \cap \mathcal{E}^c = \tilde{0}, \quad \mathcal{S}^c \cup \mathcal{E}^c = \tilde{1}.$$

Since $C_{\tilde{\tau}^{\tilde{\gamma}}, \tilde{\tau}^{\tilde{\eta}}, \tilde{\tau}^{\tilde{\mu}}}(\mathcal{S}^c, r) = \mathcal{S}^c$ and $C_{\tilde{\tau}^{\tilde{\gamma}}, \tilde{\tau}^{\tilde{\eta}}, \tilde{\tau}^{\tilde{\mu}}}(\mathcal{E}^c, r) = \mathcal{E}^c$ from Theorem 2, $\mathcal{S}^c$ and $\mathcal{E}^c$ are *r*-$\mathcal{SVNSEP}$. Suppose $\mathcal{S} = \tilde{1}$. Then, $\mathcal{E} = \mathcal{S} \cap \mathcal{E} = \tilde{0}$. It is a contradiction. Hence, $\mathcal{S}^c \in \zeta^{\tilde{X}} - \{\tilde{0}\}$. Similarly, $\mathcal{E}^c \in \zeta^{\tilde{X}} - \{\tilde{0}\}$. Furthermore, $\mathcal{S}^c \cup \mathcal{E}^c = \tilde{1}$. Thus, $\tilde{1}$ is not *r*-$\mathcal{SVNCON}$.

(2)⇒(3): It is trivial.

(3)⇒(1): Suppose that $(\tilde{X}, \tilde{\tau}^{\tilde{\gamma}}, \tilde{\tau}^{\tilde{\eta}}, \tilde{\tau}^{\tilde{\mu}})$ is not *r*-$\mathcal{SVNCON}$. Then, there exist *r*-$\mathcal{SVNSEP}$ $\mathcal{S}, \mathcal{E} \in \zeta^{\tilde{X}} - \{\tilde{0}\}$ such that $\mathcal{S} \cup \mathcal{E} = \tilde{1}$. Since $\mathcal{S} \cap \mathcal{E} \leq C_{\tilde{\tau}^{\tilde{\gamma}}, \tilde{\tau}^{\tilde{\eta}}, \tilde{\tau}^{\tilde{\mu}}}(\mathcal{S}, r) \cap \mathcal{E} = \tilde{0}$, we have $\mathcal{S} \cap \mathcal{E} = \tilde{0}$. Thus, $\mathcal{S}^c \cap \mathcal{E}^c = \tilde{0}$ implies $\mathcal{E}^c \leq \mathcal{S}$. Hence, $C_{\tilde{\tau}^{\tilde{\gamma}}, \tilde{\tau}^{\tilde{\eta}}, \tilde{\tau}^{\tilde{\mu}}}(\mathcal{S}, r) \cap \mathcal{E} = \tilde{0}$ implies, $C_{\tilde{\tau}^{\tilde{\gamma}}, \tilde{\tau}^{\tilde{\eta}}, \tilde{\tau}^{\tilde{\mu}}}(\mathcal{S}, r) \leq \mathcal{E}^c$. Thus, $C_{\tilde{\tau}^{\tilde{\gamma}}, \tilde{\tau}^{\tilde{\eta}}, \tilde{\tau}^{\tilde{\mu}}}(\mathcal{S}, r) \leq \mathcal{S}$. By Definition 9 ($C_2$), we have $C_{\tilde{\tau}^{\tilde{\gamma}}, \tilde{\tau}^{\tilde{\eta}}, \tilde{\tau}^{\tilde{\mu}}}(\mathcal{S}, r) = \mathcal{S}$. By Theorem 2, we obtain $(\tilde{\tau}^{\tilde{\gamma}}(\mathcal{S}^c) \geq r, \tilde{\tau}^{\tilde{\eta}}(\mathcal{S}^c) \leq 1-r, \tilde{\tau}^{\tilde{\mu}}(\mathcal{S}^c) \leq 1-r)$. Similarly, we have $(\tilde{\tau}^{\tilde{\gamma}}(\mathcal{E}^c) \geq r, \tilde{\tau}^{\tilde{\eta}}(\mathcal{E}^c) \leq 1-r, \tilde{\tau}^{\tilde{\mu}}(\mathcal{E}^c) \leq 1-r)$. It is a contradiction. □

Lemma 1. *Let* $(\tilde{X}, \tilde{\tau}^{\tilde{\gamma}}, \tilde{\tau}^{\tilde{\eta}}, \tilde{\tau}^{\tilde{\mu}})$ *be an* $\mathcal{SVNTS}$ *and* $\mathcal{S}, \mathcal{E}, \mathcal{R} \in \zeta^{\tilde{X}}$. *If* $\mathcal{E}$ *and* $\mathcal{R}$ *are r-*$\mathcal{SVNSEP}$, *then* $\mathcal{S} \cap \mathcal{E}$ *and* $\mathcal{S} \cap \mathcal{R}$ *are r-*$\mathcal{SVNSEP}$.

Proof. Let $\mathcal{E}$ and $\mathcal{R}$ be *r*-$\mathcal{SVNSEP}$. Thus,

$$C_{\tilde{\tau}^{\tilde{\gamma}}, \tilde{\tau}^{\tilde{\eta}}, \tilde{\tau}^{\tilde{\mu}}}(\mathcal{S} \cap \mathcal{E}, r) \cap (\mathcal{S} \cap \mathcal{R}) \leq C_{\tilde{\tau}^{\tilde{\gamma}}, \tilde{\tau}^{\tilde{\eta}}, \tilde{\tau}^{\tilde{\mu}}}(\mathcal{E}, r) \cap \mathcal{R} = \tilde{0}$$

Similarly, $C_{\tilde{\tau}^{\tilde{\gamma}}, \tilde{\tau}^{\tilde{\eta}}, \tilde{\tau}^{\tilde{\mu}}}(\mathcal{S} \cap \mathcal{R}, r) \cap (\mathcal{S} \cap \mathcal{E}) = \tilde{0}$. Thus, $\mathcal{S} \cap \mathcal{E}$ and $\mathcal{S} \cap \mathcal{R}$ are *r*-$\mathcal{SVNSEP}$. □

Theorem 4. *Let* $(\tilde{X}, \tilde{\tau}^{\tilde{\gamma}}, \tilde{\tau}^{\tilde{\eta}}, \tilde{\tau}^{\tilde{\mu}})$ *be an* $\mathcal{SVNTS}$ *and* $\mathcal{S} \in \zeta^{\tilde{X}}$. *Then, the following are equivalent.*

(1) $\mathcal{S}$ *is r-*$\mathcal{SVNCON}$,

(2) *If* $\mathcal{E}$ *and* $\mathcal{R}$ *are r-*$\mathcal{SVNSEP}$ *such that* $\mathcal{S} \leq \mathcal{E} \cup \mathcal{R}$, *then* $\mathcal{S} \cap \mathcal{E} = \tilde{0}$ *or* $\mathcal{S} \cap \mathcal{R} = \tilde{0}$,

(3) *If* $\mathcal{E}$ *and* $\mathcal{R}$ *are r-*$\mathcal{SVNSEP}$ *such that* $\mathcal{S} \leq \mathcal{E} \cup \mathcal{R}$, *then* $\mathcal{S} \leq \mathcal{E}$ *or* $\mathcal{S} \leq \mathcal{R}$.

Proof. (1) ⇒ (2): Let $\mathcal{E}, \mathcal{R} \in \zeta^{\tilde{X}}$ be *r*-$\mathcal{SVNSEP}$ such that $\mathcal{S} \leq \mathcal{E} \cup \mathcal{R}$. By Lemma 1, $\mathcal{S} \cap \mathcal{E}$ and $\mathcal{S} \cap \mathcal{R}$ are *r*-$\mathcal{SVNSEP}$. Since $\mathcal{S}$ is *r*-$\mathcal{SVNCON}$ and $\mathcal{S} = \mathcal{S} \cap (\mathcal{E} \cup \mathcal{R}) = (\mathcal{S} \cap \mathcal{E}) \cup (\mathcal{S} \cap \mathcal{R})$, then $\mathcal{S} \cap \mathcal{E} = \tilde{0}$ or $\mathcal{S} \cap \mathcal{R} = \tilde{0}$.

(2) ⇒ (3): It is easily proved.

(3) ⇒ (1): Let $\mathcal{E}$ and $\mathcal{R}$ be *r*-$\mathcal{SVNSEP}$ such that $\mathcal{S} = \mathcal{E} \cup \mathcal{R}$. By (3), $\mathcal{S} \leq \mathcal{E}$ or $\mathcal{S} \leq \mathcal{R}$. If $\mathcal{S} \leq \mathcal{E}$ and $\mathcal{E}, \mathcal{R}$ are *r*-$\mathcal{SVNSEP}$, then

$$\mathcal{R} = \mathcal{R} \cap \mathcal{S} \leq \mathcal{R} \cap \mathcal{E} \leq \mathcal{R} \cap C_{\tilde{\tau}^{\tilde{\gamma}}, \tilde{\tau}^{\tilde{\eta}}, \tilde{\tau}^{\tilde{\mu}}}(\mathcal{E}, r) = \tilde{0}.$$

Hence, $\mathcal{R} = \tilde{0}$. If $\mathcal{S} \leq \mathcal{R}$, similarly $\mathcal{E} = \tilde{0}$. □

Theorem 5. *Let* $(\tilde{X}, \tilde{\tau}^{\tilde{\gamma}}, \tilde{\tau}^{\tilde{\eta}}, \tilde{\tau}^{\tilde{\mu}})$ *be an* $\mathcal{SVNTS}$ *and* $\mathcal{S}, \mathcal{E} \in \zeta^{\tilde{X}}$.

(1) *If $\mathcal{S}$ is r-$\mathcal{SVNCON}$, $\mathcal{S} \leq \mathcal{E} \leq C_{\tilde{\tau}^{\tilde{\gamma}}\tilde{\eta}\tilde{\mu}}(\mathcal{S}, r)$, then $\mathcal{E}$ is r-$\mathcal{SVNCON}$.*

(2) *If $\mathcal{S}$ and $\mathcal{E}$ are r-$\mathcal{SVNCON}$ single-valued neutrosophic sets which are not r-$\mathcal{SVNSEP}$, then $\mathcal{S} \cup \mathcal{E}$ is r-$\mathcal{SVNCON}$.*

Proof. (1) Let $\mathcal{R}, \mathcal{D} \in \zeta^{\tilde{X}}$ be r-$\mathcal{SVNSEP}$ such that $\mathcal{E} = \mathcal{R} \cup \mathcal{D}$. Put, $\mathcal{R}_1 = \mathcal{S} \cap \mathcal{R}$ and $\mathcal{D}_1 = \mathcal{S} \cap \mathcal{D}$, then $\mathcal{R}_1$ and $\mathcal{D}_1$ are r-$\mathcal{SVNSEP}$ such that $\mathcal{S} = \mathcal{R}_1 \cup \mathcal{D}_1$. Since $\mathcal{S}$ is r-$\mathcal{SVNCON}$, $\mathcal{R}_1 = \tilde{0}$ or $\mathcal{D}_1 = \tilde{0}$. If $\mathcal{R}_1 = \tilde{0}$, then $\mathcal{S} = \mathcal{D}_1 = \mathcal{S} \cap \mathcal{D} \Rightarrow \mathcal{S} \leq \mathcal{D}$. It implies

$$\mathcal{E} \leq C_{\tilde{\tau}^{\tilde{\gamma}},\tilde{\tau}^{\tilde{\eta}},\tilde{\tau}^{\tilde{\mu}}}(\mathcal{S}, r) \leq C_{\tilde{\tau}^{\tilde{\gamma}},\tilde{\tau}^{\tilde{\eta}},\tilde{\tau}^{\tilde{\mu}}}(\mathcal{D}, r).$$

Hence, $\mathcal{R} = \mathcal{R} \cap \mathcal{E} \leq \mathcal{R} \cap C_{\tilde{\tau}^{\tilde{\gamma}},\tilde{\tau}^{\tilde{\eta}},\tilde{\tau}^{\tilde{\mu}}}(\mathcal{D}, r) = \tilde{0}$.

If $\mathcal{D}_1 = \tilde{0}$, similarly $\mathcal{D} = \tilde{0}$. Therefore, $\mathcal{E}$ is r-$\mathcal{SVNCON}$.

(2) Let $\mathcal{R}$ and $\mathcal{D}$ be r-$\mathcal{SVNSEP}$ such that $\mathcal{S} \cup \mathcal{E} = \mathcal{R} \cup \mathcal{D}$. Since $\mathcal{S}$ is r-$\mathcal{SVNCON}$, by Theorem 4 (3), $\mathcal{S} \leq \mathcal{R}$ or $\mathcal{S} \leq \mathcal{D}$. Say, $\mathcal{S} \leq \mathcal{R}$. Suppose that $\mathcal{E} \leq \mathcal{D}$. Since $(\mathcal{S} \cup \mathcal{E}) \cap \mathcal{R} = \mathcal{S}$ and $(\mathcal{S} \cup \mathcal{E}) \cap \mathcal{D} = \mathcal{E}$, by Lemma 1, $\mathcal{S}$ and $\mathcal{E}$ are r-$\mathcal{SVNSEP}$. It is a contradiction. Thus, $\mathcal{E} \leq \mathcal{R}$ Hence, $\mathcal{S} \cup \mathcal{E} \leq \mathcal{R}$, by Theorem 4 (3), $\mathcal{S} \cup \mathcal{E}$ is r-$\mathcal{SVNCON}$. □

Theorem 6. *Let $(\tilde{X}, \tilde{\tau}^{\tilde{\gamma}}, \tilde{\tau}^{\tilde{\eta}}, \tilde{\tau}^{\tilde{\mu}})$ be an $\mathcal{SVNTS}$. Let $\mathcal{B} = \{\mathcal{S}_j \in \zeta^{\tilde{X}} \mid \mathcal{S}_j$ is $r - \mathcal{SVNCON}$ sets $, j \in \Gamma\}$ be a family in $\tilde{X}$ such that no two members of $\mathcal{B}$ are r-$\mathcal{SVNSEP}$, then $\bigcup_{j \in \Gamma} \mathcal{S}_j$ is r-$\mathcal{SVNCON}$.*

Proof. Put $\mathcal{S} = \bigcup_{j \in \Gamma} \mathcal{S}_j$. Let $\mathcal{E}, \mathcal{R} \in \zeta^{\tilde{X}}$ be r-$\mathcal{SVNSEP}$ such that $\mathcal{S} = \mathcal{E} \cup \mathcal{R}$. Since any two members $\mathcal{S}_j, \mathcal{S}_i \in \mathcal{B}$ are not r-$\mathcal{SVNSEP}$, by Theorem 5 (2), $\mathcal{S}_j \cup \mathcal{S}_i$ is r-$\mathcal{SVNCON}$. From Theorem 4 (3), $\mathcal{S}_j \cup \mathcal{S}_i \leq \mathcal{E}$ or $\mathcal{S}_j \cup \mathcal{S}_i \leq \mathcal{R}$. Say, $\mathcal{S}_j \cup \mathcal{S}_i \leq \mathcal{E}$. It implies that $\mathcal{S} \leq \mathcal{E}$. Thus, $\mathcal{S}$ is r-$\mathcal{SVNCON}$. □

Corollary 1. *Let $(\tilde{X}, \tilde{\tau}^{\tilde{\gamma}}, \tilde{\tau}^{\tilde{\eta}}, \tilde{\tau}^{\tilde{\mu}})$ be an $\mathcal{SVNTS}$. Let $\mathcal{B} = \{\mathcal{S}_j \in \zeta^{\tilde{X}} \mid \mathcal{S}_j$ is $r - \mathcal{SVNCON}$ sets $, j \in \Gamma\}$ be a family in $\tilde{X}$. If $\bigcap_{j \in \Gamma} \mathcal{S}_j \neq \tilde{0}$, then $\bigcup_{j \in \Gamma} \mathcal{S}_j$ is r-$\mathcal{SVNCON}$.*

Lemma 2. *Let $f : (\tilde{X}, \tilde{\tau}_1^{\tilde{\gamma}}, \tilde{\tau}_1^{\tilde{\eta}}, \tilde{\tau}_1^{\tilde{\mu}}) \to (\tilde{Y}, \tilde{\tau}_2^{\tilde{\gamma}}, \tilde{\tau}_2^{\tilde{\eta}}, \tilde{\tau}_2^{\tilde{\mu}})$ be a mapping from an $\mathcal{SVNTS}$ $(\tilde{X}, \tilde{\tau}_1^{\tilde{\gamma}}, \tilde{\tau}_1^{\tilde{\eta}}, \tilde{\tau}_1^{\tilde{\mu}})$ to another $\mathcal{SVNTS}$ $(\tilde{Y}, \tilde{\tau}_2^{\tilde{\gamma}}.\tilde{\tau}_2^{\tilde{\eta}}, \tilde{\tau}_2^{\tilde{\mu}})$. Then, the following are equivalent, $\forall\, \mathcal{S} \in \zeta^{\tilde{X}}, \mathcal{E} \in \zeta^{\tilde{Y}}$ and $r \in \zeta_0$*

(1) *f is $\mathcal{SVN}$ − continuous.*

(2) *$f(C_{\tilde{\tau}_1^{\tilde{\gamma}},\tilde{\tau}_1^{\tilde{\eta}},\tilde{\tau}_1^{\tilde{\mu}}}(\mathcal{S}, r)) \leq C_{\tilde{\tau}_2^{\tilde{\gamma}},\tilde{\tau}_2^{\tilde{\eta}},\tilde{\tau}_2^{\tilde{\mu}}}(f(\mathcal{S}), r)$.*

(3) *$C_{\tilde{\tau}_1^{\tilde{\gamma}},\tilde{\tau}_1^{\tilde{\eta}},\tilde{\tau}_1^{\tilde{\mu}}}(f^{-1}(\mathcal{E}), r) \leq f^{-1}(C_{\tilde{\tau}_2^{\tilde{\gamma}},\tilde{\tau}_2^{\tilde{\eta}},\tilde{\tau}_2^{\tilde{\mu}}}(\mathcal{E}, r))$.*

Proof. (1)⇒(2): Suppose that f is $\mathcal{SVN}$ − *continuous*. Then, $\tilde{\tau}_1^{\tilde{\gamma}}((f^{-1}(\mathcal{S}))^c) \geq \tilde{\tau}_2^{\tilde{\gamma}}(\mathcal{S}^c)$, $\tilde{\tau}_1^{\tilde{\eta}}((f^{-1}(\mathcal{S}))^c) \leq \tilde{\tau}_2^{\tilde{\eta}}(\mathcal{S}^c)$ and $\tilde{\tau}_1^{\tilde{\mu}}((f^{-1}(\mathcal{S}))^c) \leq \tilde{\tau}_2^{\tilde{\mu}}(\mathcal{S}^c)$. Hence,

$$\begin{aligned}
C_{\tilde{\tau}_2^{\tilde{\gamma}},\tilde{\tau}_2^{\tilde{\eta}},\tilde{\tau}_2^{\tilde{\mu}}}(f(\mathcal{S}), r) &= \bigcap\{\mathcal{E} \in \zeta^{\tilde{Y}} \mid f(\mathcal{S}) \leq \mathcal{E},\ \tilde{\tau}_2^{\tilde{\gamma}}(\mathcal{E}^c) \geq r,\ \tilde{\tau}_2^{\tilde{\eta}}(\mathcal{E}^c) \leq 1 - r,\ \tilde{\tau}_2^{\tilde{\mu}}(\mathcal{E}^c) \leq 1 - r\} \\
&\geq \bigcap\{\mathcal{E} \in \zeta^{\tilde{Y}} \mid \mathcal{S} \leq f^{-1}(\mathcal{E}),\ \tilde{\tau}_1^{\tilde{\gamma}}((f^{-1}(\mathcal{E}))^c) \geq r,\ \tilde{\tau}_1^{\tilde{\eta}}((f^{-1}(\mathcal{E}))^c) \leq 1 - r, \\
&\quad \tilde{\tau}_1^{\tilde{\mu}}((f^{-1}(\mathcal{E}))^c) \leq 1 - r\} \\
&\geq \bigcap\{f(f^{-1}(\mathcal{E})) \in \zeta^{\tilde{Y}} \mid \mathcal{S} \leq f^{-1}(\mathcal{E}),\ \tilde{\tau}_1^{\tilde{\gamma}}((f^{-1}(\mathcal{E}))^c) \geq r,\ \tilde{\tau}_1^{\tilde{\eta}}((f^{-1}(\mathcal{E}))^c) \leq 1 - r, \\
&\quad \tilde{\tau}_1^{\tilde{\mu}}((f^{-1}(\mathcal{E}))^c) \leq 1 - r\} \\
&\geq f[\bigcap\{f^{-1}(\mathcal{E})) \in \zeta^{\tilde{Y}} \mid \mathcal{S} \leq f^{-1}(\mathcal{E}),\ \tilde{\tau}_1^{\tilde{\gamma}}((f^{-1}(\mathcal{E}))^c) \geq r,\ \tilde{\tau}_1^{\tilde{\eta}}((f^{-1}(\mathcal{E}))^c) \leq 1 - r, \\
&\quad \tilde{\tau}_1^{\tilde{\mu}}((f^{-1}(\mathcal{E}))^c) \leq 1 - r\}] \\
&\geq f(C_{\tilde{\tau}_1^{\tilde{\gamma}},\tilde{\tau}_1^{\tilde{\eta}},\tilde{\tau}_1^{\tilde{\mu}}}(f(\mathcal{S}), r)).
\end{aligned}$$

(2)⇒(3). For all $\mathcal{E} \in \zeta^{\tilde{Y}}$. By (2),

$$f(C_{\tilde{\tau}_1^{\tilde{\gamma}},\tilde{\tau}_1^{\tilde{\eta}},\tilde{\tau}_1^{\tilde{\mu}}}(\mathcal{S},r)) \le C_{\tilde{\tau}_2^{\tilde{\gamma}},\tilde{\tau}_2^{\tilde{\eta}},\tilde{\tau}_2^{\tilde{\mu}}}(f(\mathcal{S}),r).$$

Putting $\mathcal{S} = f^{-1}(\mathcal{E})$, we obtain

$$f(C_{\tilde{\tau}_1^{\tilde{\gamma}},\tilde{\tau}_1^{\tilde{\eta}},\tilde{\tau}_1^{\tilde{\mu}}}(f^{-1}(\mathcal{E}),r)) \le C_{\tilde{\tau}_2^{\tilde{\gamma}},\tilde{\tau}_2^{\tilde{\eta}},\tilde{\tau}_2^{\tilde{\mu}}}(f(f^{-1}(\mathcal{E})),r) \le C_{\tilde{\tau}_2^{\tilde{\gamma}},\tilde{\tau}_2^{\tilde{\eta}},\tilde{\tau}_2^{\tilde{\mu}}}(\mathcal{E},r)$$

Hence, $C_{\tilde{\tau}_1^{\tilde{\gamma}},\tilde{\tau}_1^{\tilde{\eta}},\tilde{\tau}_1^{\tilde{\mu}}}(f^{-1}(\mathcal{E}),r) \le f^{-1}(C_{\tilde{\tau}_2^{\tilde{\gamma}},\tilde{\tau}_2^{\tilde{\eta}},\tilde{\tau}_2^{\tilde{\mu}}}(\mathcal{E},r))$.

(3)⇒(1). It follows that $C_{\tilde{\tau}_1^{\tilde{\gamma}},\tilde{\tau}_1^{\tilde{\eta}},\tilde{\tau}_1^{\tilde{\mu}}}(\mathcal{E},r) = \mathcal{E}$ implies $C_{\tilde{\tau}_1^{\tilde{\gamma}},\tilde{\tau}_1^{\tilde{\eta}},\tilde{\tau}_1^{\tilde{\mu}}}(f^{-1}(\mathcal{E}),r) = f^{-1}(\mathcal{E})$. □

Theorem 7. *Let $(\tilde{X}, \tilde{\tau}_1^{\tilde{\gamma}}, \tilde{\tau}_1^{\tilde{\eta}}, \tilde{\tau}_1^{\tilde{\mu}})$, $(\tilde{Y}, \tilde{\tau}_2^{\tilde{\gamma}}, \tilde{\tau}_2^{\tilde{\eta}}, \tilde{\tau}_2^{\tilde{\mu}})$ be two $\mathcal{SVNTS}'s$ and $f : \tilde{X} \to \tilde{Y}$ is $\mathcal{SVN}$-continuous mapping. If $\mathcal{S}$ is r-$\mathcal{SVNCON}$, then $f(\mathcal{S})$ is r-$\mathcal{SVNCON}$.*

Proof. Let $\mathcal{E}, \mathcal{R} \in \zeta^{\tilde{Y}}$ be two r-$\mathcal{SVNSEP}'s$ such that $f(\mathcal{S}) = \mathcal{E} \cup \mathcal{R}$. We obtain

$$\mathcal{S} \le f(f^{-1})(\mathcal{S}) = f^{-1}(\mathcal{E} \cup \mathcal{R}) = f^{-1}(\mathcal{E}) \cup f^{-1}(\mathcal{R}).$$

Since f is $\mathcal{SVN}$-continuous, by Lemma 2,

$$C_{\tilde{\tau}_1^{\tilde{\gamma}},\tilde{\tau}_1^{\tilde{\eta}},\tilde{\tau}_1^{\tilde{\mu}}}(f^{-1}(\mathcal{E},r)) \le f^{-1}(C_{\tilde{\tau}_2^{\tilde{\gamma}},\tilde{\tau}_2^{\tilde{\eta}},\tilde{\tau}_2^{\tilde{\mu}}}(\mathcal{E},r)).$$

Thus,

$$\begin{aligned} C_{\tilde{\tau}_1^{\tilde{\gamma}},\tilde{\tau}_1^{\tilde{\eta}},\tilde{\tau}_1^{\tilde{\mu}}}(f^{-1}(\mathcal{E},r) \cap f^{-1}(\mathcal{R})) &\le f^{-1}(C_{\tilde{\tau}_2^{\tilde{\gamma}},\tilde{\tau}_2^{\tilde{\eta}},\tilde{\tau}_2^{\tilde{\mu}}}(\mathcal{E},r)) \cap f^{-1}(\mathcal{R}) \\ &= f^{-1}(C_{\tilde{\tau}_2^{\tilde{\gamma}},\tilde{\tau}_2^{\tilde{\eta}},\tilde{\tau}_2^{\tilde{\mu}}}(\mathcal{E},r) \cap \mathcal{R}) \\ &= f^{-1}(\tilde{0}) = \tilde{0}. \end{aligned}$$

Likewise, we obtain $f^{-1}(\mathcal{E}) \cap C_{\tilde{\tau}_2^{\tilde{\gamma}},\tilde{\tau}_2^{\tilde{\eta}},\tilde{\tau}_2^{\tilde{\mu}}}(\mathcal{R},r) = \tilde{0}$. It implies that $f^{-1}(\mathcal{E}), f^{-1}(\mathcal{R}) \in \zeta^{\tilde{X}}$ are r-$\mathcal{SVNSEP}'s$. Since $\mathcal{S}$ is r-$\mathcal{SVNCON}$, then we have by Theorem 4 (3), $\mathcal{S} \le f^{-1}(\mathcal{E})$ or $\mathcal{S} \le f^{-1}(\mathcal{R})$, so $\mathcal{S} \le f^{-1}(\mathcal{E})$. Thus, $f(\mathcal{S}) \le f(f^{-1}(\mathcal{E})) \le \mathcal{E}$. Hence, $f(\mathcal{S})$ is r-$\mathcal{SVNCON}$. □

Example 3. *Let $\tilde{X} = \tilde{Y} = \{a, b\}$ be a set. Define $\mathcal{E}_1, \mathcal{E}_2, \mathcal{E}_3, \mathcal{B}_1, \mathcal{B}_2, \mathcal{B}_3 \in \zeta^{\tilde{X}}$:*

$$\mathcal{E}_1 = \langle(0.5, 0.4), (0.5, 0.5), (0.9, 0.6)\rangle, \quad \mathcal{E}_2 = \langle(0.4, 0.4), (0.1, 0.1), (0.1, 0.1)\rangle,$$

$$\mathcal{E}_3 = \langle(0.3, 0.1), (0.1, 0.1), (0.1, 0.1), \quad \mathcal{B}_1 = \langle(0.4, 0.5), (0.5, 0.5), (0.6, 0.9)\rangle,$$

$$\mathcal{B}_2 = \langle(0.2, 0.2), (0.2, 0.2), (0.1, 0.1), \quad \mathcal{B}_3 = \langle(0.1, 0.1), (0.1, 0.1), (0.1, 0.1)\rangle.$$

Define $\tilde{\tau}^{\tilde{\gamma}\tilde{\eta}\tilde{\mu}}, \tilde{\sigma}^{\tilde{\gamma}\tilde{\eta}\tilde{\mu}} : \zeta^{\tilde{X}} \to \zeta^{\tilde{X}}$ as follows:

$$\tilde{\tau}^{\tilde{\gamma}}(\mathcal{S}) = \begin{cases} 1, & if\ \mathcal{S} = \tilde{0}, \\ 1, & if\ \mathcal{S} = \tilde{1}, \\ \frac{1}{2}, & if\ \mathcal{S} = \mathcal{E}_1, \\ 0, & otherwise. \end{cases} \qquad \tilde{\sigma}^{\tilde{\gamma}}(\mathcal{S}) = \begin{cases} 1, & if\ \mathcal{S} = \tilde{0}\rangle, \\ 1, & if\ \mathcal{S} = \tilde{1}, \\ \frac{1}{2}, & if\ \mathcal{S} = \mathcal{B}_1, \\ 0, & otherwise. \end{cases}$$

$$\tilde{\tau}^{\tilde{\eta}}(\mathcal{S}) = \begin{cases} 0, & if\ \mathcal{S} = \tilde{0}, \\ 0, & if\ \mathcal{S} = \tilde{1}, \\ \frac{1}{2}, & if\ \mathcal{S} = \mathcal{E}_2, \\ 1, & otherwise. \end{cases} \qquad \tilde{\sigma}^{\tilde{\eta}}(\mathcal{S}) = \begin{cases} 0, & if\ \mathcal{S} = \tilde{0}, \\ 0, & if\ \mathcal{S} = \tilde{1}, \\ \frac{1}{2}, & if\ \mathcal{S} = \mathcal{B}_2, \\ 1, & otherwise. \end{cases}$$

$$\tilde{\tau}^{\tilde{\mu}}(\mathcal{S}) = \begin{cases} 0, & if\ \mathcal{S} = \tilde{0}, \\ 0, & if\ \mathcal{S} = \tilde{1}, \\ \frac{1}{2}, & if\ \mathcal{S} = \mathcal{E}_3, \\ 1, & otherwise. \end{cases} \qquad \tilde{\sigma}^{\tilde{\mu}}(\mathcal{S}) = \begin{cases} 0, & if\ \mathcal{S} = \tilde{0}, \\ 0, & if\ \mathcal{S} = \tilde{1}, \\ \frac{1}{2}, & if\ \mathcal{S} = \mathcal{B}_3, \\ 1, & otherwise. \end{cases}$$

Define $f : (\tilde{X}, \tilde{\tau}^{\tilde{\gamma}\tilde{\eta}\tilde{\mu}})) \to (\tilde{Y}, \tilde{\sigma}^{\tilde{\gamma}\tilde{\eta}\tilde{\mu}})$ be a map as follows $f(a) = b$ and $f(b) = a$. If $\mathcal{J}^{\tilde{\gamma}}(\mathcal{B}_1) \geq \frac{1}{2}$, $\mathcal{J}^{\tilde{\eta}}(\mathcal{B}_1) \leq 1 - \frac{1}{2}$ and $\mathcal{J}^{\tilde{\mu}}(\mathcal{B}_1) \leq 1 - \frac{1}{2}$. Then, $f^{-1}(\mathcal{B}_1) = \langle (0.5, 0.4), (0.5, 0.5), (0.9, 0.6) \rangle$ is $\frac{1}{2}$-single-valued neutrosophic open set in $\tilde{X}$. Thus, f is $\mathcal{SVN}$-continuous. However, by Theorem 7, for every $\mathcal{S} \in \zeta^{\tilde{X}}$ is r-$\mathcal{SVNCON}$, then $f(\mathcal{S})$ is r-$\mathcal{SVNCON}$ in $\tilde{Y}$.

Definition 12. *Let $(\tilde{X}, \tilde{\tau}^{\tilde{\gamma}}, \tilde{\tau}^{\tilde{\eta}}, \tilde{\tau}^{\tilde{\mu}})$ be an $\mathcal{SVNTS}$. A $\mathcal{SVN}\int$ $\mathcal{S}$ is called r-single-valued neutrosophic component (r-$\mathcal{SVNCOM}$, for short) in $(\tilde{X}, \tilde{\tau}^{\tilde{\gamma}}, \tilde{\tau}^{\tilde{\eta}}, \tilde{\tau}^{\tilde{\mu}})$ if $\mathcal{S}$ is a maximal r-$\mathcal{SVNCON}$ in $(\tilde{X}, \tilde{\tau}^{\tilde{\gamma}}, \tilde{\tau}^{\tilde{\eta}}, \tilde{\tau}^{\tilde{\mu}})$, i.e., if $\mathcal{E} \geq \mathcal{S}$ and $\mathcal{E}$ is r-$\mathcal{SVNCON}$, then $\mathcal{E} = \mathcal{S}$.*

Corollary 2. *Let $(\tilde{X}, \tilde{\tau}^{\tilde{\gamma}}, \tilde{\tau}^{\tilde{\eta}}, \tilde{\tau}^{\tilde{\mu}})$ be an $\mathcal{SVNTS}$.*

(1) *If $\mathcal{S}$ is a r-$\mathcal{SVNCOM}$, $C_{\tilde{\tau}^{\tilde{\gamma}}, \tilde{\tau}^{\tilde{\eta}}, \tilde{\tau}^{\tilde{\mu}}}(\mathcal{S}, r) = \mathcal{S}$.*
(2) *If $\mathcal{S}_1, \mathcal{S}_2 \in \zeta^{\tilde{X}}$ are r-$\mathcal{SVNCOM}$ in $\tilde{X}$ such that $\mathcal{S}_1 \cap \mathcal{S}_2 = \tilde{0}$, then $\mathcal{S}_1, \mathcal{S}_2 \in \zeta^{\tilde{X}}$ are r-$\mathcal{SVNSEP}$.*
(3) *Each single-valued neutrosophic point $x_{t,s,k}$ is $\mathcal{SVNCON}$.*
(4) *Every r-$\mathcal{SVNCOM}$ is a crisp set.*

Proof. Straightforward. □

4. Stratification of Single-Valued Neutrosophic Topological Spaces

In this section, we obtain crucial results in the stratification of the single-valued neutrosophic topology as follows.

Definition 13. *The stratification of the single-valued neutrosophic topology ($\mathcal{SVNT}$) on $\tilde{X}$ is a mapping from $\zeta^{\tilde{X}}$ to ζ such that*

(SVNT1) $\tilde{\tau}^{\tilde{\gamma}}(\tilde{\alpha}) = 1$ *and* $\tilde{\tau}^{\tilde{\eta}}(\tilde{\alpha}) = \tilde{\tau}^{\tilde{\mu}}(\tilde{\alpha}) = 0, \forall\, \alpha \in \zeta$,
(SVNT2) $\tilde{\tau}^{\tilde{\gamma}}(\mathcal{S} \cap \mathcal{E}) \geq \tilde{\tau}^{\tilde{\gamma}}(\mathcal{S}) \cap \tilde{\tau}^{\tilde{\gamma}}(\mathcal{E})$, $\quad \tilde{\tau}^{\tilde{\eta}}(\mathcal{S} \cap \mathcal{E}) \leq \tau^{\tilde{\eta}}(\mathcal{S}) \cup \tilde{\tau}^{\tilde{\eta}}(\mathcal{E})$,
$\tilde{\tau}^{\tilde{\mu}}(\mathcal{S} \cap \mathcal{E}) \leq \tilde{\tau}^{\tilde{\mu}}(\mathcal{S}) \cup \tilde{\tau}^{\tilde{\mu}}(\mathcal{E})$, *for all* $\mathcal{S}, \mathcal{E} \in \zeta^{\tilde{X}}$,
(SVNT3) $\tilde{\tau}^{\tilde{\gamma}}(\cup_{j\in\Gamma}\mathcal{S}_j) \geq \cap_{j\in\Gamma}\tilde{\tau}^{\tilde{\gamma}}(\mathcal{S}_j)$, $\quad \tilde{\tau}^{\tilde{\eta}}(\cup_{i\in\Gamma}\mathcal{S}_j) \leq \cup_{j\in\Gamma}\tilde{\tau}^{\tilde{\eta}}(\mathcal{S}_j)$,
$\tilde{\tau}^{\tilde{\mu}}(\cup_{j\in\Gamma}\mathcal{S}_j) \leq \cup_{j\in\Gamma}\tilde{\tau}^{\tilde{\mu}}(\mathcal{S}_j)$, *for all* $\{\mathcal{S}_j, j \in \Gamma\} \in \zeta^{\tilde{X}}$.

The ordered pair $\mathcal{SVNTS}$ $(\tilde{X}, \tilde{\tau}^{\tilde{\gamma}}, \tilde{\tau}^{\tilde{\eta}}, \tilde{\tau}^{\tilde{\mu}})$ is called stratified. Let $(\tilde{\tau}_1^{\tilde{\gamma}}, \tilde{\tau}_1^{\tilde{\eta}}, \tilde{\tau}_1^{\tilde{\mu}})$ and $(\tilde{\tau}_2^{\tilde{\gamma}}, \tilde{\tau}_2^{\tilde{\eta}}, \tilde{\tau}_2^{\tilde{\mu}})$ be $\mathcal{SVNGO}$'s on $\tilde{X}$. We say that $(\tilde{\tau}_1^{\tilde{\gamma}}, \tilde{\tau}_1^{\tilde{\eta}}, \tilde{\tau}_1^{\tilde{\mu}})$ is finer then $(\tilde{\tau}_2^{\tilde{\gamma}}, \tilde{\tau}_2^{\tilde{\eta}}, \tilde{\tau}_2^{\tilde{\mu}})$ $[(\tilde{\tau}_2^{\tilde{\gamma}}, \tilde{\tau}_2^{\tilde{\eta}}, \tilde{\tau}_2^{\tilde{\mu}})$ is coarser then $(\tilde{\tau}_1^{\tilde{\gamma}}, \tilde{\tau}_1^{\tilde{\eta}}, \tilde{\tau}_1^{\tilde{\mu}})]$ if $\tilde{\tau}_1^{\tilde{\gamma}}(\mathcal{S}) \leq \tilde{\tau}_2^{\tilde{\gamma}}(\mathcal{S})$, $\tilde{\tau}_1^{\tilde{\eta}}(\mathcal{S}) \geq \tilde{\tau}_2^{\tilde{\eta}}(\mathcal{S})$ and $\tilde{\tau}_1^{\tilde{\mu}}(\mathcal{S}) \geq \tilde{\tau}_2^{\tilde{\mu}}(\mathcal{S})$ for all $\mathcal{S} \in \zeta^{\tilde{X}}$.

Theorem 8. *Let $(\tilde{X}, \tilde{\tau}^{\tilde{\gamma}}, \tilde{\tau}^{\tilde{\eta}}, \tilde{\tau}^{\tilde{\mu}})$ be an $\mathcal{SVNTS}$. Define the mappings $\tilde{\tau}_{st}^{\tilde{\gamma}}, \tilde{\tau}_{st}^{\tilde{\eta}}, \tilde{\tau}_{st}^{\tilde{\mu}} : \zeta^{\tilde{X}} \to \zeta$ as follows: for all $\mathcal{S} \in \zeta^{\tilde{X}}$,*

$$\tilde{\tau}_{st}^{\tilde{\gamma}}(\mathcal{S}) = \bigcup \{ \bigcap_{j\in\Gamma} \tilde{\tau}^{\tilde{\gamma}}(\mathcal{S}_j) \mid \mathcal{S} = \bigcup_{j\in\Gamma} (\mathcal{S}_j \cap \tilde{\alpha}_j) \}$$

$$\tilde{\tau}_{st}^{\tilde{\eta}}(\mathcal{S}) = \bigcap\{\bigcup_{j\in\Gamma} \tilde{\tau}^{\tilde{\eta}}(\mathcal{S}_j) \mid \mathcal{S} = \bigcup_{j\in\Gamma}(\mathcal{S}_j \cap \tilde{\alpha}_j)\}$$

$$\tilde{\tau}_{st}^{\tilde{\mu}}(\mathcal{S}) = \bigcap\{\bigcup_{j\in\Gamma} \tilde{\tau}^{\tilde{\mu}}(\mathcal{S}_j) \mid \mathcal{S} = \bigcup_{j\in\Gamma}(\mathcal{S}_j \cap \tilde{\alpha}_j)\}.$$

Then, $(\tilde{\tau}_{st}^{\tilde{\gamma}}, \tilde{\tau}_{st}^{\tilde{\eta}}, \tilde{\tau}_{st}^{\tilde{\mu}})$ is the coarsest stratified $\mathcal{SVNT}$ on $\tilde{X}$ which is finer than $(\tilde{\tau}^{\tilde{\gamma}}, \tilde{\tau}^{\tilde{\eta}}, \tilde{\tau}^{\tilde{\mu}})$.

Proof. Firstly, we will show that $(\tilde{\tau}_{st}^{\tilde{\gamma}}, \tilde{\tau}_{st}^{\tilde{\eta}}, \tilde{\tau}_{st}^{\tilde{\mu}})$ is a stratified $\mathcal{SVNT}$ on $\tilde{X}$.

(SVNT1) For every $\alpha \in \zeta$, there exists a collection $\{\tilde{1}\}$ with $\tilde{\alpha} = \tilde{\alpha} \cap \tilde{1}$, we obtain $\tilde{\tau}_{st}^{\tilde{\gamma}}(\tilde{\alpha}) \geq \tilde{\tau}^{\tilde{\gamma}}(\tilde{1}) = 1$, $\tilde{\tau}_{st}^{\tilde{\eta}}(\tilde{\alpha}) \leq \tilde{\tau}^{\tilde{\eta}}(\tilde{1}) = 0$ and $\tilde{\tau}_{st}^{\tilde{\mu}}(\tilde{\alpha}) \leq \tilde{\tau}^{\tilde{\mu}}(\tilde{1}) = 0$. Thus, $\tilde{\tau}_{st}^{\tilde{\gamma}}(\tilde{\alpha}) = 1$ and $\tilde{\tau}_{st}^{\tilde{\eta}}(\tilde{\alpha}) = \tilde{\tau}_{st}^{\tilde{\mu}}(\tilde{\alpha}) = 0$.

(SVNT2) Suppose there exists $\mathcal{E}, \mathcal{R} \in \zeta^{\tilde{X}}$ and $r \in \zeta_0$ with

$$\tilde{\tau}_{st}^{\tilde{\gamma}}(\mathcal{E} \cap \mathcal{R}) < r < \tilde{\tau}_{st}^{\tilde{\gamma}}(\mathcal{E}) \cap \tilde{\tau}_{st}^{\tilde{\gamma}}(\mathcal{R}),$$

$$\tilde{\tau}_{st}^{\tilde{\eta}}(\mathcal{E} \cap \mathcal{R}) > 1 - r > \tilde{\tau}_{st}^{\tilde{\eta}}(\mathcal{E}) \cup \tilde{\tau}_{st}^{\tilde{\eta}}(\mathcal{R}),$$

$$\tilde{\tau}_{st}^{\tilde{\mu}}(\mathcal{E} \cap \mathcal{R}) > 1 - r > \tilde{\tau}_{st}^{\tilde{\mu}}(\mathcal{E}) \cup \tilde{\tau}_{st}^{\tilde{\mu}}(\mathcal{R}).$$

Since $[\tilde{\tau}_{st}^{\tilde{\gamma}}(\mathcal{E}) > r, \tilde{\tau}_{st}^{\tilde{\gamma}}(\mathcal{R}) > r]$, $[\tilde{\tau}_{st}^{\tilde{\eta}}(\mathcal{E}) < 1 - r, \tilde{\tau}_{st}^{\tilde{\eta}}(\mathcal{R}) < 1 - r]$ and $[\tilde{\tau}_{st}^{\tilde{\gamma}}(\mathcal{E}) < 1 - r, \tilde{\tau}_{st}^{\tilde{\gamma}}(\mathcal{R}) < 1 - r]$, by the definition of $(\tilde{\tau}^{\tilde{\gamma}}, \tilde{\tau}^{\tilde{\eta}}, \tilde{\tau}^{\tilde{\mu}})$, there exist $\{\mathcal{E}_j \mid j \in \Gamma\}$ with $\mathcal{E} = \bigcup_{j\in\Gamma}(\mathcal{E}_j \cap \tilde{\alpha}_j)$ and $\{\mathcal{R}_k \mid k \in K\}$ with $\mathcal{R} = \bigcup_{k\in K}(\mathcal{R}_k \cap \tilde{\alpha}_k)$ such that

$$\tilde{\tau}_{st}^{\tilde{\gamma}}(\mathcal{E}) \geq \bigcap_{j\in\Gamma} \tilde{\tau}^{\tilde{\gamma}}(\mathcal{E}_j) > r, \quad \tilde{\tau}_{st}^{\tilde{\eta}}(\mathcal{E}) \leq \bigcup_{j\in\Gamma} \tilde{\tau}^{\tilde{\eta}}(\mathcal{E}_j) < 1 - r \text{ and } \tilde{\tau}_{st}^{\tilde{\mu}}(\mathcal{E}) \leq \bigcup_{j\in\Gamma} \tilde{\tau}^{\tilde{\mu}}(\mathcal{E}_j) < 1 - r,$$

$$\tilde{\tau}_{st}^{\tilde{\gamma}}(\mathcal{R}) \geq \bigcap_{k\in K} \tilde{\tau}^{\tilde{\gamma}}(\mathcal{R}_k) > r, \quad \tilde{\tau}_{st}^{\tilde{\mu}}(\mathcal{R}) \leq \bigcup_{k\in K} \tilde{\tau}^{\tilde{\mu}}(\mathcal{R}_k) < 1 - r \text{ and } \tilde{\tau}_{st}^{\tilde{\mu}}(\mathcal{R}) \leq \bigcup_{k\in K} \tilde{\tau}_{st}^{\tilde{\mu}}(\mathcal{R}_k) < 1 - r.$$

Since ζ is completely distributive lattice, we have

$$\begin{aligned}\mathcal{E} \cap \mathcal{R} &= [\bigcup_{j\in\Gamma}(\mathcal{E}_j \cap \tilde{\alpha}_j)] \cap [\bigcup_{k\in K}(\mathcal{R}_k \cap \tilde{\alpha}_k)] = \bigcup_{j\in\Gamma}(\mathcal{E}_j \cap \mathcal{R}_k) \cap (\tilde{\alpha}_j \cap \tilde{\alpha}_k)\\ &= \bigcup_{j\in\Gamma}(\mathcal{E}_j \cap \mathcal{R}_k) \cap \tilde{\alpha}_{jk}. \qquad (\tilde{\alpha}_{jk} = \tilde{\alpha}_j \cap \tilde{\alpha}_k).\end{aligned}$$

Moreover, since $\tilde{\tau}^{\tilde{\gamma}}(\mathcal{E}_j \cap \mathcal{R}_k) \geq \tilde{\tau}^{\tilde{\gamma}}(\mathcal{E}_j) \cap \tilde{\tau}^{\tilde{\gamma}}(\mathcal{R}_k)$, $\tilde{\tau}^{\tilde{\eta}}(\mathcal{E}_j \cap \mathcal{R}_k) \leq \tau^{\tilde{\eta}}(\mathcal{E}_j) \cup \tilde{\tau}^{\tilde{\eta}}(\mathcal{R}_k)$ and $\tilde{\tau}^{\tilde{\mu}}(\mathcal{E}_j \cap \mathcal{R}_k) \leq \tilde{\tau}^{\tilde{\mu}}(\mathcal{E}_j) \cup \tilde{\tau}^{\tilde{\mu}}(\mathcal{R}_k)$, we obtain

$$\tilde{\tau}_{st}^{\tilde{\gamma}}(\mathcal{E} \cap \mathcal{R}) \geq \bigcap_{j,k} \tilde{\tau}^{\tilde{\gamma}}(\mathcal{E}_j \cap \mathcal{R}_k) \geq \bigcap_{j,k}(\tilde{\tau}^{\tilde{\gamma}}(\mathcal{E}_j) \cap \tilde{\tau}^{\tilde{\gamma}}(\mathcal{R}_k)) = [\bigcap_{j\in\Gamma}(\tilde{\tau}^{\tilde{\gamma}}(\mathcal{E}_j)] \cap [\bigcap_{k\in K} \tilde{\tau}^{\tilde{\gamma}}(\mathcal{R}_k)] > r,$$

$$\tilde{\tau}_{st}^{\tilde{\eta}}(\mathcal{E} \cap \mathcal{R}) \leq \bigcup_{j,k} \tilde{\tau}^{\tilde{\eta}}(\mathcal{E}_j \cap \mathcal{R}_k) \leq \bigcup_{j,k}(\tilde{\tau}^{\tilde{\eta}}(\mathcal{E}_j) \cup \tilde{\tau}^{\tilde{\eta}}(\mathcal{R}_k)) = [\bigcup_{j\in\Gamma}(\tilde{\tau}^{\tilde{\eta}}(\mathcal{E}_j)] \cup [\bigcup_{k\in K} \tilde{\tau}^{\tilde{\eta}}(\mathcal{R}_k)] < 1 - r,$$

$$\tilde{\tau}_{st}^{\tilde{\gamma}}(\mathcal{E} \cap \mathcal{R}) \leq \bigcup_{j,k} \tilde{\tau}^{\tilde{\gamma}}(\mathcal{E}_j \cap \mathcal{R}_k) \leq \bigcup_{j,k}(\tilde{\tau}^{\tilde{\mu}}(\mathcal{E}_j) \cup \tilde{\tau}_{st}^{\tilde{\mu}}(\mathcal{R}_k)) = [\bigcup_{j\in\Gamma}(\tilde{\tau}^{\tilde{\mu}}(\mathcal{E}_j)] \cup [\bigcup_{k\in K} \tilde{\tau}_{st}^{\tilde{\mu}}(\mathcal{R}_k)] < 1 - r$$

It is a contradiction. Hence, for each $\mathcal{E}, \mathcal{R} \in \zeta^{\tilde{X}}$,

$$\tilde{\tau}^{\tilde{\gamma}}_{st}(\mathcal{E} \cap \mathcal{R}) \geq \tilde{\tau}^{\tilde{\gamma}}_{st}(\mathcal{E}) \cap \tilde{\tau}^{\tilde{\gamma}}_{st}(\mathcal{R}), \quad \tilde{\tau}^{\tilde{\eta}}_{st}(\mathcal{E} \cap \mathcal{R}) \leq \tau^{\tilde{\eta}}_{st}(\mathcal{E}) \cup \tilde{\tau}^{\tilde{\eta}}_{st}(\mathcal{R}), \quad \tilde{\tau}^{\tilde{\mu}}_{st}(\mathcal{E} \cap \mathcal{R}) \leq \tilde{\tau}^{\tilde{\mu}}_{st}(\mathcal{E}) \cup \tilde{\tau}^{\tilde{\mu}}_{st}(\mathcal{R}).$$

(SVNT3) Suppose there exists a family $\{\mathcal{E}_j \in \zeta^{\tilde{X}} \mid j \in \Gamma\}$ and $r \in \zeta_0$ with

$$\tilde{\tau}^{\tilde{\gamma}}_{st}(\bigcup_{j \in \Gamma} \mathcal{E}_j) < r < \bigcap_{j \in \Gamma} \tilde{\tau}^{\tilde{\gamma}}_{st}(\mathcal{E}_j),$$

$$\tilde{\tau}^{\tilde{\eta}}_{st}(\bigcup_{i \in \Gamma} \mathcal{E}_j) > 1 - r > \bigcup_{j \in \Gamma} \tilde{\tau}^{\tilde{\eta}}_{st}(\mathcal{E}_j),$$

$$\tilde{\tau}^{\tilde{\mu}}_{st}(\bigcup_{j \in \Gamma} \mathcal{E}_j) > 1 - r > \bigcup_{j \in \Gamma} \tilde{\tau}^{\tilde{\mu}}_{st}(\mathcal{E}_j)$$

Since $[\tilde{\tau}^{\tilde{\gamma}}_{st}(\mathcal{E}_j) \geq r, \tilde{\tau}^{\tilde{\eta}}_{st}(\mathcal{E}_j) \leq 1 - r$ and $\tilde{\tau}^{\tilde{\mu}}_{st}(\mathcal{E}_j) \leq 1 - r]$ for all $j \in \Gamma$, there exists a family $\{\mathcal{E}_{jk} \mid k \in K_j\}$ with $\mathcal{E}_j = \bigcup_{k \in K_j} \mathcal{E}_{jk} \cap \tilde{\alpha}_k$ such that

$$\tilde{\tau}^{\tilde{\gamma}}_{st}(\mathcal{E}_j) \geq \bigcap_{k \in K_j} \tilde{\tau}^{\tilde{\gamma}}_{st}(\mathcal{E}_{jk}) > r,$$

$$\tilde{\tau}^{\tilde{\eta}}_{st}(\mathcal{E}_j) \leq \bigcup_{k \in K_j} \tilde{\tau}^{\tilde{\eta}}_{st}(\mathcal{E}_{jk}) < 1 - r,$$

$$\tilde{\tau}^{\tilde{\mu}}_{st}(\mathcal{E}_j) \leq \bigcup_{k \in K_j} \tilde{\tau}^{\tilde{\mu}}_{st}(\mathcal{E}_{jk}) < 1 - r.$$

Since $\bigcup_{j \in \Gamma} \mathcal{E}_j = \bigcup_{j \in \Gamma}(\bigcup_{k \in K_j}(\mathcal{E}_{jk} \cap \tilde{\alpha}_k)) = \bigcup_{j,k}(\mathcal{E}_{jk} \cap \tilde{\alpha}_k)$, we obtain

$$\tilde{\tau}^{\tilde{\gamma}}_{st}(\bigcup_{j \in \Gamma} \mathcal{E}_j) \geq \bigcap_{j,k} \tilde{\tau}^{\tilde{\gamma}}(\mathcal{E}_{jk}) = \bigcap_{j \in \Gamma}(\bigcap_{k \in K_j} \tilde{\tau}^{\tilde{\gamma}}(\mathcal{E}_{jk}) \geq r,$$

$$\tilde{\tau}^{\tilde{\eta}}_{st}(\bigcup_{j \in \Gamma} \mathcal{E}_j) \leq \bigcup_{j,k} \tilde{\tau}^{\tilde{\eta}}(\mathcal{E}_{jk}) = \bigcup_{j \in \Gamma}(\bigcup_{k \in K_j} \tilde{\tau}^{\tilde{\eta}}(\mathcal{E}_{jk}) \leq 1 - r,$$

$$\tilde{\tau}^{\tilde{\mu}}_{st}(\bigcup_{j \in \Gamma} \mathcal{E}_j) \leq \bigcup_{j,k} \tilde{\tau}^{\tilde{\mu}}(\mathcal{E}_{jk}) = \bigcup_{j \in \Gamma}(\bigcup_{k \in K_j} \tilde{\tau}^{\tilde{\mu}}(\mathcal{E}_{jk}) \leq 1 - r$$

It is a contradiction. Hence, for each $\{\mathcal{E}_j\}_{j \in \Gamma} \in \zeta^{\tilde{X}}$

$$\tilde{\tau}^{\tilde{\gamma}}_{st}(\bigcup_{j \in \Gamma} \mathcal{E}_j) \geq \bigcap_{j \in \Gamma} \tilde{\tau}^{\tilde{\gamma}}_{st}(\mathcal{E}_j), \quad \tilde{\tau}^{\tilde{\eta}}_{st}(\bigcup_{i \in \Gamma} \mathcal{S}_j) \leq \bigcup_{j \in \Gamma} \tilde{\tau}^{\tilde{\eta}}_{st}(\mathcal{E}_j), \quad \tilde{\tau}^{\tilde{\mu}}_{st}(\bigcup_{i \in \Gamma} \mathcal{S}_j) \leq \bigcup_{j \in \Gamma} \tilde{\tau}^{\tilde{\mu}}_{st}(\mathcal{E}_j).$$

Secondly, for each $\mathcal{S} \in \zeta^{\tilde{X}}$, there exists a family $\{\tilde{1}\}$ with $\mathcal{S} = \mathcal{S} \cap \tilde{1}$, such that $[\tilde{\tau}^{\tilde{\gamma}}_{st}(\mathcal{S}) \geq \tilde{\tau}^{\tilde{\gamma}}(\mathcal{S})$, $\tilde{\tau}^{\tilde{\eta}}_{st}(\mathcal{S}) \leq \tilde{\tau}^{\tilde{\eta}}(\mathcal{S})$ and $\tilde{\tau}^{\tilde{\mu}}_{st}(\mathcal{S}) \leq \tilde{\tau}^{\tilde{\mu}}(\mathcal{S})]$. Hence, $(\tilde{\tau}^{\tilde{\gamma}}_{st}, \tilde{\tau}^{\tilde{\eta}}_{st}, \tilde{\tau}^{\tilde{\mu}}_{st})$ is finer than $(\tilde{\tau}^{\tilde{\gamma}}, \tilde{\tau}^{\tilde{\eta}}, \tilde{\tau}^{\tilde{\mu}})$. Finally, if a stratified $\mathcal{SVNT}$ $(\tilde{U}^{\tilde{\gamma}}, \tilde{U}^{\tilde{\eta}}, \tilde{U}^{\tilde{\mu}})$ is finer than $(\tilde{\tau}^{\tilde{\gamma}}, \tilde{\tau}^{\tilde{\eta}}, \tilde{\tau}^{\tilde{\mu}})$, we show that $[\tilde{\tau}^{\tilde{\gamma}}_{st}(\mathcal{S}) \leq \tilde{U}^{\tilde{\gamma}}(\mathcal{S})$, $\tilde{\tau}^{\tilde{\eta}}_{st}(\mathcal{S}) \geq \tilde{U}^{\tilde{\eta}}(\mathcal{S})$ and $\tilde{\tau}^{\tilde{\mu}}_{st}(\mathcal{S}) \geq \tilde{U}^{\tilde{\mu}}(\mathcal{E})]$ for each $\mathcal{S} \in \zeta^{\tilde{X}}$.

Suppose that there exist $\mathcal{E} \in \zeta^{\tilde{X}}$ and $r \in \zeta_0$ such that

$$\tilde{\tau}^{\tilde{\gamma}}_{st}(\mathcal{E}) > r > \tilde{U}^{\tilde{\gamma}}(\mathcal{E}), \quad \tilde{\tau}^{\tilde{\eta}}_{st}(\mathcal{E}) < 1 - r < \tilde{U}^{\tilde{\eta}}(\mathcal{E}), \quad \tilde{\tau}^{\tilde{\mu}}_{st}(\mathcal{E}) < 1 - r < \tilde{U}^{\tilde{\mu}}(\mathcal{E}).$$

Since $[\tilde{\tau}^{\tilde{\gamma}}_{st}(\mathcal{E}) > r, \tilde{\tau}^{\tilde{\eta}}_{st}(\mathcal{E}) < 1 - r$ and $\tilde{\tau}^{\tilde{\mu}}_{st}(\mathcal{E}) < 1 - r]$, there exists $\{\mathcal{E}_j \mid j \in \Gamma\}$ with $\mathcal{E}_j = \bigcup_{j \in \Gamma}(\mathcal{E}_j \cap \tilde{\alpha}_j)$ such that

$$\tilde{\tau}^{\tilde{\gamma}}_{st}(\mathcal{E}) \geq \bigcap_{j \in \Gamma} \tilde{\tau}^{\tilde{\gamma}}(\mathcal{E}_j) > r, \quad \tilde{\tau}^{\tilde{\eta}}_{st}(\mathcal{E}) \leq \bigcup_{j \in \Gamma} \tilde{\tau}^{\tilde{\eta}}(\mathcal{E}_j) < 1 - r, \quad \tilde{\tau}^{\tilde{\mu}}_{st}(\mathcal{E}) \leq \bigcup_{j \in \Gamma} \tilde{\tau}^{\tilde{\mu}}(\mathcal{E}_j) < 1 - r.$$

On the other hand, since $[\tilde{U}^{\tilde{\gamma}}(\mathcal{E}_j) \geq \tilde{\tau}^{\tilde{\gamma}}(\mathcal{E}_j)$, $\tilde{U}^{\tilde{\eta}}(\mathcal{E}_j) \leq \tilde{\tau}^{\tilde{\eta}}(\mathcal{E}_j)$ and $\tilde{U}^{\tilde{\mu}}(\mathcal{E}_j) \leq \tilde{\tau}^{\tilde{\mu}}(\mathcal{E}_j)]$ for all $j \in \Gamma$, we have

$$\tilde{U}^{\tilde{\gamma}}(\mathcal{E}) = \tilde{U}^{\tilde{\gamma}}(\bigcup_{j \in \Gamma}(\mathcal{E}_j \cap \tilde{\alpha}_j)) \geq \bigcap_{j \in \Gamma} \tilde{U}^{\tilde{\gamma}}(\mathcal{E}_j \cap \tilde{\alpha}_j) \geq \bigcap_{j \in \Gamma}[\tilde{U}^{\tilde{\gamma}}(\mathcal{E}_j) \cap \tilde{U}^{\tilde{\gamma}}(\tilde{\alpha}_j)] = \bigcap_{j \in \Gamma} \tilde{U}^{\tilde{\gamma}}(\mathcal{E}_j) \geq \bigcap_{j \in \Gamma} \tilde{\tau}^{\tilde{\gamma}}(\mathcal{E}_j) > r,$$

$$\tilde{U}^{\tilde{\eta}}(\mathcal{E}) = \tilde{U}^{\tilde{\eta}}(\bigcup_{j \in \Gamma}(\mathcal{E}_j \cap \tilde{\alpha}_j)) \leq \bigcup_{j \in \Gamma} \tilde{U}^{\tilde{\eta}}(\mathcal{E}_j \cap \tilde{\alpha}_j) \leq \bigcup_{j \in \Gamma}[\tilde{U}^{\tilde{\eta}}(\mathcal{E}_j) \cup \tilde{U}^{\tilde{\eta}}(\tilde{\alpha}_j)] = \bigcup_{j \in \Gamma} \tilde{U}^{\tilde{\eta}}(\mathcal{E}_j) \leq \bigcup_{j \in \Gamma} \tilde{\tau}^{\tilde{\eta}}(\mathcal{E}_j) < 1 - r$$

$$\tilde{U}^{\tilde{\mu}}(\mathcal{E}) = \tilde{U}^{\tilde{\eta}}(\bigcup_{j \in \Gamma}(\mathcal{E}_j \cap \tilde{\alpha}_j)) \leq \bigcup_{j \in \Gamma} \tilde{U}^{\tilde{\mu}}(\mathcal{E}_j \cap \tilde{\alpha}_j) \leq \bigcup_{j \in \Gamma}[\tilde{U}^{\tilde{\mu}}(\mathcal{E}_j) \cup \tilde{U}^{\tilde{\mu}}(\tilde{\alpha}_j)] = \bigcup_{j \in \Gamma} \tilde{U}^{\tilde{\mu}}(\mathcal{E}_j) \leq \bigcup_{j \in \Gamma} \tilde{\tau}^{\tilde{\mu}}(\mathcal{E}_j) < 1 - r$$

It is a contradiction. □

Remark 2. *From Defintion 13 and Theorem 8, we have* $(\tilde{\tau}^{\tilde{\gamma}}_{st}, \tilde{\tau}^{\tilde{\eta}}_{st}, \tilde{\tau}^{\tilde{\mu}}_{st})$ *as a stratification for* $\mathcal{SVNT}$ $(\tilde{\tau}^{\tilde{\gamma}}, \tilde{\tau}^{\tilde{\eta}}, \tilde{\tau}^{\tilde{\mu}})$ *on* $\tilde{X}$.

Example 4. *Let* $\tilde{X} = \{a, b, c\}$ *be a set. Define* $\mathcal{E}_1, \mathcal{E}_2 \in \zeta^{\tilde{X}}$ *as follows:*

$$\mathcal{E}_1 = \langle(0.5, 0.5, 0.5), (0.5, 0.5, 0.5), (0.5, 0.5, 0.5)\rangle; \mathcal{E}_2 = \langle(0.4, 0.4, 0.4), (0.4, 0.4, 0.4), (0.6, 0.6, 0.6)\rangle.$$

We define $\tilde{\tau}^{\tilde{\gamma}}, \tilde{\tau}^{\tilde{\eta}}, \tilde{\tau}^{\tilde{\mu}} : \zeta^{\tilde{X}} x \zeta_0 \to \zeta$ *as follows: for every* $\mathcal{S} \in \zeta^{\tilde{X}}$,

$$\tilde{\tau}^{\tilde{\gamma}}(\mathcal{S}) = \begin{cases} 1, if\ \mathcal{S} = \tilde{0}, \\ 1, if\ \mathcal{S} = \tilde{1}, \\ \frac{1}{3}, if\ \mathcal{S} = \mathcal{E}_1, \\ \frac{1}{2}, if\ \mathcal{S} = \mathcal{E}_2, \\ \frac{3}{4}, if\ \mathcal{S} = \mathcal{E}_1 \cup \mathcal{E}_2, \\ \frac{2}{3}, if\ \mathcal{S} = \mathcal{E}_1 \cap \mathcal{E}_2, \\ 0, \quad otherwise, \end{cases} \qquad \tilde{\tau}^{\tilde{\eta}}(\mathcal{S}) = \begin{cases} 0, if\ \mathcal{S} = \tilde{0}, \\ 0, if\ \mathcal{S} = \tilde{1}, \\ \frac{2}{3}, if\ \mathcal{S} = \mathcal{E}_1, \\ \frac{1}{2}, if\ \mathcal{S} = \mathcal{E}_2, \\ \frac{1}{4}, if\ \mathcal{S} = \mathcal{E}_1 \cup \mathcal{E}_2, \\ \frac{1}{3}, if\ \mathcal{S} = \mathcal{E}_1 \cap \mathcal{E}_2, \\ 1, \quad otherwise, \end{cases}$$

$$\tilde{\tau}^{\tilde{\mu}}(\mathcal{S}) = \begin{cases} 0, if\ \mathcal{S} = \tilde{0}, \\ 0, if\ \mathcal{S} = \tilde{1}, \\ \frac{2}{3}, if\ \mathcal{S} = \mathcal{E}_1, \\ \frac{1}{2}, if\ \mathcal{S} = \mathcal{E}_2, \\ \frac{1}{2}, if\ \mathcal{S} = \mathcal{E}_1 \cup \mathcal{E}_2, \\ \frac{1}{3}, if\ \mathcal{S} = \mathcal{E}_1 \cap \mathcal{E}_2, \\ 1, \quad otherwise, \end{cases}$$

If $\tilde{\gamma}_{\mathcal{S}}(\omega) = \alpha$ *for any* $0.5 < \alpha < 0.6$, $\tilde{\eta}_{\mathcal{S}}(\omega) = \alpha$ *for every* $0.5 < \alpha < 0.6$ *and* $\tilde{\gamma}_{\mathcal{S}}(\omega) = 0.6$ *for all* $\beta \geq 0.6$, *since*

$$\mathcal{S} = (\tilde{\alpha} \cap \tilde{1}) \cup (\tilde{\beta} \cap (\mathcal{E}_1 \cup \mathcal{E}_2)) = (\tilde{\alpha} \cap \tilde{1}) \cup (\tilde{\beta} \cap \mathcal{E}_2).$$

Then,

$$\tilde{\tau}^{\tilde{\gamma}}_{st}(\mathcal{S}) = [\tilde{\tau}^{\tilde{\gamma}}(\tilde{1}) \cap \tilde{\tau}^{\tilde{\gamma}}(\mathcal{E}_1 \cup \mathcal{E}_2)] \cup [\tilde{\tau}^{\tilde{\gamma}}(\tilde{1}) \cap \tilde{\tau}^{\tilde{\gamma}}(\mathcal{E}_2)] = \frac{3}{4},$$

$$\tilde{\tau}^{\tilde{\eta}}_{st}(\mathcal{S}) = [\tilde{\tau}^{\tilde{\eta}}(\tilde{1}) \cup \tilde{\tau}^{\tilde{\eta}}(\mathcal{E}_1 \cup \mathcal{E}_2)] \cap [\tilde{\tau}^{\tilde{\eta}}(\tilde{1}) \cup \tilde{\tau}^{\tilde{\eta}}(\mathcal{E}_2)] = \frac{1}{4},$$

$$\tilde{\tau}^{\tilde{\mu}}_{st}(\mathcal{S}) = [\tilde{\tau}^{\tilde{\mu}}(\tilde{1}) \cup \tilde{\tau}^{\tilde{\mu}}(\mathcal{E}_1 \cup \mathcal{E}_2)] \cap [\tilde{\tau}^{\tilde{\mu}}(\tilde{1}) \cup \tilde{\tau}^{\tilde{\mu}}(\mathcal{E}_2)] = \frac{1}{2}.$$

If $\tilde{\gamma}_{\mathcal{S}}(\omega) = \alpha \;\forall\; 0.5 < \alpha < 0.6$, $\tilde{\eta}_{\mathcal{S}}(\omega) = \alpha \;\forall\; 0.5 < \alpha < 0.6$ *and* $\tilde{\gamma}_{\mathcal{S}}(\omega) = \beta \;\forall\; 0.5 < \alpha, \beta \geq 0.6$, *we have* $\tilde{\tau}^{\tilde{\gamma}}_{st}(\mathcal{S}) = \frac{3}{4}$, $\tilde{\tau}^{\tilde{\eta}}_{st}(\mathcal{S}) = \frac{1}{4}$, $\tilde{\tau}^{\tilde{\mu}}_{st}(\mathcal{S}) = \frac{1}{2}$.

If $\tilde{\gamma}_{\mathcal{S}}(\omega) = 0.5$, $\tilde{\eta}_{\mathcal{S}}(\omega) = 0.5$ *and* $\tilde{\gamma}_{\mathcal{S}}(\omega) = 0.6$, *since* $\forall\; \beta \geq 0.6, \alpha \geq 0.5$,

$$\mathcal{S} = (\tilde{\beta} \cap (\mathcal{E}_1 \cup \mathcal{E}_2)) = (\tilde{\alpha} \cap \mathcal{E}_1) \cup (\tilde{\beta} \cap \mathcal{E}_2),$$

we obtain $\tilde{\tau}^{\tilde{\gamma}}_{st}(\mathcal{S}) = \frac{3}{4}$, $\tilde{\tau}^{\tilde{\eta}}_{st}(\mathcal{S}) = \frac{1}{4}$, $\tilde{\tau}^{\tilde{\mu}}_{st}(\mathcal{S}) = \frac{1}{2}$.

If $\tilde{\gamma}_{\mathcal{S}}(\omega) = 0.5$, $\tilde{\eta}_{\mathcal{S}}(\omega) = 0.5$, *and* $\tilde{\gamma}_{\mathcal{S}}(\omega) = \beta, \forall\; 0.5 < \beta < 0.6$, *since*

$$\mathcal{S} = (\tilde{\beta} \cap (\mathcal{E}_1 \cup \mathcal{E}_2)) = (\tilde{\beta} \cap \mathcal{E}_1) \cup (\tilde{\beta} \cap \mathcal{E}_2),$$

we obtain $\tilde{\tau}^{\tilde{\gamma}}_{st}(\mathcal{S}) = \frac{3}{4}$, $\tilde{\tau}^{\tilde{\eta}}_{st}(\mathcal{S}) = \frac{1}{4}$, $\tilde{\tau}^{\tilde{\mu}}_{st}(\mathcal{S}) = \frac{1}{2}$.

If $\tilde{\gamma}_{\mathcal{S}}(\omega) = \alpha$, $\tilde{\eta}_{\mathcal{S}}(\omega) = \alpha$, *and* $\tilde{\gamma}_{\mathcal{S}}(\omega) = \beta$, $\forall\; 0.4 < \alpha\; \beta < 0.5$ *and* $\alpha < \beta$, *since for every* $\mathcal{S}_1 = \{\tilde{1}, \mathcal{E}_1, \mathcal{E}_1 \cup \mathcal{E}_2\}$ *and* $\mathcal{S}_2 = \{\mathcal{E}_2, \mathcal{E}_1 \cap \mathcal{E}_2\}$

$$\mathcal{S} = (\tilde{\alpha} \cap \mathcal{E}_1) \cup (\tilde{\beta} \cap \mathcal{E}_2),$$

we have $\tilde{\tau}^{\tilde{\gamma}}_{st}(\mathcal{S}) = \frac{2}{3}$, $\tilde{\tau}^{\tilde{\eta}}_{st}(\mathcal{S}) = \frac{1}{3}$, $\tilde{\tau}^{\tilde{\mu}}_{st}(\mathcal{S}) = \frac{1}{3}$. *We can obtain the following:*

$$\tilde{\tau}^{\tilde{\gamma}}_{st}(\mathcal{S}) = \begin{cases} 1, \textit{if } \mathcal{S} = \tilde{\alpha}, \\ \frac{3}{4}, \textit{if } \tilde{\gamma}_{\mathcal{S}}(\omega) = \alpha, \tilde{\eta}_{\mathcal{S}}(\omega) = \alpha \textit{ and } \tilde{\mu}_{\mathcal{S}}(\omega) = \beta \textit{ for } 0.5 \leq \alpha, \beta \leq 0.6, \alpha < \beta, \\ \frac{1}{2}, \textit{if } \tilde{\gamma}_{\mathcal{S}}(\omega) = \alpha, \tilde{\eta}_{\mathcal{S}}(\omega) = \alpha \textit{ for } 0.4 \leq \alpha < 0.5 \textit{ and } \tilde{\mu}_{\mathcal{S}}(\omega) = \beta, \textit{ for } 0.5 < \beta \leq 0.6, \\ \frac{2}{3}, \textit{if } \tilde{\gamma}_{\mathcal{S}}(\omega) = \alpha, \tilde{\eta}_{\mathcal{S}}(\omega) = \alpha \textit{ and } \tilde{\mu}_{\mathcal{S}}(\omega) = \beta \textit{ for } 0.4 \leq \alpha, \beta \leq 0.5, \alpha < \beta, \\ 0, \quad \textit{otherwise,} \end{cases}$$

$$\tilde{\tau}^{\tilde{\eta}}_{st}(\mathcal{S}) = \begin{cases} 0, \textit{if } \mathcal{S} = \tilde{\alpha}, \\ \frac{1}{4}, \textit{if } \tilde{\gamma}_{\mathcal{S}}(\omega) = \alpha, \tilde{\eta}_{\mathcal{S}}(\omega) = \alpha \textit{ and } \tilde{\mu}_{\mathcal{S}}(\omega) = \beta \textit{ for } 0.5 \leq \alpha, \beta \leq 0.6, \alpha < \beta, \\ \frac{1}{2}, \textit{if } \tilde{\gamma}_{\mathcal{S}}(\omega) = \alpha, \tilde{\eta}_{\mathcal{S}}(\omega) = \alpha \textit{ for } 0.4 \leq \alpha < 0.5 \textit{ and } \tilde{\mu}_{\mathcal{S}}(\omega) = \beta, \textit{ for } 0.5 < \beta \leq 0.6, \\ \frac{1}{3}, \textit{if } \tilde{\gamma}_{\mathcal{S}}(\omega) = \alpha, \tilde{\eta}_{\mathcal{S}}(\omega) = \alpha \textit{ and } \tilde{\mu}_{\mathcal{S}}(\omega) = \beta \textit{ for } 0.4 \leq \alpha, \beta \leq 0.5, \alpha < \beta, \\ 1, \quad \textit{otherwise,} \end{cases}$$

$$\tilde{\tau}^{\tilde{\mu}}_{st}(\mathcal{S}) = \begin{cases} 0, if\ \mathcal{S} = \tilde{\alpha}, \\ \frac{1}{4}, if\ \tilde{\gamma}_{\mathcal{S}}(\omega) = \alpha, \tilde{\eta}_{\mathcal{S}}(\omega) = \alpha\ and\ \tilde{\mu}_{\mathcal{S}}(\omega) = \beta\ for\ 0.5 \leq \alpha, \beta \leq 0.6, \alpha < \beta, \\ \frac{1}{2}, if\ \tilde{\gamma}_{\mathcal{S}}(\omega) = \alpha, \tilde{\eta}_{\mathcal{S}}(\omega) = \alpha\ for\ 0.4 \leq \alpha < 0.5\ and\ \tilde{\mu}_{\mathcal{S}}(\omega) = \beta,\ for\ 0.5 < \beta \leq 0.6, \\ \frac{1}{3}, if\ \tilde{\gamma}_{\mathcal{S}}(\omega) = \alpha, \tilde{\eta}_{\mathcal{S}}(\omega) = \alpha\ and\ \tilde{\mu}_{\mathcal{S}}(\omega) = \beta\ for\ 0.4 \leq \alpha, \beta \leq 0.5,\ \alpha < \beta, \\ 1, \quad otherwise. \end{cases}$$

Theorem 9. *Let $(\tilde{X}, \tilde{\tau}^{\tilde{\gamma}}, \tilde{\tau}^{\tilde{\eta}}, \tilde{\tau}^{\tilde{\mu}})$, $(\tilde{Y}, \tilde{\mathcal{U}}^{\tilde{\gamma}}, \tilde{\mathcal{U}}^{\tilde{\eta}}, \tilde{\mathcal{U}}^{\tilde{\mu}})$ be two $\mathcal{SVNTS}$'s and let $(\tilde{\tau}^{\tilde{\gamma}}_{st}, \tilde{\tau}^{\tilde{\eta}}_{st}, \tilde{\tau}^{\tilde{\mu}}_{st})$ and $(\tilde{\mathcal{U}}^{\tilde{\gamma}}_{st}, \tilde{\mathcal{U}}^{\tilde{\eta}}_{st}, \tilde{\mathcal{U}}^{\tilde{\mu}}_{st})$ be stratification for $(\tilde{\tau}^{\tilde{\gamma}}, \tilde{\tau}^{\tilde{\eta}}, \tilde{\tau}^{\tilde{\mu}})$ and $(\tilde{\mathcal{U}}^{\tilde{\gamma}}, \tilde{\mathcal{U}}^{\tilde{\eta}}, \tilde{\mathcal{U}}^{\tilde{\mu}})$, respectively. If $f : (\tilde{X}, \tilde{\tau}^{\tilde{\gamma}}, \tilde{\tau}^{\tilde{\eta}}, \tilde{\tau}^{\tilde{\mu}}) \to (\tilde{Y}, \tilde{\mathcal{U}}^{\tilde{\gamma}}, \tilde{\mathcal{U}}^{\tilde{\eta}}, \tilde{\mathcal{U}}^{\tilde{\mu}})$ is r-$\mathcal{SVN}$-continuous, then $f : (\tilde{X}, \tilde{\tau}^{\tilde{\gamma}}_{st}, \tilde{\tau}^{\tilde{\eta}}_{st}, \tilde{\tau}^{\tilde{\mu}}_{st}) \to (\tilde{Y}, \tilde{\mathcal{U}}^{\tilde{\gamma}}_{st}, \tilde{\mathcal{U}}^{\tilde{\eta}}_{st}, \tilde{\mathcal{U}}^{\tilde{\mu}}_{st})$ is r-$\mathcal{SVN}$-continuous.*

Proof. Suppose there exist $r \in \zeta_0$, $\mathcal{R} \in \zeta^{\tilde{X}}$, such that

$$\tilde{\mathcal{U}}^{\tilde{\gamma}}_{st}(\mathcal{R}) > r > \tilde{\tau}^{\tilde{\gamma}}_{st}(f^{-1}(\mathcal{R})),$$

$$\tilde{\mathcal{U}}^{\tilde{\eta}}_{st}(\mathcal{R}) < 1 - r < \tilde{\tau}^{\tilde{\eta}}_{st}(f^{-1}(\mathcal{R}))$$

$$\tilde{\mathcal{U}}^{\tilde{\mu}}_{st}(\mathcal{R}) < 1 - r < \tilde{\tau}^{\tilde{\mu}}_{st}(f^{-1}(\mathcal{R}))$$

Since $\tilde{\mathcal{U}}^{\tilde{\gamma}}_{st}(\mathcal{R}) > r$, $\tilde{\mathcal{U}}^{\tilde{\eta}}_{st}(\mathcal{R}) < 1 - r$ and $\tilde{\mathcal{U}}^{\tilde{\mu}}_{st}(\mathcal{R}) < 1 - r$, by the definition of $(\tilde{\mathcal{U}}^{\tilde{\gamma}}_{st}, \tilde{\mathcal{U}}^{\tilde{\eta}}_{st}, \tilde{\mathcal{U}}^{\tilde{\mu}}_{st})$, there exists a family $\{\mathcal{R}_j\}_{j\in\Gamma}$ with $\mathcal{R} = \bigcup_{j\in\Gamma}(\mathcal{R}_j \cap \tilde{\alpha}_j)$ such that

$$\tilde{\mathcal{U}}^{\tilde{\gamma}}_{st}(\mathcal{R}) \geq \bigcap_{j\in\Gamma} \tilde{\mathcal{U}}^{\tilde{\gamma}}(\mathcal{R}_j) > r, \qquad \tilde{\mathcal{U}}^{\tilde{\eta}}_{st}(\mathcal{R}) \leq \bigcup_{j\in\Gamma} \tilde{\mathcal{U}}^{\tilde{\eta}}(\mathcal{R}_j) < 1 - r,$$

$$\tilde{\mathcal{U}}^{\tilde{\mu}}_{st}(\mathcal{R}) \leq \bigcup_{j\in\Gamma} \tilde{\mathcal{U}}^{\tilde{\mu}}(\mathcal{R}_j) < 1 - r.$$

Since

$$f^{-1}(\mathcal{R}) = f^{-1}(\bigcup_{j\in\Gamma}(\mathcal{R}_j \cap \tilde{\alpha}_j)) = \bigcup_{j\in\Gamma} f^{-1}(\mathcal{R}) \cap \tilde{\alpha}_j,$$

and by Remark 2 and Theorem 8, we obtain

$$\tilde{\tau}^{\tilde{\gamma}}_{st}(f^{-1}(\mathcal{R})) \geq \bigcap_{j\in\Gamma} \tilde{\tau}^{\tilde{\gamma}}(f^{-1}(\mathcal{R}_j)),$$

$$\tilde{\tau}^{\tilde{\eta}}_{st}(f^{-1}(\mathcal{R})) \leq \bigcup_{j\in\Gamma} \tilde{\tau}^{\tilde{\eta}}(f^{-1}(\mathcal{R}_j)),$$

$$\tilde{\tau}^{\tilde{\mu}}_{st}(f^{-1}(\mathcal{R})) \leq \bigcup_{j\in\Gamma} \tilde{\tau}^{\tilde{\mu}}(f^{-1}(\mathcal{R}_j)).$$

Since $f : (\tilde{X}, \tilde{\tau}^{\tilde{\gamma}}, \tilde{\tau}^{\tilde{\eta}}, \tilde{\tau}^{\tilde{\mu}}) \to (\tilde{Y}, \tilde{\mathcal{U}}^{\tilde{\gamma}}, \tilde{\mathcal{U}}^{\tilde{\eta}}, \tilde{\mathcal{U}}^{\tilde{\mu}})$ is r-$\mathcal{SVN}$-continuous, that is, $\tilde{\tau}^{\tilde{\gamma}}(f^{-1}(\mathcal{R}_j)) \geq \tilde{\mathcal{U}}^{\tilde{\gamma}}(\mathcal{R}_j)$, $\tilde{\tau}^{\tilde{\eta}}(f^{-1}(\mathcal{R}_j)) \leq \tilde{\mathcal{U}}^{\tilde{\eta}}(\mathcal{R}_j)$, $\tilde{\tau}^{\tilde{\mu}}(f^{-1}(\mathcal{R}_j)) \leq \tilde{\mathcal{U}}^{\tilde{\mu}}(\mathcal{R}_j)$ for every $j \in \Gamma$,

$$\tilde{\tau}^{\tilde{\gamma}}_{st}(f^{-1}(\mathcal{R})) \geq \bigcap_{j\in\Gamma} \tilde{\tau}^{\tilde{\gamma}}(f^{-1}(\mathcal{R}_j)) \geq \bigcap_{j\in\Gamma} \tilde{\mathcal{U}}^{\tilde{\gamma}}(\mathcal{R}_j) > r,$$

$$\tilde{\tau}^{\tilde{\eta}}_{st}(f^{-1}(\mathcal{R})) \leq \bigcup_{j\in\Gamma} \tilde{\tau}^{\tilde{\eta}}(f^{-1}(\mathcal{R}_j)) \leq \bigcup_{j\in\Gamma} \tilde{\mho}^{\tilde{\eta}}(\mathcal{R}_j) < 1-r,$$

$$\tilde{\tau}^{\tilde{\mu}}_{st}(f^{-1}(\mathcal{R})) \leq \bigcup_{j\in\Gamma} \tilde{\tau}^{\tilde{\mu}}(f^{-1}(\mathcal{R}_j)) \leq \bigcup_{j\in\Gamma} \tilde{\mho}^{\tilde{\mu}}(\mathcal{R}_j) < 1-r.$$

It is contradiction. Hence, $f : (\tilde{X}, \tilde{\tau}^{\tilde{\gamma}}_{st}, \tilde{\tau}^{\tilde{\eta}}_{st}, \tilde{\tau}^{\tilde{\mu}}_{st}) \to (\tilde{Y}, \tilde{\mho}^{\tilde{\gamma}}_{st}, \tilde{\mho}^{\tilde{\eta}}_{st}, \tilde{\mho}^{\tilde{\mu}}_{st})$ is r-$\mathcal{SVN}$-continuous. □

The converse of the previous theorem is not true in general as it will be shown by the following example.

Example 5. *Let $\tilde{X}$ be a nonempty set. Define $\mathcal{SVNT}$'s $(\tilde{\tau}^{\tilde{\gamma}}, \tilde{\tau}^{\tilde{\eta}}, \tilde{\tau}^{\tilde{\mu}})$ and $(\tilde{\mho}^{\tilde{\gamma}}, \tilde{\tau}^{\tilde{\mho}}, \tilde{\mho}^{\tilde{\mu}})$, for each $\mathcal{S} \in \zeta^{\tilde{X}}$ and define $\mathcal{R} \in \zeta^{\tilde{X}}$ as follows: $\mathcal{R} = \langle(0.5, 0.5, 0.5), (0.5, 0.5, 0.5), (0.5, 0.5, 0.5)\rangle$,*

$$\tilde{\tau}^{\tilde{\gamma}}(\mathcal{S}) = \begin{cases} 1, if\ \mathcal{S} = \tilde{0}, \tilde{1}, \\ 0, \quad otherwise, \end{cases} \quad \tilde{\tau}^{\tilde{\eta}}(\mathcal{S}) = \begin{cases} 0, if\ \mathcal{S} = \tilde{0}, \tilde{1}, \\ 1, \quad otherwise, \end{cases} \quad \tilde{\tau}^{\tilde{\mu}}(\mathcal{S}) = \begin{cases} 0, if\ \mathcal{S} = \tilde{0}, \tilde{1}, \\ 1, \quad otherwise. \end{cases}$$

$$\tilde{\mho}^{\tilde{\gamma}}(\mathcal{S}) = \begin{cases} 1, if\ \mathcal{S} = \tilde{0}, \tilde{1}, \\ \frac{1}{3}, if\ \mathcal{S} = \mathcal{R}, \\ 0, \quad otherwise, \end{cases} \quad \tilde{\mho}^{\tilde{\eta}}(\mathcal{S}) = \begin{cases} 0, if\ \mathcal{S} = \tilde{0}, \tilde{1}, \\ \frac{2}{3}, if\ \mathcal{S} = \mathcal{R}, \\ 1, \quad otherwise, \end{cases} \quad \tilde{\mho}^{\tilde{\mu}}(\mathcal{S}) = \begin{cases} 0, if\ \mathcal{S} = \tilde{0}, \tilde{1}, \\ \frac{2}{3}, if\ \mathcal{S} = \mathcal{R}, \\ 1, \quad otherwise. \end{cases}$$

Since $0 = \tilde{\tau}^{\tilde{\gamma}}_{st}(\mathcal{R}) < \tilde{\mho}^{\tilde{\gamma}}(\mathcal{R}) = \frac{1}{3}$, $0 = \tilde{\tau}^{\tilde{\eta}}_{st}(\mathcal{R}) > \tilde{\mho}^{\tilde{\eta}}(\mathcal{R}) = \frac{2}{3}$ and $0 = \tilde{\tau}^{\tilde{\mu}}_{st}(\mathcal{R}) > \tilde{\mho}^{\tilde{\mu}}(\mathcal{R}) = \frac{2}{3}$, then the identity mapping $id_x : (\tilde{X}, \tilde{\tau}^{\tilde{\gamma}}, \tilde{\tau}^{\tilde{\eta}}, \tilde{\tau}^{\tilde{\mu}}) \to (\tilde{X}, \tilde{\mho}^{\tilde{\gamma}}, \tilde{\mho}^{\tilde{\mho}}, \tilde{\mho}^{\tilde{\mu}})$ is not r-$\mathcal{SVN}$-continuous. Since for every a family $\{\tilde{1}\}$ and $\mathcal{R} = \mathcal{R} \cap \tilde{1}$, we have $\tilde{\mho}^{\tilde{\gamma}}_{st}(\mathcal{R}) \geq \tilde{\mho}^{\tilde{\gamma}}(\tilde{1}) = 1$, $\tilde{\mho}^{\tilde{\eta}}_{st}(\mathcal{R}) \leq \tilde{\mho}^{\tilde{\eta}}(\tilde{1}) = 0$ and $\tilde{\mho}^{\tilde{\mu}}_{st}(\mathcal{R}) \leq \tilde{\mho}^{\tilde{\mu}}(\tilde{1}) = 0$. Thus, $\tilde{\mho}^{\tilde{\gamma}}_{st}(\mathcal{R}) = 1$, $\tilde{\mho}^{\tilde{\eta}}_{st}(\mathcal{R}) = 0$ and $\tilde{\mho}^{\tilde{\mu}}_{st}(\mathcal{R}) = 0$. Hence,

$$\tilde{\tau}^{\tilde{\gamma}}_{st}(\mathcal{S}) = \tilde{\mho}^{\tilde{\gamma}}_{st}(\mathcal{S}) = \begin{cases} 1, \text{if } \mathcal{S} = \tilde{\alpha}, \forall \alpha \in \zeta_0, \\ 0, \quad \text{otherwise}, \end{cases} \quad \tilde{\tau}^{\tilde{\eta}}_{st}(\mathcal{S}) = \tilde{\mho}^{\tilde{\eta}}_{st}(\mathcal{S}) = \begin{cases} 0, \text{if } \mathcal{S} = \tilde{\alpha}, \forall \alpha \in \zeta_0 \\ 1, \quad \text{otherwise}, \end{cases}$$

$$\tilde{\tau}^{\tilde{\mu}}_{st}(\mathcal{S}) = \tilde{\mho}^{\tilde{\mu}}_{st}(\mathcal{S}) = \begin{cases} 0, \text{if } \mathcal{S} = \tilde{\alpha}, \forall \alpha \in \zeta_0 \\ 1, \quad \text{otherwise}. \end{cases}$$

Therefore, $id_x : (\tilde{X}, \tilde{\tau}^{\tilde{\gamma}}_{st}, \tilde{\tau}^{\tilde{\eta}}_{st}, \tilde{\tau}^{\tilde{\mu}}_{st}) \to (\tilde{X}, \tilde{\mho}^{\tilde{\gamma}}_{st}, \tilde{\mho}^{\tilde{\eta}}_{st}, \tilde{\mho}^{\tilde{\mu}}_{st})$ is r-$\mathcal{SVN}$-continuous.

In the following, we will show that every r-$\mathcal{SVNCOM}$ in the single-valued neutrosophic is r-$\mathcal{SVNCOM}$ in the stratification of it.

Theorem 10. *Let $(\tilde{X}, \tilde{\tau}^{\tilde{\gamma}}_{st}, \tilde{\tau}^{\tilde{\eta}}_{st}, \tilde{\tau}^{\tilde{\mu}}_{st})$ be a stratification of an $\mathcal{SVNTS}$ $(\tilde{X}, \tilde{\tau}^{\tilde{\gamma}}, \tilde{\tau}^{\tilde{\eta}}, \tilde{\tau}^{\tilde{\mu}})$. A $\mathcal{SVNS}$ $\mathcal{S}$ is a r-$\mathcal{SVNCOM}$ in $(\tilde{\tau}^{\tilde{\gamma}}, \tilde{\tau}^{\tilde{\eta}}, \tilde{\tau}^{\tilde{\mu}})$ iff $\mathcal{S}$ is a r-$\mathcal{SVNCOM}$ in $(\tilde{X}, \tilde{\tau}^{\tilde{\gamma}}_{st}, \tilde{\tau}^{\tilde{\eta}}_{st}, \tilde{\tau}^{\tilde{\mu}}_{st})$*

Proof. (1) Let $\mathcal{S}$ be r-$\mathcal{SVNCOM}$ in $(\tilde{X}, \tilde{\tau}^{\tilde{\gamma}}_{st}, \tilde{\tau}^{\tilde{\eta}}_{st}, \tilde{\tau}^{\tilde{\mu}}_{st})$. Suppose that $\mathcal{S}$ is not r-$\mathcal{SVNCON}$ in $(\tilde{X}, \tilde{\tau}^{\tilde{\gamma}}, \tilde{\tau}^{\tilde{\eta}}, \tilde{\tau}^{\tilde{\mu}})$. Then, $\mathcal{E} \neq \tilde{0}$ and $\mathcal{R} \neq \tilde{0}$ are r-$\mathcal{SVNSEP}$ in $(\tilde{X}, \tilde{\tau}^{\tilde{\gamma}}, \tilde{\tau}^{\tilde{\eta}}, \tilde{\tau}^{\tilde{\mu}})$ such that $\mathcal{S} = \mathcal{E} \cup \mathcal{D}$. Since $\tilde{\tau}^{\tilde{\gamma}} \leq \tilde{\tau}^{\tilde{\gamma}}_{st}$, $\tilde{\tau}^{\tilde{\eta}} \geq \tilde{\tau}^{\tilde{\eta}}_{st}$ and $\tilde{\tau}^{\tilde{\mu}} \geq \tilde{\tau}^{\tilde{\mu}}_{st}$, then, from Theorem 8, we get

$$C_{\tilde{\tau}^{\tilde{\gamma}}_{st}, \tilde{\tau}^{\tilde{\eta}}_{st}, \tilde{\tau}^{\tilde{\mu}}_{st}}(\mathcal{E}, r) \leq C_{\tilde{\tau}^{\tilde{\gamma}}, \tilde{\tau}^{\tilde{\eta}}, \tilde{\tau}^{\tilde{\mu}}}(\mathcal{E}, r), \quad C_{\tilde{\tau}^{\tilde{\gamma}}_{st}, \tilde{\tau}^{\tilde{\eta}}_{st}, \tilde{\tau}^{\tilde{\mu}}_{st}}(\mathcal{R}, r) \leq C_{\tilde{\tau}^{\tilde{\gamma}}, \tilde{\tau}^{\tilde{\eta}}, \tilde{\tau}^{\tilde{\mu}}}(\mathcal{R}, r).$$

Hence, $\mathcal{E}, \mathcal{R}$ are r-$\mathcal{SVNSEP}$ in $(\tilde{X}, \tilde{\tau}^{\tilde{\gamma}}_{st}, \tilde{\tau}^{\tilde{\eta}}_{st}, \tilde{\tau}^{\tilde{\mu}}_{st})$. Thus, $\mathcal{S}$ is not r-$\mathcal{SVNCOM}$ in $(\tilde{X}, \tilde{\tau}^{\tilde{\gamma}}_{st}, \tilde{\tau}^{\tilde{\eta}}_{st}, \tilde{\tau}^{\tilde{\mu}}_{st})$. We reach a contradiction.

(2) Now, we show that, if $\mathcal{S}$ is r-$\mathcal{SVNCOM}$ in $(\tilde{X},\tilde{\tau}^{\tilde{\gamma}},\tilde{\tau}^{\tilde{\eta}},\tilde{\tau}^{\tilde{\mu}})$, then $\mathcal{S}$ is r-$\mathcal{SVNCON}$ in $(\tilde{X},\tilde{\tau}^{\tilde{\gamma}}_{st},\tilde{\tau}^{\tilde{\eta}}_{st},\tilde{\tau}^{\tilde{\mu}}_{st})$. Let $\mathcal{S}$ be a r-$\mathcal{SVNCOM}$ in $(\tilde{X},\tilde{\tau}^{\tilde{\gamma}},\tilde{\tau}^{\tilde{\eta}},\tilde{\tau}^{\tilde{\mu}})$. Then, by Corollary 2 (1), we have $C_{\tilde{\tau}^{\tilde{\gamma}},\tilde{\tau}^{\tilde{\eta}},\tilde{\tau}^{\tilde{\mu}}}(\mathcal{S},r)=\mathcal{S}$.

Supposing that $\mathcal{S}$ is not r-$\mathcal{SVNCON}$ in $(\tilde{X},\tilde{\tau}^{\tilde{\gamma}}_{st},\tilde{\tau}^{\tilde{\eta}}_{st},\tilde{\tau}^{\tilde{\mu}}_{st})$, then $\mathcal{E}\neq\tilde{0},\mathcal{R}\neq\tilde{0}$ are r-$\mathcal{SVNSEP}$ in $(\tilde{X},\tilde{\tau}^{\tilde{\gamma}}_{st},\tilde{\tau}^{\tilde{\eta}}_{st},\tilde{\tau}^{\tilde{\mu}}_{st})$ such that $\mathcal{S}=\mathcal{E}\cup\mathcal{R}$. Since $\tilde{\tau}^{\tilde{\gamma}}\leq\tilde{\tau}^{\tilde{\gamma}}_{st}$, $\tilde{\tau}^{\tilde{\eta}}\geq\tilde{\tau}^{\tilde{\eta}}_{st}$, $\tilde{\tau}^{\tilde{\mu}}\geq\tilde{\tau}^{\tilde{\mu}}_{st}$, then $C_{\tilde{\tau}^{\tilde{\gamma}}_{st},\tilde{\tau}^{\tilde{\eta}}_{st},\tilde{\tau}^{\tilde{\mu}}_{st}}(\mathcal{S},r)\leq C_{\tilde{\tau}^{\tilde{\gamma}},\tilde{\tau}^{\tilde{\eta}},\tilde{\tau}^{\tilde{\mu}}}(\mathcal{S},r)=\mathcal{S}$. Thus, $C_{\tilde{\tau}^{\tilde{\gamma}}_{st},\tilde{\tau}^{\tilde{\eta}}_{st},\tilde{\tau}^{\tilde{\mu}}_{st}}(\mathcal{S},r)=\mathcal{S}$. Since $\mathcal{E}\leq\mathcal{S}$, we have $C_{\tilde{\tau}^{\tilde{\gamma}}_{st},\tilde{\tau}^{\tilde{\eta}}_{st},\tilde{\tau}^{\tilde{\mu}}_{st}}(\mathcal{E},r)\leq\mathcal{S}$. It implies that $\mathcal{S}=C_{\tilde{\tau}^{\tilde{\gamma}}_{st},\tilde{\tau}^{\tilde{\eta}}_{st},\tilde{\tau}^{\tilde{\mu}}_{st}}(\mathcal{E},r)\cup\mathcal{R}$. Put $C_{\tilde{\tau}^{\tilde{\gamma}}_{st},\tilde{\tau}^{\tilde{\eta}}_{st},\tilde{\tau}^{\tilde{\mu}}_{st}}(\mathcal{E},r)=\mathcal{D}$. If $x\in\text{supp}(\mathcal{E})$, then $x\in supp(\mathcal{S})$. Since $\mathcal{S}$ is a r-$\mathcal{SVNCOM}$ in $(\tilde{X},\tilde{\tau}^{\tilde{\gamma}},\tilde{\tau}^{\tilde{\eta}},\tilde{\tau}^{\tilde{\mu}})$, by Corollary 2 (4), $x_1\in\mathcal{S}=\mathcal{D}\cup\mathcal{R}$, that is, $\mathcal{D}(x)\cup\mathcal{R}(x)=1$. Since $\mathcal{D}\cap\mathcal{R}=\tilde{0}$, thus, $\mathcal{R}(x)=0$. It implies that $\mathcal{D}(x)=1$. Therefore, $\mathcal{D}$ is a crisp set. Since $\tilde{\tau}^{\tilde{\gamma}}_{st}(\tilde{1}-\mathcal{D})\geq r$, $\tilde{\tau}^{\tilde{\eta}}_{st}(\tilde{1}-\mathcal{D})\leq 1-r$, and $\tilde{\tau}^{\tilde{\mu}}_{st}(\tilde{1}-\mathcal{D})\leq 1-r$. By Theorems (1) and (2), for each family, $\{\tilde{\alpha}_j\cap\mathcal{C}_j:\tilde{1}-\mathcal{D}=\bigcup_{j\in\Gamma}\tilde{\alpha}_j\cap\mathcal{C}_j\}$,

$$\tilde{\tau}^{\tilde{\gamma}}_{st}(\tilde{1}-\mathcal{D})=\bigcup\{\bigcap_{j\in\Gamma}\tilde{\tau}^{\tilde{\gamma}}(\mathcal{C}_j)\}\geq r,\qquad \tilde{\tau}^{\tilde{\eta}}_{st}(\tilde{1}-\mathcal{D})=\bigcap\{\bigcup_{j\in\Gamma}\tilde{\tau}^{\tilde{\eta}}(\mathcal{C}_j)\}\leq 1-r,$$

$$\tilde{\tau}^{\tilde{\mu}}_{st}(\tilde{1}-\mathcal{D})=\bigcap\{\bigcup_{j\in\Gamma}\tilde{\tau}^{\tilde{\mu}}(\mathcal{C}_j)\}\leq 1-r.$$

Let $\tilde{\alpha}_j=\tilde{0}$. Since $\mathcal{D}(x)=1$ for any $x\in\text{supp}(\mathcal{D})$, we have

$$(\tilde{1}-\mathcal{D})(x)=\bigcup_{j\in\Gamma}(\tilde{\alpha}_j\cap\mathcal{C}_j)(x)\Rightarrow 1=\mathcal{D}(x)=\bigcap_{j\in\Gamma}(\tilde{1}-\tilde{\alpha}_j)(x)\cup(\tilde{1}-\mathcal{C}_j)(x).$$

Hence, $(\tilde{1}-\mathcal{D})(x)=\bigcup_{j\in\Gamma}(\mathcal{C}_j)(x)$ for $x\in supp(\mathcal{E})$. If $y\notin\text{supp}(\mathcal{D})$, then

$$1=(\tilde{1}-\mathcal{D})(y)=(\tilde{\alpha}_j\cap\mathcal{C}_j)(y)\leq\bigcup_{j\in\Gamma}\mathcal{C}_j(y).$$

Thus, for any family $\{\tilde{\alpha}_j\cap\mathcal{C}_j:\tilde{1}-\mathcal{D}=\bigcup_{j\in\Gamma}\tilde{\alpha}_j\cap\mathcal{C}_j\}$, we have $\tilde{1}-\mathcal{D}=\bigcup_{j\in\Gamma}\mathcal{C}_j$. It implies

$$\tilde{\tau}^{\tilde{\gamma}}_{st}(\tilde{1}-\mathcal{D})=\tilde{\tau}^{\tilde{\gamma}}(\tilde{1}-\mathcal{D})=\bigcap_{j\in\Gamma}\tilde{\tau}^{\tilde{\gamma}}(\mathcal{C}_j)\geq r,$$

$$\tilde{\tau}^{\tilde{\eta}}_{st}(\tilde{1}-\mathcal{D})=\tilde{\tau}^{\tilde{\eta}}(\tilde{1}-\mathcal{D})=\bigcap_{j\in\Gamma}\tilde{\tau}^{\tilde{\eta}}(\mathcal{C}_j)\leq 1-r,$$

$$\tilde{\tau}^{\tilde{\mu}}_{st}(\tilde{1}-\mathcal{D})=\tilde{\tau}^{\tilde{\mu}}(\tilde{1}-\mathcal{D})=\bigcap_{j\in\Gamma}\tilde{\tau}^{\tilde{\mu}}(\mathcal{C}_j)\leq 1-r.$$

Thus, $C_{\tilde{\tau}^{\tilde{\gamma}},\tilde{\tau}^{\tilde{\eta}},\tilde{\tau}^{\tilde{\mu}}}(\mathcal{D},r)=\mathcal{D}$. It implies

$$C_{\tilde{\tau}^{\tilde{\gamma}},\tilde{\tau}^{\tilde{\eta}},\tilde{\tau}^{\tilde{\mu}}}(C_{\tilde{\tau}^{\tilde{\gamma}}_{st},\tilde{\tau}^{\tilde{\eta}}_{st},\tilde{\tau}^{\tilde{\mu}}_{st}}(\mathcal{E},r),r)=C_{\tilde{\tau}^{\tilde{\gamma}}_{st},\tilde{\tau}^{\tilde{\eta}}_{st},\tilde{\tau}^{\tilde{\mu}}_{st}}(\mathcal{E},r).$$

Similarly, $C_{\tilde{\tau}^{\tilde{\gamma}},\tilde{\tau}^{\tilde{\eta}},\tilde{\tau}^{\tilde{\mu}}}(C_{\tilde{\tau}^{\tilde{\gamma}}_{st},\tilde{\tau}^{\tilde{\eta}}_{st},\tilde{\tau}^{\tilde{\mu}}_{st}}(\mathcal{R},r),r)=C_{\tilde{\tau}^{\tilde{\gamma}}_{st},\tilde{\tau}^{\tilde{\eta}}_{st},\tilde{\tau}^{\tilde{\mu}}_{st}}(\mathcal{R},r)$. Thus, $\mathcal{E}$ and $\mathcal{R}$ are r-$\mathcal{SVNSEP}$ in $(\tilde{X},\tilde{\tau}^{\tilde{\gamma}},\tilde{\tau}^{\tilde{\eta}},\tilde{\tau}^{\tilde{\mu}})$ from

$$C_{\tilde{\tau}^{\tilde{\gamma}},\tilde{\tau}^{\tilde{\eta}},\tilde{\tau}^{\tilde{\mu}}}(\mathcal{R},r)\cap\mathcal{E}\leq(C_{\tilde{\tau}^{\tilde{\gamma}},\tilde{\tau}^{\tilde{\eta}},\tilde{\tau}^{\tilde{\mu}}}(C_{\tilde{\tau}^{\tilde{\gamma}}_{st},\tilde{\tau}^{\tilde{\eta}}_{st},\tilde{\tau}^{\tilde{\mu}}_{st}}(\mathcal{R},r),r)\cap\mathcal{E})=(C_{\tilde{\tau}^{\tilde{\gamma}}_{st},\tilde{\tau}^{\tilde{\eta}}_{st},\tilde{\tau}^{\tilde{\mu}}_{st}}(\mathcal{R},r),r)\cap\mathcal{E})=\tilde{0},$$

$$C_{\tilde{\tau}^{\tilde{\gamma}},\tilde{\tau}^{\tilde{\eta}},\tilde{\tau}^{\tilde{\mu}}}(\mathcal{E},r)\cap\mathcal{R}\leq(C_{\tilde{\tau}^{\tilde{\gamma}},\tilde{\tau}^{\tilde{\eta}},\tilde{\tau}^{\tilde{\mu}}}(C_{\tilde{\tau}^{\tilde{\gamma}}_{st},\tilde{\tau}^{\tilde{\eta}}_{st},\tilde{\tau}^{\tilde{\mu}}_{st}}(\mathcal{D},r),r)\cap\mathcal{R})=(C_{\tilde{\tau}^{\tilde{\gamma}}_{st},\tilde{\tau}^{\tilde{\eta}}_{st},\tilde{\tau}^{\tilde{\mu}}_{st}}(\mathcal{E},r),r)\cap\mathcal{R})=\tilde{0}.$$

Thus, $\mathcal{S}$ is not r-$\mathcal{SVNCOM}$ in $(\tilde{X},\tilde{\tau}^{\tilde{\gamma}},\tilde{\tau}^{\tilde{\eta}},\tilde{\tau}^{\tilde{\mu}})$. It is a contradiction.

(3) Let $\mathcal{S}$ be a r-$\mathcal{SVNCOM}$ in $(\tilde{X},\tilde{\tau}^{\tilde{\gamma}}_{st},\tilde{\tau}^{\tilde{\eta}}_{st},\tilde{\tau}^{\tilde{\mu}}_{st})$. From (1), $\mathcal{S}$ is r-$\mathcal{SVNCON}$ in $(\tilde{X},\tilde{\tau}^{\tilde{\gamma}},\tilde{\tau}^{\tilde{\eta}},\tilde{\tau}^{\tilde{\mu}})$. There exists a r-$\mathcal{SVNCOM}$ $\mathcal{R}$ in $(\tilde{X},\tilde{\tau}^{\tilde{\gamma}},\tilde{\tau}^{\tilde{\eta}},\tilde{\tau}^{\tilde{\mu}})$ containing $\mathcal{S}$. From (2), $\mathcal{R}$ is r-$\mathcal{SVNCON}$ in $(\tilde{X},\tilde{\tau}^{\tilde{\gamma}}_{st},\tilde{\tau}^{\tilde{\eta}}_{st},\tilde{\tau}^{\tilde{\mu}}_{st})$. Thus, $\mathcal{S}=\mathcal{R}$.

Let $\mathcal{R}$ be a r-$\mathcal{SVNCOM}$ in $(\tilde{X},\tilde{\tau}^{\tilde{\gamma}},\tilde{\tau}^{\tilde{\eta}},\tilde{\tau}^{\tilde{\mu}})$. Similarly, $\mathcal{R}$ is a r-$\mathcal{SVNCOM}$ in $(\tilde{X},\tilde{\tau}^{\tilde{\gamma}}_{st},\tilde{\tau}^{\tilde{\eta}}_{st},\tilde{\tau}^{\tilde{\mu}}_{st})$. □

Example 6. *In Example 4, we proved that* $(\tilde{\tau}^{\tilde{\gamma}}_{st},\tilde{\tau}^{\tilde{\eta}}_{st},\tilde{\tau}^{\tilde{\mu}}_{st})$ *is a stratification for* $\mathcal{SVNT}$ $(\tilde{\tau}^{\tilde{\gamma}},\tilde{\tau}^{\tilde{\eta}},\tilde{\tau}^{\tilde{\mu}})$ *on* $\tilde{X}$. *By Theorem 10, we can clearly see that each r-*$\mathcal{SVNCOM}$ *in* $(\tilde{\tau}^{\tilde{\gamma}},\tilde{\tau}^{\tilde{\eta}},\tilde{\tau}^{\tilde{\mu}})$ *is r-*$\mathcal{SVNCOM}$ *in* $(\tilde{\tau}^{\tilde{\gamma}}_{st},\tilde{\tau}^{\tilde{\eta}}_{st},\tilde{\tau}^{\tilde{\mu}}_{st})$*, The opposite is also true.*

For the purpose of Symmetry, we can apply the idea in the following definition into Definition 3 and get the desired symmetry's consequences.

Note A SVNS $\mathcal{R}$ in $\tilde{X}\times\tilde{X}$ is called a single valued neutrosophic relation ($\mathcal{SVNR}$, for short) in $\tilde{X}$, denoted by $\mathcal{R}=\{\langle(x,x_1)\tilde{\gamma}_{\mathcal{R}}(x,x_1),\tilde{\eta}_{\mathcal{R}}(x,x_1),\tilde{\mu}_{\mathcal{R}}(x,x_1)\rangle:(x,x_1)\in\tilde{X}\times\tilde{X}\}$, where $\tilde{\gamma}_{\mathcal{R}}:\tilde{X}\times\tilde{X}\to[0,1]$, $\tilde{\eta}_{\mathcal{R}}:\tilde{X}\times\tilde{X}\to[0,1]$, $\tilde{\mu}_{\mathcal{R}}:\tilde{X}\times\tilde{X}\to[0,1]$ denote the truth-membership function, indeterminacy membership function and falsity-membership function of $\mathcal{R}$, respectively.

Definition 14 ([17]). *Let* $\mathcal{S},\mathcal{E}\in\zeta^{\tilde{X}}$. *If* $\mathcal{S}$ *is a single-valued neutrosophic relation on a set* $\tilde{X}$, *then* $\mathcal{S}$ *is called a single-valued neutrosophic relation on* $\mathcal{E}$ *if, for every* $x,x_1\in\tilde{X}$, $\tilde{\gamma}_{\mathcal{S}}(x,x_1)\leq\min(\tilde{\gamma}_{\mathcal{E}}(x),\tilde{\gamma}_{\mathcal{E}}(x_1))$, $\tilde{\eta}_{\mathcal{S}}(x,x_1)\geq\max(\tilde{\eta}_{\mathcal{E}}(x),\tilde{\eta}_{\mathcal{E}}(x_1))$ *and* $\tilde{\mu}_{\mathcal{S}}(x,x_1)\geq\max(\tilde{\mu}_{\mathcal{E}}(x),\tilde{\mu}_{\mathcal{E}}(x_1))$.

Moreover, a single-valued neutrosophic relation $\mathcal{S}$ *on* $\tilde{X}$ *is called symmetric if, for any* $k,k_1\in\tilde{X}$, $\tilde{\gamma}_{\mathcal{S}}(k,k_1)=\tilde{\gamma}_{\mathcal{S}}(k_1,k)$, $\tilde{\eta}_{\mathcal{S}}(k,k_1)=\tilde{\eta}_{\mathcal{S}}(k_1,k)$, $\tilde{\mu}_{\mathcal{S}}(k,k_1)=\tilde{\mu}_{\mathcal{S}}(k_1,k)$; *and* $\tilde{\gamma}_{\mathcal{E}}(k,k_1)=\tilde{\gamma}_{\mathcal{E}}(k_1,k)$ $\tilde{\eta}_{\mathcal{E}}(k,\kappa_1)=\tilde{\eta}_{\mathcal{E}}(k_1,k)$, $\tilde{\mu}_{\mathcal{E}}(k,k_1)=\tilde{\mu}_{\mathcal{E}}(k_1,k)$.

Example 7. *Let* $\tilde{X}=\{x_1,x_2,x_3,x_4,x_5\}$. *A a single valued neutrosophic relation* $\mathcal{S}$ *on* $\tilde{X}$ *is given in the following table.*

$\mathcal{S}$	x_1	x_2	x_3	x_4	x_5
x_1	(0.2, 0.6, 0.4)	(0, 0.3, 0.7)	(0.9, 0.2, 0.4)	(0.3, 0.9, 1)	(0.3, 0.9, 1)
x_2	(0.4, 0.5, 0.1)	(0.1, 0.7, 0)	(1, 1, 1)	(1, 0.3, 0)	(0.5, 0.6, 1)
x_3	(0, 1, 1)	(1, 0.5, 0)	(0, 0, 0)	(0.2, 0.8, 0.1)	(1, 0.8, 1)
x_4	(1, 0, 0)	(0, 0, 1)	(0.5, 0.7, 0.1)	(0.1, 0.4, 1)	(1, 0.8, 0.8)
x_5	(0, 1, 0)	(0.9, 0, 0)	(0, 0.1, 0.7)	(0.8, 0.9, 1)	(0.6, 1, 0)

5. Conclusions

In this paper, authors have made a study of the connectedness, the idea of component, and the stratification of single-valued neutrosophic topological spaces which are different from the study taken so far and obtained some of their basic properties. Next, the concepts of an r-$\mathcal{SVNSEP}$ and r-$\mathcal{SVNCOM}$ were introduced and studied. It has been proven that every r-$\mathcal{SVNCOM}$ in an single-valued neutrosophic topological spaces is r-$\mathcal{SVNCOM}$ in the stratification of it. We will now go into detail on some of the conclusions of the research. Firstly, a single-valued neutrosophic connected (r-$\mathcal{SVNCON}$) has the same properties in a single-valued neutrosophic topological spaces (see Theorem 3). Secondly, a single-valued neutrosophic separated (r-$\mathcal{SVNSEP}$) has the same properties in a single-valued neutrosophic topological spaces (see Theorems 4 and 5). Finally, it has been proven that every single-valued neutrosophic component (r-$\mathcal{SVNCOM}$) in single-valued neutrosophic topological spaces is r-$\mathcal{SVNCOM}$ in the stratification of it.

Discussion for Further Works

It is known that the notion of boundedness in topological spaces (see [26]) plays a significant role in topological aspects. It is also well known that the collection of bounded sets is an ideal. This concept is generalized to the concept of bornology (which is essentially an interesting ideal). There is also the corresponding generalized notion in fuzzy topics (the concept of fuzzy bornology (see [27]).

Therefore, the following ideas could be applied to the notion of single-valued neutrosophic topological spaces.

(a) The collection of bounded single-valued sets;
(b) The concept of fuzzy bornology;
(c) The notion of boundedness in topological spaces.

Author Contributions: This paper was organized by the idea of Y.S., F.S., M.A.-S., and F.A.; Writing—review and editing the original draft. All authors have read and agreed to the published version of the manuscript.

Funding: This research received no external funding.

Acknowledgments: The authors would like to express their sincere thanks to the Deanship of Scientific Research at Majmaah University for supporting this work. The authors also would like to express their sincere thanks the the referee for the useful comments and suggestion in preparing this manuscript.

Conflicts of Interest: The authors declare that there is no conflict of interest regarding the publication of this manuscript.

References

1. Zadeh, L.A. Fuzzy Sets. *Inf. Control* **1965**, *8*, 338–353. [CrossRef]
2. Atanassov, K. Intuitionistic fuzzy sets. *Fuzzy Sets Syst.* **1986**, *20*, 87–96. [CrossRef]
3. Chang, C.L. Fuzzy topological spaces. *J. Math. Anal. Appl.* **1968**, *24*, 182–190. [CrossRef]
4. Lowen, R. Fuzzy topological spaces and fuzzy compactness. *J. Math. Anal. Appl.* **1976**, *56*, 621–633. [CrossRef]
5. Lee, E.P.; Im, Y.B. Mated fuzzy topological spaces. *Int. J. Fuzzy Log. Intell. Syst.* **2001**, *11*, 161–165.
6. Liu, Y.M.; Luo, M.K. *Fuzzy Topology*; World Scientific Publishing Co.: Singapore, 1997.
7. Chattopadhyay, K.C.; Samanta, S.K. Fuzzy topology: Fuzzy closure operator, fuzzy compactness and fuzzy connectedness. *Fuzzy Sets Syst.* **1993**, *54*, 207–212. [CrossRef]
8. Chattopadhyay, K.C.; Hazra, R.N.; Samanta, S.K. Gradation of openness: Fuzzy topology. *Fuzzy Sets Syst.* **1992**, *49*, 237–242. [CrossRef]
9. Zhao, X.D. Connectedness on fuzzy topological spaces. *Fuzzy Sets Syst.* **1986**, *20*, 233–240.
10. Saber, Y.M.; Abdel-Sattar, M.A. Ideals on fuzzy topological spaces. *Appl. Math. Sci.* **2014**, *8*, 1667–1691. [CrossRef]
11. Šostak, A. On a fuzzy topological structure. In *Circolo Matematico di Palermo, Palermo; Rendiconti del Circolo Matematico di Palermo, Proceedings of the 13th Winter School on Abstract Analysis, Section of Topology, Srni, Czech Republic, 5–12 January 1985*; Circolo Matematico di Palermo: Palermo, Italy, 1985; pp. 89–103.
12. Smarandache, F. *Neutrosophy, Neutrisophic Property, Sets, and Logic*; American Research Press: Rehoboth, DE, USA, 1998.
13. Salama, A.A.; Alblowi, S.A. Neutrosophic set and neutrosophic topological spaces. *IOSR J. Math.* **2012**, *3*, 31–35. [CrossRef]
14. Salama, A.A.; Smarandache, F. *Neutrosophic Crisp Set Theory*; The Educational Publisher Columbus: Columbus, OH, USA, 2015.
15. Wang, H.; Smarandache, F.; Zhang, Y.Q.; Sunderraman, R. Single valued neutrosophic sets. *Multispace Multistruct.* **2010**, *4*, 410–413.
16. Kim, J.; Lim, P.K.; Lee, J.G.; Hur, K. Single valued neutrosophic relations. *Ann. Fuzzy Math. Inform.* **2018**, *16*, 201–221. [CrossRef]
17. Saber, Y.M.; Alsharari, F.; Smarandache, F. On single-valued neutrosophic ideals in Šostak sense. *Symmetry* **2020**, *12*, 193. [CrossRef]
18. Ubeda, T.; Egenhofer, M.J. Topological error correcting in GIS. *Adv. Spat. Databases* **1997**, *1262*, 281–297.
19. El Naschie, M.S. On the unification of heterotic strings, M theory and $\mathcal{E}^{\infty}$ theory. *Chaos Solitons Fractals* **2000**, *11*, 2397–2408. [CrossRef]

20. Shang, Y. Average consensus in multi-agent systems with uncertain topologies and multiple time-varying delays. *Linear Algebra Appl.* **2014**, *459*, 411–429. [CrossRef]
21. Smarandache, F. *A Unifying Field in Logics: Neutrosophic Logic. Neutrosophy, Neutrosophic Set, Neutrosophic Probability and Statistics*, 6th ed.; InfoLearnQuest: Ann Arbor, MI, USA, 2007.
22. Yang, H.L.; Guo, Z.L.; Liao, X. On single valued neutrosophic relations. *J. Intell. Fuzzy Syst.* **2016**, *30*, 1045–1056. [CrossRef]
23. Ye, J. A multicriteria decision-making method using aggregation operators for simplified neutrosophic sets. *J. Intell. Fuzzy Syst.* **2014**, *26*, 2450–2466. [CrossRef]
24. El-Gayyar, M. Smooth neutrosophic topological spaces. *Neutrosophic Sets Syst.* **2016**, *65*, 65–72.
25. Alsharari, F. Decomposition of single-valued neutrosophic ideal continuity via fuzzy idealization. *Neutrosophic Sets Syst.* **2020**, under revision.
26. Lambrinos, P.A. A topological notion of boundedness. *Manuscr. Math.* **1973**, *10*, 289–296. [CrossRef]
27. Yan, C.H.; Wu, C.X. Fuzzy L-bornological spaces. *Inf. Sci.* **2005**, *173*, 1–10. [CrossRef]

symmetry

MDPI

Article

A Novel Framework Using Neutrosophy for Integrated Speech and Text Sentiment Analysis

Kritika Mishra [1], Ilanthenral Kandasamy [2], Vasantha Kandasamy W. B. [2] and Florentin Smarandache [3,*]

1 Shell India Markets, RMZ Ecoworld Campus, Marathahalli, Bengaluru, Karnataka 560103, India; kritika.mishra@shell.com
2 School of Computer Science and Engineering, VIT, Vellore 632014, India; ilanthenral.k@vit.ac.in (I.K.); vasantha.wb@vit.ac.in (V.K.W.B.)
3 Department of Mathematics, University of New Mexico, 705 Gurley Avenue, Gallup, NM 87301, USA
* Correspondence: smarand@unm.edu

Received: 6 September 2020; Accepted: 9 October 2020; Published: 18 October 2020

check for updates

Abstract: With increasing data on the Internet, it is becoming difficult to analyze every bit and make sure it can be used efficiently for all the businesses. One useful technique using Natural Language Processing (NLP) is sentiment analysis. Various algorithms can be used to classify textual data based on various scales ranging from just positive-negative, positive-neutral-negative to a wide spectrum of emotions. While a lot of work has been done on text, only a lesser amount of research has been done on audio datasets. An audio file contains more features that can be extracted from its amplitude and frequency than a plain text file. The neutrosophic set is symmetric in nature, and similarly refined neutrosophic set that has the refined indeterminacies I_1 and I_2 in the middle between the extremes Truth T and False F. Neutrosophy which deals with the concept of indeterminacy is another not so explored topic in NLP. Though neutrosophy has been used in sentiment analysis of textual data, it has not been used in speech sentiment analysis. We have proposed a novel framework that performs sentiment analysis on audio files by calculating their Single-Valued Neutrosophic Sets (SVNS) and clustering them into positive-neutral-negative and combines these results with those obtained by performing sentiment analysis on the text files of those audio.

Keywords: sentiment analysis; Speech Analysis; Neutrosophic Sets; indeterminacy; Single-Valued Neutrosophic Sets (SVNS); clustering algorithm; K-means; hierarchical agglomerative clustering

1. Introduction

While many algorithms and techniques were developed for sentiment analysis in the previous years, from classification into just positive and negative categories to a wide spectrum of emotions, less attention has been paid to the concept of indeterminacy. Early stages of work were inclined towards Boolean logic which meant an absolute classification into positive or negative classes, 1 for positive and 0 for negative. Fuzzy logic uses the memberships of positive and negative that can vary in the range 0 to 1. Neutrosophy is the study of indeterminacies, meaning that not every given argument can be distinguished as positive or negative, it emphasizes the need for a neutral category. Neutrosophy theory was introduced in 1998 by Smarandache [1], and it is based on truth membership T, indeterminate membership I and false membership F that satisfies $0 \leq T + I + F \leq 3$, and the memberships are independent of each other. In case of using neutrosophy in sentiment analysis, these memberships are relabelled as positive membership, neutral membership and negative membership.

Another interesting topic is the speech sentiment analysis, it involves processing audio. Audio files cannot be directly understood by models. Machine learning algorithms do not take raw audio files as input hence it is imperative to extract features from the audio files. An audio signal is a three-dimensional signal where the three axes represent amplitude, frequency and time. Previous work on detecting the sentiment of audio files is inclined towards emotion detection as the audio datasets are mostly labelled and created in a manner to include various emotions. Then using the dataset for training classifiers are built. Speech analysis is also largely associated with speech recognition. Speech analysis is the process of analyzing and extracting information from the audio files which are more efficient than the text translation itself. Features can be extracted from audio using Librosa package in python. A total of 193 features per audio file have been retrieved including Mel-Frequency Cepstral Coefficients (MFCC), Mel spectogram, chroma, contrast, and tonnetz. The goal of this project is to establish a relationship between sentiment detected in audio and sentiment detected from the translation of the same audio to text. Work done in the domain of speech sentiment analysis is largely focused on labelled datasets because the datasets are created using actors and not collected like it is done for text where we can scrape tweets, blogs or articles. Hence the datasets are labelled as various emotions such as the Ryerson Audio-Visual Database of Emotional Speech and Song (RAVDESS) dataset which contains angry, happy, sad, calm, fearful, disgusted, and surprised classes of emotions. These datasets have no text translation provided hence no comparison can be established. With unlabelled datasets such as VoxCeleb1/2 which have been randomly collected from random YouTube videos, again the translation problem arises leading to no meaningful comparison scale. We need audio data along with the text data for comparison, so a dataset with audio translation was required. Hence LibriSpeech dataset [2] was chosen, it is a corpus of approximately 1000 h of 16 kHz read English speech.

The K-means clustering algorithm performs clustering of n values in K clusters, where each value belongs to a cluster. Since the dataset is unlabelled features extracted from the audio are clustered using the K-means clustering algorithm. Then the distance of each point from the centroid of each cluster is calculated. 1-distance implies the closeness of an audio file to every cluster. This closeness measure is used to generate Single Value Neutrosophic Sets (SVNS) for the audio. Since the data is unlabelled, we performed clustering of SVNS values using the K-means clustering.

Sentiment analysis of the text has various applications. It is used by businesses for analysing customer feedback of products and brands without having to go through all of them manually. An example of this real-life application could be social media monitoring where scraping and analysing tweets from Twitter on a certain topic or about a particular brand or personality and analysing them could very well indicate the general sentiment of the masses. Ever since internet technology started booming, data became abundant. While it is simpler to process and derive meaningful results from tabular data, it is the need for the hour to process unstructured data in the form of sentences, paragraphs or text files and PDFs. Hence NLP provides excellent sentiment analysis tools for the same. However, sentiment cannot be represented as a black and white picture with just positive and negative arguments alone. To factor in indeterminacy, we have the concept of neutrosophy which means the given argument may either be neutral or with no relation to the extremes. Work done previously related to neutrosophy will be explained in detail in the next section.

For the sentiment analysis of text part, the translation of the audio is provided as text files along with the dataset which mitigates the possibility of inefficient translation. In this paper, using Valence Aware Dictionary and Sentiment Reasoner (VADER), a lexicon and rule-based tool for sentiment analysis on the text files, SVNS values for text are generated. Then K-means clustering is applied to visualize the three clusters. The first step is the comparison of the two K-means plots indicating the formation of a cluster larger than the rest in audio SVNS implying the need for a neutral class. Then both the SVNS are combined by averaging out the two scores respectively for P_x, I_x and N_x. Again K-means clustering and hierarchical agglomerative clustering is performed on these SVNS values to get the final clusters for each file.

Neutrosophic logic uses Single Valued Neutrosophic Sets (SVNS) to implement the concept of indeterminacy in sentiment analysis. For every sentence A, its representative SVNS is generated. SVNS looks like $\langle P_A, I_A, N_A \rangle$ where 'P_A' is the positive sentiment score, 'I_A' is the indeterminacy or neutrality score and 'N_A' is the negative sentiment score. Neutrosophy was introduced to detect the paradox proposition.

In this paper, a new innovative approach is carried out in which we use unlabelled audio dataset and then generate SVNS for audio to analyse audio files from the neutrosophic logic framework. The higher-level architecture is shown in Figure 1.

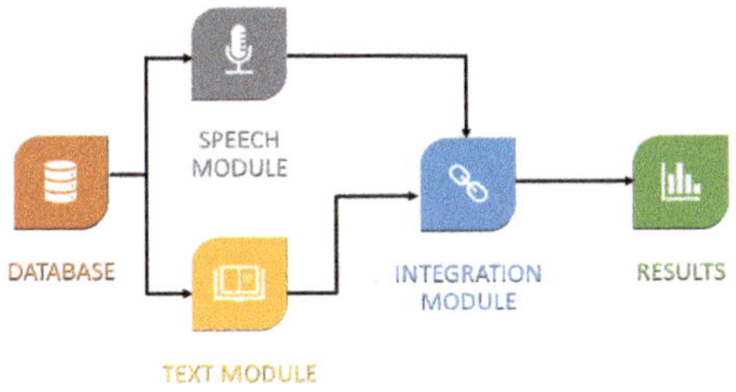

Figure 1. High level architecture.

Indeterminacy is a strong concept which has rightly indicated the importance of neutral or indeterminate class in text sentiment analysis. Coupling it with speech analysis is just an attempt to prove that not all audio can be segregated into positive and negative. There is a very good amount of neutrality present in the data that needs to be represented. We have used clustering to validate the presence of neutrality.

This paper is organized as follows: Section 1 is introductory in nature, the literature survey is provided in Section 2. In Section 3, the basic concepts related to speech sentiment analysis, text sentiment analysis and neutrosophy are recalled. The model description of the proposed framework that makes uses of neutrosophy to handle speech and text sentiment analysis is given in Section 4. In Section 5 the experimental results in terms of K-clustering and agglomerative clustering are provided. Results and discussions about combined SVNS are carried out in Section 6. The conclusions are provided in the last section.

2. Literature Survey

Emphasizing on the need and application of sentiment analysis in business and how it can play a crucial role in data monitoring on social media. The fuzzy logic model by Karen Howells and Ahmet Ertugan [3] attempts to form a five class classifier—strongly positive, positive, neutral, negative and strongly negative for tweets. It is proposed to add fuzzy logic classifier to the social bots used for data mining. It will result in the analysis of the overall positive, neutral and negative sentiments which will facilitate the companies to develop strategies to improve the customer feedback and improve the reputation of their products and brand. A study on application of sentiment analysis in the tourism industry [4] shows that most of the sentiment analysis methods perform better for positive class. One of the reasons for this could be the fact that human language is inclined towards positivity. It is even more difficult to detect neutral sentiment. Ribeiro and others have pointed out a similar observation in [5] that twelve out of twenty-four methods are better in classifying positive sentiment and neutral sentiment is harder to identify. They also concluded from their experiments that VADER tool provides consistent results for three-classes (positive, neutral, negative) classification.

Similarly, Hutto and Gilbert in [6] did an excellent job in comparing VADER tool eleven sentiment analysis techniques depending on Naïve Bayes, Support Vector Machine (SVM) and maximum entropy algorithms. They concluded that VADER is simple to understand and does not function like a black box where the internal structure of process cannot be understood as in complex machine learning and deep learning sentiment analysis techniques. VADER also performs in par with these benchmark models and is highly efficient as it only requires a fraction of second for analysis because it uses a lexicon rule-based approach, whereas its counterpart SVM can take much more time. VADER is also computationally economical as it does not need any special technical specifications such as a GPU for processing. The transparency of the tool attracts a larger audience as its users include professionals from businesses and marketing as well as it allows researchers to experiment more. Hutto and Gilbert's analysis is applied in [7] to rule out the neutral tweets. They built an election prediction model for 2016 USA elections. They used VADER to remove all the neutral tweets that were scraped to focus on positive and negative sentiments towards Donald Trump and Hilary Clinton.

Fuzzy logic gives the measure of positive and negative sentiment in decimal figures, not as absolute values 0 or 1 like Boolean logic. If truth measure is T, then F is falsehood according to the intuitionistic fuzzy set and I is the degree of indeterminacy. Neutrosophy was proposed in [1], it was taken as $0 \leq T + I + F \leq 3$. The neutrosophy theory was introduced in 1998 by Smarandache [1]. Neutrality or indeterminacy was introduced in sentiment analysis to address uncertainties. The importance of neutrosophy in sentiment analysis for the benefit of its prime users such as NLP specialists was pointed out in [8]. To mathematically apply neutrosophic logic in real world problems, Single Valued Neutrosophic Sets (SVNS) were introduced in [9]. A SVNS for sentiment analysis represented by $\langle P_A, I_A, N_A \rangle$ where 'P_A' is the positive sentiment score, 'I_A' is the indeterminacy or neutrality score and 'N_A' is the negative sentiment score.

Refined Neutrosophic sets were introduced in [10]. Furthermore, the concept of Double Valued Neutrosophic Sets (DVNS) was introduced in [11]. DVNS are an improvisation of SVNS. The indeterminacy score was split into two: one indicating indeterminacy of positive sentiment or 'T' the truth measure and the other one indicating indeterminacy of negative sentiment or 'F' the falsehood measure. DVNS are more accurate than SVNS. A minimum spanning tree clustering model was also introduced for double valued neutrosophic sets. Multi objective non-linear optimization on four-valued refined neutrosophic set was carried out in [12].

In [13] a detailed comparison between fuzzy logic and neutrosophic logic was shown by analyzing the #metoo movement. The tweets relevant to the movement are collected from Twitter. After cleaning, the tweets are then input in the VADER tool which generates SVNSs for each tweet. These SVNS are then visualized using clustering algorithms such as K-means and K-NN. Neutrosophic refined sets [10,14–16] have been developed and applied in various fields, including in sentiment analysis recently. However no one has till now attempted to do speech sentiment analysis using neutrosophy and combine it with text sentiment analysis.

A classifier with SVM in multi class mode was developed to classify a six class dataset by extracting linear prediction coefficients, derived cepstrum coefficients and mel frequency cepstral coefficients [17]. The model shows a considerable improvement and results are 91.7% accurate. After various experiments it was concluded in [18] that for emotion recognition convolutional neural networks capture rich features of the dataset when a large sized dataset is used. They also have higher accuracy compared to SVM. SVMs have certain limitations even though they can fit data with non-linearities. It was concluded that machine learning is a better solution for analysing audio. In [19] a multiple classifier system was developed for speech emotion recognition. A multimodal system was developed in [20] to analyze audio, text and visual data together. Features such as MFCC, spectral centroid, spectral flux, beat sum, and beat histogram are extracted from the audio. For text, concepts were extracted based on various rules. For visual data, facial features were incorporated. All these features were then concatenated into a single vector and classified. A similar approach was presented in [21] to build multimodal classifier using audio, textual and visual features and comparing it to its bimodal subsets (audio+text, text+visual, audio+visual). The same set of features were extracted from audio using openSMILE software whereas for text convolutional neural networks were deployed. These features were then combined using decision level fusion. From these studies it can be very well inferred that using both audio and textual features for classification will yield better or sensitive results.

3. Basic Concepts

3.1. Neutrosophy

Neutrosophy is essentially a branch of philosophy. It is based on understanding the scope and dimensions of indeterminacy. Neutrosophy forms the basis of various related fields in statistical analysis, probability, set theory, etc. In some cases, indeterminacy may require more information or in others, it may not have any linking towards either positive or negative sentiment. To represent uncertain, imprecise, incomplete, inconsistent, and indeterminate information that is present in the real world, the concept of a neutrosophic set from the philosophical point of view has been proposed.

Single Valued Neutrosophic Sets (SVNS) is an instance of a Neutrosophic set. The concept of a neutrosophic set is as follows:

Definition 1. *Consider X to be a space of points (data-points), with an element in X represented by x. A neutrosophic set A in X is denoted by a truth membership function $T_A(x)$, an indeterminacy membership function $I_A(x)$, and a falsity membership function $F_A(x)$. The functions $T_A(x)$, $I_A(x)$, and $F_A(x)$ are real standard or non-standard subsets of* $]-0, 1+[$*; that is,*

$$T_A(x) : X \leftarrow]^-0, 1^+[$$

$$I_A(x) : X \leftarrow]^-0, 1^+[,$$

$$F_A(x) : X \leftarrow]^-0, 1^+[,$$

with the condition $^-0 \leq supT_A(x) + supI_A(x) + supF_A(x) \leq 3^+$.

This definition of a neutrosophic set is difficult to apply in the real world in scientific and engineering fields. Therefore, the concept of SVNS, which is an instance of a neutrosophic set, has been introduced.

Definition 2. *Consider X be a space of points (data-points) with element in X denoted by x. An SVNS A in X is characterized by truth membership function $T_A(x)$, indeterminacy membership function $I_A(x)$, and falsity membership function $F_A(x)$. For each point $x \in X$, there are $T_A(x), I_A(x), F_A(x) \in [0,1]$, and $0 \leq T_A(x) + I_A(x) + F_A(x) \leq 3$. Therefore, an SVNS A can be represented by*

$$A = \{\langle x, TA(x), IA(x), FA(x)\rangle | x \in X\}$$

The various distance measures and clustering algorithms defined over neutrosophic sets are given in [2,11,14].

3.2. *Sentiment Analysis of Text and VADER Package*

Sentiment analysis is a very efficient tool in judging the popular sentiment revolving around any particular product, services or brand. Sentiment analysis is also known as opinion mining. It is, in all conclusive trails, a process of determining the tone behind a line of text and to get an understanding of the attitude or polarity behind that opinion. Sentiment analysis is very helpful in social media understanding, as it enables us to pick up a review of the more extensive general assessment behind specific subjects. Most of the existing sentiment analysis tools classify the arguments into positive or negative sentiment based on a set of predefined rules or 'lexicons'. This enables the tool to calculate the overall leaning polarity of the text and thus makes a decision on the overall tone of the subject.

VADER is an easy-to-use, highly accurate and consistent tool for sentiment analysis. It is fully open source with the MIT License. It has a lexicon rule-based method to detect sentiment score for three classes: positive, neutral, and negative. It provides a compound score that lies in the range $[-1, 1]$. This compound score is used to calculate the overall sentiment of the input text. If the compound score ≥ 0.05, then it is tagged as positive. If the compound score is ≤ -0.05 then it tagged as negative. The arguments with the compound score between $(-0.05, 0.05)$ is tagged as neutral. VADER uses Amazon's Mechanical Turk to acquire their ratings, which is an extremely efficient process. VADER has a built in dictionary with a list of positive and negative words. It then calculates the individual score by summing the pre-defined score for the positive and negative words present in the dictionary. VADER forms a particularly strong basis for social media texts since the tweets or comments posted on social media are often informal, with grammatical errors and contain a lot of other displays of strong emotion, such as emojis, more than one exclamation point, etc. As an example, the sentence, 'This is good!!!' will be rated as being 'more positive' than 'This is good!' by VADER. VADER was observed to be very fruitful when managing social media writings, motion picture reviews, and product reviews. This is on the grounds that VADER not just tells about the positivity and negativity score yet in addition tells us how positive or negative a text is.

VADER has a great deal of advantages over conventional strategies for sentiment analysis, including:

1. It works very well with social media content, yet promptly sums up to different areas.
2. Although it contains a human curated sentiment dictionary for analysis, it does not specifically require any training data.
3. It can be used with real time data due to its speed and efficiency.

The VADER package for Python analysis presents the negative, positive and indeterminate values for every single tweet. Every single tweet is represented as $\langle N_x, I_x, P_x \rangle$, where x belongs to the dataset.

3.3. *Speech Analysis*

An important component of this paper is speech analysis which involves processing audio. Audio files cannot be directly understood by models. Machine learning algorithms do not take raw audio files as input hence it is imperative to extract features from the audio files. An audio signal is a three-dimensional signal where the three axes represent amplitude, frequency and time. Extracting features from audio files helps in building classifiers for prediction and recommendation.

Python provides a package called librosa for the analysis of audio and music. In this work, librosa has been used to extract a total 193 features per audio file. To display an audio file as spectrogram, wave plot or colormap librosa.display is used.

Figure 2 is a wave plot of an audio file. The loudness (amplitude) of an audio file can be shown in wave plot.

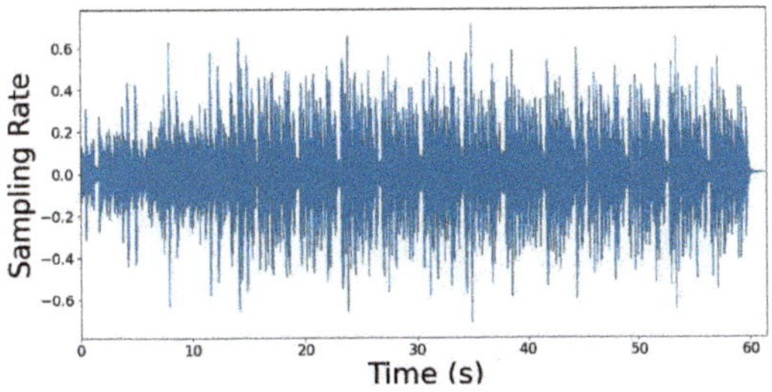

Figure 2. Wave plot of an audio file.

Figure 3 shows the spectrogram of the sample audio. Spectrogram is used to map different frequencies at a given point of time to its amplitude. It is a visual representation of the spectrum of frequencies of a sound.

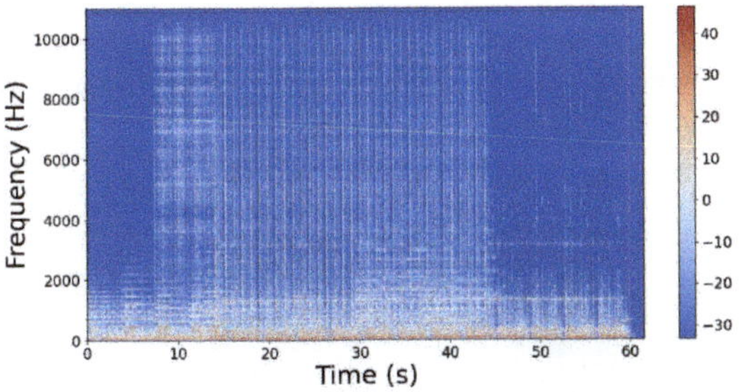

Figure 3. Spectrogram of an audio file.

The MFCC features of an audio file is shown in Figure 4. The MFCCs of a signal are a small set of features which concisely describe the overall shape of a spectral envelope. Sounds generated by a human are filtered by the shape of the vocal tract including the tongue, teeth etc. MFCCs represent the shape of the envelope that the vocal tract manifests on the short time power spectrum.

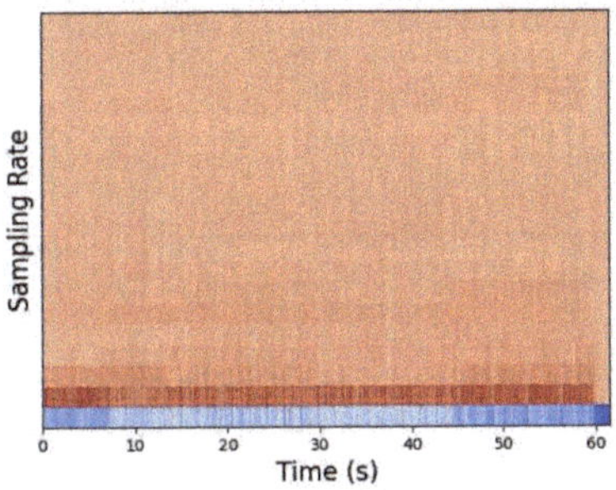

Figure 4. MFCC features of an audio file.

The chroma features of the sample audio file is represented in Figure 5. These represent the tonal content of audio files, that is the representation of pitch within the time window spread over the twelve chroma bands.

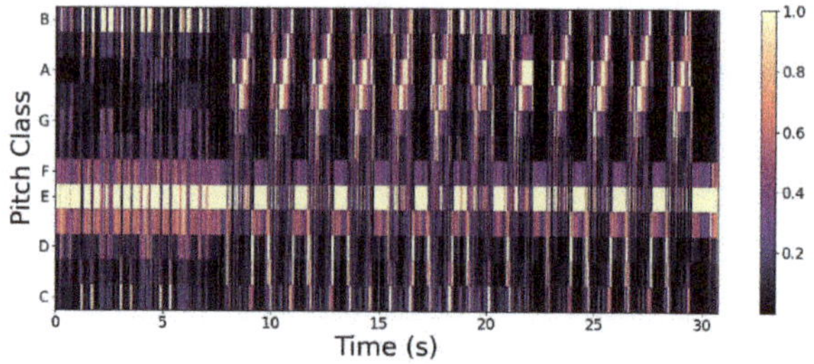

Figure 5. Chromagram of an audio file.

Figure 6 represents the mel spectrogram of the sample audio file. Mathematically, mel scale is the result of some non-linear transformation of the frequency scale. The purpose of the mel scale is that the difference in the frequencies as perceived by humans should be different for all ranges. For example, humans can easily identify the difference between 500 Hz and 1000 Hz but not between 8500 Hz and 9000 Hz.

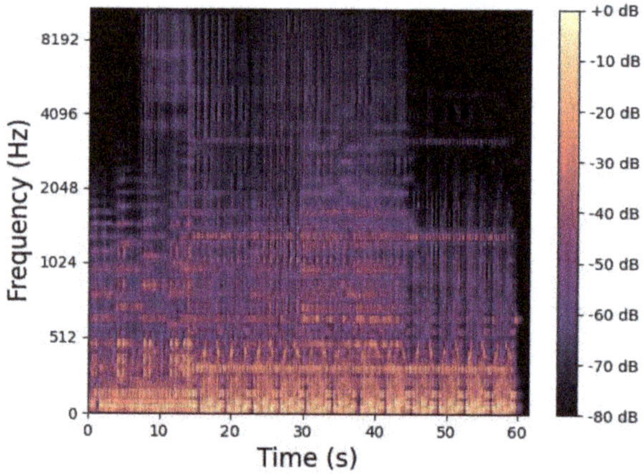

Figure 6. Mel spectrogram of an audio file.

The spectral contrast of the sample audio file is represented in Figure 7. Spectral contrast extracts the spectral peaks, valleys, and their differences in each sub-band. The spectral contrast features represent the relative spectral characteristics.

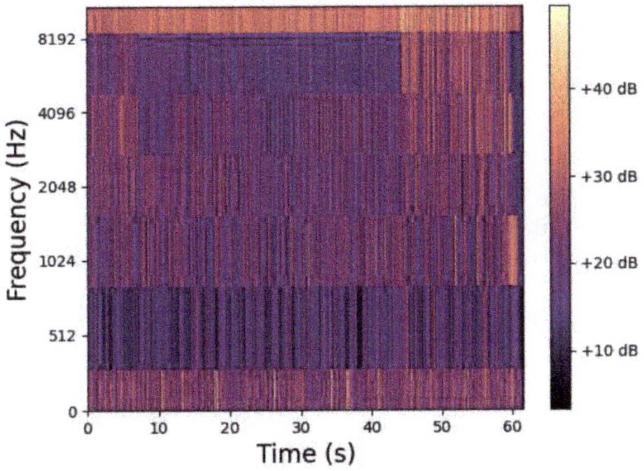

Figure 7. Spectral contrast of a sample audio file.

Figure 8 shows the tonnetz features of the sample audio file. The tonnetz is a pitch space defined by the network of relationships between musical pitches in just intonation. It estimates tonal centroids as coordinates in a six-dimensional interval space.

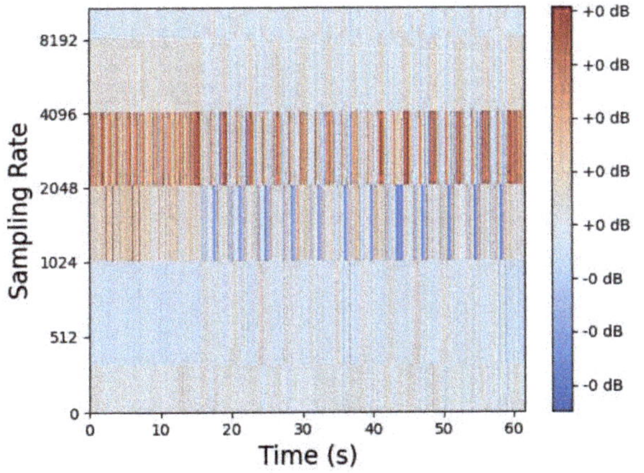

Figure 8. Tonnetz features of the sample audio file.

4. Model Description

4.1. Model Architecture

The research work follows a semi-hierarchical model where one step is followed by another but it is bifurcated into two wings one for audio and other for text and later on the SVNS are combined together in the integration module.

The overall architecture of the work is provided in Figure 9. The process begins with selecting an appropriate dataset with audio to text translations. For the audio section, convert the audio files into .wav format and extract features for further processing. Since the dataset is unlabelled the only suitable choice in the machine learning algorithms are clustering algorithms. For this module, K-means clustering was chosen. Then the Euclidean distance(x) of each point from the centre of each cluster is calculated and $1 - x$ is used as the measure of that specific class, SVNS values were obtained. Clustering was performed again to visualise the SVNS as clusters.

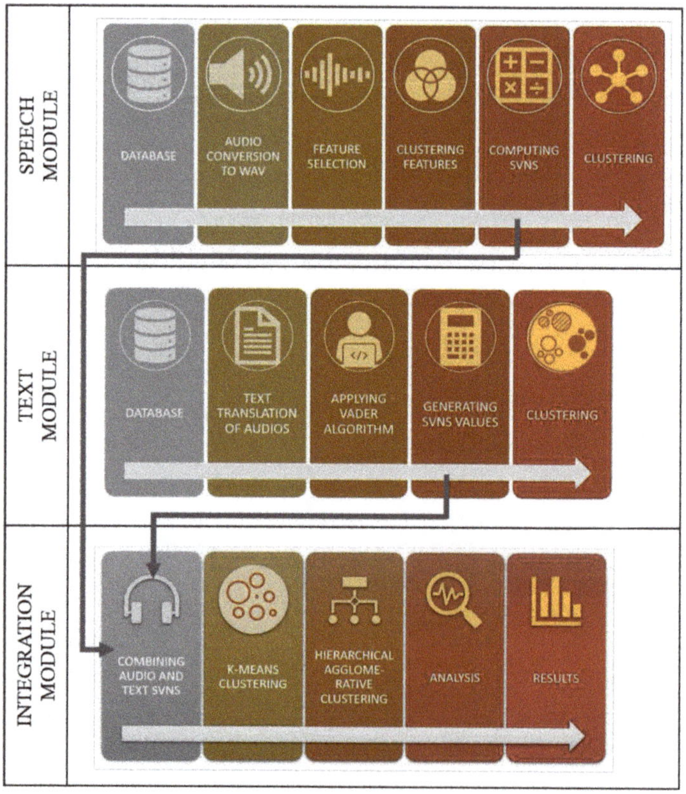

Figure 9. The model architecture.

For the text module, the text translations were considered and VADER tool was used to generate SVNS. After the generation of SVNS, it was clustered and visualized.

In the integration module the SVNS values obtained from speech module and text module was combined together, there by combining both the branches. The final SVNS are calculated by averaging the audio and text SVNS which are again clustered and visualized for comparison.

4.2. Data Processing

Dataset played a crucial role in this research work. The reason being we wanted to map audio SVNS to text SVNS for comparison so a dataset with audio translation was required. Hence LibriSpeech dataset [2] was chosen. LibriSpeech is a corpus of approximately 1000 h of 16kHz read English speech. The data is derived from read audiobooks from the LibriVox project, and has been carefully segmented and aligned. For this purpose the following folders have been used:

1. Dev-clean (337 MB with 2703 audio)
2. Train-clean (6.3 GB with 28,539 audio)

We used the dev clean (337MB) folder to test algorithms in the initial phase and then scaled up to train clean-100 (6.3 GB) to get the final results. We did not scale further due to hardware limitations. The reason for selecting the "clean" speech sets was to eliminate the more challenging audio and focus more on speech analysis. Since these are audio books, the dataset is structured in the following format. For example, 84-121123-0001.flac is present in the sub directory 121123 of directory 84, it implies that the reader ID for this audio file is 84 and the chapter is 121123. There is a separate chapters.txt which is provided along with the dataset that provides the details of the chapter. For example, 121123 is the chapter 'Maximilian' in the book 'The Count of Monte Cristo'. In the same sub directory 121123 a text file is present, 84-121123.trans.txt which contains the audio to text translation of the audio files in that directory. The reason for choosing this dataset over others is that it provides audio to text translations of the audio files.

The processing of audio file from .flac format to .wav format was carried out. The dataset was available in .flac format. It was necessary to convert these files into .wav format for further processing and extracting features. For this ffmpeg was used in shell script with bash. Ffmpeg is a free and open-source project consisting of a vast software suite of libraries and programs for handling video, audio, and other multimedia files and streams.

4.3. Feature Extraction

The audio files were then fed into the python feature extraction script which extracted 193 features per audio file. Using the Librosa package in python following features were extracted

1. MFCC (40)
2. Chroma (12)
3. Mel (128)
4. Contrast (7)
5. Tonnetz (6)

The following npy files were generated as result:

1. X_dev_clean.npy (2703 × 193)
2. X_train_clean.npy (28,539 × 193)

Then these files were normalized using sklearn. The screenshot of the normalized audio features is given in Figure 10.

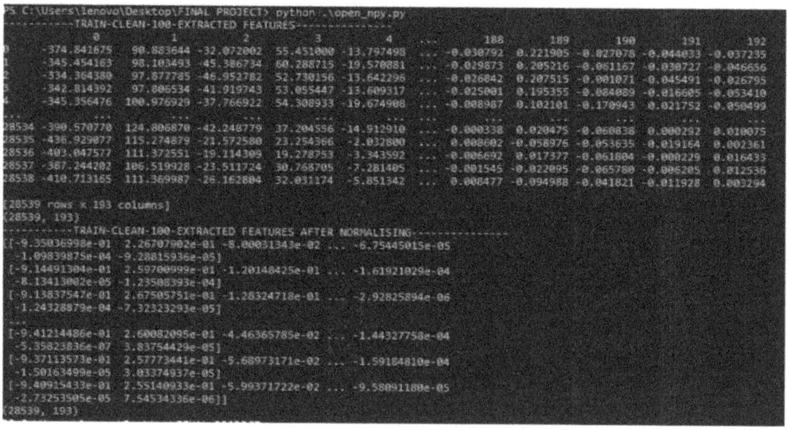

Figure 10. Normalized audio features.

4.4. Clustering and Visualization

4.4.1. K-Means

The K-means algorithms used for clustering SVNS values for sentiment analysis was proposed in [13]. It is a simple algorithm which produces the same results irrespective of the order of the dataset. The input is the SVNS values as dataset and the number of clusters (K) required. The algorithm then picks K SVNS values from the dataset randomly and assigns them as centroid. Then repeatedly the distance between other SVNS values and centroids are calculated and they are assigned to one cluster. This process continues till the centroid stops changing. Elbow method specifies what a good K (number of clusters) would be based on the sum of squared distance (SSE) between data points and their assigned clusters' centroids.

4.4.2. Hierarchical Agglomerative Clustering and Visualization

Hierarchical clustering is a machine learning algorithm used to group similar data together based on a similarity measure or the Euclidean distance between the data points. It is generally used for unlabelled data. There are two types of hierarchical clustering approaches: divisive and agglomerative. Hierarchical divisive clustering refers to top to down approach where all the data is assigned to one cluster and then partitioned further into clusters. In hierarchical agglomerative clustering all the data points are treated as individual clusters and then with every step data points closest to each other are identified and grouped together. This process is continued until all the data points are grouped into one cluster, creating a dendogram. The algorithm for hierarchical agglomerative clustering of SVNS values is given in Algorithm 1.

Algorithm 1: Hierarchical agglomerative clustering.

```
Input: N number of SVNSs {s_1, ... s_N}
Output: Cluster
begin
    Step 1: Create a distance matrix X using Euclidean distance function dist(s_i, s_j)
    for i ← 1, N do
        for j ← i + 1, N do
            x_i ← dist(s_i, s_j)
        end
    end
    Step 2: X ← {x_1, x_2, ..., x_N}
    Step 3: Perform clustering
    while X.size > 1 do
        (x_min1, x_min2) ← minimum_dist(x_a, x_b) ∀x_a, x_b ∈ X
        Remove x_min1 and x_min2 from X
        Add center{x_min1, x_min2} to X
        Alter distance matrix X accordingly
    end
    Results in cluster automatically
end
```

4.5. Generating SVNS Values

4.5.1. Speech Module

Since the dataset was unlabelled, K-means algorithm was used for clustering. With K being set to 3, the clusters were obtained. Let the cluster centres be B_1, B_2 and B_3. B_1, B_2 and B_3 were mapped as positive, neutral, and negative clusters, respectively. We randomly selected 30 samples from each cluster and mapped the maximum sentiment of the sample as the sentiment of the cluster. For every data point P, in the dataset distance was calculated to the centres of each cluster. 1-distance implied the closeness measure to each cluster or class (positive, neutral or negative). SVNS for audio were created using 1-distance and stored in a .csv file as $\langle P_A, I_A, N_A \rangle$.

4.5.2. Text Module

The next task is sentiment analysis of text translation using VADER. VADER is a tool used for sentiment analysis which provides a measure for positive, neutral and negative classes for each input sentence. Using VADER text translation for each audio was analysed and SVNS were generated and stored in .csv file as $\langle P_T, I_T, N_T \rangle$. Taking the csv file of text SVNS as input, K-means cluster with K, taken as 3, was performed.

4.5.3. Integration Module

Next, we proceed on to combine the SVNS, the audio SVNS values are represented by $\langle P_A, I_A, N_A \rangle$ and the text SVNS values are represented by $\langle P_T, I_T, N_T \rangle$ and the combined SVNS are represented by $\langle P_C, I_C, N_C \rangle$, where the component values are calculated as

$$
\begin{aligned}
P_C &= \frac{(P_T + P_A)}{2} \\
I_C &= \frac{(I_T + I_A)}{2} \\
N_C &= \frac{(N_T + N_A)}{2}
\end{aligned} \tag{1}
$$

Combined SVNS values were generated using equations given in Equation (1). The visualization of combined SVNS is carried out next. Using K-means clustering and hierarchical agglomerative clustering algorithms, the SVNS of audio, text and combined modules were visualized into 3 clusters.

5. Experimental Results and Data Visualisation

5.1. Speech Module

The elbow method specifies what a good K, the number of clusters would be based on the SSE between data points and their assigned clusters' centroids. The elbow chart of the audio were created to decide the most favourable number of clusters, they are given in Figure 11a,b for the dev-clean folder and train-clean folder, respectively.

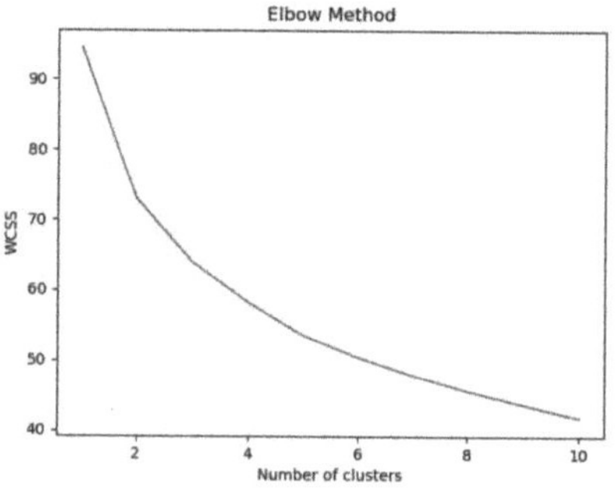

(**a**) Dev-clean

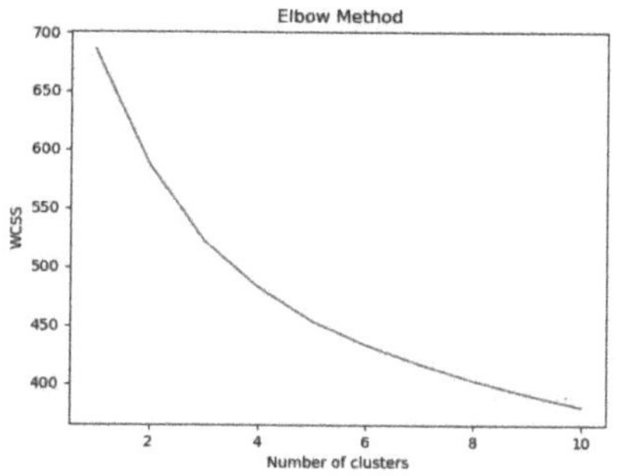

(**b**) Train-clean

Figure 11. Elbow chart for dataset.

The elbow method generates the optimum number of clusters as three as shown in Figure 11a,b. Hence, the dataset is clustered into three clusters – positive, indeterminate and negative. The results for the clustering of the dataset into three is visualised in 2D and 3D in Figures 12a,b and 13a,b. The 2D visualization of the clusters is given in Figure 12a,b for dev-clean and train-clean respectively. Figure 13a,b are the K-Means clustering in 3D for dev-clean and train-clean respectively.

Once clusters are formed, we calculate the Euclidean distance of each data point from the centre of the cluster. Let the cluster centres be B_1, B_2 and B_3. For every data point P in the dataset distance was calculated to the centres of each cluster. 1-distance implied the closeness measure to each cluster or class (positive, neutral or negative). Euclidean distance d can be calculated using the formula given Equation (2).

$$d = \sqrt{(x_2 - x_1)^2 + (y_2 - y_1)^2} \tag{2}$$

The sample SVNS values generated from the audio features is given in Figure 14a.

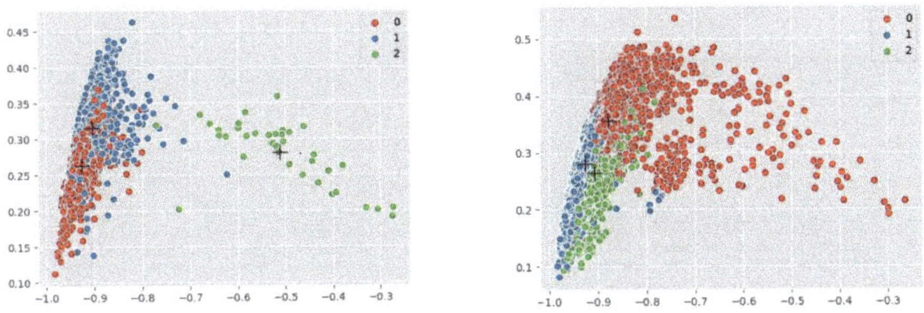

(**a**) Dev-clean (**b**) Train-clean

Figure 12. K-means clustering in 2D for audio dataset.

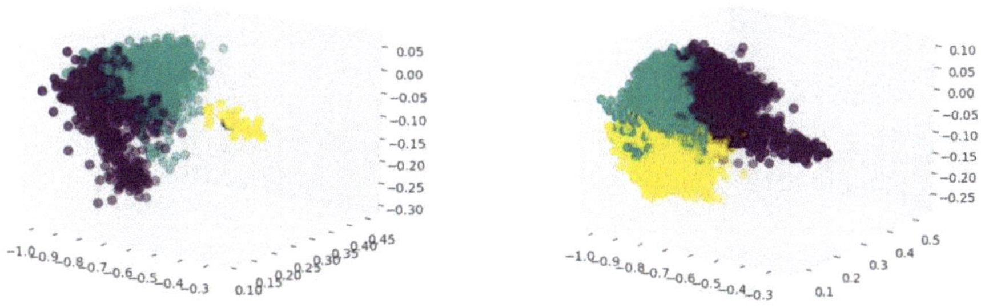

(**a**) Dev-clean (**b**) Train-clean

Figure 13. K-means clustering in 3D for audio dataset.

```
DISTANCE FROM CLUSTER CENTRE

[[0.142 0.049 0.065]
 [0.132 0.067 0.092]
 [0.155 0.055 0.083]
 [0.121 0.053 0.074]
 [0.126 0.046 0.061]
 [0.133 0.054 0.057]
 [0.129 0.058 0.042]
```

(**a**) Audio SVNS

	A	B	C
1	pos	neu	neg
18	0.125	0.835	0.04
19	0.123	0.827	0.05
20	0.047	0.953	0
21	0.1	0.9	0
22	0.116	0.767	0.117
23	0.146	0.854	0
24	0.079	0.921	0
25	0.151	0.786	0.063
26	0.09	0.849	0.061
27	0.078	0.845	0.077

(**b**) Text SVNS

Figure 14. Sample SVNS values.

5.2. Text Module

The audio to text translations are given in the dataset, a sample from the dataset is given Figure 15.

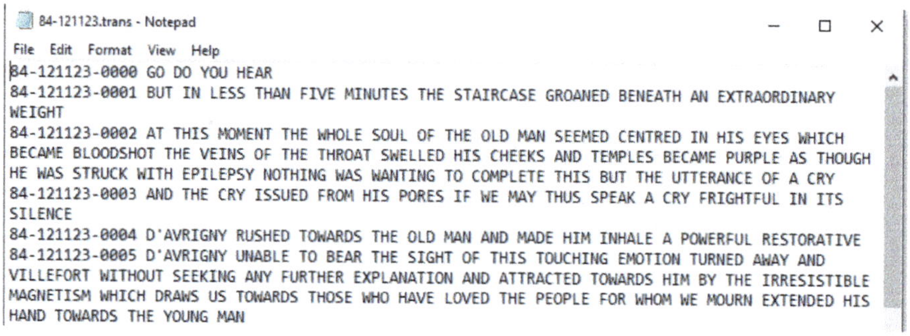

84-121123.trans - Notepad

File Edit Format View Help

84-121123-0000 GO DO YOU HEAR
84-121123-0001 BUT IN LESS THAN FIVE MINUTES THE STAIRCASE GROANED BENEATH AN EXTRAORDINARY WEIGHT
84-121123-0002 AT THIS MOMENT THE WHOLE SOUL OF THE OLD MAN SEEMED CENTRED IN HIS EYES WHICH BECAME BLOODSHOT THE VEINS OF THE THROAT SWELLED HIS CHEEKS AND TEMPLES BECAME PURPLE AS THOUGH HE WAS STRUCK WITH EPILEPSY NOTHING WAS WANTING TO COMPLETE THIS BUT THE UTTERANCE OF A CRY
84-121123-0003 AND THE CRY ISSUED FROM HIS PORES IF WE MAY THUS SPEAK A CRY FRIGHTFUL IN ITS SILENCE
84-121123-0004 D'AVRIGNY RUSHED TOWARDS THE OLD MAN AND MADE HIM INHALE A POWERFUL RESTORATIVE
84-121123-0005 D'AVRIGNY UNABLE TO BEAR THE SIGHT OF THIS TOUCHING EMOTION TURNED AWAY AND VILLEFORT WITHOUT SEEKING ANY FURTHER EXPLANATION AND ATTRACTED TOWARDS HIM BY THE IRRESISTIBLE MAGNETISM WHICH DRAWS US TOWARDS THOSE WHO HAVE LOVED THE PEOPLE FOR WHOM WE MOURN EXTENDED HIS HAND TOWARDS THE YOUNG MAN

Figure 15. Sample audio to translation.

Now the text file is processed with the VADER tool for analysis, which generates SVNS values in form of $\langle N_T, I_T, P_T \rangle$. For the sake of notational convenience, we created and populated .csv file in the order of $\langle P_T, I_T, N_T \rangle$, where P_T is positive, I_T is the indeterminate membership and N_T is the negative membership. A sample of the .csv file that contains the SVNS values is shown in Figure 14b. VADER also gives a composite score for every line, depending on which the tool also provides a class label, i.e., positive or neutral or negative. Since we were working with unlabelled data, we did not have a method to validate the labels provided by the tool.

In the case of the textual content of a novel, this is a narration, so one cannot get high values for positivity or negativity only, neutrals takes the maximum value when SVNS value is used; which is evident from Figure 14b. The obtained SVNS values are clustered using K-means algorithm and visualized in Figures 16a,b and 17a,b. Figure 16a,b are results of the K-means clustering in 2D on dev-clean and train-clean datasets respectively.

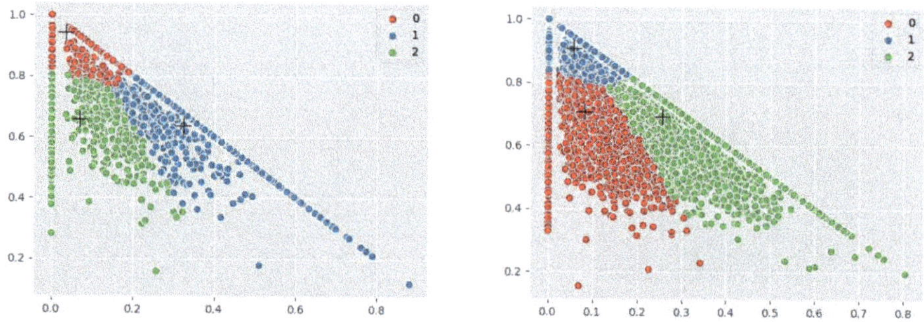

(**a**) Dev-clean results (**b**) Train-clean results

Figure 16. K-means clustering in 2D text SVNS values.

Similarly the clustering results are represented in 3D in Figure 17a,b. Dev-clean folder contains 2703 audio files and train-clean folder contains 28,539 audio files.

The clustering visualisation clearly shows the presence of 3 clusters indicating the existence of neutrality in the data.

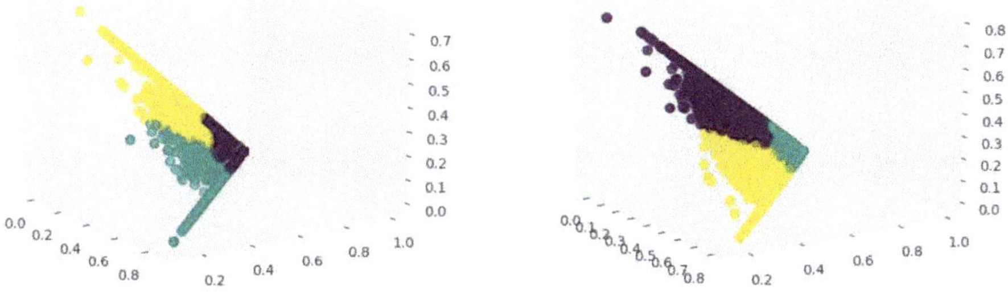

(**a**) Dev-clean (**b**) Train-clean

Figure 17. K-means clustering in 3D of text SVNS values.

5.3. Integration Module

The final SVNS are calculated by averaging the audio SVNS and text SVNS. The combined SVNS values are again clustered and visualized for comparison. We visualize the SVNS values using clustering algorithms such as K-means and hierarchical agglomerative clustering given in Algorithm 1. The K-means clustering results of combined SVNS of dev-clean and train-clean are given in Figure 18a,b respectively.

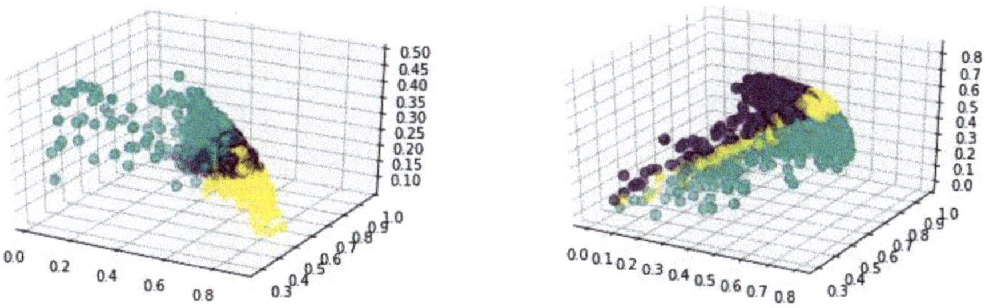

(**a**) Dev-clean (**b**) Train-clean

Figure 18. K-means clustering in 3D of combined SVNS values.

The dendograms generated while clustering the combined SVNS values of dev-clean and train-clean are given in Figure 19 and Figure 20 respectively.

The clustering results of using agglomerative clustering on the combined SVNS values of dev-clean and train-clean datasets are given in Figure 21a,b respectively.

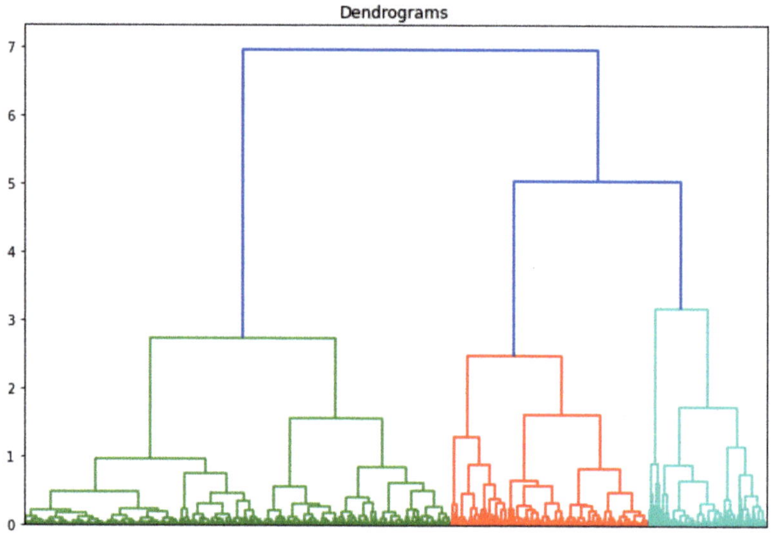

Figure 19. Dendogram of combined SVNS values of Dev-clean.

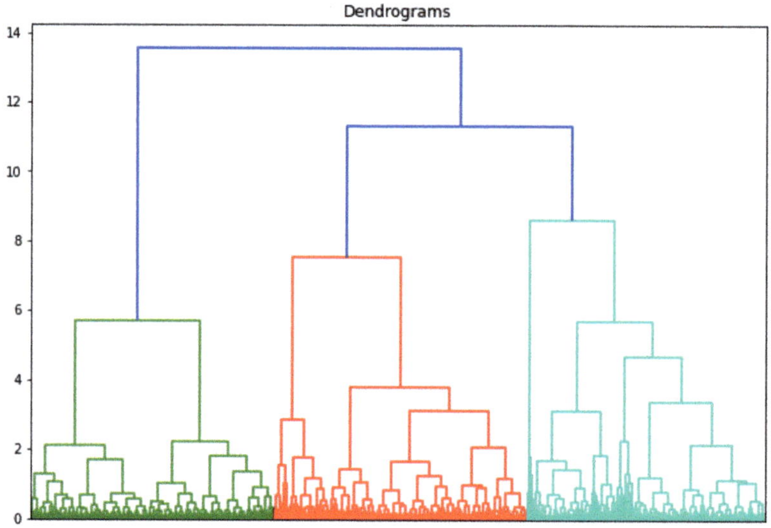

Figure 20. Dendogram of combined SVNS values of Train-clean.

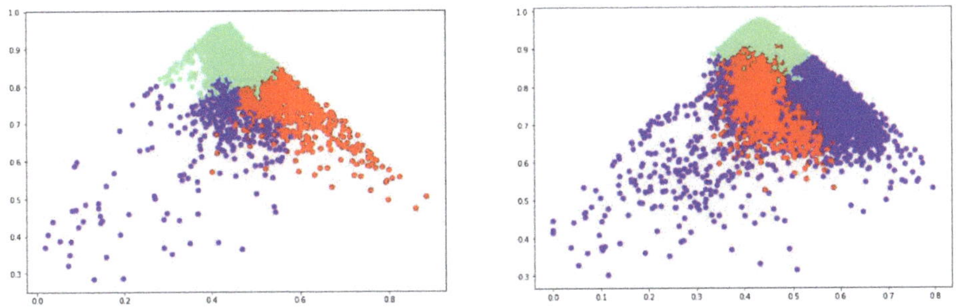

(a) Dev-clean (b) Train-clean

Figure 21. Agglomerative Clustering of combined SVNS values.

6. Result and Discussion

The visualization of clustering results and the dendogram clearly reveal the presence of neutrality in the data, which is validated by the existence of the third cluster. It is pertinent to note that, in case of sentiment analysis, data cannot be divided into positive and negative alone, the existence of neutrality needs to be acknowledged. After analysing the results of all the clustering algorithms, significant conclusions have been made. The concept of indeterminacy or neutrality has not yet been dealt with in normal or conventional and fuzzy sentiment analysis. SVNS provides a score for neutral sentiment along with positive and negative sentiments. Speech sentiment analysis using neutrosophic sets has not been done to date, whereas it can provide excellent results. The logic behind combining SVNS is to include both features related to the audio files derived from amplitude and frequency and pairing it with the analysis of text for better results. This is a much more wholesome approach than just picking either of the two.

In Table 1, the number of audio classified as cluster 1 (C1), cluster 2 (C2) and cluster 3 (C3) are shown for SVNS from audio features, text SVNS and the combined SVNS for dev-clean LibriSpeech folder which is 337 MB with 2703 audio. There is a considerable overlap in the values that are present in the cluster C1 and C2 and C3, for the three values from speech module, text module and combined module, respectively.

Table 1. Dev-clean clustering results.

SVNS	C1	C2	C3
Audio	1097	1568	38
Text	1431	675	597
Combined	1465	752	486

In Table 2, the number of audio classified as cluster 1 (C1), cluster 2 (C2) and cluster 3 (C3) are shown for SVNS from audio features, text SVNS and the combined SVNS for train-clean-100 LibriSpeech folder which is 6.3 GB with 28539 audio. Since the dataset was unlabelled there was no other choice but to cluster the features, hence the output which was received was clusters without class tags, hence it cannot be identified with these given results which cluster represents positive class, neutral class or negative class. Class tags can be obtained from VADER composite score, but since our aim was to show the presence of neutrality in the data, we did not do the mapping of the clusters to a particular class using the VADER tool provided labels.

Table 2. Train-clean-100 clustering results.

SVNS	C1	C2	C3
Audio	7830	13,234	7475
Text	9332	15,028	4179
Combined	8389	13,174	6976

Instead, if we used the max of the SVNS values present in the cluster to map the cluster to a class tag. Accordingly we obtained C1 cluster was positive class, C2 cluster was neutral and C3 cluster was the negative class. Though it can be inferred from the changing number of data points in the clusters and their ratios to one another that analysis of audio separately and text separately, and then combining the two together with neutrosophic sets is effective to address the indeterminacy and uncertainty of data.

7. Conclusions and Further Work

Work on analyzing sentiment of textual data using neutrosophic sets has been sparse and little, only [13,14] made use of SVNS and refined neutrosophic sets for sentiment analysis. Analysis of audio or speech sentiment analysis using neutrosophy has not been carried out, until now. To date, there has been no way to accommodate the neutrosophy in the sentiment analysis of audio. In the first of a kind, we used the audio features to implement the concept of neutrosophy in speech sentiment analysis. We proposed a novel framework that combines audio features, sentiment analysis, and neutrosophy to generate SVNS values. The initial phase of the work included extracting features from audio, clustering them into three clusters, and generating the SVNS. This was followed by using the VADER tool for text and generating SVNS. Now there were two SVNS for every audio file; one from the audio files and the other from the text file. These two were combined by averaging out the SVNS and the newly obtained SVNS were clustered again for final results. This is an innovative contribution to both sentiment analysis and neutrosophy. For future work, while combining the SVNS weights can be set according to priority or depending on the reliability of the data. For example, if the audio to text translations are bad then weights can be set in the ratio 4:1 for audio SVNS to text SVNS where the resulting SVNS will depend 80% on the audio SVNS and 20% on the text SVNS. Similarly, other similarity measures other than distance measures can be used for generating SVNS values for audio files.

Author Contributions: Conceptualization, V.K.W.B. and I.K.; methodology, software, validation, K.M.; formal analysis, F.S.; investigation, resources, data curation, writing—original draft preparation, V.K.W.B. and I.K.; writing—review and editing, F.S.; visualization, K.M.; supervision, project administration, I.K.; All authors have read and agreed to the published version of the manuscript.

Funding: This research received no external funding.

Conflicts of Interest: The authors declare no conflict of interest.

Abbreviations

The following abbreviations are used in this manuscript:

NLP	Natural Language Processing
SVNS	Single-Valued Neutrosophic Sets
MFCC	Mel-Frequency Cepstral Coefficients
RAVDESS	Ryerson Audio-Visual Database of Emotional Speech and Song
VADER	Valence Aware Dictionary and Sentiment Reasoner
SVM	Support Vector Machine
DVNS	Double Valued Neutrosophic Sets
SSE	Sum of Squared Distance

References

1. Smarandache, F. A unifying field in Logics: Neutrosophic Logic. In *Philosophy*; American Research Press: Rehoboth, DE, USA, 1999; pp. 1–141.
2. Panayotov, V.; Chen, G.; Povey, D.; Khudanpur, S. Librispeech: An ASR corpus based on public domain audio books. In Proceedings of the 2015 IEEE International Conference on Acoustics, Speech and Signal Processing (ICASSP), Brisbane, Australia, 19–24 April 2015; pp. 5206–5210. [CrossRef]
3. Howells, K.; Ertugan, A. Applying fuzzy logic for sentiment analysis of social media network data in marketing. In Proceedings of the 9th International Conference on Theory and Application of Soft Computing, Computing with Words and Perception, ICSCCW 2017, Budapest, Hungary, 24–25 August 2017,
4. Alaei, A.R.; Becken, S.; Stantic, B. Sentiment Analysis in Tourism: Capitalizing on Big Data. *J. Travel Res.* **2019**, *58*, 175–191. [CrossRef]
5. Ribeiro, F.N.; Araújo, M.; Gonçalves, P.; Gonçalves, M.A.; Benevenuto, F. SentiBench—A benchmark comparison of state-of-the-practice sentiment analysis methods. *EPJ Data Sci.* **2016**, *5*, 1–29. [CrossRef]
6. Gilbert, C.H.E.; Hutto, E. Vader: A parsimonious rule-based model for sentiment analysis of social media text. In Proceedings of the Eighth International AAAI Conference on Weblogs and Social Media, Ann Arbor, MI, USA, 1–4 June 2014.
7. Ramteke, J.; Shah, S.; Godhia, D.; Shaikh, A. Election result prediction using Twitter sentiment analysis. In Proceedings of the 2016 International Conference on Inventive Computation Technologies (ICICT), Coimbatore, India, 26–27 August 2016; pp. 1–5. [CrossRef]
8. Smarandache, F.; Teodorescu, M.; Gîfu, D. Neutrosophy, a Sentiment Analysis Model. In Proceedings of the RUMOR, Toronto, ON, Canada, 22 June 2017; pp. 38–41.
9. Wang, H.; Smarandache, F.; Zhang, Y.; Sunderraman, R. Single Valued Neutrosophic Sets. *Multispace Multistruct.* **2010**, *4*, 410–413.
10. Smarandache, F. n-Valued Refined Neutrosophic Logic and Its Applications in Physics. *Prog. Phys.* **2013**, *4*, 143–146.
11. Kandasamy, I. Double-valued neutrosophic sets, their minimum spanning trees, and clustering algorithm. *J. Intell. Syst.* **2018**, *27*, 163–182. [CrossRef]
12. Freen, G.; Kousar, S.; Khalil, S.; Imran, M. Multi-objective non-linear four-valued refined neutrosophic optimization. *Comput. Appl. Math.* **2020**, *39*, 35. [CrossRef]
13. Kandasamy, I.; Vasantha, W.B.; Mathur, N.; Bisht, M.; Smarandache, F. Sentiment analysis of the # MeToo movement using neutrosophy: Application of single-valued neutrosophic sets. In *Optimization Theory Based on Neutrosophic and Plithogenic Sets*; Academic Press: Cambridge, MA, USA, 2020; pp. 117–135.
14. Kandasamy, I.; Vasantha, W.B.; Obbineni, J.M.; Smarandache, F. Sentiment analysis of tweets using refined neutrosophic sets. *Comput. Ind.* **2020**, *115*, 103180. [CrossRef]
15. Kandasamy, I.; Kandasamy, W.B.V.; Obbineni, J.M.; Smarandache, F. Indeterminate Likert scale: Feedback based on neutrosophy, its distance measures and clustering algorithm. *Soft Comput.* **2020**, *24*, 7459–7468. [CrossRef]
16. Pătraşcu, V. Refined neutrosophic information based on truth, falsity, ignorance, contradiction and hesitation. *Neutrosophic Sets Syst.* **2016**, *11*, 57–66.
17. Elaiyaraja, V.; Sundaram, P.M. Audio classification using support vector machines and independent component analysis. *J. Comput. Appl.* **2012**, *5*, 34–38.
18. Huang, A.; Bao, P. Human Vocal Sentiment Analysis. *arXiv* **2019**, arXiv:1905.08632.
19. El Ayadi, M.; Kamel, M.S.; Karray, F. Survey on speech emotion recognition: Features, classification schemes, and databases. *Pattern Recogn.* **2011**, *44*, 572–587. [CrossRef]
20. Poria, S.; Cambria, E.; Howard, N.; Huang, G.B.; Hussain, A. Fusing Audio, Visual and Textual Clues for Sentiment Analysis from Multimodal Content. *Neurocomputing* **2016**, *174*, 50–59. neucom.2015.01.095. [CrossRef]

21. Poria, S.; Cambria, E.; Gelbukh, A. Deep convolutional neural network textual features and multiple kernel learning for utterance-level multimodal sentiment analysis. In Proceedings of the 2015 Conference on Empirical Methods in Natural Language Processing, Lisbon, Portugal, 13–17 September 2015; pp. 2539–2544.

Publisher's Note: MDPI stays neutral with regard to jurisdictional claims in published maps and institutional affiliations.

symmetry

MDPI

Article

A Novel MCDM Method Based on Plithogenic Hypersoft Sets under Fuzzy Neutrosophic Environment

Muhammad Rayees Ahmad [1], Muhammad Saeed [1], Usman Afzal [1] and Miin-Shen Yang [2,*]

1 Department of Mathematics, University of Management and Technology, Lahore 54770, Pakistan; F2018265003@umt.edu.pk (M.R.A.); muhammad.saeed@umt.edu.pk (M.S.); usman.afzal@umt.edu.pk (U.A.)
2 Department of Applied Mathematics, Chung Yuan Christian University, Chung-Li 32023, Taiwan
* Correspondence: msyang@math.cycu.edu.tw

Received: 18 September 2020; Accepted: 6 November 2020; Published: 10 November 2020

Abstract: In this paper, we advance the study of plithogenic hypersoft set (PHSS). We present four classifications of PHSS that are based on the number of attributes chosen for application and the nature of alternatives or that of attribute value degree of appurtenance. These four PHSS classifications cover most of the fuzzy and neutrosophic cases that can have neutrosophic applications in symmetry. We also make explanations with an illustrative example for demonstrating these four classifications. We then propose a novel multi-criteria decision making (MCDM) method that is based on PHSS, as an extension of the technique for order preference by similarity to an ideal solution (TOPSIS). A number of real MCDM problems are complicated with uncertainty that require each selection criteria or attribute to be further subdivided into attribute values and all alternatives to be evaluated separately against each attribute value. The proposed PHSS-based TOPSIS can be used in order to solve these real MCDM problems that are precisely modeled by the concept of PHSS, in which each attribute value has a neutrosophic degree of appurtenance corresponding to each alternative under consideration, in the light of some given criteria. For a real application, a parking spot choice problem is solved by the proposed PHSS-based TOPSIS under fuzzy neutrosophic environment and it is validated by considering two different sets of alternatives along with a comparison with fuzzy TOPSIS in each case. The results are highly encouraging and a MATLAB code of the algorithm of PHSS-based TOPSIS is also complied in order to extend the scope of the work to analyze time series and in developing algorithms for graph theory, machine learning, pattern recognition, and artificial intelligence.

Keywords: Soft set; hypersoft set; plithogenic hypersoft set (PHSS); multi-criteria decision making (MCDM); PHSS-based TOPSIS

1. Introduction

A strong mathematical tool is always needed in order to combat real world problems involving uncertainty in the data. This necessity has urged scholars to introduce different mathematical tools to facilitate the world for solving such problems. In 1965, the concept of fuzzy set was introduced by Zadeh [1], in which each element is assigned a membership degree in the form of a single crisp value in the interval $[0,1]$. It has been studied extensively by the researchers and a number of real life problems have been solved by fuzzy sets [2–5]. However, in some practical situations, it is seen that this membership degree is hard to be defined by a single number. The uncertainty in the membership degree became the cause to introduce the concept of interval-valued fuzzy set in which the degree of membership is an interval value in $[0,1]$. Later on, the concept of intuitionistic fuzzy set (IFS) was proposed by Atanassov [6] in 1986, which incorporates the non-membership degree. IFS had many applications [7–10].

However, IFS is unable to deal with indeterminate information, which is very common in belief systems. This inadequacy was addressed by Smarandache [11] in 2000, who introduced the concept of neutrosophic set in which membership (T), indeterminacy (I) and non-membership (F) degrees were independently quantified i.e., T, I, F $\in [0,1]$ and the sum T + I + F need not to be contained in $[0,1]$. All of these mathematical tools have been thoroughly explored and successfully applied to deal with uncertainties [12–15], yet these tools usually fail to handle uncertainty in a variety of practical situation, because these tools require all notions to be exact and do not possess a parametrization tool. Consequently, soft set was introduced by Molodstsov [16] in 1999, which can be regarded as a general mathematical tool to deal with uncertainty. Molodstsov [16] defined soft set as a parameterized family of subsets of a universe of discourse. In 2003, Maji et al. [17] introduced aggregation operations on soft sets. Soft sets and their hybrids have been successfully applied in various areas [18–21]

In a variety of real life MCDM problems, the attributes need to be further sub-divided into attribute values for a better decision. This need was fulfilled by Smarandache [22], who introduced the concept of hypersoft set as a generalization of the concept of soft set in 2018. Besides, Smarandache [22] introduced the concept of plithogenic hypersoft set with crisp, fuzzy, intuitionistic fuzzy, neutrosophic, and plithogenic sets. In 2020, Saeed et al. [23] presented a study on the fundamentals of hypersoft set theory. Smarandache [24,25] developed the aggregation operations on plithogenic set and proved that the plithogenic set is the most generalized structure that can be efficiently applied to a variety of real life problems [26–29]

A PHSS-based TOPSIS is proposed in the article to deal with MCDM problem, in which attribute may have attribute values and each attribute value has a neutrosophic degree of appurtenance of each alternative. The proposed method is authenticated by taking two different sets of alternatives. A comparison with fuzzy TOPSIS is made in each case. It shows that the results are highly inspiring. A MATLAB code of the algorithm of PHSS-based TOPSIS is also complied in order to encompass the scope of the work to analyze time series and in developing algorithms for graph theory, artificial intelligence, machine learning, pattern recognition, and neutrosophic applications in symmetry. It appears quite pertinent to point out that the article gives detailed insight on PHSS with related definitions and its implementation in MCDM process. The scope of the work can be extended in other mathematics directions as well by introducing important theorems and propositions [24]

The remainder of this article is organized, as follows. In Section 2, we briefly review some basic notions, leading to the definitions of soft sets, hypersoft sets, plithogenic sets, and plithogenic hypersoft sets (PHSSs), along with an illustrative example. Section 3 consists of the four proposed classifications of PHSSs based on different criteria. More explanations with an illustrative example for the four classifications are also made. In Section 4, the algorithm of the proposed PHSS-based TOPSIS is given, along with its application to a real life parking spot choice problem under fuzzy neutrosophic environment and its comparison with fuzzy TOPSIS. Section 5 provides the conclusion and future directions.

2. Preliminaries

This section comprises of some necessary basic concepts that are related to plithogenic hypersoft set (PHSS), which is also defined in this section along with an illustrative example for a clear understanding. Throughout the study, let $\mathcal{U}$ be a non-empty universal set, $P(\mathcal{U})$ be the power set of $\mathcal{U}$, $X \subseteq \mathcal{U}$ be a finite set of alternatives, and $\mathcal{A}$ be a finite set of n distinct parameters or attributes, as given by

$$\mathcal{A} = \{a_1, a_2, \cdots, a_n\}, \quad n \geq 1.$$

The attribute values of $a_1, a_2, \cdots, a_n$ belong to the sets $A_1, A_2, \ldots, A_n$, respectively, where $A_i \cap A_j = \phi$, for $i \neq j$, and $i, j \in \{1, 2, \ldots, n\}$. Moreover, we consider a finite number of uni-dimensional attributes and each attribute has a finite discrete set of attribute values. However, it is worth mentioning that the attributes may have an infinite number of attribute values. In such a case, every structure with non-Archimedean metrics can be dealt in depth [30,31].

2.1. Soft Sets

A soft set over $\mathcal{U}$ is a mapping $\mathcal{F} : \mathcal{B} \to P(\mathcal{U})$, $\mathcal{B} \subseteq \mathcal{A}$ with the value $\mathcal{F}_{\mathcal{B}}(\alpha) \in P(\mathcal{U})$ at $\alpha \in \mathcal{B}$ and $\mathcal{F}_{\mathcal{B}}(\alpha) = \phi$ if $\alpha \notin \mathcal{B}$. It is denoted by $(\mathcal{F}, \mathcal{B})$ and written as follows [16]:

$$(\mathcal{F}, \mathcal{B}) = \{(\alpha, \mathcal{F}_{\mathcal{B}}(\alpha)) : \alpha \in \mathcal{B},\ \mathcal{F}_{\mathcal{B}}(\alpha) \in P(\mathcal{U})\}.$$

Moreover, a soft set over $\mathcal{U}$ can be regarded as a parameterized family of the subsets of $\mathcal{U}$. For an attribute $\alpha \in \mathcal{B}$, $\mathcal{F}_{\mathcal{B}}(\alpha)$ is considered as the set of α-approximate elements of the soft set $(\mathcal{F}, \mathcal{B})$.

2.2. Hypersoft Sets

Let C denote the cartesian product of the sets $A_1, A_2, \ldots, A_n$, i.e., $C = A_1 \times A_2 \times \ldots \times A_n$, $n \geq 1$. Subsequently, a hypersoft set $(\mathcal{H}, \mathcal{C})$ over $\mathcal{U}$ is a mapping defined by $\mathcal{H} : C \to P(\mathcal{U})$ [22]. For an n-tuple $(\gamma_1, \gamma_2, \ldots, \gamma_n) \in C$,, where $\gamma_i \in A_i$, $i = 1, 2, 3, \ldots, n$, a hypersoft set is written as

$$(\mathcal{H}, \mathcal{C}) = \{(\gamma, \mathcal{H}(\gamma)) : \gamma = (\gamma_1, \gamma_2, \ldots, \gamma_n) \in \mathcal{C},\ \mathcal{H}(\gamma) \in P(\mathcal{U})\}.$$

It may be noted that hypersoft set is a generalization of soft set.

2.3. Plithogenic Sets

A set X is called a plithogenic set if all of its members are characterized by the attributes under consideration and each attribute may have any number of attribute values [24]. Each attribute value possesses a corresponding appurtenance degree of the element x, to the set X, with respect to some given criteria. Moreover, a contradiction degree function is defined between each attribute value and the dominant attribute value of an attribute in order to obtain accuracy for aggregation operations on plithogenic sets. These degrees of appurtenance and contradiction may be fuzzy, intuitionistic fuzzy or neutrosophic degrees.

Remark 1. *Plithogenic set is regarded as a generalization of crisp, fuzzy, intuitionistic fuzzy. and neutrosophic sets, since the elements of later sets are characterized by a combined single attribute value (degree of appurtenance), which has only one value for crisp and fuzzy sets i.e., membership, two values in case of intuitionistic fuzzy set i.e., membership and non-membership, and three values for neutrosophic set i.e., membership, indeterminacy, and non-membership. In the case of plithogenic set, each element is separately characterized by all attribute values under consideration in terms of degree of appurtenance.*

2.4. Plithogenic Hypersoft Set (PHSS)

Let $X \subseteq \mathcal{U}$ and $C = A_1 \times A_2 \times \ldots \times A_n$, where $n \geq 1$ and A_i is the set of all attribute values of the attribute a_i, $i = 1, 2, 3, \ldots, n$. Each attribute value γ possesses a corresponding appurtenance degree $d(x, \gamma)$ of the member $x \in X$, in accordance with some given condition or criteria. The attribute value degree of appurtenance is a function that is defined by

$$d : X \times C \to P([0,1]^j), \quad \forall \quad x \in X,$$

such that $d(x, \gamma) \in [0,1]^j$, and $P([0,1]^j)$ is the power set of $[0,1]^j$, where $j = 1, 2, 3$ are for fuzzy, intuitionistic fuzzy, and neutrosophic degree of appurtenance, respectively.

Furthermore, the degree of contradiction (dissimilarity) between any two attribute values of the same attribute is a function given by

$$c : A_i \times A_i \to P([0,1]^j), \quad 1 \leq i \leq n,\ j = 1, 2, 3.$$

For any two attribute values γ_1 and γ_2 of the same attribute, it is denoted by $c(\gamma_1, \gamma_2)$ and satisfies the following axioms:

$$c(\gamma_1, \gamma_1) = 0,$$
$$c(\gamma_1, \gamma_2) = c(\gamma_2, \gamma_1).$$

Subsequently, $(X, \mathcal{A}, C, d, c)$ is called a plithogenic hypersoft set (PHSS) [22]. For an n-tuple $(\gamma_1, \gamma_2, \ldots, \gamma_n) \in C$, $\gamma_i \in A_i$, $1 \leq i \leq n$, a plithogenic hypersoft set $F : C \rightarrow P(\mathcal{U})$ is mathematically written as

$$F(\{\gamma_1, \gamma_2, \ldots, \gamma_n\}) = \{x(d_x(\gamma_1), d_x(\gamma_2), \ldots, d_x(\gamma_n)), \quad x \in X\}.$$

Remark 2. *Plithogenic hypersoft set is a generalization of crisp hypersoft set, fuzzy hypersoft set, intuitionistic fuzzy hypersoft set, and neutrosophic hypersoft set.*

2.5. Illustrative Example

Let $\mathcal{U} = \{m_1, m_2, m_3, \ldots, m_{10}\}$ be a universe containing mobile phones. A person wants to buy a mobile phone for which the mobile phones under consideration (alternatives) are contained in $X \subseteq \mathcal{U}$, given by

$$X = \{m_2, m_3, m_5, m_8\}.$$

The characteristics or attributes of the mobile phones belong to the set $\mathcal{A} = \{a_1, a_2, a_3, a_4\}$, such that

a_1 = Processor power,
a_2 = RAM,
a_3 = Front camera resolution,
a_4 = Screen size in inches.

The attribute values of a_1, a_2, a_3, a_4 are contained in the sets A_1, A_2, A_3, A_4 given below.

$A_1 = \{$dual-core, quad-core, octa-core$\}$,
$A_2 = \{$2GB, 4GB, 8GB, 16GB$\}$,
$A_3 = \{$2MP, 5MP, 8MP, 16MP$\}$,
$A_4 = \{4, 4.5, 5, 5.5, 6\}$.

1. *Soft set*

Consider $\mathcal{B} = \{a_2, a_3\} \subseteq \mathcal{A}$. Afterwards, a soft set $(\mathcal{F}, \mathcal{B})$, defined by the mapping $\mathcal{F} : \mathcal{B} \rightarrow P(\mathcal{U})$, is given by

$$(\mathcal{F}, \mathcal{B}) = \{(a_2, \{m_2, m_5\}), (a_3, \{m_2, m_3, m_8\})\}$$

Element-wise, it may be written as

$$\mathcal{F}_{\mathcal{B}}(a_2) = \{m_2, m_5\}, \quad \mathcal{F}_{\mathcal{B}}(a_3) = \{m_2, m_3, m_8\}.$$

2. *Hypersoft set*
Let $C = A_1 \times A_2 \times A_3 \times A_4$. Then, a hypersoft set over $\mathcal{U}$ is a function $f : C \rightarrow P(\mathcal{U})$. For an element (octa-core, 8GB, 16MP, 5.5) $\in C$, it is given by

$$f(\{\text{octa-core, 8GB, 16MP, 5.5}\}) = \{m_5, m_8\}$$

3. *Plithogenic hypersoft set*

For the same tuple (octa-core, 8GB, 16MP, 5.5) $\in C$, a plithogenic hypersoft set $F : C \to P(\mathcal{U})$ is given by

$$F(\{\text{octa-core, 8GB, 16MP, 5.5}\}) = \{m_5\,(d_{m_5}(\text{octa-core}), d_{m_5}(\text{8GB}), d_{m_5}(\text{16MP}), d_{m_5}(5.5)),\\ m_8\,(d_{m_8}(\text{octa-core}), d_{m_8}(\text{8GB}), d_{m_8}(\text{16MP}), d_{m_8}(5.5))\},$$

where $d_{m_5}(\gamma)$ stands for the degree of appurtenance of the attribute value $\gamma \in$ (octa-core, 8GB, 16MP, 5.5) to the element $m_5 \in X$. A similar meaning applies to $d_{m_8}(\gamma)$.

3. The Four Classifications of PHSS

In this section, we propose the four different classifications of PHSS that are based on the number of attributes chosen for application and the characteristics of alternatives under consideration or that of the attribute value degree of appurtenance function. The same example from Section 2 is considered to each classification for a practical understanding. Figure 1 shows a diagram for these classifications.

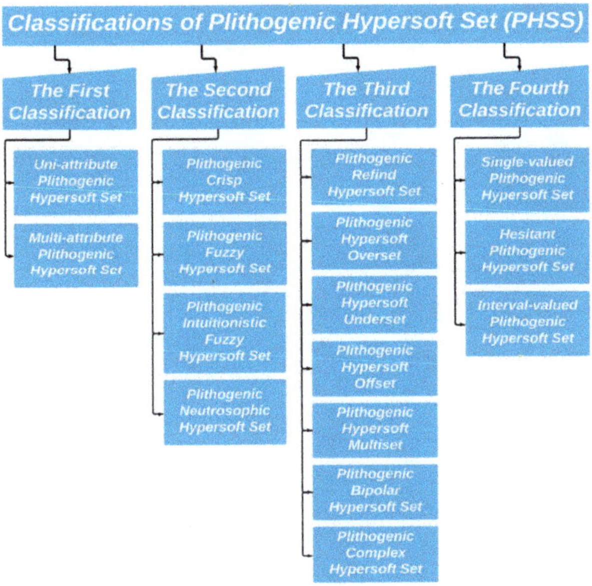

Figure 1. Flowchart of four classifications of plithogenic hypersoft sets (PHSS).

3.1. *The First Classification*

This classification is based on the number of attributes that are chosen by the decision makers for application.

3.1.1. Uni-Attribute Plithogenic Hypersoft Set

Let $\alpha \in \mathcal{A}$ be an attribute required by the experts for application purpose and the attribute values of α belong to the finite discrete set $Y = \{y_1, y_2, \ldots, y_m\}$, $m \geq 1$. Hence, the degree of appurtenance function is given by

$$d : X \times Y \to P([0,1]^j), \quad \forall\, x \in X,$$

such that $d(x,y) \subseteq [0,1]^j$, where $P([0,1]^j)$ denotes the power set of $[0,1]^j$ and $j = 1,2,3$ stands for fuzzy, intuitionistic fuzzy, or neutrosophic degree of appurtenance, respectively.

The contradiction degree function between any two attribute values of α, is given by

$$c : Y \times Y \to P([0,1]^j), \quad \forall\, y \in Y, \quad j = 1,2,3.$$

For any two attribute values $y_1, y_2 \in Y$, it is denoted by $c(y_1, y_2)$ and the following properties hold:

$$c(y_1, y_1) = 0,$$
$$c(y_1, y_2) = c(y_2, y_1).$$

Subsequently, (X, α, Y, d, c) is termed as a uni-attribute plithogenic hypersoft set. For an attribute value $y \in Y$, a uni-attribute plithogenic hypersoft set $F : Y \to P(\mathcal{U})$ is mathematically written as

$$F(y) = \{x(d_x(y)) : x \in X\}.$$

3.1.2. Multi-Attribute Plithogenic Hypersoft Set

Consider a subset $\mathcal{B}$ of $\mathcal{A}$, consisting of all attributes that were chosen by the experts, given by

$$\mathcal{B} = \{b_1, b_2, \ldots, b_m\},\ m > 1.$$

Let the attribute values of $b_1, b_2, \ldots, b_m$ belong to the sets $B_1, B_2, \ldots, B_m$, respectively, and

$$Y_m = B_1 \times B_2 \times \ldots \times B_m.$$

Afterwards, the appurtenance degree function is

$$d : X \times Y_m \to P([0,1]^j), \quad \forall \quad x \in X,$$

such that $d(x,y) \subseteq [0,1]^j$, $j = 1,2,3$. In this case, the contradiction degree function is given by

$$c : B_i \times B_i \to P([0,1]^j), \quad 1 \le i \le m,\ j = 1,2,3.$$

The degree of contradiction between any two attribute values y_1 and y_2, is denoted by $c(y_1, y_2)$ and it satisfies the following axioms:

$$c(y_1, y_1) = 0,$$
$$c(y_1, y_2) = c(y_2, y_1).$$

Subsequently, $(X, \mathcal{B}, Y_m, d, c)$ is called a multi-attribute plithogenic hypersoft set. For an m-tuple $(y_1, y_2, \ldots, y_m) \in Y_m$, $y_i \in B_i$, $1 \le i \le m$, a multi-attribute plithogenic hypersoft set $F : Y_m \to P(\mathcal{U})$ is mathematically written as

$$F(\{y_1, y_2, \ldots, y_m\}) = \{x(d_x(y_1), d_x(y_2), \ldots, d_x(y_m)), x \in X\}.$$

Example 1. *Consider the previous example in which $\mathcal{U} = \{m_1, m_2, m_3, \ldots, m_{10}\}$ and $X \subseteq \mathcal{U}$ is given by $X = \{m_2, m_3, m_5, m_8\}$. The attributes belong to the set $\mathcal{A} = \{a_1, a_2, a_3, a_4\}$, such that*

a_1 = *Processor power,*
a_2 = *RAM,*
a_3 = *Front camera resolution,*
a_4 = *Screen size in inches.*

The attribute values of a_1, a_2, a_3, a_4 are contained in the sets A_1, A_2, A_3, A_4 given below:

$A_1 = \{dual\text{-}core, quad\text{-}core, octa\text{-}core\},$
$A_2 = \{2GB, 4GB, 8GB, 16GB\},$
$A_3 = \{2MP, 5MP, 8MP, 16MP\},$
$A_4 = \{4, 4.5, 5, 5.5, 6\}.$

1. *Uni-attribute plithogenic hypersoft set*

Consider the most demanding feature of a mobile phone given by the attribute a_3 that stands for front camera resolution. The set of attribute values of a_3 is $A_3 = \{2\text{MP}, 5\text{MP}, 8\text{MP}, 16\text{MP}\}$. Then, the uni-attribute plithogenic hypersoft set $F : A_3 \to P(\mathcal{U})$ is given by

$$F(\gamma) = \{x(d_x(\gamma)), \forall\, \gamma \in A_3,\ x \in X\},$$

where $d_x(\gamma)$ denotes the degree of appurtenance of $x \in X$, to the set X, w.r.t. the attribute value $\gamma \in A_3$. For an attribute value 16MP $\in A_3$, we have

$$F(16\text{MP}) = \{m_5(d_{m_5}(16\text{MP})), m_8(d_{m_8}(16\text{MP}))\},$$

2. *Multi-attribute plithogenic hypersoft set*

Let $\mathcal{B} = \{a_3, a_4\}$ be the set of attributes required by the customer. Therefore, we need A_3 and A_4 given by

$$A_3 = \{2\text{MP}, 5\text{MP}, 8\text{MP}, 16\text{MP}\},$$
$$A_4 = \{4, 4.5, 5, 5.5, 6\}.$$

Suppose that the customer is interested to buy a mobile phone with specific requirements of 16MP front camera with 5.5 inch screen size. In this case, we take $(16\text{MP}, 5.5) \in A_3 \times A_4$ and a multi-attribute plithogenic hypersoft set $F : A_3 \times A_4 \to P(\mathcal{U})$ is given by

$$F(\{16\text{MP}, 5.5\}) = \{m_5\left(d_{m_5}(16\text{MP}), d_{m_5}(5.5)\right), m_8\left(d_{m_8}(16\text{MP}), d_{m_8}(5.5)\right)\},$$

where $d_{m_5}(\gamma)$ stands for the degree of appurtenance of m_5 to the set X w.r.t. the attribute value $\gamma \in (16\text{MP}, 5.5)$.

3.2. The Second Classification

This classification is based on the nature of the attribute value degree of appurtenance that may be crisp, fuzzy, intuitionistic fuzzy, or neutrosophic degree of appurtenance.

3.2.1. Plithogenic Crisp Hypersoft Set

A plithogenic hypersoft set X is crisp if the appurtenance degree $d_x(\gamma)$ of each member $x \in X$, w.r.t. each attribute value γ, is crisp, i.e., $d_x(\gamma)$ is either 0 or 1.

3.2.2. Plithogenic Fuzzy Hypersoft Set

If the appurtenance degree $d_x(\gamma)$ of each member $x \in X$, w.r.t. each attribute value γ, is fuzzy, then it is called the plithogenic fuzzy hypersoft set. Mathematically, $d_x(\gamma) \in P([0,1])$.

3.2.3. Plithogenic Intuitionistic Fuzzy Hypersoft Set

If the attribute value appurtenance degree $d_x(\gamma)$ of each $x \in X$, w.r.t. each attribute value, is intuitionistic fuzzy degree, then it is called the plithogenic intuitionistic fuzzy hypersoft set. Mathematically, it is written as $d_x(\gamma) \in P([0,1]^2)$.

3.2.4. Plithogenic Neutrosophic Hypersoft Set

A plithogenic hypersoft set X is called plithogenic neutrosophic hypersoft set if $d_x(\gamma) \in P([0,1]^3)$.

Example 2. *For (octa-core, 8GB, 16MP, 5.5)* $\in C$, *we have the following results:*

1. *Plithogenic crisp hypersoft set*

$$F(\{\text{octa-core, 8GB, 16MP, 5.5}\}) = \{m_5(1,1,1,1), m_8(1,1,1,1)\}.$$

2. *Plithogenic fuzzy hypersoft set*

$$F(\{\text{octa-core, 8GB, 16MP, 5.5}\}) = \{m_5(0.9,0.2,1,0.75), m_8(0.5,0.5,0.25,0.9)\}.$$

3. *Plithogenic intuitionistic fuzzy hypersoft set*

$$F(\{\text{octa-core, 8GB, 16MP, 5.5}\}) = \{m_5\left((0.9,0.1),(0.2,0.6),(1,0),(0.75,0.1)\right), m_8((0.5,0.25),(0.5,0.5),(0.25,0.1),(0.9,0))\}.$$

4. *Plithogenic neutrosophic hypersoft set*

$$F(\{\text{octa-core, 8GB, 16MP, 5.5}\}) = \{m_5((0.9,0.7,0.1),(0.2,0.3,0.6),(1,0.25,0),(0.75,0.3,0.1)), m_8((0.5,1,0.25),(0.5,0.9,0.5),(0.25,0.7,0.1),(0.9,0.8,0))\}.$$

3.3. *The Third Classification*

This classification is based on the properties of attribute values and degree of appurtenance function.

3.3.1. Plithogenic Refined Hypersoft Set

Let $(X, \mathcal{A}, C, d, c)$ be a plithogenic hypersoft set and A denote the set of attribute values of an attribute a. If an attribute value $\gamma \in A$ of the attribute a is subdivided or split into at least two or more attribute sub-values $\gamma_1, \gamma_2, \gamma_3, \ldots \in A$, such that the attribute sub-value degree of appurtenance function $d(x, \gamma_i) \in P([0,1]^j)$, for $i = 1,2,3,\ldots$ and $j = 1,2,3$ for fuzzy, intuitionistic fuzzy, neutrosophic degree of appurtenance, respectively, then X is called a refined plithogenic hypersoft set. It is represented as $(X_r, \mathcal{A}, C, d, c)$.

3.3.2. Plithogenic Hypersoft Overset

If the degree of appurtenance of any element $x \in X$ w.r.t. any attribute value $\gamma \in A$ of an attribute a is greater than 1, i.e., $d(x,\gamma) > 1$, then X is called a plithogenic hypersoft overset. It is represented as $(X_o, \mathcal{A}, C, d, c)$.

3.3.3. Plithogenic Hypersoft Underset

If the degree of appurtenance of any element $x \in X$ w.r.t. any attribute value $\gamma \in A$ of an attribute a less than 0, i.e., $d(x,\gamma) < 0$, then X is called a plithogenic hypersoft underset. It is represented as $(X_u, \mathcal{A}, C, d, c)$.

3.3.4. Plithogenic Hypersoft Offset

A plithogenic hypersoft set $(X, \mathcal{A}, C, d, c)$ is called a plithogenic hypersoft offset if it is both an overset and an underset. Mathematically, if $d(x_1, \gamma_1) > 1$ and $d(x_2, \gamma_2) < 0$ for the same or different attribute values $\gamma_1, \gamma_2 \in A$ that correspond to the same or different members $x_1, x_2 \in X$, then $(X_{\text{off}}, \mathcal{A}, C, d, c)$ is a plithogenic hypersoft offset.

3.3.5. Plithogenic Hypersoft Multiset

If an element $x \in X$ repeats itself into the set X with same plithogenic components given by

$$x(c_1, c_2, \ldots, c_n), x(c_1, c_2, \ldots, c_n),$$

or with different plithogenic components given by

$$x(c_1, c_2, \ldots, c_n), x(d_1, d_2, \ldots, d_n),$$

then $(X_n, \mathcal{A}, C, d, c)$ is called a plithogenic hypersoft multiset.

3.3.6. Plithogenic Bipolar Hypersoft Set

If the attribute value appurtenance degree function is given by

$$d : X \times C \to P([-1,0]^j) \times P([0,1]^j), \quad \forall \quad x \in X,$$

where $j = 1,2,3$, then, $(X_b, \mathcal{A}, C, d, c)$ is called plithogenic bipolar hypersoft set. It may be noted that, for an attribute value γ, $d(x,\gamma)$ allots a negative degree of appurtenance in $[-1,0]$ and a positive degree of appurtenance in $[0,1]$ to each element $x \in X$ with respect to each attribute value γ.

Remark 3. *The concept of plithogenic bipolar hypersoft set can be extended to plithogenic tripolar hypersoft set and so on up to plithogenic multipolar hypersoft set.*

3.3.7. Plithogenic Complex Hypersoft Set

If for any $x \in X$, the attribute value appurtenance degree function, with respect to any attribute value γ, is given by

$$d : X \times C \to P([0,1]^j) \times P([0,1]^j),\ j = 1,2,3,$$

such that $d(x,\gamma)$ is a complex number of the form $c_1.e^{ic_2}$, where c_1 (amplitude) and c_2 (phase) are subsets of $[0,1]$, then $(X_{com}, \mathcal{A}, C, d, c)$ is called a plithogenic complex hypersoft set.

Example 3. *Consider the same example of choosing a suitable mobile phone from the set* $X = \{m_2, m_3, m_5, m_8\}$. *The attributes are* a_1, a_2, a_3, a_4, *whose attribute values are contained in the sets* A_1, A_2, A_3, A_4.

1. *Plithogenic refined hypersoft set*

Consider an attribute a_4 = screen size in inches whose attribute values belong to the set $A_4 = \{4, 4.5, 5, 5.5, 6\}$. A refinement of A_4 is given by

$$A_4 = \{4, 4.5, 4.7, 5, 5.5, 5.8, 6\},$$

such that for all $x \in X$,

$$d(x,\gamma) \in P([0,1]^j), \forall\, \gamma \in A_4.$$

Therefore, a plithogenic refined hypersoft set $F_r : A_4 \to P(\mathcal{U})$ is given by

$$\begin{aligned} F_r(\{4, 4.5, 4.7, 5, 5.5, 5.8, 6\}) = \{ & m_5(d_{m_5}(4), d_{m_5}(4.5), d_{m_5}(4.7), d_{m_5}(5), d_{m_5}(5.5), d_{m_5}(5.8), d_{m_5}(6)), \\ & m_8(d_{m_8}(4), d_{m_8}(4.5), d_{m_8}(4.7), d_{m_8}(5), d_{m_8}(5.5), d_{m_8}(5.8), d_{m_8}(6))\}. \end{aligned}$$

2. *Plithogenic hypersoft overset*

Let each attribute value has a single-valued fuzzy degree of appurtenance to all the elements of X. Subsequently, for (octa-core, 8GB, 16MP, 5.5) $\in C$, a plithogenic hypersoft overset $F_o : C \to P(\mathcal{U})$ is given by

$$F_o(\{\text{octa-core, 8GB, 16MP, 5.5}\}) = \{m_5(0.9, 0.2, 1.3, 0.75), m_8(0.5, 0.5, 0.25, 0.9)\}.$$

It may be noted that $d_{m_5}(16\text{MP}) > 1$.

3. *Plithogenic hypersoft underset*

A plithogenic hypersoft underset defined by the function $F_u : C \to P(\mathcal{U})$ is given by

$$F_u(\{\text{octa-core, 8GB, 16MP, 5.5}\}) = \{m_5(0.9, 0.2, -0.3, 0.75), m_8(0.5, 0.5, 0.25, 0.9)\}.$$

It may be noted that $d_{m_5}(16\text{MP}) < 0$.

4. *Plithogenic hypersoft offset*

A plithogenic hypersoft offset is a function $F_{\text{off}} : C \to P(\mathcal{U})$, as given by

$$F_{\text{off}}(\{\text{octa-core, 8GB, 16MP, 5.5}\}) = \{m_5(0.9, 0.2, -0.3, 0.75), m_8(0.5, 1.5, 0.25, 0.9)\}.$$

Note that $d_{m_5}(16\text{MP}) < 0$ and $d_{m_8}(8\text{GB}) > 1$.

5. *Plithogenic hypersoft multiset*

A plithogenic hypersoft multiset $F_m : C \to P(\mathcal{U})$ is given by

$$F_m(\{\text{octa-core, 8GB, 16MP, 5.5}\}) = \{m_5(0.9, 0.2, 0.3, 0.75), m_5(0.7, 0.1, 0.9, 1), m_8(0.5, 0.5, 0.25, 0.9)\}.$$

It should be noted that the element m_5 repeats itself with different plithogenic components.

6. *Plithogenic bipolar hypersoft set*

A plithogenic bipolar hypersoft set $F_2 : C \to P(\mathcal{U})$ is given by

$$\begin{aligned} F_2(\{\text{octa-core, 8GB, 16MP, 5.5}\}) = &\{m_5(\{-0.1, 0.9\}, \{-1, 0.2\}, \{-0.9, 0.3\}, \{-0.5, 1\}), \\ &m_8(\{-0.5, 0\}, \{-0.9, 1\}, \{-0.2, 0.2\}, \{-1, 0.8\})\}. \end{aligned}$$

7. *Plithogenic complex hypersoft set*

A plithogenic complex hypersoft set $F_{\text{com}} : C \to P(\mathcal{U})$ is given by

$$\begin{aligned} F_{\text{com}}(\{\text{octa-core, 8GB, 16MP, 5.5}\}) = &\{m_5(0.9e^{0.5i}, 0.2e^{0.9i}, 0.3e^{0.25i}, 0.75e^{i}), \\ &m_8(0.5e^{0.5i}, e^{0.3i}, 0.25e^{0.75i}, 0.9e^{0.1i})\}. \end{aligned}$$

3.4. The Fourth Classification

The attribute value degree of appurtenance may be a single crisp value in $[0, 1]$, a finite discrete set or an interval value in $[0, 1]$. Therefore, we have the following classification of PHSS.

3.4.1. Single-Valued Plithogenic Hypersoft Set

A plithogenic hypersoft set is called a single-valued plithogenic hypersoft set if the attribute value appurtenance degree is a single number in $[0, 1]$.

3.4.2. Hesitant Plithogenic Hypersoft Set

If the attribute value degree of appurtenance is a finite discrete set of the form $\{m_1, m_2, \ldots, m_i\}$, $1 \leq i < \infty$, included in $[0,1]$, then such a plithogenic hypersoft set is called a hesitant plithogenic hypersoft set.

3.4.3. Interval-Valued Plithogenic Hypersoft Set

A plithogenic hypersoft set is known as an interval-valued plithogenic hypersoft set if the attribute value appurtenance degree function is an interval value in $[0,1]$. The interval value may be an open, closed, or semi open interval.

Example 4. *For (octa-core, 8GB, 16MP, 5.5) $\in C$, with each attribute value having fuzzy degree of appurtenance, we have the following results:*

1. *Single-valued plithogenic hypersoft set*

$$F(\{\text{octa-core, 8GB, 16MP, 5.5}\}) = \{m_5(0.9, 0.2, 1, 0.75), m_8(0.5, 0.5, 0.25, 0.9)\}.$$

Each attribute value is assigned a single value in $[0,1]$ as a degree of appurtenance to m_5 and m_8.

2. *Hesitant plithogenic hypersoft set*

$$\begin{aligned} F(\{\text{octa-core, 8GB, 16MP, 5.5}\}) =& \{m_5(\{0.9, 0.75\}, \{0.2, 0.7\}, \{1, 0.9\}, \{0.75, 0.5\}), \\ & m_8(\{0.5, 0.1\}, \{0.5, 0.9\}, \{0.25, 0\}, \{0.9, 1\})\}. \end{aligned}$$

3. *Interval-valued plithogenic hypersoft set*

$$\begin{aligned} F(\{\text{octa-core, 8GB, 16MP, 5.5}\}) =& \{m_5([0.25, 0.75], [0.2, 0.6], [0.1, 0.9], [0.75, 1]), \\ & m_8([0.5, 0.6], [0.3, 0.9], [0.25, 0.8], [0.9, 1])\}. \end{aligned}$$

Each attribute value has an interval value degree of appurtenance in $[0,1]$ to each element m_5 and m_8.

4. The Proposed PHSS-Based TOPSIS with Application to a Parking Problem

In this section, we use the concept of PHSS in order to construct a novel MCDM method, called PHSS-based TOPSIS, in which we extend TOPSIS based on PHSS under fuzzy neutrosophic environment. Moreover, a parking spot choice problem is constructed in order to employ the newly developed PHSS-based TOPSIS to prove its validity and efficiency. Two different sets of alternatives are considered for the application and a comparison is performed with fuzzy TOPSIS in both cases.

4.1. Proposed PHSS-Based TOPSIS Algorithm

Let $\mathcal{U}$ be a non-empty universal set, and let $X \subseteq \mathcal{U}$ be the set of alternatives under consideration, given by $X = \{x_1, x_2, \ldots, x_m\}$. Let $C = A_1 \times A_2 \times \ldots \times A_n$, where $n \geq 1$ and A_i is the set of all attribute values of the attribute a_i, $i = 1, 2, 3, \ldots, n$. Each attribute value γ has a corresponding appurtenance degree $d(x, \gamma)$ of the member $x \in X$, in accordance with some given condition or criteria. Our aim is to choose the best alternative out of the alternative set X. The construction steps for the proposed PHSS-based TOPSIS are as follows:

S1: Choose an ordered tuple $(\gamma_1, \gamma_2, \ldots, \gamma_n) \in C$ and construct a matrix of order $n \times m$, whose entries are the neutrosophic degree of appurtenance of each attribute value γ, with respect to each alternative $x \in X$ under consideration.

S2: Employ the newly developed plithogenic accuracy function A_p, to each element of the matrix obtained in S1, in order to convert each element into a single crisp value, as follows:

$$A_p(T_\gamma, I_\gamma, F_\gamma) = \frac{T_\gamma + I_\gamma + F_\gamma}{3} + \frac{T_{\gamma_d} + I_{\gamma_d} + F_{\gamma_d}}{3} \times c_F(\gamma, \gamma_d), \quad (1)$$

where $T_\gamma, I_\gamma, F_\gamma$ represent the membership, indeterminacy, and non-membership degrees of appurtenance of the attribute value γ to the set X, and $T_{\gamma_d}, I_{\gamma_d}, F_{\gamma_d}$ stand for the membership, indeterminacy, and non-membership degrees of corresponding dominant attribute value, whereas $c_F(\gamma, \gamma_d)$ denotes the fuzzy degree of contradiction between an attribute value γ and its corresponding dominant attribute value γ_d. This gives us the plithogenic accuracy matrix.

S3: Apply the transpose on the plithogenic accuracy matrix to obtain the plithogenic decision matrix $M_p = [m_{ij}]_{m\times n}$ of alternatives versus criteria.

S4: A plithogenic normalized decision matrix $N_p = [y_{ij}]_{m\times n}$ is constructed, which represents the relative performance of alternatives and whose elements are calculated as follows:

$$y_{ij} = \frac{m_{ij}}{\sqrt{\sum\limits_{i=1}^{m} m_{ij}^2}}, \quad j = 1, 2, 3, \ldots, n.$$

S5: Construct a plithogenic weighted normalized decision matrix $V_p = [v_{ij}]_{m\times n} = N_p W_n$, where $W_n = [w_1 \ w_2 \ \ldots \ w_n]$ is a row matrix of allocated weights w_k assigned to the criteria a_k, $k = 1, 2, 3, \ldots, n$ and $\sum w_k = 1, k = 1, 2, \ldots, n$. Moreover, all of the selection criteria are assigned different weights by the decision maker, depending on their importance in the decision making process.

S6: Determine the plithogenic positive ideal solution V_p^+ and plithogenic negative ideal solution V_p^- by the following formula:

$$V_p^+ = \left\{ \max_{i=1}^{m}(v_{ij}) \text{ if } a_j \in \text{benefit criteria}, \ \min_{i=1}^{m}(v_{ij}) \text{ if } a_j \in \text{cost criteria}, j = 1, 2, 3, \ldots, n \right\},$$

$$V_p^- = \left\{ \min_{i=1}^{m}(v_{ij}) \text{ if } a_j \in \text{benefit criteria}, \ \max_{i=1}^{m}(v_{ij}) \text{ if } a_j \in \text{cost criteria}, j = 1, 2, 3, \ldots, n \right\}.$$

S7: Calculate plithogenic positive distance S_i^+ and plithogenic negative distance S_i^- of each alternative from V_p^+ and V_p^-, respectively, while using the following formulas:

$$S_i^+ = \sqrt{\sum_{j=1}^{n} (v_{ij} - v_i^+)^2}, \quad i = 1, 2, 3, \ldots, m,$$

$$S_i^- = \sqrt{\sum_{j=1}^{n} (v_{ij} - v_i^-)^2}, \quad i = 1, 2, 3, \ldots, m.$$

S8: Calculate the relative closeness coefficient C_i of each alternative by the following expression:

$$C_i = \frac{S_i^-}{S_i^+ + S_i^-}, \quad i = 1, 2, 3, \ldots, m.$$

S9: The highest value from $\{C_1, C_2, \ldots, C_m\}$ belongs to the most suitable alternative. Similarly, the lowest value gives us the worst alternative.

4.2. Parking Spot Choice Problem

Based on the proposed method, a parking spot choice problem is constructed. Parking a vehicle at some suitable parking spot is an interesting real life MCDM problem. A number of questions arises in mind, for instance, how much will the parking fee be, how far is it, will it be an open or covered area, how many traffic signals will be on the way, etc. Thus, it becomes a challenging task in the presence of so many considerable criteria. This task is formulated in the form of a mathematical model in order to apply the proposed technique to choose the most suitable parking spot. Consider a person at a particular location on the road, who wants to park his car at a suitable parking place. Keeping in mind the person's various preferences, a few nearby available parking spots are considered, having different specifications in terms of parking fee, distance between the person's location and each parking spot, the number of signals between the car and the parking spot, and traffic density on the way between the car and the parking spot. Figure 2 shows the location of car to be parked at a suitable parking spot.

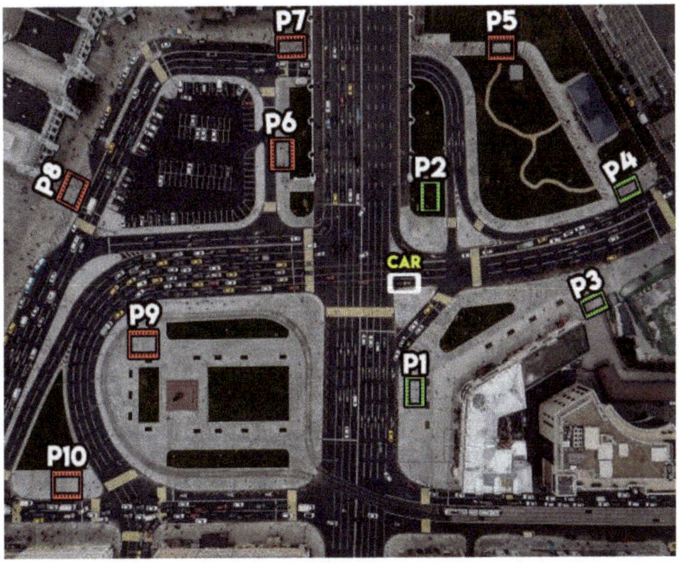

Figure 2. A real life parking spot choice problem.

Let $\mathcal{U}$ be a plithogenic universe of discourse consisting of all parking spots in the surrounding area, where

$$\mathcal{U} = \{P_1, P_2, P_3, \ldots, P_{10}\}.$$

The attributes of the parking spots, chosen for the decision, are a_1, a_2, a_3, a_4 given below:

a_1 = Parking fee,
a_2 = Distance between car and parking spot,
a_3 = Number of traffic signals between car and parking spot,
a_4 = Traffic density on the way between car and parking spot.

The attribute values of a_1, a_2, a_3, a_4 belong to the sets A_1, A_2, A_3, A_4, respectively.

$A_1 = \{\text{low fee } (f_1), \text{ medium fee } (f_2), \text{ high fee } (f_3)\}$,
$A_2 = \{\text{very near } (r_1), \text{ almost near } (r_2), \text{ near } (r_3), \text{ almost far } (r_4), \text{ far } (r_5), \text{ very far } (r_6)\}$,
$A_3 = \{\text{one signal } (s_1), \text{ two signals } (s_2)\}$,

$A_4 = \{\text{low } (d_1), \text{high } (d_2), \text{very high } (d_3)\}$.

The dominant attribute values of a_1, a_2, a_3, a_4 are chosen to be f_1, r_1, s_1 and d_1, respectively, and the single-valued fuzzy degree of contradiction between the dominant attribute value and all other attribute values is given below.

$$c_F(f_1, f_2) = \frac{1}{3}, \; c_F(f_1, f_3) = \frac{2}{3},$$
$$c_F(r_1, r_2) = \frac{1}{6}, \; c_F(r_1, r_3) = \frac{2}{6}, \; c_F(r_1, r_4) = \frac{3}{6}, \; c_F(r_1, r_5) = \frac{4}{6}, \; c_F(r_1, r_6) = \frac{5}{6},$$
$$c_F(s_1, s_2) = \frac{1}{2},$$
$$c_F(d_1, d_2) = \frac{1}{3}, \; c_F(d_1, d_3) = \frac{2}{3}.$$

Two different sets of alternatives are considered for the application of PHSS-based TOPSIS, along with a comparison with fuzzy TOPSIS in each case.

4.2.1. Case 1

In this case, the parking spots under consideration (alternatives) are contained in the set $X \subseteq \mathcal{U}$, given by

$$X = \{P_1, P_2, P_3, P_4\}.$$

The neutrosophic degree of appurtenance of each attribute value corresponding to each alternative P_1, P_2, P_3, P_4 is given in Table 1.

Let $C = A_1 \times A_2 \times A_3 \times A_4$ and consider an element $(f_2, r_1, s_2, d_1) \in C$ for which the corresponding matrix that was obtained from Table 1 is given below:

$$\begin{bmatrix} (0.7,0.9,0.1) & (0.6,0.5,0.2) & (0.2,0.3,0.6) & (0.7,0.9,0.3) \\ (0.8,0.1,0.7) & (0.9,0.4,0.5) & (0.9,0.4,0.0) & (0.8,0.4,0.2) \\ (1.0,0.8,0.6) & (0.7,0.5,0.5) & (0.4,0.4,0.7) & (0.6,0.5,0.7) \\ (0.1,0.2,1.0) & (0.3,1.0,0.6) & (0.7,0.9,0.2) & (0.9,0.7,0.5) \end{bmatrix} \tag{2}$$

Table 1. Degree of appurtenance of each attribute value w.r.t. to each alternative.

Sr.	Variables	P_1	P_2	P_3	P_4
1	f_1	(0.5, 0.1, 0.3)	(0.5, 0.0, 0.7)	(0.1, 0.4, 0.5)	(0.2, 0.1, 0.6)
2	f_2	(0.7, 0.9, 0.1)	(0.6, 0.5, 0.2)	(0.2, 0.3, 0.6)	(0.7, 0.9, 0.3)
3	f_3	(0.5, 0.5, 0.1)	(0.0, 0.1, 0.5)	(0.1, 0.1, 0.9)	(0.5, 0.7, 0.2)
4	r_1	(0.8, 0.1, 0.7)	(0.9, 0.4, 0.5)	(0.9, 0.4, 0.0)	(0.8, 0.4, 0.2)
5	r_2	(0.9, 0.3, 0.2)	(0.6, 0.1, 0.0)	(0.5, 0.2, 0.4)	(0.9, 0.1, 0.4)
6	r_3	(0.9, 0.1, 0.3)	(0.8, 0.3, 0.1)	(0.6, 0.0, 0.6)	(0.2, 0.2, 0.5)
7	r_4	(0.8, 0.3, 0.2)	(1.0, 0.1, 0.5)	(0.8, 0.5, 0.1)	(0.7, 0.3, 0.6)
8	r_5	(1.0, 0.3, 0.2)	(1.0, 0.3, 0.2)	(0.8, 0.2, 0.8)	(0.6, 0.5, 0.6)
9	r_6	(0.8, 0.1, 0.0)	(0.6, 0.8, 0.5)	(0.9, 0.7, 0.1)	(0.4, 0.8, 0.7)
10	s_1	(0.0, 0.5, 0.5)	(0.4, 0.1, 0.6)	(0.2, 0.2, 0.7)	(0.8, 0.3, 0.4)
11	s_2	(1.0, 0.8, 0.6)	(0.7, 0.5, 0.5)	(0.4, 0.4, 0.7)	(0.6, 0.5, 0.7)
12	d_1	(0.1, 0.2, 1.0)	(0.3, 1.0, 0.6)	(0.7, 0.9, 0.2)	(0.9, 0.7, 0.5)
13	d_2	(0.1, 0.4, 0.8)	(0.2, 0.2, 0.8)	(0.2, 0.6, 0.3)	(0.2, 0.8, 0.5)
14	d_3	(0.5, 0.6, 0.9)	(0.9, 0.6, 0.3)	(0.9, 0.7, 0.5)	(0.6, 0.7, 0.6)

This MCDM problem is solved by the proposed PHSS-based TOPSIS and fuzzy TOPSIS, as follows:

A. Application of PHSS-based TOPSIS for Case 1

Apply the plithogenic accuracy function (1) to the matrix (2) in order to obtain the plithogenic accuracy matrix given by:

$$\begin{bmatrix} 0.6667 & 0.5667 & 0.4778 & 0.7333 \\ 0.5333 & 0.6000 & 0.4333 & 0.4667 \\ 0.9667 & 0.7500 & 0.6833 & 0.8500 \\ 0.4333 & 0.6333 & 0.6000 & 0.7000 \end{bmatrix}.$$

The plithogenic decision matrix M_p is constructed by taking the transpose of the plithogenic accuracy matrix. It is a square matrix of order 4, given by

$$M_p = \begin{bmatrix} 0.6667 & 0.5333 & 0.9667 & 0.4333 \\ 0.5667 & 0.6000 & 0.7500 & 0.6333 \\ 0.4778 & 0.4333 & 0.6833 & 0.6000 \\ 0.7333 & 0.4667 & 0.8500 & 0.7000 \end{bmatrix}$$

A corresponding table, as shown in Table 2, of alternatives versus criteria may also be drawn to see the situation in a clear way.

Table 2. Alternatives versus criteria table.

Al/Cr	f_2	r_1	s_2	d_1
P_1	0.6667	0.5333	0.9667	0.4333
P_2	0.5667	0.6000	0.7500	0.6333
P_3	0.4778	0.4333	0.6833	0.6000
P_4	0.7333	0.4667	0.8500	0.7000

A plithogenic normalized decision matrix N_p is obtained as:

$$N_p = \begin{bmatrix} 0.5387 & 0.5205 & 0.5898 & 0.3612 \\ 0.4579 & 0.5855 & 0.4576 & 0.5280 \\ 0.3861 & 0.4229 & 0.4169 & 0.5002 \\ 0.5925 & 0.4555 & 0.5186 & 0.5836 \end{bmatrix}$$

A weighted normalized matrix W_4 is constructed as:

$$W_4 = [\,0.4, 0.22, 0.15, 0.23\,], \tag{3}$$

whereas the plithogenic weighted normalized decision matrix $V_p = [v_{ij}]_{4\times 4}$ is given, as follows:

$$V_p = \begin{bmatrix} 0.2155 & 0.1145 & 0.0885 & 0.0831 \\ 0.1832 & 0.1288 & 0.0686 & 0.1214 \\ 0.1544 & 0.0930 & 0.0625 & 0.1150 \\ 0.2370 & 0.1002 & 0.0778 & 0.1342 \end{bmatrix}$$

The plithogenic positive ideal solution V_p^+ and plithogenic negative ideal solution V_p^- are determined, as follows:

$$V_p^+ = \{0.1544, 0.0930, 0.0625, 0.0831\},$$
$$V_p^- = \{0.2370, 0.1288, 0.0885, 0.1342\}.$$

The plithogenic distance of each alternative from the V_p^+ and V_p^-, respectively, is determined as:

$$S^+ = \begin{bmatrix} 0.0697 \\ 0.0601 \\ 0.0320 \\ 0.0986 \end{bmatrix}, \quad S^- = \begin{bmatrix} 0.0573 \\ 0.0588 \\ 0.0956 \\ 0.0305 \end{bmatrix}.$$

The relative closeness coefficient $C_i, i = 1,2,3,4$, of each alternative is computed as:

$$C_1 = 0.4511,$$
$$C_2 = 0.4944,$$
$$C_3 = 0.7494,$$
$$C_4 = 0.2366.$$

The highest value corresponds to the most suitable alternative. Since $C_3 = 0.7494$ is the maximum value and it corresponds to P_3, therefore, the most suitable parking spot is P_3. The Table 3 is constructed to rank all alternatives under consideration.

Table 3. PHSS-based TOPSIS ranking table.

	S_i^+	S_i^-	C_i	**Ranking**
P_1	0.0697	0.0573	0.4511	3
P_2	0.0601	0.0588	0.4944	2
P_3	0.0320	0.0956	0.7494	1
P_4	0.0986	0.0305	0.2366	4

A bar graph presented in Figure 3 is given, in which all alternatives P_1, P_2, P_3, P_4 are ranked by PHSS-based TOPSIS.

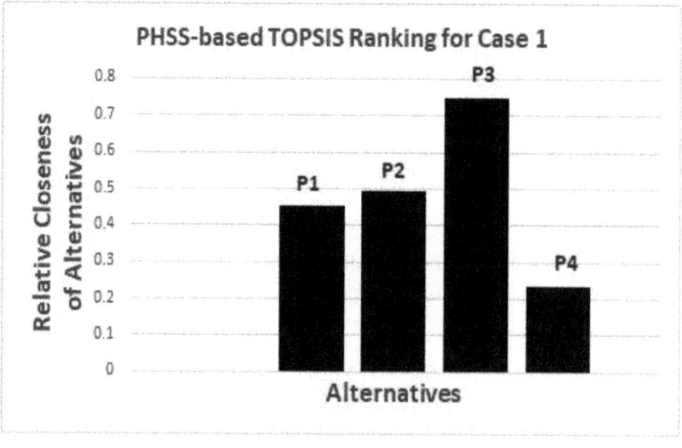

Figure 3. Ranking of Parking Spots by PHSS-based TOPSIS for Case 1.

It is evident that the parking spot P_3 is the most suitable place to park the car while P_4 is not a good choice for parking based on the selection criteria.

B. Application of Fuzzy TOPSIS for Case 1

In order to see the implementation of fuzzy TOPSIS [32–34] for the current scenario of the parking problem, we apply the average operator [27,35] to each element of the matrix 2 and take the transpose of the resulting matrix in order to obtain the decision matrix given by:

$$M = \begin{bmatrix} 0.5667 & 0.5333 & 0.8000 & 0.4333 \\ 0.4333 & 0.6000 & 0.5667 & 0.6333 \\ 0.3667 & 0.4333 & 0.5000 & 0.6000 \\ 0.6333 & 0.4667 & 0.6000 & 0.7000 \end{bmatrix}$$

Applying the fuzzy TOPSIS to the decision matrix M, along with the same weights given in matrix (3), we obtain the values of positive distance S^+, negative distance S^-, relative closeness C_i and ranking of each alternative, as given in Table 4.

Table 4. Fuzzy TOPSIS ranking table.

	S_i^+	S_i^-	C_i	**Ranking**
P_1	0.0888	0.0592	0.4000	3
P_2	0.0591	0.0841	0.5872	2
P_3	0.0320	0.1176	0.7863	1
P_4	0.1170	0.0373	0.2417	4

A bar graph in Figure 4 is given in which all alternatives P_1, P_2, P_3, P_4 are ranked by Fuzzy TOPSIS. A comparison is shown in Table 5, in which it can be seen that the result obtained by the proposed PHSS-based TOPSIS is aligned with that of fuzzy TOPSIS.

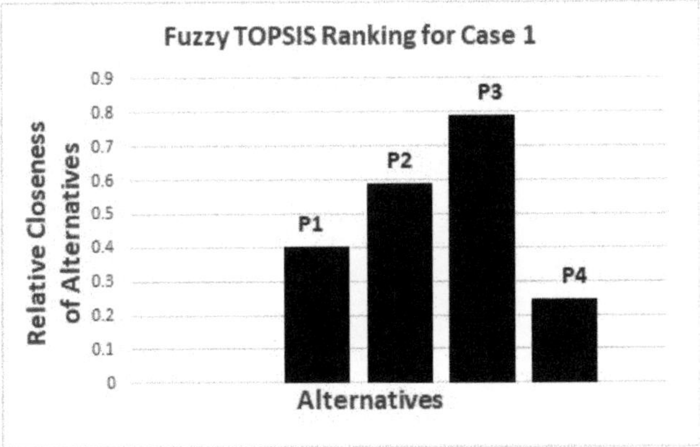

Figure 4. Ranking of Parking Spots by Fuzzy TOPSIS for Case 1.

Table 5. Comparison analysis for case 1.

Sr.	Parkings	PHSS-Based TOPSIS Ranking	Fuzzy TOPSIS Ranking
1	P_1	3rd	3rd
2	P_2	2nd	2nd
3	P_3	1st	1st
4	P_4	4th	4th

It is observed in Table 5 that the results obtained by both methods coincide in terms of the ranking of each alternative, but differ in the values of the relative closeness of each alternative. It is due to the nature of the MCDM problem in hand in which each alternative needs to be evaluated against each attribute value possessing a neutrosophic degree of appurtenance w.r.t. each alternative and a contradiction degree is defined between each attribute value and its corresponding dominant attribute value to be taken into consideration in the decision process. In such a case, the proposed PHSS-based TOPSIS produces a more reliable relative closeness of each alternative, as it can been seen in the parking spot choice problem that was chosen for the study. Therefore, it is worth noting that the proposed PHSS-based TOPSIS can be regarded as a generalization of fuzzy TOPSIS [32], because the fuzzy TOPSIS cannot be directly applied to MCDM problems in which the attribute values have a neutrosophic degree of appurtenance with respect to each alternative. In the case of the parking problem, fuzzy TOPSIS is applied after applying simple average operator to the neutrosophic elements of the matrix (2). However, it does not takes into account the degree of contradiction between the attribute values, which is the limitation of fuzzy TOPSIS. This concern is precisely addressed by the proposed PHSS-based TOPSIS.

4.2.2. Case 2

In this case, the set of parking spots under consideration is given by

$$X = \{P_1, P_5, P_6, P_7\}.$$

The neutrosophic degree of appurtenance of each attribute value that corresponds to each alternative of $\{P_1, P_5, P_6, P_7\}$ is given in Table 6.

Table 6. Degree of appurtenance of each attribute value w.r.t each alternative.

Sr.	Variables	P_1	P_5	P_6	P_7
1	f_1	(0.5, 0.1, 0.3)	(0.6, 0.6, 0.8)	(0.7, 0.2, 0.4)	(0.9, 0.5, 0.2)
2	f_2	(0.7, 0.9, 0.1)	(0.8, 0.8, 0.5)	(0.4, 0.4, 0.7)	(0.7, 0.2, 0.1)
3	f_3	(0.5, 0.5, 0.1)	(0.4, 0.2, 0.5)	(1.0, 0.5, 0.9)	(1.0, 0.7, 0.6)
4	r_1	(0.8, 0.1, 0.7)	(0.9, 0.5, 0.2)	(0.5, 0 , 0.9)	(0.8, 0.6, 0.1)
5	r_2	(0.9, 0.3, 0.2)	(0.5, 0.4, 0.2)	(0.7, 0.5, 0.4)	(0.9, 0.6, 0.8)
6	r_3	(0.9, 0.1, 0.3)	(0.5, 0.7, 0.3)	(0.9, 1.0, 0.6)	(0.2, 0 , 1.0)
7	r_4	(0.8, 0.3, 0.2)	(1.0, 0.2, 1.0)	(1.0, 0.5, 0.7)	(0.8, 0.8, 0.9)
8	r_5	(0.2, 0.3, 0.9)	(1.0, 0.1, 0.8)	(0.4, 0.6, 0.8)	(0.8, 0.6, 0.6)
9	r_6	(0.5, 0.7, 0.5)	(0.8, 0.2, 0.0)	(0.6, 0.3, 0.7)	(0.0, 0.9, 0.9)
10	s_1	(0 , 0.5, 0.5)	(0.8, 0.4, 0.6)	(0.9, 0.2, 0.2)	(0.8, 0.4, 0.7)
11	s_2	(1.0, 0.8, 0.6)	(0.7, 1.0, 0.2)	(0.2, 0.4, 0.7)	(0.9, 0 , 1.0)
12	d_1	(0.1, 0.2, 1.0)	(1.0, 0.4, 0.3)	(0.7, 0.5, 0.6)	(0.8, 0.5, 0.7)
13	d_2	(0.1, 0.4, 0.8)	(0.7, 1.0, 0.8)	(0.6, 0.6, 1.0)	(1.0, 0.8, 0.8)
14	d_3	(0.5, 0.6, 0.9)	(1.0, 0.6, 0.5)	(1.0, 1.0, 0.5)	(1.0, 0.5, 0.8)

Let $C = A_1 \times A_2 \times A_3 \times A_4$ and consider an element $(f_2, r_1, s_2, d_1) \in C$ for which the corresponding matrix obtained from Table 6, is given below:

$$\begin{bmatrix} (0.7,0.9,0.1) & (0.8,0.8,0.5) & (0.4,0.4,0.7) & (0.7,0.2,0.1) \\ (0.8,0.1,0.7) & (0.9,0.5,0.2) & (0.5,0.0,0.9) & (0.8,0.6,0.1) \\ (1.0,0.8,0.6) & (0.7,1.0,0.2) & (0.2,1.0,0.7) & (0.9,0.0,1.0) \\ (0.1,0.2,1.0) & (1.0,0.4,0.3) & (0.7,0.5,0.6) & (0.8,0.5,0.7) \end{bmatrix} \quad (4)$$

The proposed PHSS-based TOPSIS and fuzzy TOPSIS are employed, as follows:

A. Application of PHSS-Based TOPSIS for Case 2

The plithogenic accuracy matrix in this case is given by

$$\begin{bmatrix} 0.6667 & 0.9222 & 0.6444 & 0.5111 \\ 0.5333 & 0.6000 & 0.4333 & 0.4667 \\ 0.9667 & 0.9333 & 0.6500 & 0.9500 \\ 0.4333 & 0.6333 & 0.6000 & 0.7000 \end{bmatrix}.$$

Plithogenic decision matrix M_p is given by

$$M_p = \begin{bmatrix} 0.6667 & 0.5333 & 0.9667 & 0.4333 \\ 0.9222 & 0.6000 & 0.9333 & 0.6333 \\ 0.6444 & 0.4333 & 0.6500 & 0.6000 \\ 0.5111 & 0.4667 & 0.9500 & 0.7000 \end{bmatrix}$$

A plithogenic normalized decision matrix N_p is then constructed as:

$$N_p = \begin{bmatrix} 0.4748 & 0.5205 & 0.5464 & 0.3612 \\ 0.6568 & 0.5855 & 0.5275 & 0.5280 \\ 0.4590 & 0.4229 & 0.3674 & 0.5002 \\ 0.3640 & 0.4555 & 0.5369 & 0.5836 \end{bmatrix}$$

The plithogenic weighted normalized decision matrix V_p is given, as follows:

$$V_p = \begin{bmatrix} 0.1899 & 0.1145 & 0.0820 & 0.0831 \\ 0.2627 & 0.1288 & 0.0791 & 0.1214 \\ 0.1836 & 0.0930 & 0.0551 & 0.1150 \\ 0.1456 & 0.1002 & 0.0805 & 0.1342 \end{bmatrix}$$

The plithogenic positive ideal solution V_p^+ and plithogenic negative ideal solution V_p^- are determined, such that

$$V_p^+ = \{0.1456, 0.0930, 0.0551, 0.0831\},$$
$$V_p^- = \{0.2627, 0.1288, 0.0820, 0.1342\}.$$

The plithogenic positive distance S^+, plithogenic negative distance S^-, relative closeness C_i, and ranking of each alternative is shown in Table 7.

Table 7. PHSS-based TOPSIS ranking table.

	S_i^+	S_i^-	C_i	Ranking
P_1	0.0561	0.0901	0.6163	3
P_5	0.1306	0.0131	0.0912	4
P_6	0.0496	0.0929	0.6518	2
P_7	0.0576	0.1206	0.6769	1

A graphical representation of the ranking of all alternatives obtained by PHSS-based TOPSIS, is shown in Figure 5.

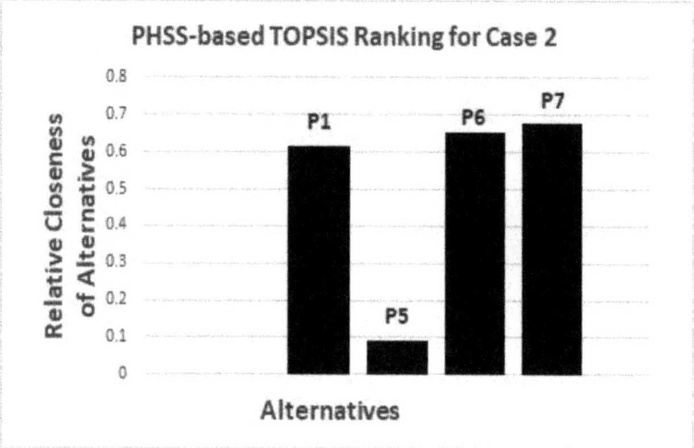

Figure 5. Ranking of Parking Spots by PHSS-based TOPSIS for Case 2.

It can be seen that the parking spot P_7 is the most suitable alternative in the light of chosen criteria.

B. Application of Fuzzy TOPSIS for Case 2

In this case, the decision matrix M for the implementation of fuzzy TOPSIS is given by

$$M = \begin{bmatrix} 0.5667 & 0.5333 & 0.8000 & 0.4333 \\ 0.7000 & 0.6000 & 0.6333 & 0.6333 \\ 0.5000 & 0.4333 & 0.6333 & 0.6000 \\ 0.3333 & 0.4667 & 0.6333 & 0.7000 \end{bmatrix}$$

By implementing the fuzzy TOPSIS to the matrix M, with the same weights given in (3), the values of positive distance S^+, negative distance S^-, relative closeness C_i, and ranking of each alternative are shown in Table 8.

Table 8. Fuzzy TOPSIS ranking table.

	S_i^+	S_i^-	C_i	Ranking
P_1	0.0908	0.0724	0.4439	3
P_5	0.1453	0.0224	0.1337	4
P_6	0.0694	0.0863	0.5543	2
P_7	0.0516	0.1397	0.7301	1

The ranking of all alternatives can also been visualized as a bar graph in Figure 6, in which all alternatives P_1, P_5, P_6, P_7 are ranked by Fuzzy TOPSIS.

The most suitable parking spot obtained by fuzzy TOPSIS is also P_7.

A comparison of rankings obtained by PHSS-based TOPSIS and fuzzy TOPSIS is shown in Table 9 for case 2.

It may be noted that similar results are obtained in case 2, with the help of proposed PHSS-based TOPSIS and fuzzy TOPSIS with exactly same ranking of each alternative, but with a considerably different values of the relative closeness of each alternative as shown in Table 9. Therefore, it is accomplished that the results that were obtained by the PHSS-based TOPSIS are valid and more reliable and PHSS-based TOPSIS can be regarded as the generalization of fuzzy TOPSIS on the basis of the study conducted in the article.

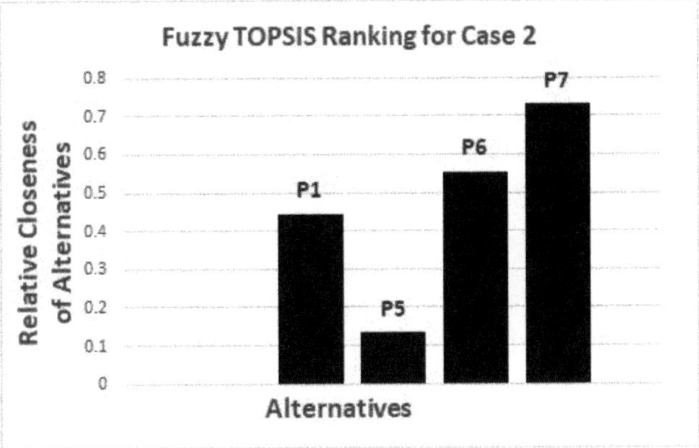

Figure 6. Ranking of Parking Spots by Fuzzy TOPSIS for Case 2.

Table 9. Comparison analysis for case 2.

Sr.	Parkings	PHSS-Based TOPSIS Ranking	Fuzzy TOPSIS Ranking
1	P_1	3rd	3rd
2	P_5	4th	4th
3	P_6	2nd	2nd
4	P_7	1st	1st

5. Conclusions

It has always been a challenging task to deal with real life MCDM problems, due to the involvement of many complexities and uncertainties. In particular, some real life MCDM problems are designed in a way that the given attributes need to be further decomposed into two or more attribute values such that each alternative is then required to be evaluated against each attribute value in order to perform a detailed analysis to reach a fair conclusion. To deal with such situations, a novel PHSS-based TOPSIS is proposed in the present study, and it is applied to a MCDM parking problem with different choices of the set of alternatives and a comparison with fuzzy TOPSIS is done to prove the validity and efficiency of the proposed method. All of the results are quite promising and graphically depicted for a clear understanding. Moreover, the algorithm of the proposed method is produced in MATLAB in order to broaden the scope of the study to other research areas, including graph theory, machine learning, pattern recognition, etc.

Author Contributions: Conceptualization, M.S. and U.A.; methodology, M.R.A. and M.S.; validation, M.R.A. and U.A.; formal analysis, M.R.A. and U.A.; investigation, M.S. and M.-S.Y.; writing—original draft preparation, M.R.A. and M.S.; writing—review and editing, U.A. and M.-S.Y.; supervision, M.S. and M.-S.Y.; funding acquisition, M.-S.Y.. All authors have read and agreed to the published version of the manuscript.

Funding: The APC was funded in part by the Ministry of Science and technology (MOST) of Taiwan under Grant MOST-109-2118-M-033-001-.

Conflicts of Interest: The authors declare no conflict of interest.

References

1. Zadeh, L.A. Fuzzy sets. *Inform. Control* **1965**, *8*, 338–353.
2. Yang, M.S.; Hung, W.L.; Chang-Chien, S.J. On a similarity measure between LR-type fuzzy numbers and its application to database acquisition. *Int. J. Intell. Syst.* **2005**, *20*, 1001–1016.
3. Meng, F.; Tang, J.; Fujita, H. Consistency-based algorithms for decision-making with interval fuzzy preference relations. *IEEE Trans. Fuzzy Syst.* **2019**, *27*, 2052–2066.

4. Ruiz-Garca, G.; Hagras, H.; Pomares, H.; Ruiz, I.R. Toward a fuzzy logic system based on general forms of interval type-2 fuzzy sets. *IEEE Trans. Fuzzy Syst.* **2019**, *27*, 2381–2395.
5. Ullah, K.; Hassan, N.; Mahmood, T.; Jan, N.; Hassan, M. Evaluation of investment policy based on multi-attribute decision-making using interval valued T-spherical fuzzy aggregation operators. *Symmetry* **2019**, *11*, 357.
6. Atanassov, K.T. Intuitionistic fuzzy sets. *Fuzzy Sets Syst.* **1986**, *20*, 87–96.
7. Hwang, C.M.; Yang, M.S.; Hung, W.L. New similarity measures of intuitionistic fuzzy sets based on the Jaccard index with its application to clustering. *Int. J. Intell. Syst.* **2018**, *33*, 1672–1688.
8. Garg, H.; Kumar, K. Linguistic interval-valued atanassov intuitionistic fuzzy sets and their applications to group decision making problems. *IEEE Trans. Fuzzy Syst.* **2019**, *27*, 2302–2311.
9. Roy, J.; Das, S.; Kar, S.; Pamucar, D. An extension of the CODAS approach using interval-valued intuitionistic fuzzy set for sustainable material selection in construction projects with incomplete weight information. *Symmetry* **2019**, *11*, 393.
10. Yang, M.S.; Hussian, Z.; Ali, M. Belief and plausibility measures on intuitionistic fuzzy sets with construction of belief-plausibility TOPSIS. *Complexity* **2020**, 1–12, doi:10.1155/2020/7849686.
11. Smarandache, F. *A Unifying Field in Logics: Neutrosophic Logic. Neutrosophy, Neutrosophic Set, Probability, and Statistics*, 2nd ed.; American Research Press: Rehoboth, DE, USA, 2000.
12. Majumdar, P.; Samanta, S.K. On similarity and entropy of neutrosophic sets. *J. Intell. Fuzzy Syst.* **2014**, *26*, 1245–1252.
13. Li, X.; Zhang, X.; Park, C. Generalized interval neutrosophic Choquet aggregation operators and their applications. *Symmetry* **2018**, *10*, 85.
14. Abdel-Basset, M.; Mohamed, M. A novel and powerful framework based on neutrosophic sets to aid patients with cancer. *Future Gener. Comput. Syst.* **2019**, *98*, 144–153.
15. Vasantha, W.B.; Kandasamy, I.; Smarandache, F. Neutrosophic components semigroups and multiset neutrosophic components semigroups. *Symmetry* **2020**, *12*, 818.
16. Molodtsov, D. Soft set theory-first results. *Comput. Math. Appl.* **1999**, *37*, 19–31.
17. Maji, P.K.; Biswas, R.; Roy, A.R. Soft set theory. *Comput. Math. Appl.* **2003**, *45*, 555–562.
18. Ali, M.I.; Feng, F.; Liu, X.; Min, W.K.; Shabir, M. On some new operations in soft set theory. *Comput. Math. Appl.* **2009**, *57*, 1547–1553.
19. Inthumathi, V.; Chitra, V.; Jayasree, S. The role of operators on soft sets in decision making problems. *Int. J. Comput. Appl. Math.* **2017**, *12*, 899–910.
20. Feng, G.; Guo, X. A novel approach to fuzzy soft set-based group decision-making. *Complexity* **2018**, *2018*, 2501489.
21. Biswas, B.; Bhattacharyya, S.; Chakrabarti, A.; Dey, K.N.; Platos, J.; Snasel, V. Colonoscopy contrast-enhanced by intuitionistic fuzzy soft sets for polyp cancer localization. *Appl. Soft Comput.* **2020**, *95*, 106492.
22. Smarandache, F. Extension of soft set to hypersoft set, and then to plithogenic hypersoft set. *Neutrosophic Sets Syst.* **2018**, *22*, 168–170.
23. Saeed, M.; Ahsan, M.; Siddique, M.K.; Ahmad, M.R. A study of the fundamentals of hypersoft set theory. *Int. Sci. Eng. Res.* **2020**, *11*, 320–329.
24. Smarandache, F. *Plithogeny, Plithogenic Set, Logic, Probobility, and Statistics*; Pons: Brussels, Belgium, 2017.
25. Smarandache, F. Plithogenic set, an extension of crisp, fuzzy, intuitionistic fuzzy, and neutrosophic sets-revisited. *Neutrosophic Sets Syst.* **2028**, *21*, 153–166.
26. Collan, M.; Luukka, P. Evaluating R & D projects as investments by using an overall ranking from four new fuzzy similarity measure-based TOPSIS variants. *IEEE Trans. Fuzzy Syst.* **2014**, *22*, 505–515.
27. Saqlain, M.; Saeed, M.; Ahmad, M.R.; Smarandache, F. Generalization of TOPSIS for neutrosophic hypersoft set using accuracy function and its application. *Neutrosophic Sets Syst.* **2019**, *27*, 131–137.
28. Khalil, A.M.; Cao, D.; Azzam, A.; Smarandache, F.; Alharbi, W.R. Combination of the single-valued neutrosophic fuzzy set and the soft set with applications in decision-making. *Symmetry* **2020**, *12*, 1361.
29. Abdel-Basset, M.; Mohamed, R. A novel plithogenic TOPSIS-CRITIC model for sustainable supply chain risk management. *J. Clean. Prod.* **2020**, *247*, 119586.
30. Schumann, A. p-Adic valued logical calculi in simulation of the slime mould behaviour. *J. Appl. Non-Class. Logics* **2015**, *25*, 125–139.

31. Schumann, A. p-Adic multiple-validity and p-adic valued logical calculi. *J. Mult.-Valued Log. Soft Comput.* **2007**, *13*, 29–60.
32. Yang, T.; Hung, C.C. Multiple-attribute decision making methods for plant layout design problem. *Robot. Comput.-Integr.* **2007**, *23*, 126–137.
33. Kabir, G.; Hasin, M. Comparative analysis of topsis and fuzzy topsis for the evaluation of travel website service quality. *Int. Qual. Res.* **2012**, *6*, 169–185.
34. Zhang, L.; Zhan, J.; Yao, Y. Intuitionistic fuzzy TOPSIS method based on CVPIFRS models: An application to biomedical problems. *Inf. Sci.* **2020**, *517*, 315–339.
35. Yager, R.R. The power average operator. *IEEE Trans. Syst. Man Cybern. Syst. Hum.* **2001**, *31*, 724–731.

symmetry

MDPI

Article

Topological Structures via Interval-Valued Neutrosophic Crisp Sets

Dongsik Jo [1], S. Saleh [2], Jeong-Gon Lee [3,*], Kul Hur [3] and Chen Xueyou [4]

1 Department of Digital Contents Engineering, Wonkwang University, Iksan 54538, Korea; dongsik1005@wku.ac.kr
2 Department of Mathematics, Faculty of Education-Zabid, Hodeidah University, Hodeidah P.O. Box 3114, Yemen; s_wosabi@hoduniv.net.ye
3 Division of Applied Mathematics, Wonkwang University, Iksan-Si 54538, Korea; kulhur@wku.ac.kr
4 School of Mathematics, Shandong University of Technology, Zibo 255049, China; xueyou-chen@163.com
* Correspondence: jukolee@wku.ac.kr

Received: 13 November 2020; Accepted: 7 December 2020; Published: 10 December 2020

check for updates

Abstract: In this paper, we introduce the new notion of interval-valued neutrosophic crisp sets providing a tool for approximating undefinable or complex concepts in real world. First, we deal with some of its algebraic structures. We also define an interval-valued neutrosophic crisp (vanishing) point and obtain some of its properties. Next, we define an interval-valued neutrosophic crisp topology, base (subbase), neighborhood, and interior (closure), respectively and investigate some of each property, and give some examples. Finally, we define an interval-valued neutrosophic crisp continuity and quotient topology and study some of each property.

Keywords: interval-valued neutrosophic crisp set; interval-valued neutrosophic crisp (vanishing) point; interval-valued neutrosophic crisp topological space; interval-valued neutrosophic crisp base (subbase); interval-valued neutrosophic crisp neighborhood; interval-valued neutrosophic crisp closure (interior); interval-valued neutrosophic crisp continuity; interval-valued neutrosophic crisp quotient topology

MSC: 54A10

1. Introduction

Numerous mathematicians have been trying to find a mathematical expression of the complexation and uncertainty in real world for a long time. For example, Zadeh [1] defined a fuzzy set as a generalization of a classical set in 1965. Zadeh [2] (1975), Pawlak [3] (1982), Atanassov [4] (1983), Atanassov and Gargov [5] (1989), Gau and Buchrer [6] (1993), Smarandache [7] (1998), Molodtsov [8] (1999), Lee [9], Torra [10], Jun et al. [11] (2012), and Lee et al. [12] (2020) introduced the concept of interval-valued fuzzy sets, rough sets, intuitionistic fuzzy sets, interval-valued intuitionistic fuzzy sets, vague sets, neutrosophic sets, soft sets, bipolar fuzzy sets, hessitant fuzzy sets, cubic sets combined by interval-valued fuzzy sets and fuzzy sets, and octahedron sets combined by interval-valued fuzzy sets, intuitionistic fuzzy sets, and fuzzy sets, in turn, in order to solve various complex and uncertain problems.

In 1996, cCoker [13] proposed the concept of an intuitionistic set as the generalization of a classical set and the special case of an intuionistic fuzzy set and he studied topological structures based on intuitionistic sets in [14]. Kim et al. [15] dealt with categorical structures based on intuitionistic sets. They also obtained further properties of intuionistic topology in [16]. In 2014, Salama et al. [17] defined neutrosophic crisp sets as the generalization of classical sets and the special case of neutrosophic sets proposed by Smarandache [7,18,19], and studied some of its properties. Moreover, they dealt with topological structures based on the neutrosophic crisp sets in [17]. Hur et al. [20] investigated categorical structures via neutrosophic crisp sets. Many researchers [21–29] have discussed topological structures via neutrosophic crisp sets. Recently, Kim et al. [30] introduced the concept of an interval-valued set as the generalization of a classical set and the specialization of an interval-valued fuzzy set, and applied it to topological structures.

This paper considers two perspectives. First, we define the interval-valued neutrosophic crisp set, a new concept that combines the interval-valued set and neutrosophic crisp set. As an example, suppose a country conducts a poll during an election that determines the highest head of administration. At this time, the preference for Candidate A is divided into three groups: Favor, neutral, and rejection among its citizens from the viewpoint of neutrosophic crisp set, but the minimum and maximum for each of a favor, neutral, and rejection from the viewpoint of interval-valued neutrosophic crisp set. The group is considered. Then, it is believed that the results of the poll by the new concept are more accurate than those by the neutrosophic crisp set. Thus, this new concept is needed. Second, since the topology can be applied to high dimensional data sets, big data, and computational evaluations (see [31–33], respectively), we study topological structures based on interval-valued neutrosophic crisp sets. In order to accomplish such research, first, we recall some definitions related to intuitionistic sets, interval-valued sets, and neutrosophic crisp sets. Secondly, we introduce the new concept of interval-valued neutrosophic crisp set and obtain some of its algebraic structures, and give some examples. We also define interval-valued neutrosophic crisp points of two types and discuss the characterizations of the inclusion, equality, intersection, and union of interval-valued neutrosophic crisp sets. Thirdly, we define an interval-valued neutrosophic crisp topology, an interval-valued neutrosophic crisp base and subbase, and study some of their properties. Fourthly, we introduce the concepts of interval-valued neutrosophic crisp neighborhoods of two types and find some of their properties. In particular, we prove that there is an IVNCT under the hypothesis satisfying some properties of interval-valued neutrosophic crisp neighborhoods. Moreover, we define an interval-valued neutrosophic crisp interior and closure and deal with some of their properties. In particular, we show that there is a unique IVNCT for interval-valued neutrosophic crisp interior [resp. closure] operators. Finally, we introduce the concepts of interval-valued neutrosophic crisp continuous [resp. open and closed] mappings and quotient topologies and obtain some of their properties.

Throughout this paper, we assume that X, Y are non-empty sets, unless otherwise stated.

2. Preliminaries

In this section, we recall the concept of an intuitionistic set proposed in [13]. We also recall some concepts and results introduced and studied in [30,34,35], respectively.

Definition 1 ([13])**.** *The form $A = (A^{\in}, A^{\notin})$ such that $A^{\in}$, $A^{\notin} \subset X$, and $A^{\in} \cap A^{\notin} = \emptyset$ is called an intuitionistic set (briefly, IS) of X, where $A^{\in}$ [resp. $A^{\notin}$] represents the set of memberships [resp. non-memberships] of elements of X to A. In fact, $A^{\in}$ [resp. $A^{\notin}$] is a subset of X agreeing or approving [resp. refusing or opposing] for a certain opinion, suggestion, or policy.*

The intuitionistic empty set [resp. the intuitionistic whole set] of X, denoted by $\bar{\emptyset}$ [resp. $\bar{X}$], is defined by $\bar{\emptyset} = (\emptyset, X)$ [resp. $\bar{X} = (X, \emptyset)$]. The set of all ISs of X will be denoted by $IS(X)$. It is also clear that for each $A \in IS(X)$, $\chi_A = (\chi_{A^{\in}}, \chi_{A^{\notin}})$ is an intuitionistic fuzzy set in X proposed by Atanassov [4]. Thus we can consider the intuitionistic set A in X as an intuitionistic fuzzy set in X.

Furthermore, we can easily check that for each $A \in IS(X)$, $A^{\in} \cup A^{\notin} \neq X$ (in fact, $A^{\in^c} \cap A^{\notin^c} \neq \emptyset$) in general (see Example 1) but if $A^{\in} \cup A^{\notin} = X$, then $A^{\in^c} \cap A^{\notin^c} = \emptyset$. We denote the family $\{A \in IS(X) : A^{\in} \cup A^{\notin} = X\}$ as $IS^*(X)$.

Example 1. *Let* $X = \{a, b, c, d, e\}$ *be a set and consider the IS A in X given by:*

$$A = (\{a, b, c\}, \{d\}).$$

Then clearly, $A^{\in} \cup A^{\notin} \neq X$. *In fact,* $A^{\in^c} \cap A^{\notin^c} \neq \emptyset$.

For the inclusion, equality, union, and intersection of intuitionistic sets, and the complement of an intuitionistic set, the operations [ ] and <> on $IS(X)$, refer to [13].

Definition 2 ([34,36])**.** *The form* $A = \langle A^T, A^I, A^F \rangle$ *such that* A^T, A^I, $A^F \subset X$ *is called a neutrosophic crisp set (briefly, NCS) in X, where* A^T, A^I, *and* A^F *represent the set of memberships, indeterminacies, and non-memberships respectively of elements of X to A.*

We consider neutrosophic crisp empty [resp. whole] sets of two types in X, denoted by $\emptyset_{1,N}$, $\emptyset_{2,N}$ *[resp.* $X_{1,N}$, $X_{2,N}$*] and defined by (see Remark 1.1.1 in [34]):*

$$\emptyset_{1,N} = \langle \emptyset, \emptyset, X \rangle, \ \emptyset_{2,N} = \langle \emptyset, X, X \rangle \ [\text{resp.} X_{1,N} = \langle X, X, \emptyset \rangle, \ X_{2,N} = \langle X, \emptyset, \emptyset \rangle].$$

We will denote the set of all NCSs in X denoted by $NC(X)$.

It is obvious that $A = \langle A, \emptyset, A^c \rangle \in NC(X)$ for each ordinary subset A of X. Then we can consider an NCS in X as the generalization of an ordinary subset of X. It is also clear that $A = \langle A^{\in}, \emptyset, A^{\notin} \rangle$ is an NCS in X for each $A \in IS(X)$. Thus an NCS in X can be considered as the generalization of an intuitionistic set in X. Furthermore, we can easily see that for each $A \in N(X)$,

$$\chi_A = \langle \chi_{A^T}, \chi_{A^I}, \chi_{A^F} \rangle$$

is a neutrosophic set in X introduced by Salama and Smarandache [7,18,19]. So an NCS is a special case of a neutrosophic set.

Definition 3 ([34])**.** *Let* $A \in NC(X)$. *Then the complement of A, denoted by* $A^{i,c}$ *(*$i = 1, 2$*) and defined by:*

$$A^{1,c} = \left\langle A^F, A^{I^c}, A^T \right\rangle, \ A^{2,c} = \left\langle A^F, A^I, A^T \right\rangle.$$

Definition 4 ([34])**.** *Let* $A, \ B \in NC(X)$. *Then A is said to be:*

(i) *A 1-type subset of B, denoted by* $A \subset_1 B$, *if it satisfies the following conditions:*

$$A^T \subset B^T, \ A^I \subset B^I, \ A^F \supset B^F,$$

(ii) *A 2-type subset of B, denoted by* $A \subset_2 B$, *if it satisfies the following conditions:*

$$A^T \subset B^T, \ A^I \supset B^I, \ A^F \supset B^F.$$

Definition 5 ([34])**.** *Let* $A, \ B \in NC(X)$.

(i) *The i-intersection of A and B, denoted by* $A \cap^i B$ *(*$i = 1, 2$*) and defined by:*

$$A \cap^1 B = \left\langle A^T \cap B^T, A^I \cap B^I, A^F \cup B^F \right\rangle, \ A \cap^2 B = \left\langle A^T \cap B^T, A^I \cup B^I, A^F \cup B^F \right\rangle.$$

(ii) *The i-union of A and B, denoted by* $A \cup^i B$ *(*$i = 1, 2$*) and defined by:*

$$A \cup^1 B = \left\langle A^T \cup B^T, A^I \cup B^I, A^F \cap B^F \right\rangle, \ A \cup^2 B = \left\langle A^T \cup B^T, A^I \cap B^I, A^F \cap B^F \right\rangle.$$

(iii) $[\,]A = \left\langle A^T, A^I, A^{T^c} \right\rangle$, $\langle\,\rangle A = \left\langle A^{F^c}, A^I, A^F \right\rangle$.

The followings are immediate results of Definitions 3, 4, and 5.

Proposition 1 (See Proposition 3.3 in [20] and also compare it with Proposition 3.5 in [15]). *Let $A, B, C \in NC(X)$ and let $i = 1, 2$. Then we have:*

(1) (See Proposition 1.1.1 in [34]) $\varnothing_{i,N} \subset_i A \subset_i X_{i,N}$,
(2) *If $A \subset_i B$ and $B \subset_i C$, then $A \subset_i C$,*
(3) $A \cap^i B \subset_i A$ *and* $A \cap^i B \subset_i B$,
(4) $A \subset_i A \cup^i B$ *and* $B \subset_i A \cup^i B$,
(5) $A \subset_i B$ *if and only if* $A \cap^i B = A$,
(6) $A \subset_i B$ *if and only if* $A \cup^i B = B$.

Proposition 2 (See Proposition 3.4 in [20] and also compare it with Proposition 3.6 in [15]). *Let $A, B, C \in NC(X)$ and let $i = 1, 2$. Then we have:*

(1) (Idempotent laws): $A \cup^i A = A$, $A \cap^i A = A$,
(2) (Commutative laws): $A \cup^i B = B \cup^i A$, $A \cap^i B = B \cap^i A$,
(3) (Associative laws): $A \cup^i (B \cup^i C) = (A \cup^i B) \cup^i C$, $A \cap^i (B \cap^i C) = (A \cap^i B) \cap^i C$,
(4) (Distributive laws): $A \cup^i (B \cap^i C) = (A \cup^i B) \cap^i (A \cup^i C)$,

$$A \cap^i (B \cup^i C) = (A \cap^i B) \cup^i (A \cap^i C),$$

(5) (Absorption laws): $A \cup^i (A \cap^i B) = A$, $A \cap^i (A \cup^i B) = A$,
(6) (DeMorgan's laws): $(A \cup^1 B)^{1,c} = A^{1,c} \cap^2 B^{1,c}$, $(A \cup^1 B)^{2,c} = A^{2,c} \cap^2 B^{2,c}$,

$$(A \cup^2 B)^{1,c} = A^{1,c} \cap^2 B^{1,c},\ (A \cup^2 B)^{2,c} = A^{2,c} \cap^1 B^{2,c},$$
$$(A \cap^1 B)^{1,c} = A^{1,c} \cup^1 B^{1,c},\ (A \cap^1 B)^{2,c} = A^{2,c} \cup^2 B^{2,c},$$
$$(A \cap^2 B)^{1,c} = A^{1,c} \cup^2 B^{1,c},\ (A \cap^2 B)^{2,c} = A^{2,c} \cup^1 B^{2,c},$$

(7) $(A^{i,c})^{i,c} = A$,
(8) *(8a)* $A \cup^i \varnothing_{i,N} = A$, $A \cap^i \varnothing_{i,N} = \varnothing_{i,N}$,
(8b) $A \cup^i X_{i,N} = X_{i,N}$, $A \cap^i X_{i,N} = A$,
(8c) ${X_{i,N}}^{i,c} = \varnothing_{i,N}$, ${\varnothing_{i,N}}^{i,c} = X_{i,N}$,
(8d) $A \cup^i A^{i,c} \neq {X_{i,N}}^{i,c}$, $A \cap^i A^{i,c} \neq \varnothing_{i,N}$, *in general.*

Definition 6 (See [34,37]). *Let $a \in X$. Then the form $a_N = \langle \{a\}, \varnothing, \{a\}^c \rangle$ [resp. $a_{NV} = \langle \varnothing, \{a\}, \{a\}^c \rangle$] is called a neutrisophic crisp [resp. vanishing] point in X.*

We denote the set of all neutrisophic crisp points and all neutrisophic crisp vanishing points in X by $N_P(X)$.

Definition 7 (See [34,37]). *Let $a \in X$ and let $A \in NC(X)$. Then,*

(i) a_N *said to belong to A, denoted by* $a_N \in A$, *if* $a \in A^T$,
(ii) a_{NV} *said to belong to A, denoted by* $a_{NV} \in A$, *if* $a \notin A^F$.

Result 1 ([34], Proposition 1.2.6). Let $A \in NC(X)$. Then,

$$A = A_N \cup^1 A_{NV},$$

where $A_N = \bigcup^1_{a_N \in A} a_N$, $A_{NV} = \bigcup^1_{a_{NV} \in A} a_{NV}$. In fact, $A_N = \left\langle A^T, \varnothing, A^{T^c} \right\rangle$ and $A_{NV} = \langle \varnothing, A^I, A^F \rangle$.

Definition 8 ([30,35]). *The form $[A^-, A^+] = \{B \subset X : A^- \subset B \subset A^+\}$ such that A^-, $A^+ \subset X$ is called an interval-valued sets (briefly, IVS) in X, where A^- [resp. A^+] represents the set of minimum [resp. maximum] memberships of elements of X to A. In fact, A^- [resp. A^+] is a minimum [resp. maximum] subset of X agreeing or approving for a certain opinion, suggestion, or policy.*

$[\varnothing, \varnothing]$ [resp. $[X, X]$] is called the interval-valued empty [resp. whole] set in X and denoted by $\widetilde{\varnothing}$ [resp. $\widetilde{X}$]. The set of all IVSs in X will be denoted by $IVS(X)$.

For any classical subset A of X, $[A, A] \in IVS(X)$ is obvious. Then we can consider an IVS in X as the generalization of a classical subset of X. Also, if $A = [A^-, A^+] \in IVS(X)$, then $\chi_A = [\chi_{A^-}, \chi_{A^+}]$ is an interval-valued fuzzy set in X introduced by Zadeh [2]. Thus an interval-valued fuzzy set can be considered as the generalization of an IVS.

Furthermore, we can easily check that for each $A \in IVS(X)$, $A^- \neq A^+$ (in fact, $A^+ \cap A^{-^c} \neq \emptyset$) in general (see Example 2) but if $A^- = A^+$, then $A^+ \cap A^{-^c} = \emptyset$. We denote the family $\{A \in IVS(X) : A^- = A^+\}$ as $IVS^*(X)$.

Example 2. *Let $X = \{a, b, c, d, e\}$ and consider the IVS A in X given by:*

$$A = [\{a, b\}, \{a,, b, c\}].$$

Then we can easily calculate that $A^- \neq A^+$ and $A^+ \cap A^{-^c} \neq \emptyset$.

For the inclusion, equality, union, and intersection of intuionistic sets, and the complement of an intuitionistic set refer to [30,35].

3. Interval-Valued Neutrosophic Crisp Sets

In this section, we introduce the concept of an interval-valued neutrosophic crisp set combined by a neutrosophic crisp set and an interval-valued set, and obtain some of its properties.

Definition 9. *The form $\langle [A^{T,-}, A^{T,+}], [A^{I,-}, A^{I,+}], [A^{F,-}, A^{F,+}] \rangle$ is called an interval-valued neutrosophic crisp set (briefly, IVNCS) in X, where $[A^{T,-}, A^{T,+}]$, $[A^{I,-}, A^{I,+}]$, $[A^{F,-}, A^{F,+}] \in IVS(X)$.*

In this case, $[A^{T,-}, A^{T,+}]$, $[A^{I,-}, A^{I,+}]$, and $[A^{F,-}, A^{F,+}]$ represent the IVS of memberships, indeterminacies, and non-memberships respectively of elements of X to A.

In particular, an IVNCS is defined as three types below.

An IVNCS $A = \langle [A^{T,-}, A^{T,+}], [A^{I,-}, A^{I,+}], [A^{F,-}, A^{F,+}] \rangle$ in X is said to be of:

(i) Type 1, if it satisfies the following conditions:

$$[A^{T,-}, A^{T,+}] \cap [A^{I,-}, A^{I,+}] = \widetilde{\emptyset},\ [A^{T,-}, A^{T,+}] \cap [A^{F,-}, A^{F,+}] = \widetilde{\emptyset},$$

$$[A^{I,-}, A^{I,+}] \cap [A^{F,-}, A^{F,+}] = \widetilde{\emptyset},$$

equivalently, $A^{T,+} \cap A^{I,+} = \emptyset$, $A^{T,+} \cap A^{F,+} = \emptyset$, $A^{I,+} \cap A^{F,+} = \emptyset$,

(ii) Type 2, if it satisfies the following conditions:

$$[A^{T,-}, A^{T,+}] \cap [A^{I,-}, A^{I,+}] = \widetilde{\emptyset},\ [A^{T,-}, A^{T,+}] \cap [A^{F,-}, A^{F,+}] = \widetilde{\emptyset},$$

$$[A^{I,-}, A^{I,+}] \cap [A^{F,-}, A^{F,+}] = \widetilde{\emptyset},\ [A^{T,-}, A^{T,+}] \cup [A^{I,-}, A^{I,+}] \cup [A^{F,-}, A^{F,+}] = \widetilde{X},$$

equivalently, $A^{T,+} \cap A^{I,+} = \emptyset$, $A^{T,+} \cap A^{F,+} = \emptyset$, $A^{I,+} \cap A^{F,+} = \emptyset$,

$A^{T,-} \cup A^{I,-} \cup A^{F,-} = X$,

(iii) Type 3, if it satisfies the following conditions:

$$[A^{T,-}, A^{T,+}] \cap [A^{I,-}, A^{I,+}] \cap [A^{F,-}, A^{F,+}] = \widetilde{\emptyset},$$

$$[A^{T,-}, A^{T,+}] \cup [A^{I,-}, A^{I,+}] \cup [A^{F,-}, A^{F,+}] = \widetilde{X},$$

equivalently, $A^{T,+} \cap A^{I,+} \cap A^{F,+} = \emptyset$, $A^{T,-} \cup A^{I,-} A^{F,-} = X$.

The set of all IVNCSs of Type 1 [resp. Type 2 and Type 3] in X is denoted by $IVN_1(X)$ [resp. $IVN_2(X)$ and $IVN_3(X)$], and $IVNCS(X) = IVN_1(X) \cup IVN_2(X) \cup IVN_3(X)$, where $IVNCS(X)$ is the set of all IVNCSs in X.

For any classical subset A of X, $\left\langle [A, A], \widetilde{\varnothing}, [A^c, A^c] \right\rangle \in IVNCS(X)$ is clear. Then we can consider an INCS in X can be considered as the generalization of a classical subset of X. Moreover, if $A = \langle [A^{T,-}, A^{T,+}], [A^{I,-}, A^{I,+}], [A^{F,-}, A^{F,+}] \rangle \in IVNCS(X)$, then:

$$\chi_A = ([\chi_{A^{T,-}}, \chi_{A^{T,+}}], [\chi_{A^{I,-}}, \chi_{A^{I,+}}], [\chi_{A^{F,-}}, \chi_{A^{F,+}}])$$

is an interval neutrosophic set in X proposed by Ye [38]. Thus we can consider an IVS as the generalization of an IVNCS.

Remark 1.

(1) $IVN_2(X) \subset IVN_1(X)$, $IVN_2(X) \subset IVN_3(X)$,
(2) $IVN_1(X) \not\subset IVN_2(X)$, $IVN_1(X) \not\subset IVN_3(X)$ *in general,*
(3) $IVN_3(X) \not\subset IVN_1(X)$, $IVN_3(X) \not\subset IVN_2(X)$ *in general.*

Example 3. *Let* $X = \{a, b, c, d, e, f, g, h, i\}$. *Consider two IVNCSs in X given by:*

$$A = \langle [\{a, b, c\}, \{a, b, c, d\}], [\{e\}, \{e, f\}], [\{g, h\}, \{g, h, i\}] \rangle,$$

$$B = \langle [\{a, b, c\}, \{a, b, c\}], [\{a, e, f\}, \{a, e, f\}], [\{g, h, i\}, \{g, h, i\}] \rangle.$$

(i) $[A^{T,-}, A^{T,+}] \cap [A^{I,-}, A^{I,+}] = \widetilde{\varnothing}$, $[A^{T,-}, A^{T,+}] \cap [A^{F,-}, A^{F,+}] = \widetilde{\varnothing}$,
$[A^{I,-}, A^{I,+}] \cap [A^{F,-}, A^{F,+}] = \widetilde{\varnothing}$. *But*
$[A^{T,-}, A^{T,+}] \cup [A^{I,-}, A^{I,+}] \cup [A^{F,-}, A^{F,+}] = [\{a, b, c, d, e, f, g, h\}, X\}] \neq \widetilde{X}$. *Then* $A \in IVN_1(X)$ *but* $A \notin IVN_2(X)$. *Moreover, we have:*

$$[A^{T,-}, A^{T,+}] \cap [A^{I,-}, A^{I,+}] \cap [A^{F,-}, A^{F,+}] = \widetilde{\varnothing}.$$

Thus $A \notin IVN_3(X)$. *So we can confirm that Remark 1 (2) holds.*

(ii) $[B^{T,-}, B^{T,+}] \cap [B^{I,-}, B^{I,+}] \cap [B^{F,-}, B^{F,+}] = \widetilde{\varnothing}$,
$[B^{T,-}, B^{T,+}] \cup BC^{I,-}, B^{I,+}] \cup [B^{F,-}, B^{F,+}] = \widetilde{X}$. *But*
$[B^{T,-}, B^{T,+}] \cap [B^{I,-}, B^{I,+}] = [\{a\}, \{a\}] \neq \widetilde{\varnothing}$.
Then $B \in IVN_3(X)$ *but* $B \notin IVN_1(X)$, $B \notin IVN_2(X)$. *Thus we can confirm that Remark 1 (3) holds.*

Definition 10. *We may define the interval-valued neutrosophic crisp empty sets and the interval-valued neutrosophic crisp whole sets, denoted by* $\varnothing_{i,IVN}$ *and* $X_{i,IVN}$ $(i = 1, 2, 3, 4)$, *respectively as follows:*

(i) $\varnothing_{1,IVN} = \left\langle \widetilde{\varnothing}, \widetilde{\varnothing}, \widetilde{X} \right\rangle$, $\varnothing_{2,IVN} = \left\langle \widetilde{\varnothing}, \widetilde{X}, \widetilde{X} \right\rangle$,
$\varnothing_{3,IVN} = \left\langle \widetilde{\varnothing}, \widetilde{X}, \widetilde{\varnothing} \right\rangle$, $\varnothing_{4,IVN} = \left\langle \widetilde{\varnothing}, \widetilde{\varnothing}, \widetilde{\varnothing} \right\rangle$,

(ii) $X_{1,IVN} = \left\langle \widetilde{X}, \widetilde{X}, \widetilde{\varnothing} \right\rangle$, $X_{2,IVN} = \left\langle \widetilde{X}, \widetilde{\varnothing}, \widetilde{\varnothing} \right\rangle$,
$X_{3,IVN} = \left\langle \widetilde{X}, \widetilde{\varnothing}, \widetilde{X} \right\rangle$, $X_{4,IVN} = \left\langle \widetilde{X}, \widetilde{X}, \widetilde{X} \right\rangle$.

Definition 11. *Let* $A \in IVNCS(X)$. *Then the complements of A, denoted by* $A^{i,c}$ $(i = 1, 2, 3)$, *is an IVNCS in X, respectively as follows:*

$$A^{1,c} = \left\langle [A^{T,-}, A^{T,+}]^c, [A^{I,-}, A^{I,+}]^c, [A^{F,-}, A^{F,+}]^c \right\rangle,$$

$$A^{2,c} = \left\langle [A^{F,-}, A^{F,+}], [A^{I,-}, A^{I,+}], [A^{T,-}, A^{T,+}] \right\rangle,$$

$$A^{3,c} = \left\langle [A^{F,-}, A^{F,+}], [A^{I,-}, A^{I,+}]^c, [A^{T,-}, A^{T,+}] \right\rangle.$$

Example 4. *Let* $A = \langle[\{a,b,c\},\{a,b,c,d\}],[\{e\},\{e,f\}],[\{g,h\},\{g,h,i\}]\rangle$ *be the IVNCS in X given in Example 3. Then we can easily check that:*

$$A^{1,c} =< [\{e,f,g,h,i\},\{d,e,f,g,h,i\}],[\{a,b,c,d,g,h,i\},\{a,b,c,d,f,g,h,i\}],$$
$$[\{a,b,c,d,e,f\},\{a,b,c,d,e,f,i\}]>,$$
$$A^{2,c} = \langle[\{g,h\},\{g,h,i\}],[\{e\},\{e,f\}],[\{a,b,c\},\{a,b,c,d\}]\rangle,$$
$$A^{3,c} =< [\{g,h\},\{g,h,i\}],[\{a,b,c,d,g,h,i\},\{a,b,c,d,f,g,h,i\}],$$
$$[\{a,b,c\},\{a,b,c,d\}]>.$$

Definition 12. *Let* $A,\ B \in IVNCS(X)$*. Then we may define the inclusions between A and B, denoted by* $A \subset_i B\ (i = 1,\ 2)$*, as follows:*

$$A \subset_1 B \text{ iff } [A^{T,-},A^{T,+}] \subset [B^{T,-},B^{T,+}],\ [A^{I,-},A^{I,+}] \subset [B^{I,-},B^{I,+}],$$
$$[A^{F,-},A^{F,+}] \supset [B^{F,-},B^{F,+}],$$
$$A \subset_2 B \text{ iff } [A^{T,-},A^{T,+}] \subset [B^{T,-},B^{T,+}],\ [A^{I,-},A^{I,+}] \supset [B^{I,-},B^{I,+}],$$
$$[A^{F,-},A^{F,+}] \supset [B^{F,-},B^{F,+}].$$

Proposition 3. *For any* $A \in IVNCS(X)$*, the followings hold:*

(1) $\varnothing_{1,IVN} \subset_1 A \subset_1 X_{1,IVN},\ \varnothing_{2,IVN} \subset_2 A \subset_2 X_{2,IVN},$
(2) $\varnothing_{i,IVN} \subset_j \varnothing_{i,IVN},\ X_{i,IVN} \subset_j X_{i,IVN},\ (i = 1,\ 2,\ 3,\ 4,\ j = 1,\ 2).$

Proof. Straightforward. □

Definition 13. *Let* $A,\ B \in IVNCS(X)$*,* $(A_j)_{j\in J} \subset IVNCS(X)$*.*

(i) *The intersection of A and B, denoted by* $A \cap^i B\ (i = 1,\ 2)$*, is an IVNCS in X defined by:*

$$A \cap^1 B =< [A^{T,-},A^{T,+}] \cap [B^{T,-},B^{T,+}],[A^{I,-},A^{I,+}] \cap [B^{I,-},B^{I,+}],$$
$$[A^{F,-},A^{F,+}] \cup [B^{F,-},B^{F,+}]>,$$
$$A \cap^2 B =< [A^{T,-},A^{T,+}] \cap [B^{T,-},B^{T,+}],[A^{I,-},A^{I,+}] \cup [B^{I,-},B^{I,+}],$$
$$[A^{F,-},A^{F,+}] \cup [B^{F,-},B^{F,+}]>.$$

(i′) *The intersection of* $(A_j)_{j\in J}$*, denoted by* $\bigcap^i_{j\in J} A_j\ (i = 1,\ 2)$*, is an IVNCS in X defined by:*

$$\bigcap^1_{j\in J} A_j = \left\langle \bigcap_{j\in J}[A_j^{T,-},A_j^{T,+}], \bigcap_{j\in J}[A_j^{I,-},A_j^{I,+}], \bigcup_{j\in J}[A_j^{F,-},A_j^{F,+}] \right\rangle,$$

$$\bigcap^2_{j\in J} A_j = \left\langle \bigcap_{j\in J}[A_j^{T,-},A_j^{T,+}], \bigcup_{j\in J}[A_j^{I,-},A_j^{I,+}], \bigcup_{j\in J}[A_j^{F,-},A_j^{F,+}] \right\rangle.$$

(ii) *The union of A and B, denoted by* $A \cup^i B\ (i = 1,\ 2)$*, is an IVNCS in X defined by:*

$$A \cup^1 B =< [A^{T,-},A^{T,+}] \cup [B^{T,-},B^{T,+}],[A^{I,-},A^{I,+}] \cup [B^{I,-},B^{I,+}],$$
$$[A^{F,-},A^{F,+}] \cap [B^{F,-},B^{F,+}]>,$$
$$A \cup^2 B =< [A^{T,-},A^{T,+}] \cup [B^{T,-},B^{T,+}],[A^{I,-},A^{I,+}] \cap [B^{I,-},B^{I,+}],$$
$$[A^{F,-},A^{F,+}] \cap [B^{F,-},B^{F,+}]>.$$

(ii′) *The union of* $(A_j)_{j\in J}$*, denoted by* $\bigcup^i_{j\in J} A_j\ (i = 1,\ 2)$*, is an IVNCS in X defined by:*

$$\bigcup^1_{j\in J} A_j = \left\langle \bigcup_{j\in J}[A_j^{T,-},A_j^{T,+}], \bigcup_{j\in J}[A_j^{I,-},A_j^{I,+}], \bigcap_{j\in J}[A_j^{F,-},A_j^{F,+}] \right\rangle,$$

$$\bigcup^2_{j\in J} A_j = \left\langle \bigcup_{j\in J}[A_j^{T,-},A_j^{T,+}], \bigcap_{j\in J}[A_j^{I,-},A_j^{I,+}], \bigcap_{j\in J}[A_j^{F,-},A_j^{F,+}] \right\rangle.$$

(iii) $[\,]A = \langle [A^{T,-}, A^{T,+}], [A^{I,-}, A^{I,+}], [A^{T,-}, A^{T,+}]^c \rangle$.
(iv) $< \ > A = \langle [A^{F,-}, A^{F,+}]^c, [A^{I,-}, A^{I,+}], [A^{F,-}, A^{F,+}] \rangle$.

From Definitions 10–13, we get similar results from Propositions 3.5 and 3.6 in [30].

Proposition 4. *Let* $A,\ B,\ C \in IVNCS(X)$, $i = 1,\ 2$. *Then,*

(1) *If* $A \subset_i B$ *and* $B \subset_i C$, *then* $A \subset_i C$,
(2) $A \subset_i A \cup^i B$ *and* $B \subset_i A \cup^i B$,
(3) $A \cap^i B \subset_i A$ *and* $A \cap^i B \subset_i B$,
(4) $A \subset_i B$ *if and only if* $A \cap^i B = A$,
(5) $A \subset_i B$ *if and only if* $A \cup^i B = B$.

Proposition 5. *Let* X $A,\ B,\ C \in IVNCS(X)$, $(A_j)_{j\in J} \subset IVNCS(X)$, *and let* $i = 1,\ 2$; $k = 1,\ 2,\ 3$. *Then*

(1) *(Idempotent laws)* $A \cup^i A = A,\ A \cap^i A = A$,
(2) *(Commutative laws)* $A \cup^i B = B \cup^i A,\ A \cap^i B = B \cap^i A$,
(3) *(Associative laws)* $A \cup^i (B \cup^i C) = (A \cup^i B) \cup^i C,\ A \cap^i (B \cap^i C) = (A \cap^i B) \cap^i C$,
(4) *(Distributive laws)* $A \cup^i (B \cap^i C) = (A \cup^i B) \cap^i (A \cup^i C)$,

$$A \cap^i (B \cup^i C) = (A \cap^i B) \cup^i (A \cap^i C),$$

(4′) *(Generalized distributive laws)* $(\bigcap^i_{j\in J} A_j) \cup^i A = \bigcap^i_{j\in J}(A_j \cup^i A)$,

$$(\bigcup^i_{j\in J} A_j) \cap^i A = \bigcup^i_{j\in J}(A_j \cap^i A),$$

(5) *(Absorption laws)* $A \cup^i (A \cap^i B) = A,\ A \cap^i (A \cup^i B) = A$,
(6) *(DeMorgan's laws)* $(A \cup^i B)^{k,c} = A^{k,c} \cap^i B^{k,c},\ (A \cap^i B)^{k,c} = A^{k,c} \cup^i B^{k,c}$,
(6′) *(Generalized DeMorgan's laws)* $(\bigcup^i_{j\in J} A_j)^{k,c} = \bigcap^i_{j\in J} A_j^{k,c}$,
(7) $(A^{k,c})^{k,c} = A$,
(8) (8a) $A \cup^i \varnothing_{i,IVN} = A,\ A \cap^i \varnothing_{i,IVN} = \varnothing_{i,IVN}$,
(8b) $A \cup^i X_{i,IVN} = X_{i,IVN},\ A \cap^i X_{i,IVN} = A$,
(8c) $X_{1,IVN}{}^{1,c} = \varnothing_{1,IVN},\ X_{1,IVN}{}^{2,c} = \varnothing_{2,IVN},\ X_{1,IVN}{}^{3,c} = \varnothing_{1,IVN}$,
$X_{2,IVN}{}^{1,c} = \varnothing_{2,IVN},\ X_{2,IVN}{}^{2,c} = \varnothing_{1,IVN},\ X_{2,IVN}{}^{3,c} = \varnothing_{2,IVN}$,
$X_{3,IVN}{}^{1,c} = \varnothing_{3,IVN},\ X_{3,IVN}{}^{2,c} = X_{3,IVN},\ X_{3,IVN}{}^{3,c} = X_{4,IVN}$,
$X_{4,IVN}{}^{1,c} = \varnothing_{4,IVN},\ X_{4,IVN}{}^{2,c} = X_{4,IVN},\ X_{4,IVN}{}^{3,c} = X_{3,IVN}$,
$\varnothing_{1,IVN}{}^{1,c} = X_{1,IVN},\ \varnothing_{1,IVN}{}^{2,c} = X_{2,IVN},\ \varnothing_{1,IVN}{}^{3,c} = X_{1,IVN}$,
$\varnothing_{2,IVN}{}^{1,c} = X_{2,IVN},\ \varnothing_{2,IVN}{}^{2,c} = X_{1,IVN},\ \varnothing_{2,IVN}{}^{3,c} = X_{2,IVN}$,
$\varnothing_{3,IVN}{}^{1,c} = X_{3,IVN},\ \varnothing_{3,IVN}{}^{2,c} = \varnothing_{3,IVN},\ \varnothing_{3,IVN}{}^{3,c} = \varnothing_{4,IVN}$,
$\varnothing_{4,IVN}{}^{1,c} = X_{4,IVN},\ \varnothing_{4,IVN}{}^{2,c} = \varnothing_{4,IVN},\ \varnothing_{4,IVN}{}^{3,c} = \varnothing_{3,IVN}$,
(8d) $A \cup^i A^{k,c} \neq X_{j,IVN},\ A \cap^i A^{k,c} \neq \varnothing_{j,IVN}$ *in general* (see Example 5), *where* $j = 1,\ 2,\ 3,\ 4$.

Example 5. *Consider the IVNCS* A *in* X *given in Example 4. Then,*

$$\begin{aligned}
&A \cap^1 A^{1,c} \\
&= \langle [\{a,b,c\},\{a,b,c,d\}], [\{e\},\{e,f\}], [\{g,h\},\{g,h,i\}] \rangle \\
&\cap^1 < [\{e,f,g,h,i\},\{d,e,f,g,h,i\}], [\{a,b,c,d,g,h,i\},\{a,b,c,d,f,g,h,i\}], \\
&\quad [\{a,b,c,d,e,f\},\{a,b,c,d,e,f,i\}] > \\
&= \langle [\varnothing,\{d\}], [\varnothing,\{f\}], [\{a,b,c,d,e,f,g,h\},X] \rangle \\
&\neq \varnothing_{j,IVN}.
\end{aligned}$$

Similarly, we can check that:

$$A \cup^1 A^{1,c} \neq X_{j,IVN},\ A \cap^1 A^{2,c} \neq \varnothing_{j,IVN},\ A \cup^1 A^{2,c} \neq X_{j,IVN}.$$

Additionally, we can easily check the remainders.

A neighborhood system of a point is very important in a classical topology. Then we propose an interval-valued neutrosophic crisp point to define the concept of an interval-valued neutrosophic crisp neighborhood. Moreover, when we deal with separation axioms in an interval-valued neutrosophic crisp topology, the notion of interval-valued neutrosophic crisp points is used. Then we define it below.

Definition 14. *Let $a \in X$, $A \in IVNCS(X)$. Then the form $\left\langle [\{a\},\{a\}], \widetilde{\emptyset}, [\{a\}^c,\{a\}^c] \right\rangle$ [resp. $\left\langle \widetilde{\emptyset}, [\{a\},\{a\}], [\{a\}^c,\{a\}^c] \right\rangle$] is called an interval-valued neutrosophic [resp. vanishing] point in X and denoted by a_{IVN} [resp. a_{IVNV}]. We will denote the set of all interval-valued neutrosophic points in X as $IVN_P(X)$.*

(i) We say that a_{IVN} belongs to A, denoted by $a_{IVN} \in A$, if $a \in A^{T,+}$.
(ii) We say that a_{IVNV} belongs to A, denoted by $a_{IVNV} \in A$, if $a \notin A^{F,+}$.

Proposition 6. *Let $A \in IVNCS(X)$. Then $A = A_{IVN} \cup^1 A_{IVNV}$,*
where $A_{IVN} = \bigcup^1_{a_{IVN} \in A} a_{IVN}$, $A_{IVNV} = \bigcup^1_{a_{IVNV} \in A} a_{IVNV}$.
In fact,

$$A_{IVN} = \left\langle [A^{T,-}, A^{T,+}], \widetilde{\emptyset}, [A^{T,-}, A^{T,+}]^c \right\rangle$$

and

$$A_{IVNV} = \left\langle \widetilde{\emptyset}, [A^{I,-}, A^{I,+}], [A^{F,-}, A^{F,+}] \right\rangle.$$

Proof.

$$\begin{aligned} A_{IVN} &= \bigcup^1_{a_{IVN} \in A} a_{IVN} = \bigcup^1_{a_{IVN} \in A} \left\langle [\{a\},\{a\}], \widetilde{\emptyset}, [\{a\}^c,\{a\}^c] \right\rangle \\ &= \left\langle \bigcup_{a_{IVN} \in A}[\{a\},\{a\}], \bigcup_{a_{IVN} \in A} \widetilde{\emptyset}, \bigcap_{a_{IVN} \in A}[\{a\}^c,\{a\}^c] \right\rangle \\ &= \left\langle [\bigcup_{a \in A^{T,-}} \{a\}, \bigcup_{a \in A^{T,+}} \{a\}], \widetilde{\emptyset}, [\bigcap_{a \in A^{T,+}} \{a\}^c, \bigcap_{a \in A^{T,-}} \{a\}^c] \right\rangle \\ &= \left\langle [A^{T,-}, A^{T,+}], \widetilde{\emptyset}, [A^{T,+c}, A^{T,-c}] \right\rangle \\ &= \left\langle [A^{T,-}, A^{T,+}], \widetilde{\emptyset}, [A^{T,-}, A^{T,+}]^c \right\rangle, \end{aligned}$$

$$\begin{aligned} A_{IVNV} &= \bigcup^1_{a_{IVNN} \in A} a_{IVNV} = \bigcup^1_{a_{IVNV} \in A} \left\langle \widetilde{\emptyset}, [\{a\},\{a\}], [\{a\}^c,\{a\}^c] \right\rangle \\ &= \left\langle \bigcup_{a_{IVNV} \in A} \widetilde{\emptyset}, \bigcup_{a_{IVNV} \in A}[\{a\},\{a\}], \bigcap_{a_{IVNV} \in A}[\{a\}^c,\{a\}^c] \right\rangle \\ &= \left\langle \widetilde{\emptyset}, [\bigcup_{a \in A^{I,-}} \{a\}, \bigcup_{a \in A^{I,+}} \{a\}, [\bigcap_{a \notin A^{F,+}} \{a\}^c, \bigcap_{a \in A^{F,-}} \{a\}^c] \right\rangle \\ &= \left\langle \widetilde{\emptyset}, [A^{I,-}, A^{I,+}], [A^{F,-}, A^{F,+}] \right\rangle. \end{aligned}$$

Then we have,

$$\begin{aligned} A_{IVN} \cup^1 A_{IVNV} &= \left\langle [A^{T,-}, A^{T,+}], \widetilde{\emptyset}, [A^{T,-}, A^{T,+}]^c \right\rangle \cup^1 \left\langle \widetilde{\emptyset}, [A^{I,-}, A^{I,+}], [A^{F,-}, A^{F,+}] \right\rangle \\ &= \left\langle [A^{T,-}, A^{T,+}] \cup \widetilde{\emptyset}, \widetilde{\emptyset} \cup [A^{I,-}, A^{I,+}], [A^{T,-}, A^{T,+}]^c \cap [A^{F,-}, A^{F,+}] \right\rangle \\ &= \left\langle [A^{T,-}, A^{T,+}], [A^{I,-}, A^{I,+}], [A^{T,+c} \cap A^{F,-}, A^{T,-c} \cap A^{F,+} \right\rangle \\ &= \langle [A^{T,-}, A^{T,+}], [A^{I,-}, A^{I,+}], [A^{F,-}, A^{F,+} \rangle \\ &= A. \end{aligned}$$

This completes the proof. □

Example 6. *Let $X = \{a,b,c,d,e,f,g,h,i\}$ and consider the IVNCS in X given by:*

$$A = \langle [\{a,b\},\{a,b,c\}], [\{d\},\{d,e\}], [\{f,g\},\{f,g,h\}] \rangle.$$

Then clearly, we have:

$$\begin{aligned} &A_{IVN} \\ &= \bigcup^1_{a_{IVI} \in A} \left\langle [\{a\},\{a\}], \widetilde{\emptyset}, [\{a\}^c,\{a\}^c] \right\rangle \\ &= \left\langle [\{a,b\},\{a,b,c\}], \widetilde{\emptyset}, [\{a\}^c \cap \{b\}^c \cap \{c\}^c, \{a\}^c \cap \{b\}^c] \right\rangle \end{aligned}$$

$= \left\langle [\{a,b\},\{a,b,c\}], \widetilde{\emptyset}, [\{d,e,f,g,h,i\},\{c,d,e,f,g,h,i\}] \right\rangle$
$= \left\langle [A^{T,-}, A^{T,+}], \widetilde{\emptyset}, [A^{T,-}, A^{T,+}]^c \right\rangle,$

A_{IVNV}
$= \bigcup^1_{a_{IVNV} \in A} \left\langle \widetilde{\emptyset}, [\{a\},\{a\}], [\{a\}^c,\{a\}^c] \right\rangle$
$=< \widetilde{\emptyset}, [\{d\},\{d,e\}], [\{a\}^c \cap \{b\}^c \cap \{c\}^c \cap \{d\}^c \cap \{e\}^c \cap \{h\}^c \cap \{i\}^c,$
$\{a\}^c \cap \{b\}^c \cap \{c\}^c \cap \{d\}^c \cap \{e\}^c \cap \{i\}^c] >$
$= \left\langle \widetilde{\emptyset}, [\{d\},\{d,e\}], [\{f,g\},\{f,g,h\}] \right\rangle$
$= \left\langle \widetilde{\emptyset}, [A^{I,-}, A^{I,+}], [A^{F,-}, A^{F,+}] \right\rangle.$

Thus $A_{IVN} \cup^1 A_{IVNV} = \langle [\{a,b\},\{a,b,c\}], [\{d\},\{d,e\}], [\{f,g\},\{f,g,h\}] \rangle = A$. *So we can confirm that Proposition 6 holds.*

Proposition 7. *Let* $(A_j)_{j \in J} \subset IVNCS(X)$ *and let* $a \in X$.
(1) $a_{IVN} \in \bigcap^1_{j \in J} A_j$ *[resp.* $a_{IVNV} \in \bigcap^1_{j \in J} A_j$*]* $\Leftrightarrow a_{IVN} \in A_j$ *[resp.* $a_{IVNV} \in A_j$*] for each* $j \in J$.
(2) $a_{IVN} \in \bigcup^1_{j \in J} A_j$ *[resp.* $a_{IVNV} \in \bigcup^1_{j \in J} A_j$*]* $\Leftrightarrow$ *there exists* $j \in J$ *such that* $a_{IVN} \in A_j$ *[resp.* $a_{IVNV} \in A_j$.

Proof. (1) Suppose $a_{IVN} \in \bigcap^1_{j \in J} A_j$ and let $A = \bigcap^1_{j \in J} A_j$. Since $A^{T,+} = \bigcap_{j \in J} A_j^{T,+}$, $a \in \bigcap_{j \in J} A_j^{T,+}$. Then $a \in A_j^{T,+}$ for each $j \in J$. Thus $a_{IVN} \in A_j$ for each $j \in J$. The converse is proved similarly. The proof of the second part is omitted.

(2) Suppose $a_{IVNV} \in \bigcup^1_{j \in J} A_j$ and let $A = \bigcup^1_{j \in J} A_j$. Since $A^{F,+} = \bigcap_{j \in J} A_j^{T,+}$, $a \notin \bigcap_{j \in J} A_j^{T,+}$. Then $a \notin A_j^{T,+}$ for some $j \in J$. Thus $a_{IVNV} \in A_j$ for some $j \in J$. The converse is shown similarly. The proof of the first part is omitted.
□

Proposition 8. *Let* $A, B \in IVNCS(X)$. *Then,*

(1) $A \subset_1 B$ *if and only if* $a_{IVN} \in A \Rightarrow a_{IVN} \in B$ *[resp.* $a_{IVNV} \in A \Rightarrow a_{IVNV} \in B$*] for each* $a \in X$.
(2) $A = B$ *if and only if* $a_{IVN} \in A \Leftrightarrow a_{IVN} \in B$ *[resp.* $a_{IVNV} \in A \Leftrightarrow a_{IVNV} \in B$*] for each* $a \in X$.

Proof. Straightforward. □

When we discuss with continuities in a classical topology, the concepts of the preimage and image of a classical subset under a mapping are used. Then we define ones of an IVNCS under a mapping as follows.

Definition 15. *Let* $f : X \to Y$ *be a mapping,* $A \in IVNCS(X)$, $B \in IVNCS(Y)$.

(i) *The image of A under f, denoted by* $f(A)$, *is an IVNCS in Y defined as:*

$$f(A) = \left\langle [f(A^{T,-}), f(A^{T,+})], [f(A^{I,-}), f(A^{I,+})], [f(A^{F,-}), f(A^{F,+})] \right\rangle.$$

(ii) *The preimage of B under f, denoted by* $f^{-1}(B)$, *is an interval set in X defined as:*

$$f^{-1}(B) = \left\langle [f^{-1}(B^{T,-}), f^{-1}(B^{T,+})], [f^{-1}(B^{I,-}), f^{-1}(B^{I,+})], [f^{-1}(B^{F,-}), f^{-1}(B^{F,+})] \right\rangle.$$

It is clear that $f(a_{IVN}) = f(a)_{IVN}$ *and* $f(a_{IVNV}) = f(a)_{IVNV}$ *for each* $a \in X$.

From the above definition, we have similar results of the image and the preimage of classical subsets under a mapping.

Proposition 9. *Let $f: X \to Y$ be a mapping, A, A_1, $A_2 \in IVNCS(X)$, $(A_j)_{j\in J} \subset IVNCS(X)$ and let B, B_1, $B_2 \in IVNCS(Y)$, $(A_j)_{j\in J} \subset IVNCS(Y)$. Let $i = 1, 2$; $k = 1, 2, 3$; $l = 1, 2, 3, 4$. Then,*

(1) If $A_1 \subset_i A_2$, then $f(A_1) \subset_i f(A_2)$,
(2) If $B_1 \subset_i B_2$, then $f^{-1}(B_1) \subset_i f^{-1}(B_1)$,
(3) $A \subset_i f^{-1}(f(A))$ and if f is injective, then $A = f^{-1}(f(A))$,
(4) $f(f^{-1}(B)) \subset_i B$ and if f is surjective, $f(f^{-1}(B)) = B$,
(5) $f^{-1}(\bigcup^i_{j\in J} B_j) = \bigcup^i_{j\in J} f^{-1}(B_j)$,
(6) $f^{-1}(\bigcap^i_{j\in J} B_j) = \bigcap^i_{j\in J} f^{-1}(B_j)$,
(7) $f(\bigcup^i_{j\in J} A_j)_i \subset_i \bigcup^i_{j\in J} f(A_j)$ and if f is surjective, then $f(\bigcup^i_{j\in J} A_j)_i = \bigcup^i_{j\in J} f(A_j)$,
(8) $f(\bigcap^i_{j\in J} A_j) \subset_i \bigcap^i_{j\in J} f(A_j)$ and if f is injective, then $f(\bigcap^i_{j\in J} A_j) = \bigcap^i_{j\in J} f(A_j)$,
(9) If f is surjective, then $f(A)^{k,c} \subset_i f(A^{k,c})$,
(10) $f^{-1}(B^{k,c}) = f^{-1}(B)^{k,c}$,
(11) $f^{-1}(\varnothing_{l,IVN}) = \varnothing_{l,IVN}$, $f^{-1}(X_{l,IVN}) = X_{l,IVN}$,
(12) $f(\varnothing_{l,IVN}) = \varnothing_{l,IVN}$ and if f is surjective, then $f(X_{l,IVN}) = X_{l,IVN}$,
(13) If $g: Y \to Z$ is a mapping, then $(g \circ f)^{-1}(C) = f^{-1}(g^{-1}(C))$, for each $C \in [Z]$.

Proof. The proofs are straightforward. □

4. Interval-Valued Topological Spaces

In this section, we define an interval-valued neutrosophic crisp topology on X and study some of its properties, and give some examples. We also introduce the concepts of an interval-valued neutrosophic crisp base and subbase, and a family of IVNCSs gets the necessary and sufficient conditions to become IVNCB and gives some examples.

From this section to the rest sections, $\subset_1$, $\cup^1$, $\cap^1$, ^{3,c}, $\varnothing_{1,IVN}$, and $X_{1,IVN}$ are denoted by $\subset, \cap, \cup, ^c$, $\varnothing_{IVN}$, and X_{IVN}, respectively.

Definition 16. *Let $\varnothing \neq \tau \subset IVNCS(X)$. Then τ is called an interval-valued neutrosophic crisp topology (briefly, IVNCT) on X, if it satisfies the following axioms:*

($IVNCO_1$) $\varnothing_{IVN}$, $X_{IVN} \in \tau$,
($IVNCO_2$) $A \cap B \in \tau$ for any A, $B \in \tau$,
($IVNCO_3$) $\bigcup_{j\in J} A_j \in \tau$ for any family $(A_j)_{j\in J}$ of members of τ.

In this case, the pair (X, τ) is called an interval-valued neutrosophic crisp topological space (briefly, IVNCTS) and each member of τ is called an interval-valued neutrosophic crisp open set (briefly, IVNCOS) in X. An IVNCS A is called an interval-valued neutrosophic crisp closed set (briefly, IVNCCS) in X, if $A^c \in \tau$.

It is obvious that $\{\varnothing_{IVN}, X_{IVN}\}$ [resp. $IVNC(X)$] is an IVNCT on X, and called the interval-valued neutrosophic crisp indiscrete topology (briefly, IVNCIT) [resp. the interval-valued neutrosophic crisp discrete topology (briefly, IVNCDT)] on X. The pair $(X, \tau_{IVN,0})$ [resp. $(X, \tau_{IVN,1})$] is called an interval-valued neutrosophic crisp indiscrete [resp. discrete] space (briefly, IVNCITS) [resp. (briefly, IVNCDTS)].

$IVNCT(X)$ represents the set of all IVNCTs on X. For an IVNCTS X, the set of all IVNCOs [resp. IVNCCSs] in X is denoted by $IVNCO(X)$ [resp. $IVNCC(X)$].

Remark 2. *(1) For each $\tau \in IVNCT(X)$, consider three families of IVSs in X:*

$$\tau^T = \{[A^{T,-}, A^{T,+}] \in IVS(X) : A \in \tau\},\ \tau^I = \{[A^{I,-}, A^{I,+}] \in IVS(X) : A \in \tau\},$$

$$\tau^F = \{[A^{F,+^c}, A^{F,-^c}] \in IVS(X) : A \in \tau\}.$$

Then we can easily check that τ^T, τ^I and τ^F are IVTs on X.

In this case, τ^T [resp. τ^I and τ^F] is called the membership [resp. indeterminacy and non-membership] topology of τ and we write $\tau = \langle \tau^T, \tau^I, \tau^F \rangle$. In fact, we can consider $(X, \tau^T, \tau^I, \tau^F)$ as an interval-valued tri-topological space on X (see the concept of bitopology introduced by Kelly [39]).

Furthermore, we can consider three intuitionistic topology on X proposed by cCoker [14]:

$$\tau_T = \{(A^{T,-}, A^{T,+^c}) \in IS(X) : A \in \tau\},\ \tau_I = \{(A^{I,-}, A^{I,+^c}] \in IS(X) : A \in \tau\},$$

$$\tau_F = \{A^{F,+^c}, A^{F,-}) \in IS(X) : A \in \tau\}.$$

Let us also consider six families of ordinary subsets of X:

$$\tau^{T,-} = \{A^{T,-} \subset X : A \in \tau\},\ \tau^{T,+} = \{A^{T,+} \subset X : A \in \tau\},$$

$$\tau^{I,-} = \{A^{I,-} \subset X : A \in \tau\},\ \tau^{I,+} = \{A^{I,+} \subset X : A \in \tau\},$$

$$\tau^{F,-} = \{A^{T,+^c} \subset X : A \in \tau\},\ \tau^{F,+} = \{A^{I,-^c} \subset X : A \in \tau\}.$$

Then clearly, $\tau^{T,-}$, $\tau^{T,+}$, $\tau^{I,+}$, $\tau^{I,-}$, $\tau^{F,-}$, $\tau^{F,+}$ *are ordinary topologies on X.*

(2) Let (X, τ_o) *be an ordinary topological space. Then there are four IVNCTs on X given by:*

$$\tau^1 = \begin{cases} \{\left\langle [G, G], \widetilde{\varnothing}, [G^c, G^c] \right\rangle \in IVNC(X) : G \in \tau_o\} & \text{if } G \neq X \\ \{\varnothing_{IVN}, X_{IVN}\} & \text{if } G = X, \end{cases}$$

$$\tau^2 = \begin{cases} \{\left\langle [G, G], \widetilde{X}, [G^c, G^c] \right\rangle \in IVNC(X) : G \in \tau_o\} & \text{if } G \neq X \\ \{\varnothing_{IVN}, X_{IVN}\} & \text{if } G = X, \end{cases}$$

$$\tau^3 = \begin{cases} \{\left\langle [\varnothing, G], \widetilde{\varnothing}, [\varnothing, G^c] \right\rangle \in IVNC(X) : G \in \tau_o\} & \text{if } G \neq \varnothing \\ \{\varnothing_{IVN}, X_{IVN}\} & \text{if } G = \varnothing, \end{cases}$$

$$\tau^4 = \begin{cases} \{\left\langle [\varnothing, G], \widetilde{X}, [\varnothing, G^c] \right\rangle \in IVNC(X) : G \in \tau_o\} & \text{if } G \neq \varnothing \\ \{\varnothing_{IVN}, X_{IVN}\} & \text{if } G = \varnothing. \end{cases}$$

(3) Let (X, τ_{IV}) *be an IVTS introduced by Kim et al. [30]. Then clearly,*

$$\tau = \{\left\langle [A^-, A^+], \widetilde{\varnothing}, [A^{+^c}, A^{-^c}] \right\rangle \in IVNC(X) : A \in \tau_{IV}\} \in IVNCT(X).$$

(4) Let (X, τ_I) *be an ITS introduced by cCoker [14]. Then clearly,*

$$\tau = \{\left\langle [A^{\in}, A^{\notin^c}], \widetilde{\varnothing}, [A^{\notin}, A^{\in^c}] \right\rangle \in IVNC(X) : A \in \tau_I\} \in IVNCT(X).$$

(5) Let (X, τ_{NC}) *be a neutrosophic crisp topological space introduced by Salama and Smarandache [34]. Then clearly,*

$$\tau = \{\left\langle [A^T, A^T], [A^I, A^I], [A^F, A^F] \right\rangle \in IVN^*(X)) : A \in \tau_{NC}\} \in IVNCT(X).$$

From Remark 2, we can easily see that an IVNCT is a generalization of a classical topology, an IVT, an IT, and neutrosophic crisp topology. Then we have the following Figure 1:

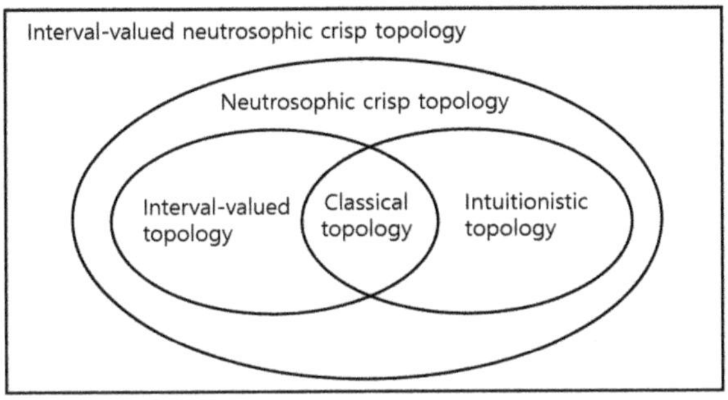

Figure 1. The relationships among five topologies.

Example 7. *(1) Let* $X = \{a, b\}$. *Then we can easily check that:*

$$\tau_{IVN,1} = \{\varnothing_{IVN}, a_{IVN}, b_{IVN}, a_{IVNV}, b_{IVNV}, \left\langle \widetilde{\varnothing}, \widetilde{\varnothing}, [\{b\}, \{b\}] \right\rangle, \langle [\{a\}, \{a\}], [\{a\}, \{a\}], [\{b\}, \{b\}] \rangle, X_{IVN}\}.$$

(2) Let $A \in IVNCS(X)$. *Then A is said to be finite, if* $A^{T,+}$, $A^{I,+}$, *and* $A^{F,+}$ *are finite. Consider the family*

$$\tau = \{U \in IVNCS(X) : U = \varnothing_{IVN} \text{ or } U^c \text{ is finite}\}.$$

Then we can easily prove that $\tau \in IVNCT(X)$.

In this case, τ *is called an interval-valued neutrosiophic crisp cofinite topology (briefly, IVNCCFT) on X.*

(3) Let $A \in IVNCS(X)$. *Then A is said to be countable, if* $A^{T,+}$, $A^{I,+}$, *and* $A^{F,+}$ *are countable. Consider the family:*

$$\tau = \{U \in IVNCS(X) : U = \varnothing_{IVN} \text{ or } U^c \text{ is countable}\}.$$

Then we can easily show that $\tau \in IVNCT(X)$.

In this case, τ *is called an interval-valued neutrosiophic crisp cocountable topology (briefly, IVNCCCT) on X.*

(4) Let $X = \{a, b, c, d, e, f, g, h, i\}$ *and the family* τ *of IVNCSs on X given by:*

$$\tau = \{\varnothing_{IVN}, A_1, A_2, A_3, A_4, X_{IVN}\},$$

where $A_1 = \langle [\{a, b\}, \{a, b, c\}], [\{e\}, \{e, f\}], [\{g\}, \{g, i\}] \rangle$,
$A_2 = \langle [\{a, d\}, \{a, c, d\}], [\{e\}, \{e\}], [\{g, h\}, \{g, h, i\}] \rangle$,
$A_3 = \langle [\{a\}, \{a, c\}], [\{e\}, \{e\}], [\{g, h\}, \{g, h, i\}] \rangle$,
$A_4 = \langle [\{a, b, d\}, \{a, b, c, d\}], [\{e\}, \{e, f\}], [\{g\}, \{g, i\}] \rangle$.

Then we can easily check that $\tau \in IVNCT(X)$.

(5) Let $X = \{0, 1\}$. *Consider the family* τ *of IVNCSs on X given by:*

$$\tau = \{\varnothing_{IVN}, \left\langle [\{0\}, \{0\}], \widetilde{\varnothing}, [\{1\}, \{1\}] \right\rangle, X_{IVN}\}.$$

Then we can easily prove that $\tau \in IVNCT(X)$. *In this case,* (X, τ) *is called the interval-valued neutrosophic crisp Sierpin'ski space.*

From Definition 16, we have the following.

Proposition 10. *Let X be an IVNCTS. Then:*

(1) $\varnothing_{IVN}, X_{IVN} \in IVNCC(X)$,

(2) $A \cup B \in IVNCC(X)$ *for any* $A,\ B \in IVNCC(X)$,
(3) $\bigcap_{j\in J} A_j \in IVNCC(X)$ *for any* $(A_j)_{j\in J} \subset IVNCC(X)$.

To discuss $IVNCT(X)$ with a view-point of lattice theory, we define an order between two IVCTs.

Definition 17. *Let* $\tau_1,\ \tau_2 \in IVNCT(X)$. *Then we say that* τ_1 *is contained in* τ_2 *or* τ_1 *is coarser than* τ_2 *or* τ_2 *is finer than* τ_1, *if* $\tau_1 \subset \tau_2$, *i.e.*, $A \in \tau_2$ *for each* $A \in \tau_1$.

For each $\tau \in IVNCT(X)$, $\tau_{IVN,0} \subset \tau \subset \tau_{IVN,1}$ is clear.

From Definitions 14 and 16, we get the following.

Proposition 11. *Let* $(\tau_j)_{j\in J} \subset IVNCT(X)$. *Then* $\bigcap_{j\in J} \tau_j \in IVNCT(X)$.
In fact, $\bigcap_{j\in J} \tau_j$ *is the coarsest IVNCT on X containing each* τ_j.

Proposition 12. *Let* $\tau,\ \gamma \in IVNCT(X)$. *We define* $\tau \wedge \gamma$ *and* $\tau \vee \gamma$ *as follows:*

$$\tau \wedge \gamma = \{W : W \in \tau,\ W \in \gamma\},$$

$$\tau \vee \gamma = \{W : W = U \cup V,\ U \in \tau,\ V \in \gamma\}.$$

Then we have:

(1) $\tau \wedge \gamma$ *is an IVNCT on X which is the finest IVNCT coarser than both* τ *and* γ,
(2) $\tau \vee \gamma$ *is an IVNCT on X which is the coarsest IVNCT finer than both* τ *and* γ,

Proof. (1) Clearly, $\tau \wedge \gamma \in IVNCT(X)$. Let η be any IVNCT on X which is coarser than both τ and γ, and let $W \in \eta$. Then $W \in \tau$ and $W \in \gamma$. Thus $W \in \tau \wedge \gamma$. So η is coarser than $\tau \wedge \gamma$.
(2) The proof is similar to (1). □

From Definition 17, Propositions 11 and 12, we can easily see that $(IVNCT(X), \subset)$ forms a complete lattice with the least element $\tau_{IVN,0}$ and the greatest element $\tau_{IVN,1}$.

A topology on a set can be a complicated collection of subsets of subsets of a set, and it can be difficult to describe the entire collection. In most cases, one describes a subcollection (called a base and a subbase) that "generates" the topology. Then we define a base and a subbase in an IVNCT. Moreover, we introduce the various intervals via IVNCSs in real line $\mathbb{R}$.

Definition 18. *Let* (X, τ) *be an IVNCTS.*
(i) A subfamily β *of* τ *is called an interval-valued neutrosophic crisp base (briefly, IVNCB) for* τ, *if for each* $A \in \tau$, $A = \varnothing_{IVN}$ *or there is* $\beta' \subset \beta$ *such that* $A = \bigcup \beta'$.
(ii) A subfamily σ *of* τ *is called an interval-valued neutrosophic crisp subbase (briefly, IVNCSB) for* τ, *if the family* $\beta = \{\bigcap \sigma' : \sigma' \text{ is a finite subset of } \sigma\}$ *is an IVNCB for* τ.

Remark 3. *(1) Let* β *be an IVNCB for an IVNCT* τ *on a non-empty set X and consider three families of IVSs in X:*

$$\beta^T = \{[A^{T,-}, A^{T,+}] \in IVS(X) : A \in \beta\},\ \beta^I = \{[A^{I,-}, A^{I,-}] \in IVS(X) : A \in \beta\},$$

$$\beta^F = \{[A^{F,+^c}, A^{F,-^c}] \in IVS(X) : A \in \beta\}.$$

Then we can easily see that β^T, β^I, *and* β^F *are an interval-valued base (see [30]) for* τ^T, τ^I, *and* τ^F, *respectively.*
Furthermore, we can consider three intuitionistic base on X defined by cCoker [14]:

$$\beta_T = \{(A^{T,-}, A^{T,+^c}) \in IS(X) : A \in \beta\},\ \beta_I = \{(A^{I,-}, A^{I,+^c}] \in IS(X) : A \in \beta\},$$

$$\beta_F = \{A^{F,+^c}, A^{F,-}) \in IS(X) : A \in \beta\}.$$

Let also us consider six families of ordinary subsets of X:

$$\beta^{T,-} = \{A^{T,-} \subset X : A \in \beta\},\ \beta^{T,+} = \{A^{T,+} \subset X : A \in \beta\},$$

$$\beta^{I,-} = \{A^{I,-} \subset X : A \in \beta\},\ \beta^{I,+} = \{A^{I,+} \subset X : A \in \beta\},$$

$$\beta^{F,-} = \{A^{T,+^c} \subset X : A \in \beta\},\ \beta^{F,+} = \{A^{I,-^c} \subset X : A \in \beta\}.$$

Then clearly, $\beta^{T,-}$, $\beta^{T,+}$, $\beta^{I,+}$, $\beta^{I,-}$, $\beta^{F,-}$, $\beta^{F,+}$ *are ordinary bases for ordinary topologies* $\tau^{T,-}$, $\tau^{T,+}$, $\tau^{I,+}$, $\tau^{I,-}$, $\tau^{F,-}$, $\tau^{F,+}$ *on X, respectively.*

(2) Let σ *be an IVNCSB for an IVNCT* τ *on a non-empty set X and consider three families of IVSs in X:*

$$\sigma^T = \{[A^{T,-}, A^{T,+}] \in IVS(X) : A \in \sigma\},\ \sigma^I = \{[A^{I,+}, A^{I,-}] \in IVS(X) : A \in \sigma\},$$

$$\sigma^F = \{[A^{F,+^c}, A^{F,-^c}] \in IVS(X) : A \in \sigma\}.$$

Then we can easily see that σ^T, σ^I, *and* σ^F *are an interval-valued subbases (see [30]) for* τ^T, τ^I, *and* τ^F, *respectively.*

Furthermore, we can consider three intuitionistic base on X defined by cCoker [14]:

$$\sigma_T = \{(A^{T,-}, A^{T,+^c}) \in IS(X) : A \in \sigma\},\ \sigma_I = \{(A^{I,-}, A^{I,+^c}] \in IS(X) : A \in \sigma\},$$

$$\sigma_F = \{A^{F,+^c}, A^{F,-}) \in IS(X) : A \in \sigma\}.$$

Let also us consider six families of ordinary subsets of X:

$$\sigma^{T,-} = \{A^{T,-} \subset X : A \in \sigma\},\ \sigma^{T,+} = \{A^{T,+} \subset X : A \in \sigma\},$$

$$\sigma^{I,-} = \{A^{I,-} \subset X : A \in \sigma\},\ \sigma^{I,+} = \{A^{I,+} \subset X : A \in \sigma\},$$

$$\sigma^{F,-} = \{A^{T,+^c} \subset X : A \in \sigma\},\ \sigma^{F,+} = \{A^{F,-^c} \subset X : A \in \sigma\}.$$

Then clearly, $\sigma^{T,-}$, $\sigma^{T,+}$, $\sigma^{I,+}$, $\sigma^{I,-}$, $\sigma^{F,-}$, $\sigma^{F,+}$ *are ordinary subbases for ordinary topologies* $\tau^{T,-}$, $\tau^{T,+}$, $\tau^{I,+}$, $\tau^{I,-}$, $\tau^{F,-}$, $\tau^{F,+}$ *on X, respectively.*

Example 8. *(1) Let* $\sigma = \{\langle[(a,b),(a,\infty)],[\varnothing,\varnothing],[\varnothing,(-\infty,a]]\rangle : a,\ b \in \mathbb{R}\}$ *be the family of IVNCs in* $\mathbb{R}$. *Then* σ *generates an IVNCT* τ *on* $\mathbb{R}$ *which is called the "usual left interval-valued neutrosophic crisp topology (briefly, ULIVNCT)" on* $\mathbb{R}$. *In fact, the IVNCB* β *for* τ *can be written in the form:*

$$\beta = \{\mathbb{R}_{IVN}\} \cup \{\cap_{\gamma \in \Gamma} S_\gamma : S_\gamma \in \sigma,\ \Gamma \text{ is finite}\}$$

and τ *consists of the following IVNCSs in* $\mathbb{R}$:

$$\tau = \{\varnothing_{IVN}, \mathbb{R}_{IVN}\} \cup \{\left\langle[\cup(a_j,b_j),(c,\infty)],\widetilde{\varnothing},\widetilde{\varnothing}\right\rangle\}$$

or

$$\tau = \{\varnothing_{IVN}, \mathbb{R}_{IVN}\} \cup \{\left\langle[\cup(a_k,b_k),\mathbb{R}],\widetilde{\varnothing},\widetilde{\varnothing}\right\rangle\},$$

where $a_j, b_j, c \in \mathbb{R}$, $\{a_j : j \in J\}$ *is bounded from below,* $c < inf\{a_j : j \in J\}$ *and* $a_k, b_k \in \mathbb{R}$, $\{a_k : k \in K\}$ *is not bounded from below.*

Similarly, one can define the "usual right interval-valued neutrosophic crisp topology (briefly, URIVNCT)" on $\mathbb{R}$ *using an analogue construction.*

(2) Consider the family σ *of IVNCSs in* $\mathbb{R}$:

$$\sigma = \{\left\langle[(a,b),(a_1,\infty)\cap(-\infty,b_1)],\widetilde{\varnothing},[\varnothing,(-\infty,a_1]\cup[b_1,\infty]\right\rangle$$

$$: a,\ b,\ a_1,\ b_1 \in \mathbb{R},\ a_1 \leq a,\ b_1 \geq b\}.$$

Then σ generates an IVNCT τ on $\mathbb{R}$ which is called the "usual interval-valued neutrosophic crisp topology (briefly, UIVNCT)" on $\mathbb{R}$. In fact, the IVNCB β for τ can be written in the form:

$$\beta = \{\mathbb{R}_{IVN}\} \cup \{\cap_{\gamma\in\Gamma} S_\gamma : S_\gamma \in \sigma,\ \Gamma \text{ is finite}\}$$

and the elements of τ can be easily written down as in (1).

(3) Consider the family $\sigma_{[0,1]}$ of IVNCSs in $\mathbb{R}$:

$$\sigma_{[0,1]} = \{\left\langle [[a,b],[a,\infty)\cap(-\infty,b]],\widetilde{\varnothing},[\varnothing,(-\infty,a]\cup[b,\infty]\right\rangle$$
$$: a,b \in \mathbb{R} \text{ and } 0 \leq a \leq b \leq 1\}.$$

Then $\sigma_{[0,1]}$ generates an IVNCT $\tau_{[0,1]}$ on $\mathbb{R}$ which is called the "usual unit closed interval interval-valued neutrosophic crisp topology" on $\mathbb{R}$. In fact, the IVNCB $\beta_{[0,1]}$ for $\tau_{[0,1]}$ can be written in the form:

$$\beta_{[0,1]} = \{\mathbb{R}_{IVN}\} \cup \{\cap_{\gamma\in\Gamma} S_\gamma : S_\gamma \in \sigma_{[0,1]},\ \Gamma \text{ is finite}\}$$

and the elements of τ can be easily written down as in (1).

In this case, $([0,1],\tau_{[0,1]})$ is called the "interval-valued neutrosophic crisp nusual unit closed interval" and denoted by $[0,1]_{IVNCI}$. In fact,

$$[0,1]_{IVNCI} = \left\langle [[0,1],[0,\infty)\cup(-\infty,1]],\widetilde{\varnothing},\widetilde{\varnothing}\right\rangle.$$

(4) Let $\beta = \{a_{IVN} : a \in X\} \cup \{a_{IVNV} : a \in X\}$. Then β is an IVNCB for the interval-valued neutrosophic crisp discrete topology τ_1 on X.

(5) Let $X = \{a,b,c,d,e,f,g,h,i\}$ and consider the family β of IVNCSs in X given by:

$$\beta = \{A, B, X_{IVN}\},$$

where $A = \langle [\{a,b\},\{a,b,c\}],[\{e\},\{e,f\}],[\{g\},\{g,i\}]\rangle,$
$B = \langle [\{a,d\},\{a,c,d\}],[\{e\},\{e\}],[\{g,h\},\{g,h,i\}]\rangle.$

Assume that β is an IVNCB for an IVNCT τ on X. Then by the definition of base, $\beta \subset \tau$. Thus $A,\ B \in \tau$. So $A\cap B = \langle [\{a\},\{a,c\}],[\{e\},\{e\}],[\{g,h\},\{g,h,i\}]\rangle \in \tau$. However for any $\beta' \subset \beta$, $A \cap B \neq \bigcup\beta'$. Hence β is not an IVNCB for an IVNCT on X.

From (1), (2), and (3) in Example 8, we can define interval-valued neutrosophic crisp intervals as following.

Definition 19. *Let $a,\ b \in \mathbb{R}$ such that $a \leq b$. Then:*

(i) (The closed interval) $[a,b]_{IVNCI} = \left\langle [[a,b],[a,-\infty)\cap(-\infty,b]],\widetilde{\varnothing},\widetilde{\varnothing}\right\rangle,$

(ii) (The open interval) $(a,b)_{IVNCI} = \left\langle [(a,b),(a,-\infty)\cap(-\infty,b)],\widetilde{\varnothing},\widetilde{\varnothing}\right\rangle,$

(iii) (The half open interval or the half closed interval)

$$(a,b]_{IVNCI} = \left\langle [(a,b],(a,-\infty)\cap(-\infty,b]],\widetilde{\varnothing},\widetilde{\varnothing}\right\rangle,$$

$$[a,b)_{IVI} = \left\langle [[a,b),[a,-\infty)\cap(-\infty,b)],\widetilde{\varnothing},\widetilde{\varnothing}\right\rangle,$$

(iv) (The half interval-valued real line)

$$(-\infty,a]_{IVNCI} = \left\langle [(-\infty,a],(-\infty,a]],\widetilde{\varnothing},\widetilde{\varnothing}\right\rangle,$$

$$(-\infty,a)_{IVNCI} = \left\langle [(-\infty,a),(-\infty,a)],\widetilde{\varnothing},\widetilde{\varnothing}\right\rangle,$$

$$[a,\infty)_{IVNCI} = \left\langle [[a,\infty),[a,\infty)], \widetilde{\emptyset}, \widetilde{\emptyset} \right\rangle,$$

$$(a,\infty)_{IVNCI} = \left\langle [(a,\infty),(a,\infty)], \widetilde{\emptyset}, \widetilde{\emptyset} \right\rangle,$$

(v) *(The interval-valued real line)*

$$(-\infty,\infty)_{IVMCI} = \left\langle [(-\infty,\infty),(-\infty,\infty)], \widetilde{\emptyset}, \widetilde{\emptyset} \right\rangle = \mathbb{R}_{IVN}.$$

The following provide a necessary and sufficient condition which a collection of IVNCSs in a set X is an IVNCB for some IVNCT on X.

Theorem 1. *Let $\beta \subset IVNCS(X)$. Then β is an IVNCB for an IVNCT τ on X if and only if it satisfies the following properties:*

(1) $X_{IVN} = \bigcup \beta$,
(2) *If $B_1, B_2 \in \beta$ and $a_{IVN} \in B_1 \cap B_2$ [resp. $a_{IVNV} \in B_1 \cap B_2$], then there exists $B \in \beta$ such that $a_{IVN} \in B \subset B_1 \cap B_2$ [resp. $a_{IVNV} \in B \subset B_1 \cap B_2$].*

Proof. The proof is the same as one in classical topological spaces. □

Example 9. *Let $X = \{a,b,c\}$ and consider the family of IVNCSs in X given by:*

$$\beta = \{A_1, A_2, A_3, A_3\},$$

where $A_1 = \langle [\{b\},\{a,b\}], [\{b\},\{b\}], [\{c\},\{c\}] \rangle$,
$A_2 = \left\langle [\{b,c\},\{b,c\}], [\{a\},\{a\}], \widetilde{\emptyset} \right\rangle$,
$A_3 = \langle [\{a\},\{a\}], [\{c\},\{c\}], [\{b\},\{b\}] \rangle$,
$A_4 = \left\langle [\{b\}, \widetilde{\emptyset}, [\{c\},\{c\}] \right\rangle$,
Then clearly, β satisfies two conditions of Theorem 1. Thus β is an IVNCB for an IVNCT τ on X. In fact, we have:

$$\tau = \{\emptyset_{IVN}, A_1, A_2, A_3, A_4, A_5, A_6, A_7, X_{IVN}\},$$

where $A_5 = \left\langle [\{b,c\}, X], [\{a,b\},\{a,b\}], \widetilde{\emptyset} \right\rangle$,
$A_6 = \left\langle [\{a,b\},\{a,b\}], [\{b,c\},\{b,c\}], \widetilde{\emptyset} \right\rangle$,
$A_7 = \left\langle \widetilde{X}, [\{a,c\},\{a,c\}], \widetilde{\emptyset} \right\rangle$.

The following provide a sufficient condition which a collection of IVNCSs in a set X is an IVNCB for some IVNCT on X.

Proposition 13. *Let $\sigma \subset IVNCS(X)$ such that $X_{IVN} = \bigcup \sigma$. Then there exists a unique IVNCT τ on X such that σ is an IVNCSB for τ.*

Proof. Let $\beta = \{B \in IVNCS(X) : B = \bigcap_{i=1}^{n} S_i$ and $S_i \in \sigma\}$. Let $\tau = \{U \in IVNCS(X) : U = \widetilde{\emptyset}$ or there is a subcollection β' of β such that $U = \bigcup \beta'\}$. Then we can show that τ is the unique IVNCT on X such that σ is an IVNCSB for τ. □

In Proposition 13, τ is called the IVNCT on X generated by σ.

Example 10. *Let $X = \{a,b,c,d,e\}$ and consider the family σ of IVNCSs in X given by:*

$$\sigma = \{A_1, A_2, A_3, A_4\},$$

where $A_1 = \langle [\{a\},\{a\}],[\{b\},\{b\}],[\{c,d\},\{c,d\}] \rangle$,
$A_2 = \langle [\{a,b,c\},\{a,b,c\}],[\{b,d\},\{b,d\}],[\{e\},\{e\}] \rangle$,
$A_3 = \langle [\{b,c,e\},\{b,c,e\}],[\{c,e\},\{c,d,e\}],[\{d\},\{d\}] \rangle$,
$A_4 = \langle [\{c,d\},\{c,d\}],[\{a,c\},\{a,c\}],[\{a,b\},\{a,b\}] \rangle$.
Then clearly, $\bigcup \sigma = X_{IVN}$. *Let* β *be the collection of all finite intersections of members of* σ. *Then we have:*

$$\beta = \{A_1, A_2, A_3, A_4, A_5, A_6, A_7, A_8, A_9, A_{10}, A_{11}, A_{12}\},$$

where $A_5 = \langle [\{a\},\{a\}],[\{b\},\{b\}],[\{c,d,e\},\{c,d,e\}] \rangle$,
$A_6 = \left\langle \tilde{\varnothing},[\{b\},\{b\}],[\{c,d\},\{c,d\}] \right\rangle$,
$A_7 = \left\langle \tilde{\varnothing},\tilde{\varnothing},[\{a,b,c,d\},\{a,b,c,d\}] \right\rangle$,
$A_8 = \langle [\{b,c\},\{b,c\}],[\varnothing,\{d\}],[\{d,e\},\{d,e\}] \rangle$,
$A_9 = \left\langle [\{c\},\{c\}],\tilde{\varnothing},[\{a,b,e\},\{a,b,e\}] \right\rangle$,
$A_{10} = \langle [\{c\},\{c\}],[\{c\},\{c\}],[\{a,b,d\},\{a,b,d\}] \rangle$,
$A_{11} = \left\langle \tilde{\varnothing},\tilde{\varnothing},[\{c,d,e\},\{c,d,e\}] \right\rangle$,
$A_{12} = \left\langle [\{c\},\{c\}],\tilde{\varnothing},[\{a,b,d,e\},\{a,b,d,e\}] \right\rangle$.
Thus we have the generated IVNCT τ *by* σ:
$\tau = \{\varnothing_{IVN}, A_1, A_2, A_3, A_4, A_5, A_6, A_7, A_8, A_9, A_{10}, A_{11}, A_{12}, A_{13}, A_{14}, A_{15}, A_{16}, A_{17}, A_{18}, X_{IVN}\}$,
where $A_{13} = \left\langle [\{a,b,c\},\{a,b,c\}],[\{b,d\},\{b,d\}],\tilde{\varnothing} \right\rangle$,
$A_{14} = \langle [\{a,b,c,e\},\{a,b,c,e\}],[\{b,c,e\},\{b,c,d,e\}],[\{d\},\{d\}] \rangle$,
$A_{15} = \left\langle [\{a,c,d\},\{a,c,d\}],[\{a,b,c\},\{a,b,c\}],\tilde{\varnothing} \right\rangle$,
$A_{16} = \left\langle [\{a,b,c,e\},\{a,b,c,e\}],[\{b,c,d,e\},\{b,c,d,e\}],\tilde{\varnothing} \right\rangle$,
$A_{17} = \left\langle [\{a,b,c,d\},\{a,b,c,d\}],[\{a,b,c,d\},\{a,b,c,d\}],\tilde{\varnothing} \right\rangle$,
$A_{18} = \left\langle \tilde{X},[\{a,c,e\},\{a,c,e\}],\tilde{\varnothing} \right\rangle$.

Remark 4. *By using "*$\subset_2$, $\cup_2$, $\cap_2$, ${}^{i,c}(i = 1,\ 2,\ 3)$, $\varnothing_{2,IN}$, $X_{2,IN}$, *and* $INC(X)$, *we can have the definitions corresponding to Definitions 16 and 18, respectively.*

5. Interval-Valued Neutrosophic Crisp Neighborhoods

In this section, we introduce the concept of interval-valued neutrosophic crisp neighborhoods of IVNPs of two types, and find their various properties and give some examples.

Definition 20. *Let X be an IVNCTS,* $a \in X$, $N \in IVNCS(X)$. *Then:*

(i) N is called an interval-valued neutrosophic crisp neighborhood (briefly, IVNCN) of a_{IVN}, *if there exists a* $U \in IVNCO(X)$ *such that:*

$$a_{IVN} \in U \subset N, \text{ i.e., } a \in U^{T,-} \subset N^{T,-},$$

(ii) N is called an interval-valued neutrosophic crisp vanishing neighborhood (briefly, IVNCVN) of a_{IVNV}, *if there exists a* $U \in IVNCO(X)$ *such that:*

$$a_{IVNV} \in U \subset N, \text{ i.e., } a \notin N^{F,+} \subset U^{F,+}.$$

The set of all IVNCNs [resp. IVNCVNs] of a_{IVN} *[resp.* a_{IVNV}*] is denoted by* $N(a_{IVN})$ *[resp.* $N(_{IVNV})$*] and will be called an IVNC neighborhood system of* a_{IVN} *[resp.* a_{IVNV}*].*

Example 11. *Let $X = \{a, b, c, d, e, f, g, h, i\}$ and let τ be the IVNCT on X given in Example 7 (4). Consider the IVNCS $N = \langle [\{a, b, d\}, \{a, b, c, d\}], [\{e\}, \{e\}], [\{g\}, \{g\}] \rangle$ in X. Then we can easily check that:*

$N \in N(a_{IVN}) \cap N(a_{IVNV}),\ N \in N(b_{IVN}) \cap N(b_{IVNV}),$
$N \in N(d_{IVN}) \cap N(d_{IVNV}),\ N \in N(c_{IVNV}).$

An IVNC neighborhood system of a_{IVN} has a similar property for a neighborhood system of a point in a classical topological space.

Proposition 14. *Let X be an IVNCTS, $a \in X$.*

[IVNCN1] If $N \in N(a_{IVN})$, then $a_{IVN} \in N$.
[IVNCN2] If $N \in N(a_{IVN})$ and $N \subset M$, then $M \in N(a_{IVN})$.
[IVNCN3] If $N, M \in N(a_{IVN})$, then $N \cap M \in N(a_{IVN})$.
[IVNCN4] If $N \in N(a_{IVN})$, then there exists $M \in N(a_{IVN})$ such that $N \in N(b_{IVN})$ for each $b_{IVN} \in M$.

Proof. The proofs of [IVNCN1], [IVNCN2], and [IVNCN4] are easy.

[IVNCN3] Suppose $N, M \in N(a_{IVN})$. Then there are $U,\ V \in IVNCO(X)$ such that $a_{IVN} \in U \subset N$ and $a_{IVN} \in V \subset M$. Let $W = U \cap V$. Then clearly, $W \in IVNCO(X)$ and $a_{IVN} \in W \subset N \cap M$. Thus $N \cap M \in N(a_{IVN})$. □

In addition, an IVNC neighborhood system of a_{IVNV} has the similar property.

Proposition 15. *Let X be an IVNCTS, $a \in X$.*

[IVNCVN1] If $N \in N(a_{IVNV})$, then $a_{IVNV} \in N$.
[IVNCVN2] If $N \in N(a_{IVNV})$ and $N \subset M$, then $M \in N(a_{IVNV})$.
[IVNCVN3] If $N, M \in N(a_{IVNV})$, then $N \cap M \in N(a_{IVNV})$.
[IVNCVN4] If $N \in N(a_{IVNV})$, then there exists $M \in N(a_{IVNV})$ such that $N \in N(b_{IVNV})$ for each $b_{IVNV} \in M$.

Proof. The proof is similar to one of Proposition 15. □

From Definition 20, we have two IVNCTs containing a given IVNCT.

Proposition 16. *Let (X, τ) be an IVNCTS and let us define two families:*

$$\tau_{IVN} = \{U \in IVNCS(X) : U \in N(a_{IVN}) \text{ for each } a_{IVN} \in U\}$$

and

$$\tau_{IVNV} = \{U \in IVNCS(X) : U \in N(a_{IVNV}) \text{ for each } a_{IVNV} \in U\}.$$

Then we have:

(1) $\tau_{IVN},\ \tau_{IVNV} \in IVNCT(X)$,
(2) $\tau \subset \tau_{IVN}$ and $\tau \subset \tau_{IVNV}$.

Proof. (1) We only prove that $\tau_{IVNV} \in IVNCT(X)$.

($IVNCO_1$) From the definition of τ_{IVNV}, we have $\varnothing_{IVN},\ X_{IVN} \in \tau_{IVNV}$.

($IVNCO_2$) Let $U,\ V \in IVN^*(X)$ such that $U, V \in \tau_{IVNV}$ and let $a_{IVNV} \in U \cap V$. Then clearly, $U,\ V \in N(a_{IVNV})$. Thus by [IVNCVN3], $U \cap V \in N(a_{IVNV})$. So $U \cap V \in \tau_{IVNV}$.

($IVNCO_3$) Let $(U_j)_{j \in J}$ be any family of IVNCSs in τ_{IVNV}, let $U = \bigcup_{j \in J} U_j$ and let $a_{IVNV} \in U$. Then by Proposition 7 (2), there is $j_0 \in J$ such that $a_{IVNV} \in U_{j_0}$. Since $U_{j_0} \in \tau_{IVNV}$, $U_{j_0} \in N(a_{IVNV})$ by the definition of τ_{IVNV}. Since $U_{j_0} \subset U$, $U \in N(a_{IVNV})$ by [IVNCVN2]. So by the definition of τ_{IVNV}, $U \in \tau_{IVNV}$.

(2) Let $U \in \tau$. Then clearly, $U \in N(a_{IVN})$ and $U \in N(a_{IVNV})$ for each $a_{IVN} \in G$ and $a_{IVNV} \in G$, respectively. Thus $U \in \tau_{IVN}$ and $U \in \tau_{IVNV}$. So the results hold. □

Remark 5. *(1) From the definitions of* τ_{IVN} *and* τ_{IVNV}*, we can easily have:*

$$\tau_{IVN} = \tau \cup \{U \in IVNCS(X) : V^{T,-} \subset U^{T,-},\ V \in \tau\}$$

and

$$\tau_{IVNV} = \tau \cup \{U \in IVNCS(X) : U^{F,+} \subset V^{F,+},\ V \in \tau\}.$$

(2) For any IVNCT τ *on a set X, we can have six IVTs on X given by:*

$$\tau^{T}_{IVN} = \{[U^{T,-}, U^{T,+}] \in IVS(X) : U \in \tau_{IVN}\},$$

$$\tau^{I}_{IVN} = \{[U^{I,-}, U^{I,+}] \in IVS(X) : U \in \tau_{IVN}$$

$$\tau^{F}_{IVN} = \{[U^{F,+^c}, U^{F,-^c}] \in IVS(X) : U \in \tau_{IVN}\},$$

$$\tau^{T}_{IVNV} = \{[U^{T,-}, U^{T,-}] \in IVS(X) : U \in \tau_{IVNV}\},$$

$$\tau^{I}_{IVNV} = \{[U^{I,-}, U^{I,+}] \in IVS(X) : U \in \tau_{IVNV}\},$$

$$\tau^{F}_{IVNV} = \{[U^{F,+^c}, U^{F,+}] \in IVS(X) : U \in \tau_{IVNV}\}.$$

Furthermore, we have 12 ordinary topologies on X:

$$\tau^{T,-}_{IVN} = \{U^{T,-} \subset X : U \in \tau_{IVN}\},\ \tau^{T,+}_{IVN} = \{U^{T,+}] \subset X : U \in \tau_{IVN}\},$$

$$\tau^{I,-}_{IVN} = \{U^{I,-} \subset X : U \in \tau_{IVN}\},\ \tau^{I,+}_{IVN} = \{U^{I,+}] \subset X : U \in \tau_{IVN}\},$$

$$\tau^{F,-}_{IVN} = \{U^{F,+^c} \subset X : U \in \tau_{IVN}\},\ \tau^{F,+}_{IVN} = \{U^{F,-^c} \subset X : U \in \tau_{IVN}\},$$

$$\tau^{T,-}_{IVNV} = \{U^{T,-} \subset X : U \in \tau_{IVNV}\},\ \tau^{T,+}_{IVNV} = \{U^{T,-} \subset X : U \in \tau_{IVNV}\},$$

$$\tau^{I,-}_{IVNV} = \{U^{I,-} \subset X : U \in \tau_{IVNV}\},\ \tau^{I,+}_{IVNV} = \{U^{I,+} \subset X : U \in \tau_{IVNV}\},$$

$$\tau^{F,-}_{IVNV} = \{U^{F,+^c} \subset X : U \in \tau_{IVNV}\},\ \tau^{F,+}_{IVNV} = \{U^{F,+} \subset X : U \in \tau_{IVNV}\}.$$

Example 12. *Let* $X = \{a, b, c, d, e, f, g, h, i\}$ *and consider IVNCT* τ *on X given in Example 7 (4). Then from Remark 5 ((1), we have:*

$$\tau_{IVN} = \tau \cup \{A_5, A_6, A_7\},$$

where $A_5 = \langle [\{a,b,c\}, \{a,b,c\}], [\{e\}, \{e,f\}], [\{g\}, \{g,i\}] \rangle$,
$A_6 = \langle [\{a,c,d\}, \{a,c,d\}], [\{e\}, \{e\}], [\{g,h\}, \{g,h,i\}] \rangle$,
$A_7 = \langle [\{a,b,c,d\}, \{a,b,c,d\}], [\{e\}, \{e,f\}], [\{g\}, \{g,i\}] \rangle$.
Additionally, we have:

$$\tau_{IVNV} = \tau \cup \{A_8, A_9, A_{10}, A_{11}\},$$

where $A_8 = \langle [\{a,b\}, \{a,b,c\}], [\{e\}, \{e,f\}], [\{g\}, \{g\}] \rangle$,
$A_9 = \langle [\{a,d\}, \{a,c,d\}], [\{e\}, \{e\}], [\{g,h\}, \{g,h\}] \rangle$,
$A_{10} = \langle [\{a\}, \{a,c\}], [\{e\}, \{e\}], [\{g\}, \{g,h\}] \rangle$,
$A_{11} = \langle [\{a,b,d\}, \{a,b,c,d\}], [\{e\}, \{e,f\}], [\{g\}, \{g\}] \rangle$.
So we can confirm that Proposition 16 holds.

Furthermore, we can obtain six IVTs on X for τ*:*

$$\tau^{T}_{IVN},\ \tau^{I}_{IVN},\ \tau^{F}_{IVN},\ \tau^{T}_{IVNV},\ \tau^{I}_{IVNV},\ \tau^{F}_{IVNV}.$$

Additionally, we have 12 ordinary topologies on X:

$$\tau^{T,-}_{IVN},\ \tau^{T,+}_{IVN},\ \tau^{I,-}_{IVN},\ \tau^{I,+}_{IVN},\ \tau^{F,-}_{IVN},\ \tau^{F,+}_{IVN},$$

$$\tau^{T,-}_{IVNV}, \tau^{T,+}_{IVNV}, \tau^{I,-}_{IVNV}, \tau^{I,+}_{IVNV}, \tau^{F,-}_{IVNV}, \tau^{F,+}_{IVNV}.$$

The following is the immediate result of Proposition 16 (2).

Corollary 1. *Let (X,τ) be an IVNCTS and let $IVNCC_\tau$ [resp. $IVNCC_{\tau_{IVN}}$ and $IVNCC_{\tau_{IVNV}}$] be the set of all IVNCCSs w.r.t. τ [resp. τ_{IVN} and τ_{IVNV}]. Then,*

$$IVNCC_\tau \subset IVNCC_{\tau_{IVN}}, \text{ and } IVNCC_\tau \subset IVNCC_{\tau_{IVNV}}.$$

Example 13. *Let (X,τ) be the IVNCTS given in Example 12. Then we have:*
$IVNCC_\tau = \{\varnothing_{IVN}, X_{IVN}, A^c_1, A^c_2, A^c_3, A^c_4\}$,
$IVNCC_{\tau_{IVN}} = IVNCC_\tau \cup \{A^c_5, A^c_6, A^c_7\}$,
$IVC_{\tau_{IVNV}} = IVC_\tau \cup \{A^c_8, A^c_9, A^c_{10}, A^c_{11}\}$,
where $A^c_1 = \langle [\{g\},\{g,i\}], [\{a,b,c,d,h\},\{a,b,c,d,f,h\}], [\{a,b\},\{a,b,c\}] \rangle$,
$A^c_2 = \langle [\{g,h\},\{g,h,i\}], [\{a,b,c,d,f\},\{a,b,c,d,f\}], [\{a,d\},\{a,c,d\}] \rangle$,
$A^c_3 = \langle [\{g,h\},\{g,h,i\}], [\{a,b,c,d,f\},\{a,b,c,d,f\}], [\{a\},\{a,c\}] \rangle$,
$A^c_4 = \langle [\{g\},\{g,i\}], [\{a,b,c,d,h\},\{a,b,c,d,f,h\}], [\{a,b,d\},\{a,b,c,d\}] \rangle$,
$A^c_5 = \langle [\{g\},\{g,i\}], [\{a,b,c,d,h\},\{a,b,c,d,f,h\}], [\{a,b,c\},\{a,b,c\}] \rangle$,
$A^c_6 = \langle [\{g,h\},\{g,h,i\}], [\{a,b,c,d,f\},\{a,b,c,d,f\}], [\{a,c,d\},\{a,c,d\}] \rangle$,
$A^c_7 = \langle [\{g\},\{g,i\}], [\{a,b,c,d,h\},\{a,b,c,d,f,h\}], [\{a,b,c,d\},\{a,b,c,d\}] \rangle$,
$A^c_8 = \langle [\{g\},\{g\}], [\{a,b,c,d,h\},\{a,b,c,d,f,h\}], [\{a,b\},\{a,b,c\}] \rangle$,
$A^c_9 = \langle [\{g,h\},\{g,h\}], [\{a,b,c,d,f\},\{a,b,c,d,f\}], [\{a,d\},\{a,c,d\}] \rangle$,
$A^c_{10} = \langle [\{g\},\{g,h\}], [\{a,b,c,d,f\},\{a,b,c,d,f\}], [\{a\},\{a,c\}] \rangle$,
$A^c_{11} = \langle [\{g\},\{g\}], [\{a,b,c,d,h\},\{a,b,c,d,f,h\}], [\{a,b,d\},\{a,b,c,d\}] \rangle$.
Thus we can confirm that Corollary 1 holds.

Now let us consider the converses of Propositions 14 and 15.

Proposition 17. *Suppose to each $a \in X$, there corresponds a set $N^*(a_{IVNV})$ of IVNCSs in X satisfying the conditions [IVNCVN1], [IVNCVN2], [IVNCVN3], and [IVNCVN4] in Proposition 15. Then there is an IVNCT on X such that $N^*(a_{IVNV})$ is the set of all IVNCVNs of a_{IVNV} in this IVNCT for each $a \in X$.*

Proof. Let,

$$\tau_{IVNV} = \{U \in IVNCS(X) : U \in N(a_{IVNV}) \text{ for each } a_{IVNV} \in U\},$$

where $N(a_{IVNV})$ denotes the set of all IVNCVNs in τ.
Then clearly, $\tau_{IVNV} \in IVNCT(X)$ by Proposition 16. We will prove that $N^*(a_{IVNV})$ is the set of all IVNCVNs of $a_{IVNV)}$ in τ_{IVNV} for each $a \in X$.

Let $V \in IVN^*(X)$ such that $V \in N^*(a_{IVNV})$ and let U be the union of all the IVNCVPs b_{IVNV} in X such that $U \in N^*(a_{IVNV})$. If we can prove that:

$$a_{IVNV} \in U \subset V \text{ and } U \in \tau_{IVNV},$$

then the proof will be complete.

Since $V \in N_*(a_{IVNV})$, $a_{IVNV} \in U$ by the definition of U. Moreover, $U \subset V$. Suppose $b_{IVNV} \in U$. Then by [IVNCVN4], there is an IVNCS $W \in N^*(b_{IVNV})$ such that $V \in N^*(c_{IVNV)})$ for each $c_{IVNV} \in W$. Thus $c_{IVNV} \in U$. By Proposition 9, $W \subset U$. So by [IVNCVN2], $U \in N^*(b_{IVNV})$ for each $b_{IVNV} \in U$. Hence by the definition of τ_{IVNV}, $U \in \tau_{IVNV}$. This completes the proof. □

Proposition 18. *Suppose to each $a \in X$, there corresponds a set $N^*(a_{IVN})$ of IVNCSs in X satisfying the conditions [IVNCN1], [IVNCN2], [IVNCN3], and [IVNCN4] in Proposition 14. Then there is an IVNCT on X such that $N^*(a_{IVN})$ is the set of all IVNCNs of $a_{IVN)}$ in this IVNCT for each $a \in X$.*

Proof. The proof is similar to Proposition 17. □

The following provide a necessary and sufficient condition which an IVNCSs is an IVNCOS in an IVNCTS.

Theorem 2. *Let (X, τ) be an IVNCTS, $A \in IVNCS(X)$. Then $A \in \tau$ if and only if $A \in N(a_{IVN})$ and $A \in N(a_{IVNV})$ for each a_{IVN}, $a_{IVNV} \in A$.*

Proof. Suppose $A \in N(a_{IVN})$ and $A \in N(a_{IVNV})$ for each a_{IVN}, $a_{IVNV} \in A$. Then there are $U_{a_{IVN}}$, $V_{a_{IVNV}} \in \tau$ such that $a_{IVN} \in U_{a_{IVN}} \subset A$ and $a_{IVNV} \in V_{a_{IVNV}} \subset A$. Thus,

$$A = (\bigcup_{a_{IVN} \in A} a_{IVN}) \cup (\bigcup_{a_{IVNV} \in A} a_{IVNV}) \subset (\bigcup_{a_{IVN} \in A} U_{a_{IVN}}) \cup (\bigcup_{a_{IVNV} \in A} V_{a_{IVNV}}) \subset A.$$

So $A = (\bigcup_{a_{IVN} \in A} U_{a_{IVN}}) \cup (\bigcup_{a_{IVNV} \in A} V_{a_{IVNV}})$. Since $U_{a_{IVN}}$, $V_{a_{IVNV}} \in \tau$, $A \in \tau$.

The proof of the necessary condition is easy. □

Now we will give the relation among three IVNCTs, τ, τ_{IVN} and τ_{IVNV}.

Proposition 19. $\tau = \tau_{IVN} \cap \tau_{IVNV}$.

Proof. From Proposition 16 (2), it is clear that $\tau \subset \tau_{IVN} \cap \tau_{IVNV}$.

Conversely, let $U \in \tau_{IVN} \cap \tau_{IVNV}$. Then clearly, $U \in \tau_{IVN}$ and $U \in \tau_{IVNV}$. Thus U is an IVNCN of each of its IVNCPs a_{IVN} and an IVNCVN of each of its IVNCVPs a_{IVNV}. Thus, there are $U_{a_{IVN}}$, $U_{a_{IVNV}} \in \tau$ such that $a_{IVN} \in U_{a_{IVN}} \subset U$ and $a_{IVNV} \in U_{a_{IVNV}} \subset U$. So we have:

$$U_{IVN} = \bigcup_{a_{IVN} \in U} a_{IVN} \subset \bigcup_{a_{IVN} \in U} U_{a_{IVN}} \subset U$$

and

$$U_{IVNV} = \bigcup_{a_{IVNV} \in U} a_{IVNV} \subset \bigcup_{a_{IVNV} \in U} U_{a_{IVNV}} \subset U.$$

By Proposition 5, we get:

$$U = U_{IVN} \cup U_{IVNV} \subset (\bigcup_{a_{IVN} \in U} U_{a_{IVN}}) \cup (\bigcup_{a_{IVNV} \in U} U_{a_{IVNV}}) \subset U, \text{ i.e.,}$$

$$U = (\bigcup_{a_{IVN} \in U} U_{a_{IVN}}) \cup (\bigcup_{a_{IVNV} \in U} U_{a_{IVNV}}).$$

It is obvious that $(\bigcup_{a_{IVN} \in U} U_{a_{IVN}}) \cup (\bigcup_{a_{IVNV} \in U} U_{a_{IVNV}}) \in \tau$. Hence $U \in \tau$. Therefore $\tau_{IVN} \cap \tau_{IVNV} \subset \tau$. This completes the proof. □

From Proposition 19, we get the following.

Corollary 2. *Let (X, τ) be an IVNCTS. Then,*

$$IVNCC_{\tau} = IVNCC_{\tau_{IVN}} \cap IVNCC_{\tau_{IVNV}}.$$

Example 14. *In Example 12, we can easily check that Corollary 2 holds.*

6. Interiors and Closures of IVNCSs

In this section, we define interval-valued neutrosophic crisp interiors and closures, and investigate some of their properties and give some examples. In particular, we will show that there is a unique IVNCT on a set X from the interval-valued neutrosophic crisp closure [resp. interior] operator.

In an IVNCTS, we can define a closure and an interior as well as two other types of closures and interiors by Proposition 16.

Definition 21. *Let* (X, τ) *be an IVNCTS,* $A \in IVNCS(X)$.

(i) *The interval-valued neutrosophic crisp closure of A w.r.t.* τ*, denoted by* $IVNcl(A)$*, is an IVNCS in X defined as:*
$$IVNcl(A) = \bigcap\{K : K^c \in \tau \text{ and } A \subset K\}.$$

(ii) *The interval-valued neutrosophic crisp interior of A w.r.t.* τ*, denoted by* $IVNint(A)$*, is an IVS in X defined as:*
$$IVNint(A) = \bigcup\{G : G \in \tau \text{ and } G \subset A\}.$$

(iii) *The interval-valued neutrosophic crisp closure of A w.r.t.* τ_{IVN}*, denoted by* $cl_{IVN}(A)$*, is an IVNCS in X defined as:*
$$cl_{IVN}(A) = \bigcap\{K : K^c \in \tau_{IVN} \text{ and } A \subset K\}.$$

(iv) *The interval-valued neutrosophic crisp interior of A w.r.t.* τ_{IVN}*, denoted by* $int_{IVN}(A)$*, is an IVS in X defined as:*
$$int_{IVN}(A) = \bigcup\{G : G \in \tau_{IVN} \text{ and } G \subset A\}.$$

(v) *The interval-valued neutrosophic crisp closure of A w.r.t.* τ_{IVNV}*, denoted by* $cl_{IVNV}(A)$*, is an IVNCS in X defined as:*
$$cl_{IVNV}(A) = \bigcap\{K : K^c \in \tau_{IVNV} \text{ and } A \subset K\}.$$

(vi) *The interval-valued neutrosophic crisp interior of A w.r.t.* τ_{IVNV}*, denoted by* $int_{IVNV}(A)$*, is an IVNCS in X defined as:*
$$int_{IVNV}(A = \bigcup\{G : G \in \tau_{IVNV} \text{ and } G \subset A\}.$$

Remark 6. *From the above definition, it is obvious that the followings hold:*
$$IVNint(A) \subset int_{IVN}(A),\ IVNint(A) \subset int_{IVNV}(A)$$
and
$$cl_{IVN}(A) \subset IVNcl(A),\ cl_{IVNV}(A) \subset IVNcl(A).$$

Example 15. *Let* (X, τ) *be the IVNCTS given in Examples 12 and 13. Consider two IVNCSs in X:*
$$A = \langle[\{a,b,c\},\{a,b,c,d\}],[\{a,e\},\{a,e,f\}],[\{g\},\{g\}]\rangle,$$
$$B = \langle[\{g,h\},\{g,h,i\}],[\{a,b,c,d,f\},\{a,b,c,d,e,f\}],[\{a\},\{a,c\}]\rangle.$$
Then,
$$\begin{aligned} IVNint(A) &= \bigcup\{G \in \tau : G \subset A\} = A_1 \cup A_3 = \langle[\{a,b\},\{a,b,c\}],[\{e\},\{e,f\}],[\{g\},\{g,i\}]\rangle, \\ int_{IVN}(A) &= \bigcup\{G \in \tau_{IVN} : G \subset A\} = A_1 \cup A_3 \cup A_5 \\ &= \langle[\{a,b,c\},\{a,b,c\}],[\{e\},\{e,f\}],[\{g\},\{g,i\}]\rangle, \\ int_{IVNV}(A) &= \bigcup\{G \in \tau_{IVNV} : G \subset A\} = A_1 \cup A_3 \cup A_8 \cup A_{10} \\ &= \langle[\{a,b\},\{a,b,c\}],[\{e\},\{e,f\}],[\{g\},\{g\}]\rangle \end{aligned}$$
and
$$\begin{aligned} IVNcl(B) &= \bigcap\{F : F^c \in \tau,\ B \subset F\} = A_2^c \cap A_3^c \\ &= \langle[\{g,h\},\{g,h,i\}],[\{a,b,c,d,f\},\{a,b,c,d,f\}],[\{a,d\},\{a,c,d\}]\rangle, \\ cl_{IVN}(B) &= \bigcap\{F : F^c \in \tau_{IVN},\ B \subset F\} = A_2^c \cap A_3^c \cap A_6^c \cap A_{10}^c \\ &= \langle[\{g,h\},\{g,h,i\}],[\{a,b,c,d,f\},\{a,b,c,d,f\}],[\{a,c,d\},\{a,c,d\}]\rangle, \end{aligned}$$

$$cl_{IVNV}(B) = \bigcap\{F : F^c \in \tau_{IVNV},\ B \subset F\} = A_2^c \cap A_3^c \cap A_9^c \cap A_{10}^c$$
$$= \langle [\{g\}, \{g,h\}], [\{a,b,c,d,f\}, \{a,b,c,d,f\}], [\{a,d\}, \{a,c,d\}] \rangle.$$

Thus we can confirm that Remark 6 holds.

Proposition 20. *Let* (X, τ) *be an IVNCTS,* $A \in IVNCS(X)$*. Then,*

$$IVNint(A^c) = (IVNcl(A))^c \text{ and } IVNcl(A^c) = (IVNint(A))^c.$$

Proof. $IVNint(A^c) = \bigcup\{U \in \tau : U \subset A^c\} = \bigcup\{U \in \tau : U \subset \left\langle A^F, A^{I^c}, A^T \right\rangle\}$
$= \bigcup\{U \in \tau : U^T \subset A^F,\ U^I \subset A^{I^c},\ U^F \supset A^T\}$
$= \bigcup\{U \in \tau : U^T \subset A^F,\ U^{I^c} \subset A^I,\ U^F \supset A^T\}$
$= (\bigcap\{U^c : U \in \tau, A \subset U^c\})^c$
$= (IVNcl(A))^c$.
Similarly, we can show that $IVNcl(A^c) = (IVNint(A))^c$. □

Proposition 21. *Let* (X, τ) *be an IVNCTS,* $A \in IVNCS(X)$*. Then,*

$$IVNint(A) = int_{IVN}(A) \cap int_{IVNV}(A).$$

Proof. The proof is straightforward from Proposition 19 and Definition 21. □

The following is the immediate result of Definition 21, and Propositions 20 and 21.

Corollary 3. *Let* (X, τ) *be an IVNCTS and let* $A \in IVNCS(X)$*. Then,*

$$IVNcl(A) = cl_{IVN}(A) \cup cl_{IVNV}(A).$$

Example 16. *Let A and B be two IVNCSs in X given in Example 15. Then we can easily check that:*

$$int_{IVN}(A) \cap int_{IVNV}(A) = IVNint(A),\ cl_{IVN}(B) \cup cl_{IVNV}(B) = IVNcl(B).$$

Theorem 3. *Let X be an IVNCTS,* $A \in IVNCS(X)$*. Then:*

(1) $A \in IVNCC(X) \Leftrightarrow$ *if* $A = IVNcl(A)$,
(2) $A \in IVNCO(X) \Leftrightarrow A = IVNint(A)$.

Proof. Straightforward. □

Proposition 22 (Kuratowski Closure Axioms). *Let X be an IVNCTS,* $A, B \in IVNCS(X)$*. Then,*

[IVNCK0] *If* $A \subset B$*, then* $IVNcl(A) \subset IVNcl(B)$,
[IVNCK1] $IVNcl(\varnothing_{IVN}) = \varnothing_{IVN}$,
[IVNCK2] $A \subset IVNcl(A)$,
[IVNCK3] $IVNcl(IVNcl(A)) = IVNcl(A)$,
[IVNCK4] $IVcl(A \cup B) = IVNcl(A) \cup IVNcl(A)$.

Proof. Straightforward. □

Let $IVNcl^* : IVNCS(X) \to IVNCS(X)$ be the mapping satisfying the properties [IVNCK1], [IVNCK2], [IVNCK3], and [IVNCK4]. Then the mapping $IVcl^*$ is called the interval-valued neutrosophic crisp closure operator (briefly, IVNCCO) on X.

Proposition 23. *Let $IVNcl^*$ be the IVNCCO on X. Then there exists a unique IVNCT τ on X such that $IVNcl^*(A) = IVNcl(A)$, for each $A \in IVNCS(X)$, where $IVNcl(A)$ denotes the interval-valued neutrosophic crisp closure of A in the IVNCTS (X, τ). In fact,*

$$\tau = \{A^c \in IVNCS(X) : IVNcl^*(A) = A\}.$$

Proof. The proof is almost similar to the case of classical topological spaces. □

Proposition 24. *⇔Let X be an IVNCTS, $A, B \in IVNCS(X)$. Then,*

[IVNCI0] *If $A \subset B$, then $IVNint(A) \subset IVNint(B)$,*
[IVNCI1] *$IVNint(X_{IVN}) = X_{IVN}$,*
[IVNCI2] *$IVNint(A) \subset A$,*
[IVNCI3] *$IVNint(IVNint(A)) = IVNint(A)$,*
[IVNCI4] *$IVNint(A \cap B) = IVNint(A) \cap IVNint(A)$.*

Proof. Straightforward. □

Let $IVNint^* : IVNCS(X) \to IVNCS(X)$ be the mapping satisfying the properties [IVNCI1], [IVNCI2], [IVNCI3], and [IVNCI4]. Then the mapping $IVNint^*$ is called the interval-valued neutrosophic crisp interior operator (briefly, IVNCIO) on X.

Proposition 25. *Let $IVNint^*$ be the IVNCIO on X. Then there exists a unique IVNCT τ on X such that $IVNint^*(A) = IVNint(A)$ for each $A \in IVNCS(X)$, where $IVNint(A)$ denotes the interval-valued neutrosophic crisp interior of A in the IVNCTS (X, τ). In fact,*

$$\tau = \{A \in IVNCS(X) : IVNint^*(A) = A\}.$$

Proof. The proof is similar to one of Proposition 23. □

7. Interval-Valued Neutrosophic Crisp Continuous Mappings

In this section, we define an interval-valued neutrosophic crisp continuity and quotient topology, and study some of their properties.

Definition 22. *Let $X, \tau)$, (Y, δ) be two IVTSs proposed in [30]. Then a mapping $f : X \to Y$ is said to be interval-valued continuous, if $f^{-1}(V) \in \tau$ for each $V \in \delta$.*

Definition 23. *Let $X, \tau)$, (Y, δ) be two IVNCTSs. Then a mapping $f : X \to Y$ is said to be interval-valued neutrosophic crisp continuous, if $f^{-1}(V) \in \tau$ for each $V \in \delta$.*

From Remark 2 (1), and Definitions 22 and 23, we can easily have the following.

Theorem 4. *Let (X, τ), (Y, δ) be two IVNCTSs and let $f : X \to Y$ be a mapping. Then f is interval-valued neutrosophic crisp continuous if and only if $f : (X, \tau^T) \to (Y, \delta^T)$, $f : (X, \tau^I) \to (Y, \delta^I)$, and $f : (X, \tau^F) \to (Y, \delta^F)$ are interval-valued continuous, respectively.*

The followings are immediate results of Proposition 9 (13) and Definition 23.

Proposition 26. *Let X, Y, Z be IVNCTSs.*

(1) *The identity mapping $id : X \to X$ is continuous.*
(2) *If $f : X \to Y$ and $g : Y \to Z$ are continuous, then $g \circ f : X \to Z$ is continuous.*

Remark 7. *From Proposition 26, we can easily see that the class of all IVNCTSs and continuous mappings, denoted by* **IVNCTop**, *forms a concrete category.*

The followings are immediate results of Definition 23.

Proposition 27. *Let* X, Y *be INCTSs.*

(1) *If* X *is an IVNCDTS, the* $f : X \to Y$ *is continuous,*
(2) *If* Y *is an IVNCITS, then* $f : X \to Y$ *is continuous.*

Theorem 5. *Let* X, Y *be IVNCTSs and let* $f : X \to Y$ *be a mapping. Then the followings are equivalent:*

(1) f *is continuous,*
(2) $f^{-1}(C) \in IVNCC(X)$ *for each* $C \in IVNCC(Y)$,
(3) $f^{-1}(S) \in IVNCO(X)$ *for each member* S *of the subbase for the IVNCT on* Y,
(4) $IVNcl(f^{-1}(B)) \subset f^{-1}(IVNcl(B))$ *for each* $B \in IVNC(Y)$,
(5) $f(IVNcl(A)) \subset IVNcl(f(A))$ *for each* $A \in IVNC(X)$.

Proof. The proofs of (1)⇒(2)⇒(3)⇒(1) are obvious.

(2)⇒(4): Suppose the condition (2) holds and let $B \in INC(Y)$. By Proposition 22 [IVNCK2], $B \subset IVNcl(B)$. Then by Proposition 9 (2), $f^{-1}(B) \subset f^{-1}(IVNcl(B))$. Thus by Proposition 22 [INCK0],

$$IVNcl(f^{-1}(B)) \subset IVNcl(f^{-1}(IVNcl(B))).$$

Since $IVNcl(B) \in IVNCC(Y)$, $f^{-1}(IVNcl(B)) \in IVNCC(X)$ by the condition (2). So by Theorem 3 (1), $IVNcl(f^{-1}(IVNcl(B))) = f^{-1}(IVNcl(B))$. Hence $IVNcl(f^{-1}(B)) \subset f^{-1}(IVNcl(B))$.

(4)⇒(5): Suppose the condition (4) holds and let $B = f(A)$ for each $A \in IVNC(X)$. Then we have $IVNcl(f^{-1}(f(A))) \subset f^{-1}(IVNcl(f(A)))$. Thus by Proposition 9 (3), $IVNcl(A) \subset f^{-1}(IVNcl(f(A)))$. So by Proposition 9 (1) and (4), $f(IVNcl(A)) \subset IVNcl(f(A))$.

(5)⇒(4): The proof is similar to (4)⇒(5). □

Theorem 6. *Let* X, Y *be IVNCTSs and let* $f : X \to Y$ *be a mapping. Then* f *is continuous if and only if* $f^{-1}(IVNint(B)) \subset IVNint(f^{-1}(B))$ *for each* $B \in INC(Y)$.

Proof. The proof is straightforward. □

Definition 24. *Let* (X, τ), (Y, δ) *be two IVNCTSs. Then a mapping* $f : X \to Y$ *is said to be:*

(i) *Interval-valued neutrosophic crisp open, if* $f(U) \in \delta$ *for each* $U \in \tau$,
(ii) *Interval-valued neutrosophic crisp closed, if* $f(C) \in IVNCC(Y)$ *for each* $C \in IVNCC(X)$.

Proposition 28. *Let* X, Y, Z *be IVNCTSs, let* $f : X \to Y$ *and* $g : Y \to Z$ *be mappings. If* f, g *are open [resp. closed], then* $g \circ g$ *is open [resp. closed].*

Proof. The proof is straightforward. □

Theorem 7. *Let* X, Y *be IVNCTSs and let* $f : X \to Y$ *be a mapping. Then* f *is open if and only if* $IVNint(f(A)) \subset f(IVNint(A))$ *for each* $A \in IVNC(X)$.

Proof. The proof is straightforward. □

Proposition 29. *Let* X, Y *be IVNCTSs and let* $f : X \to Y$ *be injective. If* f *is continuous, then* $f(IVNint(A)) \subset IVNint(f(A))$ *for each* $A \in IVNC(X)$.

Proof. The proof is straightforward. □

The following is the immediate result of Theorem 7 and Proposition 29.

Corollary 4. *Let* X, Y *be IVNCTSs and let* $f : X \to Y$ *be continuous, open, and injective. Then* $f(IVNint(A)) = IVNint(f(A))$ *for each* $A \in IVNC(X)$.

Theorem 8. *Let* X, Y *be IVNCTSs and let* $f : X \to Y$ *be a mapping. Then* f *is close if and only if* $IVNcl(f(A)) \subset f(IVNcl(A))$ *for each* $A \in IVNC(X)$.

Proof. The proof is straightforward. □

The following is the immediate result of Theorems 5 and 8.

Corollary 5. *Let* X, Y *be IVNCTSs and let* $f : X \to Y$ *be a mapping. Then* f *is continuous and closed if and only if* $f(VINcl(A)) = IVNcl(f(A))$ *for each* $A \in IVNC(X)$.

Definition 25. *Let* (X, τ), (Y, δ) *be two IVNCTSs. Then a mapping* $f : X \to Y$ *is called an interval-valued neutrosophic crisp homeomorphism, if* f *is bijective, continuous, and open.*

Theorem 9. *Let* X, Y *be IVNCDTSs and let* $f : X \to Y$ *be a mapping. Then* f *is a homeomorphism if and only if* f *is bijective.*

Proof. The proof is straightforward. □

Definition 26. *Let* (X, τ) *be an IVNCTS, let* Y *be a set and let* $f : X \to Y$ *be a surjective mapping. Let* δ *be the family of IVNCSs in* Y *given by:*

$$\delta = \{B \in IVNC(Y) : f^{-1}(B) \in \tau\}.$$

Then δ *is called the interval-valued neutrosophic crisp quotient topology (briefly, IVNCQT) on* Y.

It can easily be seen that $\delta \in IVNCT(Y)$. *It is also obvious that for each* $B \in IVNC(Y)$, B *is closed in* δ *if and only if* $f^{-1}(B)$ *is closed in* X.

Proposition 30. *Let* (X, τ), (Y, δ) *be two IVNCTSs, where* δ *is the IVNCQT on* Y. *Then a surjection* $f : X \to Y$ *is continuous and open. Moreover,* δ *is the finest topology on* Y *which* f *is continuous.*

Proof. The proof is similar to the classical case. □

The following is the immediate result of Proposition 30.

Corollary 6. *Let* (X, τ), (Y, δ) *be two IVNCTSs. If a mapping* $f : X \to Y$ *is continuous, open, and sujective, then* δ *is the IVNCQT on* Y. *But the converse does not hold in general (See Example 17).*

Example 17. *Let* $([0,1], \tau)$ *be an IVNCTS and let* $A = [\frac{1}{2}, 1]$. *Consider the characteristic function* $\chi_A : [0,1] \to \{0,1\}$, *where* $\{0,1\}$ *is the interval-valued neutrosophic crisp Sierpin'ski space (see Example 7 (5)). Then we can easily see that the topology on* $\{0,1\}$ *is the IVNCQT. On the other hand,* $(\frac{1}{2}, 1)_{IVNCI} \in \tau$ *but* $\chi_A((\frac{1}{2}, 1)_{IVNCI})$ *is not open in* $\{0,1\}$. *Thus* χ_A *is not an open mapping.*

Theorem 10. *Let* (X, τ), (Y, δ), (Z, σ) *be IVNCTSs, where* δ *is the IVNCQT on* Y. *Let* $f : X \to Y$ *and* $g : Y \to Z$ *be mappings. Then* g *is continuous if and only if* $g \circ f$ *is continuous.*

Proof. The proof is similar to the classical case. □

8. Conclusions

We obtained various properties of IVNCSs and discussed with IVNCTSs which can be considered as an interval-valued tri-opological space. Moreover, we defined an interval-valued neutrosophic crisp base and subbase and proved the characterization of an interval-valued neutrosophic crisp base. Next, we introduced the concept of interval-valued neutrosophic crisp neighborhoods and obtained some similar properties to classical neighborhoods. Furthermore, we defined interval-valued neutrosophic crisp closures and interiors, and found some properties. We also introduced the concept of interval-valued neutrosophic crisp continuities and obtained its various properties.

In future, we expect that one can apply the concept of IVNCSs to group and ring theory, BCK-algebra, and category theory, etc. We also expect that one can define the notions of interval-valued soft sets and interval-valued neutrosophic crisp soft sets. Besides, the theorems developed in this manuscript will promote future studies on the geometry calibration for multi-cameras.

Author Contributions: Create and conceptualize ideas, J.-G.L. and K.H.; Writing—original draft preparation, J.-G.L. and K.H.; Writing—review and editing, D.J.; S.S. and C.X.; Funding acquisition, D.J. All authors have read and agreed to the published version of the manuscript.

Funding: This work was supported by the Institute of Information & communications Technology Planning & Evaluation (IITP), grant funded by the Korean government(MSIT) (No. 2020-0-00226, Development of High-Definition, Unstructured Plenoptic video Acquisition Technology for Medium and Large Space).

Acknowledgments: We would like to thank the anonymous reviewers for their very careful reading and valuable comments/suggestions.

Conflicts of Interest: The authors declare no conflict of interest.

References

1. Zadeh, L.A. Fuzzy sets. *Inf. Control* **1965**, *8*, 338–353.
2. Zadeh, L.A. The concept of a linguistic variable and its application to approximate reasoning-I. *Inform. Sci.* **1975**, *8*, 199–249.
3. Pawlak, Z. Rough sets. *Int. J. Inf. Comput. Sci.* **1982**, *11*, 341–356.
4. Atanassov, K. Intuitionistic fuzzy sets. In *VII ITKR's Session*; Sgurev, V., Ed.; Academy of Sciences: Sofia, Bulgaria, 1984.
5. Atanassov, K.T.; Gargov, G. Interval-valued intuitionistic fuzzy sets. *Fuzzy Sets Syst. 31*, **1989**, 343–349.
6. Gau, W.L.; Buchrer, D.J. Vague sets. *IEEE Trans. Syst. Man Cybernet.* **1993**, *23*, 610–614.
7. Smarandache, F. *Neutrosophy Neutrisophic Property, Sets, and Logic*; Americana Research Press: Rehoboth, MA, USA, 1998.
8. Molodtsov, D. Soft set theory—First results. *Comput. Math. Appl.* **1999**, *37*, 19–31.
9. Lee, K.M. Bipolar-valued fuzzy sets and their basic operations. In Proceedings of the International Conference on Intelligent Technologies, Bangkok, Thailand, 13–15 December 2000; pp. 307–312.
10. Torra, V. Hesitant fuzzy sets. *Int. J. Intell. Syst.* **2010**, *25*, 529–539.
11. Jun, Y.B.; Kim, C.S.; Yang, K.O. Cubic sets. *Ann. Fuzzy Math. Inform.* **2012**, *4*, 83–98.
12. Lee, J.G.; Senel, G.; Lim, P.K.; Kim, J.; Hur, K. Octahedron sets. *Ann. Fuzzy Math. Inform.* **2020**, *19*, 211–238.
13. cCoker, D. A note on intuitionistic sets and intuitionistic points. *Tr. J. Math.* **1996**, *20*, 343–351.
14. cCoker, D. An introduction to intuitionistic topological spaces. *Busefal* **2000**, *81*, 51–56.
15. Kim, J.H.; Lim, P.K.; Lee, J.G.; Hur, K. The category of intuitionistic sets. *Ann. Fuzzy Math. Inform.* **2017**, *14*, 549–562.
16. Kim, J.; Lim, P.K.; Lee, J.G.; Hur, K. Intuitionistic topological spaces. *Ann. Fuzzy Math. Inform.* **2018**, *15*, 29–46.
17. Salama, A.A.; Smarandache, F.; Kroumov, V. Neutrosophic crisp sets & neutrosophic crisp topological spaces. *Neutrosophic Sets Syst.* **2014**, *2*, 25–30.
18. Smarandache, F. *A Unifying Field in Logics: Neutrosophic Logic, Neutrosophy, Neutrosophic Set, Neutrosophic Probability*; American Research Press: Rehoboth, MA, USA, 1999.

19. Smarandache, F. Neutrosophy and neutrosophic logic. In Proceedings of the First Conference on Neutrosophy, Neutrosophic Logic, Set, Probability and Statistics, Gallup, NM, USA, 5–21 January 2002.
20. Hur, K.; Lim, P.K.; Lee, J.G.; Kim, J. The category of neutrosophic crisp sets. *Ann. Fuzzy Math. Inform.* **2017**, *14*, 43–54.
21. Salama, A.A.; Broumi, S.; Smarandache, F. Neutrosophic crisp open set and neutrosophic crisp continuity via neutrosophic crisp ideals. In *Neutrosophic Theory and Its Applications*; Collected Papers; EuropaNova absl: Brussels, Belgium, 2014; Volume I, pp. 199–205. Available online: http://fs.gallup.unm.edu/NeutrosophicTheoryApplications.pdf (accessed on 10 December 2020).
22. Salama, A.A.; Smarandache, F.; Alblowi, S.A. New neutrosophic crisp topological concepts. *Neutrosophic Sets Syst.* **2014**, *4*, 50–54.
23. Salama, A.A.; Hanafy, I.M. Hewayda Elghawalby and M. S. Dabash, Neutrosophic crisp α-topological spacess. *Neutrosophic Sets Syst.* **2016**, *12*, 92–96.
24. Al-Omeri, W. Neutrosophic crisp sets via neutrosophic crisp topological spaces. *Neutrosophic Sets Syst.* **2016**, *13*, 1–9.
25. Salama, A.A.; Hanafy, I.M.; Dabash, M.S. Semi-compact and semi-Lindelof space via neutrosophic crisp set theory. *Asia Math.* **2018**, *2*, 41–48.
26. Al-Hamido, R.K. Neutrosophic crisp bi-topological spaces. *Neutrosophic Sets Syst.* **2018**, *21*, 66–73.
27. Al-Hamido, R.K.; Gharibah, T. Neutrosophic crisp tri-topological spaces, *J. New Theory* **2018**, *23*, 13–21.
28. Al-Hamido, R.K.; Gharibah, T.; Jafari, S.; Smarandache, F. On neutrosophic crisp topology via N-topology. *Neutrosophic Sets Syst.* **2018**, *23*, 96–109.
29. Al-Nafee, A.B.; Al-Hamido, R.K.; Smarandache, F. Separation axioms in neutrosophic crisp topological spaces. *Neutrosophic Sets Syst.* **2019**, *25*, 25–32.
30. Kim, J.; Jun, Y.B.; Lee, J.G.; Hur, K. Topological structures based on interval-valued sets. *Ann. Fuzzy Math. Inform.* **2020**, *20*, 273–295.
31. Singh, G.; Memoli, F.; Carlsson, G. Topological methods for the analysis of high dimensional data sets and 3D object recognition. *SPBG* **2007**, *91*, 100.
32. Snasel, V.; Nowakova, J.; Xhafa, F.; Barolli, L. Geometrical and topological approach to Big Data. *Future Genration Comput. Syst.* **2017**, *67*, 286–296.
33. Bagchi, S. On the analysis and computation of topological fuzzy measure in distributed monoid spaces. *Symmetry* **2019**, *11*, 9, doi:10.3390/sym11010009.
34. Salama, A.A.; Smarandache, F. *Neutrosophic Crisp Set Theory*; The Educational Publisher: Columbus, OH, USA, 2015.
35. Yao, Y. Interval sets and interval set algebras. In Proceedings of the 8th IEEE International Conference on Cognitive Intormatics (ICCI'09), Hong Kong, China, 15–17 June 2009; pp. 307–314.
36. Salama, A.A.; Broumi, S.; Smarandache, F. Some types of neutrosophic crisp sets and neutrosophic crisp relations. In *I. J. Information Engineering and Electronic Business*; Published Online; MECS: Hong Kong, China, 2014. Available online: http://www.mecs-press.org/ (accessed on 10 December 2020).
37. Salama, A.A. Neutrosophic crisp points & neutrosophic crisp idealss. *Neutrosophic Sets Systens* **2013**, *1*, 1–5.
38. Ye, J. Similarity measures between interval neutrosophic sets and their applications in multicriteria decision-making. *J. Intell. Fuzzy Syst.* **2014**, *26*, 165–172.
39. Kelly, J.C. Bitopological spaces. *Proc. Lond. Math. Soc.* **1963**, *13*, 71–89.

symmetry

MDPI

Article

£-Single Valued Extremally Disconnected Ideal Neutrosophic Topological Spaces

Fahad Alsharari

Department of Mathematics, College of Science and Human Studies, Hotat Sudair, Majmaah University, Majmaah 11952, Saudi Arabia; f.alsharari@mu.edu.sa

Abstract: This paper aims to mark out new concepts of r-single valued neutrosophic sets, called r-single valued neutrosophic £-closed and £-open sets. The definition of £-single valued neutrosophic irresolute mapping is provided and its characteristic properties are discussed. Moreover, the concepts of £-single valued neutrosophic extremally disconnected and £-single valued neutrosophic normal spaces are established. As a result, a useful implication diagram between the r-single valued neutrosophic ideal open sets is obtained. Finally, some kinds of separation axioms, namely r-single valued neutrosophic ideal-R_i (r-$SVNIR_i$, for short), where $i = \{0,1,2,3\}$, and r-single valued neutrosophic ideal-T_j (r-$SVNIT_j$, for short), where $j = \{1,2,2\frac{1}{2},3,4\}$, are introduced. Some of their characterizations, fundamental properties, and the relations between these notions have been studied.

Keywords: r-single valued neutrosophic £-closed; £-single valued neutrosophic irresolute mapping; £-single valued neutrosophic extremally disconnected; £-single valued neutrosophic normal; r-$SVNIR_i$; r-$SVNIT_j$

check for updates

Citation: Alsharari, F. £-Single Valued Extremally Disconnected Ideal Neutrosophic Topological Spaces. *Symmetry* **2021**, *13*, 53. http://doi.org/10.3390/sym13010053

Received: 12 November 2020
Accepted: 29 December 2020
Published: 31 December 2020

Publisher's Note: MDPI stays neutral with regard to jurisdictional claims in published maps and institutional affiliations.

1. Introduction

In 1999, Smarandache introduced the concept of a neutrosophy [1]. It has been used at various axes of mathematical theories and applications. In recent decades, the theory made an outstanding advancement in the field of topological spaces. Salama et al. and Hur et al. [2–6], for example, among many others, wrote their works in fuzzy neutrosophic topological spaces (FNTS), following Chang [7]'s discoveries in the way of fuzzy topological spaces (FTS).

Šostak, in 1985 [8], marked out a new definition of fuzzy topology as a crisp subfamily of family of fuzzy sets, which seems to be a drawback in the process of fuzzification of the concept of topological spaces. Yan, Wang, Nanjing, Liang, and Yan [9,10] developed a parallel theory in the context of intuitionistic I-fuzzy topological spaces.

The idea of "single-valued neutrosophic set" [11] was set out by Wang in 2010. Gayyar [12], in his 2016 paper, foregrounded the concept of a "smooth neutrosophic topological spaces". The ordinary single-valued neutrosophic topology was presented by Kim [13]. Recently, Saber et al. [14,15] familiarized the concepts of single-valued neutrosophic ideal open local function, single-valued neutrosophic topological space, and the connectedness and stratification of single-valued neutrosophic topological spaces.

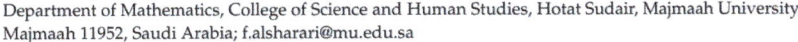

Neutrosophy, and especially neutrosophic sets, are powerful, general, and formal frameworks that generalize the concept of the ordinary sets, fuzzy sets, and intiuitionistic fuzzy sets from philosophical point of view. This paper sets out to introduce and examine a new class of sets called r-single valued £-closed in the single valued neutrosophic topological spaces in Šostak's sense. More precisely, different attributes, like £-single valued neutrosophic irresolute mapping, £-single valued neutrosophic extremally disconnected, £-single valued neutrosophic normal spaces, and some kinds of separation axioms, were developed. It can be fairly claimed that we have achieved expressive definitions, distinguished theorems, important lemmas, and counterexamples to investigate, in-depth, our consequences and to find out the best results. It is notable to say that different crucial notions in single valued neutrosophic topology were generalized in this article. Different attributes, like extremally disconnected and some kinds of separation axioms, which have a significant impact on the overall topology's notions, were also studied.

It is notable to say that the application aspects to this area of research can be further pointed to. There are many applications of neutrosophic theories in many branches of sciences. Possible applications are to control engineering and to Geographical Information Systems, and so forth, and could be secured, as mentioned by many authors, such as Reference [16–20].

In this study, $\tilde{X}$ is assumed to be a nonempty set, $\xi = [0,1]$ and $\xi_0 = (0,1]$. For $\alpha \in \xi$, $\tilde{\alpha}(\nu) = \alpha$ for all $\nu \in \tilde{X}$. The family of all single-valued neutrosophic sets on $\tilde{X}$ is denoted by $\xi^{\tilde{X}}$.

2. Preliminaries

This section is devoted to provide a complete survey and trace previous studies related to the idea of this research article.

Definition 1 ([21]). *Let $\tilde{X}$ be a non-empty set. A neutrosophic set (briefly, NS) in $\tilde{X}$ is an object having the form*

$$\sigma_n = \{\langle \nu, \tilde{\rho}_{\sigma_n}(\nu), \tilde{\varrho}_{\sigma_n}(\nu), \tilde{\eta}_{\sigma_n}(\nu)\rangle : \nu \in \tilde{X}\},$$

where

$$\tilde{\rho} : \tilde{X} \to]^{-}0, 1^{+}[, \ \tilde{\varrho} : \tilde{X} \to]^{-}0, 1^{+}[, \ \tilde{\eta} : \tilde{X} \to]^{-}0, 1^{+}[$$

and

$$^{-}0 \le \tilde{\rho}_{\sigma_n}(\nu) + \tilde{\varrho}_{\sigma_n}(\nu) + \tilde{\eta}_{\sigma_n}(\nu) \le 3^{+}$$

represent the degree of membership (namely $\tilde{\rho}_{\sigma_n}(\nu)$), the degree of indeterminacy (namely $\tilde{\varrho}_{\sigma_n}(\nu)$), and the degree of non-membership (namely $\tilde{\eta}_{\sigma_n}(\nu)$), respectively, of any $\nu \in \tilde{X}$ to the set σ_n.

Definition 2 ([11]). *Let $\tilde{X}$ be a space of points (objects), with a generic element in $\tilde{X}$ denoted by ν. Then, σ_n is called a single valued neutrosophic set (briefly, SVNS) in $\tilde{X}$, if σ_n has the form $\sigma_n = \langle \tilde{\rho}_{\sigma_n}, \tilde{\varrho}_{\sigma_n}, \tilde{\eta}_{\sigma_n}\rangle$, where $\tilde{\rho}_{\sigma_n}, \tilde{\varrho}_{\sigma_n}, \tilde{\eta}_{\sigma_n} : \tilde{X} \to [0,1]$. In this case, $\tilde{\rho}_{\sigma_n}, \tilde{\varrho}_{\sigma_n}, \tilde{\eta}_{\sigma_n}$ are called truth membership function, indeterminancy membership function, and falsity membership function, respectively.*

Let $\tilde{X}$ be a nonempty set and $\xi = [0,1]$ and $\xi_0 = (0,1]$. A single-valued neutrosophic set σ_n on $\tilde{X}$ is a mapping defined as $\sigma_n = \langle \tilde{\rho}_{\sigma_n}, \tilde{\varrho}_{\sigma_n}, \tilde{\eta}_{\sigma_n}\rangle : \tilde{X} \to \xi$ such that $0 \le \tilde{\rho}_{\sigma_n}(\nu) + \tilde{\varrho}_{\sigma_n}(\nu) + \tilde{\eta}_{\sigma_n}(\nu) \le 3$.

We denote the single-valued neutrosophic sets $\langle 0,1,1\rangle$ and $\langle 1,0,0\rangle$ by $\tilde{0}$ and $\tilde{1}$, respectively.

Definition 3 ([11]). *Let $\sigma_n = \langle \tilde{\rho}_{\sigma_n}, \tilde{\varrho}_{\sigma_n}, \tilde{\eta}_{\sigma_n}\rangle$ be an SVNS on $\tilde{X}$. The complement of the set σ_n (briefly σ_n^c) is defined as follows:*

$$\tilde{\rho}_{\sigma_n^c}(\nu) = \tilde{\eta}_{\sigma_n}(\nu), \quad \tilde{\varrho}_{\sigma_n^c}(\nu) = [\tilde{\varrho}_{\sigma_n}]^c(\nu), \quad \tilde{\eta}_{\sigma_n^c}(\nu) = \tilde{\rho}_{\sigma_n}(\nu).$$

Definition 4 ([22,23]). *Let $\tilde{X}$ be a non-empty set and let $\sigma_n, \gamma_n \in \xi^{\tilde{X}}$ be given by $\sigma_n = \langle \tilde{\rho}_{\sigma_n}, \tilde{\varrho}_{\sigma_n}, \tilde{\eta}_{\sigma_n}\rangle$ and $\gamma_n = \langle \tilde{\rho}_{\gamma_n}, \tilde{\varrho}_{\gamma_n}, \tilde{\eta}_{\gamma_n}\rangle$. Then:*

(1) *We say that $\sigma_n \subseteq \gamma_n$ if $\tilde{\rho}_{\sigma_n} \le \tilde{\rho}_{\gamma_n}$, $\tilde{\varrho}_{\sigma_n} \ge \tilde{\varrho}_{\gamma_n}$, $\tilde{\eta}_{\sigma_n} \ge \tilde{\eta}_{\gamma_n}$.*

(2) *The intersection of σ_n and γ_n denoted by $\sigma_n \cap \gamma_n$ is an SVNS and is given by*

$$\sigma_n \cap \gamma_n = \langle \tilde{\rho}_{\sigma_n} \cap \tilde{\rho}_{\gamma_n}, \ \tilde{\varrho}_{\sigma_n} \cup \tilde{\varrho}_{\gamma_n}, \ \tilde{\eta}_{\sigma_n} \cup \tilde{\eta}_{\gamma_n}\rangle.$$

(3) *The union of σ_n and γ_n denoted by $\sigma_n \cup \gamma_n$ is an SVNS and is given by*

$$\sigma_n \cup \gamma_n = \langle \tilde{\rho}_{\sigma_n} \cup \tilde{\rho}_{\gamma_n}, \ \tilde{\varrho}_{\sigma_n} \cap \tilde{\varrho}_{\gamma_n}, \ \tilde{\eta}_{\sigma_n} \cap \tilde{\eta}_{\gamma_n}\rangle.$$

For any arbitrary family $\{\sigma_n\}_{i\in j} \subseteq \xi^{\tilde{X}}$ *of SVNS, the union and intersection are given by*

(4) $\bigcap_{i\in j}[\sigma_n]_i = \langle \cap_{i\in j}\tilde{\rho}_{[\sigma_n]_i}, \cup_{i\in j}\tilde{\varrho}_{[\sigma_n]_i}, \cup_{i\in j}\tilde{\eta}_{[\sigma_n]_i}\rangle$,

(5) $\bigcup_{i\in j}[\sigma_n]_i = \langle \cup_{i\in j}\tilde{\rho}_{[\sigma_n]_i}, \cap_{i\in j}\tilde{\varrho}_{[\sigma_n]_i}, \cap_{i\in j}\tilde{\eta}_{[\sigma_n]_i}\rangle$.

Definition 5 ([12])**.** *A single-valued neutrosophic topological space is an ordered quadruple* $(\tilde{X}, \tilde{\tau}^{\tilde{\rho}}, \tilde{\tau}^{\tilde{\varrho}}, \tilde{\tau}^{\tilde{\eta}})$ *where* $\tilde{\tau}^{\tilde{\rho}}, \tilde{\tau}^{\tilde{\varrho}}, \tilde{\tau}^{\tilde{\eta}} : \xi^{\tilde{X}} \to \xi$ *are mappings satisfying the following axioms:*

(SVNT1) $\tilde{\tau}^{\tilde{\rho}}(\tilde{0}) = \tilde{\tau}^{\tilde{\rho}}(\tilde{1}) = 1$ *and* $\tilde{\tau}^{\tilde{\rho}}(\tilde{0}) = \tilde{\tau}^{\tilde{\rho}}(\tilde{1}) = \tilde{\tau}^{\tilde{\eta}}(\tilde{0}) = \tilde{\tau}^{\tilde{\eta}}(\tilde{1}) = 0$,

(SVNT2) $\tilde{\tau}^{\tilde{\rho}}(\sigma_n \cap \gamma_n) \geq \tilde{\tau}^{\tilde{\rho}}(\sigma_n) \cap \tilde{\tau}^{\tilde{\rho}}(\gamma_n)$, $\quad \tilde{\tau}^{\tilde{\varrho}}(\sigma_n \cap \gamma_n) \leq \tau^{\tilde{\varrho}}(\sigma_n) \cup \tilde{\tau}^{\tilde{\varrho}}(\gamma_n)$,
$\tilde{\tau}^{\tilde{\eta}}(\sigma_n \cap \gamma_n) \leq \tilde{\tau}^{\tilde{\eta}}(\sigma_n) \cup \tilde{\tau}^{\tilde{\eta}}(\gamma_n)$, *for all* $\sigma_n, \gamma_n \in \xi^{\tilde{X}}$,

(SVNT3) $\tilde{\tau}^{\tilde{\rho}}(\cup_{j\in\Gamma}[\sigma_n]_j) \geq \cap_{j\in\Gamma}\tilde{\tau}^{\tilde{\rho}}([\sigma_n]_j)$, $\quad \tilde{\tau}^{\tilde{\varrho}}(\cup_{i\in\Gamma}[\sigma_n]_j) \leq \cup_{j\in\Gamma}\tilde{\tau}^{\tilde{\varrho}}([\sigma_n]_j)$,
$\tilde{\tau}^{\tilde{\eta}}(\cup_{j\in\Gamma}[\sigma_n]_j) \leq \cup_{j\in\Gamma}\tilde{\tau}^{\tilde{\eta}}([\sigma_n]_j)$ *for all* $\{[\sigma_n]_j, j \in \Gamma\} \in \xi^{\tilde{X}}$.

The quadruple $(\tilde{X}, \tilde{\tau}^{\tilde{\rho}}, \tilde{\tau}^{\tilde{\varrho}}, \tilde{\tau}^{\tilde{\eta}})$ *is called a single-valued neutrosophic topological space (SVNTS, for short). We will occasionally write* $\tau^{\tilde{\rho}\tilde{\varrho}\tilde{\eta}}$ *for* $(\tau^{\tilde{\rho}}, \tau^{\tilde{\varrho}}, \tau^{\tilde{\eta}})$ *and it will cause no ambiguity*

Definition 6 ([14])**.** *Let* $(\tilde{X}, \tilde{\tau}^{\tilde{\rho}}, \tilde{\tau}^{\tilde{\varrho}}, \tilde{\tau}^{\tilde{\eta}})$ *be an SVNTS. Then, for every* $\sigma_n \in \xi^{\tilde{X}}$ *and* $r \in \xi_0$, *the single valued neutrosophic closure and the single valued neutrosophic interior of* σ_n *are defined by:*

$$C_{\tilde{\tau}^{\tilde{\rho}\tilde{\varrho}\tilde{\eta}}}(\sigma_n, s) = \bigcap\{\gamma_n \in \xi^{\tilde{X}} : \sigma_n \leq \gamma_n, \quad \tau^{\tilde{\rho}}([\gamma_n]^c) \geq r, \quad \tau^{\tilde{\varrho}}([\gamma_n]^c) \leq 1 - r, \quad \tau^{\tilde{\eta}}([\gamma_n]^c) \leq 1 - r\},$$

$$\text{int}_{\tilde{\tau}^{\tilde{\rho}\tilde{\varrho}\tilde{\eta}}}(\sigma_n, s) = \bigcup\{\gamma_n \in \xi^{\tilde{X}} : \sigma_n \geq \gamma_n, \quad \tau^{\tilde{\rho}}(\gamma_n) \geq r, \quad \tau^{\tilde{\varrho}}(\gamma_n) \leq 1 - r, \quad \tau^{\tilde{\eta}}(\gamma_n) \leq 1 - r\}.$$

Definition 7 ([24])**.** *Let* $(\tilde{X}, \tau^{\tilde{\rho}\tilde{\varrho}\tilde{\eta}})$ *be an SVNTS and* $r \in \xi_0, \sigma_n \in \xi^{\tilde{X}}$. *Then,*

(1) σ_n *is r-single valued neutrosophic semiopen (r-SVNSO, for short) iff* $\sigma_n \leq C_{\tilde{\tau}^{\tilde{\rho}\tilde{\varrho}\tilde{\eta}}}(\text{int}_{\tilde{\tau}^{\tilde{\rho}\tilde{\varrho}\tilde{\eta}}}(\sigma_n, r), r)$,

(2) σ_n *is r-single valued neutrosophic* β*-open (r-SVN*β*O, for short) iff* $\sigma_n \leq C_{\tilde{\tau}^{\tilde{\rho}\tilde{\varrho}\tilde{\eta}}}(\text{int}_{\tilde{\tau}^{\tilde{\rho}\tilde{\varrho}\tilde{\eta}}}(C_{\tilde{\tau}^{\tilde{\rho}\tilde{\varrho}\tilde{\eta}}}(\sigma_n, r), r), r)$.

The complement of $r - SVNSO$ *(resp. r-SVN*β*O) is said to be an* $r - SVNSC$ *(resp. r-SVN*β*C), respectively.*

Definition 8 ([14])**.** *Let* $\tilde{X}$ *be a nonempty set and* $v \in \tilde{X}$. *If* $s \in (0,1]$, $t \in [0,1)$ *and* $p \in [0,1)$. *Then, the single-valued neutrosophic point* $x_{s,t,p}$ *in* $\tilde{X}$ *is given by*

$$x_{s,t,p}(\kappa) = \begin{cases} (s,t,p), & \text{if } x = v, \\ (0,1,1), & \text{otherwise.} \end{cases}$$

We say $x_{s,t,p} \in \sigma_n$ *iff* $s < \tilde{\rho}_{\sigma_n}(v)$, $t \geq \tilde{\varrho}_{\sigma_n}(v)$ *and* $p \geq \tilde{\eta}_{\sigma_n}(v)$. *To avoid the ambiguity, we denote the set of all neutrosophic points by* $pt(\xi^{\tilde{X}})$.

A single-valued neutrosophic set σ_n *is said to be quasi-coincident with another single-valued neutrosophic set* γ_n, *denoted by* $\sigma_n q \gamma_n$, *if there exists an element* $v \in \tilde{X}$ *such that*

$$\tilde{\rho}_{\sigma_n}(v) + \tilde{\rho}_{\gamma_n}(v) > 1, \quad \tilde{\varrho}_{\sigma_n}(v) + \tilde{\varrho}_{\gamma_n}(v) \leq 1, \quad \tilde{\eta}_{\sigma_n}(v) + \tilde{\eta}_{\gamma_n}(v) \leq 1.$$

Definition 9 ([14])**.** *A mapping* $\mathcal{I}^{\tilde{\rho}\tilde{\varrho}\tilde{\eta}} = \mathcal{I}^{\tilde{\rho}}, \mathcal{I}^{\tilde{\varrho}}, \mathcal{I}^{\tilde{\eta}} : \xi^{\tilde{X}} \to \xi$ *is called single-valued neutrosophic ideal (SVNI) on* $\tilde{X}$ *if it satisfies the following conditions:*

(I_1) $\mathcal{I}^{\tilde{\rho}}(\tilde{0}) = 1$ *and* $\mathcal{I}^{\tilde{\varrho}}(\tilde{0}) = \mathcal{I}^{\tilde{\eta}}(\tilde{0}) = 0$.

(I_2) *If* $\sigma_n \leq \gamma_n$, *then* $\mathcal{I}^{\tilde{\rho}}(\gamma_n) \leq \mathcal{I}^{\tilde{\rho}}(\sigma_n)$, $\mathcal{I}^{\tilde{\varrho}}(\gamma_n) \geq \mathcal{I}^{\tilde{\varrho}}(\sigma_n)$, *and* $\mathcal{I}^{\tilde{\eta}}(\gamma_n) \geq \mathcal{I}^{\tilde{\eta}}(\sigma_n)$, *for* $\gamma_n, \sigma_n \in \xi^{\tilde{X}}$.

(I_3) $\mathcal{I}^{\tilde{\rho}}(\sigma_n \cup \gamma_n) \geq \mathcal{I}^{\tilde{\rho}}(\sigma_n) \cap \mathcal{I}^{\tilde{\rho}}(\gamma_n)$, $\mathcal{I}^{\tilde{\varrho}}(\sigma_n \cup \gamma_n) \leq \mathcal{I}^{\tilde{\varrho}}(\sigma_n) \cup \mathcal{I}^{\tilde{\varrho}}(\gamma_n)$ *and*
$\mathcal{I}^{\tilde{\eta}}(\sigma_n \cup \gamma_n) \leq \mathcal{I}^{\tilde{\eta}}(\sigma_n) \cup \mathcal{I}^{\tilde{\eta}}(\gamma_n)$, *for each* $\sigma_n, \gamma_n \in \xi^{\tilde{X}}$.

The triple $(\tilde{X}, \tau^{\tilde{\rho}\tilde{\varrho}\tilde{\eta}}, \mathcal{I}^{\tilde{\rho}\tilde{\varrho}\tilde{\eta}})$ *is called a single valued neutrosophic ideal topological space in Šostak's sense (SVNITS, for short).*

Definition 10 ([14]). *Let $(\tilde{X}, \tau^{\tilde{\rho}\tilde{\varrho}\tilde{\eta}}, \mathcal{I}^{\tilde{\rho}\tilde{\varrho}\tilde{\eta}})$ be an SVNITS for each $\sigma_n \in \xi^{\tilde{X}}$. Then, the single valued neutrosophic ideal open local function $[\sigma_n]_r^{£}(\tau^{\tilde{\rho}\tilde{\varrho}\tilde{\eta}}, \mathcal{I}^{\tilde{\rho}\tilde{\varrho}\tilde{\eta}})$ of σ_n is the union of all single-valued neutrosophic points $x_{s,t,k}$ such that, if $\gamma_n \in Q_{\tau^{\tilde{\rho}\tilde{\varrho}\tilde{\eta}}}(x_{s,t,k}, r)$ and $\mathcal{I}^{\tilde{\rho}}(\varsigma_n) \geq r$, $\mathcal{I}^{\tilde{\varrho}}(\varsigma_n) \leq 1 - r$, $\mathcal{I}^{\tilde{\eta}}(\varsigma_n) \leq 1 - r$, then there is at least one $v \in \tilde{X}$ for which $\tilde{\rho}_{\sigma_n}(v) + \tilde{\rho}_{\gamma_n}(v) - 1 > \tilde{\rho}_{\varsigma_n}(v)$, $\tilde{\varrho}_{\sigma_n}(v) + \tilde{\varrho}_{\gamma_n}(v) - 1 \leq \tilde{\varrho}_{\varsigma_n}(v)$, and $\tilde{\eta}_{\sigma_n}(v) + \tilde{\eta}_{\gamma_n}(v) - 1 \leq \tilde{\eta}_{\varsigma_n}(v)$.*

Occasionally, we will write $[\sigma_n]_r^{£}$ for $[\sigma_n]_r^{£}(\tau^{\tilde{\rho}\tilde{\varrho}\tilde{\eta}}, \mathcal{I}^{\tilde{\rho}\tilde{\varrho}\tilde{\eta}})$, and it will cause no ambiguity.

Remark 1 ([14]). *Let $(\tilde{X}, \tau^{\tilde{\rho}\tilde{\varrho}\tilde{\eta}}, \mathcal{I}^{\tilde{\rho}\tilde{\varrho}\tilde{\eta}})$ be an SVNITS and $\sigma_n \in \xi^{\tilde{X}}$. Then,*

$$\mathrm{CI}^{£}_{\tau^{\tilde{\rho}\tilde{\varrho}\tilde{\eta}}}(\sigma_n, r) = \sigma_n \cup [\sigma_n]_r^{£}, \qquad \mathrm{int}^{£}_{\tau^{\tilde{\rho}\tilde{\varrho}\tilde{\eta}}}(\sigma_n, r) = \sigma_n \cap [(\sigma_n^c)_r^{£}]^c.$$

It is clear that $\mathrm{CI}^{£}_{\tau^{\tilde{\rho}\tilde{\varrho}\tilde{\eta}}}$ is a single-valued neutrosophic closure operator and $(\tau^{\tilde{\rho}£}(\mathcal{I}^{\rho}, \tau^{\tilde{\varrho}£}(\mathcal{I}^{\varrho}, \tau^{\tilde{\eta}£}$ $(\mathcal{I}^{\eta})$ is the single-valued neutrosophic topology generated by $\mathrm{CI}^{£}_{\tau^{\tilde{\rho}\tilde{\varrho}\tilde{\eta}}}$, i.e.,

$$\tau^{£}(\mathcal{I})(\sigma_n) = \bigcup\{r \mid \mathrm{CI}^{£}_{\tau^{\tilde{\rho}\tilde{\varrho}\tilde{\eta}}}(\sigma_n^c, r) = \sigma^c\}.$$

Theorem 1 ([14]). *Let $\{[\sigma_n]_i\}_{i \in J} \subset \xi^{\tilde{X}}$ be a family of single-valued neutrosophic sets on $\tilde{X}$ and $(\tilde{X}, \tilde{\tau}^{\tilde{\rho}\tilde{\varrho}\tilde{\eta}}, \mathcal{I}^{\tilde{\rho}\tilde{\varrho}\tilde{\eta}})$ be an r-SVNITS. Then,*

(1) $(\bigcup([\sigma_n]_i)_r^{£} : i \in J) \leq (\bigcup[\sigma_n]_i : i \in j)_r^{£}$,
(2) $(\bigcap([\sigma_n]_i) : i \in j)_r^{£} \geq (\bigcap([\sigma_n]_i)_r^{£} : i \in J)$.

Theorem 2 ([14]). *Let $(\tilde{X}, \tilde{\tau}^{\tilde{\rho}\tilde{\varrho}\tilde{\eta}}, \mathcal{I}^{\tilde{\rho}\tilde{\varrho}\tilde{\eta}})$ be an SVNITS and $\sigma_n, \gamma_n \in \xi^{\tilde{X}}$, $r \in \zeta_0$. Then,*

(1) $\mathrm{int}^{£}_{\tilde{\tau}^{\tilde{\rho}\tilde{\varrho}\tilde{\eta}}}(\sigma_n \vee \gamma_n, r) \leq \mathrm{int}^{£}_{\tilde{\tau}^{\tilde{\rho}\tilde{\varrho}\tilde{\eta}}}(\sigma_n, r) \vee \mathrm{int}^{£}_{\tilde{\tau}^{\tilde{\rho}\tilde{\varrho}\tilde{\eta}}}(\gamma_n, r)$,
(2) $\mathrm{int}_{\tilde{\tau}^{\tilde{\rho}\tilde{\varrho}\tilde{\eta}}}(\sigma_n, r) \leq \mathrm{int}^{£}_{\tilde{\tau}^{\tilde{\rho}\tilde{\varrho}\tilde{\eta}}}(\sigma_n, r) \leq \sigma_n \leq \mathrm{CI}^{£}_{\tilde{\tau}^{\tilde{\rho}\tilde{\varrho}\tilde{\eta}}}(\sigma_n, r) \leq \mathrm{C}_{\tilde{\tau}^{\tilde{\rho}\tilde{\varrho}\tilde{\eta}}}(\sigma_n, r)$,
(3) $\mathrm{CI}^{£}_{\tilde{\tau}^{\tilde{\rho}\tilde{\varrho}\tilde{\eta}}}([\sigma_n]^c, r) = [\mathrm{int}^{£}_{\tilde{\tau}^{\tilde{\rho}\tilde{\varrho}\tilde{\eta}}}(\sigma_n, r)]^c$, *and* $[\mathrm{CI}^{£}_{\tilde{\tau}^{\tilde{\rho}\tilde{\varrho}\tilde{\eta}}}(\sigma_n, r)]^c = \mathrm{int}^{£}_{\tilde{\tau}^{\tilde{\rho}\tilde{\varrho}\tilde{\eta}}}([\sigma_n]^c, r)$,
(4) $\mathrm{int}^{£}_{\tilde{\tau}^{\tilde{\rho}\tilde{\varrho}\tilde{\eta}}}(\sigma_n \wedge \gamma_n, r) = \mathrm{int}^{£}_{\tilde{\tau}^{\tilde{\rho}\tilde{\varrho}\tilde{\eta}}}(\sigma_n, r) \wedge \mathrm{int}^{£}_{\tilde{\tau}^{\tilde{\rho}\tilde{\varrho}\tilde{\eta}}}(\gamma_n, r)$.

3. £-Single Valued Neutrosophic Ideal Irresolute Mapping

This section provides the definitions of the r-single-valued neutrosophic £-open set (SVN£O, for short), the r-single-valued neutrosophic £-closed set (SVN£C, for short) and the £-single valued neutrosophic ideal irresolute mapping (£-SVNI-irresolute, for short), in the sense of Šostak. To understand the aim of this section, it is essential to clarify its content and elucidate the context in which the definitions, theorems, and examples are performed. Some results follow.

Definition 11. *Let $(\tilde{X}, \tilde{\tau}^{\tilde{\rho}\tilde{\varrho}\tilde{\eta}}, \mathcal{I}^{\tilde{\rho}\tilde{\varrho}\tilde{\eta}})$ be an r-SVNITS for every $\sigma_n \in \xi^{\tilde{X}}$ and $r \in \zeta_0$. Then, σ_n is called r-SVN£C iff $\mathrm{CI}^{£}_{\tau^{\tilde{\rho}\tilde{\varrho}\tilde{\eta}}}(\sigma_n, r) = \sigma_n$. The complement of the r-SVN£C is called r-SVN£O.*

Proposition 1. *Let $(\tilde{X}, \tau^{\tilde{\rho}\tilde{\varrho}\tilde{\eta}}, \mathcal{I}^{\tilde{\rho}\tilde{\varrho}\tilde{\eta}})$ be an r-SVNITS and $\sigma_n \in \xi^{\tilde{X}}$. Then,*

(1) σ_n *is r-SVN£C iff* $[\sigma_n]_r^{£} \leq \sigma_n$,
(2) σ_n *is r-SVN£O iff* $([\sigma_n]_r^{£})^c \geq [\sigma_n]^c$,
(3) *If* $\tau^{\tilde{\rho}}([\sigma_n]^c) \geq r$, $\tau^{\varrho}([\sigma_n]^c) \leq 1 - r$, $\tau^{\eta}([\sigma_n]^c) \leq 1 - r$, *then* σ_n *is r-SVN£C,*
(4) *If* $\tau^{\tilde{\rho}}(\sigma_n) \geq r$, $\tau^{\varrho}(\sigma_n) \leq 1 - r$, $\tau^{\eta}(\sigma_n) \leq 1 - r$, *then* σ_n *is r-SVN£O,*
(5) *If* σ_n *is r-SVNSC (resp. r-SVNβC), then* $\mathrm{int}_{\tau^{\tilde{\rho}\tilde{\varrho}\tilde{\eta}}}([\sigma_n]_r^{£}, r) \leq \sigma_n$ *(resp.* $\mathrm{int}_{\tau^{\tilde{\rho}\tilde{\varrho}\tilde{\eta}}}([\mathrm{int}_{\tau^{\tilde{\rho}\tilde{\varrho}\tilde{\eta}}}(\sigma_n, r)]_r^{£}, r) \leq \sigma_n$).

Proof. The proof of (1) and (2) are straightforward from Definition 11.
(3) Let $\tau^{\tilde{\rho}}([\sigma_n]^c) \geq r$, $\tau^{\varrho}([\sigma_n]^c) \leq 1 - r$, $\tau^{\eta}([\sigma_n]^c) \leq 1 - r$. Then,

$$\sigma_n = \mathrm{C}_{\tilde{\tau}^{\tilde{\rho}\tilde{\varrho}\tilde{\eta}}}(\sigma_n, r) \geq \mathrm{CI}^{£}_{\tau^{\tilde{\rho}\tilde{\varrho}\tilde{\eta}}}(\sigma_n, r) = \sigma_n \cup [\sigma_n]_r^{£} \geq [\sigma_n]_r^{£}.$$

Hence, σ_n is an r-SVN£C.
(4) The proof is direct consequence of (1).
(5) Let σ_n be an r-SVNSC. Then,

$$\begin{aligned}\sigma_n \geq \text{int}_{\tilde{\tau}^{\tilde{\rho}\tilde{\varrho}\tilde{\eta}}}(C_{\tilde{\tau}^{\tilde{\rho}\tilde{\varrho}\tilde{\eta}}}(\sigma_n, r), r) &\geq \text{int}_{\tilde{\tau}^{\tilde{\rho}\tilde{\varrho}\tilde{\eta}}}(CI^{\pounds}_{\tilde{\tau}^{\tilde{\rho}\tilde{\varrho}\tilde{\eta}}}(\sigma_n, r), r) = \text{int}_{\tilde{\tau}^{\tilde{\rho}\tilde{\varrho}\tilde{\eta}}}([\sigma_n \cup [\sigma_n]^{\pounds}_r], r) \\ &\geq \text{int}_{\tilde{\tau}^{\tilde{\rho}\tilde{\varrho}\tilde{\eta}}}([\sigma_n]^{\pounds}_r, r).\end{aligned}$$

The another case is similarly proved. □

Example 1. *Suppose that* $\tilde{X} = \{a, b\}$. *Define* $\varepsilon_n, \gamma_n, \varsigma_n \in \xi^{\tilde{X}}$ *as follows:*

$$\gamma_n = \langle (0.3, 0.3), (0.3, 0.3), (0.3, 0.3) \rangle; \quad \varepsilon_n = \langle (0.7, 0.7), (0.7, 0.7), (0.7, 0.7) \rangle;$$

$$\varsigma_n = \langle (0.2, 0.2), (0.2, 0.2), (0.2, 0.2) \rangle.$$

Define $\tilde{\tau}^{\tilde{\rho}\tilde{\varrho}\tilde{\eta}}, \mathcal{I}^{\tilde{\rho}\tilde{\varrho}\tilde{\eta}} : \xi^{\tilde{X}} \to \xi$ *as follows:*

$$\tilde{\tau}^{\tilde{\rho}}(\sigma_n) = \begin{cases} 1, & \text{if } \sigma_n = \tilde{0}, \\ 1, & \text{if } \sigma_n = \tilde{1}, \\ \frac{1}{3}, & \text{if } \sigma_n = \gamma_n; \\ \frac{1}{3}, & \text{if } \sigma_n = \varepsilon_n; \\ 0, & \text{if otherwise;} \end{cases} \qquad \mathcal{I}^{\tilde{\rho}}(\sigma_n) = \begin{cases} 1, & \text{if } \sigma_n = (0,1,1), \\ \frac{1}{3}, & \text{if } \sigma_n = \varsigma_n, \\ \frac{2}{3}, & \text{if } \tilde{0} < \sigma_n < \varsigma_n; \\ 0, & \text{if otherwise;} \end{cases}$$

$$\tilde{\tau}^{\tilde{\varrho}}(\sigma_n) = \begin{cases} 0, & \text{if } \sigma_n = \tilde{0}, \\ 0, & \text{if } \sigma_n = \tilde{1}, \\ \frac{2}{3}, & \text{if } \sigma_n = \gamma_n; \\ \frac{2}{3}, & \text{if } \sigma_n = \varepsilon_n; \\ 1, & \text{if otherwise;} \end{cases} \qquad \mathcal{I}^{\tilde{\varrho}}(\sigma_n) = \begin{cases} 0, & \text{if } \sigma_n = (0,1,1), \\ \frac{2}{3}, & \text{if } \sigma_n = \varsigma_n, \\ \frac{1}{3}, & \text{if } \tilde{0} < \sigma_n < \varsigma_n; \\ 1, & \text{if otherwise;} \end{cases}$$

$$\tilde{\tau}^{\tilde{\eta}}(\sigma_n) = \begin{cases} 0, & \text{if } \sigma_n = \tilde{0}, \\ 0, & \text{if } \sigma_n = \tilde{1}, \\ \frac{2}{3}, & \text{if } \sigma_n = \gamma_n; \\ \frac{2}{3}, & \text{if } \sigma_n = \varepsilon_n; \\ 1, & \text{if otherwise;} \end{cases} \qquad \mathcal{I}^{\tilde{\eta}}(\sigma_n) = \begin{cases} 0, & \text{if } \sigma_n = (0,1,1), \\ \frac{2}{3}, & \text{if } \sigma_n = \varsigma_n, \\ \frac{1}{3}, & \text{if } \tilde{0} < \sigma_n < \varsigma_n; \\ 1, & \text{if otherwise.} \end{cases}$$

(1) $G_n = \langle (0.6, 0.6), (0.6, 0.6), (0.6, 0.6) \rangle$ *is* $\frac{1}{3}$-*SVN£C but* $\tilde{\tau}^{\tilde{\rho}}([G_n]^c) \not\geq \frac{1}{3}$, $\tilde{\tau}^{\tilde{\varrho}}([G_n]^c) \not\leq \frac{2}{3}$, *and* $\tilde{\tau}^{\tilde{\eta}}([G_n]^c) \not\leq \frac{2}{3}$,
(2) $G_n = \langle (0.6, 0.6), (0.6, 0.6), (0.6, 0.6) \rangle \geq \text{int}_{\tilde{\tau}^{\tilde{\rho}\tilde{\varrho}\tilde{\eta}}}([G_n]^{\pounds}_{\frac{1}{3}}, \frac{1}{3}) = \tilde{0}$ *but* G_n *is not is* $\frac{1}{3}$-*SVNSC.*

Lemma 1. *Let* $(\tilde{X}, \tilde{\tau}^{\tilde{\rho}\tilde{\varrho}\tilde{\eta}}, \mathcal{I}^{\tilde{\rho}\tilde{\varrho}\tilde{\eta}})$ *be an SVNITS. Then, we have the following.*
(1) *Every intersection of r-SVN£C's is r-SVN£C.*
(2) *Every union of r-SVN£O's is r-SVN£O.*

Proof. (1) Let $\{[\sigma_n]_i\}_{i \in j}$ be a family of r-SVN£C's. Then, for every $i \in j$, we obtain $[\sigma_n]_i = CI^{\pounds}_{\tilde{\tau}^{\tilde{\rho}\tilde{\varrho}\tilde{\eta}}}([\sigma_n]_i, r)$, and, by Theorem 1(2), we have

$$\begin{aligned}\bigcap_{i \in j}[\sigma_n]_i &= \bigcap_{i \in j} CI^{\pounds}_{\tilde{\tau}^{\tilde{\rho}\tilde{\varrho}\tilde{\eta}}}([\sigma_n]_i, r), r) = \bigcap_{i \in j}([\sigma_n]_i \cup ([\sigma_n]_i)^{\pounds}_r) \geq \bigcap_{i \in j}[\sigma_n]_i \cup \bigcap_{i \in j}([\sigma_n]_i)^{\pounds}_r \\ &\geq \bigcap_{i \in j}[\sigma_n]_i \cup (\bigcap_{i \in j}[\sigma_n]_i)^{\pounds}_r = CI^{\pounds}_{\tilde{\tau}^{\tilde{\rho}\tilde{\varrho}\tilde{\eta}}}(\bigcap_{i \in j}[\sigma_n]_i, r).\end{aligned}$$

Therefore, $\bigcap_{i \in \Gamma}[\sigma_n]_i$ is r-SVN£C.

(2) From Theorem 1(1). □

Lemma 2. *Let* $(\tilde{X}, \tilde{\tau}^{\tilde{\rho}\tilde{\varrho}\tilde{\eta}}, \mathcal{I}^{\tilde{\rho}\tilde{\varrho}\tilde{\eta}})$ *be an SVNITS for each* $r \in \zeta_0$. *Then,*

(1) *For each r-SVN£O* $\sigma_n \in \xi^{\tilde{X}}$, $\sigma_n q \gamma_n$ *iff* $\sigma_n q \mathrm{CI}^{£}_{\tilde{\tau}\tilde{\rho}\tilde{\varrho}\tilde{\eta}}(\gamma_n, r)$,

(2) $x_{s,t,k} q \mathrm{CI}^{£}_{\tilde{\tau}\tilde{\rho}\tilde{\varrho}\tilde{\eta}}(\gamma_n, r)$ *iff* $\sigma_n q \gamma_n$ *for every r-SVN£O* $\sigma_n \in \xi^{\tilde{X}}$ *with* $x_{s,t,k} \in \sigma_n$.

Proof. (1) Let σ_n be an r-SVN£O and $\sigma_n \bar{q} \gamma_n$. Then, for any $\nu \in \tilde{X}$, we obtain

$$\tilde{\rho}_{\sigma_n}(\nu) + \tilde{\rho}_{\gamma_n}(\nu) > 1, \quad \tilde{\varrho}_{\sigma_n}(\nu) + \tilde{\varrho}_{\gamma_n}(\nu) \leq 1, \quad \tilde{\eta}_{\sigma_n}(\nu) + \tilde{\eta}_{\gamma_n}(\nu) \leq 1.$$

This implies that $\tilde{\rho}_{\gamma_n} \leq \tilde{\rho}_{[\sigma_n]^c}$, $\tilde{\varrho}_{\gamma_n} \geq \tilde{\varrho}_{[\sigma_n]^c}$ and $\tilde{\eta}_{\gamma_n} \geq \tilde{\eta}_{[\sigma_n]^c}$; hence, $\gamma_n \leq [\sigma_n]^c$. Since σ_n is r-SVN£O, $\mathrm{CI}^{£}_{\tilde{\tau}\tilde{\rho}\tilde{\varrho}\tilde{\eta}}(\gamma_n, r) \leq \mathrm{CI}^{£}_{\tilde{\tau}\tilde{\rho}\tilde{\varrho}\tilde{\eta}}([\sigma_n]^c, r) = [\sigma_n]^c$, it follows that $\sigma_n \bar{q} \mathrm{CI}^{£}_{\tilde{\tau}\tilde{\rho}\tilde{\varrho}\tilde{\eta}}(\gamma_n, r)$.

(2) Let $x_{s,t,k} q \mathrm{CI}^{£}_{\tilde{\tau}\tilde{\rho}\tilde{\varrho}\tilde{\eta}}(\gamma_n, r)$. Then, $\sigma_n q \mathrm{CI}^{£}_{\tilde{\tau}\tilde{\rho}\tilde{\varrho}\tilde{\eta}}(\gamma_n, r)$ with $x_{s,t,k} \in \sigma_n$. By (1), we have $\gamma_n q \sigma_n$ for each r-SVN£O $\sigma_n \in \xi^{\tilde{X}}$. On the other hand, let $\sigma_n \bar{q} \gamma_n$. Then, $\gamma_n \leq [\sigma_n]^c$. Since σ_n is r-SVN£O,

$$\mathrm{CI}^{£}_{\tilde{\tau}\tilde{\rho}\tilde{\varrho}\tilde{\eta}}(\gamma_n, r) \leq \mathrm{CI}^{£}_{\tilde{\tau}\tilde{\rho}\tilde{\varrho}\tilde{\eta}}([\sigma_n]^c, r) = [\sigma_n]^c \quad and \quad \sigma_n \bar{q} \mathrm{CI}^{£}_{\tilde{\tau}\tilde{\rho}\tilde{\varrho}\tilde{\eta}}(\gamma_n, r).$$

Since $x_{s,t,k} \in \sigma_n$, we obtain $x_{s,t,k} \bar{q} \mathrm{CI}^{£}_{\tilde{\tau}\tilde{\rho}\tilde{\varrho}\tilde{\eta}}(\gamma_n, r)$ □

Definition 12. *Suppose that* $f : (\tilde{X}, \tilde{\tau}_1^{\tilde{\rho}\tilde{\varrho}\tilde{\eta}}, \mathcal{I}_1^{\tilde{\rho}\tilde{\varrho}\tilde{\eta}}) \to (\tilde{Y}, \tilde{\tau}_2^{\tilde{\rho}\tilde{\varrho}\tilde{\eta}}, \mathcal{I}_2^{\tilde{\rho}\tilde{\varrho}\tilde{\eta}})$ *is a mapping. Then,*

(1) *f is called £-SVNI-irresolute iff* $f^{-1}(\sigma_n)$ *is r-SVN£O in* $\tilde{X}$ *for any r-SVN£O* σ_n *in* $\tilde{Y}$,

(2) *f is called £-SVNI-irresolute open iff* $f(\sigma_n)$ *is r-SVN£O in* $\tilde{Y}$ *for any r-SVN£O* σ_n *in* $\tilde{X}$,

(3) *f is called £-SVNI-irresolute closed iff* $f(\sigma_n)$ *is r-SVN£C in* $\tilde{Y}$ *for any r-SVN£C* σ_n *in* $\tilde{X}$.

Theorem 3. *Let* $f : (\tilde{X}, \tilde{\tau}_1^{\tilde{\rho}\tilde{\varrho}\tilde{\eta}}, \mathcal{I}_1^{\tilde{\rho}\tilde{\varrho}\tilde{\eta}}) \to (\tilde{Y}, \tilde{\tau}_2^{\tilde{\rho}\tilde{\varrho}\tilde{\eta}}, \mathcal{I}_2^{\tilde{\rho}\tilde{\varrho}\tilde{\eta}})$ *be a mapping. Then, the following conditions are equivalent:*

(1) *f is £-SVNI-irresolute,*

(2) $f^{-1}(\sigma_n)$ *is r-SVN£C, for each r-SVN£C* $\sigma_n \in \tilde{Y}$,

(3) $f(\mathrm{CI}^{£}_{\tilde{\tau}_1^{\tilde{\rho}\tilde{\varrho}\tilde{\eta}}}(\sigma_n, r)) \leq \mathrm{CI}^{£}_{\tilde{\tau}_2^{\tilde{\rho}\tilde{\varrho}\tilde{\eta}}}(f(\sigma_n), r)$ *for each* $\sigma_n \in \xi^{\tilde{X}}$, $r \in \zeta_0$,

(4) $\mathrm{CI}^{£}_{\tilde{\tau}_1^{\tilde{\rho}\tilde{\varrho}\tilde{\eta}}}(f^{-1}(\gamma_n), r) \leq f^{-1}(\mathrm{CI}^{£}_{\tilde{\tau}_2^{\tilde{\rho}\tilde{\varrho}\tilde{\eta}}}(\gamma_n, r))$ *for each* $\gamma_n \in \xi^{\tilde{Y}}$, $r \in \zeta_0$.

Proof. (1)⇒(2): Let σ_n be an r-SVN£C in $\tilde{Y}$. Then, $[\sigma_n]^c$ is r-SVN£O in $\tilde{Y}$ by (1), we obtain $f^{-1}([\sigma_n]^c)$ is r-SVN£O. But, $f^{-1}([\sigma_n]^c) = [f^{-1}(\sigma_n)]^c$. Then, $f^{-1}(\sigma_n)$ is r-SVN£C in $\tilde{X}$.

(2)⇒(3): For each $\sigma_n \in \xi^{\tilde{X}}$ and $r \in \zeta_0$, since $\mathrm{CI}^{£}_{\tilde{\tau}_2^{\tilde{\rho}\tilde{\varrho}\tilde{\eta}}}(\mathrm{CI}^{£}_{\tilde{\tau}_2^{\tilde{\rho}\tilde{\varrho}\tilde{\eta}}}(f(\sigma_n), r) = \mathrm{CI}^{£}_{\tilde{\tau}_2^{\tilde{\rho}\tilde{\varrho}\tilde{\eta}}}(f(\sigma_n), r)$. From Definition 11, $\mathrm{CI}^{£}_{\tilde{\tau}_2^{\tilde{\rho}\tilde{\varrho}\tilde{\eta}}}(f(\sigma_n), r)$ is r-SVN£C in $\tilde{Y}$. By (2), $f^{-1}(\mathrm{CI}^{£}_{\tilde{\tau}_2^{\tilde{\rho}\tilde{\varrho}\tilde{\eta}}}(f(\sigma_n), r))$ is r-SVN£C in $\tilde{X}$. Since

$$\sigma_n \leq f^{-1}(f(\sigma_n)) \leq f^{-1}(\mathrm{CI}^{£}_{\tilde{\tau}_2^{\tilde{\rho}\tilde{\varrho}\tilde{\eta}}}(f(\sigma_n), r)),$$

by Definition 11, we get,

$$\mathrm{CI}^{£}_{\tilde{\tau}_1^{\tilde{\rho}\tilde{\varrho}\tilde{\eta}}}(\sigma_n, r) \leq \mathrm{CI}^{£}_{\tilde{\tau}_1^{\tilde{\rho}\tilde{\varrho}\tilde{\eta}}}(f^{-1}(\mathrm{CI}^{£}_{\tilde{\tau}_2^{\tilde{\rho}\tilde{\varrho}\tilde{\eta}}}(f(\sigma_n), r)), r) = f^{-1}(\mathrm{CI}^{£}_{\tilde{\tau}_2^{\tilde{\rho}\tilde{\varrho}\tilde{\eta}}}(f(\sigma_n), r)).$$

Hence,

$$f(\mathrm{CI}^{£}_{\tilde{\tau}_1^{\tilde{\rho}\tilde{\varrho}\tilde{\eta}}}(\sigma_n, r)) \leq f(f^{-1}(\mathrm{CI}^{£}_{\tilde{\tau}_2^{\tilde{\rho}\tilde{\varrho}\tilde{\eta}}}(f(\sigma_n), r))) \leq \mathrm{CI}^{£}_{\tilde{\tau}_2^{\tilde{\rho}\tilde{\varrho}\tilde{\eta}}}(f(\sigma_n), r).$$

(3)⇒(4): For each $\gamma_n \in \xi^{\tilde{Y}}$ and $r \in \zeta_0$, put $\sigma_n = f^{-1}(\gamma_n)$. By (3),

$$f(\mathrm{CI}^{£}_{\tilde{\tau}_1^{\tilde{\rho}\tilde{\varrho}\tilde{\eta}}}(f^{-1}(\gamma_n), r)) \leq \mathrm{CI}^{£}_{\tilde{\tau}_2^{\tilde{\rho}\tilde{\varrho}\tilde{\eta}}}(f(f^{-1}(\gamma_n)), r) \leq \mathrm{CI}^{£}_{\tilde{\tau}_2^{\tilde{\rho}\tilde{\varrho}\tilde{\eta}}}(\gamma_n, r).$$

It implies that $CI^{\pounds}_{\tilde{\tau}_1^{\tilde{\rho}\tilde{\varrho}\tilde{\eta}}}(f^{-1}(\gamma_n), r) \leq f^{-1}(CI^{\pounds}_{\tilde{\tau}_2^{\tilde{\rho}\tilde{\varrho}\tilde{\eta}}}(\gamma_n, r))$.

(4)⇒(1): Let γ_n be an r-SVN£O in $\tilde{Y}$. Then, $[\gamma_n]^c$ is an r-SVN£C in $\tilde{Y}$. Hence, $CI^{\pounds}_{\tilde{\tau}_2^{\tilde{\rho}\tilde{\varrho}\tilde{\eta}}}([\gamma_n]^c, r) = [\gamma_n]^c$, and, by (4), we have,

$$f^{-1}([\gamma_n]^c) = f^{-1}(CI^{\pounds}_{\tilde{\tau}_2^{\tilde{\rho}\tilde{\varrho}\tilde{\eta}}}([\gamma_n]^c, r)) \geq CI^{\pounds}_{\tilde{\tau}_1^{\tilde{\rho}\tilde{\varrho}\tilde{\eta}}}(f^{-1}([\gamma_n]^c), r).$$

On the other hand, $f^{-1}([\gamma_n]^c) \leq CI^{\pounds}_{\tilde{\tau}_1^{\tilde{\rho}\tilde{\varrho}\tilde{\eta}}}(f^{-1}([\gamma_n]^c), r)$. Thus, $f^{-1}([\gamma_n]^c) = CI^{\pounds}_{\tilde{\tau}_1^{\tilde{\rho}\tilde{\varrho}\tilde{\eta}}}(f^{-1}([\gamma_n]^c), r)$, that is $f^{-1}([\gamma_n]^c)$ is an r-SVN£C set in $\tilde{X}$. Hence, $f^{-1}(\gamma_n)$ is an r-SVN£O set in $\tilde{X}$. □

Theorem 4. *Let* $f : (\tilde{X}, \tilde{\tau}_1^{\tilde{\rho}\tilde{\varrho}\tilde{\eta}}, \mathcal{I}_1^{\tilde{\rho}\tilde{\varrho}\tilde{\eta}}) \to (\tilde{Y}, \tilde{\tau}_2^{\tilde{\rho}\tilde{\varrho}\tilde{\eta}}, \mathcal{I}_2^{\tilde{\rho}\tilde{\varrho}\tilde{\eta}})$ *be a mapping. Then, the following conditions are equivalent:*

(1) *f is £-SVNI-irresolute open,*
(2) $f(int^{\pounds}_{\tilde{\tau}_1^{\tilde{\rho}\tilde{\varrho}\tilde{\eta}}}(\sigma_n, r)) \leq int^{\pounds}_{\tilde{\tau}_2^{\tilde{\rho}\tilde{\varrho}\tilde{\eta}}}(f(\sigma_n), r)$ *for each* $\sigma_n \in \xi^{\tilde{X}}$, $r \in \zeta_0$,
(3) $int^{\pounds}_{\tilde{\tau}_1^{\tilde{\rho}\tilde{\varrho}\tilde{\eta}}}((f^{-1}(\gamma_n), r) \leq f^{-1}(int^{\pounds}_{\tilde{\tau}_2^{\tilde{\rho}\tilde{\varrho}\tilde{\eta}}}(\gamma_n, r))$ *for each* $\gamma_n \in \xi^{\tilde{Y}}$, $r \in \zeta_0$,
(4) *For any* $\gamma_n \in \xi^{\tilde{Y}}$ *and any r-SVN£C* $\sigma_n \in \xi^{\tilde{X}}$ *with* $f^{-1}(\gamma_n) \leq \sigma_n$, *there exists an r-SVN£C* $\varsigma_n \in \xi^{\tilde{Y}}$ *with* $\gamma_n \leq \varsigma_n$ *such that* $f^{-1}(\varsigma_n) \leq \sigma_n$.

Proof. (1)⇒(2): For every $\sigma_n \in \xi^{\tilde{X}}$, $r \in \zeta_0$ and $int^{\pounds}_{\tilde{\tau}_1^{\tilde{\rho}\tilde{\varrho}\tilde{\eta}}}(\sigma_n, r) \leq \sigma_n$ from Theorem 2(2), we have $f(int^{\pounds}_{\tilde{\tau}_1^{\tilde{\rho}\tilde{\varrho}\tilde{\eta}}}(\sigma_n, r)) \leq f(\sigma_n)$. By (1), $f(int^{\pounds}_{\tilde{\tau}_1^{\tilde{\rho}\tilde{\varrho}\tilde{\eta}}}(\sigma_n, r))$ is r-SVN£O in $\tilde{Y}$. Hence,

$$f(int^{\pounds}_{\tilde{\tau}_1^{\tilde{\rho}\tilde{\varrho}\tilde{\eta}}}(\sigma_n, r)) = int^{\pounds}_{\tilde{\tau}_2^{\tilde{\rho}\tilde{\varrho}\tilde{\eta}}}(f(int^{\pounds}_{\tilde{\tau}_1^{\tilde{\rho}\tilde{\varrho}\tilde{\eta}}}(\sigma_n, r))) \leq int^{\pounds}_{\tilde{\tau}_2^{\tilde{\rho}\tilde{\varrho}\tilde{\eta}}}(f(\sigma_n), r).$$

(2)⇒(3): For each $\gamma_n \in \xi^{\tilde{Y}}$ and $r \in \zeta_0$, put $\sigma_n = f^{-1}(\gamma_n)$ from (2),

$$f(int^{\pounds}_{\tilde{\tau}_1^{\tilde{\rho}\tilde{\varrho}\tilde{\eta}}}(f^{-1}(\gamma_n), r)) \leq int^{\pounds}_{\tilde{\tau}_2^{\tilde{\rho}\tilde{\varrho}\tilde{\eta}}}(f(f^{-1}(\gamma_n)), r) \leq int^{\pounds}_{\tilde{\tau}_2^{\tilde{\rho}\tilde{\varrho}\tilde{\eta}}}(\gamma_n, r).$$

It implies that

$$int^{\pounds}_{\tilde{\tau}_1^{\tilde{\rho}\tilde{\varrho}\tilde{\eta}}}(f^{-1}(\gamma_n), r) \leq f^{-1}(f(int^{\pounds}_{\tilde{\tau}_1^{\tilde{\rho}\tilde{\varrho}\tilde{\eta}}}(f^{-1}(\gamma_n), r))) \leq f^{-1}(int^{\pounds}_{\tilde{\tau}_2^{\tilde{\rho}\tilde{\varrho}\tilde{\eta}}}(\gamma_n, r)).$$

(3)⇒(4): Obvious.

(4)⇒(1): Let ε_n be an r-SVN£O in $\tilde{X}$. Put $\gamma_n = [f(\varepsilon_n)]^c$ and $\sigma_n = [\varepsilon_n]^c$ such that σ_n is r-SVN£C in $\tilde{X}$. We obtain

$$f^{-1}(\gamma_n) = f^{-1}([f(\varepsilon_n)]^c) = [f^{-1}(f(\varepsilon_n))]^c \leq [\varepsilon_n]^c = \sigma_n.$$

From (4), there exists r-SVN£O $\varsigma_n \in \xi^{\tilde{Y}}$ with $\gamma_n \leq \varsigma_n$ such that $f^{-1}(\varsigma_n) \leq \sigma_n = [\varepsilon_n]^c$. It implies $\varepsilon_n \leq [f^{-1}(\varsigma)]^c = f^{-1}([\varsigma_n]^c)$. Thus, $f(\varepsilon_n) \leq f(f^{-1}([\varsigma]^c)) \leq [\varsigma_n]^c$. On the other hand, since $\gamma_n \leq \varsigma_n$, we have

$$f(\varepsilon_n) = [\gamma]^c \geq [\varsigma_n]^c.$$

Hence, $f(\varepsilon_n) = [\varsigma_n]^c$, that is, $f(\varepsilon_n)$ is r-SVN£O in $\tilde{Y}$. □

Theorem 5. *Let* $f : (\tilde{X}, \tilde{\tau}_1^{\tilde{\rho}\tilde{\varrho}\tilde{\eta}}, \mathcal{I}_1^{\tilde{\rho}\tilde{\varrho}\tilde{\eta}}) \to (\tilde{Y}, \tilde{\tau}_2^{\tilde{\rho}\tilde{\varrho}\tilde{\eta}}, \mathcal{I}_2^{\tilde{\rho}\tilde{\varrho}\tilde{\eta}})$ *be a mapping. Then, the following conditions are equivalent:*

(1) *f is £-SVNI-irresolute closed.*
(2) $f(CI^{\pounds}_{\tilde{\tau}_1^{\tilde{\rho}\tilde{\varrho}\tilde{\eta}}}(\gamma_n, r)) \leq CI^{\pounds}_{\tilde{\tau}_2^{\tilde{\rho}\tilde{\varrho}\tilde{\eta}}}(f(\gamma_n), r)$ *for each* $\gamma_n \in \xi^{\tilde{X}}$, $r \in \zeta_0$.

Proof. Obvious. □

Theorem 6. *Let $f : (\tilde{X}, \tilde{\tau}_1^{\tilde{\rho}\tilde{\varrho}\tilde{\eta}}, \mathcal{I}_1^{\tilde{\rho}\tilde{\varrho}\tilde{\eta}}) \to (\tilde{Y}, \tilde{\tau}_2^{\tilde{\rho}\tilde{\varrho}\tilde{\eta}}, \mathcal{I}_2^{\tilde{\rho}\tilde{\varrho}\tilde{\eta}})$ be a bijective mapping. Then, the following conditions are equivalent:*

(1) *f is £-SVNI-irresolute closed,*

(2) *$\mathrm{CI}^{£}_{\tilde{\tau}_1^{\tilde{\rho}\tilde{\varrho}\tilde{\eta}}}(f^{-1}(\sigma_n), r) \leq f^{-1}(\mathrm{CI}^{£}_{\tilde{\tau}_2^{\tilde{\rho}\tilde{\varrho}\tilde{\eta}}}(\sigma_n, r))$ for each $\sigma_n \in \xi^{\tilde{Y}}$, $r \in \xi_0$.*

Proof. $(1) \Rightarrow (2)$: Suppose that f is an £-SVNI-irresolute closed. From Theorem 5(2), we claim that, for each $\gamma_n \in \xi^{\tilde{X}}$ and $r \in \xi_0$,

$$f(\mathrm{CI}^{£}_{\tilde{\tau}_1^{\tilde{\rho}\tilde{\varrho}\tilde{\eta}}}(\gamma_n, r)) \leq \mathrm{CI}^{£}_{\tilde{\tau}_2^{\tilde{\rho}\tilde{\varrho}\tilde{\eta}}}(f(\gamma_n), r).$$

Now, for all $\sigma_n \in \xi^{\tilde{Y}}$, $r \in \xi_0$, put $\gamma_n = f^{-1}(\sigma_n)$, since f is onto, it implies that $f(f^{-1}(\sigma_n)) = \sigma_n$. Thus,

$$f(\mathrm{CI}^{£}_{\tilde{\tau}_1^{\tilde{\rho}\tilde{\varrho}\tilde{\eta}}}(f^{-1}(\sigma_n), r)) \leq \mathrm{CI}^{£}_{\tilde{\tau}_2^{\tilde{\rho}\tilde{\varrho}\tilde{\eta}}}(f(f^{-1}(\sigma_n)), r) = \mathrm{CI}^{£}_{\tilde{\tau}_2^{\tilde{\rho}\tilde{\varrho}\tilde{\eta}}}(\sigma_n, r).$$

Again, since f is onto, it follows:

$$\mathrm{CI}^{£}_{\tilde{\tau}_1^{\tilde{\rho}\tilde{\varrho}\tilde{\eta}}}(f^{-1}(\sigma_n), r) = f^{-1}(f(\mathrm{CI}^{£}_{\tilde{\tau}_1^{\tilde{\rho}\tilde{\varrho}\tilde{\eta}}}(f^{-1}(\sigma_n), r))) \leq f^{-1}(\mathrm{CI}^{£}_{\tilde{\tau}_2^{\tilde{\rho}\tilde{\varrho}\tilde{\eta}}}(\sigma_n, r)).$$

$(2) \Rightarrow (1)$: Put $\sigma_n = f(\gamma_n)$. By the injection of f, we get

$$\mathrm{CI}^{£}_{\tilde{\tau}_1^{\tilde{\rho}\tilde{\varrho}\tilde{\eta}}}(\gamma_n, r) = \mathrm{CI}^{£}_{\tilde{\tau}_1^{\tilde{\rho}\tilde{\varrho}\tilde{\eta}}}(f^{-1}(f(\gamma_n)), r) \leq f^{-1}(\mathrm{CI}^{£}_{\tilde{\tau}_2^{\tilde{\rho}\tilde{\varrho}\tilde{\eta}}}(f(\gamma_n), r)),$$

for the reason that f is onto, which implies that

$$f(\mathrm{CI}^{£}_{\tilde{\tau}_1^{\tilde{\rho}\tilde{\varrho}\tilde{\eta}}}(\gamma_n, r)) \leq f(f^{-1}(\mathrm{CI}^{£}_{\tilde{\tau}_2^{\tilde{\rho}\tilde{\varrho}\tilde{\eta}}}(f(\gamma_n), r))) = \mathrm{CI}^{£}_{\tilde{\tau}_2^{\tilde{\rho}\tilde{\varrho}\tilde{\eta}}}(f(\gamma_n), r).$$

□

4. £-Single Valued Neutrosophic Extremally Disconnected and £-Single Valued Neutrosophic Normal

This section is devoted to introducing £-single valued neutrosophic extremally disconnected (£-SVNE-disconnected, for short) and £-single valued neutrosophic normal (£-SVN-normal, for short), in the sense of Šostak. These definitions and their components, together with a set of criteria for identifying the spaces, are provided to illustrate how the ideas are applied.

Definition 13. *An SVNITS $(\tilde{X}, \tilde{\tau}^{\tilde{\rho}\tilde{\varrho}\tilde{\eta}}, \mathcal{I}^{\tilde{\rho}\tilde{\varrho}\tilde{\eta}})$ is called £-SVNE-disconnected if $\tilde{\tau}^{\tilde{\rho}}(\mathrm{CI}^{£}_{\tilde{\tau}^{\tilde{\rho}}}(\sigma_n, r)) \geq r$, $\tilde{\tau}^{\tilde{\varrho}}(\mathrm{CI}^{£}_{\tilde{\tau}^{\tilde{\varrho}}}(\sigma_n, r)) \leq 1 - r$, $\tilde{\tau}^{\tilde{\eta}}(\mathrm{CI}^{£}_{\tilde{\tau}^{\tilde{\eta}}}(\sigma_n, r)) \leq 1 - r$ for each $\tilde{\tau}^{\tilde{\rho}}(\sigma_n) \geq r$, $\tilde{\tau}^{\tilde{\varrho}}(\sigma_n) \leq 1 - r$, $\tilde{\tau}^{\tilde{\eta}}(\sigma_n) \leq 1 - r$.*

Definition 14. *Let $(\tilde{X}, \tilde{\tau}^{\tilde{\rho}\tilde{\varrho}\tilde{\eta}}, \mathcal{I}^{\tilde{\rho}\tilde{\varrho}\tilde{\eta}})$ be an SVNITS and $r \in \xi_0$. Then, $\sigma_n \in \xi^{\tilde{X}}$ is said to be:*

(1) *r-single valued neutrosophic semi-ideal open set (r-SVNSIO) iff $\sigma_n \leq \mathrm{CI}^{£}_{\tilde{\tau}^{\tilde{\rho}\tilde{\varrho}\tilde{\eta}}}(\mathrm{int}_{\tilde{\tau}^{\tilde{\rho}\tilde{\varrho}\tilde{\eta}}}(\sigma_n, r), r)$,*

(2) *r-single valued neutrosophic pre-ideal open set (r-SVNPIO) iff $\sigma_n \leq \mathrm{int}_{\tilde{\tau}^{\tilde{\rho}\tilde{\varrho}\tilde{\eta}}}(\mathrm{CI}^{£}_{\tilde{\tau}^{\tilde{\rho}\tilde{\varrho}\tilde{\eta}}}(\sigma_n, r), r)$,*

(3) *r-single valued neutrosophic α-ideal open set (r-SVNαIO) iff $\sigma_n \leq \mathrm{int}_{\tilde{\tau}^{\tilde{\rho}\tilde{\varrho}\tilde{\eta}}}(\mathrm{CI}^{£}_{\tilde{\tau}^{\tilde{\rho}\tilde{\varrho}\tilde{\eta}}}(\mathrm{int}_{\tilde{\tau}^{\tilde{\rho}\tilde{\varrho}\tilde{\eta}}}\sigma_n, r), r), r)$,*

(4) *r-single valued neutrosophic β-ideal open set (r-SVNβIO) iff $\sigma_n \leq \mathrm{C}_{\tilde{\tau}^{\tilde{\rho}\tilde{\varrho}\tilde{\eta}}}(\mathrm{int}_{\tilde{\tau}^{\tilde{\rho}\tilde{\varrho}\tilde{\eta}}}(\mathrm{CI}^{£}_{\tilde{\tau}^{\tilde{\rho}\tilde{\varrho}\tilde{\eta}}}(\sigma_n, r), r), r)$,*

(5) *r-single valued neutrosophic β-ideal open (r-SVNSβIO) iff $\sigma_n \leq \mathrm{CI}^{£}_{\tilde{\tau}^{\tilde{\rho}\tilde{\varrho}\tilde{\eta}}}(\mathrm{int}_{\tilde{\tau}^{\tilde{\rho}\tilde{\varrho}\tilde{\eta}}}(\mathrm{CI}^{£}_{\tilde{\tau}^{\tilde{\rho}\tilde{\varrho}\tilde{\eta}}}(\sigma_n, r), r), r)$,*

(6) *r-single valued neutrosophic regular ideal open set (r-SVNRIO)* iff $\sigma_n = \text{int}_{\tilde{\tau}^{\tilde{\rho}\tilde{\varrho}\tilde{\eta}}}(\text{Cl}^{\pounds}_{\tilde{\tau}^{\tilde{\rho}\tilde{\varrho}\tilde{\eta}}}(\sigma_n, r), r)$.

The complement of r-SVNSIO (resp. r-SVNPIO, r-SVNαIO, r-SVNβIO, r-SVNSβIO, r-SVNRIO) are called r-SVNSIC (resp. r-SVNPIC, r-SVNαIC, r-SVNβIC, r-SVNSβIC, r-SVNRIC).

Remark 2. *The following diagram can be easily obtained from the above definition:*

$$r - SVN\alpha IO \Rightarrow r - SVNSIO \Rightarrow r - SVNSO$$
$$\Downarrow \quad\quad \Downarrow \quad\quad \Downarrow$$
$$r - SVNRIO \Rightarrow r - SVNPIO \Rightarrow r - SVN\beta IO \Rightarrow r - SVN\beta O$$
$$\Downarrow$$
$$r - SVNSIO \Rightarrow r - SVNS\beta IO \Rightarrow r - SVN\beta IO.$$

Theorem 7. *Let* $(\tilde{X}, \tilde{\tau}^{\tilde{\rho}\tilde{\varrho}\tilde{\eta}}, \mathcal{I}^{\tilde{\rho}\tilde{\varrho}\tilde{\eta}})$ *be an SVNITS and* $r \in \xi_0$. *Then, the following properties are equivalent:*

(1) $(\tilde{X}, \tilde{\tau}^{\tilde{\rho}\tilde{\varrho}\tilde{\eta}}, \mathcal{I}^{\tilde{\rho}\tilde{\varrho}\tilde{\eta}})$ *is £-SVNE-disconnected,*
(2) $\tilde{\tau}^{\tilde{\rho}}([\text{int}^{\pounds}_{\tilde{\tau}^{\tilde{\rho}}}(\sigma_n, r)]^c) \geq r$, $\tilde{\tau}^{\tilde{\varrho}}([\text{int}^{\pounds}_{\tilde{\tau}^{\tilde{\varrho}}}(\sigma_n, r)]^c) \leq 1 - r$, $\tilde{\tau}^{\tilde{\eta}}([\text{int}^{\pounds}_{\tilde{\tau}^{\tilde{\eta}}}(\sigma_n, r)]^c) \leq 1 - r$ *for each* $\tilde{\tau}^{\tilde{\rho}}([\sigma_n]^c) \geq r$, $\tilde{\tau}^{\tilde{\varrho}}([\sigma_n]^c) \leq 1 - r$, $\tilde{\tau}^{\tilde{\eta}}([\sigma_n]^c) \leq 1 - r$,
(3) $\text{Cl}^{\pounds}_{\tilde{\tau}^{\tilde{\rho}\tilde{\varrho}\tilde{\eta}}}(\text{int}_{\tilde{\tau}^{\tilde{\rho}\tilde{\varrho}\tilde{\eta}}}(\sigma_n.r), r) \leq \text{int}_{\tilde{\tau}^{\tilde{\rho}\tilde{\varrho}\tilde{\eta}}}(\text{Cl}^{\pounds}_{\tilde{\tau}^{\tilde{\rho}\tilde{\varrho}\tilde{\eta}}}(\sigma_n, r), r)$, *for each* $\sigma_n \in \xi^{\tilde{X}}$,
(4) *Every r-SVNSIO set is r-SVNPIO,*
(5) $\tilde{\tau}^{\tilde{\rho}}(\text{Cl}^{\pounds}_{\tilde{\tau}^{\tilde{\rho}}}(\sigma_n, r)) \geq r$, $\tilde{\tau}^{\tilde{\varrho}}(\text{Cl}^{\pounds}_{\tilde{\tau}^{\tilde{\varrho}}}(\sigma_n, r)) \leq 1 - r$, $\tilde{\tau}^{\tilde{\eta}}(\text{Cl}^{\pounds}_{\tilde{\tau}^{\tilde{\eta}}}(\sigma_n, r)) \leq 1 - r$ *for each r-SVNSβIO* $\sigma_n \in \xi^{\tilde{X}}$,
(6) *Every r-SVNSβIO set is r-SVNPIO,*
(7) *For each* $\sigma_n \in \xi^{\tilde{X}}$, σ_n *is r-SVNαIO set iff it is r-SVNSIO.*

Proof. (1) $\Rightarrow$ (2): The proof is direct consequence of Definition 14.

(2) $\Rightarrow$ (3): For each $\sigma_n \in \xi^{\tilde{X}}$, $\tilde{\tau}^{\tilde{\rho}}(\text{int}_{\tilde{\tau}^{\tilde{\rho}}}(\sigma_n, r)) \geq r$, $\tilde{\tau}^{\tilde{\varrho}}(\text{int}_{\tilde{\tau}^{\tilde{\varrho}}}(\sigma_n, r)) \leq 1 - r$, $\tilde{\tau}^{\tilde{\eta}}(\text{int}_{\tilde{\tau}^{\tilde{\eta}}}(\sigma_n, r)) \leq 1 - r$, and, by (2), we have

$$\tilde{\tau}^{\tilde{\rho}}([\text{int}^{\pounds}_{\tilde{\tau}^{\tilde{\rho}}}([\text{int}_{\tilde{\tau}^{\tilde{\rho}}}(\sigma_n, r)]^c, r)]^c) \geq r, \quad \tilde{\tau}^{\tilde{\varrho}}([\text{int}^{\pounds}_{\tilde{\tau}^{\tilde{\varrho}}}([\text{int}_{\tilde{\tau}^{\tilde{\varrho}}}(\sigma_n, r)]^c, r)]^c) \leq 1 - r,$$

$$\tilde{\tau}^{\tilde{\eta}}([\text{int}^{\pounds}_{\tilde{\tau}^{\tilde{\eta}}}([\text{int}_{\tilde{\tau}^{\tilde{\eta}}}(\sigma_n, r)]^c, r)]^c) \leq 1 - r.$$

Thus,

$$\tilde{\tau}^{\tilde{\rho}}(\text{Cl}^{\pounds}_{\tilde{\tau}^{\tilde{\rho}}}(\text{int}_{\tilde{\tau}^{\tilde{\rho}}}(\sigma_n, r), r)) \geq r, \quad \tilde{\tau}^{\tilde{\varrho}}([\text{Cl}^{\pounds}_{\tilde{\tau}^{\tilde{\varrho}}}(\text{int}_{\tilde{\tau}^{\tilde{\varrho}}}(\sigma_n, r), r)) \leq 1 - r, \quad \tilde{\tau}^{\tilde{\eta}}([\text{Cl}^{\pounds}_{\tilde{\tau}^{\tilde{\eta}}}(\text{int}_{\tilde{\tau}^{\tilde{\eta}}}(\sigma_n, r), r)) \leq 1 - r;$$

hence,

$$\text{Cl}^{\pounds}_{\tilde{\tau}^{\tilde{\rho}\tilde{\varrho}\tilde{\eta}}}(\text{int}_{\tilde{\tau}^{\tilde{\rho}\tilde{\varrho}\tilde{\eta}}}(\sigma_n.r), r) = \text{int}_{\tilde{\tau}^{\tilde{\rho}\tilde{\varrho}\tilde{\eta}}}(\text{Cl}^{\pounds}_{\tilde{\tau}^{\tilde{\rho}\tilde{\varrho}\tilde{\eta}}}(\text{int}_{\tilde{\tau}^{\tilde{\rho}\tilde{\varrho}\tilde{\eta}}}(\sigma_n, r), r), r) \leq \text{int}_{\tilde{\tau}^{\tilde{\rho}\tilde{\varrho}\tilde{\eta}}}(\text{Cl}^{\pounds}_{\tilde{\tau}^{\tilde{\rho}\tilde{\varrho}\tilde{\eta}}}(\sigma_n, r), r).$$

(3) $\Rightarrow$ (4): Let σ_n be an r-SVNSIO set. Then, by (4), we have

$$\sigma_n \leq \text{Cl}^{\pounds}_{\tilde{\tau}^{\tilde{\rho}\tilde{\varrho}\tilde{\eta}}}(\text{int}_{\tilde{\tau}^{\tilde{\rho}\tilde{\varrho}\tilde{\eta}}}(\sigma_n, r), r) \leq \text{int}_{\tilde{\tau}^{\tilde{\rho}\tilde{\varrho}\tilde{\eta}}}(\text{Cl}^{\pounds}_{\tilde{\tau}^{\tilde{\rho}\tilde{\varrho}\tilde{\eta}}}(\sigma_n, r), r).$$

Thus, σ_n is an r-SVNPIO set.

(4) $\Rightarrow$ (5): Since σ_n is an r-SVNSβIO set, $\sigma_n \leq \text{Cl}^{\pounds}_{\tilde{\tau}^{\tilde{\rho}\tilde{\varrho}\tilde{\eta}}}(\text{int}_{\tilde{\tau}^{\tilde{\rho}\tilde{\varrho}\tilde{\eta}}}(Cl^{\pounds}_{\tilde{\tau}^{\tilde{\rho}\tilde{\varrho}\tilde{\eta}}}(\sigma_n, r), r), r)$. Then, $\text{Cl}^{\pounds}_{\tilde{\tau}^{\tilde{\rho}\tilde{\varrho}\tilde{\eta}}}(\sigma_n, r)$ is r-SVNSIO, and, by (4), $\text{Cl}^{\pounds}_{\tilde{\tau}^{\tilde{\rho}\tilde{\varrho}\tilde{\eta}}}(\sigma_n, r) \leq \text{int}_{\tilde{\tau}^{\tilde{\rho}\tilde{\varrho}\tilde{\eta}}}(CI^{\pounds}_{\tilde{\tau}^{\tilde{\rho}\tilde{\varrho}\tilde{\eta}}}(\sigma_n, r), r)$; hence, $\tilde{\tau}^{\tilde{\rho}}\text{Cl}^{\pounds}_{\tilde{\tau}^{\tilde{\rho}}}(\sigma_n, r)) \geq r$, $\tilde{\tau}^{\tilde{\varrho}}\text{Cl}^{\pounds}_{\tilde{\tau}^{\tilde{\varrho}}}(\sigma_n, r)) \leq 1 - r$, $\tilde{\tau}^{\tilde{\eta}}\text{Cl}^{\pounds}_{\tilde{\tau}^{\tilde{\eta}}}(\sigma_n, r)) \leq 1 - r$.

(5) $\Rightarrow$ (6): Let σ_n be an r-SVNβIO set, then, by (5), $\text{Cl}^{\pounds}_{\tilde{\tau}^{\tilde{\rho}\tilde{\varrho}\tilde{\eta}}}(\sigma_n, r) \leq \text{int}_{\tilde{\tau}^{\tilde{\rho}\tilde{\varrho}\tilde{\eta}}}(Cl^{\star}(\sigma_n, r), r)$. Thus,

$$\sigma_n \leq \text{Cl}^{\pounds}_{\tilde{\tau}^{\tilde{\rho}\tilde{\varrho}\tilde{\eta}}}(\sigma_n, r) \leq \text{int}_{\tilde{\tau}^{\tilde{\rho}\tilde{\varrho}\tilde{\eta}}}(\text{Cl}^{\pounds}_{\tilde{\tau}^{\tilde{\rho}\tilde{\varrho}\tilde{\eta}}}(\sigma_n, r), r).$$

Therefore, σ_n is an r-SVNPIO set.

(6)⇒(7): Let σ_n be an r-SVNSIO. Then, σ_n is r-SVNSβIO, by (6), σ_n is an r-SVNPIO set. Since σ_n is r-SVNSIO and r-SVNPIO, σ_n is r-SVNαIO.

(7) ⇒ (1): Suppose that $\tilde{\tau}^{\tilde{\rho}}(\sigma_n) \geq r$, $\tilde{\tau}^{\tilde{\varrho}}(\sigma_n) \leq 1-r$, $\tilde{\tau}^{\tilde{\eta}}(\sigma_n) \leq 1-r$, then $\mathrm{CI}^{\pounds}_{\tilde{\tau}^{\tilde{\rho}\tilde{\varrho}\tilde{\eta}}}(\sigma_n, r)$ is r-SVNSIO, and, by (7), $\mathrm{CI}^{\pounds}_{\tilde{\tau}^{\tilde{\rho}\tilde{\varrho}\tilde{\eta}}}(\sigma_n, r)$ is r-SVNαIO. Hence,

$$\mathrm{CI}^{\pounds}_{\tilde{\tau}^{\tilde{\rho}\tilde{\varrho}\tilde{\eta}}}(\sigma_n, r) \leq \mathrm{int}_{\tilde{\tau}^{\rho\varrho\eta}}(\mathrm{CI}^{\pounds}_{\tilde{\tau}^{\tilde{\rho}\tilde{\varrho}\tilde{\eta}}}(\mathrm{int}_{\tilde{\tau}^{\rho\varrho\eta}}(Cl^{\star}(\sigma_n, r), r), r), r) = \mathrm{int}_{\tilde{\tau}^{\rho\varrho\eta}}(\mathrm{CI}^{\pounds}_{\tilde{\tau}^{\tilde{\rho}\tilde{\varrho}\tilde{\eta}}}(\sigma_n, r), r) \leq \mathrm{CI}^{\pounds}_{\tilde{\tau}^{\tilde{\rho}\tilde{\varrho}\tilde{\eta}}}(\sigma_n, r).$$

Hence,

$$\tilde{\tau}^{\tilde{\rho}}(\mathrm{CI}^{\pounds}_{\tilde{\tau}^{\rho\varrho\eta}}(\sigma_n, r)) \geq r, \quad \tilde{\tau}^{\tilde{\varrho}}(\mathrm{CI}^{\pounds}_{\tilde{\tau}^{\rho\varrho\eta}}(\sigma_n, r)) \leq 1-r, \quad \tilde{\tau}^{\tilde{\eta}}(\mathrm{CI}^{\pounds}_{\tilde{\tau}^{\rho\varrho\eta}}(\sigma_n, r)) \leq 1-r.$$

Thus, $(\tilde{X}, \tilde{\tau}^{\tilde{\rho}\tilde{\varrho}\tilde{\eta}}, \mathcal{I}^{\tilde{\rho}\tilde{\varrho}\tilde{\eta}})$ is £-SVNE-disconnected. □

Theorem 8. *Let $(\tilde{X}, \tilde{\tau}^{\tilde{\rho}\tilde{\varrho}\tilde{\eta}}, \mathcal{I}^{\tilde{\rho}\tilde{\varrho}\tilde{\eta}})$ be an SVNITS $r \in \zeta_0$ and $\sigma_n \in \zeta^{\tilde{X}}$. Then, the following are equivalent:*

(1) $(\tilde{X}, \tilde{\tau}^{\tilde{\rho}\tilde{\varrho}\tilde{\eta}}, \mathcal{I}^{\tilde{\rho}\tilde{\varrho}\tilde{\eta}})$ is £-SVNE-disconnected,

(2) $\mathrm{CI}^{\pounds}_{\tilde{\tau}^{\tilde{\rho}\tilde{\varrho}\tilde{\eta}}}(\sigma_n, r)\bar{q}\mathrm{C}_{\tilde{\tau}^{\rho\varrho\eta}}(\gamma_n, r)$, for every $\tilde{\tau}^{\tilde{\rho}}(\sigma_n) \geq r$, $\tilde{\tau}^{\tilde{\varrho}}(\sigma_n) \leq 1-r$, $\tilde{\tau}^{\tilde{\eta}}(\sigma_n) \leq 1-r$ and every r-SVN£O $\gamma_n \in \zeta^{\tilde{X}}$ with $\sigma_n\bar{q}\gamma_n$,

(3) $\mathrm{CI}^{\pounds}_{\tilde{\tau}^{\tilde{\rho}\tilde{\varrho}\tilde{\eta}}}(\mathrm{int}_{\tilde{\tau}^{\rho\varrho\eta}}(\mathrm{CI}^{\pounds}_{\tilde{\tau}^{\tilde{\rho}\tilde{\varrho}\tilde{\eta}}}(\sigma_n, r), r), r)\bar{q}\mathrm{C}_{\tilde{\tau}^{\rho\varrho\eta}}(\gamma_n, r)$, for every $\sigma_n \in \zeta^{\tilde{X}}$ and r-SVN£O $\gamma_n \in \zeta^{\tilde{X}}$ with $\sigma_n\bar{q}\gamma_n$.

Proof. (1)⇒(2): Let $\tilde{\tau}^{\tilde{\rho}}(\sigma_n) \geq r$, $\tilde{\tau}^{\tilde{\varrho}}(\sigma_n) \leq 1-r$, $\tilde{\tau}^{\tilde{\eta}}(\sigma_n) \leq 1-r$. Then, by (1),

$$\tilde{\tau}^{\tilde{\rho}}(\mathrm{CI}^{\pounds}_{\tilde{\tau}^{\rho}}(\sigma_n, r)) \geq r, \quad \tilde{\tau}^{\tilde{\varrho}}(\mathrm{CI}^{\pounds}_{\tilde{\tau}^{\varrho}}(\sigma_n, r)) \leq 1-r, \quad \tilde{\tau}^{\tilde{\eta}}(\mathrm{CI}^{\pounds}_{\tilde{\tau}^{\eta}}(\sigma_n, r)) \leq 1-r.$$

Since $[\mathrm{CI}^{\pounds}_{\tilde{\tau}^{\tilde{\rho}\tilde{\varrho}\tilde{\eta}}}(\sigma_n, r)]^c$ is an r-SVN£O and $\mathrm{CI}^{\pounds}_{\tilde{\tau}^{\tilde{\rho}\tilde{\varrho}\tilde{\eta}}}(\sigma_n, r)\bar{q}[\mathrm{CI}^{\pounds}_{\tilde{\tau}^{\tilde{\rho}\tilde{\varrho}\tilde{\eta}}}(\sigma_n, r)]^c$, it implies that

$$\mathrm{CI}^{\pounds}_{\tilde{\tau}^{\rho\varrho\eta}}(\sigma_n, r)\bar{q}\mathrm{C}_{\tilde{\tau}^{\tilde{\rho}\tilde{\varrho}\tilde{\eta}}}([\mathrm{CI}^{\pounds}_{\tilde{\tau}^{\rho\varrho\eta}}(\sigma_n, r)]^c, r).$$

(2)⇒(1): Let $\tilde{\tau}^{\tilde{\rho}}(\sigma_n) \geq r$, $\tilde{\tau}^{\tilde{\varrho}}(\sigma_n) \leq 1-r$, $\tilde{\tau}^{\tilde{\eta}}(\sigma_n) \leq 1-r$. Since $[\mathrm{CI}^{\pounds}_{\tilde{\tau}^{\tilde{\rho}\tilde{\varrho}\tilde{\eta}}}(\sigma_n, r)]^c$ is an r-SVN£O, then, by (2),

$$\mathrm{CI}^{\pounds}_{\tilde{\tau}^{\rho\varrho\eta}}(\sigma_n, r)\bar{q}\mathrm{C}_{\tilde{\tau}^{\rho\varrho\eta}}([\mathrm{CI}^{\pounds}_{\tilde{\tau}^{\rho\varrho\eta}}(\sigma_n, r)]^c, r).$$

This implies that $\mathrm{CI}^{\pounds}_{\tilde{\tau}^{\rho\varrho\eta}}(\sigma_n, r) \leq \mathrm{int}_{\tilde{\tau}^{\rho\varrho\eta}}(\mathrm{CI}^{\pounds}_{\tilde{\tau}^{\rho\varrho\eta}}(\sigma_n, r), r) \leq \mathrm{CI}^{\pounds}_{\tilde{\tau}^{\rho\varrho\eta}}(\sigma_n, r)$, so

$$\tilde{\tau}^{\tilde{\rho}}(\mathrm{CI}^{\pounds}_{\tilde{\tau}^{\rho}}(\sigma_n, r)) \geq r, \quad \tilde{\tau}^{\tilde{\varrho}}(\mathrm{CI}^{\pounds}_{\tilde{\tau}^{\varrho}}(\sigma_n, r)) \leq 1-r, \quad \tilde{\tau}^{\tilde{\eta}}(\mathrm{CI}^{\pounds}_{\tilde{\tau}^{\eta}}(\sigma_n, r)) \leq 1-r.$$

(2)⇒(3): Suppose that $\sigma_n \in \zeta^{\tilde{X}}$ and γ_n is an r-SVN£O with $\sigma_n\bar{q}\gamma_n$. Since

$$\tilde{\tau}^{\tilde{\rho}}(\mathrm{int}_{\tilde{\tau}^{\rho}}(\mathrm{CI}^{\pounds}_{\tilde{\tau}^{\rho}}(\sigma_n, r), r)) \geq r, \quad \tilde{\tau}^{\tilde{\varrho}}(\mathrm{int}_{\tilde{\tau}^{\varrho}}(\mathrm{CI}^{\pounds}_{\tilde{\tau}^{\varrho}}(\sigma_n, r), r)) \leq 1-r, \quad \tilde{\tau}^{\tilde{\eta}}(\mathrm{int}_{\tilde{\tau}^{\eta}}(\mathrm{CI}^{\pounds}_{\tilde{\tau}^{\eta}}(\sigma_n, r), r)) \leq 1-r.$$

By (2), we have $\mathrm{CI}^{\pounds}_{\tilde{\tau}^{\tilde{\rho}\tilde{\varrho}\tilde{\eta}}}(\mathrm{int}_{\tilde{\tau}^{\rho\varrho\eta}}(\mathrm{CI}^{\pounds}_{\tilde{\tau}^{\tilde{\rho}\tilde{\varrho}\tilde{\eta}}}(\sigma_n, r), r), r)\bar{q}\mathrm{C}_{\tilde{\tau}^{\rho\varrho\eta}}(\gamma_n, r)$.

(3)⇒(2): Let $\tilde{\tau}^{\tilde{\rho}}(\sigma_n) \geq r$, $\tilde{\tau}^{\tilde{\varrho}}(\sigma_n) \leq 1-r$, $\tilde{\tau}^{\tilde{\eta}}(\sigma_n) \leq 1-r$ and γ_n be an r-SVN£O with $\sigma_n\bar{q}\gamma_n$. Then, by (3), we obtain $\mathrm{CI}^{\pounds}_{\tilde{\tau}^{\tilde{\rho}\tilde{\varrho}\tilde{\eta}}}(\mathrm{int}_{\tilde{\tau}^{\rho\varrho\eta}}(\mathrm{CI}^{\pounds}_{\tilde{\tau}^{\tilde{\rho}\tilde{\varrho}\tilde{\eta}}}(\sigma_n, r), r), r)\bar{q}\mathrm{C}_{\tilde{\tau}^{\rho\varrho\eta}}(\gamma_n, r)$. Since

$$\mathrm{CI}^{\pounds}_{\tilde{\tau}^{\rho\varrho\eta}}(\sigma_n, r) \leq \mathrm{CI}^{\pounds}_{\tilde{\tau}^{\rho\varrho\eta}}(\mathrm{iny}_{\tilde{\tau}^{\rho\varrho\eta}}(\mathrm{CI}^{\pounds}_{\tilde{\tau}^{\rho\varrho\eta}}(\sigma_n, r), r), r),$$

then, we have $\mathrm{CI}^{\pounds}_{\tilde{\tau}^{\rho\varrho\eta}}(\sigma_n, r)\bar{q}\mathrm{C}_{\tilde{\tau}^{\rho\varrho\eta}}(\gamma_n, r)$. □

Definition 15. *An SVNITS $(\tilde{X}, \tilde{\tau}^{\tilde{\rho}\tilde{\varrho}\tilde{\eta}}, \mathcal{I}^{\tilde{\rho}\tilde{\varrho}\tilde{\eta}})$ is called £-SVN-normal if, for every $[\sigma_n]_1\bar{q}[\sigma_n]_2$ with $\tilde{\tau}^{\tilde{\rho}}([\sigma_n]_1) \geq r$, $\tilde{\tau}^{\tilde{\varrho}}([\sigma_n]_1) \leq 1-r$, $\tilde{\tau}^{\tilde{\eta}}([\sigma_n]_1) \leq 1-r$ and $[\sigma_n]_2$ is r-SVN£O, there exists*

$[\gamma_n]_j \in \xi^{\tilde{X}}$, for $j = \{1,2\}$ with $\tilde{\tau}^{\tilde{\rho}}([\gamma_n]_1^c) \geq r$, $\tilde{\tau}^{\tilde{\varrho}}([\gamma_n]_1^c) \leq 1-r$, $\tilde{\tau}^{\tilde{\eta}}([\gamma_n]_1^c) \leq 1-r$, $[\gamma_n]_2$ is r-SVN£C such that $[\sigma_n]_2 \leq [\gamma_n]_1$, $[\sigma_n]_1 \leq [\gamma_n]_2$ and $[\gamma_n]_1 \bar{q} [\gamma_n]_2$.

Theorem 9. *Let $(\tilde{X}, \tilde{\tau}^{\tilde{\rho}\tilde{\varrho}\tilde{\eta}}, \mathcal{I}^{\tilde{\rho}\tilde{\varrho}\tilde{\eta}})$ be an SVNITS; then, the following are equivalent:*

(1) $(\tilde{X}, \tilde{\tau}^{\tilde{\rho}\tilde{\varrho}\tilde{\eta}}, \mathcal{I}^{\tilde{\rho}\tilde{\varrho}\tilde{\eta}})$ is an £-SVN-normal.
(2) $(\tilde{X}, \tilde{\tau}^{\tilde{\rho}\tilde{\varrho}\tilde{\eta}}, \mathcal{I}^{\tilde{\rho}\tilde{\varrho}\tilde{\eta}})$ is an £-SVNE-disconnected.

Proof. (1)⇒(2): Let $\tilde{\tau}^{\tilde{\rho}}(\sigma_n) \geq r$, $\tilde{\tau}^{\tilde{\varrho}}(\sigma_n) \leq 1-r$, $\tilde{\tau}^{\tilde{\eta}}(\sigma_n) \leq 1-r$ and $[\mathrm{CI}^{\pounds}_{\tilde{\tau}^{\tilde{\rho}\tilde{\varrho}\tilde{\eta}}}(\sigma_n, r)]^c$ be an r-SVN£O. Then, $\sigma_n \bar{q} [\mathrm{CI}^{\pounds}_{\tilde{\tau}^{\tilde{\rho}\tilde{\varrho}\tilde{\eta}}}(\sigma_n, r)]^c$. By the £-SVN-normality of $(\tilde{X}, \tilde{\tau}^{\tilde{\rho}\tilde{\varrho}\tilde{\eta}}, \mathcal{I}^{\tilde{\rho}\tilde{\varrho}\tilde{\eta}})$, there exist $[\gamma_n]_i \in \xi^{\tilde{X}}$, for $i = \{1,2\}$ with

$$\tilde{\tau}^{\tilde{\rho}}([\gamma_n]_1^c) \geq r, \quad \tilde{\tau}^{\tilde{\varrho}}([\gamma_n]_1^c) \leq 1-r, \quad \tilde{\tau}^{\tilde{\eta}}([\gamma_n]_1^c) \leq 1-r,$$

and $[\gamma_n]_2^c$ r-SVN£C such that $[\mathrm{CI}^{\pounds}_{\tilde{\tau}^{\tilde{\rho}\tilde{\varrho}\tilde{\eta}}}(\sigma_n, r)]^c \leq [\gamma_n]_1$, $\sigma_n \leq [\gamma_n]_2$ and $[\gamma_n]_1 \bar{q} [\gamma_n]_2$. Since

$$\mathrm{CI}^{\pounds}_{\tilde{\tau}^{\tilde{\rho}\tilde{\varrho}\tilde{\eta}}}(\sigma_n, r) \leq \mathrm{CI}^{\pounds}_{\tilde{\tau}^{\tilde{\rho}\tilde{\varrho}\tilde{\eta}}}([\gamma_n]_2, r) = [\gamma_n]_2 \leq [\gamma_n]_1^c \leq \mathrm{CI}^{\pounds}_{\tilde{\tau}^{\tilde{\rho}\tilde{\varrho}\tilde{\eta}}}(\sigma_n, r),$$

we have $\mathrm{CI}^{\pounds}_{\tilde{\tau}^{\tilde{\rho}\tilde{\varrho}\tilde{\eta}}}(\sigma_n, r) = [\gamma_n]_2$. Since $[\mathrm{CI}^{\pounds}_{\tilde{\tau}^{\tilde{\rho}\tilde{\varrho}\tilde{\eta}}}(\sigma_n, r)]^c \leq [\gamma_n]_1 \leq [\gamma_n]_2^c = [\mathrm{CI}^{\pounds}_{\tilde{\tau}^{\tilde{\rho}\tilde{\varrho}\tilde{\eta}}}(\sigma_n, r)]^c$, so $[\mathrm{CI}^{\pounds}_{\tilde{\tau}^{\tilde{\rho}\tilde{\varrho}\tilde{\eta}}}(\sigma_n, r)]^c = [\gamma_n]_1$. Hence, $\mathrm{CI}^{\pounds}_{\tilde{\tau}^{\tilde{\rho}\tilde{\varrho}\tilde{\eta}}}(\sigma_n, r) = [\gamma_n]_1^c$ and

$$\tilde{\tau}^{\tilde{\rho}}(\mathrm{CI}^{\pounds}_{\tilde{\tau}^{\tilde{\rho}}}(\sigma_n, r)) \geq r, \quad \tilde{\tau}^{\tilde{\varrho}}(\mathrm{CI}^{\pounds}_{\tilde{\tau}^{\tilde{\varrho}}}(\sigma_n, r)) \leq 1-r, \quad \tilde{\tau}^{\tilde{\eta}}(\mathrm{CI}^{\pounds}_{\tilde{\tau}^{\tilde{\eta}}}(\sigma_n, r)) \leq 1-r.$$

Thus, $(\tilde{X}, \tilde{\tau}^{\tilde{\rho}\tilde{\varrho}\tilde{\eta}}, \mathcal{I}^{\tilde{\rho}\tilde{\varrho}\tilde{\eta}})$ is an £-SVNE-disconnected.

(2)⇒(1): Suppose that $\tilde{\tau}^{\tilde{\rho}}(\sigma_n) \geq r$, $\tilde{\tau}^{\tilde{\varrho}}(\sigma_n) \leq 1-r$, $\tilde{\tau}^{\tilde{\eta}}(\sigma_n) \leq 1-r$ and γ_n is an r-SVN£O with $\sigma_n \bar{q} \gamma_n$. By the £-SVNE-disconnected of $(\tilde{X}, \tilde{\tau}^{\tilde{\rho}\tilde{\varrho}\tilde{\eta}}, \mathcal{I}^{\tilde{\rho}\tilde{\varrho}\tilde{\eta}})$, we have

$$\tilde{\tau}^{\tilde{\rho}}(\mathrm{CI}^{\pounds}_{\tilde{\tau}^{\tilde{\rho}}}(\sigma_n, r)) \geq r, \quad \tilde{\tau}^{\tilde{\varrho}}(\mathrm{CI}^{\pounds}_{\tilde{\tau}^{\tilde{\varrho}}}(\sigma_n, r)) \leq 1-r, \quad \tilde{\tau}^{\tilde{\eta}}(\mathrm{CI}^{\pounds}_{\tilde{\tau}^{\tilde{\eta}}}(\sigma_n, r)) \leq 1-r,$$

and $[\mathrm{CI}^{\pounds}_{\tilde{\tau}^{\tilde{\rho}\tilde{\varrho}\tilde{\eta}}}(\sigma_n, r)]^c$ is r-SVN£O. Since $\sigma_n \bar{q} \gamma_n$, $\sigma_n \leq \mathrm{CI}^{\pounds}_{\tilde{\tau}^{\tilde{\rho}\tilde{\varrho}\tilde{\eta}}}(\sigma_n, r)$ and $\gamma_n \leq [\mathrm{CI}^{\pounds}_{\tilde{\tau}^{\tilde{\rho}\tilde{\varrho}\tilde{\eta}}}(\sigma_n, r)]^c$. Thus, $(\tilde{X}, \tilde{\tau}^{\tilde{\rho}\tilde{\varrho}\tilde{\eta}}, \mathcal{I}^{\tilde{\rho}\tilde{\varrho}\tilde{\eta}})$ is an £-SVN-normal. □

Theorem 10. *Let $(\tilde{X}, \tilde{\tau}^{\tilde{\rho}\tilde{\varrho}\tilde{\eta}}, \mathcal{I}^{\tilde{\rho}\tilde{\varrho}\tilde{\eta}})$ be an SVNITS, $\sigma_n, \sigma_n \xi^{\tilde{X}}$ and $r \in \xi_0$. Then, the following properties are equivalent:*

(1) $(\tilde{X}, \tilde{\tau}^{\tilde{\rho}\tilde{\varrho}\tilde{\eta}}, \mathcal{I}^{\tilde{\rho}\tilde{\varrho}\tilde{\eta}})$ is £-SVNE-disconnected.
(2) If σ_n is r-SVNRIO, then σ_n is r-SVN£C.
(3) If σ_n is r-SVNRIC, then σ_n is r-SVN£O.

Proof. (1)⇒(2): Let σ_n be an r-SVNRIO. Then, $\sigma_n = \mathrm{int}_{\tilde{\tau}^{\tilde{\rho}\tilde{\varrho}\tilde{\eta}}}(\mathrm{CI}^{\pounds}_{\tilde{\tau}^{\tilde{\rho}\tilde{\varrho}\tilde{\eta}}}(\sigma_n, r), r)$ and $\tilde{\tau}^{\tilde{\rho}}(\sigma_n) \geq r$, $\tilde{\tau}^{\tilde{\varrho}}(\sigma_n) \leq 1-r$, $\tilde{\tau}^{\tilde{\eta}}(\sigma_n) \leq 1-r$. By (1),

$$\tilde{\tau}^{\tilde{\rho}}(\mathrm{CI}^{\pounds}_{\tilde{\tau}^{\tilde{\rho}}}(\sigma_n, r)) \geq r, \quad \tilde{\tau}^{\tilde{\varrho}}(\mathrm{CI}^{\pounds}_{\tilde{\tau}^{\tilde{\varrho}}}(\sigma_n, r)) \leq 1-r, \quad \tilde{\tau}^{\tilde{\eta}}(\mathrm{CI}^{\pounds}_{\tilde{\tau}^{\tilde{\eta}}}(\sigma_n, r)) \leq 1-r.$$

Hence $\sigma_n = \mathrm{int}_{\tilde{\tau}^{\tilde{\rho}\tilde{\varrho}\tilde{\eta}}}(\mathrm{CI}^{\pounds}_{\tilde{\tau}^{\tilde{\rho}\tilde{\varrho}\tilde{\eta}}}(\sigma_n, r), r) = \mathrm{CI}^{\pounds}_{\tilde{\tau}^{\tilde{\rho}\tilde{\varrho}\tilde{\eta}}}(\sigma_n, r)$.

(2)⇒(1): Suppose that $\sigma_n = \mathrm{int}_{\tilde{\tau}^{\tilde{\rho}\tilde{\varrho}\tilde{\eta}}}(\mathrm{CI}^{\pounds}_{\tilde{\tau}^{\tilde{\rho}\tilde{\varrho}\tilde{\eta}}}(\sigma_n, r), r)$, then $\tilde{\tau}^{\tilde{\rho}}(\sigma_n) \geq r$, $\tilde{\tau}^{\tilde{\varrho}}(\sigma_n) \leq 1-r$, $\tilde{\tau}^{\tilde{\eta}}(\sigma_n) \leq 1-r$, by (2), σ_n is r-SVN£C. This implies that

$$\mathrm{CI}^{\pounds}_{\tilde{\tau}^{\tilde{\rho}\tilde{\varrho}\tilde{\eta}}}(\sigma_n, r) \leq \mathrm{CI}^{\pounds}_{\tilde{\tau}^{\tilde{\rho}\tilde{\varrho}\tilde{\eta}}}(\mathrm{int}_{\tilde{\tau}^{\tilde{\rho}\tilde{\varrho}\tilde{\eta}}}(\mathrm{CI}^{\pounds}_{\tilde{\tau}^{\tilde{\rho}\tilde{\varrho}\tilde{\eta}}}(\sigma_n, r), r), r) = \mathrm{int}_{\tilde{\tau}^{\tilde{\rho}\tilde{\varrho}\tilde{\eta}}}(\mathrm{CI}^{\pounds}_{\tilde{\tau}^{\tilde{\rho}\tilde{\varrho}\tilde{\eta}}}(\sigma_n, r), r) \leq \mathrm{CI}^{\pounds}_{\tilde{\tau}^{\tilde{\rho}\tilde{\varrho}\tilde{\eta}}}(\sigma_n, r).$$

Thus,

$$\tilde{\tau}^{\tilde{\rho}}(\mathrm{CI}^{\pounds}_{\tilde{\tau}^{\tilde{\rho}}}(\sigma_n, r)) \geq r, \quad \tilde{\tau}^{\tilde{\varrho}}(\mathrm{CI}^{\pounds}_{\tilde{\tau}^{\tilde{\varrho}}}(\sigma_n, r)) \leq 1-r, \quad \tilde{\tau}^{\tilde{\eta}}(\mathrm{CI}^{\pounds}_{\tilde{\tau}^{\tilde{\eta}}}(\sigma_n, r)) \leq 1-r,$$

then $(\tilde{X}, \tilde{\tau}^{\tilde{\rho}\tilde{\varrho}\tilde{\eta}}, \mathcal{I}^{\tilde{\rho}\tilde{\varrho}\tilde{\eta}})$ is an £-SVNE-disconnected.

(2) ⇔ (3): Obvious. □

Remark 3. *The union of two r-SVNRIO sets need not to be an r-SVNRIO.*

Theorem 11. *If $(\tilde{X}, \tilde{\tau}^{\tilde{\rho}\tilde{\varrho}\tilde{\eta}}, \mathcal{I}^{\tilde{\rho}\tilde{\varrho}\tilde{\eta}})$ is £-SVNE-disconnected and $\sigma_n, \gamma_n \in \xi^{\tilde{X}}$, $r \in \xi_0$. Then, the following properties hold:*

(1) *If σ_n and γ_n are r-SVNRIC, then $\sigma_n \wedge \gamma_n$ is r-SVNRIC.*
(2) *If σ_n and γ_n are r-SVNRIO, then $\sigma_n \vee \gamma_n$ is r-SVNRIO.*

Proof. Let σ_n and γ_n be r-SVNRIC. Then, $\tilde{\tau}^{\tilde{\rho}}([\sigma_n]^c) \geq r, \tilde{\tau}^{\tilde{\varrho}}([\sigma_n]^c) \leq 1-r, \tilde{\tau}^{\tilde{\eta}}([\sigma_n]^c) \leq 1-r$ and $\tilde{\tau}^{\tilde{\rho}}([\gamma_n]^c) \geq r, \tilde{\tau}^{\tilde{\varrho}}([\gamma_n]^c) \leq 1-r, \tilde{\tau}^{\tilde{\eta}}([\gamma_n]^c) \leq 1-r$, by Theorem 7, we have

$$\tilde{\tau}^{\tilde{\rho}}([\mathrm{int}^{£}_{\tilde{\tau}^{\tilde{\rho}}}(\sigma_n, r)]^c) \geq r, \quad \tilde{\tau}^{\tilde{\varrho}}([\mathrm{int}^{£}_{\tilde{\tau}^{\tilde{\varrho}}}(\sigma_n, r)]^c) \leq 1-r, \quad \tilde{\tau}^{\tilde{\eta}}([\mathrm{int}^{£}_{\tilde{\tau}^{\tilde{\eta}}}(\sigma_n, r)]^c) \leq 1-r,$$

and

$$\tilde{\tau}^{\tilde{\rho}}([\mathrm{int}^{£}_{\tilde{\tau}^{\tilde{\rho}}}(\gamma_n, r)]^c) \geq r, \quad \tilde{\tau}^{\tilde{\varrho}}([\mathrm{int}^{£}_{\tilde{\tau}^{\tilde{\varrho}}}(\gamma_n, r)]^c) \leq 1-r, \quad \tilde{\tau}^{\tilde{\eta}}([\mathrm{int}^{£}_{\tilde{\tau}^{\tilde{\eta}}}(\gamma_n, r)]^c) \leq 1-r.$$

This implies that

$$\begin{aligned} \sigma_n \wedge \gamma_n &= C_{\tilde{\tau}^{\tilde{\rho}\tilde{\varrho}\tilde{\eta}}}(\mathrm{int}^{£}_{\tilde{\tau}^{\tilde{\rho}\tilde{\varrho}\tilde{\eta}}}(\sigma_n, r), r) \wedge C_{\tilde{\tau}^{\tilde{\rho}\tilde{\varrho}\tilde{\eta}}}(\mathrm{int}^{£}_{\tilde{\tau}^{\tilde{\rho}\tilde{\varrho}\tilde{\eta}}}(\gamma_n, r), r) \\ &= \mathrm{int}^{£}_{\tilde{\tau}^{\tilde{\rho}\tilde{\varrho}\tilde{\eta}}}(\sigma_n, r) \wedge \mathrm{int}^{£}_{\tilde{\tau}^{\tilde{\rho}\tilde{\varrho}\tilde{\eta}}}(\gamma_n, r) = \mathrm{int}^{£}_{\tilde{\tau}^{\tilde{\rho}\tilde{\varrho}\tilde{\eta}}}(\sigma_n \wedge \gamma_n, r) \\ &\leq C_{\tilde{\tau}^{\tilde{\rho}\tilde{\varrho}\tilde{\eta}}}(\mathrm{int}^{£}_{\tilde{\tau}^{\tilde{\rho}\tilde{\varrho}\tilde{\eta}}}(\sigma_n \wedge \gamma_n, r), r). \end{aligned}$$

On the other hand,

$$\begin{aligned} C_{\tilde{\tau}^{\tilde{\rho}\tilde{\varrho}\tilde{\eta}}}(\mathrm{int}^{£}_{\tilde{\tau}^{\tilde{\rho}\tilde{\varrho}\tilde{\eta}}}(\sigma_n \wedge \gamma_n, r), r) &= C_{\tilde{\tau}^{\tilde{\rho}\tilde{\varrho}\tilde{\eta}}}(\mathrm{int}^{£}_{\tilde{\tau}^{\tilde{\rho}\tilde{\varrho}\tilde{\eta}}}(\sigma_n, r) \wedge \mathrm{int}^{£}_{\tilde{\tau}^{\tilde{\rho}\tilde{\varrho}\tilde{\eta}}}(\gamma_n, r), r) \\ &\leq C_{\tilde{\tau}^{\tilde{\rho}\tilde{\varrho}\tilde{\eta}}}(\mathrm{int}^{£}_{\tilde{\tau}^{\tilde{\rho}\tilde{\varrho}\tilde{\eta}}}(\sigma_n, r), r) \wedge C_{\tilde{\tau}^{\tilde{\rho}\tilde{\varrho}\tilde{\eta}}}(\mathrm{int}^{£}_{\tilde{\tau}^{\tilde{\rho}\tilde{\varrho}\tilde{\eta}}}(\gamma_n, r), r) \\ &= \sigma_n \wedge \gamma_n. \end{aligned}$$

Thus, $C_{\tilde{\tau}^{\tilde{\rho}\tilde{\varrho}\tilde{\eta}}}(\mathrm{int}^{£}_{\tilde{\tau}^{\tilde{\rho}\tilde{\varrho}\tilde{\eta}}}(\sigma_n \wedge \gamma_n, r), r) = \sigma_n \wedge \gamma_n$. Therefore, $\sigma_n \wedge \gamma_n$ is an r-SVNRIC.
(2) The proof is similar to that of (1). □

Theorem 12. *Let $(\tilde{X}, \tilde{\tau}^{\tilde{\rho}\tilde{\varrho}\tilde{\eta}}, \mathcal{I}^{\tilde{\rho}\tilde{\varrho}\tilde{\eta}})$ be an SVNITS and $r \in \xi_0$. Then, the following properties are equivalent:*

(1) *$(\tilde{X}, \tilde{\tau}^{\tilde{\rho}\tilde{\varrho}\tilde{\eta}}, \mathcal{I}^{\tilde{\rho}\tilde{\varrho}\tilde{\eta}})$ is £-SVNE-disconnected,*
(2) *$\tilde{\tau}^{\tilde{\rho}}(\mathrm{CI}^{£}_{\tilde{\tau}^{\tilde{\rho}}}(\sigma_n, r)) \geq r, \tilde{\tau}^{\tilde{\varrho}}(\mathrm{CI}^{£}_{\tilde{\tau}^{\tilde{\varrho}}}(\sigma_n, r)) \leq 1-r, \tilde{\tau}^{\tilde{\eta}}(\mathrm{CI}^{£}_{\tilde{\tau}^{\tilde{\eta}}}(\sigma_n, r)) \leq 1-r$, for every r-SVNSIO $\sigma_n \in \xi^{\tilde{X}}$,*
(3) *$\tilde{\tau}^{\tilde{\rho}}(\mathrm{CI}^{£}_{\tilde{\tau}^{\tilde{\rho}}}(\sigma_n, r)) \geq r, \tilde{\tau}^{\tilde{\varrho}}(\mathrm{CI}^{£}_{\tilde{\tau}^{\tilde{\varrho}}}(\sigma_n, r)) \leq 1-r, \tilde{\tau}^{\tilde{\eta}}(\mathrm{CI}^{£}_{\tilde{\tau}^{\tilde{\eta}}}(\sigma_n, r)) \leq 1-r$, for every r-SVNPIO $\sigma_n \in \xi^{\tilde{X}}$,*
(4) *$\tilde{\tau}^{\tilde{\rho}}(\mathrm{CI}^{£}_{\tilde{\tau}^{\tilde{\rho}}}(\sigma_n, r)) \geq r, \tilde{\tau}^{\tilde{\varrho}}(\mathrm{CI}^{£}_{\tilde{\tau}^{\tilde{\varrho}}}(\sigma_n, r)) \leq 1-r, \tilde{\tau}^{\tilde{\eta}}(\mathrm{CI}^{£}_{\tilde{\tau}^{\tilde{\eta}}}(\sigma_n, r)) \leq 1-r$, for every r-SVNRIO $\sigma_n \in \xi^{\tilde{X}}$.*

Proof. (1) ⇒ (2) and (1) ⇒ (3). Let σ_n be an r-SVNSIO (r-SVNPIO). Then, σ_n is r-SVNSβIO, and, by Theorem 7, we have,

$$\tilde{\tau}^{\tilde{\rho}}(\mathrm{CI}^{£}_{\tilde{\tau}^{\tilde{\rho}}}(\sigma_n, r)) \geq r, \tilde{\tau}^{\tilde{\varrho}}(\mathrm{CI}^{£}_{\tilde{\tau}^{\tilde{\varrho}}}(\sigma_n, r)) \leq 1-r, \tilde{\tau}^{\tilde{\eta}}(\mathrm{CI}^{£}_{\tilde{\tau}^{\tilde{\eta}}}(\sigma_n, r)) \leq 1-r.$$

(2)⇒(4) and (3)⇒(4). Let σ_n be an r-SVNRIO. Then, σ_n is r-SVNPIO and r-SVNSIO. Thus,

$$\tilde{\tau}^{\tilde{\rho}}(\mathrm{CI}^{£}_{\tilde{\tau}^{\tilde{\rho}}}(\sigma_n, r)) \geq r, \quad \tilde{\tau}^{\tilde{\varrho}}(\mathrm{CI}^{£}_{\tilde{\tau}^{\tilde{\varrho}}}(\sigma_n, r)) \leq 1-r, \quad \tilde{\tau}^{\tilde{\eta}}(\mathrm{CI}^{£}_{\tilde{\tau}^{\tilde{\eta}}}(\sigma_n, r)) \leq 1-r.$$

(4)⇒(1). Suppose that

$$\tilde{\tau}^{\tilde{\rho}}(\mathrm{int}_{\tilde{\tau}^{\tilde{\rho}}}(\mathrm{CI}^{£}_{\tilde{\tau}^{\tilde{\rho}}}(\sigma_n, r), r)) \geq r, \quad \tilde{\tau}^{\tilde{\varrho}}(\mathrm{int}_{\tilde{\tau}^{\tilde{\varrho}}}(\mathrm{CI}^{£}_{\tilde{\tau}^{\tilde{\varrho}}}(\sigma_n, r), r)) \geq r, \quad \tilde{\tau}^{\tilde{\eta}}(\mathrm{int}_{\tilde{\tau}^{\tilde{\eta}}}(\mathrm{CI}^{£}_{\tilde{\tau}^{\tilde{\eta}}}(\sigma_n, r), r)) \geq r.$$

Then, by (4), we have

$$\tilde{\tau}^{\tilde{\rho}}(\mathrm{CI}^{\pounds}_{\tilde{\tau}\tilde{\rho}}(\mathrm{int}_{\tilde{\tau}\tilde{\rho}}(\mathrm{CI}^{\pounds}_{\tilde{\tau}\tilde{\rho}}(\sigma_n,r),r),r)) \geq r, \qquad \tilde{\tau}^{\tilde{\varrho}}(\mathrm{CI}^{\pounds}_{\tilde{\tau}\tilde{\varrho}}(\mathrm{int}_{\tilde{\tau}\tilde{\varrho}}(\mathrm{CI}^{\pounds}_{\tilde{\tau}\tilde{\varrho}}(\sigma_n,r),r),r)) \geq r,$$
$$\tilde{\tau}^{\tilde{\eta}}(\mathrm{CI}^{\pounds}_{\tilde{\tau}\tilde{\eta}}(\mathrm{int}_{\tilde{\tau}\tilde{\eta}}(\mathrm{CI}^{\pounds}_{\tilde{\tau}\tilde{\eta}}(\sigma_n,r),r),r)) \geq r.$$

Hence,

$$\begin{aligned}\mathrm{CI}^{\pounds}_{\tilde{\tau}\tilde{\rho}\tilde{\varrho}\tilde{\eta}}(\sigma_n,r) &\leq \mathrm{CI}^{\pounds}_{\tilde{\tau}\tilde{\rho}\tilde{\varrho}\tilde{\eta}}(\mathrm{int}_{\tilde{\tau}\tilde{\rho}\tilde{\varrho}\tilde{\eta}}(\mathrm{CI}^{\pounds}_{\tilde{\tau}\tilde{\rho}\tilde{\varrho}\tilde{\eta}}(\sigma_n,r),r),r)\\ &= \mathrm{int}_{\tilde{\tau}\tilde{\rho}\tilde{\varrho}\tilde{\eta}}(\mathrm{CI}^{\pounds}_{\tilde{\tau}\tilde{\rho}\tilde{\varrho}\tilde{\eta}}(\mathrm{int}_{\tilde{\tau}\tilde{\rho}\tilde{\varrho}\tilde{\eta}}(\mathrm{CI}^{\pounds}_{\tilde{\tau}\tilde{\rho}\tilde{\varrho}\tilde{\eta}}(\sigma_n,r),r),r),r)\\ &= \mathrm{int}_{\tilde{\tau}\tilde{\rho}\tilde{\varrho}\tilde{\eta}}(\mathrm{CI}^{\pounds}_{\tilde{\tau}\tilde{\rho}\tilde{\varrho}\tilde{\eta}}(\sigma_n,r),r) \leq \mathrm{CI}^{\pounds}_{\tilde{\tau}\tilde{\rho}\tilde{\varrho}\tilde{\eta}}(\sigma_n,r).\end{aligned}$$

Thus, $\tilde{\tau}^{\tilde{\rho}}(\mathrm{CI}^{\pounds}_{\tilde{\tau}\tilde{\rho}}(\sigma_n,r)) \geq r, \tilde{\tau}^{\tilde{\varrho}}(\mathrm{CI}^{\pounds}_{\tilde{\tau}\tilde{\varrho}}(\sigma_n,r)) \leq 1-r, \tilde{\tau}^{\tilde{\eta}}(\mathrm{CI}^{\pounds}_{\tilde{\tau}\tilde{\eta}}(\sigma_n,r)) \leq 1-r$; hence, $(\tilde{X}, \tilde{\tau}^{\tilde{\rho}\tilde{\varrho}\tilde{\eta}}, \mathcal{I}^{\tilde{\rho}\tilde{\varrho}\tilde{\eta}})$ is an £-SVNE-disconnected. □

Definition 16. *Let $(\tilde{X}, \tilde{\tau}^{\tilde{\rho}\tilde{\varrho}\tilde{\eta}}, \mathcal{I}^{\tilde{\rho}\tilde{\varrho}\tilde{\eta}})$ be an SVNITS. Then, σ_n is said to be an r-SVN£SO if $\sigma_n \leq C_{\tilde{\tau}\tilde{\rho}\tilde{\varrho}\tilde{\eta}}(\mathrm{int}^{\pounds}_{\tilde{\tau}\tilde{\rho}\tilde{\varrho}\tilde{\eta}}(\sigma_n,r),r)$.*

Definition 17. *Let $(\tilde{X}, \tilde{\tau}^{\tilde{\rho}\tilde{\varrho}\tilde{\eta}}, \mathcal{I}^{\tilde{\rho}\tilde{\varrho}\tilde{\eta}})$ be an SVNITS for each $r \in \zeta_0$, $\sigma_n \in \zeta^{\tilde{X}}$ and $x_{s,t,p} \in Pt(\zeta^{\tilde{X}})$. Then, $x_{s,t,p}$ is called an r-SVN$\delta\mathcal{I}$-cluster point of σ_n if, for every $\gamma_n \in Q_{\tilde{\tau}\tilde{\rho}\tilde{\varrho}\tilde{\eta}}(x_{s,t,p},r)$, we have $\sigma_n q \mathrm{int}_{\tilde{\tau}\tilde{\rho}\tilde{\varrho}\tilde{\eta}}(\mathrm{CI}^{\pounds}_{\tilde{\tau}\tilde{\rho}\tilde{\varrho}\tilde{\eta}}(\gamma_n,r),r)$.*

Definition 18. *Let $(\tilde{X}, \tilde{\tau}^{\tilde{\rho}\tilde{\varrho}\tilde{\eta}}, \mathcal{I}^{\tilde{\rho}\tilde{\varrho}\tilde{\eta}})$ be an SVNITS for each $r \in \zeta_0$, $\sigma_n \in \zeta^{\tilde{X}}$ and $x_{s,t,p} \in Pt(\zeta^{\tilde{X}})$. Then, the single-valued neutrosophic $\delta\mathcal{I}$-closure operator is a mapping $C_{\delta I\tilde{\tau}\tilde{\rho}\tilde{\varrho}\tilde{\eta}} : \zeta^{\tilde{X}} \times \zeta_0 \to \zeta^{\tilde{X}}$ that is defined as: $C_{\delta I\tilde{\tau}\tilde{\rho}\tilde{\varrho}\tilde{\eta}}(\sigma_n,r) = \bigvee\{x_{s,t,p} \in Pt(\zeta^{\tilde{X}})$ is r-SVN$\delta\mathcal{I}$-cluster point of $\sigma_n\}$.*

Lemma 3. *Let $(\tilde{X}, \tilde{\tau}^{\tilde{\rho}\tilde{\varrho}\tilde{\eta}}, \mathcal{I}^{\tilde{\rho}\tilde{\varrho}\tilde{\eta}})$ be an SVNITS. Then, σ_n is r-SVN£SO iff $C_{\tilde{\tau}\tilde{\rho}\tilde{\varrho}\tilde{\eta}}(\sigma_n,r) = C_{\tilde{\tau}\tilde{\rho}\tilde{\varrho}\tilde{\eta}}$ $(\mathrm{int}^{\pounds}_{\tilde{\tau}\tilde{\rho}\tilde{\varrho}\tilde{\eta}}(\sigma_n,r),r)$.*

Proof. Obvious. □

Lemma 4. *Let $(\tilde{X}, \tilde{\tau}^{\tilde{\rho}\tilde{\varrho}\tilde{\eta}}, \mathcal{I}^{\tilde{\rho}\tilde{\varrho}\tilde{\eta}})$ be an SVNITS for each $\sigma_n \in \zeta^{\tilde{x}}$ and $r \in \zeta_0$. Then, $C_{\tilde{\tau}\tilde{\rho}\tilde{\varrho}\tilde{\eta}}(\sigma_n,r) \leq C_{\delta I\tilde{\tau}\tilde{\rho}\tilde{\varrho}\tilde{\eta}}(\sigma_n,r)$.*

Proof. Obvious. □

Lemma 5. *Let $(\tilde{X}, \tilde{\tau}^{\tilde{\rho}\tilde{\varrho}\tilde{\eta}}, \mathcal{I}^{\tilde{\rho}\tilde{\varrho}\tilde{\eta}})$ be an SVNITS and σ_n be an r-SVN£SO. Then, $C_{\tilde{\tau}\tilde{\rho}\tilde{\varrho}\tilde{\eta}}(\sigma_n,r) = C_{\delta I\tilde{\tau}\tilde{\rho}\tilde{\varrho}\tilde{\eta}}(\sigma_n,r)$.*

Proof. We show that $C_{\tilde{\tau}\tilde{\rho}\tilde{\varrho}\tilde{\eta}}(\sigma_n,r) \leq C_{\delta I\tilde{\tau}\tilde{\rho}\tilde{\varrho}\tilde{\eta}}(\sigma_n,r)$. Suppose that $C_{\tilde{\tau}\tilde{\rho}\tilde{\varrho}\tilde{\eta}}(\sigma_n,r) \not\geq C_{\delta I\tilde{\tau}\tilde{\rho}\tilde{\varrho}\tilde{\eta}}(\sigma_n,r)$,; then, there exist $\nu \in \tilde{X}$ and $s,t,p \in \zeta_0$ such that

$$\tilde{\rho}_{C_{\tilde{\tau}\tilde{\rho}}(\sigma_n,r)}(\nu) < s \leq \tilde{\rho}_{C_{\delta I\tilde{\tau}\tilde{\rho}}(\sigma_n,r)}(\nu), \qquad \tilde{\varrho}_{C_{\tilde{\tau}\tilde{\varrho}}(\sigma_n,r)}(\nu) \geq t > \tilde{\varrho}_{C_{\delta I\tilde{\tau}\tilde{\varrho}}(\sigma_n,r)}(\nu), \tag{1}$$
$$\tilde{\eta}_{C_{\tilde{\tau}\tilde{\eta}}(\sigma_n,r)}(\nu) \geq p > \tilde{\eta}_{C_{\delta I\tilde{\tau}\tilde{\eta}}(\sigma_n,r)}(\nu).$$

By the definition of $C_{\tilde{\tau}\tilde{\rho}\tilde{\varrho}\tilde{\eta}}$, there exists $\tilde{\tau}^{\tilde{\rho}}([\gamma_n]^c) \geq r$, $\tilde{\tau}^{\tilde{\varrho}}([\gamma_n]^c) \leq 1-r$, $\tilde{\tau}^{\tilde{\eta}}([\gamma_n]^c) \leq 1-r$ with $\sigma_n \leq \gamma_n$ such that

$$\tilde{\rho}_{C_{\tilde{\tau}\tilde{\rho}}(\sigma_n,r)}(\nu) \leq \tilde{\rho}_{\gamma_n}(\nu) < s < \tilde{\rho}_{C_{\delta I\tilde{\tau}\tilde{\rho}}(\sigma_n,r)}(\nu), \qquad \tilde{\varrho}_{C_{\tilde{\tau}\tilde{\varrho}}(\sigma_n,r)}(\nu) \geq \tilde{\varrho}_{\gamma_n}(\nu) > t > \tilde{\varrho}_{C_{\delta I\tilde{\tau}\tilde{\rho}}(\sigma_n,r)}(\nu),$$
$$\tilde{\eta}_{C_{\tilde{\tau}\tilde{\eta}}(\sigma_n,r)}(\nu) \geq \tilde{\rho}_{\gamma_n}(\nu) > p > \tilde{\eta}_{C_{\delta I\tilde{\tau}\tilde{\eta}}(\sigma_n,r)}(\nu).$$

Then, $[\gamma_n]^c \in Q_{\tilde{\tau}^{\tilde{\rho}\tilde{\varrho}\tilde{\eta}}}(x_{s,t,p}, r)$ and

$$[\sigma_n]^c \geq [\gamma_n]^c \quad \Rightarrow \mathrm{CI}^{\pounds}_{\tilde{\tau}^{\tilde{\rho}\tilde{\varrho}\tilde{\eta}}}([\sigma_n]^c, r) \geq \mathrm{CI}^{\pounds}_{\tilde{\tau}^{\tilde{\rho}\tilde{\varrho}\tilde{\eta}}}([\gamma_n]^c, r)$$
$$\Rightarrow \mathrm{CI}^{\pounds}_{\tilde{\tau}^{\tilde{\rho}\tilde{\varrho}\tilde{\eta}}}([\sigma_n]^c, r) \geq \mathrm{int}_{\tilde{\tau}^{\tilde{\rho}\tilde{\varrho}\tilde{\eta}}}([\gamma_n]^c, r)$$
$$\Rightarrow [\mathrm{int}^{\pounds}_{\tilde{\tau}^{\tilde{\rho}\tilde{\varrho}\tilde{\eta}}}(\sigma_n, r)]^c \geq [\gamma_n]^c.$$

Thus, $\mathrm{int}^{\pounds}_{\tilde{\tau}^{\tilde{\rho}\tilde{\varrho}\tilde{\eta}}}(\sigma_n, r)\overline{q}[\gamma_n]^c$. Hence, $\mathrm{int}_{\tilde{\tau}^{\tilde{\rho}\tilde{\varrho}\tilde{\eta}}}(\mathrm{CI}^{\pounds}_{\tilde{\tau}^{\tilde{\rho}\tilde{\varrho}\tilde{\eta}}}([\gamma_n]^c, r), r)\overline{q}\mathrm{C}_{\tilde{\tau}^{\tilde{\rho}\tilde{\varrho}\tilde{\eta}}}(\mathrm{int}^{\pounds}_{\tilde{\tau}^{\tilde{\rho}\tilde{\varrho}\tilde{\eta}}}(\sigma_n, r), r)$. Since σ_n is an r-SVN£SO, we have $\mathrm{int}_{\tilde{\tau}^{\tilde{\rho}\tilde{\varrho}\tilde{\eta}}}(\mathrm{CI}^{\pounds}_{\tilde{\tau}^{\tilde{\rho}\tilde{\varrho}\tilde{\eta}}}(\gamma_n, r), r)\overline{q}\sigma_n$. So, $x_{s,t,p}$ is not an r-SVN$\delta\mathcal{I}$-cluster point of σ_n. It is a contradiction for equation 3. Thus, $\mathrm{C}_{\tilde{\tau}^{\tilde{\rho}\tilde{\varrho}\tilde{\eta}}}(\sigma_n, r) \geq \mathrm{C}_{\delta\mathbf{I}\tilde{\tau}^{\tilde{\rho}\tilde{\varrho}\tilde{\eta}}}(\sigma_n, r)$. By Lemma 4, we have $\mathrm{C}_{\tilde{\tau}^{\tilde{\rho}\tilde{\varrho}\tilde{\eta}}}(\sigma_n, r) = \mathrm{C}_{\delta\mathbf{I}\tilde{\tau}^{\tilde{\rho}\tilde{\varrho}\tilde{\eta}}}(\sigma_n, r)$. □

Theorem 13. *Let $(\tilde{X}, \tilde{\tau}^{\tilde{\rho}\tilde{\varrho}\tilde{\eta}}, \mathcal{I}^{\tilde{\rho}\tilde{\varrho}\tilde{\eta}})$ be an SVNITS. Then, the following properties are equivalent:*

(1) *$(\tilde{X}, \tilde{\tau}^{\tilde{\rho}\tilde{\varrho}\tilde{\eta}}, \mathcal{I}^{\tilde{\rho}\tilde{\varrho}\tilde{\eta}})$ is £-SVNE-disconnected,*
(2) *If σ_n is r-SVNSβIO and γ_n is r-SVN£SO, then $\mathrm{CI}^{\pounds}_{\tilde{\tau}^{\tilde{\rho}\tilde{\varrho}\tilde{\eta}}}(\sigma_n, r) \wedge \mathrm{C}_{\tilde{\tau}^{\tilde{\rho}\tilde{\varrho}\tilde{\eta}}}(\gamma_n, r) \leq \mathrm{C}_{\tilde{\tau}^{\tilde{\rho}\tilde{\varrho}\tilde{\eta}}}(\sigma_n \wedge \gamma_n)$,*
(3) *If σ_n is r-SVNSIO and γ_n is r-SVN£SO, then $\mathrm{CI}^{\pounds}_{\tilde{\tau}^{\tilde{\rho}\tilde{\varrho}\tilde{\eta}}}(\sigma_n, r) \wedge \mathrm{C}_{\tilde{\tau}^{\tilde{\rho}\tilde{\varrho}\tilde{\eta}}}(\gamma_n, r) \leq \mathrm{C}_{\tilde{\tau}^{\tilde{\rho}\tilde{\varrho}\tilde{\eta}}}(\sigma_n \wedge \gamma_n)$,*
(4) *$\mathrm{CI}^{\pounds}_{\tilde{\tau}^{\tilde{\rho}\tilde{\varrho}\tilde{\eta}}}(\sigma_n, r)\overline{q}\mathrm{C}_{\tilde{\tau}^{\tilde{\rho}\tilde{\varrho}\tilde{\eta}}}(\gamma_n, r)$, for every r-SVNSIO set $\sigma_n \in \xi^{\tilde{X}}$ and every r-SVN£SO $\gamma_n \in \xi^{\tilde{X}}$ with $\sigma_n\overline{q}\gamma_n$,*
(5) *If σ_n is an r-SVNPIO and γ_n is an r-SVN£SO, then $\mathrm{CI}^{\pounds}_{\tilde{\tau}^{\tilde{\rho}\tilde{\varrho}\tilde{\eta}}}(\sigma_n, r) \wedge \mathrm{C}_{\tilde{\tau}^{\tilde{\rho}\tilde{\varrho}\tilde{\eta}}}(\gamma_n, r) \leq \mathrm{C}_{\tilde{\tau}^{\tilde{\rho}\tilde{\varrho}\tilde{\eta}}}(\sigma_n \wedge \gamma_n)$.*

Proof. (1)⇒(2): Let σ_n be an r-SVNSβIO and γ_n be an r-SVN£SO, by Theorem 7, $\tilde{\tau}^{\tilde{\rho}}(\mathrm{CI}^{\pounds}_{\tilde{\tau}^{\tilde{\rho}}}(\sigma_n, r))$
$\geq r$, $\tilde{\tau}^{\tilde{\varrho}}(\mathrm{CI}^{\pounds}_{\tilde{\tau}^{\tilde{\varrho}}}(\sigma_n, r)) \leq 1 - r$, $\tilde{\tau}^{\tilde{\eta}}(\mathrm{CI}^{\pounds}_{\tilde{\tau}^{\tilde{\eta}}}(\sigma_n, r)) \leq 1 - r$. Then,

$$\mathrm{CI}^{\pounds}_{\tilde{\tau}^{\tilde{\rho}\tilde{\varrho}\tilde{\eta}}}(\sigma_n, r) \wedge \quad \mathrm{C}_{\tilde{\tau}^{\tilde{\rho}\tilde{\varrho}\tilde{\eta}}}(\gamma_n, r) \leq \mathrm{C}_{\tilde{\tau}^{\tilde{\rho}\tilde{\varrho}\tilde{\eta}}}(\mathrm{int}^{\pounds}_{\tilde{\tau}^{\tilde{\rho}\tilde{\varrho}\tilde{\eta}}}(\gamma_n, r), r) \leq \mathrm{C}_{\tilde{\tau}^{\tilde{\rho}\tilde{\varrho}\tilde{\eta}}}[\mathrm{CI}^{\pounds}_{\tilde{\tau}^{\tilde{\rho}\tilde{\varrho}\tilde{\eta}}}(\gamma_n, r) \wedge \mathrm{int}^{\pounds}_{\tilde{\tau}^{\tilde{\rho}\tilde{\varrho}\tilde{\eta}}}(\gamma_n, r), r]$$
$$\leq \mathrm{C}_{\tilde{\tau}^{\tilde{\rho}\tilde{\varrho}\tilde{\eta}}}[\mathrm{CI}^{\pounds}_{\tilde{\tau}^{\tilde{\rho}\tilde{\varrho}\tilde{\eta}}}[\gamma_n \wedge \mathrm{int}^{\pounds}_{\tilde{\tau}^{\tilde{\rho}\tilde{\varrho}\tilde{\eta}}}(\gamma_n, r), r], r] \leq \mathrm{C}_{\tilde{\tau}^{\tilde{\rho}\tilde{\varrho}\tilde{\eta}}}[\mathrm{C}_{\tilde{\tau}^{\tilde{\rho}\tilde{\varrho}\tilde{\eta}}}[\gamma_n \wedge \mathrm{int}^{\pounds}_{\tilde{\tau}^{\tilde{\rho}\tilde{\varrho}\tilde{\eta}}}(\gamma_n, r), r], r]$$
$$\leq \mathrm{C}_{\tilde{\tau}^{\tilde{\rho}\tilde{\varrho}\tilde{\eta}}}[\gamma_n \wedge \mathrm{int}^{\pounds}_{\tilde{\tau}^{\tilde{\rho}\tilde{\varrho}\tilde{\eta}}}(\gamma_n, r), r] \leq \mathrm{C}_{\tilde{\tau}^{\tilde{\rho}\tilde{\varrho}\tilde{\eta}}}[\gamma_n \wedge \gamma_n, r].$$

Hence, $\mathrm{CI}^{\pounds}_{\tilde{\tau}^{\tilde{\rho}\tilde{\varrho}\tilde{\eta}}}(\sigma_n, r) \wedge \mathrm{C}_{\tilde{\tau}^{\tilde{\rho}\tilde{\varrho}\tilde{\eta}}}(\gamma_n, r) \leq \mathrm{C}_{\tilde{\tau}^{\tilde{\rho}\tilde{\varrho}\tilde{\eta}}}(\sigma_n \wedge \gamma_n)$.

(2)⇒(3): It follows from the fact that every r-SVNSIO set is an r-SVNSβIO.

(3)⇒(4): Clear.

(4)⇒(1): Let σ_n be an r-SVNSIO. Since $[\mathrm{CI}^{\pounds}_{\tilde{\tau}^{\tilde{\rho}\tilde{\varrho}\tilde{\eta}}}(\sigma_n, r)]^c \leq \mathrm{C}_{\tilde{\tau}^{\tilde{\rho}\tilde{\varrho}\tilde{\eta}}}(\mathrm{int}^{\pounds}_{\tilde{\tau}^{\tilde{\rho}\tilde{\varrho}\tilde{\eta}}}(\mathrm{CI}^{\pounds}_{\tilde{\tau}^{\tilde{\rho}\tilde{\varrho}\tilde{\eta}}}([\sigma_n]^c, r), r), r)$ we have, $[\mathrm{CI}^{\pounds}_{\tilde{\tau}^{\tilde{\rho}\tilde{\varrho}\tilde{\eta}}}(\sigma_n, r)]^c$ is an r-SVN£SO. Then, by (4), $\mathrm{CI}^{\pounds}_{\tilde{\tau}^{\tilde{\rho}\tilde{\varrho}\tilde{\eta}}}(\sigma_n, r)\overline{q}\mathrm{C}_{\tilde{\tau}^{\tilde{\rho}\tilde{\varrho}\tilde{\eta}}}([\mathrm{CI}^{\pounds}_{\tilde{\tau}^{\tilde{\rho}\tilde{\varrho}\tilde{\eta}}}(\sigma_n, r)]^c, r)$. Thus, $\mathrm{CI}^{\pounds}_{\tilde{\tau}^{\tilde{\rho}\tilde{\varrho}\tilde{\eta}}}(\sigma_n, r) \leq [\mathrm{C}_{\tilde{\tau}^{\tilde{\rho}\tilde{\varrho}\tilde{\eta}}}(\mathrm{CI}^{\pounds}_{\tilde{\tau}^{\tilde{\rho}\tilde{\varrho}\tilde{\eta}}}(\sigma_n, r)]^c, r)]^c = \mathrm{int}_{\tilde{\tau}^{\tilde{\rho}\tilde{\varrho}\tilde{\eta}}}(\mathrm{CI}^{\pounds}_{\tilde{\tau}^{\tilde{\rho}\tilde{\varrho}\tilde{\eta}}}(\sigma_n, r), r)$. Therefore, $\tilde{\tau}^{\tilde{\rho}}(\mathrm{CI}^{\pounds}_{\tilde{\tau}^{\tilde{\rho}}}(\sigma_n, r)) \geq r$, $\tilde{\tau}^{\tilde{\varrho}}(\mathrm{CI}^{\pounds}_{\tilde{\tau}^{\tilde{\varrho}}}(\sigma_n, r)) \leq 1 - r$, $\tilde{\tau}^{\tilde{\eta}}(\mathrm{CI}^{\pounds}_{\tilde{\tau}^{\tilde{\eta}}}(\sigma_n, r)) \leq 1 - r$. Thus, by Theorem 12, $(\tilde{X}, \tilde{\tau}^{\tilde{\rho}\tilde{\varrho}\tilde{\eta}}, \mathcal{I}^{\tilde{\rho}\tilde{\varrho}\tilde{\eta}})$ is £-SVNE-disconnected.

(2)⇒(5): It follows from the fact that every r-SVNPIO is an r-SVNSβIO. □

Corollary 1. *Let $(\tilde{X}, \tilde{\tau}^{\tilde{\rho}\tilde{\varrho}\tilde{\eta}}, \mathcal{I}^{\tilde{\rho}\tilde{\varrho}\tilde{\eta}})$ be an SVNITS. Then, the following properties are equivalent:*

(1) *$(\tilde{X}, \tilde{\tau}^{\tilde{\rho}\tilde{\varrho}\tilde{\eta}}, \mathcal{I}^{\tilde{\rho}\tilde{\varrho}\tilde{\eta}})$ is £-SVNE-disconnected.*
(2) *If σ_n is an r-SVNSβIO and γ_n is an r-SVN£SO, then $\mathrm{CI}^{\pounds}_{\tilde{\tau}^{\tilde{\rho}\tilde{\varrho}\tilde{\eta}}}(\sigma_n, r) \wedge \mathrm{C}_{\delta\mathbf{I}\tilde{\tau}^{\tilde{\rho}\tilde{\varrho}\tilde{\eta}}}(\gamma_n, r) \leq \mathrm{C}_{\tilde{\tau}^{\tilde{\rho}\tilde{\varrho}\tilde{\eta}}}(\sigma_n \wedge \gamma_n)$.*
(3) *If σ_n is an r-SVNSIO and γ_n is an r-SVN£SO, then $\mathrm{CI}^{\pounds}_{\tilde{\tau}^{\tilde{\rho}\tilde{\varrho}\tilde{\eta}}}(\sigma_n, r) \wedge \mathrm{C}_{\delta\mathbf{I}\tilde{\tau}^{\tilde{\rho}\tilde{\varrho}\tilde{\eta}}}(\gamma_n, r) \leq \mathrm{C}_{\tilde{\tau}^{\tilde{\rho}\tilde{\varrho}\tilde{\eta}}}(\sigma_n \wedge \gamma_n)$.*
(4) *If σ_n is an r-SVNPIO and γ_n is an r-SVN£SO, then $\mathrm{CI}^{\pounds}_{\tilde{\tau}^{\tilde{\rho}\tilde{\varrho}\tilde{\eta}}}(\sigma_n, r) \wedge \mathrm{C}_{\delta\mathbf{I}\tilde{\tau}^{\tilde{\rho}\tilde{\varrho}\tilde{\eta}}}(\gamma_n, r) \leq \mathrm{C}_{\tilde{\tau}^{\tilde{\rho}\tilde{\varrho}\tilde{\eta}}}(\sigma_n \wedge \gamma_n)$.*

Proof. It follows directly from Lemma 3 and 5. □

5. Some Types of Separation Axioms

In this section, some kinds of separation axioms, namely r-single valued neutrosophic ideal-R_i (r-$SVNIR_i$, for short), where $i = \{0,1,2,3\}$, and r-single valued neutrosophic ideal-T_j (r-$SVNIT_j$, for short), where $j = \{1,2,2\frac{1}{2},3,4\}$, in the sense of Šostak are defined. Some of their characterizations, fundamental properties, and the relations between these notions have been studied.

Definition 19. *Let $(\tilde{X}, \tilde{\tau}^{\tilde{\rho}\tilde{\varrho}\tilde{\eta}}, \mathcal{I}^{\tilde{\rho}\tilde{\varrho}\tilde{\eta}})$ be an SVNITS and $r \in \zeta_0$. Then, $\tilde{X}$ is called:*

(1) *r-$SVNIR_0$ iff $x_{s,t,p}\overline{q}CI^{\pounds}_{\tau\tilde{\rho}\tilde{\varrho}\tilde{\eta}}(y_{s_1,t_1,p_1}, r)$ implies $y_{s_1,t_1,p_1}\overline{q}CI^{\pounds}_{\tau\tilde{\rho}\tilde{\varrho}\tilde{\eta}}(x_{s,t,p}, r)$ for any $x_{s,t,p} \neq y_{s_1,t_1,p_1}$.*

(2) *r-$SVNIR_1$ iff $x_{s,t,p}\overline{q}CI^{\pounds}_{\tau\tilde{\rho}\tilde{\varrho}\tilde{\eta}}(y_{s_1,t_1,p_1}, r)$ implies that there exist r-SVN£O sets $\sigma_n, \gamma_n \in \xi^{\tilde{X}}$ such that $x_{s,t,p} \in \sigma_n$, $y_{s_1,t_1,p_1} \in \gamma_n$ and $\sigma_n\overline{q}\gamma_n$.*

(3) *r-$SVNIR_2$ iff $x_{s,t,p}\overline{q}\varsigma_n = CI^{\pounds}_{\tau\tilde{\rho}\tilde{\varrho}\tilde{\eta}}(\varsigma_n, r)$ implies there exist r-SVN£O sets $\sigma_n, \gamma_n \in \xi^{\tilde{X}}$ such that $x_{s,t,p} \in \sigma_n$, $\varsigma_n \leq \gamma_n$ and $\sigma_n\overline{q}\gamma_n$.*

(4) *r-$SVNIR_3$ iff $[\varsigma_n]_1 = CI^{\pounds}_{\tau\tilde{\rho}\tilde{\varrho}\tilde{\eta}}([\varsigma_n]_1, r)\overline{q}[\varsigma_n]_2 = CI^{\pounds}_{\tau\tilde{\rho}\tilde{\varrho}\tilde{\eta}}([\varsigma_n]_2, r)$ implies that there exist r-SVN£O sets $\sigma_n, \gamma_n \in \xi^{\tilde{X}}$ such that $[\varsigma_n]_1 \leq \sigma_n$, $[\varsigma_n]_2 \leq \gamma_n$ and $\sigma_n\overline{q}\gamma_n$.*

(5) *r-$SVNIT_1$ iff $x_{s,t,p}\overline{q}y_{s_1,t_1,p_1}$ implies that there exists r-SVN£O $\sigma_n \in \xi^{\tilde{X}}$ such that $x_{s,t,p} \in \sigma_n$ and $y_{s_1,t_1,p_1}\overline{q}\sigma_n$.*

(6) *r-$SVNIT_2$ iff $x_{s,t,p}\overline{q}y_{s_1,t_1,p_1}$ implies that there exist r-SVN£O sets $\sigma_n, \gamma_n \in \xi^{\tilde{X}}$ such that $x_{s,t,p} \in \sigma_n$, $y_{s_1,t_1,p_1} \in \gamma_n$ and $\sigma_n\overline{q}\gamma_n$.*

(7) *r-$SVNIT_{2\frac{1}{2}}$ iff $x_{s,t,p}\overline{q}y_{s_1,t_1,p_1}$ implies that there exist r-SVN£O sets $\sigma_n, \gamma_n \in \xi^{\tilde{X}}$ such that $x_{s,t,p} \in \sigma_n$, $y_{s_1,t_1,p_1} \in \gamma_n$ and $CI^{\pounds}_{\tau\tilde{\rho}\tilde{\varrho}\tilde{\eta}}(\sigma_n, r)\overline{q}CI^{\pounds}_{\tau\tilde{\rho}\tilde{\varrho}\tilde{\eta}}(\gamma_n, r)$.*

(8) *r-$SVNIT_3$ iff it is r-$SVNITR_2$ and r-$SVNIT_1$.*

(9) *r-$SVNIT_4$ iff it is r-$SVNITR_3$ and r-$SVNIT_1$.*

Theorem 14. *Let $(\tilde{X}, \tilde{\tau}^{\tilde{\rho}\tilde{\varrho}\tilde{\eta}}, \mathcal{I}^{\tilde{\rho}\tilde{\varrho}\tilde{\eta}})$ be an SVNITS and $r \in \zeta_0$. Then, the following statements are equivalent:*

(1) *$(\tilde{X}, \tilde{\tau}^{\tilde{\rho}\tilde{\varrho}\tilde{\eta}}, \mathcal{I}^{\tilde{\rho}\tilde{\varrho}\tilde{\eta}})$ is r-$SVNIR_0$.*

(2) *If $x_{s,t,p}\overline{q}\sigma_n = CI^{\pounds}_{\tau\tilde{\rho}\tilde{\varrho}\tilde{\eta}}(\sigma_n, r)$, then there exists r-SVN£O $\gamma_n \in \xi^{\tilde{X}}$ such that $x_{s,t,p}\overline{q}\gamma_n$ and $\sigma_n \leq \gamma_n$.*

(3) *If $x_{s,t,p}\overline{q}\sigma_n = CI^{\pounds}_{\tau\tilde{\rho}\tilde{\varrho}\tilde{\eta}}(\sigma_n, r)$, then $CI^{\pounds}_{\tau\tilde{\rho}\tilde{\varrho}\tilde{\eta}}(x_{s,t,p}, r)\overline{q}\sigma_n = CI^{\pounds}_{\tau\tilde{\rho}\tilde{\varrho}\tilde{\eta}}(\sigma_n, r)$.*

(4) *If $x_{s,t,p}\overline{q}CI^{\pounds}_{\tau\tilde{\rho}\tilde{\varrho}\tilde{\eta}}(y_{s_1,t_1,p_1}, r)$, then $CI^{\pounds}_{\tau\tilde{\rho}\tilde{\varrho}\tilde{\eta}}(x_{s,t,p}, r)\overline{q}CI^{\pounds}_{\tau\tilde{\rho}\tilde{\varrho}\tilde{\eta}}(y_{s_1,t_1,p_1}, r)$.*

Proof. (1)⇒(2): Let $x_{s,t,p}\overline{q}\sigma_n = CI^{\pounds}_{\tau\tilde{\rho}\tilde{\varrho}\tilde{\eta}}(\sigma_n, r)$. Then,

$$s + \tilde{\rho}_{\sigma_n}(\nu) < 1, \quad t + \tilde{\varrho}_{\sigma_n}(\nu) \geq 1, \quad p + \tilde{\eta}_{\sigma_n}(\nu) \geq 1,$$

for every $y_{s_1,t_1,p_1} \in \sigma_n$, we have $s_1 < \tilde{\rho}_{\sigma_n}(\nu)$, $t_1 \geq \tilde{\varrho}_{\sigma_n}(\nu)$ and $p_1 \geq \tilde{\eta}_{\sigma_n}(\nu)$. Thus, $x_{s,t,p}\overline{q}CI^{\pounds}_{\tau\tilde{\rho}\tilde{\varrho}\tilde{\eta}}(y_{s_1,t_1,p_1}, r)$. Since $(\tilde{X}, \tilde{\tau}^{\tilde{\rho}\tilde{\varrho}\tilde{\eta}}, \mathcal{I}^{\tilde{\rho}\tilde{\varrho}\tilde{\eta}})$ is an r-$SVNIR_0$, we obtain $y_{s_1,t_1,p_1}\overline{q}CI^{\pounds}_{\tau\tilde{\rho}\tilde{\varrho}\tilde{\eta}}(x_{s,t,p}, r)$. By Lemma 2(2), there exists an r-SVN£O $\varsigma_n \in \xi^{\tilde{X}}$ such that $x_{s,t,p}\overline{q}\varsigma_n$ and $y_{s_1,t_1,p_1} \leq \varsigma_n$. Let

$$\gamma_n = \bigvee_{y_{s_1,t_1,p_1} \in \sigma_n} \{\varsigma_n : x_{s,t,p}\overline{q}\varsigma_n,\ y_{s_1,t_1,p_1} \in \varsigma_n\}.$$

From Lemma 1(1), γ_n is an r-SVN£O. Then, $x_{s,t,p}\overline{q}\gamma_n$, $\sigma_n \leq \gamma_n$.

(2)⇒(3): Let $x_{s,t,p}\overline{q}\sigma_n = CI^{\pounds}_{\tau\tilde{\rho}\tilde{\varrho}\tilde{\eta}}(\sigma_n, r)$. Then, there exists an r-SVN£O $\gamma_n \in \xi^{\tilde{X}}$ such that $x_{s,t,p}\overline{q}\gamma_n$ and $\sigma_n \leq \gamma_n$. Since for every $\nu \in \tilde{X}$,

$$s < 1 - \tilde{\rho}_{\gamma_n}(\nu), \quad t \geq 1 - \tilde{\varrho}_{\gamma_n}(\nu), \quad p \geq 1 - \tilde{\eta}_{\gamma_n}(\nu),$$

we obtain

$$CI^{\pounds}_{\tau\tilde{\rho}\tilde{\varrho}\tilde{\eta}}(x_{s,t,p}, r) \leq CI^{\pounds}_{\tau\tilde{\rho}\tilde{\varrho}\tilde{\eta}}([\gamma_n]^c, r) = [\gamma_n]^c \leq [\sigma_n]^c.$$

Therefore, $\mathrm{CI}^{\pounds}_{\tau^{\tilde{\rho}\tilde{\varrho}\tilde{\eta}}}(x_{s,t,p}, r)\overline{q}\sigma_n = \mathrm{CI}^{\pounds}_{\tau^{\tilde{\rho}\tilde{\varrho}\tilde{\eta}}}(\sigma_n, r)$.

(3)⇒(4): Let $x_{s,t,p}\overline{q}\mathrm{CI}^{\pounds}_{\tau^{\tilde{\rho}\tilde{\varrho}\tilde{\eta}}}(y_{s_1,t_1,p_1}, r)$. Then, $x_{s,t,p}\overline{q}\mathrm{CI}^{\pounds}_{\tau^{\tilde{\rho}\tilde{\varrho}\tilde{\eta}}}(y_{s_1,t_1,p_1}, r) = \mathrm{CI}^{\pounds}_{\tau^{\tilde{\rho}\tilde{\varrho}\tilde{\eta}}}(\mathrm{CI}^{\pounds}_{\tau^{\tilde{\rho}\tilde{\varrho}\tilde{\eta}}}(y_{s_1,t_1,p_1}, r), r)$. By (3), $s_1, t_1, p_1(x_{s,t,p}, r)\overline{q}\mathrm{CI}^{\pounds}_{\tau^{\tilde{\rho}\tilde{\varrho}\tilde{\eta}}}(y_{s_1,t_1,p_1}, r)$.

(4)⇒(1): Clear. □

Theorem 15. *Let $(\tilde{X}, \tilde{\tau}^{\tilde{\rho}\tilde{\varrho}\tilde{\eta}}, \mathcal{I}^{\tilde{\rho}\tilde{\varrho}\tilde{\eta}})$ be an SVNITS and $r \in \zeta_0$. Then, if $\tilde{X}$ is*

(1) [r-$SVNIR_3$ and r-$SVNIR_0$] $\Rightarrow^{(a)}$ r-$SVNIR_2 \Rightarrow^{(b)}$ r-$SVNIR_1 \Rightarrow^{(c)}$ r-$SVNIR_0$.
(2) r-$SVNIT_2 \Rightarrow r$-$SVNIR_1$.
(3) r-$SVNIT_3 \Rightarrow r$-$SVNIR_2$.
(4) r-$SVNIT_4 \Rightarrow r$-$SVNIR_3$.
(5) r-$SVNIT_4 \Rightarrow^{(a)} r$-$SVNIT_3 \Rightarrow^{(b)} r$-$SVNIT_{2\frac{1}{2}} \Rightarrow^{(c)} r$-$SVNIT_2 \Rightarrow^{(d)} r$-$SVNIT_1$.

Proof. (1_a). Let $x_{s,t.p}\overline{q}\varsigma_n = \mathrm{CI}^{\pounds}_{\tau^{\tilde{\rho}\tilde{\varrho}\tilde{\eta}}}(\varsigma_n, r)$, by Theorem 14(3), $\mathrm{CI}^{\pounds}_{\tau^{\tilde{\rho}\tilde{\varrho}\tilde{\eta}}}(x_{s,t,p}, r)\overline{q}\varsigma_n = \mathrm{CI}^{\pounds}_{\tau^{\tilde{\rho}\tilde{\varrho}\tilde{\eta}}}(\varsigma_n, r)$. Since $(\tilde{X}, \tilde{\tau}^{\tilde{\rho}\tilde{\varrho}\tilde{\eta}}, \mathcal{I}^{\tilde{\rho}\tilde{\varrho}\tilde{\eta}})$ is r-$SVNIR_3$ and $\mathrm{CI}^{\pounds}_{\tau^{\tilde{\rho}\tilde{\varrho}\tilde{\eta}}}(x_{s,t,p}, r) = \mathrm{CI}^{\pounds}_{\tau^{\tilde{\rho}\tilde{\varrho}\tilde{\eta}}}(\mathrm{CI}^{\pounds}_{\tau^{\tilde{\rho}\tilde{\varrho}\tilde{\eta}}}(x_{s,t,p}, r), r)$, there exist r-SVN£O sets $\sigma_n, \gamma_n \in \zeta^{\tilde{X}}$ such that $x_{s,t,p} \in \mathrm{CI}^{\pounds}_{\tau^{\tilde{\rho}\tilde{\varrho}\tilde{\eta}}}(x_{s,t,p}, r) \leq \sigma_n$, $\varsigma_n \leq \gamma_n$ and $\sigma_n\overline{q}\gamma_n$. Hence, $(\tilde{X}, \tilde{\tau}^{\tilde{\rho}\tilde{\varrho}\tilde{\eta}}, \mathcal{I}^{\tilde{\rho}\tilde{\varrho}\tilde{\eta}})$ is r-$SVNIR_2$.

(1_b). For each $x_{s,t,p}\overline{q}\mathrm{CI}^{\pounds}_{\tau^{\tilde{\rho}\tilde{\varrho}\tilde{\eta}}}(y_{s_1,t_1,p_1}, r)$, by r-$SVNIR_2$ of $\tilde{X}$, there exist r-SVN£O sets $\sigma_n, \gamma_n \in \zeta^{\tilde{X}}$ such that $x_{s,t,p} \in \sigma_n$, $y_{s_1,t_1,p_1}, r \in \mathrm{CI}^{\pounds}_{\tau^{\tilde{\rho}\tilde{\varrho}\tilde{\eta}}}(y_{s_1,t_1,p_1}, r, r) \leq \gamma_n$ and $\sigma_n\overline{q}\gamma_n$. Thus, $(\tilde{X}, \tilde{\tau}^{\tilde{\rho}\tilde{\varrho}\tilde{\eta}}, \mathcal{I}^{\tilde{\rho}\tilde{\varrho}\tilde{\eta}})$ is r-$SVNIR_1$.

(1_c). Let $(\tilde{X}, \tilde{\tau}^{\tilde{\rho}\tilde{\varrho}\tilde{\eta}}, \mathcal{I}^{\tilde{\rho}\tilde{\varrho}\tilde{\eta}})$ be r-$SVNIR_1$. Then, for every $x_{s,t,p}\overline{q}\mathrm{CI}^{\pounds}_{\tau^{\tilde{\rho}\tilde{\varrho}\tilde{\eta}}}(y_{s_1,t_1,p_1}, r, r)$ and $x_{s,t,p} \neq y_{s_1,t_1,p_1}$, there exist r-SVN£O sets $\sigma_n, \gamma_n \in \zeta^{\tilde{X}}$ such that $x_{s,t,p} \in \sigma_n$, $y_{s_1,t_1,p_1} \in \gamma_n$ and $\sigma_n\overline{q}\gamma_n$. Hence, $x_{s,t,p} \in \sigma_n \leq [\gamma_n]^c$. Since γ_n is an r-SVN£O set, we obtain $\mathrm{CI}^{\pounds}_{\tau^{\tilde{\rho}\tilde{\varrho}\tilde{\eta}}}(x_{s,t,p}, r) \leq \mathrm{CI}^{\pounds}_{\tau^{\tilde{\rho}\tilde{\varrho}\tilde{\eta}}}([\gamma_n]^c, r) = [\gamma_n]^c \leq [y_{s_1,t_1,p_1}]^c$. Thus, $y_{s_1,t_1,p_1}\overline{q}\mathrm{CI}^{\pounds}_{\tau^{\tilde{\rho}\tilde{\varrho}\tilde{\eta}}}(x_{s,t,p}, r)$ and $(\tilde{X}, \tilde{\tau}^{\tilde{\rho}\tilde{\varrho}\tilde{\eta}}, \mathcal{I}^{\tilde{\rho}\tilde{\varrho}\tilde{\eta}})$ is r-$SVNIR_0$.

(2). Let $x_{s,t,p}\overline{q}\mathrm{CI}^{\pounds}_{\tau^{\tilde{\rho}\tilde{\varrho}\tilde{\eta}}}(y_{s_1,t_2,p_1}, r)$. Then, $x_{s,t,p}\overline{q}y_{s_1,t_1,p_1}$. By r-$SVNIT_2$ of $\tilde{X}$, there exist r-SVN£O sets $\sigma_n, \gamma_n \in \zeta^{\tilde{X}}$ such that $x_{st,p} \in \sigma_n$, $y_{s_1,t_1,p_1} \in \gamma_n$ and $\sigma_n\overline{q}\gamma_n$. Hence, $(\tilde{X}, \tilde{\tau}^{\tilde{\rho}\tilde{\varrho}\tilde{\eta}}, \mathcal{I}^{\tilde{\rho}\tilde{\varrho}\tilde{\eta}})$ is r-$SVNIR_1$.

(3) and (4) The proofs are direct consequence of (2) .

(5_a). The proof is direct consequence of (1).

(5_b). For each $x_{s,t,p}\overline{q}y_{s_1,t_1,p_1}$, since $\tilde{X}$ is both r-$SVNIR_2$ and r-$SVNIT_1$, then, there exists an r-SVN£O set $\varsigma_n \in \zeta^{\tilde{X}}$ such that $x_{s,t,p} \in \varsigma_n$ and $y_{s_1,t_1,p_1}\overline{q}\varsigma_n$. Then,

$$x_t \in \varsigma_n = \mathrm{int}^{\pounds}_{\tau^{\tilde{\rho}\tilde{\varrho}\tilde{\eta}}}(\varsigma_n, r) \leq \mathrm{int}^{\pounds}_{\tau^{\tilde{\rho}\tilde{\varrho}\tilde{\eta}}}([y_{s_1,t_1,p_1}]^c, r) = [\mathrm{CI}^{\pounds}_{\tau^{\tilde{\rho}\tilde{\varrho}\tilde{\eta}}}(y_{s_1,t_1,p_1}, r)]^c.$$

Hence, $x_{s,t,p}\overline{q}\mathrm{CI}^{\pounds}_{\tau^{\tilde{\rho}\tilde{\varrho}\tilde{\eta}}}(y_{s_1,t_1,p_1}, r)$. By r-$SVNIR_2$ of $\tilde{X}$, there exist r-SVN£O sets $\sigma_n, \gamma_n \in \zeta^{\tilde{X}}$ such that $x_{s,t,p} \in \sigma_n$, $\mathrm{CI}^{\pounds}_{\tau^{\tilde{\rho}\tilde{\varrho}\tilde{\eta}}}(y_{s_1,t_1,p_1}, r) \leq \gamma_n$ and $\sigma_n\overline{q}\gamma_n$. Thus, $\sigma_n \leq [\gamma_n]^c$, so

$$\mathrm{CI}^{\pounds}_{\tau^{\tilde{\rho}\tilde{\varrho}\tilde{\eta}}}(\sigma_n, r) \leq \mathrm{CI}^{\pounds}_{\tau^{\tilde{\rho}\tilde{\varrho}\tilde{\eta}}}([\gamma_n]^c, r) = [\gamma_n]^c \leq [\mathrm{CI}^{\pounds}_{\tau^{\tilde{\rho}\tilde{\varrho}\tilde{\eta}}}(y_{s_1,t_1,p_1}, r)]^c.$$

It implies $\mathrm{CI}^{\pounds}_{\tau^{\tilde{\rho}\tilde{\varrho}\tilde{\eta}}}(\sigma_n, r)\overline{q}\mathrm{CI}^{\pounds}_{\tau^{\tilde{\rho}\tilde{\varrho}\tilde{\eta}}}(y_{s_1,t_1,p_1}, r)$ with $x_{s,t,p} \in \sigma_n$ and $y_{s_1,t_1,p_1} \in \mathrm{CI}^{\pounds}_{\tau^{\tilde{\rho}\tilde{\varrho}\tilde{\eta}}}(y_{s_1,t_1,p_1}, r)$. Thus, $(\tilde{X}, \tilde{\tau}^{\tilde{\rho}\tilde{\varrho}\tilde{\eta}}, \mathcal{I}^{\tilde{\rho}\tilde{\varrho}\tilde{\eta}})$ is r-$SVNIT_{2\frac{1}{2}}$.

(5_c). Let $x_{s,t,p}\overline{q}y_{s_1,t_1,p_1}$. Then, by r-$SVNIT_{2\frac{1}{2}}$ of $\tilde{X}$, there exist r-SVN£O sets $\sigma_n, \gamma_n \in \zeta^{\tilde{X}}$ such that $x_{s,t,p} \in \sigma_n$, $y_{s_1,t_1,p_1} \in \gamma_n$ and $\mathrm{CI}^{\pounds}_{\tau^{\tilde{\rho}\tilde{\varrho}\tilde{\eta}}}(\sigma_n, r)\overline{q}\mathrm{CI}^{\pounds}_{\tau^{\tilde{\rho}\tilde{\varrho}\tilde{\eta}}}(\gamma_n, r)$, which implies that $\sigma_n\overline{q}\gamma_n$. Thus, $(\tilde{X}, \tilde{\tau}^{\tilde{\rho}\tilde{\varrho}\tilde{\eta}}, \mathcal{I}^{\tilde{\rho}\tilde{\varrho}\tilde{\eta}})$ is r-$SVNIT_2$.

(5_d). Similar to the proof of (5_c). □

Theorem 16. *Let $(\tilde{X}, \tilde{\tau}^{\tilde{\rho}\tilde{\varrho}\tilde{\eta}}, \mathcal{I}^{\tilde{\rho}\tilde{\varrho}\tilde{\eta}})$ be an SVNITS and $r \in \zeta_0$. Then, the following statements are equivalent:*

(1) $(\tilde{X}, \tilde{\tau}^{\tilde{\rho}\tilde{\varrho}\tilde{\eta}}, \mathcal{I}^{\tilde{\rho}\tilde{\varrho}\tilde{\eta}})$ is r-$SVNIR_2$.

(2) *If $x_{s,t,p} \in \sigma_n$ and σ_n is r-SVN£O set, then there exists r-SVN£O set $\gamma_n \in \xi^{\tilde{X}}$ such that $x_{s,t,p} \in \gamma_n \leq \mathrm{CI}^{\pounds}_{\tau^{\tilde{\rho}\tilde{\varrho}\tilde{\eta}}}(\gamma_n, r) \leq \sigma_n$.*

(3) *If $x_{s,t,p}\overline{q}\sigma_n = \mathrm{CI}^{\pounds}_{\tau^{\tilde{\rho}\tilde{\varrho}\tilde{\eta}}}(\sigma_n, r)$, then there exists r-SVN£O set $[\gamma_n]_j \in \xi^{\tilde{X}}$, $j = \{1,2\}$ such that $x_{s,t,p} \in [\gamma_n]_1$, $\sigma_n \leq [\gamma_n]_2$ and $\mathrm{CI}^{\pounds}_{\tau^{\tilde{\rho}\tilde{\varrho}\tilde{\eta}}}([\gamma_n]_1, r)\overline{q}\mathrm{CI}^{\pounds}_{\tau^{\tilde{\rho}\tilde{\varrho}\tilde{\eta}}}([\gamma_n]_2, r)$.*

Proof. Similar to the proof of Theorem 14. □

Theorem 17. *Let $(\tilde{X}, \tilde{\tau}^{\tilde{\rho}\tilde{\varrho}\tilde{\eta}}, \mathcal{I}^{\tilde{\rho}\tilde{\varrho}\tilde{\eta}})$ be an SVNITS and $r \in \zeta_0$. Then, the following statements are equivalent:*

(1) *$(\tilde{X}, \tilde{\tau}^{\tilde{\rho}\tilde{\varrho}\tilde{\eta}}, \mathcal{I}^{\tilde{\rho}\tilde{\varrho}\tilde{\eta}})$ is r-$SVNIR_3$.*

(2) *If $[\sigma_n]_1\overline{q}[\sigma_n]_2$ and $[\sigma_n]_1, [\sigma_n]_2$ are r-SVN£C sets, then there exists r-SVN£O set $\gamma_n \in \xi^{\tilde{X}}$ such that $[\sigma_n]_1 \leq \gamma_n$ and $\mathrm{CI}^{\pounds}_{\tau^{\tilde{\rho}\tilde{\varrho}\tilde{\eta}}}(\gamma_n, r) \leq [\sigma_n]_2$.*

(3) *For any $[\sigma_n]_1 \leq [\sigma_n]_2$, where $[\sigma_n]_1$ is an r-SVN£O set, and $[\sigma_n]_2$ is an r-SVN£C set, then, there exists an r-SVN£O set $\gamma_n \in \xi^{\tilde{X}}$ such that $[\sigma_n]_1 \leq \gamma_n \leq \mathrm{CI}^{\pounds}_{\tau^{\tilde{\rho}\tilde{\varrho}\tilde{\eta}}}(\gamma_n, r) \leq [\sigma_n]_2$.*

Proof. Similar to the proof of Theorem 15. □

Theorem 18. *Let $f : (\tilde{X}, \tilde{\tau}_1^{\tilde{\rho}\tilde{\varrho}\tilde{\eta}}, \mathcal{I}_1^{\tilde{\rho}\tilde{\varrho}\tilde{\eta}}) \to (\tilde{Y}, \tilde{\tau}_2^{\tilde{\rho}\tilde{\varrho}\tilde{\eta}}, \mathcal{I}_2^{\tilde{\rho}\tilde{\varrho}\tilde{\eta}})$ be a £-SVNI-irresolute, bijective, £-SVNI-irresolute open mapping and $(\tilde{X}, \tilde{\tau}_1^{\tilde{\rho}\tilde{\varrho}\tilde{\eta}}, \mathcal{I}_1^{\tilde{\rho}\tilde{\varrho}\tilde{\eta}})$ is r-$SVNIR_2$. Then, $(\tilde{Y}, \tilde{\tau}_2^{\tilde{\rho}\tilde{\varrho}\tilde{\eta}}, \mathcal{I}_2^{\tilde{\rho}\tilde{\varrho}\tilde{\eta}})$ is r-$SVNIR_2$.*

Proof. Let $y_{s,t,p}\overline{q}\varsigma_n = Cl^{\star}(\varsigma_n, r)$. Then, by Definition 11, ς_n is an r-SVN£C set in $\tilde{Y}$. By Theorem 3(2), $f^{-1}(\varsigma_n)$ is an r-SVN£C set in $\tilde{X}$. Put $y_{s,t,p} = f(x_{s,t,p})$. Then, $x_{s,t,p}\overline{q}f^{-1}(\varsigma_n)$. By r-$SVNIR_2$ of $\tilde{X}$, there exist r-SVN£O sets $\sigma_n, \gamma_n \in \xi^{\tilde{X}}$ such that $x_{s,t,p} \in \sigma_n$, $f^{-1}(\varsigma_n) \leq \gamma_n$ and $\sigma_n\overline{q}\gamma_n$. Since f is bijective and £-SVNI-irresolute open, $y_{s,t,p} \in f(\sigma_n)$, $\varsigma_n \leq f(f^{-1}(\varsigma_n)) \leq f(\gamma_n)$ and $f(\sigma_n)\overline{q}f(\gamma_n)$. Thus, $(\tilde{Y}, \tilde{\tau}_2^{\tilde{\rho}\tilde{\varrho}\tilde{\eta}}, \mathcal{I}_2^{\tilde{\rho}\tilde{\varrho}\tilde{\eta}})$ is r-$SVNIR_2$. □

Theorem 19. *Let $f : (\tilde{X}, \tilde{\tau}_1^{\tilde{\rho}\tilde{\varrho}\tilde{\eta}}, \mathcal{I}_1^{\tilde{\rho}\tilde{\varrho}\tilde{\eta}}) \to (\tilde{Y}, \tilde{\tau}_2^{\tilde{\rho}\tilde{\varrho}\tilde{\eta}}, \mathcal{I}_2^{\tilde{\rho}\tilde{\varrho}\tilde{\eta}})$ be an £-SVNI-irresolute, bijective, £-SVNI-irresolute open mapping and $(\tilde{X}, \tilde{\tau}_1^{\tilde{\rho}\tilde{\varrho}\tilde{\eta}}, \mathcal{I}_1^{\tilde{\rho}\tilde{\varrho}\tilde{\eta}})$ be an r-$SVNIR_3$. Then, $(\tilde{Y}, \tilde{\tau}_2^{\tilde{\rho}\tilde{\varrho}\tilde{\eta}}, \mathcal{I}_2^{\tilde{\rho}\tilde{\varrho}\tilde{\eta}})$ is r-$SVNIR_3$.*

Proof. Similar to the proof of Theorem 18. □

6. Conclusions

In summary, we have introduced the definition of the r-single valued neutrosophic £-closed and r-single valued neutrosophic £-open sets over single valued neutrosophic ideal topology space in Šostak's sense. Many consequences have been arisen up to show that how far topological structures are preserved by these r-single valued neutrosophic £-closed. We also have provided some counterexamples where such properties fail to be preserved. The most important contribution to this area of research is that we have introduced the notion of £-single valued neutrosophic irresolute mapping, £-single valued neutrosophic extremally disconnected spaces, £-single valued neutrosophic normal spaces and that we defined some kinds of separation axioms, namely r-$SVNIR_i$, where $i = \{0,1,2,3\}$, and r-$SVNIT_j$, where $j = \{1,2,2\frac{1}{2},3,4\}$, in the sense of Šostak. Some of their characterizations, fundamental properties, and the relations between these notions have been studied.

Funding: This research received no external funding.

Institutional Review Board Statement: Not applicable.

Informed Consent Statement: Not applicable.

Acknowledgments: The author would like to express his sincere thanks to Majmaah University for supporting this work. The author is also grateful to the reviewers for their valuable comments and suggestions which led to the improvement of this research.

Conflicts of Interest: The author declares that there is no conflict of interest regarding the publication of this manuscript.

Discussion for Further Works: The theory in this article can be extended in the following natural ways. One may study the properties of neutrosophic metric topological spaces using the concepts defined through this paper.

References

1. Smarandache, F. *Unifying A Field in Logics, Neutrosophy: Neutrosophic Probability*; Set and Logic; American Research Press: Rehoboth, NM, USA, 1999.
2. Hur, K.; Lim, P.K.; Lee, J.G.; Kim, J. The category of neutrosophic sets. *Neutrosophic Sets Syst.* **2016**, *14*, 12–20.
3. Hur, K.; Lim, P.K.; Lee, J.G.; Kim, J. The category of neutrosophic crisp sets. *Ann. Fuzzy Math. Inform.* **2017**, *14*, 43–54. [CrossRef]
4. Salama, A.A.; Alblowi, S.A. Neutrosophic set and neutrosophic topological spaces. *IOSR J. Math.* **2012**, *3*, 31–35. [CrossRef]
5. Salama, A.A.; Smarandache, F. *Neutrosophic Crisp Set Theory*; Educational Publisher: Columbus, OH, USA, 2015.
6. Salama, A.A.; Smarandache, F.; Kroumov, V. Neutrosophic crisp sets and neutrosophic crisp topological spaces. *Neutrosophic Sets Syst.* **2014**, *2*, 25–30.
7. Chang, C.L. Fuzzy topological spaces. *J. Math. Anal. Appl.* **1968**, *24* , 182–190. [CrossRef]
8. Šostak, A. On a fuzzy topological structure. In *Circolo Matematico di Palermo, Palermo; Rendiconti del Circolo Matematico di Palermo, Proceedings of the 13th Winter School on Abstract Analysis, Section of Topology, Srni, Czech Republic, 5–12 January 1985*; Circolo Matematico di Palermo: Palermo, Italy, 1985; pp. 89–103.
9. Liang, C.; Yan, C. Base and subbase in intuitionistic *I*-fuzzy topological spaces. *Hacet. J. Math. Stat.* **2014**, *43*, 231–247.
10. Yan, C.H.; Wang, X.K. Intuitionistic *I*-fuzzy topological spaces. *Czechoslov. Math. J.* **2010**, *60*, 233–252. [CrossRef]
11. Wang, H.; Smarandache, F.; Zhang, Y.Q.; Sunderraman, R. Single valued neutrosophic sets. *Multispace Multistruct.* **2010**, *4*, 410–413.
12. El-Gayyar, M. Smooth neutrosophic topological spaces. *Neutrosophic Sets Syst.* **2016**, *65*, 65–72.
13. Kim, J.; Smarandache, F.; Lee J.G.; Hur, K. Ordinary Single Valued Neutrosophic Topological Spaces. *Symmetry* **2019**, *11*, 1075. [CrossRef]
14. Saber, Y.; Alsharari, F.; Smarandache, F. On single-valued neutrsophic ideals in Šostak's sense. *Symmetry* **2020**, *12*, 193. [CrossRef]
15. Saber, Y.; Alsharari, F.; Smarandache, F. Connectedness and Stratification of Single-Valued Neutrosophic Topological Spaces. *Symmetry* **2020**, *12*, 1464. [CrossRef]
16. Abdel-Basset, M.; Ali, M.; Atef, A. Uncertainty assessments of linear time-cost tradeoffs using neutrosophic set. *Comput. Ind. Eng.* **2020**, *141*, 106–286 [CrossRef]
17. Abdel-Basset, M.; Ali, M.; Atef, A. Resource levelling problem in construction projects under neutrosophic environment. *J. Supercomput.* **2020** *76*, 964–988. [CrossRef]
18. Abdel-Basset, M.; Gamal, A.; Son, L. H.; Smarandache, F. A Bipolar Neutrosophic Multi Criteria Decision Making Framework for Professional Selection. *Appl. Sci.* **2020** *10*, 12–20. [CrossRef]
19. Ubeda, T.; Egenhofer M. J. Topological error correcting in GIS. *Adv. Spat. Databases* **2005**, 281–297.
20. Shang, Y. Average consensus in multi-agent systems with uncertain topologies and multiple time-varying 443 delays. *Linear Algebra Its Appl.* **2014**, 459, 411–429. [CrossRef]
21. Smarandache, F. *A Unifying Field in Logics: Neutrosophic Logic. Neutrosophy, Neutrosophic Set, Neutrosophic Probability and Statistics*, 6th ed.; InfoLearnQuest: Ann Arbor, MI, USA, 2007.
22. Ye, J. A multicriteria decision-making method using aggregation operators for simplified neutrosophic sets. *J. Intell. Fuzzy Syst.* **2014**, *26*, 2450–2466. [CrossRef]
23. Yang, H.L.; Guo, Z.L.; Liao, X. On single valued neutrosophic relations. *J. Intell. Fuzzy Syst.* **2016**, *30*, 1045–1056. [CrossRef]
24. Alsharari, F. Decomposition of single-valued neutrosophic ideal continuity via fuzzy idealization. *Neutrosophic Sets Syst.* **2020**, *38*, 145–163.

symmetry

Article

A Study on Some Properties of Neutrosophic Multi Topological Group

Bhimraj Basumatary [1,*], Nijwm Wary [1], Dimacha Dwibrang Mwchahary [2], Ashoke Kumar Brahma [3], Jwngsar Moshahary [4], Usha Rani Basumatary [1] and Jili Basumatary [1]

1 Department of Mathematical Sciences, Bodoland University, Kokrajhar 783370, India; nijwmwr0@gmail.com (N.W.); usharanibsty@gmail.com (U.R.B.); jilibasumatary@gmail.com (J.B.)
2 Department of Mathematics, Kokrajhar Government College, Kokrajhar 783370, India; ddmwchahary@kgc.edu.in
3 Department of Mathematics, Science College, Kokrajhar 783370, India; ashoke_brahma@rediffmail.com
4 Department of Agricultural Statistics, SCS College of Agriculture, AAU, Dhubri 783376, India; jwngsar.moshahary@aau.ac.in
* Correspondence: brbasumatary14@gmail.com

Abstract: In this paper, we studied some properties of the neutrosophic multi topological group. For this, we introduced the definition of semi-open neutrosophic multiset, semi-closed neutrosophic multiset, neutrosophic multi regularly open set, neutrosophic multi regularly closed set, neutrosophic multi continuous mapping, and then studied the definition of a neutrosophic multi topological group and some of their properties. Moreover, since the concept of the almost topological group is very new, we introduced the definition of neutrosophic multi almost topological group. Finally, for the purpose of symmetry, we used the definition of neutrosophic multi almost continuous mapping to define neutrosophic multi almost topological group and study some of its properties.

Keywords: neutrosophic multi continuous mapping; neutrosophic multi topological group; neutrosophic multi almost continuous mapping; neutrosophic multi almost topological group

Citation: Basumatary, B.; Wary, N.; Mwchahary, D.D.; Brahma, A.K.; Moshahary, J.; Basumatary, U.R.; Basumatary, J. A Study on Some Properties of Neutrosophic Multi Topological Group. *Symmetry* **2021**, *13*, 1689. https://doi.org/10.3390/sym13091689

Academic Editors: Florentin Smarandache, Yanhui Guo and Alexei Kanel-Belov

Received: 3 July 2021
Accepted: 6 September 2021
Published: 13 September 2021

1. Introduction

Following the introduction of the fuzzy set (FS) [1], a variety of studies on generalisations of FS concepts were performed. In the sense that the theory of sets should have been a particular case of the theory of FSs, the theory of FSs is a generalisation of the classical theory of sets. Following the generalisation of FSs, many scholars used the theory of generalised FSs in a variety of fields in science and technology. Fuzzy topology (FT) was first introduced by Chang [2], and Intuitionistic fuzzy topological space (FITS) was defined by Coker [3]. Many researchers studied topology based on neutrosophic sets (NS), such as Lupianez [4–7] and Salama et al. [8]. Kelly [9] defined the concept of bitopological space (BTS) in 1963. Kandil et al. [10] studied the topic of fuzzy bitopological space (FBTS). Some characteristics of Intuitionistic Fuzzy Bitopological Space (IFBTS) were addressed by Lee et al. [11]. Garg [12] investigated how to rank interval-valued Pythagorean FSs using a modified score function. A Pythagorean fuzzy method for order of preference by similarity to ideal solution (TOPSIS) method based on Pythagorean FSs was discussed, which took the experts' preferences in the form of interval-valued Pythagorean fuzzy decision matrices. Moreover, different explorations of the theory of Pythagorean FSs can be seen in [13–19]. Yager [20] proposed the q-rung orthopair FSs, in which the sum of the qth powers of the membership (MS) and non-MS degrees is restricted to one [21]. Peng and Liu [22] studied the systematic transformation for information measures for q-rung orthopair FSs. Pinar and Boran [23] applied a q-rung orthopair fuzzy multi-criteria group decision-making method for supplier selection based on a novel distance measure.

Cuong et al. [24] proposed a picture FS as an extension of FS and Intuitionistic fuzzy set (IFS) that contains the concept of an element's positive, negative, and neutral MS degree. Cuong [25] investigated several picture FS characteristics and proposed distance measurements between picture FS. Phong et al. [26] investigated some picture fuzzy relation compositions. Cuong et al. [27] examined the basic fuzzy logic operators: negations, conjunctions, and disjunctions, as well as their implications on picture FSs, and also developed main operations for fuzzy inference processes in picture fuzzy systems. For picture FSs, Cuong et al. [28] demonstrated properties of an involutive picture negator and some related De Morgan fuzzy triples. Viet et al. [29] presented a picture fuzzy inference system based on MS graph, and Singh [30] studied correlation coefficients of picture FS. Garg [31] studied some picture fuzzy aggregation operations and their applications to multi-criteria decision-making. Quek et al. [32] used T-spherical fuzzy weighted aggregation operators to investigate the MADM problem. Garg [33] suggested interactive aggregation operators for T-spherical FSs and used the proposed operators to solve the MADM problem. Zeng et al. [34] studied on multi-attribute decision-making process with immediate probabilistic interactive averaging aggregation operators of T-spherical FSs and its application in the selection of solar cells. Munir et al. [35] investigated T-spherical fuzzy Einstein hybrid aggregation operators and how they could be applied in multi-attribute decision-making issues. Mahmood et al. [36] proposed the idea of a spherical FS and consequently a T-spherical FS.

Many researchers also studied FT and then generalised it in the IFS and then to the neutrosophic topology. Warren [37] studied the boundary of an FS in FT. Warren [37] studied some properties of the boundary of an FS and found that some properties are not the same as the properties of the crisp boundary of a set. Later, many authors studied the properties of the boundary of an FS. Tang [38] made heavy use of the notion of fuzzy boundary. Kharal [39] studied Frontier and Semifrontier in IFTSs. Salama et al. [40] studied generalised neutrosophic topological space (NTS), where they have discussed on properties of generalised closed sets. Azad [41] introduced the concepts of fuzzy semi-continuity (FSC), fuzzy almost continuity (FAC), and fuzzy weakly continuity (FWC) (FWC). Smarandache [42,43] suggested neutrosophic set (NS) theory, which generalised FST and IFST and incorporated a degree of indeterminacy as an independent component. Mwchahary et al. [44] studied on properties of the boundary of neutrosophic bitopological space (NBTS). Many authors studied the properties of the boundary of an FS by several methods (FS, IFS, and NS), but some of its properties are not the same as the properties of the crisp boundary of a set.

Blizard [45] traced multisets back to the very origin of numbers, arguing that in ancient times, the number was often represented by a collection of n strokes, tally marks, or units. The idea of fuzzy multiset (FMS) was introduced by Yager [46] as fuzzy bags. In the interest of brevity, we consider our attention to the basic concepts such as an open FMS, closed FMS, interior, closure, and continuity of FMSs. Yager, in [46], generalised the FS by introducing the concept of FMS (fuzzy bag), and he discussed a calculus for them in [47]. An element of an FMS can occur more than once with possibly the same or different MS values. If every element of an FMS can occur at most once, we go back to FSs [48]. In [49], Onasanya et al. defined the multi-fuzzy group (FMG), and in [50,51], the authors defined fuzzy multi-polygroups and fuzzy multi-Hv-ideals and studied their properties. In [52], Neutrosophic Multigroup (NMG) and their applications are observed. A new type of FS (FMS) was studied by Sebastian et al. [53]. This set makes use of ordered sequences of MS functions to express problems that are not covered by other extensions of FS theory, such as pixel colour. Dey et al. [54] were the first to establish the concept of multi-fuzzy complex numbers and multi-fuzzy complex sets. Over a distributive lattice, the authors [54] proposed multi fuzzy complex nilpotent matrices. Yong et al. [55] recently proposed the notion of the multi-fuzzy soft set, which is a more general fuzzy soft set, for its application to decision making.

Motivation

There is a lot of ambiguity information in the real world that crisp values cannot manage. The FS theory [1], proposed by Zadeh, is an age-old and excellent tool for dealing with uncertain information; however, it can only be used on random processes. As a result, Sebastian et al. [56] introduced FMSs, Atanassov [57] suggested the IFS theory, and Shinoj et al. [58] launched intuitionistic FMSs, all based on FS theory. The theories mentioned above have expanded in a variety of ways and have applications in a variety of fields, including algebraic structures. Some of the selected papers are those on FSs [59–61], FMSs [62–64], IFSs [65–72], and intuitionistic FMSs [73]. However, these theories are incapable of dealing with all forms of uncertainty, such as indeterminate and inconsistent data in various decision-making situations. To address this shortfall, Smarandache [74] proposed the NS theory, which makes Atanassov's [57] theory very practical and easy to apply. In this current decade, neutrosophic environments are mainly interested by different fields of researchers. In Mathematics, much theoretical research has also been observed in the sense of neutrosophic environment. A more theoretical study will be required to build a broad framework for decision-making and to define patterns for the conception and implementation of complex networks. Deli et al. [75] and Ye [76,77] proposed the notion of neutrosophic multiset (NMS) for modelling vagueness and uncertainty in order to improve the NS theory further. From the literature survey, it was noticed that precisely the properties of the neutrosophic multi topological group (NMTG) are not performed. Now, as an update for the research in NMS, we introduced the definition of a neutrosophic semi-open set, neutrosophic semi-closed set, neutrosophic regularly open set, neutrosophic regularly closed set, neutrosophic continuous mapping, neutrosophic open mapping, neutrosophic closed mapping, neutrosophic semi-continuous mapping, neutrosophic semi-open mapping, neutrosophic semi-closed mapping. Moreover, we tried to prove some of their properties and also cited some examples. We defined the neutrosophic multi almost topological group by using the definition of neutrosophic multi almost continuous mapping and investigate some properties and theorems of a neutrosophic multi almost topological group.

2. Materials and Methods

Definition 1 ([42]). *Let X be a non-empty fixed set. A neutrosophic set (NS)* A *is an object with the form* $\mathrm{A} = \{< x,\ \mu_A\ ,\sigma_A,\ \gamma_A >\ : x \in \mathrm{X}\}$, *where* $T, I, F\ : \mathrm{X} \longrightarrow [0,\ 1]$ *and* $0 \leq \mu_A + \sigma_A + \gamma_A \leq 3$ *and* $\mu_A\ (x)$, $\sigma_A(x)$, *and* $\gamma_A(x)$ *represents the degree of MS function, the degree indeterminacy, and the degree of non-MS function, respectively, of each element* $x \in X$ *to set* A.

Definition 2 ([78]). *A neutrosophic multiset (NMS) is a type of neutrosophic set (NS) in which one or more elements are repeated with the same or different neutrosophic components.*

Example 1. *Let* $X = \{a, b, c\}$ *then*

$$\mathcal{A} = \left\{ \begin{array}{c} < a, 0.6, 0.1, 0.2 >, < a, 0.5, 0.1, 0.3 >, < a, 0.4, 0.2, 0.4 >, \\ < b, 0.3, 0.5, 0.4 >, < b, 0.2, 0.5, 0.6 >, < b, 0.1, 0.5, 0.7 >, \\ < c, 0.4, 0.5, 0.6 >, < c, 0.3, 0.5, 0.7 >, < c, 0.2, 0.6, 0.8 > \end{array} \right\}$$

is an NMS, as the elementsa , b, care repeated.

However,$B = \{< a, 0.8, 0.3, 0.1 >,\ < b, 0.5, 0.3, 0.4 >, < c, 0.4, 0.4, 0.6 >\}$*is an NS and not an NMS.*

Definition 3 ([52]). *The Empty NMS is defined as* $0_{\mathrm{NM}} = \left\{ \mathrm{m} \in \mathrm{X}; < m_{(0,1,1)} > \right\}$, *where* m *can be repeated.*

Definition 4 ([52]). *The Whole NMS is defined as* $1_{NM} = \left\{ \mathrm{m} \in \mathrm{X}; < m_{(1,0,0)} > \right\}$, *where* m *can be repeated.*

Definition 5 ([52]). *Let* $X \neq \phi$, *and a neutrosophic multiset (NMS) A on X can be expressed as* $A = \left\{ m \in X; \left(m_{<\mathfrak{T}_{A(m)}, \mathfrak{I}_{A(m)}, \mathfrak{F}_{A(m)}>} \right) \right\}$, *then the complement of A is defined as* $A^C = \left\{ m \in X; \left(m_{<\mathfrak{F}_{A(m)}, 1-\mathfrak{I}_{A(m)}, \mathfrak{T}_{A(m)}>} \right) \right\}$. *where m can be repeated depending on its multiplicity, and the* $\mathfrak{T}, \mathfrak{I}, \mathfrak{F}$ *values may or may not be equal.*

Definition 6 ([52]). *Let* $X \neq \phi$ *and* $A = \left\{ m \in X; \left(m_{<\mathfrak{T}_{A(m)}, \mathfrak{I}_{A(m)}, \mathfrak{F}_{A(m)}>} \right) \right\}$ *and* $B = \left\{ m \in X; \left(m_{<\mathfrak{T}_{B(m)}, \mathfrak{I}_{B(m)}, \mathfrak{F}_{B(m)}>} \right) \right\}$ *are NMSs. Then*

(i) $A \Cap B = \left\{ m \in X; m_{<\min(\mathfrak{T}_A(m), \mathfrak{T}_B(m)), \max(\mathfrak{I}_A(m), \mathfrak{I}_B(m)), \max(\mathfrak{F}_A(m), \mathfrak{F}_B(m))>} \right\}$;

(ii) $A \Cup B = \left\{ m \in X; m_{<\max(\mathfrak{T}_A(m), \mathfrak{T}_B(m)), \min(\mathfrak{I}_A(m), \mathfrak{I}_B(m)), \min(\mathfrak{F}_A(m), \mathfrak{F}_B(m))>} \right\}$.

Definition 7 ([78]). *Let* $X \neq \phi$, *and a neutrosophic multiset topology (NMT) on X is a family* τ_X *of neutrosophic multi subsets of X if the following conditions hold:*

(i) $0_{NM}, 1_{NM} \in \tau_X$;

(ii) $G_1 \Cap G_2 \in \tau_X$ for $G_1, G_2 \in \tau_X$;

(iii) $\Cup G_i \in \tau_X, \forall \{G_{N_i} : i \in J\} \preccurlyeq \tau_X$.

Then (X, τ_X) *is known as a neutrosophic multi topological space (NMTS), and any NMS in* τ_X *is called a neutrosophic multi-open set (NMOS). The element of* τ_X *are said to be NMOSs, an NMS F is neutrosophic multi closed set (NMCoS) if* F^c *is NMOS.*

Definition 8 ([52]). *Let X be a classical group and A be a neutrosophic multiset (NMS) on X. Then A is said to be neutrosophic multi groupoid over X if*

(i) $T_i^G(mn) \geq T_i^G(m) \longrightarrow T_i^G(n)$;

(ii) $I_i^G(mn) \leq I_i^G(m) \longrightarrow I_i^G(n)$;

(iii) $F_i^G(mn) \leq F_i^G(m) \longrightarrow F_i^G(n), \ \forall\, m, n \in X$ *and* $i = 1, 2, \ldots, P$.

Moreover, A is said to be neutrosophic multi-group (NMG) over X if the neutrosophic multi groupoid satisfies the following:

(i) $T_i^G(m^{-1}) \geq T_i^G(m)$;

(ii) $I_i^G(m^{-1}) \leq I_i^G(m)$;

(iii) $F_i^G(m^{-1}) \leq F_i^G(m), \forall\, m \in X$ *and* $i = 1, 2, \ldots, P$.

Definition 9 ([52]). *Let* $\mathcal{G}$ *be an NMG in a group X, and e be the identity of X. We define the NMS* $\mathbb{G}_e$ *by*

$$\mathbb{G}_e = \{m \in X : \mathfrak{T}_{\mathbb{G}}(m) = \mathfrak{T}_{\mathbb{G}}(e),\ \mathfrak{I}_{\mathbb{G}}(m) = \mathfrak{I}_{\mathbb{G}}(e),\ \mathfrak{F}_{\mathbb{G}}(m) = \mathfrak{F}_{\mathbb{G}}(e)\}$$

We note for an NMG $\mathbb{G}$ *in a group X, for every* $m \in X: \ \mathfrak{T}_{\mathbb{G}}(m^{-1}) = \mathfrak{T}_{\mathbb{G}}(m),\ \mathfrak{I}_{\mathbb{G}}(m^{-1}) = \mathfrak{I}_{\mathbb{G}}(m)$ *and* $\mathfrak{F}_{\mathbb{G}}(m^{-1}) = \mathfrak{F}_{\mathbb{G}}(m)$. *Moreover, for the identity* $e \in X: \ \mathfrak{T}_{\mathbb{G}}(e) \succcurlyeq \mathfrak{T}_{\mathbb{G}}(m),\ \mathfrak{I}_{\mathbb{G}}(e) \succcurlyeq \mathfrak{I}_{\mathbb{G}}(m)$ *and* $\mathfrak{F}_{\mathbb{G}}(e) \preccurlyeq \mathfrak{F}_{\mathbb{G}}(m)$.

3. Results

Definition 10. *Let* (X, τ_X) *be NMTS. Then for an NMS* $A = \{< x, \mu_{N_i}, \sigma_{N_i}, \delta_{N_i} > : x \in X\}$, *the neutrosophic interior of A can be defined as* $NM \backsim Int\ (A) = \{< x, \Cup\mu_{N_i}, \Cap\sigma_{N_i}, \Cap\delta_{N_i} > : x \in X\}$.

Definition 11. *Let* (X, τ_X) *be NMTS. Then for an NMS* $A = \{< x, \mu_{N_i}, \sigma_{N_i}, \delta_{N_i} > : x \in X\}$, *the neutrosophic closure of A can be defined as* $NM \backsim Cl\ (A) = \{< x, \Cap\mu_{N_i}, \Cup\sigma_{N_i}, \Cup\delta_{N_i} > : x \in X\}$.

Definition 12. *Let $\mathbb{G}$ be an NMG on a group X. Let τ_X be a NMT on $\mathbb{G}$, then $(\mathbb{G}, \tau_X)$ is known as a neutrosophic multi topological group (NMTG) if it satisfies the given conditions:*

(i) $\alpha : (\mathbb{G}, \tau_X) \times (\mathbb{G}, \tau_X) \longrightarrow (\mathbb{G}, \tau_X)$ *defined by* $\alpha(m, n) = mn, \forall\, m, n \in X$, *is relatively neutrosophic multi continuous;*

(ii) $\beta : (\mathbb{G}, \tau_X) \longrightarrow (\mathbb{G}, \tau_X)$ *defined by* $\beta(m) = m^{-1}, \forall\, m \in X$, *is relatively neutrosophic multi continuous.*

Definition 13. *Let $\mathcal{A}$ be an NMS of an NMTS (X, τ_X), then $\mathcal{A}$ is called a neutrosophic multi semi-open set (NMSOS) of X if $\exists$ a $\mathcal{B} \in \tau_X$, such that $\mathcal{A} \preccurlyeq MN \backsim Int(MN \sim Cl(\mathcal{B}))$.*

Example 2. *Let* $X = \{a, b\}$:

$$\mathcal{A} = \left\{ \begin{array}{l} < a, 0.8, 0.1, 0.2 >, < a, 0.7, 0.1, 0.3 >, < a, 0.6, 0.2, 0.4 >, \\ < b, 0.7, 0.2, 0.3 >, < b, 0.6, 0.3, 0.4 >, < b, 0.4, 0.2, 0.5 > \end{array} \right\};$$

$$\mathcal{B} = \left\{ \begin{array}{l} < a, 0.9, 0.1, 0.1 >, < a, 0.8, 0.1, 0.2 >, < a, 0.7, 0.2, 0.3 >, \\ < b, 0.8, 0.2, 0.2 >, < b, 0.7, 0.2, 0.3 >, < b, 0.5, 0.2, 0.4 > \end{array} \right\}.$$

Then $\tau = \{0_X, 1_X, \mathcal{B}\}$ is neutrosophic multi topological space.
Then $Cl(\mathcal{B}) = 1_X,\ Int(Cl(\mathcal{B})) = 1_X$.
Hence, $\mathcal{B}$ is NMSOS.

Definition 14. *Let $\mathcal{A}$ be an NMS of an NMTS (X, τ_X), then $\mathcal{A}$ is called a neutrosophic multi semi-closed set (NMSCoS) of X if $\exists$ a $\mathcal{B}^c \in \tau_X$, such that $MN \backsim Cl(MN \sim Int(\mathcal{B})) \preccurlyeq \mathcal{A}$.*

Lemma 1. *Let $\phi : X \longrightarrow Y$ be a mapping and $\{\mathcal{A}_\alpha\}$ be a family of NMSs of Y, then (1) $\phi^{-1}(\Cup \mathcal{A}_\alpha) = \Cup\, \phi^{-1}(\mathcal{A}_\alpha)$ and (ii) $\phi^{-1}(\Cap \mathcal{A}_\alpha) = \Cap \phi^{-1}(\mathcal{A}_\alpha)$.*

Proof. Proof is straightforward. □

Lemma 2. *Let $\mathcal{A}$, $\mathcal{B}$ be NMSs of X and Y, then $1_X - \mathcal{A} \times \mathcal{B} = (\mathcal{A}^c \times 1_X) \Cup (1_X \times \mathcal{B}^c)$.*

Proof. Let (p, q) be any element of $X \times Y$, $(1_X - \mathcal{A} \times \mathcal{B})(p, q) = \max(1_X - \mathcal{A}(p), 1_X - \mathcal{B}(q)) = \max\{(\mathcal{A}^c \times 1_X)(p, q), (\mathcal{B}^c \times 1_X)(p, q)\} = \{(\mathcal{A}^c \times 1_X) \Cup (1_X \times \mathcal{B}^c)\}(p, q)$, for each $(p, q) \in X \times Y$. □

Lemma 3. *Let $\phi_i : X_i \longrightarrow Y_i$ and $\mathcal{A}_i$ be NMSs of Y_i, $i = 1, 2$; we have $(\phi_1 \times \phi_2)^{-1}(\mathcal{A}_1 \times \mathcal{A}_2) = {\phi_1}^{-1}(\mathcal{A}_1) \times {\phi_2}^{-1}(\mathcal{A}_2)$.*

Proof. For each $(p_1, p_2) \in X_1 \times X_2$, we have

$$\begin{aligned} (\phi_1 \times \phi_2)^{-1}(\mathcal{A}_1 \times \mathcal{A}_2)(p_1, p_2) &= (\mathcal{A}_1 \times \mathcal{A}_2)((\phi_1(p_1), \phi_2(p_2)) \\ &= \min\{\mathcal{A}_1\phi_1(p_1), \mathcal{A}_2\phi_2(p_2)\} \\ &= \min\{{\phi_1}^{-1}(\mathcal{A}_1)(p_1),\ {\phi_2}^{-1}(\mathcal{A}_2)(p_2)\} \\ &= ({\phi_1}^{-1}(\mathcal{A}_1) \times {\phi_2}^{-1}(\mathcal{A}_2))(p_1, p_2). \end{aligned}$$

□

Lemma 4. *Let $\psi : X \longrightarrow X \times Y$ be the graph of a mapping $\phi : X \longrightarrow Y$. Then, if $\mathcal{A}, \mathcal{B}$ is NMSs of X and Y, $\psi^{-1}(\mathcal{A} \times \mathcal{B}) = \mathcal{A} \Cap \phi^{-1}(\mathcal{B})$.*

Proof. For each $p \in X$, we have

$$\begin{aligned} \psi^{-1}(\mathcal{A} \times \mathcal{B})(p) = (\mathcal{A} \times \mathcal{B})\psi(p) &= (\mathcal{A} \times \mathcal{B})(p, \phi(p)) \\ &= \min\{\mathcal{A}(p), \mathcal{B}(\phi(p))\} \\ &= (\mathcal{A} \Cap \phi^{-1}(\mathcal{B}))(p). \end{aligned}$$

□

Lemma 5. *For a family* $\{\mathcal{A}\}_\alpha$ *of NMSs of NMTS* (X, τ_X), $\Cup NM \backsim Cl(\mathcal{A}_\alpha) \preccurlyeq NM \backsim Cl(\Cup(\mathcal{A}_\alpha))$. *In the case that* $\mathcal{B}$ *is a finite set,* $\Cup NM \backsim Cl(\mathcal{A}_\alpha) \preccurlyeq NM \backsim Cl(\Cup(\mathcal{A}_\alpha))$. *Moreover,* $\Cup NM \backsim Int(\mathcal{A}_\alpha) \preccurlyeq NM \backsim Int(\Cup(\mathcal{A}_\alpha))$, *where a subfamily* $\mathcal{B}$ *of* (X, τ_X) *is said to be subbase for* (X, τ_X) *if the collection of all intersections of members of* $\mathcal{B}$ *forms a base for* (X, τ_X).

Lemma 6. *For an NMS* $\mathcal{A}$ *of an NMTS* (X, τ_X), *(a)* $1_{NM} - NM \backsim Int(\mathcal{A}) = NM \backsim Cl(1_{NM} - \mathcal{A})$, *and (b)* $1_{NM} - NM \backsim Cl(\mathcal{A}) = NM \backsim Int(1_{NM} - \mathcal{A})$.

Proof. Proof is straightforward. □

Theorem 1. *The statements below are equivalent:*

(i) $\mathcal{A}$ *is an NMCoS;*
(ii) $\mathcal{A}^c$ *is an NMOS;*
(iii) $NM \backsim Int(NM \backsim Cl(\mathcal{A})) \preccurlyeq \mathcal{A}$;
(iv) $NM \backsim Cl(NM \backsim Int(\mathcal{A}^c)) \succcurlyeq \mathcal{A}^c$.

Proof. (i) and (ii) are equivalent follows from Lemma 6, since for an NMS $\mathcal{A}$ of an NMTS (X, τ_X) such that $1_{NM} - NM \backsim Int(\mathcal{A}) = NM \backsim Cl(1_{NM} - \mathcal{A})$ and $1_{NM} - NM \backsim Cl(\mathcal{A}) = NM \backsim Int(1_{NM} - \mathcal{A})$.

(i)⇒(iii). By definition, ∃ an NMCoS $\mathcal{B}$ such that $NM \backsim Int(\mathcal{B}) \preccurlyeq \mathcal{A} \preccurlyeq \mathcal{B}$; hence, $NM \backsim Int(\mathcal{B}) \preccurlyeq \mathcal{A} \preccurlyeq NM \backsim Cl(\mathcal{A}) \preccurlyeq \mathcal{B}$. Since $NM \backsim Int(\mathcal{B})$ is the largest NMOS contained in $\mathcal{B}$, we have $NM \backsim Int(NM \backsim Cl(\mathcal{B})) \preccurlyeq NM \backsim Int(\mathcal{B}) \preccurlyeq \mathcal{A}$;

(iii)⇒(i) follows by taking $\mathcal{B} = NM \backsim Cl(\mathcal{A})$;

(ii)⇔(iv) can similarly be proved. □

Theorem 2. *(i) Arbitrary union of NMSOSs is an NMSOS;*
(ii) Arbitrary intersection of NMSCoSs is an NMSCoS.

Proof. (i) Let $\{\mathcal{A}_\alpha\}$ be a collection of NMSOSs of an NMTS (X, τ_X). Then ∃ a $\mathcal{B}_\alpha \in \tau_X$ such that $\mathcal{B}_\alpha \preccurlyeq \mathcal{A}_\alpha \preccurlyeq NM \backsim Cl(\mathcal{B}_\alpha)$ for each α. Thus, $\Cap \mathcal{B}_\alpha \preccurlyeq \Cup\mathcal{A}_\alpha \preccurlyeq \Cup NM \backsim Cl(\mathcal{B}_\alpha) \preccurlyeq NM \backsim Cl(\Cup(\mathcal{B}_\alpha))$ (Lemma 5), and $\Cup\mathcal{B}_\alpha \in \tau_X$, this shows that $\Cup\mathcal{B}_\alpha$ is an NMSOS;

(ii) Let $\{\mathcal{A}_\alpha\}$ be a collection of NMSCoSs of an NMTS (X, τ_X). Then ∃ a $\mathcal{B}_\alpha \in \tau_X$ such that $NM \backsim Int(\mathcal{B}_\alpha) \preccurlyeq \mathcal{A}_\alpha \preccurlyeq \mathcal{B}_\alpha$ for each α. Thus, $NM \backsim Int(\Cap(\mathcal{B}_\alpha)) \preccurlyeq \Cap NM \backsim Int(\mathcal{B}_\alpha) \preccurlyeq \Cap\mathcal{A}_\alpha \preccurlyeq \Cap\mathcal{B}_\alpha$ (Lemma 5), and $\Cup\mathcal{B}_\alpha \in \tau_X$, this shows that $\Cap\mathcal{B}_\alpha$ is an NMSCoS. □

Remark 1. *It is clear that every NMOS (NMCoS) is an NMSOS (NMSCoS). The converse is not true.*

Example 3. *From Example 2, it is clear that* $\mathcal{B}$ *is a neutrosophic multi semi-open set, but* $\mathcal{B}$ *is not NMOS.*

Theorem 3. *If* (X, τ_X) *and* (Y, τ_Y) *are NMTSs, and X is a product related to Y. Then the product* $\mathcal{A} \times \mathcal{B}$ *of an NMSOS* $\mathcal{A}$ *of X and an NMSOS* $\mathcal{B}$ *of Y is an NMSOS of the neutrosophic multi-product space* $X \times Y$.

Proof. Let $\mathcal{P} \preccurlyeq \mathcal{A} \preccurlyeq NM \backsim Cl(\mathcal{P})$ and $\mathcal{Q} \preccurlyeq \mathcal{B} \preccurlyeq NM \backsim Cl(\mathcal{Q})$, where $\mathcal{P} \in \tau_X$ and $\mathcal{Q} \in \tau_Y$. Then $\mathcal{P} \times \mathcal{Q} \preccurlyeq \mathcal{A} \times \mathcal{B} \preccurlyeq NM \backsim Cl(\mathcal{P}) \times NM \backsim Cl(\mathcal{Q})$. For NMSs $\mathcal{P}$'s of X and $\mathcal{Q}$'s of Y, we have:

(a) $\inf\{\mathcal{P}, \mathcal{Q}\} = \min\{\inf \mathcal{P}, \inf \mathcal{Q}\}$;
(b) $\inf\{\mathcal{P} \times 1_{NM}\} = (\inf \mathcal{P}) \times 1_{NM}$;
(c) $\inf\{1_{NM} \times \mathcal{Q}\} = 1_{NM} \times (\inf \mathcal{Q})$.

It is sufficient to prove $Nm \backsim Cl(\mathcal{A} \times \mathcal{B}) \succcurlyeq NM \backsim Cl(\mathcal{A}) \times NM \backsim Cl(\mathcal{B})$. Let $\mathcal{P} \in \tau_X$ and $\mathcal{Q} \in \tau_Y$. Then

$$\begin{aligned} NM \backsim Cl(\mathcal{A} \times \mathcal{B}) &= \inf\{(\mathcal{P} \times \mathcal{Q})^c | (\mathcal{P} \times \mathcal{Q})^c \succcurlyeq \mathcal{A} \times \mathcal{B} \\ &= \inf\{(\mathcal{P}^c \times 1_{NM}) \Cup (1_{NM} \times \mathcal{Q}^c) | (\mathcal{P}^c \times 1_{NM}) \Cup (1_{NM} \times \mathcal{Q}^c) \\ &\succcurlyeq \mathcal{A} \times \mathcal{B}\} \\ &= \inf\{(\mathcal{P}^c \times 1_{NM}) \Cup (1_{NM} \times \mathcal{Q}^c) | \ \mathcal{P}^c \succcurlyeq \mathcal{A} \ or \ \mathcal{Q}^c \succcurlyeq \mathcal{B}\} \\ &= \min\begin{bmatrix} inf\ \{(\mathcal{P}^c \times 1_{NM}) \Cup (1_{NM} \times \mathcal{Q}^c) | \mathcal{P}^c \succcurlyeq \mathcal{A}\}, \\ inf\ \{(\mathcal{P}^c \times 1_{NM}) \Cup (1_{NM} \times \mathcal{Q}^c) | \ \mathcal{Q}^c \succcurlyeq \mathcal{B}\} \end{bmatrix} \end{aligned}$$

Since, $\inf\{(\mathcal{P}^c \times 1_{NM}) \Cup (1_{NM} \times \mathcal{Q}^c) \ |\mathcal{P}^c \succcurlyeq \mathcal{A}\} \succcurlyeq \inf\{(\mathcal{P}^c \times 1_{NM}) \ |\mathcal{P}^c \succcurlyeq \mathcal{A}\}$

$$= \inf\{\mathcal{P}^c | \ \mathcal{P}^c \succcurlyeq \mathcal{A}\} \times 1_{NM} = NM \backsim Cl(\mathcal{A}) \times 1_{NM}$$

and $\inf\{(\mathcal{P}^c \times 1_{NM}) \Cup (1_{NM} \times \mathcal{Q}^c) \mid \mathcal{Q}^c \succcurlyeq \mathcal{B}\} \succcurlyeq \inf\{(1_{NM} \times \mathcal{Q}^c) \mid \mathcal{Q}^c \succcurlyeq \mathcal{B}\}$

$$= 1_{NM} \times \inf\{\mathcal{Q}^c | \mathcal{Q}^c \succcurlyeq \mathcal{B}\} = 1_{NM} \times NM \backsim Cl(\mathcal{B})$$

we have, $NM \backsim Cl(\mathcal{A} \times \mathcal{B}) \succcurlyeq \min\{NM \backsim Cl(\mathcal{A}) \times 1_{NM}, 1_{NM} \times NM \backsim Cl(\mathcal{B})\} = NM \backsim Cl(\mathcal{A}) \times NM \backsim Cl(\mathcal{B})$, hence the result. □

Definition 15. *An NMS $\mathcal{A}$ of an NMTS (X, τ_X) is called a neutrosophic multi regularly open set (NMROS) of (X, τ_X) if $NM \backsim Int(NM \backsim Cl(\mathcal{A})) = \mathcal{A}$.*

Example 4. *Let $X = \{a, b\}$* and

$$\mathcal{A} = \left\{ \begin{array}{l} <a, 0.4, 0.5, 0.5>, <a, 0.3, 0.5, 0.6>, <a, 0.2, 0.6, 0.7>, \\ <b, 0.5, 0.7, 0.6>, <b, 0.4, 0.5, 0.7>, <b, 0.3, 0.5, 0.8> \end{array} . \right\}$$

Then $\tau = \{0_X, 1_X, \mathcal{A}\}$ is neutrosophic multi topological space.
Clearly, $Cl(\mathcal{A}) = \mathcal{A}^C$, $Int(Cl(\mathcal{A})) = \mathcal{A}$.
Hence, $\mathcal{A}$ is NMROS.

Definition 16. *An NMS $\mathcal{A}$ of an NMTS (X, τ_X) is called a neutrosophic multi regularly closed set (NMRCoS) of (X, τ_X) if $NM \backsim Cl(NM \backsim Int(\mathcal{A})) = \mathcal{A}$.*

Theorem 4. *An NMS $\mathcal{A}$ of NMTS (X, τ_X) is an NMRO if $\mathcal{A}^c$ is NMRCo.*

Proof. It follows from Lemma 3. □

Remark 2. *It is obvious that every NMROS (NMRCoS) is an NMOS (NMCoS). The converse need not be true.*

Example 5. *Let $X = \{a, b\}$ and*

$$\mathcal{A} = \left\{ \begin{array}{l} <a, 0.8, 0.1, 0.2>, <a, 0.7, 0.1, 0.3>, <a, 0.6, 0.2, 0.4>, \\ <b, 0.7, 0.2, 0.3>, <b, 0.6, 0.3, 0.4>, <b, 0.4, 0.2, 0.5> \end{array} \right\};$$

$$\mathcal{B} = \left\{ \begin{array}{l} <a, 0.9, 0.1, 0.1>, <a, 0.8, 0.1, 0.2>, <a, 0.7, 0.2, 0.3>, \\ <b, 0.8, 0.2, 0.2>, <b, 0.7, 0.2, 0.3>, <b, 0.5, 0.2, 0.4> \end{array} \right\}.$$

Then $\tau = \{0_X, 1_X, \mathcal{B}\}$ is a neutrosophic multi topological space.
Then $Cl(\mathcal{B}) = 1_X$, $Int(Cl(\mathcal{B})) = 1_X$, which is not NMROS.

Remark 3. *The union (intersection) of any two NMROSs (NMRCoS) need not be an NMROS (NMRCoS).*

Example 6. *Let* $X = \{a, b\}$ *and*
$\tau = \{0_X, 1_X, \mathcal{A}, \mathcal{B}, \mathcal{A} \longrightarrow \mathcal{B}\}$ is a neutrosophic multi topological space, where

$$\mathcal{A} = \left\{ \begin{array}{c} < a, 0.4, 0.5, 0.6 >, < a, 0.3, 0.5, 0.7 >, < a, 0.2, 0.6, 0.8 >, \\ < b, 0.7, 0.5, 0.3 >, < b, 0.6, 0.5, 0.4 >, < b, 0.4, 0.5, 0.6 > \end{array} \right\};$$

$$\mathcal{B} = \left\{ \begin{array}{c} < a, 0.6, 0.5, 0.4 >, < a, 0.7, 0.5, 0.3 >, < a, 0.8, 0.4, 0.2 >, \\ < b, 0.3, 0.5, 0.7 >, < b, 0.4, 0.5, 0.6 >, < b, 0.6, 0.5, 0.4 > \end{array} \right\};$$

$$\mathcal{A} \bigcup \mathcal{B} = \left\{ \begin{array}{c} < a, 0.6, 0.5, 0.4 >, < a, 0.7, 0.5, 0.3 >, < a, 0.8, 0.4, 0.2 >, \\ < b, 0.7, 0.5, 0.3 >, < b, 0.6, 0.5, 0.4 >, < b, 0.4, 0.5, 0.6 > \end{array} \right\}.$$

Here, $Cl(\mathcal{A}) = \mathcal{B}^C$, $Int(Cl(\mathcal{A})) = \mathcal{A}$, and $Cl(\mathcal{B}) = \mathcal{A}^C$, $Int(Cl(\mathcal{B})) = \mathcal{B}$.
Then $Cl(\mathcal{A} \bigcup \mathcal{B}) = 1_X$.
Thus, $Int(Cl(\mathcal{A} \bigcup \mathcal{B})) = 1_X$.
Hence, $\mathcal{A}$ and $\mathcal{B}$ is NROS, but $\mathcal{A} \bigcup \mathcal{B}$ is not NROS.

Theorem 5. *(i) The intersection of any two NMROSs is an NMROS;*
(ii) The union of any two NMRCoSs is an NMRCoS.

Proof. (i) Let $\mathcal{A}_1$ and $\mathcal{A}_2$ be any two NMROSs of an NMTS (X, τ_X). Since $\mathcal{A}_1 \Cap \mathcal{A}_2$ is NMOS (from Remark 3), we have $\mathcal{A}_1 \Cap \mathcal{A}_2 \preccurlyeq NM \backsim Int(NM \backsim Cl(\mathcal{A}_1 \Cap \mathcal{A}_2))$. Now, $NM \backsim Int(NM \backsim Cl(\mathcal{A}_1 \Cap \mathcal{A}_2)) \preccurlyeq NM \backsim Int(NM \backsim Cl(\mathcal{A}_1)) = \mathcal{A}_1$ and $NM \backsim Int(NM \backsim Cl(\mathcal{A}_1 \Cap \mathcal{A}_2)) \preccurlyeq NM \backsim Int(NM \backsim Cl(\mathcal{A}_2)) = \mathcal{A}_2$ implies that $NM \backsim Int(NM \backsim Cl(\mathcal{A}_1 \Cap \mathcal{A}_2)) \preccurlyeq \mathcal{A}_1 \Cap \mathcal{A}_2$, hence the theorem;

(ii) Let $\mathcal{A}_1$ and $\mathcal{A}_2$ be any two NMROSs of an NMTS (X, τ_X). Since $\mathcal{A}_1 \Cup \mathcal{A}_2$ is NMOS (from Remark 3), we have $\mathcal{A}_1 \Cup \mathcal{A}_2 \succcurlyeq NM \backsim Cl(NM \backsim Int(\mathcal{A}_1 \Cup \mathcal{A}_2))$. Now, $NM \backsim Cl(NM \backsim Int(\mathcal{A}_1 \Cup \mathcal{A}_2)) \succcurlyeq NM \backsim Cl(NM \backsim Int(\mathcal{A}_1)) = \mathcal{A}_1$ and $NM \backsim Cl(NM \backsim Int(\mathcal{A}_1 \Cup \mathcal{A}_2)) \succcurlyeq NM \backsim Cl(NM \backsim Int(\mathcal{A}_2)) = \mathcal{A}_2$ implies that $\mathcal{A}_1 \Cup \mathcal{A}_2 \preccurlyeq NM \backsim Cl(NM \backsim Int(\mathcal{A}_1 \Cup \mathcal{A}_2))$, hence the theorem. □

Theorem 6. *(i) The closure of an NMOS is an NMRCoS;*
(ii) The interior of an NMCoS is an NMROS.

Proof. (i) Let $\mathcal{A}$ be an NMOS of an NMTS (X, τ_X), clearly, $NM \backsim Int(NM \backsim Cl(\mathcal{A})) \preccurlyeq NM \backsim Cl(\mathcal{A}) \Rightarrow NM \backsim Cl(NM \backsim Int(NM \backsim Cl(\mathcal{A}))) \preccurlyeq NM \backsim Cl(\mathcal{A})$. Now, $\mathcal{A}$ is NMOS implies that $\mathcal{A} \preccurlyeq NM \backsim Int(NM \backsim Cl(\mathcal{A}))$, and hence, $NM \backsim Cl(\mathcal{A}) \preccurlyeq NM \backsim Cl(NM \backsim Int(NM \backsim Cl(\mathcal{A})))$. Thus, $NM \backsim Cl(\mathcal{A})$ is NMRCoS;

(ii) Let $\mathcal{A}$ be an NMCoS of an NMTS (X, τ_X), clearly, $NM \backsim Cl(NM \backsim Int(\mathcal{A})) \succcurlyeq NM \backsim Int(\mathcal{A}) \Rightarrow NM \backsim Int(NM \backsim Cl(NM \backsim Int(\mathcal{A}))) \succcurlyeq NM \backsim Int(\mathcal{A})$. Now, $\mathcal{A}$ is NMCoS implies that $\mathcal{A} \succcurlyeq NM \backsim Cl(NM \backsim Int(\mathcal{A}))$, and hence, $NM \backsim Int(\mathcal{A}) \succcurlyeq NM \backsim Int(NM \backsim Cl(NM \backsim Int(\mathcal{A})))$. Thus, $NM \backsim Int(\mathcal{A})$ is NMROS. □

Definition 17. *Let* $\phi : (X, \tau_X) \longrightarrow (Y, \tau_Y)$ *be a mapping from an NMTS* (X, τ_X) *to another NMTS* (Y, τ_Y)*, then* ϕ *is known as a neutrosophic multi continuous mapping (NMCM), if* $\phi^{-1}(\mathcal{A}) \in \tau_X$ *for each* $\mathcal{A} \in \tau_Y$*, or equivalently* $\phi^{-1}(\mathcal{B})$ *is an NMCoS of* X *for each CoNMS* $\mathcal{B}$ *of* Y.

Example 7. *Let* $X = Y = \{a, b, c\}$ *and*

$$\mathcal{A} = \left\{ \begin{array}{c} < a, 0.4, 0.5, 0.6 >, < a, 0.3, 0.5, 0.7 >, < a, 0.2, 0.6, 0.8 >, \\ < b, 0.3, 0.5, 0.4 >, < b, 0.2, 0.5, 0.6 >, < b, 0.1, 0.5, 0.7 >, \\ < c, 0.4, 0.5, 0.6 >, < c, 0.3, 0.5, 0.7 >, < c, 0.2, 0.6, 0.8 > \end{array} \right\};$$

$$\mathcal{B} = \left\{ \begin{array}{l} < a, 0.6, 0.1, 0.2 >, < a, 0.5, 0.1, 0.3 >, < a, 0.4, 0.2, 0.4 >, \\ < b, 0.3, 0.5, 0.4 >, < b, 0.2, 0.5, 0.6 >, < b, 0.1, 0.5, 0.7 >, \\ < c, 0.4, 0.5, 0.6 >, < c, 0.3, 0.5, 0.7 >, < c, 0.2, 0.6, 0.8 > \end{array} \right\}.$$

Then $\tau_X = \{0_X, 1_X, \mathcal{A}\}$ and $\tau_Y = \{0_Y, 1_Y, \mathcal{B}\}$ are neutrosophic multi topological spaces.
Now, define a mapping $f : (X, \tau_X) \longrightarrow (Y, \tau_Y)$ by $f(a) = f(c) = c$ and $f(b) = b$.
Thus, f is NMCM.

Definition 18. *Let $\phi : (X, \tau_X) \longrightarrow (Y, \tau_Y)$ be a mapping from an NMTS (X, τ_X) to another NMTS (Y, τ_Y), then ϕ is called a neutrosophic multi open mapping (NMOM) if $\phi(\mathcal{A}) \in \tau_Y$ for each $\mathcal{A} \in \tau_X$.*

Definition 19. *Let $\phi : (X, \tau_X) \longrightarrow (Y, \tau_Y)$ be a mapping from an NMTS (X, τ_X) to another NMTS (Y, τ_Y), then ϕ is said to be a neutrosophic multi-closed mapping (NMCoM) if $\phi(\mathcal{B})$ is an NMCoS of Y for each NMCoS $\mathcal{B}$ of X.*

Definition 20. *Let $\phi : (X, \tau_X) \longrightarrow (Y, \tau_Y)$ be a mapping from an NMTS (X, τ_X) to another NMTS (Y, τ_Y), then ϕ is called a neutrosophic multi semi-continuous mapping (NMSCM), if $\phi^{-1}(\mathcal{A})$ is the NMSOS of X, for each $\mathcal{A} \in \tau_Y$.*

Definition 21. *Let $\phi : (X, \tau_X) \longrightarrow (Y, \tau_Y)$ be a mapping from an NMTS (X, τ_X) to another NMTS (Y, τ_Y), then ϕ is called a neutrosophic multi semi-open mapping (NMSOM) if $\phi(\mathcal{A})$ is a SONMS for each $\mathcal{A} \in \tau_X$.*

Example 8. *Let $X = Y = \{a, b, c\}$ and*

$$\mathcal{A} = \left\{ \begin{array}{l} < a, 0.6, 0.1, 0.2 >, < a, 0.5, 0.1, 0.3 >, < a, 0.4, 0.2, 0.4 >, \\ < b, 0.3, 0.5, 0.4 >, < b, 0.2, 0.5, 0.6 >, < b, 0.1, 0.5, 0.7 >, \\ < c, 0.4, 0.5, 0.6 >, < c, 0.3, 0.5, 0.7 >, < c, 0.2, 0.6, 0.8 > \end{array} \right\};$$

$$\mathcal{B} = \left\{ \begin{array}{l} < a, 0.3, 0.5, 0.4 >, < a, 0.2, 0.5, 0.6 >, < a, 0.1, 0.5, 0.7 >, \\ < b, 0.6, 0.1, 0.2 >, < b, 0.5, 0.1, 0.3 >, < b, 0.4, 0.2, 0.4 >, \\ < c, 0.4, 0.5, 0.6 >, < c, 0.3, 0.5, 0.7 >, < c, 0.2, 0.6, 0.8 > \end{array} \right\}.$$

Then $\tau_X = \{0_X, 1_X, \mathcal{A}\}$ and $\tau_Y = \{0_Y, 1_Y, \mathcal{B}\}$ are neutrosophic multi topological spaces.
Clearly, $\mathcal{A}$ is a semi-open set.
Then a mapping $f : (X, \tau_X) \longrightarrow (Y, \tau_Y)$ defined by $f(a) = b, f(b) = a$ and $f(c) = c$.
Hence, f is NMSOM.

Definition 22. *Let $\phi : (X, \tau_X) \longrightarrow (Y, \tau_Y)$ be a mapping from an NMTS (X, τ_X) to another NMTS (Y, τ_Y), then ϕ is called a neutrosophic multi semi-closed mapping (NMSCoM) if $\phi(\mathcal{B})$ is an NMSCoS for each NMCoS $\mathcal{B}$ of X.*

Remark 4. *From Remark 1, an NMCM (NMOM, NMCoM) is also an NMSCM (NMSOM, NMSCoM).*

Example 9. *Let $X = Y = \{a, b, c\}$ and*

$$\mathcal{A} = \left\{ \begin{array}{l} < a, 0.4, 0.5, 0.6 >, < a, 0.3, 0.5, 0.7 >, < a, 0.2, 0.6, 0.8 >, \\ < b, 0.3, 0.5, 0.4 >, < b, 0.2, 0.5, 0.6 >, < b, 0.1, 0.5, 0.7 >, \\ < c, 0.4, 0.5, 0.6 >, < c, 0.3, 0.5, 0.7 >, < c, 0.2, 0.6, 0.8 > \end{array} \right\};$$

$$\mathcal{B} = \left\{ \begin{array}{l} < a, 0.4, 0.5, 0.6 >, < a, 0.3, 0.5, 0.7 >, < a, 0.2, 0.6, 0.8 >, \\ < b, 0.4, 0.6, 0.4 >, < b, 0.3, 0.5, 0.5 >, < b, 0.2, 0.5, 0.6 >, \\ < c, 0.6, 0.5, 0.5 >, < c, 0.4, 0.5, 0.6 >, < c, 0.2, 0.6, 0.9 > \end{array} \right\}.$$

Then $\tau_X = \{0_X, 1_X, \mathcal{A}\}$ and $\tau_Y = \{0_Y, 1_Y, \mathcal{B}\}$ are neutrosophic multi topological spaces.
Let us define a mapping $f : (X, \tau_X) \longrightarrow (Y, \tau_Y)$ by $f(a) = f(c) = c$ and $f(b) = b$.
Thus, f is NMSCM, which is not an NMCM.

Theorem 7. *Let X_1, X_2, Y_1 and Y_2 be NMTSs such that X_1 is product related to X_2. Then, the product $\phi_1 \times \phi_2 : X_1 \times X_2 \longrightarrow Y_1 \times Y_2$ of NMSCMs $\phi_1 : X_1 \longrightarrow Y_1$ and $\phi_2 : X_2 \longrightarrow Y_2$ is NMSCM.*

Proof. Let $\mathcal{A} \equiv \Cup(\mathcal{A}_\alpha \times \mathcal{B}_\beta)$, where $\mathcal{A}_\alpha$'s and $\mathcal{B}_\beta$'s are NMOSs of Y_1 and Y_2, respectively, be an NMOS of $Y_1 \times Y_2$. By using Lemma 1(i) and Lemma 3, we have

$$(\phi_1 \times \phi_2)^{-1}(\mathcal{A}) = \Cup\left[\phi_1{}^{-1}(\mathcal{A}_\alpha) \times \phi_2{}^{-1}(\mathcal{A}_\beta)\right]$$

where $(\phi_1 \times \phi_2)^{-1}(\mathcal{A})$ is an NMSOS follows from Theorem 3 and Theorem 2 (i). □

Theorem 8. *Let X, X_1 and X_2 be NMTSs and $p_i : X_1 \times X_2 \longrightarrow X_i$ $(i = 1,\ 2)$ be the projection of $X_1 \times X_2$ onto X_i. Then, if $\phi : X \longrightarrow X_1 \times X_2$ is an NMSCM, $p_i\phi$ is also NMSCM.*

Proof. For an NMOS $\mathcal{A}$ of X_i, we have $(p_i\phi)^{-1}(\mathcal{A}) = \phi^{-1}(p_i{}^{-1}(\mathcal{A}))$. p_i is an NMCM and ϕ is an NMSCM, which implies that $(p_i\phi)^{-1}(\mathcal{A})$ is an NMSOS of X. □

Theorem 9. *Let $\phi : X \longrightarrow Y$ be a mapping from an NMTS X to another NMTS Y. Then if the graph $\psi : X \longrightarrow X \times Y$ of ϕ is NMSCM, ϕ is also NMSCM.*

Proof. From Lemma 4, $\phi^{-1}(\mathcal{A}) = 1_{NM} \Cap \phi^{-1}(\mathcal{A}) = \psi^{-1}(1_{NM} \times \mathcal{A})$, for each NMOS $\mathcal{A}$ of Y. Since ψ is an NMSCM and $1_{NM} \times \mathcal{A}$ is an NMOS $X \times Y$, $\phi^{-1}(\mathcal{A})$ is an NMSOS of X and hence ϕ is an NMSCM. □

Remark 5. *The converse of Theorem 9 is not true.*

Definition 23. *A mapping $\phi : (X, \tau_X) \longrightarrow (Y, \tau_Y)$ from an NMTS X to another NMTS Y is known as a neutrosophic multi almost continuous mapping (NMACM), if $\phi^{-1}(\mathcal{A}) \in \tau_X$ for each NMROS $\mathcal{A}$ of Y.*

Example 10. *Let $X = Y = \{a, b\}$ and*

$$\mathcal{A} = \left\{ \begin{array}{l} < a, 0.4, 0.5, 0.5 >, < a, 0.3, 0.5, 0.6 >, < a, 0.2, 0.6, 0.7 >, \\ < b, 0.5, 0.7, 0.6 >, < b, 0.4, 0.5, 0.7 >, < b, 0.3, 0.5, 0.8 > \end{array} \right\};$$

$$\mathcal{B} = \left\{ \begin{array}{l} < a, 0.5, 0.7, 0.6 >, < a, 0.4, 0.5, 0.7 >, < a, 0.3, 0.5, 0.8 >, \\ < b, 0.4, 0.5, 0.5 >, < b, 0.3, 0.5, 0.6 >, < b, 0.2, 0.6, 0.7 > \end{array} \right\}.$$

Then $\tau_X = \{0_X, 1_X, \mathcal{A}\}$ and $\tau_Y = \{0_Y, 1_Y, \mathcal{B}\}$ are neutrosophic multi topological spaces.
Clearly, $Cl(\mathcal{B}) = \mathcal{B}^C$, $Int(Cl(\mathcal{B})) = \mathcal{B}$.
Hence, $\mathcal{B}$ is NMROS.
Now, let us define a mapping $f : (X, \tau_X) \rightarrow (Y, \tau_Y)$ by $f(a) = b, f(b) = a$.
Thus, f is NMACM.

Theorem 10. *Let $\phi : (X, \tau_X) \rightarrow (Y, \tau_Y)$ be a mapping. Then the below statements are equivalent:*

(a) ϕ *is an NMACM;*
(b) $\phi^{-1}(\mathcal{F})$ *is an NMCoS, for each NMRCoS* $\mathcal{F}$ *of Y;*
(c) $\phi^{-1}(\mathcal{A}) \preccurlyeq NM \backsim Int(\phi^{-1}(NM \backsim Int(NM \backsim Cl(\mathcal{A}))))$*, for each NMOS* $\mathcal{A}$ *of Y;*
(d) $NM \backsim Cl\ (\phi^{-1}(NM \backsim Cl(NM \backsim Int(\mathcal{F})))) \preccurlyeq \phi^{-1}(\mathcal{F})$*, for each NMCoS* $\mathcal{F}$ *of Y.*

Proof. Consider that $\phi^{-1}(\mathcal{A}^c) = \left(\phi^{-1}(A)\right)^c$, for any NMS $\mathcal{A}$ of Y, (a) $\Leftrightarrow$ (b) follows from Theorem 4.

(a) $\Rightarrow$ (c). Since $\mathcal{A}$ is an NMOS of Y, $\mathcal{A} \preccurlyeq NM \backsim Int(Cl(\mathcal{A}))$, hence, $\phi^{-1}(\mathcal{A}) \preccurlyeq \phi^{-1}(NM \backsim Int(NM \backsim Cl(\mathcal{A})))$. From Theorem 6 (ii), $NM \backsim Int(NM \backsim Cl(\mathcal{A}))$ is an NMROS of Y, hence $\phi^{-1}(NM \backsim Int(NM \backsim Cl(\mathcal{A})))$ is an NMOS of X. Thus, $\phi^{-1}(\mathcal{A}) \preccurlyeq \phi^{-1}(NM \backsim Int(NM \backsim Cl(\mathcal{A}))) = NM \backsim Int(\phi^{-1}(NM \backsim Int(NM \backsim Cl(\mathcal{A}))).$

(c) $\Rightarrow$ (a). Let $\mathcal{A}$ be an NMROS of Y, then we have $\phi^{-1}(\mathcal{A}) \preccurlyeq NM \backsim Int(\phi^{-1}(NM \backsim Int(NM \backsim Cl(\mathcal{A})))) = NM \backsim Int(\phi^{-1}(\mathcal{A}))$. Thus, $\phi^{-1}(\mathcal{A}) = NM \backsim Int(\phi^{-1}(\mathcal{A}))$. This shows that $\phi^{-1}(\mathcal{A})$ is an NMOS of X.

(b) $\Leftrightarrow$ (d) similarly can be proved. □

Remark 6. *Clearly, an NMCM is an NMACM. The converse need not be true.*

Example 11. *Let* $X = Y = \{a, b\}$ *and*

$$\mathcal{A} = \left\{ \begin{array}{c} < a, 0.4, 0.5, 0.5 >, < a, 0.3, 0.5, 0.6 >, < a, 0.2, 0.6, 0.7 >, \\ < b, 0.5, 0.7, 0.6 >, < b, 0.4, 0.5, 0.7 >, < b, 0.3, 0.5, 0.8 > \end{array} \right\};$$

$$\mathcal{B} = \left\{ \begin{array}{c} < a, 0.5, 0.5, 0.6 >, < a, 0.6, 0.5, 0.7 >, < a, 0.2, 0.6, 0.9 >, \\ < b, 0.4, 0.4, 0.7 >, < b, 0.3, 0.5, 0.5 >, < b, 0.4, 0.5, 0.6 > \end{array} \right\}.$$

Then, $\tau_X = \{0_X, 1_X, \mathcal{A}\}$ and $\tau_Y = \{0_Y, 1_Y, \mathcal{B}\}$ are neutrosophic multi topological spaces.
Clearly, $Cl(\mathcal{B}) = \mathcal{B}^C$, $Int(Cl(\mathcal{B})) = \mathcal{B}$.
Hence, $\mathcal{B}$ is NMROS in τ_Y.
Now, a mapping $f : (X, \tau_X) \to (Y, \tau_Y)$ defined by $f(a) = a, f(b) = b$.
Then clearly, f is NMACM but not NMCM.

Theorem 11. *Neutrosophic multi semi-continuity and neutrosophic multi almost continuity are independent notions.*

Definition 24. *AN NMTS* (X, τ_X) *is called a neutrosophic multi semi-regularly space (NMSRS) if and only if the collection of all NMROSs of X forms a base for NMT* τ_X.

Theorem 12. *Let* $\phi : (X, \tau_X) \to (Y, \tau_Y)$ *be a mapping from an NMTS X to an NMSRS Y. Then* ϕ *is NMACM iff* ϕ *is NMCM.*

Proof. From Remark 6, it suffices to prove that if ϕ is NMACM, then it is NMCM. Let $\mathcal{A} \in \tau_Y$, then $\mathcal{A} = \Cup \mathcal{A}_\alpha$, where $\mathcal{A}_\alpha$'s are NMROSs of Y. Now, from Lemma 1(i), 5, and Theorem 10 (c), we obtain

$$\phi^{-1}(\mathcal{A}) = \Cup \phi^{-1}(\mathcal{A}_\alpha) \preccurlyeq \Cup NM \backsim Int\Big(\phi^{-1}(NM \backsim Cl(\mathcal{A}_\alpha))\Big) = \Cup NM \backsim Int\Big(\phi^{-1}(\mathcal{A}_\alpha)\Big).$$

$$\preccurlyeq NM \backsim Int \Cup \Big(\phi^{-1}(\mathcal{A}_\alpha)\Big) = NM \backsim Int\Big(\phi^{-1}(\mathcal{A}_\alpha)\Big).$$

which shows that $\phi^{-1}(\mathcal{A}_\alpha) \in \tau_X$. □

Theorem 13. *Let* X_1, X_2, Y_1 *and* Y_2 *be the NMTSs, such that* Y_1 *is product related to* Y_2. *Then the product* $\phi_1 \times \phi_2 : X_1 \times X_2 \to Y_1 \times Y_2$ *of NMACMs* $\phi_1 : X_1 \to Y_1$ *and* $\phi_2 : X_2 \to Y_2$ *is NMACM.*

Proof. Let $\mathcal{A} = \Cup (\mathcal{A}_\alpha \times \mathcal{B}_\beta)$, where $\mathcal{A}_\alpha$'s and $\mathcal{B}_\beta$'s are NMOSs of Y_1 and Y_2, respectively, be an NMOS of $Y_1 \times Y_2$. From Lemma 1(i), 3, 5, and *Theorems* 6, and 10 (c), we have

$$(\phi_1 \times \phi_2)^{-1}(\mathcal{A}) = \Cup \left\{ \phi_1{}^{-1}(\mathcal{A}_\alpha) \times \phi_2{}^{-1}(\mathcal{B}_\beta) \right\}$$

$$\preccurlyeq \Cup \begin{bmatrix} NM \backsim Int(\phi_1{}^{-1}(NM \backsim Int(NM \backsim Cl(\mathcal{A}_\alpha)))) \\ \times NM \backsim Int(\phi_2{}^{-1}(NM \backsim Int(NM \backsim Cl(\mathcal{B}_\beta)))) \end{bmatrix}$$

$$\preccurlyeq \Cup \left[NM \backsim Int\left\{\phi_1{}^{-1}(NM \backsim Int(NM \backsim Cl(\mathcal{A}_\alpha))) \times \phi_2{}^{-1}(NM \backsim Int(NM \backsim Cl(\mathcal{B}_\beta)))\right\}\right]$$

$$\preccurlyeq NM \backsim Int\left[\Cup\ (\phi_1 \times \phi_2)^{-1}\{NM \backsim Int(NM \backsim Cl(\mathcal{A}_\alpha)) \times NM \backsim Int(NM \backsim Cl(\mathcal{B}_\beta))\}\right]$$

$$= NM \backsim Int\left[\Cup\ (\phi_1 \times \phi_2)^{-1}\{NM \backsim Int(NM \backsim Cl(\mathcal{A}_\alpha \times \mathcal{B}_\beta))\}\right]$$

$$\preccurlyeq NM \backsim Int\left[(\phi_1 \times \phi_2)^{-1}\{NM \backsim Int(NM \backsim Cl(\Cup\ (\mathcal{A}_\alpha \times \mathcal{B}_\beta)))\}\right]$$

$$= NM \backsim Int\left[(\phi_1 \times \phi_2)^{-1}(NM \backsim Int(NM \backsim Cl(\mathcal{A})))\right]$$

Thus, by Theorem 10 (c), $\phi_1 \times \phi_2$ is NMACM. □

Theorem 14. *Let X, X_1 and X_2 be an NMTSs and $p_i : X_1 \times X_2 \to X_i (i = 1, 2)$ be the projection of $X_1 \times X_2$ onto X_i. Then if $\phi : X \to X_1 \times X_2$ is an NMACM, $p_i\phi$ is also an NMACM.*

Proof. Since p_i is NMCM Definition 16, for any NMS $\mathcal{A}$ of X_i, we have (i) $NM \backsim Cl(p_i{}^{-1}(\mathcal{A})) \preccurlyeq p_i{}^{-1}(\mathrm{NM} \backsim \mathrm{Cl}(\mathcal{A}))$ and (ii) $NM \backsim Int(p_i{}^{-1}(\mathcal{A})) \succcurlyeq p_i{}^{-1}(\mathrm{NM} \backsim Int(\mathcal{A}))$. Again, since (i) each p_i is an NMOS, and (ii) for any NMS $\mathcal{A}$ of X_i (a) $\mathcal{A} \preccurlyeq p_i{}^{-1}p_i(\mathcal{A})$ and (b) $p_i{}^{-1}p_i(\mathcal{A}) \preccurlyeq \mathcal{A}$, we have $p_i(NM \backsim Int(p_i{}^{-1}(\mathcal{A}))) \preccurlyeq p_ip_i{}^{-1}(\mathcal{A}) \preccurlyeq \mathcal{A}$, and hence, $p_i(NM \backsim Int(p_i{}^{-1}(\mathcal{A}))) \preccurlyeq NM \backsim Int(\mathcal{A})$. □

Thus, $NM \backsim Int(p_i{}^{-1}(\mathcal{A})) \preccurlyeq p_i{}^{-1}p_i(NM \backsim Int(p_i{}^{-1}(\mathcal{A}))) \preccurlyeq (p_i{}^{-1}(NM \backsim Int(\mathcal{A}))$ establishes that $NM \backsim Int(p_i{}^{-1}(\mathcal{A})) \preccurlyeq p_i{}^{-1}(NM \backsim Int(\mathcal{A}))$. Now, for any NMOS $\mathcal{A}$ of X_i,

$$\begin{aligned}(p_i\phi)^{-1}(\mathcal{A}) = \ & \phi^{-1}(p_i{}^{-1}(\mathcal{A})) \\ & \preccurlyeq NM \backsim Int\{\phi^{-1}(NM \backsim Int(NM \backsim Cl(p_i{}^{-1}(\mathcal{A}))))\} \\ & \preccurlyeq NM \backsim Int\{\phi^{-1}(NM \backsim Int(p_i{}^{-1}(NM \backsim Cl(\mathcal{A}))))\} \\ & = NM \backsim Int\{\phi^{-1}(p_i{}^{-1}(NM \backsim Int(NM \backsim Cl(\mathcal{A}))))\} \\ & = NM \backsim Int(p_i\phi)^{-1}(NM \backsim Int(NM \backsim Cl(\mathcal{A})))\end{aligned}$$

Theorem 15. *Let X and Y be NMTSs such that X is product related to Y and let $\phi : X \to Y$ be a mapping. Then, the graph $\psi : X \to X \times Y$ of ϕ is NMACM if ϕ is NMACM.*

Proof. Consider that ψ is an NMACM and $\mathcal{A}$ is an NMOS of Y. Then, using Lemma 4 and Theorems 10 (c), we have

$$\begin{aligned}\phi^{-1}(\mathcal{A}) & = 1_{NM} \Cap \phi^{-1}(\mathcal{A}) \\ & = \psi^{-1}(1_{NM} \times \mathcal{A}) \preccurlyeq NM \backsim Int(\psi^{-1}(NM \backsim Int(NM \backsim Cl(1_{NM} \times \mathcal{A})))) \\ & = NM \backsim Int(\psi^{-1}(1_{NM} \times NM \backsim Int(NM \backsim Cl(\mathcal{A})))) \\ & = NM \backsim Int(\psi^{-1}(NM \backsim Int(1_{NM} \times NM \backsim Cl(\mathcal{A})))) \\ & = NM \backsim Int(\psi^{-1}(NM \backsim Int(NM \backsim Cl(\mathcal{A}))))\end{aligned}$$

Thus, by Theorem 10 (c), ϕ is NMACM.

Conversely, let ϕ be an NMACM and $\mathcal{B} = \Cup\ (\mathcal{B}_\alpha \times \mathcal{A}_\beta)$, where $\mathcal{B}_\alpha$'s and $\mathcal{A}_\beta$'s are NMOSs of X and Y, respectively, be an NMOS of $X \times Y$.

Since $\mathcal{B}_\alpha \Cap NM \backsim Int(\phi^{-1}(NM \backsim Int(NM \backsim Cl(\mathcal{A}_\beta))))$ is an NMOSs of X contained in

$$NM \backsim Int(NM \backsim Cl(\mathcal{B}_\alpha)) \Cap \phi^{-1}(NM \backsim Int(NM \backsim Cl(\mathcal{A}_\beta))),$$

$$\mathcal{B}_\alpha \Cap NM \backsim Int\left(\phi^{-1}(NM \backsim Int(NM \backsim Cl(\mathcal{A}_\beta)))\right)$$

$$\preccurlyeq NM \backsim Int\Big[NM \backsim Int(NM \backsim Cl(\mathcal{B}_\alpha)) \Cap \phi^{-1}(NM \backsim Int(NM \backsim Cl(\mathcal{A}_\beta)))\Big]$$

and hence, using Lemmas 1(i), 4 and 5, and Theorems 10 (c), we have

$$\begin{aligned}
\phi^{-1}(\mathcal{B}) &= \phi^{-1}(\Cup (\mathcal{B}_\alpha \times \mathcal{A}_\beta)) \\
&= \Cup \left[\mathcal{B}_\alpha \Cap \phi^{-1}(\mathcal{A}_\beta)\right] \\
&\preccurlyeq \Cup \left[\mathcal{B}_\alpha \Cap NM \backsim Int(\phi^{-1}(NM \backsim Int(NM \backsim Cl(\mathcal{A}_\beta))))\right] \\
&\preccurlyeq \Cup \left[NM \backsim Int(NM \backsim Int(NM \backsim Cl(\mathcal{B}_\alpha))) \Cap \phi^{-1}(NM \backsim Int(NM \backsim Cl(\mathcal{A}_\beta)))\right] \\
&\preccurlyeq NM \backsim Int[\Cup \ \psi^{-1}(NM \backsim Int(NM \backsim Cl(\mathcal{B}_\alpha))) \times NM \backsim Int(NM \backsim Cl(\mathcal{A}_\beta))] \\
&= NM \backsim Int\left[\psi^{-1}(\Cup (NM \backsim Int(NM \backsim Cl(\mathcal{B}_\alpha \times \mathcal{A}_\beta))))\right] \\
&\preccurlyeq NM \backsim Int\left[\psi^{-1}(NM \backsim Int(NM \backsim Cl(\Cup (\mathcal{B}_\alpha \times \mathcal{A}_\beta))))\right] \\
&= NM \backsim Int\left[\psi^{-1}(NM \backsim Int(NM \backsim Cl(\mathcal{B})))\right]
\end{aligned}$$

Thus, by Theorem 10(c), ψ is NMACM. □

Definition 25. *Let $\mathbb{G}$ be an NMG on a group X. Now, if τ_X is an NMT on $\mathbb{G}$, then $(\mathbb{G}, \tau_X)$ is said to be a neutrosophic multi almost topological group (NMATG) if the given conditions are satisfied:*

(i) $\alpha : (\mathbb{G}, \tau_X) \times (\mathbb{G}, \tau_X) \to (\mathbb{G}, \tau_X) : \alpha(m,n) = mn$ *is NMACM;*

(ii) $\beta : (\mathbb{G}, \tau_X) \to (\mathbb{G}, \tau_X) : \beta(m) = m^{-1}$ *is NMACM.*

Then $(\mathbb{G}, \tau_X)$ is known as an NMATG.

Remark 7. *$(\mathbb{G}, \tau_X)$ is an NMATG if the below conditions hold good:*

(i) *For $g_1, g_2 \in \mathbb{G}$ and every NMROS $\mathcal{P}$ containing g_1g_2 in $\mathbb{G}$, $\exists$ open neighborhoods $\mathcal{R}$ and $\mathcal{S}$ of g_1 and g_2 in $\mathbb{G}$ such that $\mathcal{R} * \mathcal{S} \preccurlyeq \mathcal{P}$;*

(ii) *For $g \in \mathbb{G}$ and every N in $\mathbb{G}$ containing g^{-1}, $\exists$ open neighborhood $\mathcal{R}$ of g in $\mathbb{G}$ so that $\mathcal{R}^{-1} \preccurlyeq \mathcal{S}$.*

Remark 8. *For any $\mathcal{P}, \mathcal{Q} \preccurlyeq \mathbb{G}$, we denote $\mathcal{P} * \mathcal{Q}$ by $\mathcal{PQ}$ and defined as $\mathcal{PQ} = \{gh : g \in \mathcal{P},\ h \in \mathcal{Q}\}$ and $\mathcal{P}^{-1} = \{g^{-1} : g \in \mathcal{P}\}$. If $\mathcal{P} = \{a\}$ for each $a \in \mathbb{G}$, we denote $\mathcal{P} * \mathcal{Q}$ by $a\mathcal{Q}$ and $\mathcal{Q} * \mathcal{P}$ by $\mathcal{P}a$.*

Example 12. *Let, $\mathbb{G} = (\mathbb{Z}_3, +)$ be a classical group and*

$$\mathcal{A} = \left\{ \begin{array}{l} <0,\ 0.4, 0.5, 0.6>, <0,\ 0.3, 0.5, 0.7>, <0,\ 0.2, 0.6, 0.8>, \\ <1,\ 0.3, 0.5, 0.4>, <1,\ 0.2, 0.5, 0.6>, <1,\ 0.1, 0.5, 0.7>, \\ <2,\ 0.4, 0.5, 0.6>, <2,\ 0.3, 0.5, 0.7>, <2,\ 0.2, 0.6, 0.8> \end{array} \right\}$$

Then $\tau_{\mathbb{G}} = \{0_G, 1_G, \mathcal{A}\}$ is NTS and the mapping $\alpha : (\mathbb{G}, \tau_{\mathbb{G}}) \times (\mathbb{G}, \tau_{\mathbb{G}}) \to (\mathbb{G}, \tau_{\mathbb{G}}) : \alpha(m,n) = mn$ and $\beta : (\mathbb{G}, \tau_{\mathbb{G}}) \to (\mathbb{G}, \tau_{\mathbb{G}}) : \beta(m) = m^{-1}$ are NMACM. Hence, $(\mathbb{G}, \tau_{\mathbb{G}})$ is NMATG.

Theorem 16. *Let $(\mathbb{G}, \tau_X)$ be an NMATG and let a be any element of $\mathbb{G}$. Then*

(a) $\mu_a : (\mathbb{G}, \tau_X) \to (\mathbb{G}, \tau_X) : \mu_a(x) = ax, \forall x \in \mathbb{G}$, *is NMACM;*

(b) $\lambda_a : (\mathbb{G}, \tau_X) \to (\mathbb{G}, \tau_X) : \lambda_a(x) = xa, \forall x \in \mathbb{G}$, *is NMACM.*

Proof. (a) Let $p \in \mathbb{G}$ and let $\mathcal{R}$ be an NMROS containing ap in $\mathbb{G}$. By Definition 25, $\exists$ open neighborhoods $\mathcal{P}, \mathcal{Q}$ of a, p in $\mathbb{G}$ such that $\mathcal{PQ} \preccurlyeq \mathcal{R}$. Especially, $a\mathcal{Q} \preccurlyeq \mathcal{R}$, i.e., $\mu_a(\mathcal{Q}) \preccurlyeq \mathcal{R}$. This proves that μ_a is NMACM at p, and hence, μ_a is NMACM.

(b) Suppose $p \in \mathbb{G}$ and $\mathcal{R} \in NMRO(\mathbb{G})$ contain pa. Then $\exists$ open sets $p \in \mathcal{P}$ and $a \in \mathcal{Q}$ in $\mathbb{G}$ such that $\mathcal{PQ} \preccurlyeq \mathcal{R}$. This proves $\mathcal{P}a \preccurlyeq \mathcal{R}$. This shows that λ_a is NMACM at p. Since arbitrary element p is in $\mathbb{G}$, hence, λ_a is NMACM. □

Theorem 17. *Let $\mathcal{U}$ be NMROS in a NMATG $(\mathbb{G}, \tau_X)$. The below conditions hold good:*

(a) $m\mathcal{U} \in NMROS(\mathbb{G}), \forall m \in \mathbb{G}$;

(b) $\mathcal{U}m \in NMROS(\mathbb{G}), \forall m \in \mathbb{G}$;

(c) $\mathcal{U}^{-1} \in NMROS(\mathbb{G})$.

Proof. (a) We first show that $m\mathcal{U} \in \tau_X$. Let $p \in m\mathcal{U}$. Then by Definition 25 of NMATGs, $\exists$ NMOSs $m^{-1} \in W_1$ and $p \in W_2$ in $\mathbb{G}$ such that $W_1W_2 \preccurlyeq \mathcal{U}$. Especially, $m^{-1}W_2 \preccurlyeq \mathcal{U}$. That is, equivalently, $W_2 \preccurlyeq m\mathcal{U}$. This indicates that $p \in NM \backsim Int(m\mathcal{U})$ and thus, $NM \backsim Int(m\mathcal{U}) = m\mathcal{U}$. That is $m\mathcal{U} \in \tau_X$. Consequently, $m\mathcal{U} \preccurlyeq NM \backsim Int(NM \backsim Cl(m\mathcal{U}))$.

Now, we have to prove that $NM \backsim Int(NM \backsim Cl(m\mathcal{U})) \preccurlyeq m\mathcal{U}$. As $\mathcal{U}$ is NMOS, $NM \backsim Cl(\mathcal{U}) \in NMRCS(\mathbb{G})$. By Theorem 16, $\mu_{m^{-1}} : (\mathbb{G}, \tau_X) \to (\mathbb{G}, \tau_X)$ is NMACM, and therefore, $mNM \backsim Cl(\mathcal{U})$ is NMCoS. Thus, $NM \backsim Int(NM \backsim Cl(m\mathcal{U})) \preccurlyeq NM \backsim Cl(m\mathcal{U}) \preccurlyeq mNM \backsim Cl(\mathcal{U})$, i.e., $m^{-1}NM \backsim Int(NM \backsim Cl(m\mathcal{U})) \preccurlyeq NM \backsim Cl(\mathcal{U})$. Since $NM \backsim Int(NM \backsim Cl(m\mathcal{U}))$ is NMROS, it follows that $m^{-1}NM \backsim Int(NM \backsim Cl(m\mathcal{U})) \preccurlyeq NM \backsim Int(NM \backsim Cl(\mathcal{U})) = \mathcal{U}$, i.e., $NM \backsim Int(NM \backsim Cl(m\mathcal{U})) \preccurlyeq m\mathcal{U}$. Thus $m\mathcal{U} = NM \backsim Int(NM \backsim Cl(m\mathcal{U}))$. This proves that $m\mathcal{U} \in NMROS(\mathbb{G})$.

(b) Following the same steps as in part (1) above, we can prove that $\mathcal{U}m \in NMROS(\mathbb{G}), \forall\, m \in \mathbb{G}$.

(c) Let $p \in \mathcal{U}^{-1}$, then $\exists$ open set $p \in W$ in $\mathbb{G}$ such that $W^{-1} \preccurlyeq \mathcal{U} \Rightarrow W \preccurlyeq \mathcal{U}^{-1}$. Thus, $\mathcal{U}^{-1}$ has interior point p. Thus, $\mathcal{U}^{-1}$ is NMOS. That is, $\mathcal{U}^{-1} \preccurlyeq NM \backsim Int(NM \backsim Cl(\mathcal{U}^{-1}))$. Now we have to prove that $NM \backsim Int(NM \backsim Cl(\mathcal{U}^{-1})) \preccurlyeq \mathcal{U}^{-1}$. Since $\mathcal{U}$ is NMOS, $NM \backsim Cl(\mathcal{U})$ is NMRCoS and thus $NM \backsim Cl(\mathcal{U})^{-1}$ is CoNMS in $\mathbb{G}$. Thus, $NM \backsim Int(NM \backsim Cl(\mathcal{U}^{-1})) \preccurlyeq NM \backsim Cl(\mathcal{U}^{-1}) \preccurlyeq NM \backsim Cl(\mathcal{U})^{-1} \Rightarrow NM \backsim Int(NM \backsim Cl(\mathcal{U}^{-1})) \preccurlyeq (NM \backsim Cl(\mathcal{U}))^{-1} \preccurlyeq \mathcal{U}^{-1}$. Thus, $\mathcal{U}^{-1} = NM \backsim Int(NM \backsim Cl(\mathcal{U}^{-1}))$. This proves that $\mathcal{U}^{-1} \in NMROS(\mathbb{G})$. □

Corollary 1. *Let $\mathcal{Q}$ be any NMRCoS in an NMATG in $\mathbb{G}$. Then*

(a) $m\mathcal{Q} \in NMRCS(\mathbb{G})$*, for each* $m \in \mathbb{G}$;

(b) $\mathcal{Q}^{-1} \in NMRCS(\mathbb{G})$.

Theorem 18. *Let $\mathcal{U}$ be any NMROS in an NMATG $\mathbb{G}$. Then*

(a) $NM \backsim Cl(\mathcal{U}m) = NM \backsim Cl(\mathcal{U})m$*, for each* $m \in \mathbb{G}$;

(b) $NM \backsim Cl(m\mathcal{U}) = mNM \backsim Cl(\mathcal{U})$*, for each* $m \in \mathbb{G}$;

(c) $NM \backsim Cl(\mathcal{U}^{-1}) = NM \backsim Cl(\mathcal{U})^{-1}$.

Proof. (a) Assume $p \in NM \backsim Cl(\mathcal{U}m)$ and consider $q = pm^{-1}$. Let $q \in W$ be NMOS in $\mathbb{G}$. Then $\exists$ NMOSs $m^{-1} \in V_1$ and $p \in V_2$ in $\mathbb{G}$, such that $V_1V_2 \preccurlyeq NM \backsim Int(NM \backsim Cl(W))$. By hypothesis, there is $g \in \mathcal{U}m \Cap V_2 \Rightarrow gm^{-1} \in \mathcal{U} \Cap V_1V_2 \preccurlyeq \mathcal{U} \Cap NM \backsim Int(NM \backsim Cl(W)) \Rightarrow \mathcal{U} \Cap NM \backsim Int(NM \backsim Cl(W)) \neq 0_{NM} \Rightarrow \mathcal{U} \Cap (NM \backsim Cl(W)) \neq 0_{NM}$. Since $\mathcal{U}$ is NMOS, $\mathcal{U} \Cap W \neq 0_{NM}$. That is, $m \in NM \backsim Cl(\mathcal{U})m$.

Conversely, let $q \in NM \backsim Cl(\mathcal{U})m$. Then $q = pg$ for some $p \in NM \backsim Cl(\mathcal{U})$.

To prove $NM \backsim Cl(\mathcal{U})m \preccurlyeq NM \backsim Cl(\mathcal{U}m)$.

Let $pg \in W$ be an NMOS in $\mathbb{G}$. Then $\exists$ NMOSs $m \in V_1$ in $\mathcal{G}$ and $p \in V_2$ in $\mathbb{G}$ so that $V_1V_2 \preccurlyeq NM \backsim Int(NM \backsim Cl(W))$. Since $p \in NM \backsim Cl(\mathcal{U})$, $\mathcal{U} \Cap V_2 \neq 0_{NM}$. There is $g \in \mathcal{U} \Cap V_2$. This implies $gm \in (\mathcal{U}m) \Cap NM \backsim Int(NM \backsim Cl(W)) \Rightarrow (\mathcal{U}m) \Cap (NM \backsim Cl(W)) \neq 0_{NM}$. From Theorem 17, $\mathcal{U}m$ is NMOS and thus $(\mathcal{U}m) \Cap W \neq 0_{NM}$, therefore $q \in NM \backsim Cl(\mathcal{U}m)$. Therefore $NM \backsim Cl(\mathcal{U}m) = NM \backsim Cl(\mathcal{U})m$.

(b) Following the same steps as in part (1) above, we can prove that $NM \backsim Cl(m\mathcal{U}) = mNM \backsim Cl(\mathcal{U})$.

(c) Since $NM \backsim Cl(\mathcal{U})$ is NMRCoS, $NM \backsim Cl(\mathcal{U})^{-1}$ is NMCoS in $\mathbb{G}$. Therefore, $\mathcal{U}^{-1} \preccurlyeq NM \backsim Cl(\mathcal{U})^{-1}$ this gives $NM \backsim Cl(\mathcal{U}^{-1}) \preccurlyeq NM \backsim Cl(\mathcal{U})^{-1}$. Next, let $q \in NM \backsim Cl(\mathcal{U})^{-1}$. Then $q = p^{-1}$, for some $p \in NM \backsim Cl(\mathcal{U})$. Let $q \in V$ be any NMOS in $\mathbb{G}$. Then $\exists$ open set $\mathcal{U}$ in $\mathbb{G}$ such that $p \in \mathcal{U}$ with $\mathcal{U}^{-1} \preccurlyeq NM \backsim Int(NM \backsim Cl(V))$. Moreover, there is $m \in \mathcal{A} \Cap \mathcal{U}$ which implies $m^{-1} \in \mathcal{U}^{-1} \Cap NM \backsim Int(NM \backsim Cl(V))$. That is, $\mathcal{U}^{-1} \Cap NM \backsim$

$Int(NM \backsim Cl(V)) \neq 0_{NM} \Rightarrow \mathcal{U}^{-1} \Cap NM \backsim Cl(V) \neq 0_{NM} \Rightarrow \mathcal{U}^{-1} \Cap V \neq 0_{NM}$, since $\mathcal{U}^{-1}$ is NMOS. Therefore, $q \in NM \backsim Cl(\mathcal{U})^{-1}$. Hence, $NM \backsim Cl(\mathcal{U}^{-1}) \preccurlyeq NM \backsim Cl(\mathcal{U})^{-1}$. □

Theorem 19. *Let $\mathcal{Q}$ be NMRCo subset in an NMATG $\mathbb{G}$. Then the below assertions are true:*

(a) $NM \backsim Int(m\mathcal{Q}) = aNM \backsim Int(\mathcal{Q}), \forall m \in \mathbb{G}$;
(b) $NM \backsim Int(\mathcal{Q}m) = NM \backsim Int(\mathcal{Q})a, \forall m \in \mathbb{G}$;
(c) $NM \backsim Int(\mathcal{Q}^{-1}) = NM \backsim Int(\mathcal{Q})^{-1}$.

Proof. (a) Since $\mathcal{Q}$ is NMRCoS, $NM \backsim Int(\mathcal{Q})$ is NMROS in $\mathbb{G}$. Consequently, $mNM \backsim Int(\mathcal{Q}) \preccurlyeq NM \backsim Int(m\mathcal{Q})$. Conversely, let $q \in NM \backsim Int(m\mathcal{Q})$ be an arbitrary element. Suppose $q = mp$, for some $p \in \mathcal{Q}$. By hypothesis, this proves $m\mathcal{Q}$ is NMCoS, and that is $NM \backsim Int(m\mathcal{Q})$ is NMROS in $\mathbb{G}$. Assume that $m \in U$ and $p \in V$ be NMOSs in $\mathbb{G}$, such that $UV \preccurlyeq NM \backsim Int(m\mathcal{Q})$. Then $mV \preccurlyeq m\mathcal{Q}$, which means that $mV \preccurlyeq mNM \backsim Int(\mathcal{Q})$. Thus, $NM \backsim Int(m\mathcal{Q}) \preccurlyeq mNM \backsim Int(\mathcal{Q})$.

(b) Following the same steps as in part (1) above, we can prove that $NM \backsim Int(\mathcal{Q}m) \preccurlyeq NM \backsim Int(\mathcal{Q})m$.

(c) Since $NM \backsim Int(\mathcal{Q})$ is NMROS, $NM \backsim Int(\mathcal{Q})^{-1}$ is NMOS in $\mathbb{G}$. Therefore, $\mathcal{Q}^{-1} \preccurlyeq NM \backsim Int(\mathcal{Q})^{-1}$ implies that $NM \backsim Int(\mathcal{Q}^{-1}) \preccurlyeq NM \backsim Int(\mathcal{Q})^{-1}$. Next, let q be an arbitrary element of $NM \backsim Int(\mathcal{Q})^{-1}$. Then $q = p^{-1}$, for some $p \in NM \backsim Int(\mathcal{Q})$. Let $q \in V$ be NMOS in $\mathbb{G}$. Then $\exists$ NMOS U is in $\mathbb{G}$, such that $p \in U$ with $U^{-1} \preccurlyeq NM \backsim Cl(NM \backsim Int(V))$. Moreover, there is $g \in \mathcal{Q} \Cap U$, which implies $g^{-1} \in \mathcal{Q}^{-1} \Cap NM \backsim Cl(NM \backsim Int(V))$. That is $\mathcal{Q}^{-1} \Cap NM \backsim Cl(NM \backsim Int(V)) \neq 0_{NM} \Rightarrow \mathcal{Q}^{-1} \Cap NM \backsim Int(V) \neq 0_{NM} \Rightarrow \mathcal{Q}^{-1} \Cap V \neq 0_{NM}$, since $\mathcal{Q}^{-1}$ is NMCoS. Hence, $NM \backsim Int(\mathcal{Q}^{-1}) = NM \backsim Int(\mathcal{Q})^{-1}$. □

Theorem 20. *Let $\mathcal{U}$ be any NMSOS in an NMATG $\mathbb{G}$. Then*

(a) $NM \backsim Cl(m\mathcal{U}) \preccurlyeq mNM \backsim Cl(\mathcal{U}), \forall\, m \in \mathbb{G}$;
(b) $NM \backsim Cl(\mathcal{U}m) \preccurlyeq NM \backsim Cl(\mathcal{U})m, \forall\, m \in \mathbb{G}$;
(c) $NM \backsim Cl(\mathcal{U}^{-1}) \preccurlyeq NM \backsim Cl(\mathcal{U})^{-1}$.

Proof. (a) As $\mathcal{U}$ is NMSOS, $NM \backsim Cl(\mathcal{U})$ is NMRCoS. From Theorem 16, $\mu_{m^{-1}} : (\mathbb{G}, \tau_X) \longrightarrow (\mathbb{G}, \tau_X)$ is NMACM. Thus, $mNM \backsim Cl(\mathcal{U})$ is NMCoS. Hence, $NM \backsim Cl(m\mathcal{U}) \preccurlyeq mNM \backsim Cl(\mathcal{U})$.

(b) As $\mathcal{U}$ is NMSOS, $NM \backsim Cl(\mathcal{U})$ is NMRCoS. From Theorem 16, $\lambda_{m^{-1}} : (\mathbb{G}, \tau_X) \longrightarrow (\mathbb{G}, \tau_X)$ is NMACM. Thus, $NM \backsim Cl(\mathcal{U})m$ is NMCoS. Therefore, $NM \backsim Cl(\mathcal{U}m) \preccurlyeq NM \backsim Cl(\mathcal{U})m$.

(c) Since $\mathcal{U}$ is NMSOS, $NM \backsim Cl(\mathcal{U})$ is NMRCoS, and hence, $NM \backsim Cl(\mathcal{U})^{-1}$ is NMCoS. Consequently, $NM \backsim Cl(\mathcal{U}) \preccurlyeq NM \backsim Cl(\mathcal{U})^{-1}$. □

Theorem 21. *Let $\mathcal{U}$ be both NMSO and NMSCo subset of an NMATG $\mathbb{G}$. Then the below statements hold:*

(a) $NM \backsim Cl(m\mathcal{U}) = mNM \backsim Cl(\mathcal{U})$, *for each* $m \in \mathbb{G}$;
(b) $NM \backsim Cl(\mathcal{U}m) = NM \backsim Cl(\mathcal{U})m$, *for each* $m \in \mathbb{G}$;
(c) $NM \backsim Cl(\mathcal{U}^{-1}) = NM \backsim Cl(\mathcal{U})^{-1}$.

Proof. (a) Since $\mathcal{U}$ is NMSOS, $NM \backsim Cl(\mathcal{U})$ is NMRCoS, from which it implies that $NM \backsim Cl(m\mathcal{U}) \preccurlyeq mNM \backsim Cl(\mathcal{U})$. Further, neutrosophic multi semi-openness of $\mathcal{U}$ gives $NM \backsim Cl(\mathcal{U}) = NM \backsim Cl(NM \backsim Int(\mathcal{U})) \Rightarrow mNM \backsim Cl(\mathcal{U}) = mNM \backsim Cl(NM \backsim Int(\mathcal{U})$. As $\mathcal{U}$ is NMSCoS, $NM \backsim Int(\mathcal{U})$ is NMROS in $\mathbb{G}$. From Theorem 20, $mNM \backsim Cl(\mathcal{U}) = mNM \backsim Cl(NM \backsim Int(\mathcal{U})) = NM \backsim Cl(mNM \backsim Int(\mathcal{U})) \preccurlyeq NM \backsim Cl(m\mathcal{U})$. Hence, NM $\backsim Cl(m\mathcal{U}) = mNM \backsim Cl(\mathcal{U})$.

(b) Following the same steps as in part (1) above, we can prove that $NM \backsim Cl(\mathcal{U}m) = NM \backsim Cl(\mathcal{U})m$.

(c) By hypothesis, this proves $NM \backsim Cl(\mathcal{U})$ is NMRCoS and therefore $NM \backsim Cl(\mathcal{U})^{-1}$ is NMCoS. Consequently, $NM \backsim Cl(\mathcal{U}^{-1}) \preccurlyeq NM \backsim Cl(\mathcal{U})^{-1}$. Next, since $\mathcal{U}$ is NMSOS, $NM \backsim Cl(\mathcal{U}) = NM \backsim Cl(NM \backsim Int(\mathcal{U})) \Rightarrow NM \backsim Cl(\mathcal{U})^{-1} = NM \backsim Cl(NM \backsim Int(\mathcal{U})$. Moreover, as $\mathcal{U}$ is NMSCoS, $NM \backsim Int(\mathcal{U})$ is NMROS. From Theorem 18, $NM \backsim Cl(\mathcal{U})^{-1} = NM \backsim Cl\left(NM \backsim Int(\mathcal{U})^{-1}\right) \preccurlyeq NM \backsim Cl(\mathcal{U}^{-1})$. This shows that $NM \backsim Cl(\mathcal{U}^{-1}) = NM \backsim Cl(\mathcal{U})^{-1}$. □

Theorem 22. *From Theorem 21, the following statements hold:*

(a) $NM \backsim Int(m\mathcal{U}) = mNM \backsim Int(\mathcal{U})$, *for each* $m \in \mathbb{G}$;
(b) $NM \backsim Int(\mathcal{U}m) = NM \backsim Int(\mathcal{U})m$, *for each* $m \in \mathbb{G}$;
(c) $NM \backsim Int(\mathcal{U}^{-1}) = NM \backsim Int(\mathcal{U})^{-1}$.

Proof. (a) As $\mathcal{U}$ is NMSCoS, $NM \backsim Int(\mathcal{U})$ is NMROS. From Theorem 16, $\mu_{m^{-1}} : (\mathbb{G}, \tau_X) \longrightarrow (\mathbb{G}, \tau_X)$ is NMACM. Therefore, ${\mu^{-1}}_{m^{-1}}(NM \backsim Int(\mathcal{U})) = mNM \backsim Int(\mathcal{U})$ is NMOS. Thus, $mNM \backsim Int(\mathcal{U}) \preccurlyeq NM \backsim Int(m\mathcal{U})$. Next, by assumption, it implies that $NM \backsim Int(\mathcal{U}) = NM \backsim Int(NM \backsim Cl(\mathcal{U})) \Rightarrow mNM \backsim Int(\mathcal{U}) = mNM \backsim Int(NM \backsim Cl(\mathcal{U}))$. As $\mathcal{U}$ is NMSOS, $NM \backsim Cl(\mathcal{U})$ is NMRCoS. From Theorem 19, $mNM \backsim Int(NM \backsim Cl(\mathcal{U})) = NM \backsim Int(mNM \backsim Cl(\mathcal{U})) \succcurlyeq NM \backsim Int(m\mathcal{U})$. That is, $NM \backsim Int(m\mathcal{U}) \preccurlyeq mNM \backsim Int(\mathcal{U})$. Therefore, we have, $NM \backsim Int(m\mathcal{U}) = mNM \backsim Int(\mathcal{U})$. Hence, it was proved.

(b) As $\mathcal{U}$ is NMSCoS, $NM \backsim Int(\mathcal{U})$ is NMROS. From Theorem 16, $\mu_{m^{-1}} : (\mathbb{G}, \tau_X) \longrightarrow (\mathbb{G}, \tau_X)$ is NMACM. Thus, ${\lambda^{-1}}_{m^{-1}}(NM \backsim Int(\mathcal{U})) = mNM \backsim Int(\mathcal{U})$ is NMOS. Therefore, $NM \backsim Int(\mathcal{U})m \preccurlyeq NM \backsim Int(\mathcal{U}m)$. Next, by assumption, this proves that $NM \backsim Int(\mathcal{U}) = NM \backsim Int(NM \backsim Cl(\mathcal{U})) \Rightarrow NM \backsim Int(\mathcal{U})m = NM \backsim Int(NM \backsim Cl(\mathcal{U}))m$. As $\mathcal{U}$ is NMSOS, $NM \backsim Cl(\mathcal{U})$ is NMRCoS. From Theorem 19, $NM \backsim Int(NM \backsim Cl(\mathcal{U}))m = NM \backsim Int(NM \backsim Cl(\mathcal{U})m) \succcurlyeq NM \backsim Int(\mathcal{U}m)$. That is, $NM \backsim Int(\mathcal{U}m) \preccurlyeq NM \backsim Int(\mathcal{U})m$. Therefore, $NM \backsim Int(\mathcal{U}m) = NM \backsim Int(\mathcal{U})m$. Hence, it was proved.

(c) From assumption, this proves that $NM \backsim Int(\mathcal{U})$ is NMROS and therefore $NM \backsim Int(\mathcal{U})^{-1}$ is NMOS. Consequently, $NM \backsim Int(\mathcal{U}^{-1}) \preccurlyeq NM \backsim Int(\mathcal{U})^{-1}$. Next, as $\mathcal{U}$ is NMSCoS, $NM \backsim Int(\mathcal{U}) = NM \backsim Int(NM \backsim Cl(\mathcal{U})) \Rightarrow NM \backsim Int(\mathcal{U})^{-1} = NM \backsim Int(NM \backsim Cl(\mathcal{U}))^{-1}$. Moreover, as $\mathcal{U}$ is NMSOS, $NM \backsim Cl(\mathcal{U})$ is NMRCoS. From Theorem 19, $NM \backsim Int(\mathcal{U})^{-1} = NM \backsim Int(NM \backsim Cl(\mathcal{U})^{-1}) \preccurlyeq NM \backsim Int(\mathcal{U}^{-1})$. This proves that $NM \backsim Int(\mathcal{U}^{-1}) = NM \backsim Int(\mathcal{U})^{-1}$. □

Theorem 23. *Let* $\mathcal{A}$ *be NMOS in an NMATG* $\mathbb{G}$. *Then* $a\mathcal{A} \preccurlyeq NM \backsim Int(aNM \backsim Int(NM \backsim Cl(\mathcal{A})))$ *for* $a \in \mathbb{G}$.

Proof. Since $\mathcal{A}$ is NMOS, so $\mathcal{A} \preccurlyeq NM \backsim Int(NM \backsim Cl(\mathcal{A})) \Rightarrow a\mathcal{A} \preccurlyeq aNM \backsim Int(NM \backsim Cl(\mathcal{A}))$. From Theorem 17, $aNM \backsim Int(NM \backsim Cl(\mathcal{A}))$ is NMOS (in fact, NMROS). Hence, $a\mathcal{A} \preccurlyeq NM \backsim Int(aNM \backsim Int(NM \backsim Cl(\mathcal{A})))$. □

Theorem 24. *Let* $\mathcal{Q}$ *be any neutrosophic multi-closed subset in an NMATG* $\mathbb{G}$. *Then* $NM \backsim Cl(aNM \backsim Cl(NM \backsim Int(\mathcal{A}))) \preccurlyeq a\mathcal{Q}$ *for each* $a \in \mathbb{G}$.

Proof. Since $\mathcal{Q}$ is NMCoS, so $\mathcal{Q} \succcurlyeq NM \backsim Cl(NM \backsim Int(\mathcal{Q})) \Rightarrow a\mathcal{Q} \succcurlyeq aNM \backsim Cl(NM \backsim Int(\mathcal{Q}))$. From Theorem 17, $aNM \backsim Cl(NM \backsim Int(\mathcal{Q}))$ is NMCoS (in fact, NMRCoS). Therefore, $a\mathcal{Q} \succcurlyeq NM \backsim Cl(aNM \backsim Cl(NM \backsim Int(\mathcal{A})))$. Hence, $NM \backsim Cl(aNM \backsim Cl(NM \backsim Int(\mathcal{A}))) \preccurlyeq a\mathcal{Q}$. □

4. Conclusions

To deal with uncertainty, the NS uses the truth membership function, indeterminacy membership function, and falsity membership function. By discovering this concept, we were able to generalise the idea of an almost topological group to an NMATG. First, we developed the definitions of NMSOS, NMSCoS, NMROS, NMRCoS, NMCM, NMOM, NMCoM, NMSCM, NMSOM, NMSCoM to propose the definition of NMATG. Some properties of NMACM were demonstrated. Finally, we defined NMATG and demonstrated some of their properties using the definition of NMACM. In this study, an NMATG is conceptualised for the environments of the NS along with some of their elementary properties and theoretic operations. Novel numerical examples are given for definitions and remarks to study NMATG. We expect that our study may spark some new ideas for the construction of the NMATG. Future work may include the extension of this work for:

(1) The development of the NMATG of the neutrosophic multi-vector spaces, etc.;
(2) Dealing NMATG with multi-criteria decision-making techniques.

Author Contributions: Conceptualization: B.B., N.W., D.D.M., A.K.B., J.M., U.R.B. and J.B. All authors have contributed equally to this paper in all aspects. All authors have read and agreed to the published version of the manuscript.

Funding: This research received no external funding.

Institutional Review Board Statement: Not applicable.

Informed Consent Statement: Not applicable.

Data Availability Statement: Not applicable.

Acknowledgments: We would like to thank the anonymous reviewers for their very careful reading and valuable suggestions.

Conflicts of Interest: The authors declare no conflict of interest.

References

1. Zadeh, L.A. Fuzzy sets. *Inf. Control* **1965**, *8*, 338–353. [CrossRef]
2. Chang, C.L. Fuzzy Topological Space. *J. Math. Anal. Appl.* **1968**, *24*, 182–190. [CrossRef]
3. Coker, D. An introduction to intuitionistic fuzzy topological spaces. *Fuzzy Sets Syst.* **1997**, *88*, 81–89. [CrossRef]
4. Lupianez, F.G. On neutrosophic topology. *Int. J. Syst. Cybern.* **2008**, *37*, 797–800. [CrossRef]
5. Lupianez, F.G. Interval neutrosophic sets and topology. *Int. J. Syst. Cybern.* **2009**, *38*, 621–624. [CrossRef]
6. Lupianez, F.G. On various neutrosophic topologies. *Int. J. Syst. Cybern.* **2009**, *38*, 1009–1013.
7. Lupianez, F.G. On neutrosophic paraconsistent topology. *Int. J. Syst. Cybern.* **2010**, *39*, 598–601. [CrossRef]
8. Salama, A.A.; Smarandache, F.; Kroumov, V. Closed sets and Neutrosophic Continuous Functions. *Neutrosophic Sets Syst.* **2014**, *4*, 4–8.
9. Kelly, J.C. Bitopological spaces. *Proc. Lond. Math. Soc.* **1963**, *3*, 71–89. [CrossRef]
10. Kandil, A.; Nouth, A.A.; El-Sheikh, S.A. On fuzzy bitopological spaces. *Fuzzy Sets Syst.* **1995**, *74*, 353–363. [CrossRef]
11. Lee, S.J.; Kim, J.T. Some Properties of Intuitionistic Fuzzy Bitopological Spaces. In Proceedings of the 6th International Conference on Soft Computing and Intelligent Systems, and The 13th IEEE International Symposium on Advanced Intelligence Systems, Kobe, Japan, 20–24 November 2012; pp. 20–24. [CrossRef]
12. Garg, H. A new improved score function of an interval-valued Pythagorean fuzzy set based TOPSIS method. *Int. J. Uncertain. Quantif.* **2017**, *7*, 463–474. [CrossRef]
13. Peng, X.; Yang, Y. Some results for Pythagorean fuzzy sets. *Int. J. Intell. Syst.* **2015**, *30*, 1133–1160. [CrossRef]
14. Peng, X.; Selvachandran, G. Pythagorean fuzzy set: State of the art and future directions. *Artif. Intell. Rev.* **2017**, *52*, 1873–1927. [CrossRef]
15. Beliakov, G.; James, S. Averaging aggregation functions for preferences expressed as Pythagorean membership grades and fuzzy orthopairs. In Proceedings of the IEEE International Conference on Fuzzy Systems (FUZZ-IEEE), Beijing, China, 6–11 July 2014; pp. 298–305.
16. Dick, S.; Yager, R.R.; Yazdanbakhsh, O. On Pythagorean and complex fuzzy set operations. *IEEE Trans. Fuzzy Syst.* **2016**, *24*, 1009–1021. [CrossRef]
17. Gou, X.J.; Xu, Z.S.; Ren, P.J. The properties of continuous Pythagorean fuzzy Information. *Int. J. Intell. Syst.* **2016**, *31*, 401–424. [CrossRef]
18. He, X.; Du, Y.; Liu, W. Pythagorean fuzzy power average operators. *Fuzzy Syst. Math.* **2016**, *30*, 116–124.

19. Ejegwa, P.A. Distance and similarity measures of Pythagorean fuzzy sets. *Granul. Comput.* **2018**, *5*, 225–238. [CrossRef]
20. Yager, R.R. Generalized orthopair fuzzy sets. *IEEE Trans. Fuzzy Syst.* **2017**, *25*, 1222–1230. [CrossRef]
21. Yager, R.R.; Alajlan, N. Approximate reasoning with generalized orthopair fuzzy sets. *Inf. Fusion* **2017**, *38*, 65–73. [CrossRef]
22. Peng, X.; Liu, L. Information measures for q-rung orthopair fuzzy sets. *Int. J. Intell. Syst.* **2019**, *34*, 1795–1834. [CrossRef]
23. Pinar, A.; Boran, F.E. A q-rung orthopair fuzzy multi-criteria group decision making method for supplier selection based on a novel distance measure. *Int. J. Mach. Learn. Cybern.* **2020**, *11*, 1749–1780. [CrossRef]
24. Cuong, B.C.; Kreinovich, V. Picture Fuzzy Sets—A new concept for computational intelligence problems. In Proceedings of the 2013 Third World Congress on Information and Communication Technologies (WICT 2013), Hanoi, Vietnam, 15–18 December 2013; pp. 1–6. [CrossRef]
25. Cuong, B.C. Picture fuzzy sets. *J. Comput. Sci. Cybern.* **2014**, *30*, 409–420.
26. Phong, P.H.; Hieu, D.T.; Ngan, R.T.H.; Them, P.T. Some compositions of picture fuzzy relations. In Proceedings of the 7th National Conferenceon Fundamental and Applied Information Technology Research, FAIR'7, Thai Nguyen, Vietnam, 19–20 June 2014; pp. 19–20.
27. Cuong, B.C.; Hai, P.V. Some fuzzy logic operators for picture fuzzy sets. In Proceedings of the Seventh International Conference on Knowledge and Systems Engineering, Ho Chi Minh City, Vietnam, 8–10 October 2015; pp. 132–137.
28. Cuong, B.C.; Ngan, R.T.; Hai, B.D. An involutive picture fuzzy negator on picture fuzzy sets and some De Morgan triples. In Proceedings of the Seventh International Conference on Knowledge and Systems Engineering, Ho Chi Minh City, Vietnam, 8–10 October 2015; pp. 126–131.
29. Viet, P.V.; Chau, H.T.M.; Hai, P.V. Some extensions of membership graphs for picture inference systems. In Proceedings of the 2015 Seventh International Conference on Knowledge and Systems Engineering, (KSE), Ho Chi Minh City, Vietnam, 8–10 October 2015; IEEE: Piscataway, NJ, USA, 2015; pp. 192–197.
30. Singh, P. Correlation coefficients for picture fuzzy sets. *J. Intell. Fuzzy Syst.* **2015**, *28*, 591–604. [CrossRef]
31. Garg, H. Some picture fuzzy aggregation operators and their applications to multicriteria decision-making. *Arab. J. Sci. Eng.* **2017**, *42*, 5275–5290. [CrossRef]
32. Quek, S.G.; Selvachandran, G.; Munir, M.; Mahmood, T.; Ullah, K.; Son, L.H.; Pham, T.H.; Kumar, R.; Priyadarshini, I. Multi-attribute multi-perception decision-making based on generalized T-spherical fuzzy weighted aggregation operators on neutrosophic sets. *Mathematics* **2019**, *7*, 780. [CrossRef]
33. Garg, H.; Munir, M.; Ullah, K.; Mahmood, T.; Jan, N. Algorithm for T-spherical fuzzy multi-attribute decision making based on improved interactive aggregation operators. *Symmetry* **2018**, *10*, 670. [CrossRef]
34. Zeng, S.; Garg, H.; Munir, M.; Mahmood, T.; Hussain, A. A multi-attribute decision making process with immediate probabilistic interactive averaging aggregation operators of T-spherical fuzzy sets and its application in the selection of solar cells. *Energies* **2019**, *12*, 4436. [CrossRef]
35. Munir, M.; Kalsoom, H.; Ullah, K.; Mahmood, T.; Chu, Y.M. T-spherical fuzzy Einstein hybrid aggregation operators and their applications in multi-attribute decision making problems. *Symmetry* **2020**, *12*, 365. [CrossRef]
36. Mahmood, T.; Ullah, K.; Khan, Q.; Jan, N. An approach towards decision making and medical diagnosis problems using the concept of spherical fuzzy sets. *Neural Comput. Appl.* **2018**, *31*, 7041–7053. [CrossRef]
37. Warren, R.H. Boundary of a fuzzy set. *Indiana Univ. Math. J.* **1977**, *26*, 191–197. [CrossRef]
38. Tang, X. Spatial Object Modeling in Fuzzy Topological Spaces with Applications to Land Cover Change in China. Ph.D. Thesis, University of Twente, Enschede, The Netherlands, 2004.
39. Kharal, A. A Study of Frontier and Semifrontier in Intuitionistic Fuzzy Topological Spaces. *Sci. World J.* **2014**, *2014*, 674171. [CrossRef]
40. Salama, A.A.; Alblowi, S. Generalized neutrosophic set and generalized neutrosophic topological spaces. *Comp. Sci. Eng.* **2012**, *2*, 129–132. [CrossRef]
41. Azad, K.K. On Fuzzy Semi-continuity, Fuzzy Almost Continuity and Fuzzy Weakly Continuity. *J. Math. Anal. Appl.* **1981**, *82*, 14–32. [CrossRef]
42. Smarandache, F. *Neutrosophy and Neutrosophic Logic, First International Conference on Neutrosophy, Neutrosophic Logic, Set, Probability, and Statistics*; University of New Mexico: Gallup, NM, USA, 2002.
43. Smarandache, F. Neutrosophic set—A generalization of the intuitionistic fuzzy set. *Int. J. Pure Appl. Math.* **2005**, *24*, 287–297.
44. Mwchahary, D.D.; Basumatary, B. A note on Neutrosophic Bitopological Space. *Neutrosophic Sets Syst.* **2020**, *33*, 134–144.
45. Blizard, W. Multiset theory. *Notre Dame J. Form. Logic* **1989**, *30*, 36–66. [CrossRef]
46. Yager, R.R. On the theory of bags. *Int. J. Gen Syst.* **1986**, *13*, 23–37. [CrossRef]
47. Miyamoto, S. Fuzzy Multisets and Their Generalizations. In *Multiset Processing*; Springer: Berlin, Germany, 2001; pp. 225–235.
48. Onasanya, B.O.; Hoskova-Mayerova, S. Some Topological and Algebraic Properties of alpha-level Subsets Topology of a Fuzzy Subset. *An. St. Univ. Ovidius Constanta* **2018**, *26*, 213–227.
49. Onasanya, B.O.; Hoskova-Mayerova, S. Multi-fuzzy group induced by multisets. *Ital. J. Pure Appl. Math.* **2019**, *41*, 597–604.
50. Al Tahan, M.; Hoskova-Mayerova, S.; Davvaz, B. Fuzzy multi-polygroups. *J. Intell. Fuzzy Syst.* **2020**, *38*, 2337–2345. [CrossRef]
51. Al Tahan, M.; Hoskova-Mayerova, S.; Davvaz, B. Some results on (generalized) fuzzy multi-Hv-ideals of Hv-rings. *Symmetry* **2019**, *11*, 1376. [CrossRef]
52. Bakbak, D.; Uluçay, V.; Sahin, M. Neutrosophic Multigroups and Applications. *Mathemaics* **2019**, *7*, 95. [CrossRef]

53. Sebastian, S.; Ramakrishnan, T.V. Multi-fuzzy sets: An extension of fuzzy sets. *Fuzzy Inf. Eng.* **2011**, *1*, 35–43. [CrossRef]
54. Dey, A.; Pal, M. Multi-fuzzy complex numbers and multi-fuzzy complex sets. *Int. J. Fuzzy Syst. Appl.* **2014**, *4*, 15–27. [CrossRef]
55. Yong, Y.; Xia, T.; Congcong, M. The multi-fuzzy soft set and its application in decision making. *Appl. Math. Model* **2013**, *37*, 4915–4923. [CrossRef]
56. Sebastian, S.; Ramakrishnan, T.T. Multi-Fuzzy Sets. *Int. Math. Forum* **2010**, *5*, 2471–2476.
57. Atanassov, K.T. Intuitionistic fuzzy sets. *Fuzzy Sets Syst.* **1986**, *20*, 87–96. [CrossRef]
58. Shinoj, T.K.; John, S.S. Intuitionistic fuzzy multisets and its application in medical diagnosis. *World Acad. Sci. Eng. Technol.* **2012**, *6*, 1418–1421.
59. Abdullah, S.; Naeem, M. A new type of interval valued fuzzy normal subgroups of groups. *New Trends Math. Sci.* **2015**, *3*, 62–77.
60. Mordeson, J.N.; Bhutani, K.R.; Rosenfeld, A. *Fuzzy Group Theory*; Springer: Berlin/Heidelberg, Germany, 2005.
61. Liu, Y.L. Quotient groups induced by fuzzy subgroups. *Quasigroups Related Syst.* **2004**, *11*, 71–78.
62. Baby, A.; Shinoj, T.K.; John, S.J. On Abelian Fuzzy Multi Groups and Orders of Fuzzy Multi Groups. *J. New Theory* **2015**, *5*, 80–93.
63. Muthuraj, R.; Balamurugan, S. Multi-Fuzzy Group and its Level Subgroups. *Gen. Math. Notes* **2013**, *17*, 74–81.
64. Tella, Y. On Algebraic Properties of Fuzzy Membership Sequenced Multisets. *Br. J. Math. Comput. Sci.* **2015**, *6*, 146–164. [CrossRef]
65. Fathi, M.; Salleh, A.R. Intuitionistic fuzzy groups. *Asian J. Algebra* **2009**, *2*, 1–10. [CrossRef]
66. Li, X.P.; Wang, G.J. (l, a)-Homomorphisms of Intuitionistic Fuzzy Groups. *Hacettepe J. Math. Stat.* **2011**, *40*, 663–672.
67. Palaniappan, N.; Naganathan, S.; Arjunan, K. A study on Intuitionistic L-fuzzy Subgroups. *Appl. Math. Sci.* **2009**, *3*, 2619–2624.
68. Sharma, P.K. Homomorphism of Intuitionistic fuzzy groups. *Int. Math. Forum* **2011**, *6*, 3169–3178.
69. Sharma, P.K. On the direct product of Intuitionistic fuzzy subgroups. *Int. Math. Forum* **2012**, *7*, 523–530.
70. Sharma, P.K. On intuitionistic fuzzy abelian subgroups. Adv. *Fuzzy Sets Syst.* **2012**, *12*, 1–16.
71. Xu, C.Y. Homomorphism of Intuitionistic Fuzzy Groups. In Proceedings of the 2007 International Conference on Machine Learning and Cybernetics, Hong Kong, China, 19–22 August 2007; pp. 178–183. [CrossRef]
72. Yuan, X.H.; Li, H.X.; Lee, E.S. On the definition of the intuitionistic fuzzy subgroups. *Comput. Math. Appl.* **2010**, *59*, 3117–3129. [CrossRef]
73. Shinoj, T.K.; John, S.S. Intuitionistic fuzzy multigroups. *Ann. Pure Appl. Math.* **2015**, *9*, 131–143.
74. Smarandache, F. *A Unifying Field in Logics*; Infinite Study: Conshohocken, PA, USA, 1998.
75. Deli, I.; Broumi, S.; Smarandache, F. On neutrosophic multisets and its application in medical diagnosis. *J. New Theory* **2015**, *6*, 88–98.
76. Ye, S.; Ye, J. Dice Similarity Measure between Single Valued Neutrosophic Multisets and Its Application in Medical Diagnosis. *Neutrosophic Sets Syst.* **2014**, *6*, 49–54.
77. Ye, S.; Fu, J.; Ye, J. Medical Diagnosis Using Distance-Based Similarity Measures of Single Valued Neutrosophic Multisets. *Neutrosophic Sets Syst.* **2015**, *7*, 47–52.
78. Smarandache, F. *Neutrosophic Perspectives: Triplets, Duplets, Multisets, Hybrid Operators, Modal Logic, Hedge Algebras and Applications*; Pons Publishing House Brussels: Brussels, Belgium, 2017; p. 323.

MDPI
St. Alban-Anlage 66
4052 Basel
Switzerland
Tel. +41 61 683 77 34
Fax +41 61 302 89 18
www.mdpi.com

Symmetry Editorial Office
E-mail: symmetry@mdpi.com
www.mdpi.com/journal/symmetry

www.ingramcontent.com/pod-product-compliance
Lightning Source LLC
LaVergne TN
LVHW070250100526
838202LV00015B/2203